Lecture Notes in Artificial Intell

Edited by J. G. Carbonell and J. Siekmann

Subseries of Lecture Notes in Computer Science

Jean-François Boulicaut
Floriana Esposito Fosca Giannotti
Dino Pedreschi (Eds.)

Knowledge Discovery in Databases: PKDD 2004

8th European Conference on Principles and Practice of Knowledge Discovery in Databases
Pisa, Italy, September 20-24, 2004
Proceedings

Series Editors

Jaime G. Carbonell, Carnegie Mellon University, Pittsburgh, PA, USA
Jörg Siekmann, University of Saarland, Saarbrücken, Germany

Volume Editors

Jean-François Boulicaut
INSA Lyon
LIRIS CNRS FRE 2672, 69621 Villeurbanne Cedex, France
E-mail: jean-francois.boulicaut@insa-lyon.fr

Floriana Esposito
University of Bari
Department of Computer Science
Via Orabona 4, 70126 Bari, Italy
E-mail: esposito@di.uniba.it

Fosca Giannotti
Science and Technology Institute
Knowledge Discovery and Delivery (KDD)
Via G. Moruzzi 1, 56124 Pisa, Italy
E-mail: Fosca.Giannotti@isti.cnr.it

Dino Pedreschi
University of Pisa
Department of Computer Science
Via F. Buonarroti 2, 56125 Pisa, Italy
E-mail: pedre@di.unipi.it

Library of Congress Control Number: 2004111516

CR Subject Classification (1998): I.2, H.2, J.1, H.3, G.3, I.7, F.4.1

ISSN 0302-9743
ISBN 3-540-23108-0 Springer Berlin Heidelberg New York

Springer is a part of Springer Science+Business Media

springeronline.com

Printed in Germany

Typesetting: Camera-ready by author, data conversion by Olgun Computergrafik
Printed on acid-free paper SPIN: 11322801 06/3142 5 4 3 2 1 0

Preface

The proceedings of ECML/PKDD 2004 are published in two separate, albeit intertwined, volumes: the Proceedings of the 15th European Conference on Machine Learning (LNAI 3201) and the Proceedings of the 8th European Conferences on Principles and Practice of Knowledge Discovery in Databases (LNAI 3202). The two conferences were co-located in Pisa, Tuscany, Italy during September 20–24, 2004.

It was the fourth time in a row that ECML and PKDD were co-located. After the successful co-locations in Freiburg (2001), Helsinki (2002), and Cavtat-Dubrovnik (2003), it became clear that researchers strongly supported the organization of a major scientific event about machine learning and data mining in Europe.

We are happy to provide some statistics about the conferences. 581 different papers were submitted to ECML/PKDD (about a 75% increase over 2003); 280 were submitted to ECML 2004 only, 194 were submitted to PKDD 2004 only, and 107 were submitted to both. Around half of the authors for submitted papers are from outside Europe, which is a clear indicator of the increasing attractiveness of ECML/PKDD.

The Program Committee members were deeply involved in what turned out to be a highly competitive selection process. We assigned each paper to 3 reviewers, deciding on the appropriate PC for papers submitted to both ECML and PKDD. As a result, ECML PC members reviewed 312 papers and PKDD PC members reviewed 269 papers. We accepted for publication regular papers (45 for ECML 2004 and 39 for PKDD 2004) and short papers that were associated with poster presentations (6 for ECML 2004 and 9 for PKDD 2004). The global acceptance rate was 14.5% for regular papers (17% if we include the short papers).

The scientific program of ECML/PKDD 2004 also included 5 invited talks, a wide workshop and tutorial program (10 workshops plus a Discovery Challenge workshop, and seven tutorials) and a demo session.

We wish to express our gratitude to:

- the authors of the submitted papers;
- the program committee members and the additional referees for their exceptional contribution to a tough but crucial selection process;
- the invited speakers: Dimitris Achlioptas (Microsoft Research, Redmond), Rakesh Agrawal (IBM Almaden Research Center), Soumen Chakrabarti (Indian Institute of Technology, Bombay), Pedro Domingos (University of Washington, Seattle), and David J. Hand (Imperial College, London);
- the workshop chairs Donato Malerba and Mohammed J. Zaki;
- the tutorial chairs Katharina Morik and Franco Turini;
- the discovery challenge chairs Petr Berka and Bruno Crémilleux;

- the publicity chair Salvatore Ruggieri;
- the demonstration chairs Rosa Meo, Elena Baralis, and Codrina Lauth;
- the members of the ECML/PKDD Steering Committee Peter Flach, Luc De Raedt, Arno Siebes, Nada Lavrač, Dragan Gamberger, Ljupčo Todorovski, Hendrik Blockeel, Tapio Elomaa, Heikki Mannila, and Hannu T.T. Toivonen;
- the members of the Award Committee, Michael May and Foster Provost;
- the workshops organizers and the tutorialists;
- the extremely efficient Organization Committee members, Maurizio Atzori, Miriam Baglioni, Sergio Barsocchi, Jérémy Besson, Francesco Bonchi, Stefano Ferilli, Tiziana Mazzone, Mirco Nanni, Ruggero Pensa, Simone Puntoni, Chiara Renso, Salvatore Rinzivillo, as well as all the other members of the KDD Lab in Pisa, Laura Balbarini and Cristina Rosamilia of L&B Studio, Elena Perini and Elena Tonsini of the University of Pisa;
- the great Web masters Mirco Nanni, Chiara Renso and Salvatore Rinzivillo;
- the directors of the two research institutions in Pisa that jointly made this event possible, Piero Maestrini (ISTI-CNR) and Ugo Montanari (Dipartimento di Informatica);
- the administration staff of the two research institutions in Pisa, in particular Massimiliano Farnesi (ISTI-CNR), Paola Fabiani and Letizia Petrellese (Dipartimento di Informatica);
- Richard van de Stadt (www.borbala.com) for his efficient support to the management of the whole submission and evaluation process by means of the CyberChairPRO software;
- Alfred Hofmann of Springer for co-operation in publishing the proceedings.

We gratefully acknowledge the financial support of KDNet, the Pascal Network, Kluwer and the Machine Learning journal, Springer, the Province of Lucca, the Province of Pisa, the Municipality of Pisa, Microsoft Research, COOP, Exeura, Intel, Talent, INSA-Lyon, ISTI-CNR Pisa, the University of Pisa, the University of Bari, and the patronage of Regione Toscana.

There is no doubt that the impressive scientific activities in machine learning and data mining world-wide were well demonstrated in Pisa. We had an exciting week in Tuscany, enhancing further co-operations between the many researchers who are pushing knowledge discovery into becoming a mature scientific discipline.

July 2004

Jean-François Boulicaut,
Floriana Esposito,
Fosca Giannotti,
and Dino Pedreschi

ECML/PKDD 2004 Organization

Executive Committee

Program Chairs	Jean-François Boulicaut (INSA Lyon) Floriana Esposito (Università di Bari) Fosca Giannotti (ISTI-CNR) Dino Pedreschi (Università di Pisa)
Workshop Chairs	Donato Malerba (University of Bari) Mohammed J. Zaki (Rensselaer Polytechnic Institute)
Tutorial Chairs	Katharina Morik (University of Dortmund) Franco Turini (University of Pisa)
Discovery Challenge Chairs	Petr Berka (University of Economics, Prague) Bruno Crémilleux (University of Caen)
Publicity Chair	Salvatore Ruggieri (University of Pisa)
Demonstration Chairs	Rosa Meo (University of Turin) Elena Baralis (Politecnico of Turin) Ina Lauth (Fraunhofer Institute for Autonomous Intelligent Systems)
Steering Committee	Peter Flach (University of Bristol) Luc De Raedt (Albert-Ludwigs University, Freiburg) Arno Siebes (Utrecht University) Nada Lavrač (Jozef Stefan Institute) Dragan Gamberger (Rudjer Boskovic Institute) Ljupčo Todorovski (Jozef Stefan Institute) Hendrik Blockeel (Katholieke Universiteit Leuven) Tapio Elomaa (Tampere University of Technology) Heikki Mannila (Helsinki Institute for Information Technology) Hannu T.T. Toivonen (University of Helsinki)
Awards Committee	Michael May (Fraunhofer Institute for Autonomous Intelligent Systems, KDNet representative) Floriana Esposito (PC representative) Foster Provost (Editor-in-Chief of Machine Learning Journal, Kluwer)
Organizing Committee	Maurizio Atzori (KDDLab, ISTI-CNR) Miriam Baglioni (KDDLab, University of Pisa) Sergio Barsocchi (KDDLab, ISTI-CNR) Jérémy Besson (INSA Lyon) Francesco Bonchi (KDDLab, ISTI-CNR) Stefano Ferilli (University of Bari) Tiziana Mazzone (KDDLab) Mirco Nanni (KDDLab, ISTI-CNR) Ruggero Pensa (INSA Lyon) Chiara Renso (KDDLab, ISTI-CNR) Salvatore Rinzivillo (KDDLab, University of Pisa)

ECML 2004 Program Committee

Hendrik Blockeel, Belgium
Marco Botta, Italy
Henrik Boström, Sweden
Jean-François Boulicaut, France
Ivan Bratko, Slovenia
Pavel Brazdil, Portugal
Nello Cristianini, USA
James Cussens, UK
Ramon Lopes de Mantaras, Spain
Luc De Raedt, Germany
Luc Dehaspe, Belgium
Josè del R. Millan, Switzerland
Sašo Džeroski, Slovenia
Tapio Elomaa, Finland
Floriana Esposito, Italy
Peter Flach, UK
Johannes Fürnkranz, Germany
Joao Gama, Portugal
Dragan Gamberger, Croatia
Jean-Gabriel Ganascia, France
Fosca Giannotti, Italy
Attilio Giordana, Italy
Haym Hirsh, USA
Thomas Hofmann, USA
Tamas Horvath, Germany
Thorsten Joachims, USA
Dimitar Kazakov, UK
Roni Khardon, USA
Joerg Kindermann, Germany
Yves Kodratoff, France
Igor Kononenko, Slovenia
Stefan Kramer, Germany
Miroslav Kubat, USA
Stephen Kwek, USA
Nada Lavrač, Slovenia
Charles Ling, Canada
Donato Malerba, Italy
Heikki Mannila, Finland
Stan Matwin, Canada
Dunja Mladenic, Slovenia
Katharina Morik, Germany
Hiroshi Motoda, Japan
Remi Munos, France
Richard Nock, France
David Page, USA
Georgios Paliouras, Greece
Dino Pedreschi, Italy
Bernhard Pfahringer, New Zealand
Enric Plaza, Spain
Juho Rousu, UK
Celine Rouveirol, France
Tobias Scheffer, Germany
Michele Sebag, France
Giovanni Semeraro, Italy
Arno Siebes, The Netherlands
Robert Sloan, USA
Gerd Stumme, Germany
Henry Tirri, Finland
Ljupčo Todorovski, Slovenia
Luis Torgo, Portugal
Peter Turney, Canada
Maarten van Someren, The Netherlands
Paul Vitanyi, The Netherlands
Sholom Weiss, USA
Dietrich Wettschereck, UK
Gerhard Widmer, Austria
Marco Wiering, The Netherlands
Ruediger Wirth, Germany
Stefan Wrobel, Germany
Thomas Zeugmann, Germany
Tong Zhang, USA
Blaž Zupan, Slovenia

PKDD 2004 Program Committee

Elena Baralis, Italy
Michael Berthold, Germany
Elisa Bertino, USA
Hendrik Blockeel, Belgium
Jean-François Boulicaut, France
Christopher W. Clifton, USA
Bruno Cremilleux, France
Luc De Raedt, Germany
Luc Dehaspe, Belgium
Sašo Džeroski, Slovenia
Tapio Elomaa, Finland
Floriana Esposito, Italy
Martin Ester, Canada
Ad Feelders, The Netherlands
Ronen Feldman, IL
Peter Flach, UK
Eibe Frank, New Zealand
Alex Freitas, UK
Johannes Fürnkranz, Germany
Dragan Gamberger, Croatia
Minos Garofalakis, USA
Fosca Giannotti, Italy
Christophe Giraud-Carrier, Switzerland
Bart Goethals, Finland
Howard Hamilton, Canada
Robert Hilderman, Canada
Haym Hirsh, USA
Frank Hoeppner, Germany
Se Hong, USA
Samuel Kaski, Finland
Daniel Keim, Germany
Jorg-Uwe Kietz, Switzerland
Ross King, UK
Yves Kodratoff, France
Joost Kok, The Netherlands
Stefan Kramer, Germany
Laks Lakshmanan, Canada
Nada Lavrač, Slovenia
Donato Malerba, Italy
Giuseppe Manco, Italy
Heikki Mannila, Finland
Stan Matwin, Canada
Michael May, Germany
Rosa Meo, Italy
Dunja Mladenic, Slovenia
Katharina Morik, Germany
Shinichi Morishita, Japan
Hiroshi Motoda, Japan
Gholamreza Nakhaeizadeh, Germany
Claire Nedellec, France
David Page, USA
Dino Pedreschi, Italy
Zbigniew Ras, USA
Jan Rauch, Czech Rebuclic
Christophe Rigotti, France
Gilbert Ritschard, Switzerland
John Roddick, Australia
Yucel Saygin, Turkey
Michele Sebag, France
Marc Sebban, France
Arno Siebes, The Netherlands
Andrzej Skowron, Poland
Myra Spiliopoulou, Germany
Nicolas Spyratos, France
Reinhard Stolle, USA
Gerd Stumme, Germany
Einoshin Suzuki, Japan
Ah-Hwee Tan, Singapore
Ljupčo Todorovski, Slovenia
Hannu Toivonen, Finland
Luis Torgo, Portugal
Shusaku Tsumoto, Japan
Franco Turini, Italy
Maarten van Someren, The Netherlands
Ke Wang, Canada
Louis Wehenkel, Belgium
Dietrich Wettschereck, UK
Gerhard Widmer, Austria
Ruediger Wirth, Germany
Stefan Wrobel, Germany
Osmar R. Zaiane, Canada
Mohammed Zaki, USA
Carlo Zaniolo, USA
Djamel Zighed, France
Blaž Zupan, Slovenia

ECML/PKDD 2004 Additional Reviewers

Fabio Abbattista
Markus Ackermann
Erick Alphonse
Oronzo Altamura
Massih Amini
Ahmed Amrani
Anastasia Analiti
Nicos Angelopoulos
Fabrizio Angiulli
Luiza Antonie
Annalisa Appice
Josep-Lluis Arcos
Eva Armengol
Thierry Artieres
Maurizio Atzori
Anne Auger
Ilkka Autio
Jérôme Azé
Vincenzo Bacarella
Miriam Baglioni
Yijian Bai
Cristina Baroglio
Teresa Basile
Ganesan Bathumalai
Fadila Bentayeb
Margherita Berardi
Bettina Berendt
Petr Berka
Guillaume Beslon
Philippe Bessières
Matjaz Bevk
Steffen Bickel
Gilles Bisson
Avrim Blum
Axel Blumenstock
Damjan Bojadžiev
Francesco Bonchi
Toufik Boudellal
Omar Boussaid
Janez Brank
Nicolas Bredeche
Ulf Brefeld
Wray Buntine
Christoph Büscher
Benjamin Bustos
Niccolo Capanni
Amedeo Cappelli
Martin R.J. Carpenter
Costantina Caruso
Ciro Castiello
Barbara Catania
Davide Cavagnino
Michelangelo Ceci
Alessio Ceroni
Jesús Cerquides
Eugenio Cesario
Silvia Chiusano
Fang Chu
Antoine Cornuéjols
Fabrizio Costa
Gianni Costa
Tom Croonenborghs
Tomaz Curk
Maria Damiani
Agnieszka Dardzinska
Tijl De Bie
Edwin D. De Jong
Kurt De Grave
Marco Degemmis
Janez Demšar
Damjan Demšar
Michel de Rougemont
Nicola Di Mauro
Christos Dimitrakakis
Simon Dixon
Kurt Driessens
Isabel Drost
Chris Drummond
Wenliang Du
Nicolas Durand
Michael Egmont-Petersen
Craig Eldershaw
Mohammed El-Hajj
Roberto Esposito
Timm Euler
Theodoros Evgeniou
Anna Maria Fanelli
Nicola Fanizzi
Ayman Farahat
Sebastien Ferre
Stefano Ferilli
Daan Fierens
Thomas Finley
Sergio Flesca
François Fleuret
Francesco Folino
Francesco Fornasari
Blaz Fortuna
Andrew Foss
Keith Frikken
Barbara Furletti
Thomas Gärtner
Ugo Galassi
Arianna Gallo
Byron Gao
Paolo Garza
Liqiang Geng
Claudio Gentile
Pierre Geurts
Zoubin Ghahramani
Arnaud Giacometti
Emiliano Giovannetti
Piotr Gmytrasiewicz
Judy Goldsmith
Anna Gomolinska
Udo Grimmer
Matthew Grounds
Antonella Guzzo
Amaury Habrard
Stephan ten Hagen
Jörg Hakenberg
Mark Hall
Greg Hamerly
Ji He
Jaana Heino
Thomas Heitz
Frank Herrmann
Haitham Hindi
Ayca Azgin Hintoglu
Joachim Hipp
Susanne Hoche
Pieter Jan 't Hoen
Andreas Hotho
Tomas Hrycej
Luigi Iannone
Inaki Inza
François Jacquenet
Aleks Jakulin
Jean-Christophe Janodet
Nathalie Japkowicz
Tony Jebara

Tao-Yuan Jen
Tao Jiang
Xing Jiang
Yuelong Jiang
Alípio Jorge
Pierre-Emmanuel Jouve
Matti Kääriäinen
Spiros Kapetanakis
Vangelis Karkaletsis
Andreas Karwath
Branko Kavšek
Steffen Kempe
Kristian Kersting
Jahwan Kim
Minsoo Kim
Svetlana Kiritchenko
Richard Kirkby
Jyrki Kivinen
Willi Kloesgen
Gabriella Kókai
Petri Kontkanen
Dimitrios Kosmopoulos
Mark-A. Krogel
Jussi Kujala
Matjaž Kukar
Kari Laasonen
Krista Lagus
Lotfi Lakhal
Stéphane Lallich
Gert Lanckriet
John Langford
Carsten Lanquillon
Antonietta Lanza
Michele Lapi
Dominique Laurent
Yan-Nei Law
Neil Lawrence
Gregor Leban
Sau Dan Lee
Gaëlle Legrand
Edda Leopold
Claire Leschi
Guichong Li
Oriana Licchelli
Per Lidén
Jussi T. Lindgren
Francesca A. Lisi
Bing Liu
Zhenyu Liu
Peter Ljubič
Marco Locatelli
Huma Lodhi
Ricardo Lopes
Pasquale Lops
Robert Lothian
Claudio Lucchese
Jack Lutz
Tuomo Malinen
Michael Maltrud
Suresh Manandhar
Alain-Pierre Manine
Raphael Marée
Berardi Margherita
Elio Masciari
Cyrille Masson
Nicolas Méger
Carlo Meghini
Corrado Mencar
Amar-Djalil Mezaour
Tatiana Miazhynskaia
Alessio Micheli
Taneli Mielikäinen
Ingo Mierswa
Tommi Mononen
Martin Možina
Thierry Murgue
Mirco Nanni
Phu Chien Nguyen
Tuan Trung Nguyen
Alexandru Niculescu-Mizil
Siegfried Nijssen
Janne Nikkilä
Blaž Novak
Alexandros Ntoulas
William O'Neill
Kouzou Ohara
Arlindo L. Oliveira
Santiago Ontañón
Riccardo Ortale
Martijn van Otterlo
Gerhard Paass
Ignazio Palmisano
Christian Panse
Andrea Passerini
Jaakko Peltonen
Lourdes Pena
Raffaele Perego
José Ramón Quevedo Pérez
Fernando Perez-Cruz
Georgios Petasis
Johann Petrak
Sergios Petridis
Viet Phan-Luong
Dimitris Pierrakos
Joël Plisson
Neoklis Polyzotis
Luboš Popelínský
Roland Priemer
Kai Puolamäki
Sabine Rabaseda
Filip Radlinski
Mika Raento
Jan Ramon
Ari Rantanen
Pierre Renaux
Chiara Renso
Rita Ribeiro
Lothar Richter
Salvatore Rinzivillo
François Rioult
Stefano Rizzi
Céline Robardet
Mathieu Roche
Pedro Rodrigues
Teemu Roos
Benjamin Rosenfeld
Roman Rosipal
Fabrice Rossi
Olga Roudenko
Antonin Rozsypal
Ulrich Rückert
Salvatore Ruggieri
Stefan Rüping
Nicolas Sabouret
Aleksander Sadikov
Taro L. Saito
Lorenza Saitta
Luka Šajn
Apkar Salatian
Marko Salmenkivi
Craig Saunders
Alexandr Savinov
Jelber Sayyad Shirabad
Francesco Scarcello
Christoph Schmitz
Joern Schneidewind
Martin Scholz
Tobias Schreck
Ingo Schwab
Mihaela Scuturici

Vasile-Marian Scuturici
Alexander K. Seewald
Jouni K. Seppänen
Jun Sese
Georgios Sigletos
Marko Robnik-Šikonja
Fabrizio Silvestri
Janne Sinkkonen
Mike Sips
Dominik Slezak
Giovanni Soda
Larisa Soldatova
Arnaud Soulet
Alessandro Sperduti
Jaroslaw Stepaniuk
Olga Stepankova
Umberto Straccia
Alexander L. Strehl
Thomas Strohmann
Jan Struyf
Dorian Šuc
Henri-Maxime Suchier
Johan Suykens
Piotr Synak
Marcin Szczuka
Prasad Tadepalli
Andrea Tagarelli
Julien Tane
Alexandre Termier
Evimaria Terzi
Franck Thollard
Andrea Torsello
Alain Trubuil
Athanasios Tsakonas
Chrisa Tsinaraki
Ville Tuulos
Yannis Tzitzikas
Jaideep S. Vaidya
Pascal Vaillant
Alexandros Valarakos
Anneleen Van Assche
Antonio Varlaro
Guillaume Vauvert
Julien Velcin
Celine Vens
Naval K. Verma
Ricardo Vilalta
Alexei Vinokourov
Daniel Vladusic
Nikos Vlassis
Alessandro Vullo
Bernard Ženko
Martin Žnidaršič
Haixun Wang
Xin Wang
Yizhou Wang
Hannes Wettig
Nirmalie Wiratunga
Jakub Wroblewski
Michael Wurst
Dan Xiao
Tomoyuki Yamada
Robert J. Yan
Hong Yao
Ghim-Eng Yap
Kihoon Yoon
Bianca Zadrozny
Fabio Zambetta
Farida Zehraoui
Bernard Zenko
Xiang Zhang
Alexander Zien
Albrecht Zimmermann

ECML/PKDD 2004 Tutorials

Evaluation in Web Mining
Bettina Berendt, Ernestina Menasalvas, Myra Spiliopoulou

Symbolic Data Analysis
Edwin Diday, Carlos Marcelo

Radial Basis Functions: An Algebraic Approach (with Data Mining Applications)
Amrit L. Goel, Miyoung Shin

Mining Unstructured Data
Ronen Feldman

Statistical Approaches Used in Machine Learning
Bruno Apolloni, Dario Malchiodi

Rule-Based Data Mining Methods for Classification Problems in the Biomedical Domain
Jinyan Li, Limsoon Wong

Distributed Data Mining for Sensor Networks
Hillol Kargupta

ECML/PKDD 2004 Workshops

Statistical Approaches for Web Mining (SAWM)
Marco Gori, Michelangelo Ceci, Mirco Nanni

Symbolic and Spatial Data Analysis: Mining Complex Data Structures
Paula Brito, Monique Noirhomme

Third International Workshop on Knowledge Discovery in Inductive Databases (KDID 2004)
Bart Goethals, Arno Siebes

Data Mining and Adaptive Modelling Methods for Economics and Management (IWAMEM 2004)
Pavel Brazdil, Fernando S. Oliveira, Giulio Bottazzi

Privacy and Security Issues in Data Mining
Yücel Saygin

Knowledge Discovery and Ontologies
Paul Buitelaar, Jürgen Franke, Marko Grobelnik, Gerhard Paaß, Vojtech Svátek

Mining Graphs, Trees and Sequences (MGTS 2004)
Joost Kok, Takashi Washio

Advances in Inductive Rule Learning
Johannes Fürnkranz

Data Mining and Text Mining for Bioinformatics
Tobias Scheffer

Knowledge Discovery in Data Streams
Jesus Aguilar-Ruiz, Joao Gama

Table of Contents

Invited Papers

Contributed Papers

Posters

Demonstration Papers

Random Matrices in Data Analysis

Dimitris Achlioptas

Microsoft Research, Redmond, WA 98052, USA
optas@microsoft.com

Abstract. We show how carefully crafted random matrices can achieve distance-preserving dimensionality reduction, accelerate spectral computations, and reduce the sample complexity of certain kernel methods.

1 Introduction

Given a collection of n data points (vectors) in high-dimensional Euclidean space it is natural to ask whether they can be projected into a lower dimensional Euclidean space without suffering great distortion. Two particularly interesting classes of projections are: i) projections that tend to preserve the interpoint distances, and ii) projections that maximize the average projected vector length.

In the last few years, distance-preserving projections have had great impact in theoretical computer science where they have been useful in a variety of algorithmic settings, such as approximate nearest neighbor search, clustering, learning mixtures of distributions, and computing statistics of streamed data.

The general idea is that by providing a low dimensional representation of the data, distance-preserving embeddings dramatically speed up algorithms whose run-time depends exponentially in the dimension of the working space. At the same time, the provided guarantee regarding pairwise distances often allows one to show that the solution found by working in the low dimensional space is a good approximation to the solution in the original space.

Perhaps the most commonly used projections aim at maximizing the average projected vector length, thus retaining most of the variance in the data. This involves representing the data as a matrix A, diagonalizing $A = UDV$, and projecting A onto subspaces spanned by the vectors in U or V corresponding to the largest entries in D. Variants of this idea are known as Karhunen-Loève transform, Principal Component Analysis, Singular Value Decomposition and others.

In this paper we examine different applications of random matrices to both kinds of projections, all stemming from variations of the following basic fact: if R is an $n \times n$ random matrix whose entries are i.i.d. Normal random variables, $N(0,1)$, then the matrix $\frac{1}{\sqrt{n}} R$ is very close to being orthonormal.

2 Euclidean Distance Preservation

A classic result of Johnson and Lindenstrauss [7] asserts that any set of n points in $\mathbb{R}^d$ can be embedded into $\mathbb{R}^k$, with $k = O(\log n)$, so that all pairwise distances are maintained within an arbitrarily small factor. More precisely,

J.-F. Boulicaut et al. (Eds.): PKDD 2004, LNAI 3202, pp. 1–7, 2004.

Lemma 1 ([7]). *Given $0 < \epsilon \leq 1$ and an integer n, let k be a positive integer such that $k \geq k_0 = (12/\epsilon^2)\log n$. For every set P of n points in $\mathbb{R}^d$ there exists $f : \mathbb{R}^d \to \mathbb{R}^k$ such that for all $u, v \in P$*

$$(1-\epsilon)||u-v||^2 \leq ||f(u)-f(v)||^2 \leq (1+\epsilon)||u-v||^2 \ .$$

Perhaps, a naive attempt to construct an embedding as above would be to pick a random set of k coordinates from the original space. Unfortunately, two points can be very far apart while differing only along one original dimension, dooming this approach. On the other hand, if (somehow) for all pairs of points, all coordinates contributed "roughly equally" to their distance, such a sampling scheme would be very natural. This consideration motives the following idea: first apply a random *rotation* to the n points, and then pick the first k coordinates as the new coordinates. The random rotation can be viewed as a form of insurance against axis alignment, analogous to applying a random permutation before running Quicksort.

Of course, applying a random rotation and then taking the first k coordinates is equivalent to projecting the n points on a uniformly random k-dimensional subspace. Indeed, this is exactly how the original proof of Lemma 1 by Johnson and Lindenstrauss proceeds: to implement the embedding, multiply the $n \times d$ data matrix A with a random $d \times k$ orthonormal matrix. Dasgupta and Gupta [5] and, independently, Indyk and Motwani [6] more recently gave a simpler proof of Lemma 1 by taking the following more relaxed approach towards orthonormality.

The key idea is to consider what happens if we multiply A with a random $d \times k$ matrix R whose entries are independent Normal random variables with mean 0 and variance 1, i.e., $N(0,1)$. It turns out that while we do not explicitly enforce either orthogonality or normality in R, its columns will come very close to having both of these properties. This is because, as d increases: (i) the length of each column-vector concentrates around its expectation as the sum of d independent random variables; (ii) by the spherical symmetry of the Gaussian distribution, each column-vector points in a uniformly random direction in $\mathbb{R}^d$, making the $k \leq d$ independent column-vectors nearly orthogonal with high probability.

More generally, let R be a random matrix whose entries are independent random variables with $\mathbf{E}(r_{ij}) = 0$ and $\text{Var}(r_{ij}) = 1$. If $f : \mathbb{R}^d \to \mathbb{R}^k$ is given by

$$f(x) = \frac{1}{\sqrt{k}}\, x\, R \ ,$$

it is easy to check that for any vector $x \in \mathbb{R}^d$ we have $\mathbf{E}\left(||f(x)||\right) = ||x||$. Effectively, the squared inner product of x with each column of R acts as an independent estimate of $||x||^2$, making $||f(x)||^2$ the consensus estimate (sum) of the k estimators. Seen from this angle, requiring the k vectors to be orthonormal simply maximizes the mutual information of the k estimators. For good dimensionality reduction, we also need to minimize the variance of the estimators.

In [1], it was shown that taking $r_{ij} = \pm 1$ with equal probability, in fact, slightly reduces the number of required dimensions k (as the variance of each column-estimator is slightly smaller). At the same time, and more importantly, this choice of r_{ij} makes f a lot easier to work with in practice.

3 Computing Low Rank Approximations

Given n points in $\mathbb{R}^d$ represented as an $n \times d$ matrix A, one of the most common tasks in data analysis is to find the "top k" singular vectors of A and then project A onto the subspace they span. Such low rank approximations are used widely in areas such as computer vision, information retrieval, and machine learning to extract correlations and remove noise from matrix-structured data.

Recall that the top singular vector of a matrix A is the maximizer of $\|Ax\|_2$ over all unit vectors x. This maximum is known as the L_2 norm of A and the maximizer captures the dominant linear trend in A. Remarkably, this maximizer can be discovered by starting with a random unit vector $x \in \mathbb{R}^d$ and repeating the following "voting process" until it reaches a fixpoint, i.e., until x stops rotating:

- Have each of the n rows in A vote on candidate x, i.e., compute $y = Ax \in \mathbb{R}^n$.
- Compose a new candidate by combining the rows of A, weighing each row by its enthusiasm for x, i.e., update $x \leftarrow \dfrac{A^T y}{\|A^T y\|} \in \mathbb{R}^d$.

The above idea extends to $k > 1$. To find the k-dimensional invariant subspace of A, one starts with a random subspace, i.e., a random $d \times k$ orthonormal matrix, and repeatedly multiplies by $A^T A$ (orthonormalizing after each multiplication). Computing the singular row-vectors of A, i.e., the eigenvectors of $B = A^T A$, is often referred to as Principal Component Analysis (PCA). The following process achieves the exact same goal, by extracting the dominant trends in A sequentially, in order of strength: let A_0 be the all zeros matrix; for $i = 1, \ldots, k$:

- Find the top singular vector, x_i, of $A - A_{i-1}$, via the voting process above.
- Let $A_i = A_{i-1} + Ax_i x_i^T$, i.e., A_i is the optimal rank i approximation to A.

To get an idea of how low rank approximations can remove noise, let G be an $n \times d$ random matrix whose entries are i.i.d. $N(0, \sigma^2)$ random variables. We saw earlier that each column of G points in an independent, uniformly random direction in $\mathbb{R}^n$. As a result, when n is large, with high probability the $d \leq n$ columns of G are nearly orthogonal and there is *no* low-dimensional subspace that simultaneously accommodates many of them. This means that when we compute a low rank approximation of $A + G$, as long as σ is "not too large" (in a sense we will make precise), the columns of G will exert little influence as they do not strongly favor any particular low-dimensional subspace. Assuming that A contains strong linear trends, it is its columns that will command and receive accommodation.

To make this intuition more precise, we first state a general bound on the impact that a matrix N can have on the optimal rank k approximation of a matrix A, denoted by A_k, as a function of $\|N_k\|$. Recall that $\|A\|_F = \sqrt{\sum_{i,j} A_{ij}^2}$.

Lemma 2. *For any matrices A and N, if $\widehat{A} = A + N$ then*

$$\|A - \widehat{A}_k\|_2 \le \|A - A_k\|_2 + 2\|N_k\|_2 \quad \text{and}$$

$$\|A - \widehat{A}_k\|_F \le \|A - A_k\|_F + \|N_k\|_F + 2\sqrt{\|N_k\|_F \|A_k\|_F} \ .$$

Notice that all error terms above scale with $\|N_k\|$. As a result, whenever N is poorly approximated in k dimensions, i.e., $\|N_k\|$ is small, the error caused by adding N to a matrix A is also small.

Let us consider the norms of our Gaussian perturbation matrix.

Fact 1 *Let G be a random $n \times d$ matrix, where $d \le n$, whose entries are i.i.d. random variables $N(0, \sigma^2)$. For any $\epsilon > 0$, with probability $1 - 1/\mathrm{poly}(n, \epsilon)$,*

$$\|G\|_2 = \|G_k\|_2 < (2+\epsilon)\,\sigma\sqrt{n} \ \text{ and } \ \|G_k\|_F < (2+\epsilon)\,\sigma\sqrt{kn} \ .$$

Remarkably, the upper bound above for $\|G\|_2$ is within a factor of 2 of the *lower bound* $\sigma\sqrt{n}$ on the L_2 norm of *any* $n \times d$ matrix with mean squared entry σ^2. In other words, a random Gaussian matrix is nearly as unstructured as possible, resembling white noise in the flatness of its spectrum. On the other hand, $\|A\|_2$ can be as large as $\sigma\sqrt{dn}$ for an $n \times d$ matrix A with mean squared entry σ^2.

This capacity of spectral techniques to remove Gaussian noise is by now very well-understood. We will see that the above geometric explanation of this fact can actually accommodate much more general noise models, e.g. N_{ij} that are not identically distributed and, in fact, whose distribution depends on A_{ij}. In the next section, this generality will enable the notion of "computation-friendly noise", i.e., noise that enhances (rather than hinders) spectral computations.

Fact 1 also suggests a criterion for choosing a good value of k when seeking low rank approximations of a $n \times d$ data matrix A:

$$\|A - A_k\|_2 \sim \sigma\sqrt{n}, \text{ where } \sigma^2 \text{ is the mean squared entry in } A - A_k.$$

In words: we should stop when, after projecting onto the top k singular vectors, we are left with a matrix, $A - A_k$, whose strongest linear trend is comparable to that of a random matrix of similar scale.

3.1 Co-opting the Noise Process

Computing optimal low rank approximations of large matrices often runs against practical computational limits since the algorithms for this task generally require superlinear time and a large working set. On the other hand, in many applications it is perfectly acceptable just to find a rank k matrix C satisfying

$$\|A - C\| \le \|A - A_k\| + \delta \ ,$$

where A_k is the optimal rank k approximation of the input matrix A, and δ captures an appropriate notion of "error tolerance" for the domain at hand.

In [2], it was shown that with the aid of randomization one can exploit such an "error allotment" to aid spectral computations. The main idea is as follows.

Imagine, first, that we squander the error allotment by obliviously adding to A a Gaussian matrix G, as in the previous section. While this is not likely to yield a computationally advantageous matrix, we saw that at least it is rather harmless. The first step in using noise to aid computation is realizing that G is innocuous due precisely to the following three properties of its entries:

independence, zero mean, small variance.

The fact that the G_{ij} are Gaussian is *not* essential: a fundamental result of Füredi and Komlós [4] shows that Fact 1 generalizes to random matrices where the entries can have different, in fact arbitrary, distributions as long as all N_{ij} are zero-mean, independent, and their variance is bounded by σ^2.

To exploit this fact for computational gain, given a matrix A, we will create a distribution of noise matrices N that *depends on* A, yet is such that the random variables N_{ij} still enjoy independence, zero mean, and small variance. In particular, we will be able to choose N so that $\widehat{A} = A + N$ has computationally useful properties, such as sparsity, yet N is sufficiently random for $\|N_k\|$ to be small with high probability.

Example: Set $N_{ij} = \pm A_{ij}$ with equal probability, independently for all i, j.

In this example, the random variables N_{ij} are independent, $\mathbf{E}[N_{ij}] = 0$ for all i, j, and the standard deviation of N_{ij} equals A_{ij}. On the other hand, the matrix $\widehat{A} = A + N$ will have about half as many non-zero entries as A, i.e., it will be about twice as sparse. Therefore, while $\|A\|_2$ can be proportional to $\sqrt{dn}$, the error term $\|N\|_2$, i.e., the price for the sparsification, is only proportional to $\sqrt{n}$.

The rather innocent example above can be greatly generalized. To simplify exposition, in the following, we assume that $A_{ij} \in [-1, +1]$.

- **Quantization:** For all i, j, independently, set $\widehat{A}_{ij}$ to $+1$ with probability $(1 + A_{ij})/2$, and to -1 with probability $(1 - A_{ij})/2$. Clearly, for all i, j, we have $\mathbf{E}[N_{ij}] = \mathbf{E}[\widehat{A}_{ij} - A_{ij}] = 0$, while $\mathrm{Var}(N_{ij}) \le N_{ij}^2 \le 4$.
- **Uniform sampling:** For any desired fraction $p \in (0, 1]$, set $\widehat{A}_{ij} = A_{ij}/p$ with probability p, and 0 otherwise. Now, $\mathrm{Var}(N_{ij}) = A_{ij}^2(1 - p)/p \le 1/p$, so that the error grows only as $1/\sqrt{p}$ as we retain a p-fraction of all entries.
- **Weighted sampling:** For all i, j, independently, set $\widehat{A}_{ij} = A_{ij}/p_{ij}$ with probability p_{ij}, and 0 otherwise, where $p_{ij} = pA_{ij}^2$. This way we retain even fewer small entries, while maintaining $\mathrm{Var}(N_{ij}) = 1/p - A_{ij}^2 \le 1/p$.

Reducing the number of non-zero entries and their representation length causes standard eigenvalue algorithms to work faster. Moreover, the reduced memory footprint of the matrix $\widehat{A}$ enables the handling of larger data sets. At a high level, we perform data reduction by randomly perturbing each data vector so as to simplify its representation, i.e., sparsify and quantize. The point is that the perturbation vectors we use, by virtue of their independence, do not fit in a small subspace, acting effectively as "white noise" that is largely filtered out.

4 Kernel Principal Component Analysis

Given a collection $\mathcal{X}$ of training data $x_1, \ldots, x_n \in \mathbb{R}^d$, techniques such as linear SVMs and PCA extract features from $\mathcal{X}$ by computing linear functions of $\mathcal{X}$. However, often the structure present in the training data is not a linear function of the data representation. Worse, many data sets do not readily support linear operations such as addition and scalar multiplication (text, for example).

In a "kernel method" the idea is to map $\mathcal{X}$ into a space $\mathcal{H}$ equipped with inner product. The dimension of $\mathcal{H}$ can be very large, even infinite, and therefore it may not be practical (or possible) to work with the mapped data explicitly by applying $\Phi : \mathcal{X} \to \mathcal{H}$. Nevertheless, in many interesting cases it is possible to efficiently evaluate the dot products $\langle \Phi(x_i), \Phi(x_j) \rangle$ via a positive definite kernel k for Φ, i.e., a function k so that $k(x_i, x_j) = \langle \Phi(x_i), \Phi(x_j) \rangle$. Algorithms whose operations can be expressed in terms of inner products can thus operate on $\Phi(\mathcal{X})$ implicitly, given only the *Gram* matrix

$$K_{ij} := k(x_i, x_j) \ .$$

Given n training data points, the Kernel PCA (KPCA) method [8] begins by forming the Gram matrix K above and computing the ℓ largest eigenvalues, $\lambda_1, \ldots, \lambda_\ell$, and corresponding eigenvectors, $e_1, \ldots, e_\ell$ of K, for some appropriate choice of $\ell \leq n$. Then, given an input point x, the method computes the value of the ℓ nonlinear feature extractors, corresponding to the inner product of the vector $k(x) = (k(x, x_1), k(x, x_2), \ldots, k(x, x_n))$ with each of the eigenvectors. These feature-values can be used for clustering, classification etc.

While Kernel PCA is very powerful the matrix K, in general, is dense making the input size scale as n^2, where n is the number of training points. As kernel functions become increasingly more sophisticated, e.g. invoking dynamic programming to evaluate the similarity $k(x_i, x_j)$ of two strings x_i, x_j, just the cost of $\Theta(n^2)$ kernel evaluations to construct K rapidly becomes prohibitive.

The uniform sparsification and quantization techniques of the previous section are ideally suited for speeding up KPCA. In particular, "sparsification" here means that we actually only construct a matrix $\widehat{K}$ by computing $k(x_i, x_j)$ for a uniformly random subset of all input pairs x_i, x_j and filling in 0 for the remaining pairs. In [3], it was proven that as long as K has strong linear structure (which is what justifies KPCA in the first place), with high probability, the invariant subspaces of $\widehat{K}$ will be very close to those of K.

Also, akin to quantization, we can replace each exact evaluation of $k(x_i, x_j)$ with a more easily computable unbiased estimate for it. In [3], it was shown that for kernels where: i) $\mathcal{X} \subseteq \mathbb{R}^d$, and, ii) $k(x_i, x_j)$ depends only on $\|x_i - x_j\|$ and/or $x_i \cdot x_j$, one can use random projections, as described in Section 2 for this purpose. Note that this covers some of the most popular kernels, e.g., radial basis functions (RBF) and polynomial kernels.

5 Future Work

Geometric and spectral properties of random matrices with zero-mean, independent entries are the key ingredients in all three examples we considered [1–3]. More general ensembles of random matrices hold great promise for algorithm design and call for a random matrix theory motivated from a computational perspective. Two natural directions are the investigation of matrices with limited independence, and the development of concentration inequalities for non-linear functionals of random matrices.

We saw that sampling and quantizing matrices can be viewed as injecting "noise" into them to endow useful properties such as sparsity and succinctness. The distinguishing feature of this viewpoint is that the effect of randomization is established without an explicit analysis of the interaction between randomness and computation. Instead, matrix norms act as an interface between the two domains: (i) matrix perturbation theory asserts that matrices of small spectral norm cannot have a large effect in eigencomputations, while (ii) random matrix theory asserts that matrices of zero-mean, independent random variables with small variance have small spectral norm. Is it possible to extend this style of analysis to other machine-learning settings, e.g. Support Vector Machines?

Acknowledgments

Many thanks to Robert Kleinberg, Heikki Mannila, and Frank McSherry for reading earlier drafts and providing helpful suggestions.

References

1. Dimitris Achlioptas, *Database-friendly random projections: Johnson-Lindenstrauss with binary coins*, JCSS **66** (2003), no. 4, 671–687.
2. Dimitris Achlioptas and Frank McSherry, *Fast computation of low rank matrix approximations*, JACM, to appear.
3. Dimitris Achlioptas, Frank McSherry and Bernhard Schölkopf, *Sampling techniques for kernel methods*, NIPS 2002, pp. 335–342.
4. Zoltán Füredi and János Komlós, *The eigenvalues of random symmetric matrices*, Combinatorica **1** (1981), no. 3, 233–241.
5. Sanjoy Dasgupta and Anupam Gupta, *An elementary proof of the Johnson-Lindenstrauss lemma*, Technical report 99-006, UC Berkeley, March 1999.
6. Piotr Indyk and Rajeev Motwani, *Approximate nearest neighbors: towards removing the curse of dimensionality*, STOC 1998, pp. 604–613.
7. William B. Johnson and Joram Lindenstrauss, *Extensions of Lipschitz mappings into a Hilbert space*, Amer. Math. Soc., Providence, R.I., 1984, pp. 189–206.
8. Bernhard Schölkopf, Alex J. Smola and Klaus-Robert Müller, *Nonlinear component analysis as a kernel Eigenvalue problem*, Neural Computation **10** (1998), no. 5, 1299–1319.

Data Privacy

Rakesh Agrawal

IBM Almaden Research Center, San Jose, CA 95120, USA

There is increasing need to build information systems that protect the privacy and ownership of data without impeding the flow of information. We will present some of our current work to demonstrate the technical feasibility of building such systems:

Privacy-preserving data mining. The conventional wisdom held that data mining and privacy were adversaries, and the use of data mining must be restricted to protect privacy. Privacy-preserving data mining cleverly finesses this conflict by exploting the difference between the level where we care about privacy, i.e., individual data, and the level where we run data mining algorithms, i.e., aggregated data. User data is randomized such that it is impossible to recover anything meaningful at the individual level, while still allowing the data mining algorithms to recover aggregate information, build mining models, and provide actionable insights.

Hippocratic databases. Unlike the current systems, Hippocratic databases include responsibility for the privacy of data they manage as a founding tenet. Their core capabilities have been distilled from the principles behind current privacy legislations and guidelines. We identify the technical challenges and problems in designing Hippocratic databases, and also outline some solutions.

Sovereign information sharing. Current information integration approaches are based on the assumption that the data in each database can be revealed completely to the other databases. Trends such as end-to-end integration, outsourcing, and security are creating the need for integrating information across autonomous entities. In such cases, the enterprises do not wish to completely reveal their data. In fact, they would like to reveal minimal information apart from the answer to the query. We have formalized the problem, identified key operations, and designed algorithms for these operations, thereby enabling a new class of applications, including information exchange between security agencies, intellectual property licensing, crime prevention, and medical research.

References

1. R. Agrawal, R. Srikant: Privacy Preserving Data Mining. ACM Int'l Conf. on Management of Data (SIGMOD), Dallas, Texas, May 2000.
2. R. Agrawal, J. Kiernan, R. Srikant, Y. Xu: Hippocratic Databases. 28th Int'l Conf. on Very Large Data Bases (VLDB), Hong Kong, August 2002.
3. R. Agrawal, A. Evfimievski, R. Srikant: Information Sharing Across Private Databases. ACM Int'l Conf. on Management of Data (SIGMOD), San Diego, California, June 2003.

J.-F. Boulicaut et al. (Eds.): PKDD 2004, LNAI 3202, p. 8, 2004.

Breaking Through the Syntax Barrier: Searching with Entities and Relations

Soumen Chakrabarti

IIT Bombay
soumen@cse.iitb.ac.in

Abstract. The next wave in search technology will be driven by the identification, extraction, and exploitation of real-world entities represented in unstructured textual sources. Search systems will either let users express information needs naturally and analyze them more intelligently, or allow simple enhancements that add more user control on the search process. The data model will exploit graph structure where available, but not impose structure by fiat. First generation Web search, which uses graph information at the macroscopic level of inter-page hyperlinks, will be enhanced to use fine-grained graph models involving page regions, tables, sentences, phrases, and real-world-entities. New algorithms will combine probabilistic evidence from diverse features to produce responses that are not URLs or pages, but entities and their relationships, or explanations of how multiple entities are related.

1 Toward More Expressive Search

Search systems for unstructured textual data have improved enormously since the days of boolean queries over title and abstract catalogs in libraries. Web search engines index much of the full text from billions of Web pages and serve hundreds of millions of users per day. They use rich features extracted from the graph structure and markups in hypertext corpora.

Despite these advances, even the most popular search engines make us feel that we are searching with mere strings: we do not find direct expression of the entities involved in our information need, leave alone relations that must hold between those entities in a proper response. In a plenary talk at the 2004 World-wide Web Conference, Udi Manber commented:

> If music had been invented ten years ago along with the Web, we would all be playing one-string instruments (and not making great music).

referring to the one-line text boxes in which users type in 1–2 keywords and expect perfect gratification with the responses.

Apart from classical Information Retrieval (IR), several communities are coming together in the quest of expressive search, but they are coming from very different origins.

Databases and XML: To be sure, the large gap between the user's information need and the expressed query is well-known. The database community has been traditionally uncomfortable with the imprecise nature of queries inherent in IR.

J.-F. Boulicaut et al. (Eds.): PKDD 2004, LNAI 3202, pp. 9–16, 2004.

The preference for precise semantics has persisted from SQL to XQuery (the query language proposed for XML data). The rigor, while useful for system-building, has little appeal for the end-user, who will not type SQL, leave alone XQuery.

Two communities are situated somewhere between "uninterpreted" keyword search systems and the rigor of database query engines. Various sub-communities of natural language processing (NLP) researchers are concerned with NL interfaces to query systems. The other community, which has broad overlaps with the NLP community, deals with information extraction (IE).

NLP: Classical NLP is concerned with annotating grammatical natural language with parts of speech (POS), chunking phrases and clauses, disambiguating polysemous words, extracting a syntactic parse, resolving pronoun and other references, analyze roles (eating with a spoon vs. with a friend), prepare a complete computer-usable representation of the knowledge embedded in the original text, and perform automatic inference with this knowledge representation. Outside controlled domains, most of these, especially the latter ones, are very ambitious goals. Over the last decade, NLP research has gradually moved toward building robust tools for the simpler tasks [19].

IE: Relatively simple NLP tasks, such as POS tagging, named entity tagging, and word sense disambiguation (WSD) share many techniques from machine learning and data mining. Many such tasks model unstructured text as a sequence of tokens generated from a finite state machine, and solve the reverse problem: given the output token sequence, estimate the state sequence. E.g., if we are interested in extracting dates from text, we can have a positive and a negative state, and identify the text spans generated by the positive state. IE is commonly set up as a supervised learning problem, which requires training text with labeled spans.

Obviously, to improve the search experience, we need that

- Users express their information need in some more detail, while minimizing additional cognitive burden
- The system makes intelligent use of said detail, thus rewarding the burden the user agrees to undertake

This new contract will work only if the combination of social engineering and technological advances work efficiently in concert.

2 The New Contract: Query Syntax

Suitable user interfaces, social engineering, and reward must urge the user to express their information need in some more detail. Relevance feedback, offering query refinements, and encouraging the user to drill down into response clusters are some ways in which systems collect additional information about the user's information need. But there are many situations where direct input from the user can be useful. I will discuss two kinds of query augmentation.

Fragments of types: If the token *2000* appears in grammatical text, current technology can usually disambiguate between the year and some other number, say a money amount. There is no reason why search interfaces cannot accept a query with a type hint so as to avoid spurious matches. There is also no reason a user cannot look for persons related to SVMs using the query `PersonType NEAR "SVM"`, where `PersonType` is the anticipated response type and *SVM* a word to match. To look for a book in SVMs published around year 2000, one might type `BookType (NEAR "SVM" year~2000)`. I believe that the person composing the query, being the stakeholder in response quality, can be encouraged to provide such elementary additional information, provided the reward is quickly tangible. Moreover, reasonably deep processing power can be spent on the query, and this may even be delegated to the client computer.

Attributes, roles and relations: Beyond annotating query tokens with type information, the user may want to express that they are looking for "a business that repairs iMacs," "the transfer bandwidth of USB2.0," and "papers written in 1985 by C. Mohan." It should be possible to express broad relations between entities in the query, possibly the placeholder entity that must be instantiated into the answer. The user may constrain the placeholder entity using attributes (e.g. MacOS-compliant software), roles and relations (e.g., a student advised by X). The challenge will be to support an ever-widening set of attribute types, roles and relations while ensuring ongoing isolation and compatibility between knowledge bases, features, and algorithms.

Compared to query syntax and preprocessing, whose success depends largely on human factors, we have more to say about executing the internal form of the query on a preprocessed corpus.

3 The New Contract: Corpus and Query Processing

While modest changes may be possible in users' query behavior, there is far too much inertia to expect content creators to actively assist mediation in the immediate future. Besides, questions preprocessing can be distributed economically, but corpus processing usually cannot.

The situation calls for relatively light processing of the corpus, at least until query time. During large scale use, however, a sizable fraction of the corpus may undergo complex processing. It would be desirable but possibly challenging to cache the intermediate results in a way that can be reused efficiently.

3.1 Supervised Entity Extraction

Information extraction (IE), also called named entity tagging, annotates spans of unstructured text with markers for instances of specified types, such as people, organizations, places, dates, and quantities.

A popular framework [11] models the text as a linear sequence of tokens being generated from a Markov state machine. A parametric model for state transition and symbol emission is learned from labeled training data. Then the model is evaluated on test data, and spans of tokens likely to be generated by desired states are picked off as extracted entities.

Generative models such as hidden Markov models (HMMs) have been used for IE for a while [7]. If $\mathbf{s}$ is the (unknown) sequence of states and $\mathbf{x}$ the sequence of output features, HMMs seek to optimize the joint likelihood $\Pr(\mathbf{s}, \mathbf{x})$.

In general, $\mathbf{x}$ is a sequence of feature vectors. Apart from the tokens themselves, some derived features found beneficial in IE are of the form: Does the token

- Contain a digit, or digits and commas?
- Contain patterns like DD:DD or DDDD or DD's where D is a digit?
- Follow a preposition?
- Look like a proper noun (as flagged by a part-of-speech tagger[1])?
- Start with an uppercase letter?
- Start with an uppercase letter and continue with lowercase letters?
- Look like an abbreviation (e.g., uppercase letters alternating with periods)?

The large dimensionality of the feature vectors usually corners us into naive independence assumptions about $\Pr(\mathbf{s}, \mathbf{x})$, and the large redundancy across features then lead to poor estimates of the joint distribution.

Recent advances in modeling conditional distributions [18] directly optimize $\Pr(\mathbf{s}|\mathbf{x})$, allowing the use of many redundant features without attempting to model the distribution over $\mathbf{x}$ itself.

3.2 Linkage Analysis and Alias Resolution

After the IE step, spans of characters and tokens are marked with type identifiers. However, many string spans (called *aliases*) may refer to a single entity (e.g., *IBM*, *International Business Machines*, *Big Blue*, *the computer giant* or www.ibm.com). The variations may be based on abbreviations, pronouns, anaphora, hyperlinks and other creative ways to create shared references to entities. Some of these aliases are syntactically similar to each other but others are not.

In general, detecting aliases from unstructured text, also called *coreferent resolution*, in a complete and correct manner is considered "NLP complete," i.e., requires deep language understanding and vast amounts of world knowledge. Alias resolution is an active and difficult area of NLP research. In the IE community, more tangible success has been achieved within the relatively limited scope of **record linkage**.

In record linkage, the first IE step results in structured tables of entities, each having attributes and relations to other entities. E.g., we may apply IE techniques to bibliographies at the end of research papers to populate a table of papers, authors, conferences/journals, etc. Multiple rows in each table may refer to the same object. Similar problems may arise in Web search involving names of people, products, and organizations.

The goal of record linkage is to partition rows in each table into equivalence classes, all rows in a class being references to one real-world entity. Obviously, knowing that two different rows in the author table refer to the same person (e.g., one may abbreviate the first name) may help us infer that two rows in the paper table refer to the same real-world paper.

A veriety of new techniques are being brought to bear on record linkage [10] and coreferent resolution [20], and this is an exciting area of current research.

[1] Many modern part-of-speech taggers are in turn driven by state transition models.

3.3 Bootstrapping Ontologies from the Web

The set of entity types of interest to a search system keeps growing and changing. A fixed set of types and entities may not keep up. The system may need to actively explore the corpus to propose new types and extract entities for old and new types. Eventually, we would like the system to learn how to learn.

Suppose we want to discover instances of some type of entity (city, say) on the Web. We can exploit the massive redundancy of the Web and use some very simple patterns [16,8,1,13]:

"cities" {","} "such as" NPList2
NP1 {","} "and other cities"
"cities" {","} "including" NPList2
NP1 "is a city"

Here { } denotes an optional pattern and NP is a noun phrase. These patterns are fired off as queries to a number of search engines. A set of *rules* test the response Web pages for the existence of valid instantiations of the patterns. A rule may look like this:

NP1 "such as" NPList2 AND
head(NP1)="cities" AND
properNoun(head(each(NPList2)))
$\Rightarrow$ instanceOf(City,head(each(NPList2)))

KnowItAll [13] makes a probabilistic assessment of the quality of the extraction by collecting co-occurrence statistics on the Web of terms carefully chosen from the extracted candidates and pre-defined *discriminator phrases.* E.g., if X is a candidate actor, "X starred in" or "starring X" would be good discriminator phrases. KnowItAll uses the *pointwise mutual information* (PMI) formulation by Turney [24] to measure the association between the candidate instance I and the discriminator phrase D: $\text{PMI}(I, D) = |\text{Hits}(D + I)|/|\text{Hits}(I)|$.

Apart from finding instances of types, it is possible to discover subtypes. E.g., if we wish to find instances of *scientists*, and we have a seed set of instances, we can discover that physicists and biologists are scientists, make up new patterns from the old ones (e.g. "scientist X" to "physicist X") and improve our harvest of new instances.

In Sections 3.5 and 3.6 we will see how automatic extraction of ontologies can assist next-generation search.

3.4 Searching Relational Data with NL Queries

In this section and the next (§3.5), we will assume that information extraction and alias analysis have led to a reasonably clean entity-relationship (ER) graph. The graphs formed by nodes corresponding to authors, papers, conferences and journal in DBLP, and actors/actresses, movies, awards, genres, ratings, producers and music directors in the Internet Movie Database (IMDB) are examples of reasonably clean entity-relationship data graphs. Other real-life examples involve e-commerce product catalogs and personal information management data, with organizations, people, locations, emails, papers, projects, seminars, etc.

There is a long history of systems that give a natural language interface (NLI) to relational engines [4], but, as in general NLP research, recent work has moved from highly engineered solutions to arbitrarily complex problems to less knowledge-intensive and more robust solutions for limited domains [21]. E.g., for a table `JOB(description,platform,company)` and the NL query *What are the HP jobs on a UNIX system?*, the translation to SQL might be `select distinct description from JOB WHERE company = 'HP' and platform = 'UNIX'`. The main challenge is to agree on a perimeter of NL questions within which an algorithm is required to find a correct translation, and to reliably detect when this is not possible.

3.5 Searching Entity-Relationship Graphs

NLI systems take advantage of the precise schema information available with the "corpus" as well the well-formed nature of the query, even if it is framed in uncontrolled natural language. The output of IE systems has less elaborate type information, the relations are shallower, and the questions are most often a small set of keywords, from users who are used to Web search and do not wish to learn about any schema information in framing their queries.

Free-form keyword search in ER graphs raises many interesting issues, including the query language, the definition of a "document" in response to a query, how to score a document which may be distributed in the graph, and how to search for these subgraphs efficiently.

Multiple words in a query may not all match within a single row in a single table, because ER graphs are typically highly normalized using foreign key constraints. In an ER version of DBLP, paper titles and author names are in different tables, connected by a relation `wrote(author,paper)`. In such cases, what is the appropriate unit of response? Recent systems [6,3,17] adopt the view that the response should be some minimal graph that connects at least one node containing each query keyword.

Apart from type-free keyword queries, one may look for a single node of a specified type (say, a paper) with high proximity to nodes satisfying various predicates, e.g., keyword match ("indexing", "SIGIR") or conditions on numeric fields (`year<1995`). Resetting random walks [5] are a simple way to answer such queries. These techniques are broadly similar to Pagerank [9], except that the random surfer teleports only to nodes that satisfy the predicates. Biased random walks with restarts are also related to effective conductance in resistive networks. In a large ER graph, it is also nontrivial to *explain* to the user why/how entities are related; this is important for diagnostics and eliciting user confidence. Conductance-based approaches work well [14]: we can connect +1 V to one node, ground the other, penalize high-fanout nodes using a grounded sink connected to every node, and report subgraphs that conduct the largest current out of the source node.

Recent years have seen an explosion of analysis and search systems for ER graphs, and I expect the important issues regarding meaningfulness of results and system scalability to be resolved in the next few years.

3.6 Open-Domain Question Answering

Finally, the Web at large will continue to be an "open-domain" system where comprehensive and accurate entity and relation extraction will remain elusive. No schema of entities and relationships can be complete at any time, even if they become more comprehensive over time. Moreover, even a cooperative user will not be able to remember and exploit a universal "type system" in asking questions. Instead, search systems will provide some basic set of *roles* [15] that apply broadly. Questions will express roles or refinements of roles, and will be matched to probabilistic role annotations in the corpus.

In open-domain QA, question analysis and response scoring will necessarily be far more tentative. Some basic machine learning will reveal that the question *When was television invented?* expects the type of the answer (atype) to be a *date*, and that the answer is almost certainly only a few tokens from the word *television* or its synonym. In effect, current technology [22,2,12,23] can translate questions into the general form

```
find x from corpus where x InstanceOf(Atype(question))
  and x RelatedTo GroundConstants(question)
```

Here `Atype(question)` represents the concept of time, and we are looking for a reference to an entity `x` which is an instance of time. (This is where a system like KnowItAll comes into play.) In the example above, *television* or *TV* would be in `GroundConstants(question)`.

Checking the predicate `RelatedTo` is next to impossible in general. QA systems employ a variety of approximations. These may be as crude as linear proximity (the number of of tokens separating `x` from `GroundConstants(question)`. Linear proximity is already surprisingly effective [23]. More sophisticated systems[2] attempt a parse of the question and the passage, and verify that `x` and `GroundConstants(question)` are related in a way specified by (a parse of) the question. As might be expected, there is a trade-off beteen speed and robustness on one hand and accuracy and brittleness on the other.

4 Conclusion

Many of the pieces required for better searching are coming together. Current an upcoming research will introduce synergy as well as build large, robust applications. The applications will need to embrace bootstrapping and life-long learning better than before. The architecture must isolate feature extraction, models, and algorithms for estimation and inferencing. The interplay between processing stages makes this goal very challenging. The applications must be able to share models and parameters across different tasks and across time.

References

1. E. Agichtein and L. Gravano. Snowball: Extracting relations from large plain-text collections. In *International Conference on Digital Libraries (DL)*, volume 5. ACM, 2000.
2. E. Agichtein, S. Lawrence, and L. Gravano. Learning search engine specific query transformations for question answering. In *WWW Conference*, pages 169–178, 2001.
3. S. Agrawal, S. Chaudhuri, and G. Das. DBXplorer: A system for keyword-based search over relational databases. In *ICDE*, San Jose, CA, 2002. IEEE.

[2] Visit, e.g., http://www.languagecomputer.com/

4. I. Androutsopoulos, G. D. Ritchie, and P. Thanisch. Natural language interfaces to databases–an introduction. *Journal of Language Engineering*, 1(1):29–81, 1995.
5. A. Balmin, V. Hristidis, and Y. Papakonstantinou. Authority-based keyword queries in databases using ObjectRank. In *VLDB*, Toronto, 2004.
6. G. Bhalotia, A. Hulgeri, C. Nakhe, S. Chakrabarti, and S. Sudarshan. Keyword searching and browsing in databases using BANKS. In *ICDE*, San Jose, CA, 2002. IEEE.
7. D. M. Bikel, R. L. Schwartz, and R. M. Weischedel. An algorithm that learns what's in a name. *Machine Learning*, 34(1–3):211–231, 1999.
8. S. Brin. Extracting patterns and relations from the World Wide Web. In P. Atzeni, A. O. Mendelzon, and G. Mecca, editors, *WebDB Workshop*, volume 1590 of *LNCS*, pages 172–183, Valencia, Spain, Mar. 1998. Springer.
9. S. Brin and L. Page. The anatomy of a large-scale hypertextual web search engine. In *Proceedings of the 7th World-Wide Web Conference (WWW7)*, 1998.
10. W. Cohen and J. Richman. Learning to match and cluster entity names. In *SIGKDD*, volume 8, 2002.
11. T. G. Dietterich. Machine learning for sequential data: A review. In T. Caelli, editor, *Structural, Syntactic, and Statistical Pattern Recognition*, volume 2396 of *Lecture Notes in Computer Science*, pages 15–30. Springer-Verlag, 2002.
12. S. Dumais, M. Banko, E. Brill, J. Lin, and A. Ng. Web question answering: Is more always better? In *SIGIR*, pages 291–298, 2002.
13. O. Etzioni, M. Cafarella, D. Downey, S. Kok, A.-M. Popescu, T. Shaked, S. Soderland, D. S. Weld, and A. Yates. Web-scale information extraction in KnowItAll. In *WWW Conference*, New York, 2004. ACM.
14. C. Faloutsos, K. S. McCurley, and A. Tomkins. Connection subgraphs in social networks. In *Workshop on Link Analysis, Counterterrorism, and Privacy*, SIAM International Conference on Data Mining, 2004.
15. D. Gildea and D. Jurafsky. Automatic labeling of semantic roles. *Computational Linguistics*, 28(3):245–288, 2002.
16. M. Hearst. Automatic acquisition of hyponyms from large text corpora. In *International Conference on Computational Linguistics*, volume 14, pages 539–545, 1992.
17. V. Hristidis, L. Gravano, and Y. Papakonstantinou. Efficient IR-style keyword search over relational databases. In *VLDB*, pages 850–861, 2003.
18. J. Lafferty, A. McCallum, and F. Pereira. Conditional random fields: Probabilistic models for segmenting and labeling sequence data. In *ICML*, 2001.
19. R. J. Mooney. Learning semantic parsers: An important but under-studied problem. In *AAAI Spring Symposium on Language Learning: An Interdisciplinary Perspective*, pages 39–44, Mar. 2004.
20. V. Ng and C. Cardie. Improving machine learning approaches to coreference resolution. In *ACL*, volume 40, 2002.
21. A. Popescu, O. Etzioni, and H. Kautz. Towards a theory of natural language interfaces to databases. In *Intelligent User Interfaces*, pages 149–157, Miami, 2003. ACM.
22. J. Prager, E. Brown, A. Coden, and D. Radev. Question-answering by predictive annotation. In *SIGIR*, pages 184–191. ACM, 2000.
23. G. Ramakrishnan, S. Chakrabarti, D. A. Paranjpe, and P. Bhattacharyya. Is question answering an acquired skill? In *WWW Conference*, pages 111–120, New York, 2004.
24. P. D. Turney. Mining the Web for synonyms: PMI-IR versus LSA on TOEFL. In *ECML*, 2001.

Real-World Learning with Markov Logic Networks

Pedro Domingos

Department of Computer Science and Engineering
University of Washington
Seattle, WA 98195, USA
pedrod@cs.washington.edu
http://www.cs.washington.edu/homes/pedrod

Machine learning and data mining systems have achieved many impressive successes, but to become truly widespread they must be able to work with less help from people. This requires automating the data cleaning and integration process, handling multiple types of objects and relations at once, and easily incorporating domain knowledge. In this talk, I describe how we are pursuing these aims using Markov logic networks, a representation that combines first-order logic and probabilistic graphical models. Data from multiple sources is integrated by automatically learning mappings between the objects and terms in them. Rich relational structure is learned using a combination of ILP and statistical techniques. Knowledge is incorporated by viewing logic statements as soft constraints on the models to be learned. Application to a real-world university domain shows our approach to be accurate, efficient, and less labor-intensive than traditional ones.

This work, joint with Parag and Matthew Richardson, is described in further detail in Richardson and Domingos [1], Richardson and Domingos [2], and Parag and Domingos [3].

References

1. Richardson, M., & Domingos, P.: *Markov Logic Networks.* Technical Report, Department of Computer Science and Engineering, University of Washington, Seattle, Washington, U.S.A. (2004). http://www.cs.washington.edu/homes/pedrod/mln.pdf.
2. Richardson, M., & Domingos, P.: Markov logic: A unifying framework for statistical relational learning. In *Proceedings of the ICML-2004 Workshop on Statistical Relational Learning and its Connections to Other Fields*, Banff, Alberta, Canada (2004). http://www.cs.washington.edu/homes/pedrod/mus.pdf.
3. Parag, & Domingos, P.: Multi-relational record linkage. In *Proceedings of the KDD-2004 Workshop on Multi-Relational Data Mining*, Seattle, Washington, U.S.A. (2004). http://www.cs.washington.edu/homes/pedrod/mrrl.pdf

J.-F. Boulicaut et al. (Eds.): PKDD 2004, LNAI 3202, p. 17, 2004.

Strength in Diversity: The Advance of Data Analysis

David J. Hand

Department of Mathematics, Imperial College, 180 Queen's Gate,
London SW7 2AZ, UK
d.j.hand@imperial.ac.uk

Abstract. The scientific analysis of data is only around a century old. For most of that century, data analysis was the realm of only one discipline - statistics. As a consequence of the development of the computer, things have changed dramatically and now there are several such disciplines, including machine learning, pattern recognition, and data mining. This paper looks at some of the similarities and some of the differences between these disciplines, noting where they intersect and, perhaps of more interest, where they do not. Particular issues examined include the nature of the data with which they are concerned, the role of mathematics, differences in the objectives, how the different areas of application have led to different aims, and how the different disciplines have led sometimes to the same analytic tools being developed, but also sometimes to different tools being developed. Some conjectures about likely future developments are given.

1 Introduction

This paper gives a personal view of the state of affairs in data analysis. That means that inevitably I will be making general statements, so that most of you will be able to disagree on some details. But I am trying to paint a broad picture, and I hope that you will agree with the overall picture.

We live in very exciting times. In fact, from the perspective of a professional data analyst, I would say we live in the *most* exciting of times. Not so long ago, analysing data was characterised by drudgery, by manual arithmetic, and the need to take great care over numerical trivia. Nowadays, all that has been swept aside, with the burden of tedium having been taken over by the computer. What we are left with are the high-level interpretations and strategic decisions; we look at the summary values derived by the computers and make our statements and draw conclusions and base our actions on these. It is clear from this that the computer has become *the* essential tool for data analysis.

But there is more. The computer has not merely swept aside the tedium. The awesome speed of numerical manipulation has permitted the development of entirely new kinds of data analytic tools, being applied in entirely new ways, to entirely new kinds, and indeed sizes, of data sets. The computer has given us new ways to look at things. The old image, that data analysis was the realm of the boring obsessive, is now so diametrically opposite to the new truth as to be laughable.

J.-F. Boulicaut et al. (Eds.): PKDD 2004, LNAI 3202, pp. 18–26, 2004.

This paper describes some of the history, some of the tools, and something of how I see the present status of data analysis. So perhaps I should begin with a definition. *Data analysis is the science of discovery in data, and of processing data to extract evidence so that one can make properly informed decisions.* In brief, data analysis is *applied philosophy of science*: the theory and methods, not of any particular scientific discipline itself, but of *how to find things out.*

2 The Evolution of Data Analytic Disciplines

The origins of data analysis can be traced back as far back as one likes. Think of Kepler and Gauss analysing astronomical data, of Florence Nightingale using plots to demonstrate that soldiers were dying because of poor hygiene rather than military action, of Quetelet's development of 'social mechanics', and the fact that world's oldest statistical society, the Royal Statistical Society, was established in 1834. But these 'origins' really only represent the initial stirrings: it wasn't until the start of the 20th century that a proper scientific discipline of data analysis really began to be formed. That discipline was statistics, and for the first half of the 20th century statistics was the only data analytic game in town. Until around 1950, statistics *was* the science of data analysis. (You will have to permit me some poetic leeway in my choice of dates: 1960 might be more realistic.)

Then, around the middle of the 20th century, the computer arrived and a revolution began. Statistics began to change rapidly in response to the awesome possibilities the computer provided. There is no doubt that, had statistics been born now, at the start of the 21st century, rather than 100 years ago at the start of the 20th, it would be a very different kind of animal. (Would we have the *t*-test?.) Moreover, although statistics was the intellectual owner of data *analysis* up until about 1950, it was never the intellectual owner of *data* per se, and in the following decades other changes occurred which were to challenge the position assumed by statistics. In particular, another discipline grew up, whose primary responsibility was, initially, the storage and manipulation of data. From data manipulation to data analysis was then hardly a large step. Statistics was no longer the only player.

Nowadays, of course, computer science has grown into a vast edifice, and different subdisciplines of it have developed as specialised areas of data analysis, all overlapping with each other and overlapping with their intellectual parent, statistics. These subdisciplines include machine learning, pattern recognition, and data mining, and one could arguably include image processing, neural networks, and perhaps even computational learning theory and other areas also. I cannot avoid remarking that Charles Babbage, typically regarded as one of the fathers of computing with his *analytical engine*, would have been fascinated by these developments: he was also one of the founders of the Royal Statistical Society. Of course, these various data analytic disciplines are not carbon copies of each other. They have subtly different aims and emphases, and often deal with rather different kinds of data (e.g. in terms of data set size, correlations, complexities, etc.). One of my aims in this talk is to examine some of these differences. Moreover, if the computer has been the strongest influence leading to the development of new data analytic technologies, application areas have always been and continue to have a similar effect. Thus we have psychometrics, bioinformatics, chemometrics, technometrics, and other areas, all addressing the same sorts of problems, but in different areas. I shall say more about this below.

3 Data

I toyed briefly with the idea of calling this talk 'analysing tomorrow's data' since one of the striking things about the modern world of data analysis is that the data with which we now have to deal could not have been imagined 100 years ago. Then the data had to be painstakingly collected by hand since there was no alternative, but nowadays much data acquisition is automatic. This has various consequences.

Firstly, astronomically vast data sets are readily acquired. Books on data mining (e.g. [2],[3]), which is that particular data analytic discipline especially concerned with analysing large data sets, illustrate the sorts of sizes which are now being encountered. The word *terabyte* is no longer unusual. When I was taught statistical data analysis, I was taught that first one must familiarise oneself with one's data: plot it this way and that, look for outliers and anomalies, fit simple models and examine diagnostics. With a billion data points (one of the banking data sets I was presented with) this is clearly infeasible. Other problems involve huge numbers of variables, and perhaps relatively few data points, posing complex theoretical as well as practical questions: bioinformatics, genomics, and proteomics are important sources of such problems.

Secondly, one might have thought that automatic data acquisition would mean better data quality, since there would be no subjective human intervention. Unfortunately, this has not turned out to be the case. New ways of collecting data has meant new ways for the data collection process to go wrong. Worse, large data sets can make it more difficult to detect many of the data anomalies.

Data can be of low quality in many ways: individual values may be distorted or absent, entire records may be missing, measurement error may be large, and so on. As discussed below, much of statistics is concerned with *inference* - with making statements about objects or values not seen or measured, on the basis of those which have been. Thus we might want to make statements about other objects from a population, or about the future behaviour of objects. Accurate inferences can only be made if one has accurate information on how the data were collected. Statisticians have therefore predicated their analyses on the assumption that the available observations were drawn in well-specified ways, or that the departures from these ways were understood and could be modelled. Unfortunately, with many of today's data sets, such assumptions often cannot be made. This has sometimes made statisticians (quite properly) wary of analysing such data. But the data still have to be analysed: the questions still need answers. This is one reason why data mining has been so successful, at least at first glance. Data miners have been prepared to examine distorted data, and to attempt to draw conclusions about it. It has to be said, however, that often that willingness has arisen from a position of ignorance, rather than one of awareness of the risks that were being taken. Relatively few reports of the conclusions extracted from a data mining exercise, for example, qualify those conclusions with a discussion of the possible impact of selectivity bias on the data being analysed. This is interesting because, almost by definition, data mining is secondary data analysis: the analysis of data collected for some other purpose. The data may be of perfect quality for its original purpose (e.g. calculating your grocery bill in the store), but of poor quality for subsequent mining (e.g. because some items were grouped together in the bill).

A third difference between many modern data analysis problems and those of the past is that nowadays they are often dynamic. Electronics permit data to be collected as things happen, and this opens the possibility of of making decisions as the data are collected. An example is in commercial transactions, where a customer can supply information and expects an immediate decision. In such circumstances one does not have the luxury of taking the data back to one's laboratory and analysing it at leisure. Speech recognition is another example. This issue has led to new kinds of analytic tools, with an emphasis on speed and not merely accuracy. No particular area of data analysis seems to have precedence for such problems, but the computer science side, perhaps especially machine learning clearly regards such problems as important.

Although every kind of data analytic discipline must contend with all kinds of data, there is no doubt that different kinds are more familiar in different areas. Computational areas probably place more emphasis on categorical data than on continuous data, and this is reflected in the types of data analytic tools (e.g. methods for extracting association rules) which have been developed.

4 The Role of Mathematics

Modern statistics is often regarded as a branch of mathematics. This is entirely inappropriate. Indeed, the qualitative change induced by the advent of the computer means that statistics could equally be regarded as a branch of computer science.

In a sense statistics, and data analysis more generally, is the opposite of mathematics. Mathematics begins with assumptions about the structure of the universe of discourse (the axioms) and seeks to deduce the consequences. Data analysis, on the other hand, begins with observations of the consequences (the data) and seeks to infer something about the structure of the universe. One consequence of this is that one can be a good mathematician without understanding anything about any area to which the mathematics will be applied – one primarily needs facility with mathematical symbol manipulation – but one cannot be a good statistician without being able to relate the analysis to the world from which the data arose. This is why one hears of mathematics prodigies, but never statistics prodigies. Analysis requires understanding.

There are other differences as well. Nowadays a computer is an essential and indispensable tool for statistics, but one can still do much mathematics without a computer. This is brought home to our undergraduate students, taking mathematics degrees, with substantial statistical components, when they come to use software: statistical software packages such as Splus, R, SAS, SPSS, Stata, etc., are very different from mathematical packages such as Maple and Mathematica. Carrying out even fairly basic statistical analyses using the latter can be a non-trivial exercise.

David Finney has commented that it is no more true to describe statistics as a branch of mathematics than it would be to describe engineering as a branch of mathematics, and John Nelder has said '*The main danger, I believe, in allowing the ethos of mathematics to gain too much influence in statistics is that statisticians will be tempted into types of abstraction that they believe will be thought respectable by mathematicians rather than pursuing ideas of value to statistics.*'

There is no doubt that the misconception of statistics as mathematics has been detrimental in the past, especially in commercial and business applications. Data mining, in particular took advantage of this - its very name spells glamour and excitement, the possibility of gaining a market edge for free. But there are also other examples where the image of statistics slowed its uptake. For example, experimental design (that branch of statistics concerned with efficient and cost effective ways to collect data) was used in only relatively few sectors (mostly manufacturing). Reformulations of experimental design ideas under names such as the Taguchi method and Six Sigma, however, have had a big impact. If anything ought to convince my academic colleagues of the power of packaging and presentation, then it should be these examples.

5 Several Cultures Separated by a Common Language

The writer George Bernard Shaw once described England and America as '*two cultures divided by a common language*', and I sometimes feel that the same applies to the various data analytic disciplines. Over the years, I have seen several intense debates between proponents of the different disciplines. Part of the reason for this lies in the different philosophical approaches to investigation. Statistics, perhaps because of its mathematical links, places a premium on proof and mathematical demonstration of the properties of data analytic tools. For example, demonstrating mathematically that an algorithm will always converge. Machine learning, on the other hand, places more emphasis on empirical testing. Of course there is overlap. Most methodological statistics papers include at least one example of the methods applied to real problems, and most machine learning papers describe the ideas in mathematical terms, but there is a clear difference in what is regarded as of central importance.

Another reason for the debates has been that many of the ideas were developed in parallel, by researchers naturally keen to establish their priority and reputation. This led to claims to the effect that 'we developed it first' or 'we demonstrated that property years ago.' This was certainly evident in the debates on recursive partitioning tree classifiers, which were developed in parallel by the machine learning and statistics communities.

Misunderstandings can also arise because different schools place emphasis on different things. Early computer science perspectives on data mining stressed the finding of patterns in databases. This is perfectly natural: it is something often required (e.g. what percentage of my employees earn more than €x p.a.?). However, this is of limited interest to a statistician, who will normally want to make an inference to a wider population or to the future (e.g. what percentage of my employees are likely to earn more than €x p.a. next year?). Much work on association analysis has ignored this inferential aspect. Moreover, much work has also made a false causal assumption:

while it is *interesting* to know that ten times as many people who bought A also bought B, it is *valuable* to know that if people can be induced to buy A they will also buy B, and the two are not the same.

While there have been tensions between the different areas when they develop similar models, each from their own perspective, there is no doubt that these tensions can be immensely beneficial from a scientific perspective. A nice example of this is the work on feedforward neural networks. These originally came from the computer (or, one might argue, the cybernetics, electrical engineering, or even biological) side of things. The perspective of a set of fairly simple interacting processors dominated. Later, however, statisticians became involved and translated the ideas into mathematical terms: such models can be written as nested sequences of nonlinear transformations of linear combinations of variables. Once written in fairly standard terms, one can apply the statistical results of a century of theoretical work. In particular, one could explain that the early neural network claims of very substantial improvement in predictive power were likely to be in large part due to overfitting the design data, and to present ideas and tools for avoiding this problem. Of course, nowadays all these are well understood by the neural network community, but this was certainly not the case in the early days (I can remember papers presenting absurdly overoptimistic claims), even though statisticians had known about the issues for decades.

If the computer is leading to a unification of the data analytic schools, so also are some theoretical developments. The prime examples here, of course, are Bayesian ideas. Bayes's theorem tells us how we should update our knowledge in the light of new information. This is the very essence of learning, so it is not surprising that machine learning uses these ideas. With the advent of practical computational tools for evaluating high dimensional integrals, such as MCMC, statistics has also undergone a dramatic Bayesian revolution, not only in terms of dynamic updating models but also in terms of model averaging. Indeed, model averaging, like the understanding of overfitting (indeed, closely connected to it), has led to deep theoretical advances. Tools such as boosting and bagging are based on these sorts of principles. Boosting, in particular, is interesting from our perspective because it illustrates the potential synergy which can arise from the disparate emphases of the different disciplines. Originally developed by the machine learning community, who proposed it on fairly intuitive grounds and showed that it worked in practical applications, it was then explored theoretically by statisticians, who showed its strong links to generalised additive models, a well-understood class of statistical tools. The most recent tool to experience this initial development, followed by an exposure to the ideas and viewpoints of other data analytic disciplines, is that of support vector machines.

In fact, perceptrons (the progenitor of support vector machines) and logistic discrimination provide a very nice illustration of the difference in emphasis between, in this case, statistical and machine learning models for classification. Logistic discrimination fits a model to the probability that an object with given features $\mathbf{x}$ will belong to class 0 rather than class 1. Typically, the model is fitted by finding the parameters which maximise the design set log likelihood:

$$\log L \propto \sum_{i=1}^{n} \log \hat{p}(0 | \mathbf{x}_i). \tag{1}$$

Classification is then effected by comparing an estimated probability with a threshold. It is immediately clear from (1) that all design set data points contribute - it is really an average of contributions. This is fine if one's model $\hat{p}(0|\mathbf{x})$ has the form of the 'true' function $p(0|\mathbf{x})$. But this is a brave assumption. It is likely that the model is not perfect. If so, one must question the wisdom of letting data points with estimated probability far from the classification threshold contribute the same amount to the fit criterion (1) as do those near to it (see [4]). In contrast, perceptron models focus attention on whether or not the design set points are correctly classified: quality of fit of a model far from the decision surface, which is broadly irrelevant to classification performance, does not come into it.

An example of another area which has been developed in rather different ways by different disciplines is the area I call *pattern discovery*. This is the search for, identification of, and description of anomalously high local densities of data points. The computer science literature has focused on algorithms for finding such configurations. In particular, a great deal of work has occurred when the data are character strings, in, especially text search (e.g. web search engines) and nucleotide sequences. In contrast, the statistical work has concentrated on the inference problem, developing scan statistics for deciding whether a local anomaly represents a real underlying structure is just random variation of a background model. Ideas of this kind have been developed in many application areas, including bioinformatics, technical stock chart analysis, astronomy, market basket analysis, and others, but the realisation that they are all tackling very similar problems appears to be only recent.

Implicit in the last two paragraphs is one of the fundamental differences in emphasis between computational and statistical approaches to data analysis - again an understandable difference in view of their origins. This is the emphasis of the computational approaches on algorithms (e.g. the perceptron error-correcting algorithm) and the emphasis of the statistical approaches on models (e.g. the logistic discrimination model). Both algorithms and models are, of course, important when tackling real problems.

It is my own personal view that one can also characterise the difference between the two perspectives, at least to some extent, in terms of degree of risk. The computational schools seem often prepared to try something without the assurance that it will work, or that it will always work, but in the hope (or knowledge from previous analyses) that it will sometimes work. The statistical schools seem more risk averse, requiring more assurance before carrying out an analysis. Perhaps this is illustrated by the approaches to pattern discovery mentioned above: the data mining community develops algorithms with which to detect possible patterns, while the statistical community develops tools to tell whether they are real or merely chance. Once again, both perspectives are valuable, especially in tandem: adventurous risk-taking offers the possibility of major breakthroughs, while careful analysis shows one that the method gives reliable results.

6 Future Tools and Application Areas?

Of course, the various data analytic disciplines are constantly evolving. We live in very exciting times because of the tools which have been developed over the past few decades, but that development has not stopped. If anything, it has accelerated and will continue to do so as the computational infrastructure continues to develop. This means faster and larger (in terms of all dimensions of datasets). Judging from the past, this will translate into analytic tools about which one previously could only have dreamt, and, further, into tools one could not even have imagined.

If the computer is one force driving the development of new data analytic tools, I can see at least two others.

The first of these are application areas, mentioned above. Certainly, the growth of statistics over the 20th century was strongly directed by the applications. Thus agricultural requirements led to the early development of experimental design, psychology motivated the development of factor analysis and other latent variable models, medicine led to survival analysis, and so on. In other areas, speech recognition stimulated work on hidden Markov models, robotics stimulated work on reinforcement learning, etc. Of course, once developments have been started, and the power of the tools being developed has been recognised, other application areas rapidly adopt the tools.

As with the impact of developing computational infrastructure, I see no reason to expect this influence of application areas to stop. We are currently witnessing the particular requirements of genomic, proteomic, and related data leading to the development of new analytic tools; for example, methods for handling *fat data* - data involving many (perhaps tens of thousands of) variables, but few (perhaps a few tens of) data points. Mathematical finance is likewise an area which is shifting its centre of gravity towards analysis. Until recently characterised by mathematical areas such as stochastic calculus, it is increasingly recognised that data analysis is also needed - the values of the model parameters must come from somewhere. More generally, the area of personal finance is beginning to provide a rich source of novel problems, requiring novel solutions. The world wide web, of course, is another source of new types of data, and new problems. This area, in particular, is a source of data which is characterised by its dynamic properties, and I expect the analysis of dynamic data to play an even more crucial role in future developments. Decisions in telecoms systems, even in day-to-day purchasing transactions, are needed *now*, not after a leisurely three months' analysis of a customer's track record and characteristics. Delay loses business.

The second additional driving force I can see is also not really a new one. It has always been with us, but it will lead to the development of new kinds of tools, in response to new demands and also enabled by the advancing computational infrastructure. This is the need to model finer and finer aspects of the presenting problems. A recent example of this is in the analysis of repeated measures data. The last two decades have witnessed a very exciting efflorescence of ideas for tackling such data. The essential problem is to recognise and take account of the fact that repeated measurements data are likely to be correlated (with the (multiple) series being too short to use time series ideas). Classical assumptions of independence are all very well, but more accurate models and predictions result when the dependence is modelled. Another example of such 'finer aspects' of the presenting problem, which has typically been ignored up until now, is the fact that predictive models are likely to be applied to data drawn from distributions different from that from which the design data were drawn (perhaps a case for dynamic models). There are many other examples.

There is, however, a cautionary comment to be made in connection with this driving force. It is easy to go too far. There is little point is developing a method to cope with some aspect of the data if the inaccuracies induced by that aspect are trivial in comparison with those arising from other causes. Data analysis is not a merely mathematical exercise of data manipulation.

If we data analysts live in exciting times, I think it is clear that the future will be even more exciting. Looking back on the past it is obvious that the tensions between the different data analytic disciplines have, in the end, been beneficial: we can learn from the perspectives and emphases of the other approaches. In particular, we should learn that the other disciplines can almost certainly shed light on and help each of us gain greater understanding of what we are trying to do. We should look for the *synergies*, not the *antagonisms*.

I'd like to conclude with two quotations. The first is from John Chambers, the computational statistician who developed Splus and who won the 1998 ACM Software System Award for that work. He wrote: *'Greater statistics can be defined simply, if loosely, as everything related to learning from data, from the first planning or collection to the last presentation or report. Lesser statistics is the body of specifically statistical methodology that has evolved within the profession - roughly, statistics as defined by texts, journals, and doctoral dissertations. Greater statistics tends to be inclusive, eclectic with respect to methodology, closely associated with other disciplines, and practiced by many outside of academia and often outside of professional statistics. Lesser statistics tends to be exclusive, oriented to mathematical techniques, less frequently collaborative with other disciplines, and primarily practiced by members of university departments of statistics.'* [1]

John has called the discipline of data analysis 'greater statistics', but I am sure we can all recognise what we do in his description. What we call it is not important. As Juliet puts it in Act II, Scene ii of Shakespeare's *Romeo and Juliet*:

> *'What's in a name? that which we call a rose*
> *By any other name would smell as sweet.'*

References

1. Chambers J.M. Greater or lesser statistics: a choice for future research. *Statistics and Computing*, **3**, (1993) 182-184.
2. Giudici P. *Applied Data Mining*. Chichester: Wiley. (2003)
3. Hand D.J., Mannila H., and Smyth P. *Principles of Data Mining*, Cambridge, Massachusetts: MIT Press. (2001)
4. Hand D.J. and Vinciotti V. Local versus global models for classification problems: fitting models where it matters. *The American Statistician*. **57**, (2003) 124-131.

Mining Positive and Negative Association Rules: An Approach for Confined Rules

Maria-Luiza Antonie and Osmar R. Zaïane

Department of Computing Science, University of Alberta
Edmonton, Alberta, Canada
{luiza,zaiane}@cs.ualberta.ca

Abstract. Typical association rules consider only items enumerated in transactions. Such rules are referred to as *positive association rules. Negative association rules* also consider the same items, but in addition consider negated items (i.e. absent from transactions). Negative association rules are useful in market-basket analysis to identify products that conflict with each other or products that complement each other. They are also very convenient for associative classifiers, classifiers that build their classification model based on association rules. Many other applications would benefit from negative association rules if it was not for the expensive process to discover them. Indeed, mining for such rules necessitates the examination of an exponentially large search space. Despite their usefulness, and while they were referred to in many publications, very few algorithms to mine them have been proposed to date. In this paper we propose an algorithm that extends the support-confidence framework with a sliding correlation coefficient threshold. In addition to finding confident positive rules that have a strong correlation, the algorithm discovers negative association rules with strong negative correlation between the antecedents and consequents.

1 Introduction

Association rule mining is a data mining task that discovers relationships among items in a transactional database. Association rules have been extensively studied in the literature for their usefulness in many application domains such as recommender systems, diagnosis decisions support, telecommunication, intrusion detection, etc. The efficient discovery of such rules has been a major focus in the data mining research community. From the original *apriori* algorithm [1] there have been a remarkable number of variants and improvements of association rule mining algorithms [2].

Association rule analysis is the task of discovering association rules that occur frequently in a given data set. A typical example of association rule mining application is the market basket analysis. In this process, the behaviour of the customers is studied when buying different products in a shopping store. The discovery of interesting patterns in this collection of data can lead to important marketing and management strategic decisions. For instance, if a customer buys

J.-F. Boulicaut et al. (Eds.): PKDD 2004, LNAI 3202, pp. 27–38, 2004.

bread, what is the probability that he/she buys milk as well? Depending on the probability of such an association, marketing personnel can develop better planning of the shelf space in the store or can base their discount strategies on such associations/correlations found in the data.

All the traditional association rule mining algorithms were developed to find positive associations between items. By positive associations we refer to associations between items existing in transactions (i.e. items bought). What about associations of the type: "customers that buy Coke do not buy Pepsi" or "customers that buy juice do not buy bottled water"? In addition to the positive associations, the negative association can provide valuable information, in devising marketing strategies. Interestingly, very few have focused on negative association rules due to the difficulty in discovering these rules.

Although some researchers pointed out the importance of negative associations [3], only few groups of researchers [4], [5], [6] proposed an algorithm to mine these types of associations. This not only illustrates the novelty of negative association rules, but also the challenge in discovering them.

1.1 Contributions of This Paper

The main contributions of this work are as follows:

1. We devise a new algorithm to generate both positive and negative association rules. There are very few papers to discuss and discover negative association rules. Our algorithm differs from those in the sense that it uses a different interestingness measure and it generates the association rules from a different candidate set.
2. To avoid adding new parameters that would make tuning difficult and thus impractical, we introduce an automatic thresholding on the correlation coefficient. We automatically and progressively slide the threshold to find strong correlations.
3. We compare our algorithm with other existing algorithms that can generate negative association rules and discuss their performances.

The remainder of the paper is organized as follows: Section 2 gives an overview of the basic concepts involved in association rule mining. In Section 3 we introduce our approach for positive and negative rule generation based on correlation measure. Section 4 presents related work for comparison with our approach. Experimental results are described in Section 5 along with the performance of our system compared to known algorithms. We summarize our research and discuss some future work directions in Section 6.

2 Basic Concepts and Terminology

This section introduces association rules terminology and some related work on negative association rules.

2.1 Association Rules

Formally, association rules are defined as follows: Let $\mathcal{I} = \{i_1, i_2, ...i_n\}$ be a set of items. Let $\mathcal{D}$ be a set of transactions, where each transaction T is a set of items such that $T \subseteq \mathcal{I}$. Each transaction is associated with a unique identifier TID. A transaction T is said to contain X, a set of items in $\mathcal{I}$, if $X \subseteq T$. An *association rule* is an implication of the form "$X \Rightarrow Y$", where $X \subseteq \mathcal{I}, Y \subseteq \mathcal{I}$, and $X \cap Y = \emptyset$. The rule $X \Rightarrow Y$ has a *support* s in the transaction set $\mathcal{D}$ if $s\%$ of the transactions in $\mathcal{D}$ contain $X \cup Y$. In other words, the support of the rule is the probability that X and Y hold together among all the possible presented cases. It is said that the rule $X \Rightarrow Y$ holds in the transaction set $\mathcal{D}$ with *confidence* c if $c\%$ of transactions in $\mathcal{D}$ that contain X also contain Y. In other words, the confidence of the rule is the conditional probability that the consequent Y is true under the condition of the antecedent X. The problem of discovering all association rules from a set of transactions $\mathcal{D}$ consists of generating the rules that have a *support* and *confidence* greater than given thresholds. These rules are called *strong rules*, and the framework is known as the *support-confidence framework* for association rule mining.

2.2 Negative Association Rules

Example 1. Suppose we have an example from the market basket data. In this example we want to study the purchase of organic versus non-organic vegetables in a grocery store. Table 1 gives us the data collected from 100 baskets in the store. In Table 1 "organic" means the basket contains organic vegetables and "¬ organic" means the basket does not contain organic vegetables. The same applies for non-organic. On this data, let us find the positive association rules in the "support-confidence" framework. The association rule "non-organic → organic" has 20% support and 25% confidence (supp(non-organic ∧ organic)/supp(non-organic)). The association rule "organic → non-organic" has 20% support and 50% confidence (supp(non-organic ∧ organic)/supp(organic)). The support is considered fairly high for both rules. Although we may reject the first rule on the confidence basis, the second rule seems a valid rule and may be considered in the data analysis. Now, let us compute the statistical correlation between the *non-organic* and *organic* items. A more elaborated discussion on the correlation measure is given in Section 3.1. The correlation coefficient between these two items is -0.61. This means that the two items are negatively correlated. This measure sheds a new light on the data analysis on these specific items. The rule "organic → non-organic" is misleading. The correlation brings new information that can help in devising better marketing strategies.

The example above illustrates some weaknesses in the "support-confidence" framework and the need for the discovery of more interesting rules. The interestingness of an association rule can be defined in terms of the measure associated with it, as well as in the form an association can be found.

Brin *et. al* [3] mentioned for the first time in the literature the notion of negative relationships. Their model is chi-square based. They use the statistical

Table 1. Example 1 data

	organic	¬organic	$\sum_{row}$
non-organic	20	60	80
¬non-organic	20	0	20
$\sum_{col}$	40	60	100

Table 2. 2x2 contingency table

	Y	¬Y	$\sum_{row}$
X	f_{11}	f_{10}	f_{1+}
¬X	f_{01}	f_{00}	f_{0+}
$\sum_{col}$	f_{+1}	f_{+0}	N

test to verify the independence between two variables. To determine the nature (positive or negative) of the relationship, a correlation metric was used. In [6] the authors present a new idea to mine strong negative rules. They combine positive frequent itemsets with domain knowledge in the form of a taxonomy to mine negative associations. However, their algorithm is hard to generalize since it is domain dependant and requires a predefined taxonomy. A similar approach is described in [7]. Wu *et. al* [4] derived a new algorithm for generating both positive and negative association rules. They add on top of the support-confidence framework another measure called *mininterest* for a better pruning of the frequent itemsets generated. In [5] the authors use only negative associations of the type $X \rightarrow \neg Y$ to substitute items in market basket analysis.

We define as *generalized negative association rule*, a rule that contains a negation of an item (i.e a rule for which its antecedent or its consequent can be formed by a conjunction of presence or absence of terms). An example for such association would be as follows: $A \wedge \neg B \wedge \neg C \wedge D \rightarrow E \wedge \neg F$. To the best of our knowledge there is no algorithm that can determine such type of associations. Deriving such an algorithm is not an easy problem, since it is well known that the itemset generation in the association rule mining process is an expensive one. It would be necessary not only to consider all items in a transaction, but also all possible items absent from the transaction. There could be a considerable exponential growth in the candidate generation phase. This is especially true in datasets with highly correlated attributes. That is why it is not feasible to extend the attribute space by adding the negated attributes and use the existing association rule algorithms. Although we are currently investigating this problem, in this paper we generate a subset of the generalized negative association rules. We refer to them as *confined negative association rules.* A confined negative association rule is one of the follows: $\neg X \rightarrow Y$, $X \rightarrow \neg Y$ or $\neg X \rightarrow \neg Y$, where the entire antecedent or consequent must be a conjunction of negated attributes or a conjunction of non-negated attributes.

3 Discovering Positive and Negative Association Rules

The most common framework in the association rules generation is the "support-confidence" one. Although these two parameters allow the pruning of many associations that are discovered in data, there are cases when many uninteresting rules may be produced. In this paper we consider another framework that adds to the support-confidence some measures based on correlation analysis. Next section introduces the correlation coefficient, which we add to the support-confidence framework in this work.

3.1 Correlation Coefficient

Correlation coefficient measures the strength of the linear relationship between a pair of two variables. It is discussed in the context of association patterns in [8]. For two variables X and Y, the correlation coefficient is given by the following formula:

$$\rho = \frac{Cov(X,Y)}{\sigma_X \sigma_Y} \quad . \tag{1}$$

In Equation 1, $Cov(X,Y)$ represents the covariance of the two variables and σ_X stands for the standard deviation. The range of values for ρ is between -1 and +1. If the two variables are independent then ρ equals 0. When $\rho = +1$ the variables considered are perfectly positive correlated. Similarly, When $\rho = -1$ the variables considered are perfectly negative correlated. A positive correlation is evidence of a general tendency that when the value of X increases/decreases so does the value of Y. A negative correlation occurs when for the increase/decrease of X value we discover a decrease/increase in the value of Y.

Let X and Y be two binary variables. Table 2 summarizes the information about X and Y variables in a dataset in a 2x2 contingency table. The cells of this table represent the possible combinations of X and Y and give the frequency associated with each combination. N is the size of the dataset considered.

Given the values in the contingency table for binary variables, Pearson introduced the ϕ *correlation coefficient* which is given in the equation 2:

$$\phi = \frac{f_{11} f_{00} - f_{10} f_{01}}{\sqrt{f_{+0} f_{+1} f_{1+} f_{0+}}} \quad . \tag{2}$$

We can transform this equation by replacing f_{00}, f_{01}, f_{10}, f_{0+} and f_{+0} as follows:

$$\phi = \frac{f_{11}(N - f_{10} - f_{01} - f_{11}) - f_{10} f_{01}}{\sqrt{f_{+0} f_{+1} f_{1+} f_{0+}}} \tag{3}$$

$$\phi = \frac{f_{11} N - f_{11} f_{10} - f_{11} f_{01} - f_{11}^2 - f_{10} f_{01}}{\sqrt{f_{+0} f_{+1} f_{1+} f_{0+}}} \tag{4}$$

$$\phi = \frac{f_{11} N - (f_{11} + f_{10})(f_{11} + f_{01})}{\sqrt{f_{+0} f_{+1} f_{1+} f_{0+}}} \tag{5}$$

$$\phi = \frac{N f 11 - f_{1+} * f_{f+1}}{\sqrt{f_{1+}(N - f_{1+}) f_{+1}(N - f_{+1})}} \quad . \tag{6}$$

The measure given in Equation 6 is the measure that we use in the association rule generation.

Cohen [9] discusses about the correlation coefficient and its strength. In his book, he considers that a correlation of 0.5 is large, 0.3 is moderate, and 0.1 is small. The interpretation of this statement is that anything greater than 0.5 is large, 0.5-0.3 is moderate, 0.3-0.1 is small, and anything smaller than 0.1 is insubstantial, trivial, or otherwise not worth worrying about as described in [10].

We use these arguments to introduce an automatic progressive thresholding process. We start by setting our correlation threshold to 0.5. If no strong correlated rules are found the threshold slides progressively to 0.4, 0.3 and so on until some rules are found with moderate correlations. This progressive process eliminates the need for manually adjusted thresholds. It is well known that the more parameters a user is given, the more difficult it becomes to tune the system. Association rule mining is certainly not immune to this phenomenon.

3.2 Our Algorithm

Traditionally, the process of mining for association rules has two phases: first, mining for frequent itemsets; and second, generating strong association rules from the discovered frequent itemsets. In our algorithm, we combine the two phases and generate the relevant rules on-the-fly while analyzing the correlations within each candidate itemset. This avoids evaluating item combinations redundantly. Indeed, for each generated candidate itemset, we compute all possible combinations of items to analyze their correlations. At the end, we keep only those rules generated from item combinations with strong correlation. The strength of the correlation is indicated by a correlation threshold, either given as input or by default set to 0.5 (see above for rational). If the correlation between item combinations X and Y of an itemset XY, where X and Y are itemsets, is negative, negative association rules are generated when their confidence is high enough. The produced rules have either the antecedent or the consequent negated: ($\neg X \rightarrow Y$ and $X \rightarrow \neg Y$), even if the support is not higher than the support threshold. However, if the correlation is positive, a positive association rule with the classical support-confidence idea is generated. If the support is not adequate, a negative association rule that negates both the antecedent and the consequent is generated when its confidence and support are high.

The algorithm generates all positive and negative association rules that have a strong correlation. If no rule is found, either positive or negative, the correlation threshold is automatically lowered to ease the constraint on the strength of the correlation and the process is redone. Figure 1 gives the detailed pseudo-code for our algorithm.

Initially both sets of negative and positive association rules are set to empty (line 1). After generating all the frequent 1-itemsets (line 2) we iterate to generate all frequent k-itemsets, stored in F_k (line 8). F_k is verified from a set of candidate C_k computed in line 4. The iteration from line 2 stops when no longer frequent itemsets are possible. Unlike the join made in the traditional *Apriori* algorithm, to generate candidates at level k, instead of joining frequent $(k-1)$-itemsets, we join the frequent itemsets at level $k-1$ with the frequent 1-itemsets (line 4). This is because we want to extend the set of candidate itemsets and have the possibility to analyze the correlation of more item combinations. The rational will be explained later. Every candidate itemset generated this way is on one hand tested for support (line 7), and on the other hand used to analyze possible correlations even if its support is below the minimum support (loop from line 9 to 22). Correlations for all possible pair combinations for each candidate itemset are computed. For an itemset i and a pair combination (X, Y) such that $i = X \cup Y$, the correlation coefficient is calculated (line 10). If the correlation is positive and strong enough, a positive association rule of the type $X \rightarrow Y$ is generated, if the $supp(X \cup Y)$ is above the minimum support threshold and the confidence of the

```
Algorithm Positive and Negative Association Rules Generation
Input TD, minsupp, minconf, and ρmin, respectively Transactional Database,
minimum support, minimum confidence, and correlation threshold.
Output AR: Positive and Negative Association Rules.
Method:
(0)  if ρmin is undefined then ρmin = 0.5
(1)  positiveAR ← ∅; negativeAR ← ∅ /*positive and negative AR sets*/
(2)  scan the database and find the set of frequent 1-itemsets (F1)
(3)  for (k = 2, Fk−1 ≠ ∅, k + +){
(4)      Ck= Fk−1 ⋈ F1
(5)      foreach i ∈ Ck {
(6)          s=support(TD,i) /*support of item i is computed*/
(7)          if s≥ minsupp then
(8)              Fk ← Fk ∪ {i} /*item i is added to Fk*/
(9)          foreach X,Y (i = X ∪ Y) {
(10)             ρ=correlation(X,Y) /*correlation btw X and Y is computed*/
(11)             if ρ ≥ ρmin then
(12)               if s≥ minsupp then
(13)                 if confidence(X → Y) ≥ minconf then
(14)                   positiveAR ← positiveAR ∪ {X → Y}
(15)               else if confidence(¬X → ¬Y) ≥ minconf and
                           supp(¬X¬Y) ≥ minsupp then
(16)                   negativeAR ← negativeAR ∪ {¬X → ¬Y}
(17)             if ρ ≤ −ρmin then /*ρ < 0 and |ρ| ≥ ρmin */
(18)               if confidence(X → ¬Y) ≥ minconf then
(19)                 negativeAR ← negativeAR ∪ {X → ¬Y}
(20)               if confidence(¬X → Y) ≥ minconf then
(21)                 negativeAR ← negativeAR ∪ {¬X → Y}
(22)         }
(23)     }
(24) }
(25) AR ← positiveAR ∪ negativeAR
(26) if AR = ∅ then {
(27)   ρmin = ρmin − 0.1
(28)   if ρmin ≥ 0 then go to step (3)
(29) }
(30) return AR
```

Fig. 1. Discovering positive and negative confined association rules

rule is strong. Otherwise, if we still have a positive and strong correlation but the support is below the minimum support, a negative association rule of the type $\neg X \rightarrow \neg Y$ is generated if its confidence is above the minimum confidence threshold (lines 15-16). On the other hand, if the correlation test gives a strong negative correlation, association rules of the types $X \rightarrow \neg Y$ and $\neg X \rightarrow Y$ are generated and appended to the set of association rules if their confidence is adequate. The result is compiled by combining all discovered positive and negative association rules. Lines 26 onward, illustrate the automatic progressive thresholding for the correlation coefficient. If no rules are generated at a given correlation level, the threshold is lowered by 0.1 (line 27) and the process reiterated.

4 Related Work in Negative Association Rule Mining

In this section, we discuss two known algorithms that generate negative association rules. We compare our approach with them later in the experiments section.

4.1 Negative Association Rule Algorithms

We give a short description of the existing algorithms that can generate positive and negative association rules. For more details, please refer to [4] and [5].

First, we discuss the algorithm proposed by Wu *et. al* [4]. They add on top of the support-confidence framework another measure called *mininterest* (the argument is that a rule $A \rightarrow B$ is of interest only if $supp(A \cup B) - supp(A)supp(B) \geq mininterest$). The authors consider as itemsets of interest those itemsets (positive or negative) that exceed minimum support and minimum interest thresholds. Although, [4] introduces the "mininterest" parameter, the authors do not discuss how to set it and what would be the impact on the results when changing this parameter. The approach differs from our algorithm in that in our algorithm we use the correlation coefficient as measure of interestingness, which was thoroughly studied in the statistics community. In addition, the value of our parameter is well defined and it is not as sensitive to the dataset as the *mininterest* parameter. In our algorithm (line 9) we compute the correlation coefficient for every pair X,Y of an item i where $i = X \cup Y$. As described earlier, when such a pair is found correlated an association rule is generated from it. In [4], they compute the interest for every pair X,Y of the item i where $i = X \cup Y$. However, they extract rules from itemset i only if any expression $i = X \cup Y$ exceeds the minimum interest threshold. We claim that by adding this condition they are loosing some potential interesting association rules. In addition, in our algorithm the candidate set C_k is generated as a join between F_{k-1} *and* F_1. In [4] the candidate set C_k is generated as a union of two frequent itemsets in F_i *for* $1 \leq i \leq k-1$. This turns out to be expensive. Since we all make the assumption that a k-itemset must have all its subsets in F_{k-1} we prove in the next theorem that our join generates the same itemsets as in [4].

Theorem. All candidate items $c \in C_k$ generated by $F_i \bowtie F_j, 1 \leq i,j \leq k-1$ for which $\exists t \in c$ such that $t \in F_{k-1}$, can be discovered by $F_{k-1} \bowtie F_1$.

Proof. Let us suppose $\exists c \in C_k$ such that $c \in F_i \bowtie F_j, 1 \leq i,j \leq k-1$ and $c \notin F_{k-1} \bowtie F_1$. Given the condition stated in theorem $\exists t \in c$ such that $t \in F_{k-1}$. Since $c \notin F_{k-1} \bowtie F_1$ and $t \in F_{k-1}$ it follows that $c - t \notin F_1$. This is false as $c - t$ is of length one and $c \in C_k$ was generated from frequent itemsets. Thus $\forall c \in C_k, c \in F_{k-1} \bowtie F_1$. Q.E.D

Second, we present the algorithm proposed in [5]. The algorithm is named by the authors SRM (substitution rule mining). We refer to it in the same way throughout the paper. The authors develop an algorithm to discover negative associations of the type $X \rightarrow \neg Y$. These association rules can be used to discover to which items are substitutes for others in market basket analysis. Their algorithm discovers first what they call *concrete items*, which are those itemsets that have a high chi-square value and exceed the expected support. Once these itemsets are discovered, they compute the correlation coefficient for each pair of them. From those pairs that are negatively correlated, they extract the desired rules (of the type $X \rightarrow \neg Y$). This paper, although interesting for the substitution items application, it is limited in the kind of rules that can discover.

Table 3. TD (a)

TID	Items
1	A,C,D
2	B,C
3	C
4	A,B,F
5	A,C,D
6	E
7	B,F
8	B,C,F
9	A,B,E
10	A,D

Table 4. TD (b)

TID	Items	Equivalent bit vector
1	$A, \neg B, C, D, \neg E, \neg F$	(101100)
2	$\neg A, B, C, \neg D, \neg E, \neg F$	(011000)
3	$\neg A, \neg B, C, \neg D, \neg E, \neg F$	(001000)
4	$A, B, \neg C, \neg D, \neg E, F$	(110001)
5	$A, \neg B, C, D. \neg E, \neg F$	(101100)
6	$\neg A, \neg B, \neg C, \neg D, E, \neg F$	(000010)
7	$\neg A, B, \neg C, \neg D, \neg E, F$	(010001)
8	$\neg A, B, C, \neg D, \neg E, F$	(011001)
9	$A, B, \neg C, \neg D, E, \neg F$	(110010)
10	$A, \neg B, \neg C, D, \neg E, \neg F$	(100100)

Using the next example, which is an extension of the example presented in [5], we present some of the differences among the three algorithms.

Example 2. Let us consider a small transactional table with 10 transactions and 6 items. In Table 3 a small transactional database is given. To illustrate the challenges in mining negative association rules we create another transactional database where for each transaction, the complement of each missing item is appended to it. The new created dataset is shown in Table 4. This new database can be mined with the existing association rule mining algorithms. However, there are a few drawbacks of this naive approach. In practice, the data collections are very large, thus adding all the complemented items to the original database requires a large storage space. Not only the storage space has to increase considerably, but the execution times as well, in particular when the number of unique items in the database is very large. In addition, many association rules would be generated, many of them being of no interest to the applications at hand.

Using a minimum support of 0.2, the following itemsets are discovered using the three discussed algorithms. For this example the *correlation coefficient* was set to 0.5, and the *minimum interest* to 0.07.

Table 5. 2-itemsets

Correlation	Interest	Concrete
AD	AD	AD
BF	BF	BF
BD	BD	BD
CE	CE	
	DF	

Table 6. 3-itemsets

Correlation	Interest	Concrete
ACD		ACD
ABC	ABC	
ABD		ABD
BCD		

In Table 5 and Table 6, the first column presents the results when our approach was used. The second column uses the algorithm from[4], while in the third one the results are obtained using the approach in [5]. In both tables the positive itemsets are separated by the negative ones by a double horizontal line. The positive itemsets are in the upper part of the tables. As it can be seen, for the 2-itemsets all three algorithms find the same positive ones. The differences occur for the negative itemsets. The itemset DF has a *minimum interest* of 0.09, but it has a *correlation* of only 0.42. That is why it is not found by our approach or by the SRM algorithm [5]. The itemset CE is not found by SRM because their condition is that the itemset should have higher correlation than the minimum value. In our approach the condition is to be greater or equal. Since the itemset CE has a *correlation* of 0.5 it is discovered by our algorithm, but not by SRM.

In Table 6 there are differences for both, the positive and the negative ones. The algorithm that uses the *minimum interest* parameter discovers only the ABC itemset because it is the only one that has all the pairs X,Y of the item ABC where $ABC = X \cup Y$ above the parameter. Although. all the other itemsets discovered by the other algorithms have at least two strong pairs they are not considered of interest. Our approach and SRM generate the same positive 3-itemset. The itemsets BCD and ABC are not discovered by SRM because none of its subsets of two items are generated as concrete during the process.

From the itemsets that were shown in Table 5 and Table 6 a set of association rules can be generated. Here we show, some of the rules that were generated from the itemsets that were discovered by one algorithm, but not by others. From itemset CE, the association rule $negE \to C$ can be found with support 0.5 and confidence of 62%. This rule seems to be strong, but it is missed by the SRM algorithm. From itemset DF, which is discovered only by the minimum interest algorithm, the association rules $negD \to F$ and $D \to \neg F$ can be discovered. However, both rules have support 0.3 and confidence of 42%. These rules could have been eliminated when the confidence threshold is set to 50%, thus our approach and SRM do not miss much by not generating them. In addition, our approach generates the 3-itemset BCD. From this itemset the rule $B \to \neg C \neg D$ is discovered and it has support of 0.2 and confidence of 60%.

5 Experimental Results

We conducted our experiments on a real dataset to study the behaviour of the algorithms compared. We used the Reuters-21578 text collection [11]. Reuters dataset had 6488 transaction, when only the ten largest categories were kept.

We compare the three algorithms discussed in the sections above. For each algorithm a set of values for their main interestingness measure was used in the experiments. Our algorithm and SRM [5] had the correlation coefficient set to 0.5, 0.4 and 0.3. In [4] the authors used the value 0.07 in their examples. We used this value and two others in its vicinity (0.05, 0.07 and 0.09). Each algorithm was run to generate a set of association rules. For lack of space the results are reported only for correlation coefficient 0.4 and minimum interest 0.07. For all the results, please see [12].

For these association rules a number of measures were computed: support (supp), confidence (conf), Piatetsky-Shapiro measure (PS), Yule's Q (Q), co-

Table 7. Results for Reuters text collection

		#rules	supp	conf	PS	Q	IS	J
(a) Results for rules of type $X \rightarrow Y$								
corr	0.4	235	**0.23±0.03**	**0.79±0.16**	**0.14±0.02**	0.84±0.28	**0.78±0.08**	**0.63±0.12**
int	0.07	219	**0.23±0.03**	0.79±0.18	0.13±0.02	**0.85±0.27**	0.76±0.09	0.61±0.14
SRM	0.4	297	0.22±0.03	0.76±0.20	0.12±0.03	0.82±0.27	0.73±0.10	0.57±0.15
(b) Results for rules of type $X \rightarrow \neg Y$								
		#rules	supp	conf	PS	Q	IS	J
corr	0.4	6	**0.33±0.10**	**0.99±0.0**	**0.11±0.01**	**0.99±0.0**	**0.72±0.05**	**0.52±0.08**
int	0.07	4	0.25±0.01	0.98±0.02	0.08±0.01	0.70±0.47	0.62±0.03	0.39±0.03
SRM	0.4	6	**0.33±0.10**	**0.99±0.0**	**0.11±0.01**	**0.99±0.0**	**0.72±0.05**	**0.52±0.08**
(c) Results for rules of type $\neg X \rightarrow Y$								
		#rules	supp	conf	PS	Q	IS	J
corr	0.4	6	0.33±0.10	**0.49±0.08**	**0.11±0.01**	**0.99±0.0**	**0.72±0.05**	**0.52±0.08**
int	0.07	4	**0.34±0.06**	0.46±0.09	0.08±0.01	0.70±0.47	0.67±0.06	0.45±0.08
(d) Results for rules of type $\neg X \rightarrow \neg Y$								
		#rules	supp	conf	PS	Q	IS	J
corr	0.4	1474	0.31±0.09	0.41±0.13	0.15±0.02	**0.84±0.20**	0.80±0.06	0.66±0.10
int	0.07	148	**0.49±0.09**	**0.67±0.13**	**0.16±0.05**	0.81±0.39	**0.87±0.08**	**0.77±0.11**

sine measure (IS) and the Jaccard measure (J). These measures evaluate the interestingness of the discovered pattern. For more details on these measures for frequent patterns see [13]. In [13] a set of measures are compared and discussed. The measures are clustered with respect to their similarity. We chose to compute a few measures from different clusters to ensure the diversity of evaluation.

Tables 7 presents the results obtained for the Reuters dataset. We conducted the experiments with support 20% and confidence 0%. In each table a subset of the obtained rules are compared. Table 7 (a) compares rules of the type $X \rightarrow Y$, Table 7 (b) rules of the type $X \rightarrow \neg Y$, Table 7 (c) rules of the type $\neg X \rightarrow Y$ and Table 7 (d) rules of the type $\neg X \rightarrow \neg Y$. In each table the average of the measurement and the standard deviation are reported. The value in bold represents the best value for each measure.

Table 7 (a) shows that for positive association rules our approach tends to generate a more interesting set of rules compared to the other methods.

For rules of type $X \rightarrow \neg Y$ (Table 7 (b)) our approach and SRM perform best. They produce the same set of rules for correlation values of 0.4.

Table 7 (c) and Table 7 (d) compare our approach with the one in [4] only, since SRM algorithm does not generate this kind of rules.

In Table 7 (c) the symmetric rules of the ones in Table 7 (b) are generated, since the confidence is set to 0% and the correlation and minimum interest are computed for XY itemset.

However, for the rules of type $\neg X \rightarrow \neg Y$ (Table 7 (d)) the method in [4] generates a smaller set of rules, but with higher values for the measures.

6 Conclusions and Future Research Directions

In this paper we introduced a new algorithm to generate both positive and negative association rules. Our method adds to the support-confidence framework the correlation coefficient to generate stronger positive and negative rules. We compared our algorithm with other existing algorithms on a real dataset. We discussed their performances on a small example for a better illustration of the algorithms and we presented and analyze experimental results for a text collection. The results prove that our algorithm can discover strong patterns. In addition, our method generates all types of confined rules, thus allowing to be used in different applications where all these types of rules could be needed or just a subset of them.

Acknowledgements

This work was partially supported by Alberta Ingenuity Fund, iCORE and NSERC Canada.

References

1. Agrawal, R., Imielinski, T., Swami, A.: Mining association rules between sets of items in large databases. In: Proc. of SIGMOD. (1993) 207–216
2. Goethals, B., Zaki, M., eds.: FIMI'03: Workshop on Frequent Itemset Mining Implementations. Volume 90 of CEUR Workshop Proceedings series. (2003) http://CEUR-WS.org/Vol-90/.
3. Brin, S., Motwani, R., Silverstein, C.: Beyond market basket: Generalizing association rules to correlations. In: Proc. of SIGMOD. (1997) 265–276
4. Wu, X., Zhang, C., Zhang, S.: Mining both positive and negative association rules. In: Proc. of ICML. (2002) 658–665
5. Teng, W., Hsieh, M., Chen, M.: On the mining of substitution rules for statistically dependent items. In: Proc. of ICDM. (2002) 442–449
6. Savasere, A., Omiecinski, E., Navathe, S.: Mining for strong negative associations in a large database of customer transactions. In: Proc. of ICDE. (1998) 494–502
7. Yuan, X., Buckles, B., Yuan, Z., Zhang, J.: Mining negative association rules. In: Proc. of ISCC. (2002) 623–629
8. Tan, P., Kumar, V.: Interestingness measures for association patterns: A perspective. In: Proc. of Workshop on Postprocessing in Machine Learning and Data Mining. (2000)
9. Cohen, J.: Statistical power analysis for the behavioral sciences (2nd ed.). Lawrence Erlbaum, New Jersey (1988)
10. Hopkins, W.: A new view of statistics. http://www.sportsci.org/resource/stats/ (2002)
11. Reuters-21578: (The Reuters-21578 text categorization test collection) http://www.research.att.com/~lewis/reuters21578.html.
12. Antonie, M.L., Zaïane, O.: Mining positive and negative association rules: An approach for confined rules. Technical Report TR04-07, Dept. of Computing Science, University of Alberta (2004) ftp://ftp.cs.ualberta.ca/pub/TechReports/2004/TR04-07/TR04-07.ps.
13. Tan, P., Kumar, V., Srivastava, J.: Selecting the right interestingness measure for association patterns. In: Proc. of SIGKDD. (2002) 32–41

An Experiment on Knowledge Discovery in Chemical Databases

Sandra Berasaluce[1,2,3], Claude Laurenço[1,2], Amedeo Napoli[3], and Gilles Niel[1]

[1] LSIC – ENSCM, 8, rue de l'Ecole Normale, 34296 Montpellier,
[2] LIRMM, 161, rue Ada, 34392 Montpellier,
[3] LORIA, BP 239, 54506 Vandœuvre-lès-Nancy

Abstract. In this paper, we present an experiment on knowledge discovery in chemical reaction databases. Chemical reactions are the main elements on which relies synthesis in organic chemistry, and this is why chemical reactions databases are of first importance. From a problem-solving process point of view, synthesis in organic chemistry must be considered at several levels of abstraction: mainly a strategic level where general synthesis methods are involved, and a tactic level where actual chemical reactions are applied. The research work presented in this paper is aimed at discovering general synthesis methods from chemical reaction databases in order to design generic and reusable synthesis plans. The knowledge discovery process relies on frequent levelwise itemset search and association rule extraction, but also on chemical knowledge involved within every step of the knowledge discovery process. Moreover, the overall process is supervised by an expert of the domain. The principles of this original experiment on mining chemical reaction databases and its results are detailed and discussed.

Keywords: knowledge discovery, data mining, frequent level-wise itemset search, association rule, knowledge-based system.

1 Introduction

In this paper, we present an experiment on the application of knowledge discovery algorithms for mining chemical reaction databases. Chemical reactions are the main elements on which relies synthesis in organic chemistry, and this is why chemical reaction databases are of first importance. From a problem-solving process point of view, synthesis in organic chemistry must be considered at several levels of abstraction: mainly a strategic level where general synthesis methods are involved, and a tactic level where actual chemical reactions are applied. The research work presented in this paper is aimed at discovering general synthesis methods from chemical reaction databases in order to design generic and reusable synthesis plans. This can be understood in the following way: mining reaction databases at the tactic level for finding synthesis methods at the strategic level. This knowledge discovery process relies on the one hand on mining algorithms, i.e. frequent levelwise itemset search and association rule extraction, and, on the other hand, on domain knowledge, that is involved at every step of the knowledge discovery process.

J.-F. Boulicaut et al. (Eds.): PKDD 2004, LNAI 3202, pp. 39–51, 2004.

This research work is carried out within a long-term project for designing chemical information systems whose goal is to help a chemist building a synthesis plan [14, 19]. Actually, the general problem of synthesis relies on the design of a synthesis plan followed by an experimentation of this synthesis plan. Synthesis planning is mainly based on an analytical reasoning process, called *retrosynthesis*, where the first element of the plan is the *target* molecule, i.e. the molecule that has to be built (see fig. 1). This process can be likened to a goal-directed problem-solving approach: the target molecule is iteratively transformed by applying reactions for obtaining simpler fragments, until finding starting materials that are easy to build or to obtain (this constitutes a synthesis pathway). For a given target molecule, a huge number of starting materials and reactions may exist, e.g. thousands of commercially available chemical compounds. Thus, exploring all the possible pathways issued from a target molecule leads to a combinatorial explosion. Therefore the choice of reaction sequences to be used within the planning process is of first importance, and strategies are needed for efficiently solving the synthesis planning problem.

At present, reaction database management systems are the most useful tools for helping the chemist in synthesis planning. Other knowledge systems have been developed since the 70s for helping synthesis planning based on a retrosynthetic approach. The main problem in this kind of system is the constitution of the knowledge base. In our research work, we are designing a new kind of knowledge system for synthesis planning, combining the principles of knowledge systems, database systems, and knowledge discovery [4, 3, 19]. One aspect of this research is to study how data mining techniques may contribute to knowledge extraction from reaction databases, and beyond that, to the structuring of these databases and the improvement in their querying.

This paper presents a preliminary experiment carried on two commercial reaction databases[1] using frequent itemset search and association rule extraction [2, 16]. This study is original and novel within the domain of organic synthesis planning, and is of first importance, with respect to chemical researches. Regarding the knowledge discovery research, we stress the fact that knowledge extraction in an application domain has to be guided by knowledge domain if substantial results have to be obtained. Indeed, the knowledge extraction process is performed under the supervision of a domain expert, but the computing process itself has to be guided by domain knowledge, at every step, i.e. cleaning and transforming data, and interpreting results. We claim that the role of knowledge within the knowledge extraction process is most of the time underestimated, and one of the goal of this paper is to show that taking advantage of the functionalities of a knowledge system within the knowledge discovery process may be of first importance for obtaining accurate and realistic results.

The paper is organized as follows. First, we introduce the chemical context, describing the synthesis problem. Then, we detail the selection and the preprocessing of data, i.e. organic synthesis reactions from reaction databases, and then, the application of data mining techniques to these data, namely frequent

[1] Supplied by Molecular Design Ltd – MDL (`http://www.mdli.com`).

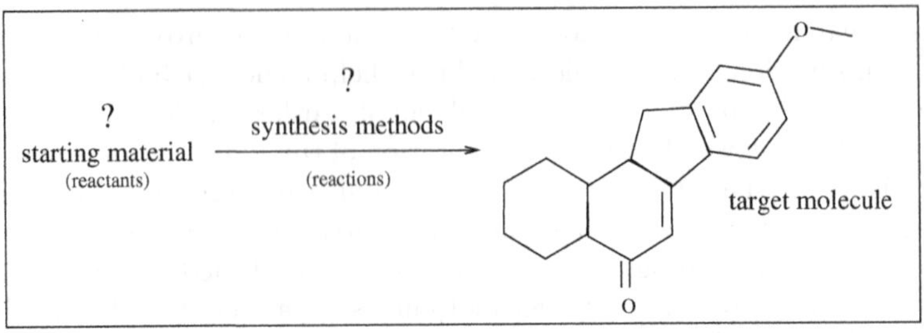

Fig. 1. The general schema of a synthesis problem.

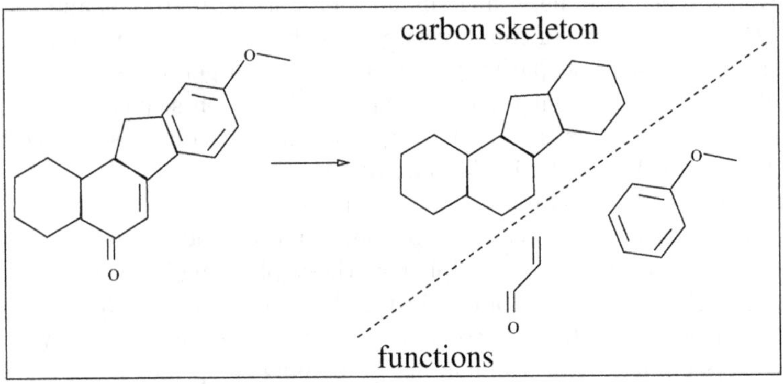

Fig. 2. Skeleton and functional groups of a target molecule.

itemset search and association rules extraction. We show how the results of such a knowledge discovery process give insights for information organization and retrieval within reaction databases. Moreover, the extracted knowledge units after been validated by a chemist may be useful in the search for efficient reactions in association with a synthesis problem. Then we conclude the paper with a discussion regarding the present research work and the research perspectives.

2 The Chemical Context

2.1 The Synthesis Problem

The information needs for a chemist solving a synthesis problem is related to a search in the literature for specific reactions solving synthesis problems considered to be similar to the current one. There is a very huge number of specific reactions described within articles in the literature, certainly more than 10 millions. Reaction documentation is complex and not yet standardized: many classification systems have been proposed, based on reaction mechanism, or electron properties, but they are not really useful for studying synthesis in the large. Actually, the main questions for the synthesis chemist are related to chemical families to which a target molecule belongs, and to the synthesis methods, i.e. a reaction or a sequence of reactions building structural patterns, to be used for building these families. For the sake of simplicity, we will use hereafter only the term "reaction" for mentioning a basic reaction or a synthesis method as well.

Two main categories of reactions may be distinguished: reactions building the skeleton of a molecule –the arrangement of carbon atoms on which relies an organic molecule–, and reactions changing the functionality of a molecule, i.e. changing a function into another function (see fig. 2). In our framework, a function is mainly used for recognizing a given molecule as a member of a chemical family, for predicting and explaining the molecule reactivity. It is defined as a connected molecular substructure composed of multiple carbon-carbon bonds, carbon-heteroatom bonds –an heteroatom is an atom that is not a carbon atom–, and heteroatom-heteroatom bonds. Here, we are mainly interested in reactions changing the functionality, and in the following questions: (i) what is the starting function F_i for a given formed function F_j? (ii) what are the reactions allowing the transformation of a function F_i into a function F_j? (iii) what are the functions F_i remaining unchanged during the application of a reaction?

2.2 The Reaction Databases: Data Selection and Preprocessing

The experiment reported hereafter has been carried out on two reaction databases, namely the "Organic Syntheses" database ORGSYN-2000 including 5486 records, and the "Journal of Synthetic Methods" database JSM-2002 including 75291 records. The selection of these databases relies on size and quality criteria. In these databases, the filtering of the data related to functional transformations has been performed within a data preprocessing step, where only structural information about the reaction has been considered (details are given in 3.2).

The purpose of the preprocessing step of data mining is to improve the quality of the selected data by cleaning and normalizing the data. Reaction databases such as ORGSYN-2000 and JSM-2002 may be seen as a collection of records, where every record contains one chemical equation involving structural information, that can be read, according to the reaction model, as the transformation of an *initial state* –or the set of *reactants*– into a *final state* –or the set of *products*– associated with an atom-to-atom mapping between the initial and final states (see fig. 3).

In our framework, data preprocessing has mainly consisted in exporting and analyzing the structural information recorded in the databases for extracting and for representing the functional transformations in a target format that has been processed afterwards. The considered transformations are functional modifications, functional addition and deletion, i.e. adding or deleting a function. Moreover, no distinction has been made between one-step or multi-steps reactions. Errors in the atom-to-atom mapping have been neglected, as a reaction is considered at an abstract level, the so-called *block level*, as explained hereafter. The abstraction of a reaction from the atom level into the block level is carried out using the RESYN-ASSISTANT system [19, 3] (some details on RESYN-ASSISTANT are given in § 3.2).

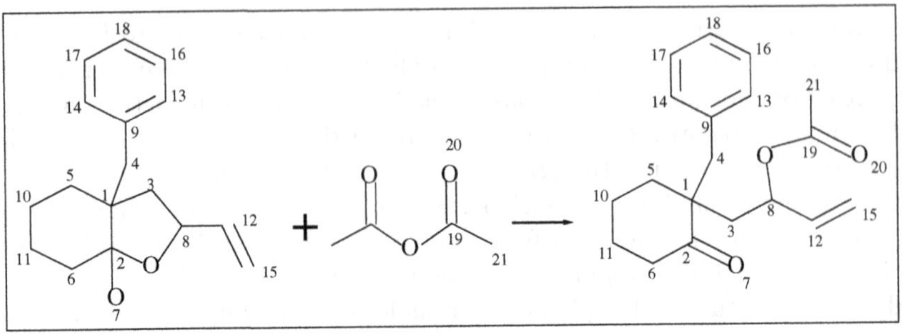

Fig. 3. The structural information on a reaction with the associated atom-to-atom mapping (reaction #13426 in the JSM-2002 database).

In the following, we discuss the whole process of chemical reaction databases manipulation, involving data transformation and data mining, for retrieving and organizing chemical reaction databases.

3 Knowledge Discovery in Reaction Databases

3.1 An Overview of the Knowledge Discovery Process

The knowledge discovery process in chemical reaction databases is considered as an interactive and iterative experimental process. An expert of the data domain, called hereafter the *analyst*, plays a central role in this process since he is in charge of controlling all the steps of the process (as discussed e.g. in [9, 5]). According to given synthesis objectives, the analyst selects first the data to be analyzed, applies data mining modules for extracting knowledge units from data, and finally interprets and validates the units having a sufficient plausibility for being reused. For carrying out this process with benefits, the analyst may take advantage of his own knowledge of the domain –he is an expert–, and, as well, of a set of modules including a knowledge system, ontologies, molecule and reaction databases[2].

Hence, in our approach, the knowledge discovery process is, first of all, guided by the analyst and domain knowledge. The knowledge discovery process itself is based on frequent itemsets search and association rules extraction. Practically, the Close and the Pascal algorithms have been used for data processing [16, 15]. Their application and the results that have been obtained are discussed in the next sections.

3.2 The Modeling of Reactions for Knowledge Discovery

Data on organic reactions are generally recorded in databases within structural and textual entries: the former describes the structural formulae of substances

[2] More generally, the Web in the large could be taken into account if necessary.

in terms of "molecular graphs" while the latter refers to reaction conditions, names and roles of implied substances, bibliographical references, keywords and comments. In our experiment, we have been mainly interested in the so-called functionality changes –or interchanges– occurring during a reaction. These interchanges can be recognized and represented by comparing the functionality of the reactants with that of the products: the removal of some (old) functions and the creation of some (new) functions can be made explicit. The comparison relies on the atom-to-atom mapping where functionality interchanges correspond to the substitution of an atom from one function to another.

Formally, the representation of a reaction equation relies on the atom-to-atom mapping relation between the graphs of the reactants and the graphs of the products, defining three bond sets: the set of the *broken* or *destroyed* bonds, of *formed* bonds and of *unchanged* bonds. Actually these three function modifications correspond to subgoals that are achieved during the synthesis, preparing a main objective [7].

The knowledge system RESYN-ASSISTANT has been designed for assisting the chemist in the design of organic synthesis problems. In particular, the RESYN-ASSISTANT system is able to recognize the building blocks of a molecule, and among these blocks, the functional blocks, or more simply functions. Every function is defined by a name, and is represented as a graph, called functional graph, modeling the structure of the function. The set of functional graphs is partially ordered by a subsumption relation based on a typed subgraph relation. In this way, the set of functional graphs constitutes a concept hierarchy, called $\mathcal{H}_f$, that is part of the knowledge base of the RESYN-ASSISTANT system. Recognizing a function $\mathtt{F_k}$ within a molecule say `M` means that the structure of `M` includes the functional graph $\mathtt{F_k}$ as a subgraph. This recognition process is based on the classification of `M` within the function hierarchy $\mathcal{H}_f$. At present, the $\mathcal{H}_f$ hierarchy includes about five hundred named functions.

The RESYN-ASSISTANT system has been extended to recognize the building blocks of reactions. Based on the atom-to-atom mapping, the system establishes the correspondence between the recognized blocks of the same nature, and determines their role in the reaction. The abstraction of the reaction introduced in figure 3 from the atom-to-atom level into the block level is shown in figure 4. A function may be present in a reactant, in a product, or in both. In the last case, the function is unchanged. In the two other cases, the function in the reactant is destroyed, or the function in the product is formed. During a reaction, either one or more reactant functions may contribute to form the functions in

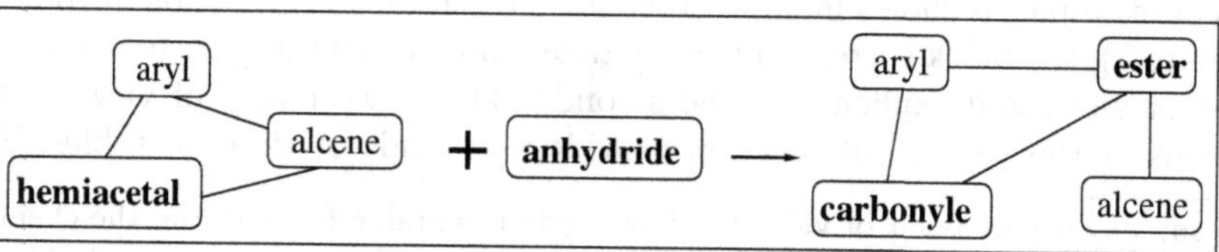

Fig. 4. The analysis of reaction #13426 in the JSM-2002 database in terms of blocks.

the products. At the end of the preprocessing step, the information obtained by the recognition process is incorporated into the representation of the reaction.

For allowing the application of the algorithms Close and Pascal for frequent itemsets search, the data on reaction have to be transformed into a Boolean table. Thus, the representation of a molecule as a composition of functional blocks cannot be used in a straightforward way. Moreover, a reaction can be considered from two main points of view, depending on the fact that the atom-to-atom mapping is taken into account or not (see fig. 5):

- a global point of view on the functionality interchanges leads to consider a single entry R corresponding to an analyzed reaction, to which is associated a list of properties, i.e. formed and/or destroyed and/or unchanged functions,
- a specific point of view on the functionality transformations that is based on the consideration of a number of different entries R_k corresponding to the different functions being formed, i.e. the atom-to-atom mapping gives the explicit correspondence between the blocks that are formed, destroyed, and unchanged (see figure 3).

Entries/Blocks	**Destroyed blocks**		**Formed blocks**		**Unchanged blocks**	
	anhydride	hemiacetal	carbonyle	ester	alcene	aryle
without correspondence entry **R**	×	×	×	×	×	×
with correspondence entry **R_1**	×	×		×	×	×
entry **R_2**		×	×		×	×

Fig. 5. The original data are prepared for the mining task: the Boolean transformation of the data can be done by not taking into account the atom mapping, i.e. one single line in the Boolean table, or by taking into account the atom mapping, i.e. two lines in the table.

For example, as shown in figure 5 (in association with the reaction introduced in figure 3), the block correspondence is taken into account implicitly in the first point of view, and explicitly in the second[3]. These two points of view on the analysis of the content of reaction provide two kinds of Boolean tables. The

[3] From a synthesis point of view, the first mode is suitable for studying the chemoselectivity of functionality interchanges, and the second mode is more suitable for comparing the relative reactivities of the studied functions.

rows correspond to the entries related to a single reaction, one row for the global (or implicit) point of view, and two or more for the specific (or explicit) point of view. The columns correspond to the three families of functions, destroyed, formed and unchanged (the same functions are repeated three times, one time per type of columns). Two remarks can be done: firstly, both correspondence have been used during the experiment, and, secondly, in both cases, spatial information on the graph structure of the molecules is lost.

3.3 The Search for Itemsets and the Extraction of Association Rules

The Close and the Pascal algorithms have been applied to Boolean tables (built as indicated just above) for generating first itemsets, i.e. sets of functions (with an associated support), and then association rules. The study of the extracted frequent itemsets may be done with different points of view. Firstly, studying frequent itemsets of length 2 or 3 enables the analyst to determine basic relations between functions. For example searching for a formed functions F_f ($_{-f}$ for formed) deriving from a broken function F_d ($_{-d}$ for destroyed) leads to the study of the itemsets $F_d \sqcap F_f$, where the symbol $\sqcap$ stands for the conjunction of functions. In some cases, a reaction may depend on functions present in both reactants and products that remain unchanged ($_{-u}$ for unchanged) during the reaction application, leading to the study of frequent itemsets such as $F_f \sqcap F_u \sqcap F_d$. This kind of itemsets can be searched for extracting a "protection function" supposed to be stable under given experimental conditions.

The extraction of association rules gives a complementary perspective on the knowledge extraction process. For example, searching for the more frequent ways to form a function F_f from a function F_d leads to the study of rules such as $F_f \longrightarrow F_d$: indeed, this rule has to be read in a retrosynthetic way, i.e. if the function F_f is formed then this means that the function F_d is destroyed. Again, this rule can be generalized in the following way: determining how a function F_f is formed from two destroyed functions F_{d1} and F_{d2}, knowing say that the function F_{d1} is actually destroyed, leads to the study of the association rules such as $F_f \sqcap F_{d1} \longrightarrow F_{d2}$. It must be noticed that for the sake of simplicity, the examples have been kept formal here. Concrete examples can be found either in [3] or in [4].

As usual, the number of itemsets and of association rules to be considered depends on:

- the way of considering the block correspondence, either implicit (one entry per reaction) or explicit (two or more entries per reaction),
- the minimal value of the support of the itemsets to be considered,
- the confidence level chosen for considering and interpreting the association rules.

The results obtained by the application of the data mining algorithms are discussed in the two next sections, firstly from a chemical point of view, and then from a knowledge discovery point of view.

4 Chemical Interpretation of the Knowledge Extraction Results

A whole set of results of the application of the data mining process on the ORGSYN-2000 and JSM-2002 databases is given in [4]. These results show that both reaction databases share many common points though they differ in terms of size and data coverage, i.e. among 500 functions included in the $\mathcal{H}_f$ hierarchy, only 170 are retrieved from ORGSYN-2000 while 300 functions are retrieved from JSM-2002. The same five functions are ranked at the first places in both databases with the highest occurrence frequency. However, some significant differences can be observed: a given function may be much more frequent in the ORGSYN-2000 database than in JSM-2002 database, and reciprocally. These differences can be roughly explained by different data selection criteria and editor motivations for both databases.

A qualitative and statistical study of the results has shown the following behaviors. Some functions have a high stability, i.e. they mostly remain unchanged, and, in the contrary, some others functions are very reactive, i.e. they are mostly destroyed. All the reactive functions are more present in reactants than in products, and some functions are more often formed. Some functions, that are among the most widely used functions in organic synthesis, are more often present and destroyed in reactants, e.g. `alcohol` and `carboxylic acid`. For example, among the standard reactions involving functions, it is well-known –for chemists– that the `ester` function derives from a combination of two functions, one of them being mostly an `alcohol`. The search for a second function relies on the study of rules such as $\mathtt{ester_f} \sqcap \mathtt{alcohol_d} \longrightarrow \mathtt{F_d}$. The main functions that are retrieved are `anhydride`, `carboxylic acid`, `ester`, and `acyl chloride`. If the chemist is interested in the unchanged functions, then the analysis of the rule $\mathtt{ester_f} \sqcap \mathtt{alcohol_d} \sqcap \mathtt{anhydride_d} \longrightarrow \mathtt{F_u}$ gives functions such as `acetal`, `phenyl`, `alkene`, and `carboxylic acid`.

These first results provide a good overview on the function stability and reactivity. They also give partial answers to the questions that have been posed in section 2.1. However, some questions remain open, such as the classification of reactions with respect to a given point of view, e.g. reactivity, stereochemistry,...

Working on functionality interchanges within organic synthesis is a complex problem, and the data mining experiment presented in this paper raises the question of the selection of the reaction databases. The choice of the ORGSYN-2000 and the JSM-2002 databases has been guided by their coverage relevance. The ORGSYN-2000 database provides an electronic version of the entire series of Organic Syntheses, and offers an access to new general synthesis methods. The principle followed by the editors of Organic Syntheses is particularly interesting since each synthesis method has been checked by experts in laboratories. This practice confers to the data a high confidence value, because they have been verified in compound preparations. On the other hand, the JSM-2002 database is a document-based organic reaction database presenting a high coverage in organic synthesis (from 1975). To be selected and recorded, a reaction must be novel or have a particular advantage over an existing one. In addition the reaction must have a clear experimental method, and must be repeatable. For these reasons, the ORGSYN-2000 and the JSM-2002 databases appear to be suitable for a global study of chemical functionality. Both databases contain fine-grained selected data, and thus the information retrieval process is necessarily focused. The ORGSYN-2000 and the JSM-2002 databases have proven to be useful sources for exploring organic synthesis knowledge rather than for providing exhaustive information about particular reactions.

5 Discussion: Frequent Itemsets, Rules and Chemical Reactions

First of all, it can be mentioned that only a few research works hold on the application of data mining methods on reaction databases (see for example [8, 6, 12, 13]). Moreover, these studies have different objectives, and are mainly concerned with molecular graph manipulation rather than reaction database mining. Another study on the lattice-based classification of dynamic knowledge units has been a valuable source of inspiration for the present work [10], leading to the division of functions in three categories, formed, destroyed, and unchanged. The work in [10] is more focused on formal concept analysis and lattice construction rather than on data mining concerns. A number of topics can be discussed here regarding the experiment presented in this paper:

- The abstraction of reactions within blocks and the separation in three kinds of blocks, namely formed, destroyed, and unchanged blocks. Indeed, this is one of the most original idea in that research work, that is responsible of the good results that have been obtained. This idea of the separation into three families may be reused in other contexts involving dynamic data. However, the transformation into a Boolean table has led to a loss of information, e.g. the connection information on reactions and blocks. This loss of information on the connection of the entries introduces a bias in the data mining process, that is quite difficult to take into account.
- Frequent items or association rules are generic elements that can be used either to index (and thus organize) reactions or to retrieve reactions. Termed in another way, this means that frequent itemsets or extracted association rules may be in certain cases considered as a kind of meta-data giving meta-information on the bases that are under study. For example, questions that the chemist wants to be answered are the following: if `A` $\longrightarrow$ `B` is true, and `B` $\longrightarrow$ `C` is also true, then it can be deduced that `A` $\longrightarrow$ `C` is also true, meaning that we can have access to three reactions if needed (or the access to two reactions allows the access to a third inferred reaction).
- Knowledge is used at every step of the knowledge extraction process, e.g. the coupling of the knowledge extraction process with the RESYN-ASSISTANT system, and domain ontologies such as the function ontologies, the role of the analyst,... Indeed, and this is one of the major lesson of this experiment: the knowledge discovery process in a specific domain such as organic synthesis

has to be *knowledge-intensive*, and has to be guided by domain knowledge, and an analyst as well, for obtaining substantial results.

- The role of the analyst includes fixing the thresholds, and interpreting of the results. The thresholds must be chosen in function of the objectives of the analyst, and in function of the content of the databases. A threshold of 1% for an item support means that for a thousand of reactions, ten reactions may form a family: this is not a bad hypothesis. Moreover, if ten thousand reactions are considered, then 1% means that a hundred reactions may form a family, and in this case, this is a very realistic hypothesis. This shows that the thresholds are linked in a very close way to the knowledge of the domain. Here, we see again the influence of the domain: the value of a threshold here is very different from the values that can be used for a threshold in marketing analysis.
  Another remark may be done on what could be called "exceptions", i.e. a reaction that appear only once in a database; this means that there is no other reaction of the same kind in the database, or, that the item associated with this reaction is a unique one. The notion of exception has no substantial meaning here: one unique reaction in one database may be found under several examples in another database. A unique exemplar is rather a matter of point of view taken by the editors of the considered database.
- Other research directions have to investigated, namely sequential patterns [1, 17], or working with closed itemsets and icebergs [18]. Regarding closed itemsets and icebergs, it must be noticed that closed itemsets are the longer itemsets, and those that potentially bring the most of information for the current mining problem. Thus, it could be interesting to consider only these closed itemsets, and to work more in the spirit of formal concept analysis, where a concept lattice –based on closed sets of properties– is built [11].
  Moreover, the use of data mining methods such as frequent itemsets search or association rule extraction has proven to be useful, and has provided encouraging results. It could be interesting to test other (symbolic) data mining methods, e.g. OLAP technology, relational mining, cluster analysis, or Bayesian network classification, knowing that numerical methods such as hidden Markov models or neural networks are not really adapted to the kind of data that are considered in our experiment.

6 Conclusion

In this paper, we have presented an experiment on knowledge discovery in chemical reaction databases. Two databases have been deeply studied and mined, namely the ORGSYN-2000 and the JSM-2002 databases, using frequent levelwise itemset search and association rule extraction. The main topic of interest in the reactions is related with functionality interchanges. Thus, the reactions in the databases have been abstracted in terms of three kinds of building blocks for molecules involved in the reactions, namely formed, destroyed and unchanged blocks. This categorization of blocks has been the basis for building the Boolean tables on which data mining algorithms such as Close and Pascal have been applied. From a chemical point of view, the results are very encouraging, and provide a set of meta-data for organizing and retrieving chemical reaction according to given synthesis objectives. From a knowledge discovery point of view, a number of questions can be discussed, such as the value of the thresholds (usually lower than in marketing analysis), on the processing of the data, and on the interpretation of the results. Moreover, two major elements have to be pointed out, and can be reused in other contexts : the categorization of dynamic data such as reactions into three families, here, formed, destroyed and unchanged functions, and the use of knowledge at every stage of the knowledge discovery process. Indeed, in a domain such as organic synthesis, the knowledge discovery process has to be fully guided by domain knowledge, and the analyst, an expert of the domain, as well. There are a number of research perspectives following the present work, including the adaptation of sequential pattern algorithms to chemical reactions, taking actually into account the structures of the molecules involved in reactions, and working in the spirit of concept analysis for lattice-based classification of the data.

References

1. R. Agrawal and R. Srikant. Mining sequential patterns. In P.S. Yu and A.L. P. Chen, editors, *Proceedings of the Eleventh International Conference on Data Engineering, (ICDE-95), Taipei, Taiwan*, pages 3–14. IEEE Computer Society, 1995.
2. Y. Bastide, R. Taouil, N. Pasquier, G. Stumme, and L. Lakhal. Mining frequent patterns with counting inference. *ACM SIGKDD Explorations*, 2(2):66–75, 2000.
3. S. Berasaluce. *Fouille de données at acquisition de connaissances à partir de bases de données de réactions chimiques.* Thèse de chimie informatique et théorique, Université Henri Poincaré Nancy 1, 2002.
4. S. Berasaluce, C. Laurenço, A. Napoli, and G. Niel. Data mining in reaction databases: extraction of knowledge on chemical functionality transformations. Technical Report A04-R-049, LORIA, Nancy, 2004.
5. R.J. Brachman and T. Anand. The Process of Knowledge Discovery in Databases. In U.M. Fayyad, G. Piatetsky-Shapiro, P. Smyth, and R. Uthurusamy, editors, *Advances in Knowledge Discovery and Data Mining*, pages 37–57, Menlo Park, California, 1996. AAAI Press / MIT Press.
6. R. Chittimoori, L. B. Holder, and D. J. Cook. Applying the Subdue substructure discovery system to the chemical toxicity domain. In *Proceedings of the Florida AI Research Symposium*, pages 90–94, 1999.
7. E.J. Corey and X.M. Cheng. *The Logic of Chemical Synthesis.* John Wiley & Sons, New York, 1989.
8. L. Dehaspe, H. Toivonen, and R.D. King. Finding frequent substructures in chemical compounds. In *Proceedings of the 4th International Conference on Knowledge Discovery and Data Mining*, pages 30–36, 1998.
9. U. Fayyad, G. Piatetsky-Shapiro, and P. Smyth. Knowledge Discovery and Data Mining: Towards a Unifying Framework. In *Proceedings of the Second International Conference on Knowledge Discovery & Data Mining (KDD-96), Portland, Oregon*, pages 82–88, 1996.

10. B. Ganter and S. Rudolph. Formal Concept Analysis Methods for Dynamic Conceptual Graphs. In H.S.Delugach and G. Stumme, editors, *Conceptual Structures: Broadening the Base – 9th International Conference on Conceptual Structures, ICCS-2001, Stanford*, Lecture Notes in Artificial Intelligence 2120, pages 143–156. Springer, Berlin, 2001.
11. B. Ganter and R. Wille. *Formal Concept Analysis*. Springer, Berlin, 1999.
12. A. Inokuchi, T. Washio, and H. Motoda. An Apriori-based algorithm for mining frequent substructures from graph data. In D. Zighed, J. Komorowski, and J.M. Zytkow, editors, *Principles of Data Mining and Knowledge Discovery, 4th European Conference, PKDD-2000, Lyon, France, 2000, Proceedings*, Lecture Notes in Computer Science 1910, pages 13–23. Springer, 2000.
13. M. Kuramochi and G. Karypis. An efficient algorithm for discovering frequent subgraphs. Technical Report 02–026, Department of Computer Science, University of Minnesota, 2002. To be published in IEEE Transactions on Knowledge and Data Engineering.
14. A. Napoli, C. Laurenço, and R. Ducournau. An object-based representation system for organic synthesis planning. *International Journal of Human-Computer Studies*, 41(1/2):5–32, 1994.
15. N. Pasquier, Y. Bastide, R. Taouil, and L. Lakhal. Discovering frequent closed itemsets for association rules. In C. Beeri and P. Buneman, editors, *Database Theory - ICDT'99 Proceedings, 7th International Conference, Jerusalem, Israel*, Lecture Notes in Computer Science 1540, pages 398–416. Springer, 1999.
16. N. Pasquier, Y. Bastide, R. Taouil, and L. Lakhal. Pruning closed itemset lattices for association rules. *International Journal of Information Systems*, 24(1):25–46, 1999.
17. M. Sena and G. Karypis. SLPMiner: An algorithm for finding frequent sequential patterns using length-decreasing support constraint. Technical Report 02–023, Department of Computer Science, University of Minnesota, 2002.
18. G. Stumme, R. Taouil, Y. Bastide, N. Pasquier, and L. Lakhal. Computing iceberg concept lattices with titanic. *Journal of Data and Knowledge Engineering*, 42(2):189–222, 2002.
19. P. Vismara and C. Laurenço. An abstract representation for molecular graphs. *DIMACS Series in Discrete Mathematics and Theoretical Computer Science*, 51:343–366, 2000.

Shape and Size Regularization in Expectation Maximization and Fuzzy Clustering

Christian Borgelt and Rudolf Kruse

Dept. of Knowledge Processing and Language Engineering
Otto-von-Guericke-University of Magdeburg
Universitätsplatz 2, D-39106 Magdeburg, Germany
{borgelt,kruse}@iws.cs.uni-magdeburg.de

Abstract. The more sophisticated fuzzy clustering algorithms, like the Gustafson–Kessel algorithm [11] and the fuzzy maximum likelihood estimation (FMLE) algorithm [10] offer the possibility of inducing clusters of ellipsoidal shape and different sizes. The same holds for the EM algorithm for a mixture of Gaussians. However, these additional degrees of freedom often reduce the robustness of the algorithm, thus sometimes rendering their application problematic. In this paper we suggest shape and size regularization methods that handle this problem effectively.

1 Introduction

Prototype-based clustering methods, like fuzzy clustering [1, 2, 12], expectation maximization (EM) [6] of a mixture of Gaussians [9], or learning vector quantization [15, 16], often employ a distance function to measure the similarity of two data points. If this distance function is the *Euclidean distance*, all clusters are (hyper-)spherical. However, more sophisticated approaches rely on a cluster-specific *Mahalanobis distance*, making it possible to find clusters of (hyper-) ellipsoidal shape. In addition, they relax the restriction (as it is present, e.g., in the fuzzy c-means algorithm) that all clusters have the same size [13]. Unfortunately, these additional degrees of freedom often reduce the robustness of the clustering algorithm, thus sometimes rendering their application problematic.

In this paper we consider how shape and size parameters of a cluster can be regularized, that is, modified in such a way that extreme cases are ruled out and/or a bias against extreme cases is introduced, which effectively improves robustness. The basic idea of shape regularization is the same as that of Tikhonov regularization for linear optimization problems [18, 8], while size and weight regularization is based on a bias towards equality as it is well-known from Laplace correction or Bayesian approaches to the estimation of probabilities.

This paper is organized as follows: in Sections 2 and 3 we briefly review some basics of mixture models and the expectation maximization algorithm as well as fuzzy clustering. In Section 4 we discuss our shape, size, and weight regularization schemes. In Section 5 we present experimental results on well-known data sets and finally, in Section 6, we draw conclusions from our discussion.

J.-F. Boulicaut et al. (Eds.): PKDD 2004, LNAI 3202, pp. 52–62, 2004.

2 Mixture Models and the EM Algorithm

In a mixture model [9] it is assumed that a given data set $\mathcal{X} = \{\boldsymbol{x}_j \mid j = 1, \ldots, n\}$ has been drawn from a population of c clusters. Each cluster is characterized by a probability distribution, specified as a prior probability and a conditional probability density function (cpdf). The data generation process may then be imagined as follows: first a cluster i, $i \in \{1, \ldots, c\}$, is chosen for a datum, indicating the cpdf to be used, and then the datum is sampled from this cpdf. Consequently the probability of a data point $\boldsymbol{x}$ can be computed as

$$p_{\boldsymbol{X}}(\boldsymbol{x}; \Theta) = \sum_{i=1}^{c} p_C(i; \Theta_i) \cdot f_{\boldsymbol{X}|C}(\boldsymbol{x}|i; \Theta_i),$$

where C is a random variable describing the cluster i chosen in the first step, $\boldsymbol{X}$ is a random vector describing the attribute values of the data point, and $\Theta = \{\Theta_1, \ldots, \Theta_c\}$ with each Θ_i, containing the parameters for one cluster (that is, its prior probability $\theta_i = p_C(i; \Theta_i)$ and the parameters of the cpdf).

Assuming that the data points are drawn independently from the same distribution (i.e., that the probability distributions of their underlying random vectors $\boldsymbol{X}_j$ are identical), we can compute the probability of a data set $\mathcal{X}$ as

$$P(\mathcal{X}; \Theta) = \prod_{j=1}^{n} \sum_{i=1}^{c} p_{C_j}(i; \Theta_i) \cdot f_{\boldsymbol{X}_j|C_j}(\boldsymbol{x}_j|i; \Theta_i),$$

Note, however, that we do not know which value the random variable C_j, which indicates the cluster, has for each example case $\boldsymbol{x}_j$. Fortunately, though, given the data point, we can compute the posterior probability that a data point $\boldsymbol{x}$ has been sampled from the cpdf of the i-th cluster using Bayes' rule as

$$p_{C|\boldsymbol{X}}(i|\boldsymbol{x}; \Theta) = \frac{p_C(i; \Theta_i) \cdot f_{\boldsymbol{X}|C}(\boldsymbol{x}|i; \Theta_i)}{f_{\boldsymbol{X}}(\boldsymbol{x}; \Theta)} = \frac{p_C(i; \Theta_i) \cdot f_{\boldsymbol{X}|C}(\boldsymbol{x}|i; \Theta_i)}{\sum_{k=1}^{c} p_C(k; \Theta_k) \cdot f_{\boldsymbol{X}|C}(\boldsymbol{x}|k; \Theta_k)}.$$

This posterior probability may be used to complete the data set w.r.t. the cluster, namely by splitting each datum $\boldsymbol{x}_j$ into c data points, one for each cluster, which are weighted with the posterior probability $p_{C_j|\boldsymbol{X}_j}(i|\boldsymbol{x}_j; \Theta)$. This idea is used in the well known expectation maximization (EM) algorithm [6], which consists in alternately computing these posterior probabilities and estimating the cluster parameters from the completed data set by maximum likelihood estimation.

For clustering numeric data it is usually assumed that the cpdf of each cluster is an m-variate normal distribution (Gaussian mixture [9, 3]), i.e.

$$\begin{aligned} f_{\boldsymbol{X}|C}(\boldsymbol{x}|i; \Theta_i) &= N(\boldsymbol{x}; \boldsymbol{\mu}_i, \boldsymbol{\Sigma}_i) \\ &= \frac{1}{\sqrt{(2\pi)^m |\boldsymbol{\Sigma}_i|}} \exp\left(-\frac{1}{2}(\boldsymbol{x} - \boldsymbol{\mu})^\top \boldsymbol{\Sigma}_i^{-1} (\boldsymbol{x} - \boldsymbol{\mu})\right), \end{aligned}$$

where $\boldsymbol{\mu}_i$ is the mean vector and $\boldsymbol{\Sigma}_i$ the covariance matrix of the normal distribution, $i = 1, \ldots, c$, and m is the number of dimensions of the data space.

In this case the maximum likelihood estimation formulae are

$$\theta_i = \frac{1}{n}\sum_{j=1}^{n} p_{C|\mathbf{X}_j}(i|\boldsymbol{x}_j;\Theta) \quad \text{and} \quad \boldsymbol{\mu}_i = \frac{\sum_{j=1}^{n} p_{C|\mathbf{X}_j}(i|\boldsymbol{x}_j;\Theta)\cdot \boldsymbol{x}_j}{\sum_{j=1}^{n} p_{C|\mathbf{X}_j}(i|\boldsymbol{x}_j;\Theta)}$$

for the prior probability θ_i and the mean vector $\boldsymbol{\mu}_i$ and

$$\boldsymbol{\Sigma}_i = \frac{\sum_{j=1}^{n} p_{C|\mathbf{X}_j}(i|\boldsymbol{x}_j;\Theta)\cdot (\boldsymbol{x}_j - \boldsymbol{\mu}_i)(\boldsymbol{x}_j - \boldsymbol{\mu}_i)^\top}{\sum_{j=1}^{n} p_{C|\mathbf{X}_j}(i|\boldsymbol{x}_j;\Theta)}$$

for the covariance matrix $\boldsymbol{\Sigma}_i$ of the i-th cluster, $i = 1, \ldots, c$.

3 Fuzzy Clustering

While most classical clustering algorithms assign each datum to exactly one cluster, thus forming a crisp partition of the given data, fuzzy clustering allows for *degrees of membership*, to which a datum belongs to different cluster [1, 2, 12]. Most fuzzy clustering algorithms are objective function based: they determine an optimal (fuzzy) partition of a given data set $\mathbf{X} = \{\boldsymbol{x}_j \mid j = 1, \ldots, n\}$ into c clusters by minimizing an objective function

$$J(\mathbf{X}, \mathbf{U}, \mathbf{C}) = \sum_{i=1}^{c}\sum_{j=1}^{n} u_{ij}^w d_{ij}^2$$

subject to the constraints

$$\sum_{j=1}^{n} u_{ij} > 0, \quad \text{for all } i \in \{1, \ldots, c\}, \qquad \text{and} \tag{1}$$

$$\sum_{i=1}^{c} u_{ij} = 1, \quad \text{for all } j \in \{1, \ldots, n\}, \tag{2}$$

where $u_{ij} \in [0, 1]$ is the membership degree of datum $\boldsymbol{x}_j$ to cluster i and d_{ij} is the distance between datum $\boldsymbol{x}_j$ and cluster i. The $c \times n$ matrix $\mathbf{U} = (u_{ij})$ is called the *fuzzy partition matrix* and $\mathbf{C}$ describes the set of clusters by stating location parameters (i.e. the cluster center) and maybe size and shape parameters for each cluster. The parameter w, $w > 1$, is called the *fuzzifier* or *weighting exponent*. It determines the "fuzziness" of the classification: with higher values for w the boundaries between the clusters become softer, with lower values they get harder. Usually $w = 2$ is chosen. Hard clustering results in the limit for $w \to 1$. However, a hard assignment may also be determined from a fuzzy result by assigning each data point to the cluster to which it has the highest degree of membership.

Constraint (1) guarantees that no cluster is empty and constraint (2) ensures that each datum has the same total influence by requiring that the membership degrees of a datum must add up to 1. Because of the second constraint this approach is usually called *probabilistic fuzzy clustering*, because with it the membership degrees for a datum formally resemble the probabilities of its being a member of the corresponding clusters. The partitioning property of a probabilistic clustering algorithm, which "distributes" the weight of a datum to the different clusters, is due to this constraint.

Unfortunately, the objective function J cannot be minimized directly. Therefore an iterative algorithm is used, which alternately optimizes the membership degrees and the cluster parameters [1, 2, 12]. That is, first the membership degrees are optimized for fixed cluster parameters, then the cluster parameters are optimized for fixed membership degrees. The main advantage of this scheme is that in each of the two steps the optimum can be computed directly. By iterating the two steps the joint optimum is approached (although, of course, it cannot be guaranteed that the global optimum will be reached – the algorithm may get stuck in a local minimum of the objective function J).

The update formulae are derived by simply setting the derivative of the objective function J w.r.t. the parameters to optimize equal to zero (necessary condition for a minimum). Independent of the chosen distance measure we thus obtain the following update formula for the membership degrees [12]:

$$u_{ij} = \frac{d_{ij}^{-\frac{2}{w-1}}}{\sum_{k=1}^{c} d_{kj}^{-\frac{2}{w-1}}}, \qquad (3)$$

that is, the membership degrees represent the relative inverse squared distances of a data point to the different cluster centers, which is a very intuitive result.

The update formulae for the cluster parameters, however, depend on what parameters are used to describe a cluster (location, shape, size) and on the chosen distance measure. Therefore a general update formula cannot be given. Here we briefly review the three most common cases: The best-known fuzzy clustering algorithm is the fuzzy c-means algorithm, which is a straightforward generalization of the classical crisp c-means algorithm. It uses only cluster centers for the cluster prototypes and relies on the *Euclidean distance*, i.e.,

$$d_{ij}^2 = (\boldsymbol{x}_j - \boldsymbol{\mu}_i)^\top (\boldsymbol{x}_j - \boldsymbol{\mu}_i),$$

where $\boldsymbol{\mu}_i$ is the center of the i-th cluster. Consequently it is restricted to finding spherical clusters of equal size. The resulting update rule is

$$\boldsymbol{\mu}_i = \frac{\sum_{j=1}^{n} u_{ij} \boldsymbol{x}_j}{\sum_{j=1}^{n} u_{ij}}, \qquad (4)$$

that is, the new cluster center is the weighted mean of the data points assigned to it, which is again a very intuitive result.

The Gustafson–Kessel algorithm [11] uses the *Mahalanobis distance*, i.e.,

$$d_{ij}^2 = (\boldsymbol{x}_j - \boldsymbol{\mu}_i)^\top \boldsymbol{\Sigma}_i^{-1} (\boldsymbol{x}_j - \boldsymbol{\mu}_i),$$

where $\boldsymbol{\mu}_i$ is the cluster center and $\boldsymbol{\Sigma}_i$ is a cluster-specific covariance matrix with determinant 1 that describes the shape of the cluster, thus allowing for ellipsoidal clusters of equal size. This distance function leads to same update rule (4) for the clusters centers. The covariance matrices are updated according to

$$\boldsymbol{\Sigma}_i = |\boldsymbol{\Sigma}_i^*|^{-\frac{1}{m}} \boldsymbol{\Sigma}_i^* \qquad \text{where} \qquad \boldsymbol{\Sigma}_i^* = \frac{\sum_{j=1}^n u_{ij} (\boldsymbol{x}_j - \boldsymbol{\mu}_i)(\boldsymbol{x}_j - \boldsymbol{\mu}_i)^\top}{\sum_{j=1}^n u_{ij}} \tag{5}$$

and m is the number of dimensions of the data space. $\boldsymbol{\Sigma}_i^*$ is called the *fuzzy covariance matrix*, which is simply normalized to determinant 1 to meet the abovementioned constraint. Compared to standard statistical estimation procedures, this is also a very intuitive result. It should be noted that the restriction to cluster of equal size may be relaxed by simply allowing general covariance matrices. However, depending on the characteristics of the data, this additional degree of freedom can deteriorate the robustness of the algorithm.

Finally, the fuzzy maximum likelihood estimation (FMLE) algorithm [10] is based on the assumption that the data was sampled from a mixture of c multivariate normal distributions as in the statistical approach of mixture models (cf. Section 2). It uses a (squared) distance that is inversely proportional to the probability that a datum was generated by the normal distribution associated with a cluster and also incorporates the prior probability of the cluster, i.e.,

$$d_{ij}^2 = \left(\frac{\theta_i}{\sqrt{(2\pi)^m |\boldsymbol{\Sigma}_i|}} \exp\left(-\frac{1}{2} (\boldsymbol{x}_j - \boldsymbol{\mu}_i)^\top \boldsymbol{\Sigma}_i^{-1} (\boldsymbol{x}_j - \boldsymbol{\mu}_i) \right) \right)^{-1},$$

where θ_i is the prior probability of the cluster, $\boldsymbol{\mu}_i$ is the cluster center, $\boldsymbol{\Sigma}_i$ a cluster-specific covariance matrix, which in this case is not required to be normalized to determinant 1, and m the number of dimensions of the data space (cf. Section 2). For the FMLE algorithm the update rules are not derived from the objective function due to technical obstacles, but by comparing it to the expectation maximization (EM) algorithm for a mixture of normal distributions (cf. Section 2), which, by analogy, leads to the same update rules for the cluster center and the cluster-specific covariance matrix as for the Gustafson–Kessel algorithm [12], that is, equations (4) and (5). The prior probability θ_i is, in direct analogy to statistical estimation (cf. Section 2), computed as

$$\theta_i = \frac{1}{n} \sum_{j=1}^n u_{ij}. \tag{6}$$

Note that the difference to the expectation maximization algorithm consists in the different ways in which the membership degrees (equation (3)) and the posterior probabilities in the EM algorithm are computed.

Since the high number of free parameters of the FMLE algorithm renders it unstable on certain data sets, it is usually recommended [12] to initialize it with a few steps of the very robust fuzzy c-means algorithm. The same holds, though to a somewhat lesser degree, for the Gustafson–Kessel algorithm.

It is worth noting that of both the Gustafson–Kessel as well as the FMLE algorithm there exist so-called *axes-parallel* version, which restrict the covariance matrices $\boldsymbol{\Sigma}_i$ to diagonal matrices and thus allow only axes-parallel ellipsoids [14]. These variants have certain advantages w.r.t. robustness and execution time.

4 Regularization

Regularization, as we use this term in this paper, means to modify the parameters of a cluster in such a way that certain conditions are satisfied or at least that a tendency (of varying strength, as specified by a user) towards satisfying these conditions is introduced. In particular we consider regularizing the (ellipsoidal) shape, the (relative) size, and the (relative) weight of a cluster.

4.1 Shape Regularization

The shape of a cluster is represented by its covariance matrix $\mathbf{\Sigma}_i$. Intuitively, $\mathbf{\Sigma}_i$ describes a general (hyper-)ellipsoidal shape, which can be obtained, for example, by computing the Cholesky decomposition or the eigenvalue decomposition of $\mathbf{\Sigma}_i$ and mapping the unit (hyper-)sphere with it.

Shape regularization means to modify the covariance matrix, so that a certain relation of the lengths of the major axes of the represented (hyper-)ellipsoid is obtained or that at least a tendency towards this relation is introduced. Since the lengths of the major axes are the roots of the eigenvalues of the covariance matrix, regularizing it means shifting the eigenvalues of $\mathbf{\Sigma}_i$. Note that such a shift leaves the eigenvectors unchanged, i.e., the orientation of the represented (hyper-)ellipsoid is preserved. Note also that such a shift of the eigenvalues is the basis of the Tikhonov regularization of linear optimization problems [18, 8], which inspired our approach. We suggest two methods:

Method 1: The covariance matrices $\mathbf{\Sigma}_i$, $i = 1, \ldots, c$, are adapted according to

$$\mathbf{\Sigma}_i^{(\text{adap})} = \sigma_i^2 \cdot \frac{\mathbf{S}_i + h^2\mathbf{1}}{\sqrt[m]{|\mathbf{S}_i + h^2\mathbf{1}|}} = \sigma_i^2 \cdot \frac{\mathbf{\Sigma}_i + \sigma_i^2 h^2\mathbf{1}}{\sqrt[m]{|\mathbf{\Sigma}_i + \sigma_i^2 h^2\mathbf{1}|}},$$

where m is the dimension of the data space, $\mathbf{1}$ is a unit matrix, $\sigma_i^2 = \sqrt[m]{|\mathbf{\Sigma}_i|}$ is the equivalent isotropic variance (equivalent in the sense that it leads to the same (hyper-)volume, i.e., $|\mathbf{\Sigma}_i| = |\sigma_i^2\mathbf{1}|$), $\mathbf{S}_i = \sigma_i^{-2}\mathbf{\Sigma}_i$ is the covariance matrix scaled to determinant 1, and h is the regularization parameter.

This regularization shifts up all eigenvalues by the value of $\sigma_i^2 h^2$ and then renormalizes the resulting matrix so that the determinant of the old covariance matrix is preserved (i.e., the (hyper-)volume is kept constant). This regularization tends to equalize the lengths of the major axes of the represented (hyper-) ellipsoid and thus introduces a tendency towards (hyper-)spherical clusters. This tendency is the stronger, the greater the value of h. In the limit, for $h \to \infty$, the clusters are forced to be exactly spherical; for $h = 0$ the shape is left unchanged.

Method 2: The above method always changes the length ratios of the major axes and thus introduces a general tendency towards (hyper-)spherical clusters. In this method, however, a limit r, $r > 1$, for the length ratio of the longest to the shortest major axis of the (hyper-)ellipsoid is used and only if this limit is exceeded, the eigenvalues are shifted in such a way that the limit is satisfied.

Formally: let λ_k, $k = 1, \ldots m$, be the eigenvalues of the matrix $\mathbf{\Sigma}_i$. Set

$$h^2 = \begin{cases} 0, & \text{if } \dfrac{\max_{k=1}^m \lambda_k}{\min_{k=1}^m \lambda_k} \le r^2, \\ \dfrac{\max_{k=1}^m \lambda_k - r^2 \min_{k=1}^m \lambda_k}{\sigma_i^2 (r^2 - 1)}, & \text{otherwise,} \end{cases}$$

and execute Method 1 with this value of h^2.

4.2 Size Regularization

The size of a cluster can be described in different ways, for example, by the determinant of its covariance matrix $\mathbf{\Sigma}_i$, which is a measure of the clusters squared (hyper-)volume, an equivalent isotropic variance σ_i^2 or an equivalent isotropic radius (standard deviation) σ_i (equivalent in the sense that they lead to the same (hyper-)volume, see above). The latter two measures are defined as

$$\sigma_i^2 = \sqrt[m]{|\mathbf{\Sigma}_i|} \quad \text{and} \quad \sigma_i = \sqrt{\sigma_i^2} = \sqrt[2m]{|\mathbf{\Sigma}_i|}$$

and thus the (hyper-)volume may also be written as $\sigma_i^m = \sqrt{|\mathbf{\Sigma}_i|}$.

Size regularization means to ensure a certain relation between the cluster sizes or at least to introduce a tendency into this direction. We suggest three different versions of size regularization, in each of which the measure that is used to describe the cluster size is specified by an exponent a of the equivalent isotropic radius σ_i, with the special cases:

$a = 1$: equivalent isotropic radius,
$a = 2$: equivalent isotropic variance,
$a = m$: (hyper-)volume.

Method 1: The equivalent isotropic radii σ_i are adapted according to

$$\sigma_i^{(\text{adap})} = \sqrt[a]{s \cdot \frac{\sum_{k=1}^c \sigma_k^a}{\sum_{k=1}^c (\sigma_k^a + b)} \cdot (\sigma_i^a + b)} = \sqrt[a]{s \cdot \frac{\sum_{k=1}^c \sigma_k^a}{cb + \sum_{k=1}^c \sigma_k^a} \cdot (\sigma_i^a + b)}.$$

That is, each cluster size is increased by the value of the regularization parameter b and then the sizes are renormalized so that the sum of the cluster sizes is preserved. However, the parameter s may be used to scale the sum of the sizes up or down (by default $s = 1$). For $b \to \infty$ the cluster sizes are equalized completely, for $b = 0$ only the parameter s has an effect. This method is inspired by Laplace correction or Bayesian estimation with an uniformative prior (see below).

Method 2: This method does not renormalize the sizes, so that the size sum increases by cb. However, this missing renormalization may be mitigated to some degree by specifying a value of the scaling parameter s that is smaller than 1. The equivalent isotropic radii σ_i are adapted according to

$$\sigma_i^{(\text{adap})} = \sqrt[a]{s \cdot (\sigma_i^a + b)}.$$

Method 3: The above methods always change the relation of the cluster sizes and thus introduce a general tendency towards clusters of equal size. In this method, however, a limit r, $r > 1$, for the size ratio of the largest to the smallest cluster is used and only if this limit is exceeded, the sizes are changed in such a way that the limit is satisfied. To achieve this, b is set according to

$$b = \begin{cases} 0, & \text{if } \dfrac{\max_{k=1}^{c} \sigma_k^a}{\min_{k=1}^{c} \sigma_k^a} \le r, \\ \dfrac{\max_{k=1}^{c} \sigma_k^a - r \min_{k=1}^{c} \sigma_k^a}{r-1}, & \text{otherwise,} \end{cases}$$

and then Method 1 is executed with this value of b.

4.3 Weight Regularization

A cluster weight θ_i only appears in the mixture model approach and the FMLE algorithm, where it describes the prior probability of a cluster. For the cluster weight we may use basically the same regularization methods as for the cluster size, with the exception of the scaling parameter s, since the θ_i are probabilities, i.e., we must ensure $\sum_{i=1}^{c} \theta_i = 1$. Therefore we have:

Method 1: The cluster weights θ_i

$$\theta_i^{(\text{adap})} = \frac{\sum_{k=1}^{c} \theta_k}{\sum_{k=1}^{c} (\theta_k + b)} \cdot (\theta_i + b) = \frac{\sum_{k=1}^{c} \theta_k}{cb + \sum_{k=1}^{c} \theta_k} \cdot (\theta_i + b),$$

where b is the regularization parameter. Note that this method is equivalent to a Laplace corrected estimation of the prior probabilities or a Bayesian estimation with an uninformative (uniform) prior.

Method 2: The value of the regularization parameter b is computed as

$$b = \begin{cases} 0, & \text{if } \dfrac{\max_{k=1}^{c} \theta_k}{\min_{k=1}^{c} \theta_k} \le r, \\ \dfrac{\max_{k=1}^{c} \theta_k - r \min_{k=1}^{c} \theta_k}{r-1}, & \text{otherwise,} \end{cases}$$

with a user-specified maximum weight ratio r, $r > 1$, and then Method 1 is executed with this value of b.

5 Experiments

We implemented our regularization methods as part of an expectation maximization and fuzzy clustering program written by the first author of this paper and applied it to several different data sets from the UCI machine learning repository [4]. In all data sets each dimension was normalized to mean value 0 and standard deviation 1 in order to avoid any distortions that may result from different scaling of the coordinate axes.

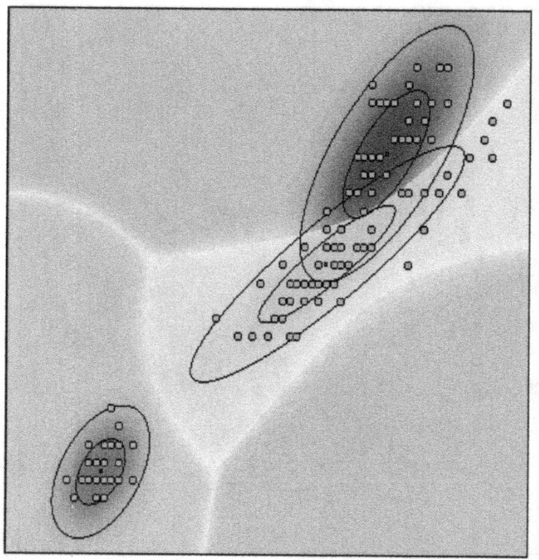
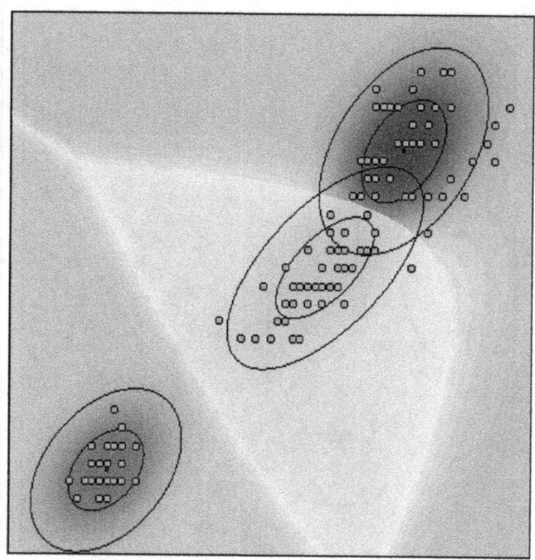

Fig. 1. Result of Gustafson-Kessel algorithm on the iris data with fixed cluster size without (left) and with shape regularization (right, method 2 with $r = 4$). Both images show the petal length (horizontal) and the petal width (vertical). Clustering was done on all four attributes (sepal length and sepal width in addition to the above).

As one illustrative example, we present here the result of clustering the iris data (excluding, of course, the class attribute) with the Gustafson–Kessel algorithm using three cluster of fixed size (measured as the isotropic radius) of 0.4 (since all dimensions are normalized to mean 0 and standard deviation 1, 0.4 is a good size of a cluster if three clusters are to be found). The result without shape regularization is shown in Figure 1 on the left. Due to the few data points located in a thin diagonal cloud on the right border on the figure, the middle cluster is drawn into a very long ellipsoid. Although this shape minimizes the objective function, it may not be a desirable result, because the cluster structure is not compact enough. Using shape regularization method 2 with $r = 4$ the cluster structure shown on the right in Figure 1 is obtained. In this result the clusters are more compact and resemble the class structure of the data set.

As another example let us consider the result of clustering the wine data with the fuzzy maximum likelihood estimation (FMLE) algorithm using three clusters of variable size. We used attributes 7, 10, and 13, which are the most informative w.r.t. the class assignments. One result without size regularization is shown in Figure 2 on the left. However, the algorithm is much too unstable to present a unique result. Often enough clustering fails completely, because one cluster collapses to a single data point – an effect that is due to the steepness of the Gaussian probability density function. This situation is considerably improved with size regularization, a result of which (which sometimes, with a fortunate initialization, can also be achieved without) is shown on the right in Figure 2. It was obtained with method 3 with $r = 2$. Although the result is still not unique and sometimes clusters still focus on very few data points, the algorithm is considerably more stable and reasonable results are obtained much more of-

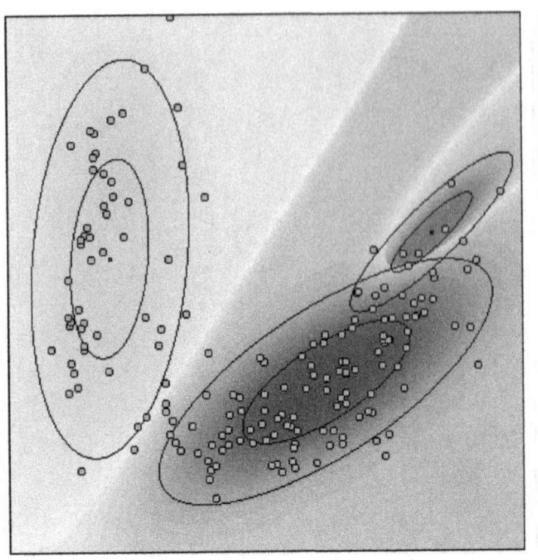
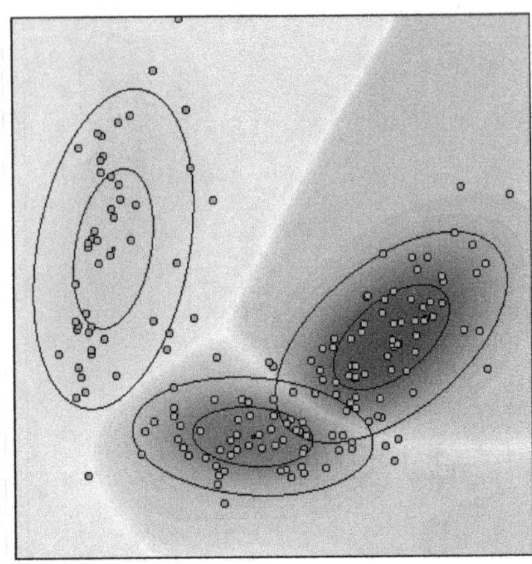

Fig. 2. Result of fuzzy maximum likelihood estimation (FMLE) algorithm on the wine data with fixed cluster weight without (left) and with size regularization (right, method 3 with $r = 2$). Both images show attribute 7 (horizontal) and attribute 10 (vertical). Clustering was done on attributes 7, 10, and 13.

ten than without regularization. Hence we can conclude that size regularization considerably improves the robustness of the algorithm.

6 Conclusions

In this paper we suggested shape and size regularization methods for clustering algorithms that use a cluster-specific Mahalanobis distance to describe the shape and the size of a cluster. The basic idea is to introduce a tendency towards equal length of the major axes of the represented (hyper-)ellipsoid and towards equal cluster sizes. As the experiments show, these methods improve the robustness of the more sophisticated fuzzy clustering algorithms, which without them suffer from instabilities even on fairly simple data sets. Regularized clustering can even be used without an initialization by the fuzzy c-means algorithm. It should be noted that with a time-dependent shape regularization parameter one may obtain a soft transition from the fuzzy c-means algorithm (spherical clusters) to the Gustafson-Kessel algorithm (general ellipsoidal clusters).

Software. A free implementation of the described methods as command line programs for expectation maximization and fuzzy clustering can be found at

`http://fuzzy.cs.uni-magdeburg.de/~borgelt/software.html#cluster`

References

1. J.C. Bezdek. *Pattern Recognition with Fuzzy Objective Function Algorithms.* Plenum Press, New York, NY, USA 1981
2. J.C. Bezdek, J. Keller, R. Krishnapuram, and N. Pal. *Fuzzy Models and Algorithms for Pattern Recognition and Image Processing.* Kluwer, Dordrecht, Netherlands 1999
3. J. Bilmes. A Gentle Tutorial on the EM Algorithm and Its Application to Parameter Estimation for Gaussian Mixture and Hidden Markov Models. University of Berkeley, Tech. Rep. ICSI-TR-97-021, 1997
4. C.L. Blake and C.J. Merz. UCI Repository of Machine Learning Databases. `http://www.ics.uci.edu/~mlearn/MLRepository.html`
5. H.H. Bock. *Automatische Klassifikation.* Vandenhoeck & Ruprecht, Göttingen, Germany 1974
6. A.P. Dempster, N. Laird, and D. Rubin. Maximum Likelihood from Incomplete Data via the EM Algorithm. *Journal of the Royal Statistical Society (Series B)* 39:1–38. Blackwell, Oxford, United Kingdom 1977
7. R.O. Duda and P.E. Hart. *Pattern Classification and Scene Analysis.* J. Wiley & Sons, New York, NY, USA 1973
8. H. Engl, M. Hanke, and A. Neubauer. *Regularization of Inverse Problems.* Kluwer, Dordrecht, Netherlands 1996
9. B.S. Everitt and D.J. Hand. *Finite Mixture Distributions.* Chapman & Hall, London, UK 1981
10. I. Gath and A.B. Geva. Unsupervised Optimal Fuzzy Clustering. *IEEE Trans. Pattern Analysis & Machine Intelligence* 11:773–781. IEEE Press, Piscataway, NJ, USA, 1989
11. E.E. Gustafson and W.C. Kessel. Fuzzy Clustering with a Fuzzy Covariance Matrix. *Proc. 18th IEEE Conference on Decision and Control (IEEE CDC, San Diego, CA)*, 761–766, IEEE Press, Piscataway, NJ, USA 1979
12. F. Höppner, F. Klawonn, R. Kruse, and T. Runkler. *Fuzzy Cluster Analysis.* J. Wiley & Sons, Chichester, England 1999
13. A. Keller and F. Klawonn. Adaptation of Cluster Sizes in Objective Function Based Fuzzy Clustering. In: C.T. Leondes, ed. *Database and Learning Systems IV*, 181–199. CRC Press, Boca Raton, FL, USA 2003
14. F. Klawonn and R. Kruse. Constructing a Fuzzy Controller from Data. *Fuzzy Sets and Systems* 85:177-193. North-Holland, Amsterdam, Netherlands 1997
15. T. Kohonen. *Learning Vector Quantization for Pattern Recognition.* Technical Report TKK-F-A601. Helsinki University of Technology, Finland 1986
16. T. Kohonen. *Self-Organizing Maps.* Springer-Verlag, Heidelberg, Germany 1995 (3rd ext. edition 2001)
17. R. Krishnapuram and J. Keller. A Possibilistic Approach to Clustering, *IEEE Transactions on Fuzzy Systems*, 1:98-110. IEEE Press, Piscataway, NJ, USA 1993
18. A.N. Tikhonov and V.Y. Arsenin. Solutions of Ill-Posed Problems. J. Wiley & Sons, New York, NY, USA 1977
19. H. Timm, C. Borgelt, and R. Kruse. A Modification to Improve Possibilistic Cluster Analysis. *Proc. IEEE Int. Conf. on Fuzzy Systems (FUZZ-IEEE 2002, Honolulu, Hawaii).* IEEE Press, Piscataway, NJ, USA 2002

Combining Multiple Clustering Systems

Constantinos Boulis and Mari Ostendorf

Department of Electrical Engineering,
University of Washington, Seattle, WA 98195, USA
{boulis,mo}@ee.washington.edu

Abstract. Three methods for combining multiple clustering systems are presented and evaluated, focusing on the problem of finding the correspondence between clusters of different systems. In this work, the clusters of individual systems are represented in a common space and their correspondence estimated by either "clustering clusters" or with Singular Value Decomposition. The approaches are evaluated for the task of topic discovery on three major corpora and eight different clustering algorithms and it is shown experimentally that combination schemes almost always offer gains compared to single systems, but gains from using a combination scheme depend on the underlying clustering systems.

1 Introduction

Clustering has an important role in a number of diverse fields, such as genomics [1], lexical semantics [2], information retrieval [3] and automatic speech recognition [4], to name a few. A number of different clustering approaches have been suggested [5] such as agglomerative clustering, mixture densities and graph partitioning. Most clustering methods focus on individual criteria or models and do not address issues of combining multiple different systems. The problem of combining multiple clustering systems is analogous to the classifier combination problem, that has received increased attention over the last years [6]. Unlike the classifier combination problem, though, the correspondence between clusters of different systems is unknown. For example, consider two clustering systems applied to nine data points and clustered in three groups. System A's output is $\boldsymbol{o}_A = [1, 1, 2, 3, 2, 2, 1, 3, 3]$ and system B's output is $\boldsymbol{o}_B = [2, 2, 3, 1, 1, 3, 2, 1, 1]$, where the i-th element of $\boldsymbol{o}$ is the group to which data point i is assigned. Although the two systems appear to be making different decisions, they are in fact very similar. Cluster 1 of system A and cluster 2 of system B are identical, and cluster 2 of system A and cluster 3 of system B agree 2 out of 3 times, as cluster 3 of system A and cluster 1 of system B. If the correspondence problem is solved then a number of system combination schemes can be applied.

Finding the optimum correspondence requires a criterion and a method for optimization. The criterion used here is maximum agreement, i.e. find the correspondence where clusters of different systems make the maximum number of the same decisions. Second, we must optimize the selected criterion. Even if we assume a 0 or 1 correspondence between clusters with only two systems of M

J.-F. Boulicaut et al. (Eds.): PKDD 2004, LNAI 3202, pp. 63–74, 2004.

topics each, a brute-force approach would require the evaluation of $M!$ possible solutions. In this work, three novel methods are presented for determining the correspondence of clusters and combining them. Two of the three methods are formulated and solved with linear optimization and the third uses singular value decomposition.

Another contribution of this work is the empirical result that the combination schemes are not independent of the underlying clustering systems. Most of the past work has focused on combining systems generated from a single clustering algorithm (using resampling or different initial conditions), usually k-means. In this work, we experimentally show that the relative gains of applying a combination scheme are not the same across eight different clustering algorithms. For example, although the mixture of multinomials was one of the worse performing clustering algorithms, it is shown that when different runs were combined it achieved the best performance of all eight clustering algorithms in two out of three corpora. The results suggest that an algorithm should not be evaluated solely on the basis of its individual performance, but also on the combination of multiple runs.

2 Related Work

Combining multiple clustering systems has recently attracted the interest of several researchers in the machine learning community. In [7], three different approaches for combining clusters based on graph-partitioning are proposed and evaluated. The first approach avoids the correspondence problem by defining a pairwise similarity matrix between data points. Each system is represented by a $D \times D$ matrix (D is the total number of observations) where the (i, j) position is either 1 if observations i and j belong to the same cluster and 0 otherwise. The average of all matrices is used as the input to a final similarity-based clustering algorithm. The core of this idea also appears in [8–12]. A disadvantage of this approach is that it has quadratic memory and computational requirements. Even by exploiting the fact that each of the $D \times D$ matrices is symmetric and sparse, this approach is impractical for high D.

The second approach taken in [7], is that of a hypergraph cutting problem. Each one of the clusters of each system is assumed to be a hyperedge in a hypergraph. The problem of finding consensus among systems is formulated as partitioning a hypergraph by cutting a minimum number of hyperedges. This approach is linear with the number of data points, but requires fairly balanced data sets and all hyperedges having the same weight. A similar approach is presented in [13], where each data point is represented with a set of meta-features. Each meta-feature is the cluster membership for each system, and the data points are clustered using a mixture model. An advantage of [13] is that it can handle missing meta-features, i.e. a system failing to cluster some data points. Algorithms of this type, avoid the cluster correspondence problem by clustering directly the data points.

The third approach presented in [7], is to deal with the cluster correspondence problem directly. As stated in [7], the objective is to *"cluster clusters"*, where each cluster of a system is a hyperedge and the objective is to combine similar hyperedges. The data points will be assigned to the combined hyperedge they most strongly belong to. Clustering hyperedges is performed by using graph-partitioning algorithms. The same core idea can also be found in [10, 14–16]. In [10], different clustering solutions are obtained by resampling and are aligned with the clusters estimated on all the data. In both [14, 15], the different clustering solutions are obtained by multiple runs of the k-means algorithm with different initial conditions. An agglomerative pairwise cluster merging scheme is used, with a heuristic to determine the corresponding clusters. In [16], a two-stage clustering procedure is proposed. Resampling is used to obtain multiple solutions of k-means. The output centroids from multiple runs are clustered with a new k-means run. A disadvantage of [16] is that it requires access to the original features of the data points, while all other schemes do not. Our work falls in the third approach, i.e. attempts to first find a correspondence between clusters and then combine clusters without requiring the original observations.

3 Finding Cluster Correspondence

In this paper, three novel methods to address the cluster correspondence problem are presented. The first two cast the correspondence problem as an optimization problem, and the third method is based on singular value decomposition.

3.1 Constrained and Unconstrained Search

We want to find the assignment of clusters to entities (metaclusters) such that the overall agreement among clusters is maximized. Suppose $\boldsymbol{R}_{\{c,s\}}$ is the $D \times 1$ vector representation of cluster c of system s (with D being the total number of documents). The k-th element of $\boldsymbol{R}_{\{c,s\}}$ is $p(cluster = c|observation = k, system = s)$. The agreement between clusters $\{c, s\}$ and $\{c', s'\}$ is defined as:

$$g_{\{c,s\},\{c',s'\}} = \boldsymbol{R}^T_{\{c,s\}} \cdot \boldsymbol{R}_{\{c',s'\}} \tag{1}$$

In addition, suppose that $\lambda^{\{m\}}_{\{c,s\}} = 1$ if cluster c of system s is assigned to metacluster m and 0 otherwise, and $r^{\{m\}}_{\{c,s\}}$ is the "reward" of assigning cluster c of system s to metacluster m, defined as:

$$r^{\{m\}}_{\{c,s\}} = \frac{1}{|I(m)|} \sum_{\{c',s'\} \in I(m)} g_{\{c,s\},\{c',s'\}} \quad , \{c', s'\} \in I(m) \iff \lambda^{\{m\}}_{\{c',s'\}} \neq 0 \tag{2}$$

We seek to find the argument that maximizes:

$$\boldsymbol{\lambda}^* = arg \max_{\boldsymbol{\lambda}} \sum_{m=1}^{M} \sum_{s=1}^{S} \sum_{c=1}^{C_s} \lambda^{\{m\}}_{\{c,s\}} r^{\{m\}}_{\{c,s\}} \tag{3}$$

$$\text{subject to the constraints} \sum_{m=1}^{M} \lambda^{\{m\}}_{\{c,s\}} = 1, \quad \forall c, s \tag{4}$$

Optionally, we may want to add the following constraint:

$$\sum_{c=1}^{C_s} \lambda_{\{c,s\}}^{\{m\}} = 1, \quad \forall s, m \tag{5}$$

This is a linear optimization problem and efficient techniques exist for maximizing the objective function. In our implementation, the GNU Linear Programming library was used[1]. The scheme that results from omitting the constraints of equation (5) is referred to as *unconstrained*, while including them results in the *constrained* combination scheme. The added constraints ensure that exactly one cluster from each system is assigned to each metacluster and are useful when $C_s = C \; \forall s$. The entire procedure is iterative, starting from an initial assignment of clusters to metaclusters and alternating between equations (2) and (3).

The output of the clustering procedure is matrix $\mathbf{F}$ of size $D \times M$, where each column is the centroid of each metacluster. The $\boldsymbol{F}_m$ column is given by:

$$\boldsymbol{F}_m = \frac{1}{|I(m)|} \sum_{\{c,s\} \in I(m)} \boldsymbol{R}_{\{c,s\}}^T \tag{6}$$

This can be the final output or a clustering stage can be applied using the $\mathbf{F}$ matrix as the observation representations. Note that the assignments can be continuous numbers between 0 and 1 (soft decisions) and that the systems do not need to have the same number of clusters, nor do the final number of metaclusters need to be the same as the number of clusters. To simplify the experiments, here we have assumed that the number of clusters is known and equal to the number of topics, i.e. $C_s = M = \#$of topics $\forall s$. The methodology presented here does not assume access to the original features and therefore it can be applied in cases irrespective of whether the original features were continuous or discrete.

The optimization procedure is very similar to any partition-based clustering procedure trained with the Expectation-Maximization algorithm, like k-means. In fact, this scheme is "clustering clusters", i.e. expressing clusters in a common vector space and grouping them into similar sets. Although the problem is formulated from the optimization perspective, any clustering methodology can be applied (statistical, graph-partitioning). However, there are two reasons that favor the optimization approach. First, it directly links the correspondence problem to an objective function that can be maximized. Second, it allows us to easily integrate constraints during clustering such as equation (5). As it is shown in section 5, the constrained clustering scheme offers gains over the unconstrained case, when it is appropriate for the task.

3.2 Singular Value Decomposition Combination

The third combination approach we introduce is based on Singular Value Decomposition (SVD). As before, we will assume that all systems have the same number of clusters for notational simplicity, though it is not required of the

[1] http://www.gnu.org/software/glpk/glpk.html

algorithm. Just as before, we construct matrix $\mathbf{R}$ of size $D \times SC$ (D is the number of observations, S is the number of systems, C the number of clusters), where each row contains the cluster posteriors of all systems for a given observation. $\mathbf{R}$ can be approximated as $\mathbf{R} \approx \mathbf{U} * \mathbf{S} * \mathbf{\Lambda}^{\mathbf{t}}$ where $\mathbf{U}$ is orthogonal and of size $D \times C$, $\mathbf{S}$ is diagonal and of size $C \times C$ and $\mathbf{\Lambda}$ is orthogonal and of size $(SC) \times C$. The final metaspace is $\mathbf{R} * \mathbf{\Lambda}$ of size $D \times C$. If we define $p_s(c|d) = p(cluster = c|observation = d, system = s)$, $c = 1 \ldots C, s = 1 \ldots S, d = 1 \ldots D$ and $h_C(l) = l - C\lfloor l/C \rfloor$ (remainder of division), then the $\phi_{d,c}$ element of $\mathbf{R} * \mathbf{\Lambda}$ is given by:

$$\phi_{d,c} = \sum_{k=1}^{S} \lambda_{g_c(k),c} p_k \left(h_C(g_c(k)) | d \right) \tag{7}$$

where $g_c(\cdot)$ is a function that aligns clusters of different systems and is estimated by SVD. In essence, SVD identifies the most correlated clusters, i.e. finds $g_c(\cdot)$ and combines them with linear interpolation. The λ weights provide a soft alignment of clusters. After SVD, a final clustering is performed using the $\phi_{d,c}$ representation.

4 Evaluating Clustering Systems

There is no consensus in the literature on how to evaluate clustering decisions. In this work, we used two measures to evaluate the clustering output. The first is the classification accuracy of a one-to-one mapping between clusters and true classes. The problem of finding the optimum assignment of M clusters to M classes can be formulated and solved with linear programming. If $r_{i,j}$ is the "reward" of assigning cluster i to class j (which can be the number of observations they agree), $\lambda_{i,j}$=1 if cluster i is assigned to class j and 0 otherwise are the parameters to estimate, then we seek to find: $\max_{\lambda_{i,j}} \sum_{i,j} r_{i,j} \lambda_{i,j}$ under the constraints $\sum_i \lambda_{i,j} = 1$ and $\sum_j \lambda_{i,j} = 1$. The constraints will ensure a one-to-one mapping.

The second measure we used is the normalized mutual information (NMI) between clusters and classes, introduced in [7]. The measure does not assume a fixed cluster-to-class mapping but rather takes the average mutual information between every pair of cluster and class. It is given by:

$$NMI = \frac{\sum_{i=1}^{M} \sum_{j=1}^{M} n_{i,j} \log \left(\frac{n_{i,j} D}{n_i m_j} \right)}{\sqrt{\sum_{i=1}^{M} n_i \log \frac{n_i}{D} \sum_{j=1}^{M} m_j \log \frac{m_j}{D}}} \tag{8}$$

where $n_{i,j}$ is the number of observations cluster i and class j agree, n_i is the number of observations assigned to cluster i, m_j the number of observation of class j and D the total number of observations. It can be shown that $0 < NMI \leq 1$ with NMI=1 corresponding to perfect classification accuracy.

5 Experiments

The multiple clustering system combination schemes that are introduced in this paper are general and can, in principle, be applied to any clustering problem. The task we have chosen to evaluate our metaclustering schemes is topic discovery, i.e. clustering documents according to their topic. Topic discovery is an especially hard clustering problem because of the high dimensionality of the data points and the redundancy of many features. To simplify our experiments, the number of topics is assumed to be known. This is an assumption that is not true in many practical cases, but standard techniques such as Bayesian Information Criterion [17] can be used to select the number of topics. It should be noted that the unconstrained and SVD combination schemes do not require the same number of clusters for all systems. On the other hand, the constrained clustering scheme was proposed based on this assumption.

5.1 Corpora

The techniques proposed in this work are applied on three main corpora with different characteristics. The first corpus is 20Newsgroups[2], a collection of 18828 postings into one of 20 categories (newsgroups). The second corpus is a subset of Reuters-21578[3], consisting of 1000 documents equally distributed among 20 topics. The third corpus is Switchboard-I release 2.0 [18], a collection of 2263 5-minute telephone conversations on 67 possible topics. Switchboard-I and to a smaller extent 20Newsgroups, are characterized with a spontaneous, less structured style. On the other hand, Reuters-21578 contains carefully prepared news stories for broadcasting. 20Newsgroups and the subset of Reuters are balanced, i.e. documents are equally divided by topics, but Switchboard-I is not. Also, the median length of a document varies significantly across corpora (155 words for 20Newsgroups, 80 for the subset of Reuters-21578 and 1328 for Switchboard-I). Standard processing was applied in all corpora. Words in the default stoplist of CLUTO (total 427 words) are removed, the remaining stemmed and only tokens with T or more occurrences (T=5 for 20Newsgroups, T=2 for Reuters-21578 and Switchboard-I) are retained. These operations result in 26857 unique tokens and 1.4M total tokens in 20Newsgroups, 4128 unique tokens and 50.5K total tokens in Reuters, and 11550 unique and 0.4M total tokens in Switchboard.

5.2 Clustering Algorithms

A number of different clustering systems were used, including the mixture of multinomials (MixMulti) and the optimization-based clustering algorithms and criteria described in [19]. The MixMulti algorithm clusters documents by estimating a mixture of multinomial distributions. The assumption is that each topic is characterized by a different multinomial distribution, i.e. different counts of each word given a topic. The probability of a document d is given by: $p(d) \propto \sum_{c=1}^{M} p(c) \prod_{w \in W_d} p(w|c)^{n(w,d)}$ where M is the number of topics, W_d is the set of unique words that appear in document d, $p(w|c)$ is the probability of word w given cluster c and $n(w,d)$ is the count of word w in document d. The cluster

[2] http://www.ai.mit.edu/~jrennie/20Newsgroups/

[3] http://www.daviddlewis.com/resources/testcollections/

Table 1. Performance of different combination schemes on various clustering algorithms for 20Newsgroups.

	Single Run	Best of 100 runs	SVD Combin.	Constr. Combin.	Unconstr. Combin.	No Combin.
I_1						
Accuracy	.422	.412	.418	.417	.408	.459
NMI	.486	.485	.481	.480	.463	.500
I_2						
Accuracy	.575	.603	.634	.615	.639	.624
NMI	.601	.621	.637	.628	.640	.637
E_1						
Accuracy	.579	.604	.648	.641	.610	.635
NMI	.588	.606	.639	.631	.628	.633
G_1						
Accuracy	.535	.561	.581	.562	.578	.576
NMI	.561	.585	.593	.581	.582	.589
G_1'						
Accuracy	.576	.608	.642	.630	.563	.644
NMI	.584	.603	.631	.622	.620	.632
H_1						
Accuracy	.570	.584	.636	.641	.549	.642
NMI	.593	.610	.629	.627	.592	.628
H_2						
Accuracy	.586	.611	.656	.639	.602	.641
NMI	.598	.616	.646	.634	.628	.638
MixMulti						
Accuracy	.534	.620	.679	.677	.621	.651
NMI	.587	.625	.662	.656	.651	.662

c that each document is generated from is assumed to be hidden. Training such a model is carried out using the Expectation-Maximization algorithm [20]. In practice, smoothing the multinomial distributions is necessary. The mixture of multinomials algorithm is the unsupervised analogue of the Naive Bayes algorithm and has been successfully used in the past for document clustering [21]. Mixture models, in general, have been extensively used for data mining and pattern discovery [22].

The software package CLUTO[4] was used for the optimization-based algorithms. Using CLUTO, a number of different clustering methods (hierarchical, partitional and graph-partitioning) and criteria can be used. For example, the I_2 criterion maximizes the function $\sum_{k=1}^{M}\sqrt{\sum_{\boldsymbol{u},\boldsymbol{v}\in c_k}\cos(\boldsymbol{u},\boldsymbol{v})}$, where c_k is the set of documents in cluster k and $\boldsymbol{u},\boldsymbol{v}$ are the tfidf vector representations of documents u, v respectively. The I_2 criterion attempts to maximize intra-cluster similarity. Other criteria, like E_1, attempt to minimize inter-cluster similarity

[4] http://www-users.cs.umn.edu/~karypis/cluto/

and yet other criteria, like H_2, attempt to optimize a combination of both. For more information on the optimization criteria and methods, see [19].

Having determined the clustering algorithms to use, the next question is how to generate the systems to be combined. We may combine systems from different clustering algorithms, pick a single algorithm and generate different systems by resampling, or pick a single algorithm and use different initial conditions for each system. In this work we chose the last option.

5.3 Results

On all results reported in this work the *direct* clustering method was used for the CLUTO algorithms. For the *single run* case, the number reported is the average of 100 independent runs. For the *best of 100 runs* case, the number is the average of 10 runs where each run selects the system with the highest objective function out of 100 trials. A trial is an execution of a clustering algorithm with a different initial condition. For the metaclustering schemes, the final clustering is performed, with the default values of CLUTO. 100 runs of the CLUTO algorithm are performed and the one with the highest objective function selected.

In Table 1, the performance of the three combination schemes applied on eight different clustering algorithms on 20Newsgroups is shown. For every clustering algorithm except I_1, we can observe significant gains of the combination schemes compared to a single run or selecting the system with the highest objective function. The results show that the SVD combination outperforms the constrained combination which in turn outperforms the unconstrained combination. This suggests that the constraints introduced are meaningful and lead to improved performance over the unconstrained scheme. Also shown in Table 1 are the results from not using any combination scheme. This means that the clusters of different systems are not combined but rather the cluster posteriors for all systems are used as a new document representation. This corresponds to using matrix $\mathbf{R}$ from subsection 3.2 without any dimensionality reduction. This is the approach taken in [13]. From Table 1, we see that for the MixMulti case there are gains from using SVD combination rather than using no combination of clusters at all. For other systems, gains are small or differences are insignificant, except for I_1 again where accuracy degrades significantly.

In Table 2, the performance of the three combination schemes over the same eight algorithms on a 1000-document subset of Reuters-21578 is shown. The same trends as in Table 1 seem to hold. Combination appears to offer significant improvements for all clustering algorithms, with SVD combination having a lead over the other two combination schemes. In most cases, SVD combination is better than the best individual clustering system. As in Table 1, the constrained scheme is superior to unconstrained but not as good as SVD combination.

In Table 3 the experiments are repeated for the Switchboard corpus. In contrast to previous tables, the combination schemes do not offer an improvement for the CLUTO algorithms and for the unconstrained scheme there is even a degradation compared to the single run case. However, the mixture of multinomials records a very big improvement of about 40% on classification accuracy.

Table 2. Performance of different combination schemes on various clustering algorithms for a 1000-document subset of Reuters-21578.

	Single Run	**Best of 100 runs**	**SVD Combin.**	**Constr. Combin.**	**Unconstr. Combin.**	**No Combin.**
I_1						
Accuracy	.636	.644	.696	.669	.673	.686
NMI	.697	.697	.735	.711	.725	.726
I_2						
Accuracy	.709	.797	.838	.838	.764	.808
NMI	.760	.805	.821	.819	.797	.814
E_1						
Accuracy	.710	.797	.855	.837	.773	.849
NMI	.745	.790	.830	.819	.799	.822
G_1						
Accuracy	.652	.660	.707	.721	.705	.709
NMI	.699	.716	.723	.727	.723	.727
G'_1						
Accuracy	.692	.771	.814	.816	.782	.827
NMI	.730	.771	.797	.800	.790	.804
H_1						
Accuracy	.709	.822	.844	.834	.789	.835
NMI	.758	.820	.821	.819	.801	.817
H_2						
Accuracy	.719	.814	.854	.849	.799	.828
NMI	.761	.812	.837	.833	.813	.833
MixMulti						
Accuracy	.502	.525	.582	.543	.542	.586
NMI	.597	.609	.658	.644	.633	.651

It is interesting to note that for the Switchboard corpus, although the mixture of multinomials method was by far the worse clustering algorithm, after SVD combination it clearly became the best method. The same happened for the 20Newsgroups corpus where the mixture of multinomials was among one of the worse-performing methods and after SVD combination it became the best. These results suggest that when developing clustering algorithms, issues of the performance of metaclustering are distinct than issues of performance of single systems.

5.4 Factor Analysis of Results

In this subsection we try to determine the relative importance of two factors in the combination schemes: the mean and variance of the classification accuracy of individual systems. Comparing Table 1 or 2 with Table 3 the gains in 20Newsgroups or Reuters are higher than Switchboard and the variance of individual systems is higher in 20Newsgroups and Reuters than Switchboard. To assess the effect of each one of these two factors (mean and variance of individual systems)

Table 3. Performance of different combination schemes on various clustering algorithms for Switchboard.

	Single Run	Best of 100 runs	SVD Combin.	Constr. Combin.	Unconstr. Combin.	No Combin.
I_1						
Accuracy	.819	.848	.826	.820	.789	.836
NMI	.908	.914	.913	.907	.898	.915
I_2						
Accuracy	.831	.863	.841	.837	.807	.845
NMI	.913	.920	.920	.918	.910	.922
E_1						
Accuracy	.798	.819	.819	.777	.736	.818
NMI	.882	.886	.890	.883	.863	.891
G_1						
Accuracy	.711	.711	.765	.751	.741	.762
NMI	.868	.870	.887	.877	.875	.888
G_1'						
Accuracy	.789	.808	.811	.801	.749	.803
NMI	.875	.878	.880	.877	.859	.878
H_1						
Accuracy	.826	.861	.842	.811	.757	.841
NMI	.910	.918	.918	.899	.895	.918
H_2						
Accuracy	.814	.845	.840	.817	.773	.830
NMI	.897	.903	.905	.900	.886	.901
MixMulti						
Accuracy	.635	.699	.888	.756	.739	.876
NMI	.787	.818	.924	.899	.892	.921

we generated 300 systems and chose a set of 100 for metaclustering depending on high/medium/low variance and similar mean (Table 4) or high/medium/low mean and similar variance (Table 5). The results of Table 4 do not show a significant impact of variance on the combination results. The results of Table 5 show a clear impact of the mean on the combination results. However, from Tables 1, 2 and 3 we know that the performance of the combined system does not depend simply on the performance of the individual systems: the MixMulti result for Switchboard compared with the CLUTO results is a counterexample. It appears that there are unexplained interactions of mean, variance and clustering algorithms that will make the combination more successful in some cases and less successful in other cases.

6 Summary

We have presented three new methods for the combination of multiple clustering systems and evaluated them on three major corpora and on eight different clustering algorithms. Identifying the correspondence between clusters of differ-

Table 4. Effect of combining sets of 100 systems with approximately the same mean and different levels of variance. The (stdev,acc) cells contain the standard deviation and mean of classification accuracy for each set. Systems are generated with the E_1 criterion on 20Newsgroups and combined with SVD.

	Low Variance	**Medium Variance**	**High Variance**
(stdev,acc)	(.010,.577)	(.023,.578)	(.056,.580)
Accuracy	.640	.631	.635
NMI	.630	.629	.633

Table 5. Effect of combining sets of 100 systems with approximately the same variance and different levels of mean. The (stdev,acc) cells contain the standard deviation and mean of classification accuracy for each set. Systems are generated with the E_1 criterion on 20Newsgroups and combined with SVD.

	Low Mean	**Medium Mean**	**High Mean**
(stdev,acc)	(.018,.538)	(.010,.577)	(.019,.617)
Accuracy	.581	.641	.669
NMI	.616	.632	.647

ent systems was achieved by "clustering clusters", using constrained or unconstrained clustering or by applying SVD. We have empirically demonstrated that the combination schemes can offer gains in most cases. Issues of combination of multiple runs of an algorithm can be important. The combination of different runs of mixture of multinomials algorithm was shown to outperform seven state-of-the-art clustering algorithms on two out of three corpora. In the future we will attempt to gain a better understanding of the conditions under which poor individual systems can lead to improved performance when combined.

References

1. Golub, T., Slonim, D., Tamayo, P., Huard, C., Gaasenbeek, M., Mesirov, J., Coller, H., Loh, M., Downing, J., Caligiuri, M.: Molecular classification of cancer: class discovery and class prediction by gene expression monitoring. Science **286** (1999) 531–537
2. Schütze, H.: Automatic word sense discrimination. Computational Linguistics **24** (1998) 97–124
3. Zamir, O., Etzioni, O.: Grouper: a dynamic clustering interface to Web search results. Computer Networks **31** (1999) 1361–1374
4. Bellegarda, J.: Large vocabulary speech recognition with multispan statistical language models. IEEE Trans. on Speech and Audio Processing **8** (2000) 76–84
5. Jain, A.K., Murty, M.N., Flynn, P.J.: Data clustering: a review. ACM Computing Surveys **31** (1999) 264–323

6. Bauer, E., Kohavi, R.: An empirical comparison of voting classification algorithms: Bagging, boosting, and variants. Machine Learning **36** (1999) 105–139
7. Strehl, A., Ghosh, J.: Cluster ensembles – a knowledge reuse framework for combining multiple partitions. Machine Learning Research **3** (2002) 583–617
8. Monti, S., Tamayo, P., Mesirov, J., Golub, T.: Consensus clustering: a resampling-based method for class discovery and visualization of gene-expression microaray data. Machine Learning **52** (2003) 91–118
9. Fred, A., Jain, A.: Data clustering using evidence accumulation. In: Proc. of the International Conference on Pattern Recognition. (2002) 276–280
10. Dudoit, S., Fridlyand, J.: Bagging to improve the accuracy of a clustering procedure. Bioinformatics **19** (2003) 1090–1099
11. Zeng, Y., Tang, J., Garcia-Frias, J., Gao, G.: An adaptive meta-clustering approach: Combining the information from different clustering results. In: Proc. IEEE Computer Society Bioinformatics Conference. (2002) 276–281
12. Fern, X., Brodley, C.: Random projection for high dimensional data: A cluster ensemble approach. In: Proc. of the 20th International Conf. on Machine Learning, (ICML). (2003) 186–193
13. Topchy, A., Jain, A., Punch, W.: A mixture model for clustering ensembles. In: Proc. of SIAM Conference on Data Mining. (2004)
14. Dimitriadou, E., Weingessel, A., Hornik, K.: A combination scheme for fuzzy clustering. Inter. J. of Pattern Recognition and Artificial Intelligence **16** (2002) 901–912
15. Frossyniotis, D., Pertselakis, M., Stafylopatis, M.: A multi-clustering fusion algorithm. In: Proc. of the 2nd Hellenic Conference on Artificial Intelligence. (2002) 225–236
16. Bradley, P., Fayyad, U.: Refining initial points for K-Means clustering. In: Proc. 15th International Conf. on Machine Learning, (ICML). (1998) 91–99
17. Schwartz, G.: Estimating the dimension of a model. The Annals of Statistics **6(2)** (1978) 461–464
18. Godfrey, J., Holliman, E., McDaniel, J.: Switchboard: Telephone speech corpus for research development. In: Proc. of ICASSP. (1992) 517–520
19. Zhao, Y., Karypis, G.: Empirical and theoretical comparisons of selected criterion functions for document clustering. Machine Learning (2004) 311–331
20. Dempster, A., Laird, N., Rubin, D.: Maximum likelihood from incomplete data via the EM algorithm. Journal of the Royal Statistical Society **Series B, 39(1)** (1977) 1–38
21. Nigam, K., McCallum, A., Thrun, S., Mitchell, T.: Learning to classify text from labeled and unlabeled documents. In: Proc. of AAAI. (1998) 792–799
22. Cheeseman, P., Stutz, J.: Bayesian classification (AutoClass): Theory and results. In: Advances in Knowledge Discovery and Data Mining, AAAI Press/MIT Press (1996)

Reducing Data Stream Sliding Windows by Cyclic Tree-Like Histograms*

Francesco Buccafurri and Gianluca Lax

DIMET, Università degli Studi Mediterranea di Reggio Calabria
Via Graziella, Località Feo di Vito, 89060 Reggio Calabria, Italy
bucca@unirc.it,lax@unirc.it

Abstract. Data reduction is a basic step in a KDD process useful for delivering to successive stages more concise and meaningful data. When mining is applied to data streams, that are continuous data flows, the issue of suitably reducing them is highly interesting, in order to arrange effective approaches requiring multiple scans on data, that, in such a way, may be performed over one or more reduced sliding windows. A class of queries, whose importance in the context of KDD is widely accepted, corresponds to *sum range queries*. In this paper we propose a histogram-based technique for reducing sliding windows supporting approximate arbitrary (i.e., non biased) sum range queries. The histogram, based on a hierarchical structure (opposed to the flat structure of traditional ones), results suitable for directly supporting hierarchical queries, and, thus, drill-down and roll-up operations. In addition, the structure well supports sliding window shifting and quick query answering (both these operations are logarithmic in the sliding window size). Experimental analysis shows the superiority of our method in terms of accuracy w.r.t. the state-of-the-art approaches in the context of histogram-based sliding window reduction techniques.

1 Introduction

It is well known that data pre-processing techniques (data cleaning and data reduction), when applied prior to mining, may significantly improve the overall data mining results. This is particularly true in the context of data stream mining, where data comes continuously and mining may be done on the basis of sliding windows including only the most recent data [1]. Indeed, in order to give significance to the sliding window itself, the size, that is the number of most recent data we keep in each instant, should be as large as possible. As a consequence any technique capable of reducing (i.e., compressing) sliding windows by maintaining a good approximate representation of data distribution inside it,

* This work was partially funded by the Italian National Council Research under the "Reti Internet: efficienza, integrazione e sicurezza" project and by the European Union under the "SESTANTE - Strumenti Telematici per la Sicurezza e l'Efficienza Documentale della Catena Logistica di Porti e Interporti" Interreg III-B Mediterranee Occidentale project

J.-F. Boulicaut et al. (Eds.): PKDD 2004, LNAI 3202, pp. 75–86, 2004.

and, at the same time, by smoothing possible outliers, is certainly relevant in the field of data stream mining. Observe that, reducing sliding windows allows us also to keep simultaneously more than just one approximate sliding window, in order to implement similarity queries and other analysis, like *change mining queries* [12], useful for trend analysis and, in general, for understanding the dynamics of the data stream. In sum, since in a typical streaming environment, only limited memory resources are available [14], reduction is a key factor allowing us query processing also requiring multiple scans on data. But, which properties a sliding window reduction technique has to satisfy? Necessarily, the reduced sliding window should maintain in a certain measure the *semantic nature* of original data, in such a way that meaningful queries for mining activities can be submitted to reduced data in place of original ones. Then, for a given kind of query, accuracy of the reduced structure should be enough independent of the position where the query is applied. Indeed, mining needs the possibility of freely querying data. In addition, the reduction technique should not to limit too much the capability of drilling-down and rolling-up data.

In this paper we propose a histogram-based technique for reducing sliding windows supporting approximate arbitrary range-sum queries satisfying all the above properties. Observe that range-sum queries represents a class of queries very frequent in the field of data stream mining. Our histogram, called *c-tree*, differently from traditional ones, is based on a hierarchical structure. Its nodes contain, hierarchically, pre-computed range-sum queries, stored by approximate (via bit saving) encoding. For this reason, the structure directly supports the estimation of arbitrary range-sum queries (indeed, range-sum queries are either embedded in the histogram or derivable by linear interpolation by the latter ones). Reduction derives both from aggregation implemented by leaves of the tree (discretization), and from the saving of bits obtained by representing range queries with less than 32 bits (assumed enough for an exact representation). The number of bits used for representing range queries decreases as the level of the tree increases. The structure is designed as dynamic, in the sense that each update, for maintaining the c-tree on the sliding window, can be applied in logarithmic time in worst case (w.r.t. the window size). Moreover, answering to a range query requires at most logarithmic time too. Observe that hierarchical structure directly supports querying at different abstraction levels, thus allowing drill-down and roll-up operations. Finally, bucket summarization smoothes each data value by consulting the "neighborhood" or values around it. This works to remove the noise from data. But the main feature we have to remark for our histogram concerns its accuracy. Indeed, in order the reduction technique to have significance, error should be either guaranteed or heuristically shown to be low (and this is our case), compared with that of the state-of-the-art techniques. There is no large literature about the important issue of evaluating approximate arbitrary range queries on sliding windows. Most of the recent approaches are based on histograms [17, 16] and Wavelet [15, 21]. Other approaches use sampling [2, 19, 8] and sketches [7, 13]. Histograms are a lossy compression technique widely applied in various application contexts, like query optimization, statistical and temporal databases, and OLAP applications. [11] deals with the problem of reducing sliding windows by error guaranteed histograms (called *exponential histograms*), by solving the problem only in case of biased range queries (i.e., queries involving the last q data of the sliding window) that are significant queries in the context of data stream processing. For arbitrary queries, error may increase dramatically, especially if queries are distant from the most recent data of the sliding window. The proposal of [6] presents the same characteristics. Unfortunately, limiting range queries to a particular case is not acceptable in the KDD context, as observed earlier. Thus, having a technique not focused on analytic error guarantee, but experimentally shown to be uniformly accurate w.r.t. arbitrary range queries is strongly preferable to biased (even error guaranteed) methods.

Validation of our method is conducted experimentally by comparing the c-tree with the V-optimal histogram [18]. We have chosen such a histogram since its superiority (in terms of accuracy) w.r.t. the state-of-the-art proposals is proven in a recent paper [16]. Actually, [16] concerns a more efficient version of V-optimal (called *ϵ-approximate V-optimal histogram*) defined in order to have an effective method (since V-optimal updating is polylinear). But, in the same paper, it is shown that the ϵ-approximate V-optimal histogram is (slightly) less accurate than the classical V-optimal defined in [18]. This guarantees the significance of our comparison. In addition, we observe that while our method requires $O(\log w)$ time for answering to a range query (where w is the size of the sliding window), V-Optimal requires $O(w)$ time (by assuming that the number of bucket is linear on the sliding window size). [16] shows that the ϵ-approximate approach is also definitely superior w.r.t. Wavelet approach [21]. Anyway, we perform comparisons also with Wavelet histograms [20]. Observe that, in order to increase significance of our experiments, we have used exact V-optimal [18] and Wavelet [20] histograms, that are more accurate than histograms defined in [16] and [21], respectively. Indeed, the latter two approaches were introduced since both [18] and [20] are not effective in the context of data streams due to their high maintenance computational cost.

The plan of the paper is the following. The c-tree histogram is introduced in Section 2 where we describe also the dynamics of the c-tree, that is how it is updated while the sliding window is moved. Section 3 shows how the c-tree histogram is represented; moreover some considerations about the proposed approach are remarked. In Section 4 we show how to use the c-tree for evaluating range queries. Section 5 analyzes experimental results validating the method. In Section 6 we draw our conclusions.

2 The c-Tree Histogram

In this section we describe the core of our proposal, consisting of a tree-like histogram, named *c-tree*, used for managing data streams under sliding windows. A *sliding window* of size w[1] (on a data stream D) is the tuple $\langle x_w, x_{w-1}, \dots, x_1\rangle$

[1] For simplicity, we assume that $w = 2^z$ for a given positive integer $z > 0$

containing the w-most recent values of D in arrival ordering (observe that x_w represents the oldest value whereas x_1 is the most recent one). Clearly, at each new time instant the sliding window is updated by inserting from the right hand the new value of D and deleting the left-most one (corresponding to the oldest value of the sliding window).

The histogram is built on top of the sliding window, by hierarchically summarizing the values occurring in it. In order to describe the c-tree we chose a constructive fashion. In particular we define the initial configuration (at the time instant 0 – coinciding with the origin of the data flow) and we show, at a generic instant coinciding with the arrival of a new data, how the c-tree is updated.

Initial Configuration. The c-tree histogram consists of:

1. A full binary tree T with n levels, where n is a parameter set according to the required data reduction (this issue will be treated in Section 5). Each leaf node N of T is associated with a range $\langle l(N), u(N)\rangle$ of size $d = \frac{w}{2^{n-1}}$ and the set of such ranges produces an equi-width partition of the array $\langle 1, w\rangle$. In addition, we require that adjacent leaves correspond to adjacent ranges of $\langle 1, w\rangle$ and the left-most leaf corresponds to the range $\langle 1, d\rangle$. We denote by $Val(N)$ the value of a node N. In the initial state, all nodes of T contains the value 0.

2. A buffer (of size 2) $B = \langle e, s\rangle$, where $0 \leq e < d$ and $s \geq 0$. s represents the sum of the e most recent elements of the sliding window. Initially, $e = s = 0$.

3. An index P, with $1 \leq P \leq 2^{n-1}$, identifying a leaf node of T. P is initially set to 1, and thus it identifies the left-most leaf of T.

We denote by H the above data structure. Now we describe how H is updated when new data arrive.

State Transition. Let x_t be the data coming at the instant $t > 0$. Then, $e := (e+1)$ **mod** d and $s := s + x_t$. Now, if $e \neq 0$ (i.e., the buffer B is not full), then the updating of H halts. Otherwise (i.e., $e = 0$), the value s (which summarizes the last d data) has to be stored in T and, then, the buffer has to be emptied. We explain now how the insertion of s in T is implemented. Let $\delta = s - val(N_P)$, where N_P is the leaf of T identified by P. Then, $val(N_P) := val(N_P) + s$ and δ is also added to all nodes belonging to the path from N_P to the root of H. Finally, e and s of B are reset (i.e., they assume value 0) and $P := P$ **mod** $2^{n-1} + 1$ (this way, leaf nodes of the tree are managed as a cyclic array). Observe that P points to the leaf node containing the less recent data, and such data are replaced by new data incoming. Each update operation requires $O(\log w)$ time, where w is the size of the sliding window. Now we show an example of 3-levels c-tree building and updating.

Example 1. Let $\langle 35, 51, 40, 118, 132, 21, 15, 16, 18, 29, ...\rangle$ be the data stream order by arrival time increasing and let the sliding window size be 8; moreover, let $d = w/2^{n-1} = 2$. Initially, $e = 0$, $s = 0$, $P = 1$ and the value of all nodes of T is 0. The first data coming from the stream is 35, thus $e = 1$ and $s = 35$. Since $e \neq 0$ no other updating operation has to be done. Then, the data coming from the stream is 51, thus $e = 0$ and $s = 35 + 51 = 86$. Since $e = 0$, the first leaf node of T is set to the value s, and all nodes belonging to the path between

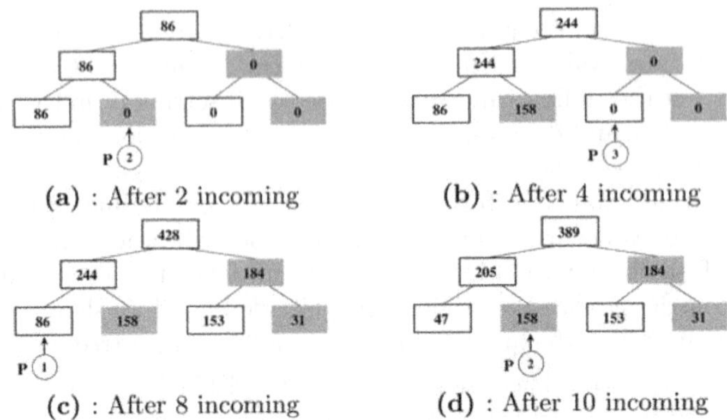

(a) : After 2 incoming (b) : After 4 incoming

(c) : After 8 incoming (d) : After 10 incoming

Fig. 1. The c-tree of Example 1

such leaf and the root are increased by $\delta = 86$. Finally, $P = 2$, and e and s are reset. In Figure 1.(a) the resulting c-tree is reported. Therein (as well as in the other figures of the example), we have omitted buffer values since they are null. Moreover, right-hand child nodes are represented with the color grey. This is because, as we will explain in Section 3, these nodes have not to be saved since they can be computed from white nodes. For the first 8 data arrivals, updates proceeds as before. In Figure 1.(b) and 1.(c) we report just the snapshot after 4 and 8 updates, respectively. The pointer P is now changed assuming the value 1. Now, the data 18 arrives. Thus, $e = 1$ and $s = 18$. At the next time instant, $e = 0$ and $s = 47$. Since $e = 0$, $\delta = 47 - 86 = -39$ is added to the leaf node pointed by P (that is the first leaf) determining its new value 47. Moreover, nodes belonging to the path between such leaf and the root are increased by δ. At this point, P assumes the value 2. The final c-tree is shown in Figure 1.(d).

3 c-Tree Representation

In this section we describe how the c-tree histogram is represented. Beside storing just necessary nodes, we use a bit-saving based encoding in order to reduce storing space. As already sketched in Example 1, each right-hand child node can be derived as a difference between the value of the parent node and the value of the sibling node. As a consequence, right-hand child node values have not to be stored. In addition, we encode node values through length-variable representations. In particular: (1) The root is encoded by 32 bits (we assume that anyway the overall sum of the sliding window data can be represented by 32 bits with no scaling error). (2) The root left-child is represented by k bits (where k is a parameter suitably set – we will discuss about such an issue in Section 5). (3) All nodes which belong to the same level are represented by the same number of bits. (4) Nodes belonging to a level, say l, with $2 \leq l \leq n-1$, are represented by a bit less than nodes belonging to the level $l-1$.

Substantially, the approach is based on the assumption that, in the average, the sum of occurrences of a given interval of the frequency vector, is twice than the sum of the occurrences of each half of such an interval. This assumption is chosen as a heuristic criterion for designing c-tree, and this explains the choice of reducing by 1 per level the number of bits used for representing numbers. Clearly, the sum contained in a given node is represented as a fraction of the sum contained in the parent node. Observe that, in principle, it could be used also a representation allowing possibly different number of bits for nodes belonging to the same level, depending on the actual value contained into nodes. However, we should deal with the spatial *overhead* due to these variable codes. The reduction of 1 bit per level appears as a reasonable compromise. Our approach is validated by previous results shown in [3, 4] for histograms on persistent data and in [5] for improving estimation inside histogram buckets.

Remark. We remark that this bit-saving approach is not applicable to non-indexed histograms (by a tree). Indeed, for a "flat" histogram, the scaling size used for representing numbers would be related to the overall sliding window sum value, that is, bucket values would be represented as a fraction of this overall sum, with a considerable increasing of the scaling error. One could argue that also data-distribution-driven histograms, like V-Optimal [18], whose accuracy has been widely proven in the literature, could be improved by building a tree index on top, and by reducing the storage space by trivially applying our bit-saving approach. However, such indexed histograms, induce a *non* equi-width partition, and, as a consequence, the reduction of 1 bit per level in the index of our approach would be not well founded.

Encoding a given node N with a certain number of bits, say i, is done in a standard fashion. Let denote by P the parent node of N. The value $val(N)$ of the node N will be recovered not exactly, in general. It will be affected by a certain scaling approximation. We denote by $\widetilde{val^i}(N)$ the encoding of $val(N)$ done with i bits and by $\overline{val^i}(N)$ the approximation of $val(N)$ obtained by $\widetilde{val^i}(N)$.
We have that: $\widetilde{val^i}(N) = Round(\frac{val(N)}{val(P)} \cdot (2^i - 1))$. Clearly, $0 \leq \widetilde{val^i}(N) \leq 2^i - 1$.

Concerning the approximation of $val(N)$ it results: $\overline{val^i}(N) = (\frac{\widetilde{val^i}(N)}{2^i-1} \cdot val(P))$ The *absolute error* due to the i-bit encoding of the node N, with parent node P, is: $\epsilon_a(val(N), val(P), i) = |val(N) - \overline{val^i}(N)|$. It can be easily verified that: $0 \leq \epsilon_a(val(N), val(P), i) \leq \frac{val(P)}{2^{i+1}}$.

We conclude this section by analyzing both overall scaling error and storage space required by the c-tree, once the two input parameters are fixed, that is: n, i.e., the number of levels n and k, i.e., the number of bits uses for encoding the left-hand child of the root. Concerning scaling error we have to understand how it is propagated over the path from the root to the leaves of the tree. Indeed, the error for a stand-alone node is analyzed above. We may determine an upper bound of the worst-case error by considering the sum of the maximum scaling error at each stage. Assume that R is the maximum value appearing in the data stream and w is the sliding window size. According to considerations above, since at the first level we use k bits for encoding numbers, the maximum absolute error at this level is $\frac{R \cdot w}{2^{k+1}}$. Going down to the second level cannot increase the maximum error. Indeed, we double the scale granularity (since coding is reduced by 1 bit) but the maximum allowed value is halved. More precisely, the maximum absolute error at the second level is $\frac{R \cdot \frac{w}{2}}{2^k}$. Clearly, the same reasoning can be applied to lower levels, so that the above claim is easily verified. In sum, the maximum absolute scaling error of the c-tree is $\frac{R \cdot w}{2^{k+1}}$; interestingly, observe that the error is independent of the tree depth n.

Concerning the storage space (in bits) required by the c-tree, we have:

$$(n-1) + \lceil \log(R \cdot d) \rceil + \lceil \log(d) \rceil + 32 + \sum_{h=0}^{n-2} (k-h) \cdot 2^h \tag{1}$$

where $d = \frac{w}{2^{n-1}}$ and the first three components of the sum takes account of P, s and e, respectively, while $32 + \sum_{h=0}^{n-2}(k-h) \cdot 2^h$ is the space required for saved nodes of T (recall that only left child nodes are stored). In Section 5 we will discuss about the setting of the parameters n and k.

4 Evaluation of a Range-Sum Query

In this section we describe the algorithm used for evaluating the answer to a range query $Q(t_1, t_2)$, where $0 \leq t_1 < t_2 < w$, that computes the sum of data arrived between time instants $t - t_1$ and $t - t_2$, respectively, where t denotes the current time instant. For example, if $t_1 = 0$ and $t_2 = 5$ it represents the sum of the 5 most recent data. C-tree allows us to reduce the storage space required for storing data of the sliding windows and, at the same time, to give fast yet approximate answers to range queries. As usual in this context, this approximation is the price we have to pay for having a small data structure to manage and for obtaining fast query answering. We now introduce some definitions that will be use in the algorithm.

Notations: Given a range query $Q(t_1, t_2)$, with $t_1 > e$: (1) Let η be the set of leaf nodes containing at least one data involved in the range query. Let $l_i = (P - \mathbf{ceil}\langle \frac{t_i - e}{d} \rangle)/2^{n-1}$, with $i = 1, 2$ be the indexes of the two leaf nodes L_1 and L_2. η consists of all leaf nodes succeeding L_2 and preceding L_1 (including both L_1 and L_2) in the ordering obtained by considering leaf nodes as a cyclic array. (2) Given a non leaf node of N, let *L(N)* be the set of leaf nodes descending from N. (3) Given a leaf node N, we define $I(N, Q) = \frac{i}{d} \cdot val(N)$, where i is the number of data stored by N involved in Q. $I(N, Q)$ computes the contribution of a leaf node to a range query (by linear interpolation). (4) Let $\overline{Q}$ be the estimation of the range query computed by means the c-tree.

First, suppose that $t_1 > e$ (recall e is the number of data in the buffer B) in such a way that the range query doesn't involve data in the buffer B. The algorithm for the range query evaluation is performed by calling the function *contribution* on the root of the c-tree.

The function *contribution* is shown below:

function *contribution (N)*
  if (N is a leaf)
    $\overline{Q} = \overline{Q} + I(N, Q)$
    return $\overline{Q}$ `//the function halts.`
  endif else
    for each N_x child of N
      if ($L(N_x) \subseteq \eta$)
        $\overline{Q} = \overline{Q} + val(N_x)$ **endif**
      if ($L(N_x) \cap \eta \neq \emptyset$)
        contribution (N_x) **endif**
endfunction

The first test checks if N is a leaf node, and in such a case the function, before halting, computes the contribution of N to Q by linear interpolation. In case N is not a leaf node, it is tested if all nodes descending from N_x (denoting a child of N) are involved in the query. If this is the case, their contribution to the range query coincides with the value of N_x. In case not all nodes descending from N_x are involved in the query, but only some of them, their contribution is obtained by recursively calling the function on N_x. The algorithm performs, in the worst case, two descents from the root to two leaves. Thus, asymptotic computational cost of answering a range query is $O(\log w)$ where, w is the window size. Note that the exact cost is upper bounded by n, where $n = \lceil \log \frac{w}{d} \rceil + 1$ is the number of levels of the c-tree and d is the size of leaf nodes.
In case $t_1 < t_2 \leq e$, the range query involves only data in the buffer B and $Q(t_1, t_2) = (t_2 - t_1) \cdot \frac{s}{e}$ (recall that s represents the sum of data buffered in B). Finally, in case $t_1 \leq e < t_2$, we have that $Q(t_1, t_2) = Q(t_1, e) + Q(e, t_2)$ that can be computed by exploiting the two above cases.

5 Experiments

We start this section by describing the test bed used for our experiments.
Available Storage: For experiments conducted we have used 22 four-byte numbers for all techniques. According to (1) (given in Section 3), the above constraint has to be taken into account when the two basic parameters n and k of the c-tree are set. We have chosen to fix these parameters to values $n = 7$ and $k = 14$ (we will motivate such a choice next in this section).

Techniques: We compare our technique with (the motivations of such a choice are given in the Introduction): (1) *V-Optimal* (VO) [18], which produces 11 bucket; for each bucket both upper bound and value are stored; (2) *Wavelet* (WA) [20], which are constructed using the *bi-orthogonal* 2.2 decomposition of the MATLAB 5.3 with 11 four-byte Wavelet coefficients plus another 11 four-byte numbers for storing coefficient positions.

Synthetic Data Streams: Synthetic data streams are obtained by randomly generating 10000 data values belonging to the range $[0, 100]$.

Real-Life Data Streams: Real-life data have been retrieved from [10] and represent the daily maximum air temperature stored by the station STBARBRA.A in the County of Santa Barbara from 1994 to 2001. Its size is 2922 and the range is from 10.6 to 38.3 degree Celsius.

Query Set and Error Metric: In our experiments, we use two different query sets for evaluating the effectiveness of the various methods: (1) QS_1 consists of all range queries from 1 to q with $1 \leq q \leq w$ and (2) QS_2 is the set of all range queries having size $round(\frac{w}{10})$, where, we recall, w is the size of the sliding window. At each time we measure the error $E(t)$ produced by techniques on the above query set by using the average of the relative error $\frac{1}{Q}\sum_{i=1}^{Q} e_i^{rel}$, where Q is the cardinality of the query set, and e_i^{rel} is the *relative error* , i.e., $e_i^{rel} = \frac{|S_i - \widetilde{S}_i|}{S_i}$, where S_i and $\widetilde{S}_i$ are, respectively, the actual answer and the estimated answer of the query i-th of the considered query set. Then we compute the average of the error $E(t)$ over the entire data stream duration. After a suitable initial delay sufficient to fill the sliding window, queries are applied at each new arrival.

Sliding Window Size: In our experiments, we use sliding windows of size 64, 128, 256, 512, 1024, that are dimensions frequently used for experiments in this context (e.g., see [6, 9, 16]).

Now we consider the problem of the choice of a suitable value for n and k, that are, we recall, number of levels of the c-tree and number of bits used for encoding the left child node of the root (for the successive levels, as already mentioned, we drop 1 bit per level) respectively. Observe that, according to the result about the error given in Section 3, setting the parameter k means fixing also the error due to scaling approximation. We have performed some experiments on synthetic data in order to test the error dependence on the parameters n and k for different window sizes by using the average relative error on query set 1. Experiments produce similar curves (not reported here for space limitations) showing that the error decreases as n increases and decreases as k increases until $k = 11$ and then it remains near constant. Indeed, the error consists of two components: (1) the error due to the interpolation inside the leaves nodes partially involved in the query, and (2) the scaling approximation. For $k > 11$, the last component is negligible, and the error keeps a quasi-constant behavior since the first component depends only on n. Therefore, in order to reduce the error, we should set k to a value as large as possible allowing us to represent leaves with a sufficient number of bits (not to much lower than the threshold heuristically determined above). However, for a fixed compression ratio, this may limit the depth of the tree and, thus, the resolution determined by the leaves. As a consequence, the error arising from linear interpolation done inside leaf nodes increases. In sum, the choice of k plays the role of solving the above trade-off. These criteria are employed in experiments in order to choose the value of n and k, respectively, on the basis of the storage space amount.

Now we present results obtained by experiments. For each data set we have calculated the average relative error on both query set 1 and query set 2.

In Figures 2.(a) and 2.(b) we have reported results obtained on real and synthetic data sets, respectively varying the sliding window size (we have considered sizes: 64, 128, 256, 512) and using query set 1. C-tree shows the best accuracy, with significant gaps especially respect to Wavelet. Note that c-tree in case of sliding window of size 64 does not produce error since there is no discretization and, furthermore, a leaf node is encoded by 9 bits which are sufficient to represent exactly a single data value.

In Figures 3.(a) and 3.(b) we have replicated the previous experiment considering the behavior of techniques on query set 2. Observe that accuracy of techniques becames worse on query set 2 since the range query size is very small (indeed, the range query involves only the 10% of sliding window data). Also this comparison shows the superiority of the c-tree over other histogram methods. In Figure 4.(a) and 4.(b) we have studied the accuracy of c-tree versus the number levels, fixing $k = 14$, with sliding windows of size 256, 512 and 1024. In this experiment we have used the query set 1. Finally, we observe that, thanks to experiments conducted with query set 2, we have verified that the behaviour of the c-tree is "macroscopically" independent of the position of the range query in the window. Macroscopically here means that even though some queries can be privileged (for instance those involving only entire buckets), it happens that both average and variability of the query answer error is not biased. This basically reflects the equi-width nature of the c-tree histogram.

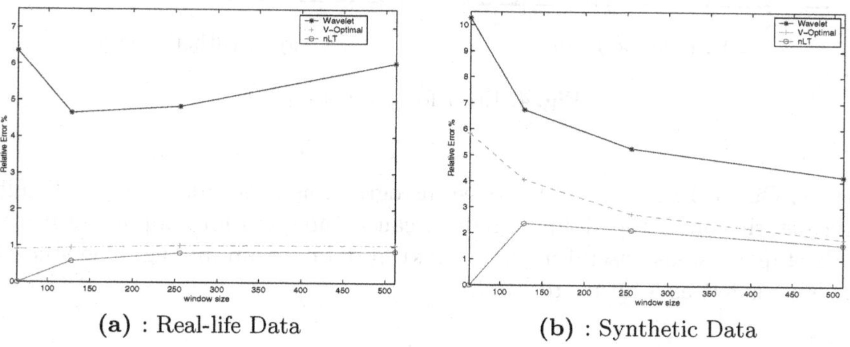

(a) : Real-life Data

(b) : Synthetic Data

Fig. 2. Error for query set 1

6 Conclusions and Future Work

In this paper we have presented a tree-like histogram used for reducing sliding windows and supporting fast approximate answers to arbitrary range-sum queries on them. Through a large set of experiments, the method is successfully compared with the most relevant proposals among the related histogram-based approaches. The histogram is designed for implementing data stream preprocessing in a KDD process which exploits arbitrary hierarchical range-sum

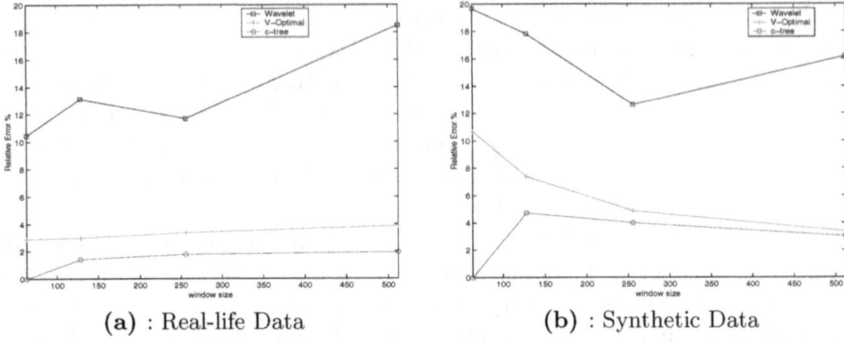

(a) : Real-life Data (b) : Synthetic Data

Fig. 3. Error for query set 2

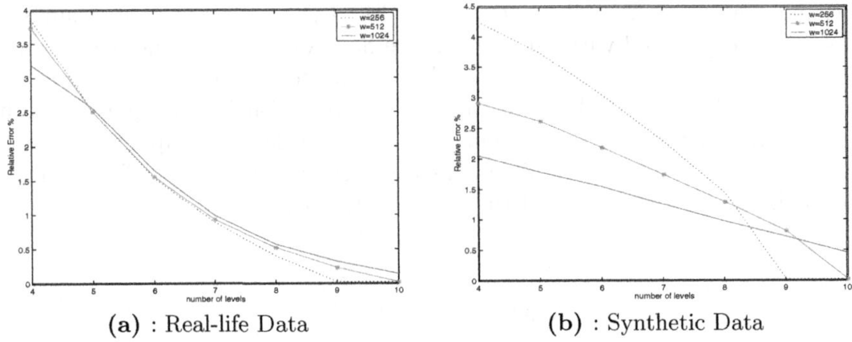

(a) : Real-life Data (b) : Synthetic Data

Fig. 4. Error for query set 1

queries. Our feeling is that the c-tree histogram, as it encodes a concise multi-layer description of the sliding window, can be adapted for supporting further kinds of queries, also useful in the context of data stream mining. The study of this issue is left as a future research.

References

1. B. Babcock, S. Babu, M. Datar, R. Motwani, and J. Widom. Models and issues in data stream system. In *PODS*, pages 1–16, 2002.
2. B. Babcock, M. Datar, and R. Motwani. Sampling from a moving window over streaming data. In *Proc. of the thirteenth annual ACM-SIAM Symposium on Discrete algorithms*, pages 633–634, 2002.
3. F. Buccafurri and G. Lax. Fast range query estimation by n-level tree histograms. *Data & Knowledge Engineering Journal.* To appear.
4. F. Buccafurri and G. Lax. Pre-computing approximate hierarchical range queries in a tree-like histogram. In *Proc. of the Int. Conf. on Data Warehousing and Knowledge Discovery, DaWak*, pages 350–359, 2003.

5. F. Buccafurri, L. Pontieri, D. Rosaci, and D. Saccà. Improving range query estimation on histograms. In *Proc. of the Int. Conf. on Data Engineering, ICDE*, pages 628–638, 2002.
6. A. Bulut and A. K. Singh. SWAT: Hierarchical stream summarization in large networks. In *ICDE*, pages 303–314, 2003.
7. M. Charikar, K. Chen, and M. Farach-Colton. Finding frequent items in data streams. In *Proc. of the 29th Int. Colloquium on Automata, Languages and Programming*, pages 693–703. Springer-Verlag, 2002.
8. S. Chaudhuri, R. Motwani, and V. Narasayya. On random sampling over joins. In *Proc. of the 1999 ACM SIGMOD Int. Conf. on Management of data*, pages 263–274. ACM Press, 1999.
9. A. Das, J. Gehrke, and M. Riedewald. Approximate join processing over data streams. In *Proc. of the ACM SIGMOD Int. Conf. on Management of data*, pages 40–51. ACM Press, 2003.
10. California Weather Databases. http://www.ipm.ucdavis.edu/calludt.cgi/wxstationdata?map=santabarbara.html&stn=stbarbra.a.
11. M. Datar, A. Gionis, P. Indyk, and R. Motwani. Maintaining stream statistics over sliding windows: (extended abstract). In *Proc. of the thirteenth annual ACM-SIAM symposium on Discrete algorithms*, pages 635–644, 2002.
12. G. Dong, J. Han, L. V. S. Lakshmanan, J. Pei, H. Wang, and P. S. Yu. Online mining of changes from data streams: Research problems and preliminary results. In *Proc. of the 2003 ACM SIGMOD Workshop on Management and Processing of Data Streams*, 2003.
13. J. Feigenbaum, S. Kannan, M. Strauss, and M. Viswanathan. An approximate l1-difference algorithm for massive data streams. In *Proc. of the 40th Annual Symposium on Foundations of Computer Science*, page 501. IEEE Computer Society, 1999.
14. M.N. Garofalakis, J. Gehrke, and R. Rastogi. Querying and mining data streams: You only get one look (tutorial). In *Proc. of the Int. Conf. on Management of Data ACM SIGMOD*, page 635, 2002.
15. A. C. Gilbert, Y. Kotidis, S. Muthukrishnan, and M. Strauss. Surfing wavelets on streams: One-pass summaries for approximate aggregate queries. In *The VLDB Journal*, pages 79–88, 2001.
16. S. Guha and N. Koudas. Approximating a data streams for querying and estimation: Algorithms and performance evaluation. In *ICDE*, pages 567–576, 2002.
17. S. Guha, N. Koudas, and K. Shim. Data-streams and histograms. In *Proc. of the thirty-third annual ACM symposium on Theory of computing*, pages 471–475. ACM Press, 2001.
18. H. V. Jagadish, N. Koudas, S. Muthukrishnan, V. Poosala, K. C. Sevcik, and T. Suel. Optimal histograms with quality guarantees. In *Proc. 24th Int. Conf. Very Large Data Bases, VLDB*, pages 275–286, 24–27 1998.
19. G. S. Manku and R. Motwani. Approximate frequency counts over data streams. In *Proc. Int. Conf. on Very Large Data Bases*, pages 346–357, 2002.
20. Y. Matias, J. S. Vitter, and M. Wang. Wavelet-based histograms for selectivity estimation. In *Proc. of the 1998 ACM SIGMOD Int. Conf. on Management of data*, pages 448–459. ACM Press, 1998.
21. Y. Matias, J. S. Vitter, and M. Wang. Dynamic maintenance of wavelet-based histograms. In *The VLDB Journal*, pages 101–110, 2000.

A Framework for Data Mining Pattern Management

Barbara Catania[1], Anna Maddalena[1], Maurizio Mazza[1], Elisa Bertino[2], and Stefano Rizzi[3]

[1] University of Genova (Italy)
{catania,maddalena,mazza}@disi.unige.it
[2] Purdue University (IL)
bertino@cerias.purdue.edu
[3] University of Bologna (Italy)
srizzi@deis.unibo.it

Abstract. To represent and manage data mining patterns, several aspects have to be taken into account: (i) patterns are heterogeneous in nature; (ii) patterns can be extracted from raw data by using data mining tools (a-posteriori patterns) but also defined by the users and used for example to check how well they represent some input data source (a-priori patterns); (iii) since source data change frequently, issues concerning pattern validity and synchronization are very important; (iv) patterns have to be manipulated and queried according to specific languages. Several approaches have been proposed so far to deal with patterns, however all of them lack some of the previous characteristics. The aim of this paper is to present an overall framework to cope with all these features.

1 Introduction

In many different modern contexts, a huge quantity of raw data is collected. An usual approach to analyze such data is to generate some compact knowledge artifacts (i.e., clusters, association rules, frequent itemsets, etc.) through data processing methods, to make them manageable from humans while preserving as much as possible their hidden information or discovering new interesting correlations. Those knowledge artifacts, which can be very heterogeneous and complex, are also called *patterns*. Although a large variety of techniques for pattern mining exist, we still miss comprehensive environments supporting the development of *knowledge intensive* applications. Such an environment goes much beyond the use of pattern mining techniques; it has to provide support for combining heterogeneous patterns, for characterizing their temporal behavior, and for querying and manipulating them. In what follows we elaborate on these requirements.

Heterogeneity. There are many different application contexts from which various types of patterns can be generated and need to be managed. For example, in the market-basket analysis, common patterns are association rules, which identify sets of items usually sold together, or clusters, used to realize a market segmentation analysis. Moreover, we may be interested not only in patterns generated from raw data by using some data mining tools (a-posteriori patterns)

J.-F. Boulicaut et al. (Eds.): PKDD 2004, LNAI 3202, pp. 87–98, 2004.

but also in patterns known by the users and used for example to check how well some data source is represented by them (a-priori patterns).

Temporal information. Since source data change with high frequency, an important issue consists in determining whether existing patterns, after a certain time, still represent the data source from which they have been generated, possibly being able to change pattern information when the quality of the representation changes. Two different time information can be considered: (i) *transaction time*, i.e. the time the pattern "starts to live" in the system. For a-priori patterns, it is the instant when the user inserts the pattern in the system; for a-posteriori patterns, it is the instant when the pattern is extracted from raw data and inserted in the system; (ii) *validity period*, i.e., the time interval in which the pattern is assumed to be reliable with respect to its data source. The validity period can be either assigned by the user or by the system, depending on the quality of raw data representation (*semantic validity*) achieved by the pattern.

Pattern languages. Patterns should be manipulated (e.g. extracted, synchronized, deleted) and queried through a Pattern Manipulation Language (PML) and a Pattern Query Language (PQL). PML must support the management of a-posteriori and a-priori patterns. PQL must support both operations against patterns and operations combining patterns with raw data (cross-over queries).

Several approaches have been proposed so far to deal with patterns, however all of them lack some of the previous characteristics. Most of them deal with specific types of a-posteriori patterns, often stored together with raw data, and do not consider temporal information [6, 8, 12–15]. However, as it has been recognized [5], due to the quite different characteristics of raw data and patterns, to ensure an efficient handling of both, it could be better to use two dedicated systems: a traditional Data Base Management System (DBMS) for raw data and a specific Pattern Based Management System (PBMS) for patterns.

In this paper we propose a comprehensive framework to deal with patterns within a PBMS, addressing the above requirements, and we develop in details some key notions of the framework, such as: (i) a temporal pattern representation model, allowing one to associate time and validity with patterns; (ii) a temporal pattern manipulation language (TPML) and a temporal pattern query language (TPQL), supporting specialized predicates and operators to deal with temporal information. To the best of our knowledge this is the first proposal dealing with temporal aspects of pattern representation and management.

The remainder of the paper is organized as follows. In Section 2, the basic architecture and the pattern model are introduced. In Sections 3 and 4, the TPML and TPQL are discussed, respectively. Related work is then discussed in Section 5. Finally, Section 6 presents some conclusions and outlines future work.

2 The Pattern Model

Pattern-Base Management System. A *Pattern-Base Management System* (PBMS), first introduced in the context of the PANDA project [5], is a system for handling patterns defined over raw data.

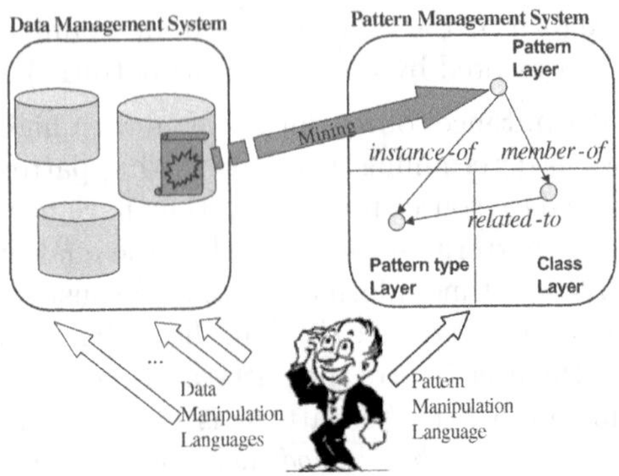

Fig. 1. PBMS Architecture

The overall architecture of the system is shown in Fig. 1. The *Data Management System* on the left-hand side of the figure deals with data collections, whereas the *Pattern Management System*, on the right-hand side, deals with patterns. The user may interact with both systems by mean of dedicated manipulation languages. Within the PBMS, we distinguish three different layers: (i) the *pattern layer*, which is populated with patterns (pattern-base); (ii) the *pattern type layer*, which holds built-in and user-defined types for patterns; (iii) the *class layer*, which holds definitions of pattern classes, i.e., collections of patterns. End-users may directly interact with the PBMS: to this end, the PBMS adopts ad-hoc techniques not only for representing and storing patterns, but also for querying patterns or recalculating them from raw data.

Basic Model Concepts. Based on the proposed architecture, the concepts at the basis of the pattern model are: pattern types, patterns, and classes (for additional details, see [15]). A *pattern type* is the intensional form of patterns, giving a formal description of their structure and relationship with source data. It is a record with five elements: (i) the *pattern name* n; (ii) the *structure schema* s, which defines the pattern space by describing the structure of the patterns instances of the pattern type; (iii) the *source schema* d, which defines the related source space by describing the dataset from which patterns are constructed; (iv) the *measure schema* m, which is a tuple describing the measures which quantify the quality of the source data representation achieved by the pattern; (v) the *formula* f, which describes the relationship between the source space and the pattern space, thus representing the semantics of the pattern. Inside f, attributes are interpreted as free variables ranging over the components of either the source or the pattern space. Note that, though in some particular domains f may exactly express the inter-space relationship, in most cases it will describe it only approximatively.

Given a pattern type pt, a *mining function* μ for pt takes as input a data source, applies a certain computation to it, and returns a set of patterns, instances of pt. We then call *measure function* the function computing the measures

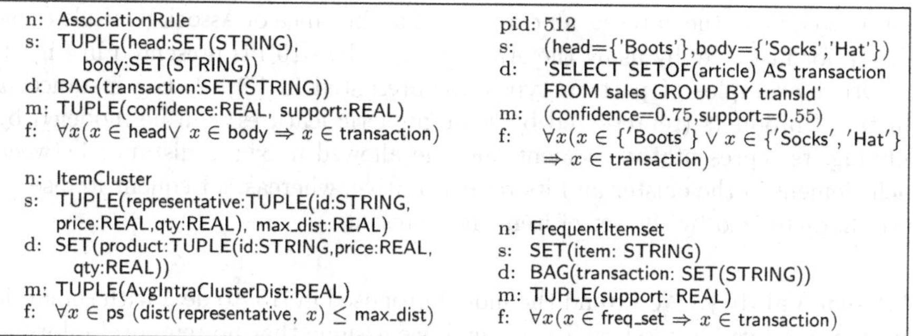

n: AssociationRule
s: TUPLE(head:SET(STRING), body:SET(STRING))
d: BAG(transaction:SET(STRING))
m: TUPLE(confidence:REAL, support:REAL)
f: ∀x(x ∈ head∨ x ∈ body ⇒ x ∈ transaction)
pid: 512
s: (head={'Boots'},body={'Socks','Hat'})
d: 'SELECT SETOF(article) AS transaction FROM sales GROUP BY transId'
m: (confidence=0.75,support=0.55)
f: ∀x(x ∈ {'Boots'} ∨ x ∈ {'Socks', 'Hat'} ⇒ x ∈ transaction)
n: ItemCluster
s: TUPLE(representative:TUPLE(id:STRING, price:REAL,qty:REAL), max_dist:REAL)
d: SET(product:TUPLE(id:STRING,price:REAL, qty:REAL))
m: TUPLE(AvgIntraClusterDist:REAL)
f: ∀x ∈ ps (dist(representative, x) ≤ max_dist)
n: FrequentItemset
s: SET(item: STRING)
d: BAG(transaction: SET(STRING))
m: TUPLE(support: REAL)
f: ∀x(x ∈ freq_set ⇒ x ∈ transaction)

Fig. 2. Examples of pattern types and patterns

of patterns over a certain dataset. We store such information in some catalog of the pattern layer.

Patterns are instances of a specific pattern type containing: (i) a pattern identifier *pid*; (ii) a structure that positions the pattern within the pattern space; (iii) a source that identifies the specific dataset the pattern relates to[1]; (iv) a measure that estimates the quality of the raw data representation achieved by the pattern; (v) an instantiated formula, obtained from the one in the pattern type by instantiating each attribute appearing in s with the corresponding value, and letting the attributes appearing in d range over the source space. Dot notation and path expressions can be used to denote pattern components.

A *class* is a set of semantically related patterns of a certain pattern type and constitutes the key concept in defining a pattern query language.

Example 1. Consider the following scenario. A commercial vendor traces shop transactions and he applies data mining techniques to determine how he can further increase his sales. To this purpose, the vendor deals with several kinds of patterns: (i) *association rules*, representing correlations between items sold; (ii) *clusters of products*, grouping sold products with respect to their price and sold quantity; (iii) *frequent itemsets*, recording items most frequently sold together.

As an example, consider the pattern type for association rules in Fig. 2. The structure schema is a tuple modeling the head and the body as sets of strings representing products. The source schema specifies that association rules are constructed from a bag of transactions, each defined as a set of products. The measure schema includes two common measures to assess the rule relevance: its confidence (what percentage of transactions including the head also include the body) and its support (what percentage of the whole set of transactions include both the head and the body). The formula represents (exactly, in this case) the pattern/source data relationship by associating each rule with the set of transactions which support it. Now suppose that data related to sales transactions are stored in a relational table sales(transId, article, qty). Using an extended SQL

[1] When no otherwise stated, data sources are intensionally described as queries over the raw data.

syntax to denote the dataset, an example of an instance of AssociationRule (generated, for instance, by using the *Apriori* algorithm [10]) is presented in Fig. 2.

Other examples of pattern types are presented in Fig. 2. For instance, a cluster of products (represented by some numeric features) can be modeled by defining its representative element and the allowed maximal distance between each element in the cluster and its representative, whereas, a frequent itemset is just characterized by the set of items it represents. □

Pattern Validity. We extend the model proposed in [15] to deal with semantic and temporal validity issues. To this end, we assume that no temporal information is available from raw data and we associate each pattern with a *transaction time* and a *validity period*. *Transaction time* is automatically computed by the PBMS and points out when a pattern has been inserted in the system. This information cannot be changed by the user, thus it is just recorded in system catalogs. On the other hand, the *validity period* is the interval $[StartTime, EndTime)$ in which the pattern can be considered reliable with respect to raw data, and therefore usable. The validity period can be queried by the user, thus it must be inserted in the model. Moreover, we suppose that the validity period can be either assigned and managed by the user or by the system, depending on the operations performed over patterns (see Section 3). For the sake of simplicity, we deal with a fixed time granularity tg, chosen by the PBMS administrator. Thus, the validity period schema is always of type $[StartTime\colon tg, EndTime\colon tg)$ where tg is fixed. Each pattern is then extended with a new component vt, representing the actual pattern validity period according to the chosen granularity.

Temporal validity specifies that the pattern is *assumed* to be valid in that period. However, since raw data change with a high frequency, the pattern, in its validity period, may not correctly represent raw data it is associated with. To this end, we also introduce the concept of *semantic validity* with respect to a data source and the notion of *safety*, for patterns that are both temporally and semantically valid in a certain instant of time. Note that semantic validity (and thus safety) can only be checked at an instant-by-instant base, since we do not know how the data source will change in the future and how it was in the past.

Definition 1. *Let p be a pattern, with $p.m =< m_1 : v_1, ..., m_n : v_n >$, and t an instant of time. Suppose each measure m_i is associated with a boolean operator θ_i such that $v_1 \theta_i v_2$ means that v_1 is "better than" v_2. p is temporally valid at t if $t \in p.vt$. p is semantically valid at t with respect to a data source D with thresholds $v_1, ..., v_n$ if and only if, $D \models_t p$ and $p.m.m_i\ \theta_i\ v_i$, $i = 1, ..., n$. $D \models_t p$ means that p can be extracted from D at time t. p is safe at t with respect to a data source D with thresholds $v_1, ..., v_n$ if it is both temporally and semantically valid at t with respect to D and $v_1, ..., v_n$.* □

Semantic validity can be seen as a function of time. By checking semantic validity periodically, we may plot how measures change in the time.

Example 2. Consider the pattern in Fig. 2 (say p). Suppose $p.vt$ = [1-APR-04,31-MAR-05), thus p is valid from 1-APR-04 to 31-MAR-05. For instance, p is temporally valid on 22-MAY-04. However, since raw data is continuously changing, it may be possible that, on 22-MAY-04, no transaction in the p data source (say D) contains both 'Hat' and 'Boots'. Thus, the support and the confidence of p in D on 22-MAY-04 are 0. Thus, on 22-MAY-04, p is not semantically valid with respect to D, for any threshold values, and it is not safe. □

3 Temporal Pattern Manipulation Language (TPML)

TPML must support primitives to generate patterns from raw data, to insert them in the PBMS, to delete, and to update patterns. These operations are defined by taking into account validity issues and differences between a-posteriori and a-priori patterns.

Insertion Operations. To cope with both a-posteriori and a-priori patterns, three different types of insertions are supported: extraction, direct insertion, and recomputation. Insertion operators must set the validity period of the interested patterns, which is an additional parameter of the operator. If the user does not specify any validity period, it is set by default to $[Current\_time, +\infty)$.

Extraction $\mathcal{E}(pt, d, cond, \mu, pr)$ extracts patterns of pattern type *pt* from a raw dataset d by applying mining function μ. To these (a-posteriori) patterns the validity period *pr* is assigned and they are inserted in the PBMS if they satisfy condition *cond*, defined by using predicates that will be presented in Section 4.1.

Direct Insertion $\mathcal{I}(pt, d, s, m, pr)$ allows the user to insert in the PBMS patterns from scratch (a-priori patterns) by taking as input a pattern type *pt*, a source d, a structure s, a tuple of measure values m, and a validity period *pr*.

Recomputation $\mathcal{R}(pt, cond, d, \mu_m, pr)$ generates new patterns from old ones, by recomputing their measures over a given raw dataset. More precisely, given the instances of a pattern type *pt* satisfying a given condition *cond*, the measures of those patterns over a raw dataset D are computed, accordingly to some input measure function μ_m. New patterns are created and inserted into the system.

Example 3. Consider Example 1. At the end of every month, the vendor mines his transaction data to extract association rules and frequent sold item sets. A validity period is assigned to each extracted pattern, from the first day till the last day of the month. In order to generate such patterns, the extraction operator can be used (Fig. 3 lines 2-3 and 5) against the relational view $JuneSales$, storing information concerning sales in June, with mining functions $\mu_{aPriori}$ and $\mu_{FreqSeq}$. The extracted patterns are then inserted in classes $MinedAssociationRules$ and $UsedFrequentItemsets$, respectively (Fig. 3 lines 4 and 6) (see below for class-based operators). Furthermore, to specialize his advertising campaign, he mines clusters of products, based on numerical information concerning price and sold quantity of each product. Since the vendor does not know the expiration time of those clusters, he assumes they are always valid. To implement this behavior, the extraction operator is used against the relational view *Products*, storing information concerning sold products (Fig. 3 line 7), by using mining function μ_{SLink}. Such patterns are then inserted in class $SoldItemClusters$ (Fig. 3 lines 8). □

1: /* **Pattern generation** */
2: $AR = \mathcal{E}(AssociationRules, JuneSales, \langle confidence \geq 0.30, support \geq 0.25 \rangle,$
3: $\mu_{aPriori},$ [1-JUN-04, 30-JUN-04])
4: $FORALL\ i \in AR\ DO\ \mathcal{I}_C(i, MinedAssociationRules)$
5: $FS = \mathcal{E}(FrequentItemset, JuneSales, support \geq 0.30, \mu_{FreqSeq},$ [1-JUN-04, 30-JUN-04])
6: $FORALL\ j \in FS\ DO\ \mathcal{I}_C(j, UsedFrequentItemsets)$
7: $CS = \mathcal{E}(ItemCluster, Products, AvgIntraClusterDist \leq 0.20, \mu_{SLink})$
8: $FORALL\ k \in CS\ DO\ \mathcal{I}_C(k, SoldItemClusters)$
9: /* **Pattern deletion** */
10: $\delta(AssociationRule, support \leq 0.50)$
11: /* **A-priori pattern management** */
12: $v = \mathcal{I}\ (FrequentItemset, DS, \{A, B\}, <0>)$
13: $\mathcal{I}_C(v, UsedFrequentItemset)$
14: $\mathcal{S}(v, true, \mu_{FreqSeq})$
15: /* **Restoring temporal validity** */
16: $\mathcal{D}_C$(Current_time $after\ vt$, c)
17: /* **Promotion of a new product** */
18: $W = \pi_{(<representative>,<>)}\ (\sigma_{f(P)}\ (SoldItemClusters))$
19: $W \bowtie_{s.representative \in s.head, cf}\ (MinedAssociationRules)$
20: $\mathcal{S}(AssociationRule, true, \mu_{aPriori})$
21: $\mathcal{S}(FrequentItemset, true, \mu_{FreqSeq})$
22: $\mathcal{S}(ItemCluster, true, \mu_{SLink})$

Fig. 3. Pattern manipulation session

Deletion Operator. $\delta(pt, cond)$ removes the instances of pattern type *pt* satisfying condition *cond* from the pattern layer if they belong to no class.

Example 4. Consider Example 1. Suppose the vendor is no more interested in association rules with support lower than 0.50. Thus he performs a deletion (Fig. 3 line 10). Note that association rules extracted at lines 2-3 are not deleted since they have already been inserted in a class. □

Update Operators. We assume that only measures and the validity period can be changed, by using the following operators.
New_Period $\mathcal{N}_P(pt, cond, pr)$ updates a set of patterns, instances of a pattern type *pt* and satisfying a condition *cond*, by setting the validity period to *pr*. Note that this operator does not recompute the measure values of the patterns.
Synchronize $\mathcal{S}(pt, cond, \mu_m)$ makes patterns safe *without changing* the validity period. More precisely, it re-computes the measure values associated with temporally valid patterns, instances of a pattern type *pt* and satisfying a condition *cond*, to reflect data source modifications, by using the input measure function μ_m. Only temporally valid patterns are synchronized since all the others, by definition of safety, cannot become safe.
Validate $\mathcal{V}(pt, cond, \mu_m)$ makes patterns safe *by changing* their validity period. More precisely, it first recomputes the measure values associated with temporally valid patterns, instances of a pattern type *pt* and satisfying a condition *cond*. If such measure values are better than the ones associated with patterns before validation, patterns are semantically valid. Thus, similarly to synchronization, measures are modified, and the validity period is left unchanged. On the other hand, if measures are worst than before, the validity period of the pattern is changed, setting the end time to *Current_time* and a new pattern is created, with the same structure and dataset than the previous one, but with the new measures

and validity period $[Current\_time, +\infty)$. Since, after validation, patterns are semantically valid at the starting and ending points of their validity period, it is possible to use the validity period as an approximation of the periods in which a pattern is semantically valid.

Example 5. Consider Example 1. After a certain period of time, the vendor receives information about sale transactions of a new shop (suppose those data are stored in a dataset DS). Suppose he wants to trace information concerning how often a product A (e.g. milk) and another product B (e.g. cookies) are sold together in this shop. To this purpose, he first inserts the pattern representing such itemset in the system with measure value equal to 0. To this purpose, he uses a direct insertion operation (Fig. 3 lines 12-13) and then he synchronizes the pattern with raw data to get the right frequency (Fig. 3 line 14). □

Operators for Classes. According to our model, a pattern must be inserted in at least one class in order to be queried. Thus, two TPML operations are provided: insertion, $\mathcal{I}_C(p, c)$, of a pattern p into class c, and deletion, $\mathcal{D}_C(cond, c)$, of all the patterns in class c satisfying condition *cond*. Note that the deletion operator just removes patterns from a class but leaves them in the system.

Example 6. Consider Example 1. In general, the user may be interested in restoring the temporal validity of a certain class c, i.e. he may want to delete from c all patterns that are not temporally valid at the current time. Such behavior can be achieved by using the $\mathcal{D}_C$ operator, by using a temporal predicate in its condition (Fig. 3 line 16). □

4 Temporal Pattern Query Language (TPQL)

TPQL supports the retrieval of patterns from the PBMS, taking temporal issues into account. Each operator of TPQL takes classes as input and returns a set of patterns as output. Moreover, *cross-over operators*, binding patterns with raw data, are provided. In the following, before presenting the TPQL operators, some useful predicates are identified.

4.1 Pattern Predicates

Predicates over Pattern Components. Let p_1 and p_2 be two patterns. The general forms of a predicate over pattern components are $t_1 \theta t_2$ and $t_1 \theta o$, where t_1 and t_2 are path expressions that denote components of patterns p_1 and p_2, of *compatible* type, o is a constant suitable for the type of t_1, and θ is an operator, suitable for the type of t_1, t_2, and o. We consider the following special cases:

– If t_1 and t_2 are data sources, then $\theta \in \{=^i, \subseteq^i, =^e, \subseteq^e\}$. Constants o in this case are queries characterizing a dataset. $=^i$ stands for equivalence and $\subseteq^i$ for containment between intensional data source descriptions (i.e., between queries). These predicates do not require accessing raw data and can be

checked by using results obtained in the literature for queries. On the other hand, $=^e$ and $\subseteq^e$ are checked by accessing raw data (thus, they are cross-over predicates). More precisely, $t_1 =^e t_2$ if and only if $\forall x\ (x \in t_1 \Leftrightarrow x \in t_2)$ and $t_1 \subseteq^e t_2$ if and only if $\forall x\ (x \in t_1 \Rightarrow x \in t_2)$.
- If t_1 and t_2 are pattern formulas, then $\theta \in \{\equiv, \preceq\}$. $t_1 \equiv t_2$ is true if and only if t_1 and t_2 are equivalent formulas; $t_1 \preceq t_2$ is true if and only if t_1 logically implies t_2. Given a tuple o, containing one value for each free variable in t_1, $t_1(o)$ is true if and only if t_1 instantiated with the values in o is true.
- If t_1 and t_2 are validity periods, then $\theta \in \{equals, before, meets, overlaps, during, starts, finishes\}$. The meaning of such predicates is defined in [16]. o in this case is a temporal value, according to the chosen granularity.

Predicates over Patterns. In the following, p_1 and p_2 are patterns.

- *Identity* ($=$). $p_1 = p_2$ if $p_1.pid = p_2.pid$.
- *Shallow equality* ($=^s$). $p_1 =^s p_2$ if their corresponding components, except for pid and the validity period v, are equal. For the data source, we consider intensional equality.
- *Intensional subsumption* ($\preceq^i$). $p_1 \preceq^i p_2$ if they have the same structure but p_1 represents a smaller set of raw data, i.e. $p_1.s = p_2.s$, $p_1.d \subseteq^i p_2.d$ and $p_1.f \preceq p_2.f$.
- *Extensional subsumption* ($\preceq^e$). $p_1 \preceq^e p_2$ if they have the same structure but p_1 represents a smaller set of raw data through the considered formula, i.e. $p_1.s = p_2.s$ and $p_1.d_{\lceil p_1.f} \subseteq p_2.d_{\lceil p_2.f}$, where $d_{\lceil f}$ represents the set of source data items satisfying the formula.
- *Goodness* ($\nearrow$). $p_1 \nearrow p_2$ if they have the same pattern type, $p_1 \preceq^e p_2$, and p_1 measures are better than p_2 measures, i.e., assuming that $pt.m = \langle m_1, ..., m_n \rangle$, $p_1.m_i \theta_i p_2.m_i$, $i = 1, ..., n$ [2].
- *Temporal validity* (ω_T). Given a pattern p_1 and a temporal value t, $\omega_T(p_1, t)$ is true if and only if p_1 is temporally valid at time t.
- *Semantic validity* (ω_S). Given a pattern p of type pt, a data source D, a measure function μ_m for pt, and some thresholds $v_1, ..., v_n$, $\omega_S(p, D, \mu_m, < v_1, ..., v_n >)$ is true if and only if p is semantically valid with respect to D and $v_1, ..., v_n$, assuming to compute measure values by using μ_m.

Note that $\preceq^e$, $\nearrow$, and ω_S are cross-over predicates.

4.2 Query Operators

Basic Operators. In the PBMS framework, queries are executed against classes. Besides typical relational operators (such as renaming, set-based operators), several other query operators are proposed (see Table 1). For example, projection is revisited to project out structure and measure components. The selection operator allows one to select patterns belonging to a certain class

[2] According to Def. 1, θ_i is a predicate expressing that $p_1.m_i$ is "better than" $p_2.m_i$.

Table 1. TPQL basic operators

Name	Operator	Description
Projection	$\pi_{(l\_s,l\_m)}(c)$ where: c is a pattern class, l_s is a non empty list of attributes appearing the pattern structure, and l_m a list of attributes appearing in the pattern measure	it reduces the structure and the measures of the patterns in c by projecting out components not appearing in l_s and l_m
Selection	$\sigma_F(c)$ where: c is a class and F is a selection predicate	it selects the patterns in c satisfying F
Join	$c_1 \bowtie_{F,cf} c_2$ where: c_1 and c_2 are two classes, F: join predicate , and cf: composition function	it combines patterns belonging to c_1 and c_2, if they satisfy the join predicate F; each new pattern is generated by using the composition function cf

satisfying a certain condition, using any predicate introduced in Section 4.1. When using cross-over predicates, it becomes a cross-over operator. Finally, the join operator combines patterns belonging to two different classes, with possibly different pattern types. It requires the specification of a join predicate and a composition function, which defines the pattern type of the result.

Temporal Operators. Since we deal with temporal information associated with patterns, the need arises of querying such information. By using the proposed query operators (especially selection and join) and the predicates defined over validity periods (see Section 4), several interesting temporal queries can be specified. For instance, the user may be interested in retrieving from a certain class c, at a fixed instant of time (e.g. 'now'), all safe patterns. To this purpose, selection can be used as follows: $\sigma_{\omega_S(p,p.d,\mu_m,v)\wedge\omega_T(p,'now')}(c)$ [3]. As another example, retrieval of the patterns belonging to a certain class c, which are temporally valid in a given interval of time (e.g. a certain year), can be specified as follows: $\sigma_{vt\ during\ [01-JAN-03,31-DEC-03)}(c)$.

Cross-over Operators. They correlate patterns with raw data, providing a way for navigating from the pattern layer to the raw data layer and vice versa. *Drill-Through* γ. It allows one to retrieve the subset of source data associated with at least one pattern in a class c, satisfying condition *cond*:

$\gamma(c, cond) = \{x | \exists p \in c, cond(c) = true, x \in p.d\}$.

Data Covering θ_d. Let p be a pattern of type pt, D a data source, μ a mining function for patterns of type pt, and $v =< v_1, ..., v_n >$ some user-specified thresholds. Data covering allows us to determine the subset of source data represented by at least one pattern in the class. To this purpose, the formula is used:

$\theta_d(c, D, \mu) = \{x | x \in D, \exists p \in c, p.f(x) = true\}$.

Cross-over selection. When using a cross-over predicate within a selection, we need to access raw data to execute the query. For example, suppose that c is a class of association rules and D a dataset suitable for patterns in c. The query $\sigma_{D\subseteq^e d\wedge support>0.6}(c)$ returns all rules in c representing a superset of D, with a support greater than 0.6.

[3] p denotes a generic pattern in c.

Example 7. Consider Example 1. Suppose that from April 2005, the vendor will start to sell a certain product P and he wants to know how P can be promoted. To do that, he looks for a correlation between P and some other items sold. With such information, he may activate an advertising campaign to promote some other product he already sells in order to stimulate the demand for P. In this way, when he starts to sell P, probably customers will start to buy it without the need for a dedicated advertising campaign. A possible approach could be the following. First of all it determines in which cluster of products P belongs and gets the representative R, by using a selection and a projection operator (Fig. 3 line 18). Then, he determines which products stimulate the sale of R by considering bodies of association rules having R in their head. This result can be achieved by performing a join operation (Fig. 3 line 19) between patterns just retrieved and association rules already mined. We assume that the used composition function returns patterns representing the bodies of the selected association rules. Products in association rule bodies are such that whenever a customer buys one of them, with a high probability he buys also R. Since P is in the cluster represented by R, P and R are similar with respect to customer preferences, thus it is most likely that when the vendor starts to sell P, customers will behave as for R. When, on April 1 2005, the vendor starts to sell P, new data are collected and patterns previously extracted may become unreliable. Thus, a synchronization is required between data and patterns (Fig. 3 lines 20-22). □

5 Related Work

Several approaches have been proposed to model patterns. Among standardization efforts for modeling patterns, we recall the Predictive Model Markup Language (PMML) [4], the ISO SQL/MM standard [3], and the Common Warehouse Model (CWM) framework [2]. Although these approaches represent a wide range of data mining results, they do not provide a generic model to handle arbitrary pattern types. Furthermore, their main purpose is to enable an easy interchange of metadata not their effective manipulation.

In inductive databases, data and patterns are stored and queried together [1, 7, 8, 12]. They rely on specific (but extensible) types of patterns and are primarily focused on a-posteriori patterns. Moreover, validity is not considered. Within this framework, the entire knowledge discovering process is a querying process [12–14]. However, new SQL-based operators do not allow the user to specify specific mining functions [9]. Moreover, none of the proposed languages deals with pattern validity and synchronization aspects.

In [11], the authors propose a unified algebraic framework for multi-step knowledge discovery. Similarly to our approach, they model different types of patterns and maintain data and patterns separated. However, temporal information is not considered.

Previous work strictly related to the work presented here has been reported in two previous papers by us and other authors [6, 15], where a model for patterns and a pattern query language have been proposed. The major differences between that work and the one presented here is the extension with temporal features. We believe that this is relevant extension from both a theoretical and architectural point of view.

6 Concluding Remarks

In this paper, we presented a general framework for patterns representation and management, taking into account validity information, a-priori and a-posteriori patterns. The resulting framework seems general enough to cope with real data mining applications. We are currently working on the development of a prototype of the proposed framework. Future work includes the definition of a pattern calculus, equivalent to the proposed algebra, the analysis of their expressive power and complexity, and the comparison of the expressive power of existing approaches to deal with patterns. We also plan to further investigate semantic validity to extend temporal analysis capabilities for patterns.

References

1. The CINQ project. `http://www.cinq-project.org`, 1998-2002
2. Common Warehouse Metamodel (CWM). `http://www.omg.org/cwm`, 2001.
3. ISO SQL/MM Part 6. `http://www.sql-99.org/SC32/WG4/Progression_Documents/FCD/fcd-datamining-2001-05.pdf`, 2001.
4. Predictive Model Markup Language (PMML). `http://www.dmg.org/pmmlspecs_v2/pmml_v2_0.html`, 2003.
5. The PANDA Project. `http://dke.cti.gr/panda/`, 2002.
6. E. Bertino, B. Catania, and A. Maddalena. Towards a Language for Pattern Manipulation and Querying. In *Proc. of the 1st Int. Workshop on Pattern Representation and Management (PaRMa'04)*, 2004.
7. J. F. Boulicaut, M. Klemettinen, and H. Mannila. Modeling KDD Processes within the Inductive Database Framework. In *Proc. of the Data Warehousing and Knowledge Discovery*, pages 293–302, 1999.
8. L. De Raedt. A Perspective on Inductive Databases. *ACM SIGKDD Explorations Newsletter*, 4(2):69–77, 2002.
9. B. Goethals and J. Van den Bussche. A Priori versus a Posteriori Filtering of Association Rules. In *Proc. of the ACM SIGMOD Workshop on Research Issues in Data Mining and Knowledge Discovery*, 1999.
10. J. Han and M. Kamber. *Data Mining: Concepts and Techniques.* Academic Press, 2001.
11. T. Johnson, L.V.S Lakshmanan, and R.T. Ng. *The 3W Model and Algebra for Unified Data Mining.* In *Proc. of the 26th Int. Conf. on Very Large Data Bases*, 2000.
12. T. Imielinski and H. Mannila. A Database Perspective on Knowledge Discovery. *Communications of the ACM*, 39(11):58–64, 1996.
13. T. Imielinski and A. Virmani. MSQL: A Query Language for Database Mining. *Data Mining and Knowledge Discovery*, 2(4):373–408, 1999.
14. R. Meo, G. Psaila, and S. Ceri. An Extension to SQL for Mining Association Rules. *Data Mining and Knowledge Discovery*, 2(2):195–224, 1999.
15. S. Rizzi et Al. Towards a Logical Model for Patterns. In *Proc. of the 22nd Int. Conf. on Conceptual Modeling (ER 2003)*, 2003.
16. R. T. Snodgrass, editor. *The TSQL2 Temporal Query Language.* Kluwer, 1995.

Spatial Associative Classification at Different Levels of Granularity: A Probabilistic Approach

Michelangelo Ceci, Annalisa Appice, and Donato Malerba

Dipartimento di Informatica, Università degli Studi
via Orabona, 4, 70126 Bari, Italy
{ceci,appice,malerba}@di.uniba.it

Abstract. In this paper we propose a novel spatial associative classifier method based on a multi-relational approach that takes spatial relations into account. Classification is driven by spatial association rules discovered at multiple granularity levels. Classification is probabilistic and is based on an extension of naïve Bayes classifiers to multi-relational data. The method is implemented in a Data Mining system tightly integrated with an object relational spatial database. It performs the classification at different granularity levels and takes advantage from domain specific knowledge in form of rules that support qualitative spatial reasoning. An application to real-world spatial data is reported. Results show that the use of different levels of granularity is beneficial.

1 Introduction

The rapidly expanding amount of spatial data gathered by collection tools, such as satellite systems or remote sensing systems have paved the way for advances in spatial data structures [12], spatial reasoning [8] and computational geometry [23] to serve multiple tasks including storage and sophisticated treatment of real-world geometry in a spatial database. A spatial database contains (spatial) objects that are characterized by a geometrical representation (e.g. point, line, and region in a 2D context) as well as several non-spatial attributes. The widespread use of spatial databases in real-world applications (e.g geo-marketing or environmental analysis) is leading to an increasing interest in Spatial Data Mining, i.e. in mining interesting and useful but implicit knowledge. Classification of spatial objects is a fundamental task in Spatial Data Mining, where training data consists of multiple target spatial objects (primary data), possibly spatially-related with other non-target spatial objects (secondary data). The goal is to learn the concept associated with each class on the basis of the interaction of two or more spatially-referenced objects or space-dependent attributes, according to a particular spacing or set of arrangements [15].

While a lot of research has been conducted, both in propositional and multi-relational setting, on mining classification models from data eventually stored in multiple tables of a relational database, only a few works deal with classification models to be discovered in spatial database. Indeed, mining spatial classification models presents two main sources of complexity, that is, the implicit definition of spatial relations and the granularity of the spatial objects. The former is due to the fact that the geometrical representation (e.g. point, line, and region in a 2D context) and the relative positioning of spatial objects with respect to some reference system, define implicitly spatial relations of different nature, such as directional and topological.

J.-F. Boulicaut et al. (Eds.): PKDD 2004, LNAI 3202, pp. 99–111, 2004.

Modeling these spatial relations is a key challenge in classification problems that arise in spatial domains [24]. Indeed, both the attribute values of the object to be classified and the attribute values of spatially related objects may be relevant for assigning an object to a class from a given set of classes. The second source of complexity refers to the fact that spatial objects can be described at multiple levels of granularity. For instance, UK census data can be geo-referenced with respect to the hierarchy of areal objects:

$$\text{ED} \rightarrow \text{Ward} \rightarrow \text{District} \rightarrow \text{County},$$

based on the inside relationship between locations. Therefore, some kind of *taxonomic knowledge* of task-relevant geographic layers may also be taken into account to obtain descriptions at different granularity levels (*multiple-level classification*).

In this paper we propose a novel spatial classification method based on a multi-relational approach that takes spatial relations into account. Classification is probabilistic and is based on the extension of naive Bayes classifiers to multi-relational data. Classification rules are automatically generated by means of a spatial association rule discovery system characterized by the capability of generating association rules at multiple levels of granularity. In this way, the proposed method can deal with both sources of complexity presented above. The proposed method has been implemented in a Data Mining system tightly integrated with an object-relational spatial database. It can perform the classification at different levels of granularity and takes advantage from domain specific knowledge expressed in form of rules to support qualitative spatial reasoning. Finally, it handles categorical as well as numerical data through a contextual discretization method.

The paper is organized as follows. In the next section we discuss the background of this research and some related works. The mining of multi-level spatial association rules for classification purpose is presented in Section 3 while the multi-relational Naïve Bayes classification is described in Section 4. Section 5 describes the system architecture. Finally, an application is presented in Section 6 and some conclusions are drawn.

2 Background and Motivations

The problem of classifying spatial objects has been investigated by some researchers. Ester et al. [10] proposed a neighbourhood graph based extension of decision trees that considers both non-spatial attributes of the classified objects and relations with neighbouring objects. However, the proposed method does not take into account hierarchical relations defined on spatial objects as well as non-spatial attributes (e.g. number of residents) of neighbouring objects. In contrast, Kopersky [15] described an efficient method that classifies spatial objects by considering both spatial and hierarchical relations between spatial objects and takes into account non-spatial attributes for neighbouring objects. However this method suffers from severe limitations due to the restrictive representation formalism known as *single-table assumption* [26]. More specifically, it is assumed that data to be mined are represented in a single table of a relational database, such that each row (or tuple) represents an independent unit of the sample population and columns correspond to properties of units. This requires that non-spatial properties of neighboring objects be represented in aggregated form causing a consequent loss of information and a change in the units of analysis.

In [20], the authors proposed to exploit the expressive power of predicate logic to represent both spatial relations and background knowledge, such as spatial hierarchies. In addition the logical notions of generality order and of downward refinement operator on the space of patterns may be profitably used to define both the search space and the search strategy. For this purpose, the ILP system ATRE [21] has been integrated in the data mining server of a prototypical Geographical Information System (GIS), named INGENS, which allows, among other things, to mine classification rules for geographical objects stored in an object-oriented database. Training is based on a set of examples and counterexamples of geographic concepts of interest to the user (e.g., ravine or steep slopes). The first-order logic representation of the training examples is automatically extracted from maps, although it is still controlled by the user who can select a suitable level of abstraction and/or aggregation of data by means of a data mining query language [19].

Similarly, the discovery of spatial association rules, that is spatial and a-spatial relationships among spatial objects, has been investigated both in propositional and multi-relational setting. A *spatial association rule* is a rule of the form "$P \rightarrow Q$ (s, c)" such that both P (body) and Q (head) are sets of literals, some of which refer to spatial properties, and $P \cap Q = \varnothing$. $P \cup Q$ is named *pattern*. The support s estimates the probability $p(P \cup Q)$, while the confidence c estimates the probability $p(Q|P)$.

Koperski and Han [14] implemented the module Geo-associator of the spatial data mining system GeoMiner that mines rules from data represented in a single relation (table) of a relational database. In contrast, in [16], the authors proposed an ILP approach to spatial association rules discovery. The algorithm SPADA (Spatial Pattern Discovery Algorithm), reported in their work, allows the extraction of multi-level spatial association rules, that is, association rules involving spatial objects at different granularity levels. SPADA has been implemented as a module of the system ARES (Association Rules Extractor from Spatial data) [2], which also supports users in the complex processes of extracting spatial objects from the spatial database, specifying the background knowledge on the application domain and defining a search bias.

Despite the fact that spatial association rule mining is a descriptive task, while classification of spatial objects is a predictive task, recent studies in Data Mining and Machine Learning have investigated the opportunity of combining association rules discovery and classification, by taking advantage of employing association rules for classification purpose [6, 3]. This approach is named associative classification [17] and several advantages are reported in the literature for this approach. First, differently from most of classifiers as decision trees, association rules consider the simultaneous correspondence of values of different attributes, hence allowing to achieve better accuracy [3]. Second, it makes association rule mining techniques applicable to classification tasks. Third, the user can decide to mine both association rules and a classification model in the same data mining process [17]. Fourth, the associative classification approach helps to solve *understandability* problems [4, 25] that may occur with some classification methods. Indeed, many rules produced by standard classification systems are difficult to understand because these systems often use only domain independent biases and heuristics, which may not fulfil user's expectation. With the associative classification approach, the problem of finding understandable rules is reduced to a post-processing task [17]; filtering based on user-defined rule template may help in extracting understandable rules.

Although associative classification methods present several interesting aspects, they also suffer from some limitations. First, most of methods reported in the literature work under the *single-table assumption*, which is a strong limitation in those application domains characterized by a spatial dimension. Second, they have a categorical output which convey no information on the potential uncertainty in classification. Small changes in the attribute values of an object being classified may result in sudden and inappropriate changes to the assigned class. Missing or imprecise information may prevent a new object from being classified at all. In alternative, to overcome these deficiencies, we propose to use a probabilistic classifier that returns, in addition to the result of the classification, the confidence of the classification. This is an important aspect because of the increasing attention on the ROC curve analysis [11] that defines an evaluation measure to take into account the confidence of the classification. Third, reported methods require additional heuristics to identify the most effective rule at classifying a new object. Alternatively, in the proposed approach, the evaluation of the class is based on the computation of probabilities taking into account all the rules.

3 Multi-level Spatial Association Rules

In [2] the problem of mining spatial association rules has been formalized as follows:

Given a spatial database (SDB), a set S of *reference objects* tagged with a class label $c_j \in \{C_1, C_2, \ldots, C_L\}$, some sets R_k, $1 \le k \le m$, of *task-relevant objects*, a background knowledge BK including some *spatial hierarchies* H_k on objects in R_k, M *granularity levels* in the descriptions (1 is the highest while M is the lowest), a set of *granularity assignments* ψ_k which associate each object in H_k with a granularity level, a couple of thresholds *minsup*[l] and *minconf*[l] for each granularity level, a language bias LB that constrains the search space;

Find strong multi-level spatial association rules, that is, association rules involving spatial objects at different granularity levels.

The reference objects are the main subject of the description, that is, the observation units, while the task relevant objects are spatial objects that are relevant for the task in hand and are spatially related to the former. The sets R_k typically correspond to layers of the spatial database, while hierarchies H_k define *is-a* (i.e., taxonomical) relations of spatial objects in the same layer (e.g. river *is-a* water body). Objects of each hierarchy are mapped to one or more of the M user-defined description granularity levels in order to deal uniformly with several hierarchies at once. Both frequency of patterns and strength of rules depend on the granularity level l at which patterns/rules describe data. Therefore, a pattern P ($s\%$) at level l is *frequent* if $s \ge minsup[l]$ and all ancestors of P with respect to H_k are frequent at their corresponding levels. An association rule $Q \rightarrow R$ ($s\%$, $c\%$) at level l is *strong* if the pattern $Q \cup R$ ($s\%$) is frequent and $c \ge minconf[l]$.

The problem above is solved by the algorithm SPADA [16] that operates in three steps for each granularity level: i) pattern generation; ii) pattern evaluation; iii) rule generation and evaluation. SPADA takes advantage of statistics computed at granularity level l when computing the supports of patterns at granularity level l+1.

In the system ARES (http://www.di.uniba.it/~malerba/software/ARES/index.htm) SPADA has been loosely coupled with a spatial database, since data stored in the SDB Oracle Spatial are pre-processed and then represented in a deductive database (DDB). For instance, spatial intersection between two objects X and Y is represented by the extensional predicate *crosses(X,Y)*. In this way, the expressive power of first-order logic in databases is exploited to specify both the background knowledge BK, such as spatial hierarchies and domain specific knowledge, and the language bias LB. Spatial hierarchies allow to face with one of the main issues of spatial data mining, that is, the representation and management of spatial objects at different levels of granularity, while the domain specific knowledge stored as a set of rules in the intensional part of the DDB supports qualitative spatial reasoning. On the other hand, the LB is relevant to allow the user to specify his/her bias for interesting solutions, and then to exploit this bias to improve both the efficiency of the mining process and the quality of the discovered rules. In SPADA, the language bias is expressed as a set of constraint specifications for either patterns or association rules. Pattern constrains allow to specify a literal or a set of literals that should occur one or more times in discovered patterns. During the *rule generation* phase, patterns that do not satisfy a pattern constraint are filtered out. Similarly, rule constraints are used do specify literals that should occur in the head or body of discovered rules.

In a more recent release of SPADA (3.1) a new rule constraint has been introduced in order to specify the maximum number of literal that should occur in the head of a rule. In this way users may define the head structure of a rule requiring the presence of exactly a specific literal and nothing more. In the case this literal describes the class label, multi-level spatial association rules discovered by ARES may be used for classification.

4 Naïve Bayes Classification

Once a set of rules has been extracted for each level, it is used in the construction of a naïve Bayesian classifier [5], which aims to classify any target object $o \in S$ by maximizing the *posterior probability* $P(C_i|o)$ that o is of class C_i, that is:

$$class(o) = arg\ max_i\ P(C_i|o)$$

By applying the Bayes theorem, $P(C_i|o)$ can be reformulated as follows:

$$P(C_i|o) = \frac{P(C_i)P(o|C_i)}{P(o)} \qquad (1)$$

The term $P(o|C_i)$ is estimated by means of the *naïve Bayes assumption*:

$$P(o|C_i)=P(o_1,o_2,\ldots,o_m|C_i)=P(o_i|C_i) \times P(o_2|C_i) \times \ldots \times P(o_m|C_i)$$

where $o_1,o_2,\ldots,o_m$ represent the set of the properties, different from the class, used to describe the object. This assumption is clearly false if the predictor variables are statistically dependent. However, even in this case, the naïve Bayesian classifier can give good results [5].

In (1) the value $P(C_i)$ is the prior probability of the class C_i. Since $P(o)$ is independent of the class C_i, it does not affect $f(o)$, that is,

$$\text{class(o)}= \arg\max_i \; P(C_i)P(o|C_i) \tag{2}$$

However, this formulation of the problem holds in the *single-table assumption* data representation formalism, where an object represents an independent unit of the sample population described by means of a set of properties. In the multi-relational setting [7], the target object is related to other non-target objects. In order to take into account the relations of the target object, a modification of the problem formulation is necessary. For this purpose, a key role is played by the extracted association rules. In particular, the idea is to consider the set of rules to guide the computation of $P(o|C_i)$.

Given the object $o \in S$, we consider the subset of the extracted rules that can be used to classify o. More formally, we consider the subset R of rules whose body is satisfied by the object to be classified both in terms of the values of properties of involved spatial objects and in terms of the spatial relations between objects. For example, if S is the set of wards in a district, a ward w satisfies the rule:

mortality_rate(A, low) ← *wards_relatedTo_waters(A, B), waters_typewater(B, river), cars_per_person(A, high)*

when w is spatially related (intersects) to a river and is characterized by a high average number of cars per person.

We use R to estimate $P(o|C_i)$. In particular, we estimate $P(o|C_i)$ by means of the probabilities associated to both spatial relations (e.g. *wards_relatedTo_waters(A,B)*) and properties (e.g. *waters__typewater(B,RIVER)*, *cars_per_person(A,high)*) associated to each rule in R.

For instance, if $R = \{R_1, R_2\}$, where R_1 and R_2 are two association rules of class C_i extracted by SPADA:

$$R_1\text{: } \beta_{1,0} :- \beta_{1,1}, \beta_{1,2} \qquad R_2\text{: } \beta_{2,0} :- \beta_{2,1}, \beta_{2,2}$$

where $\beta_{1,1}$ and $\beta_{1,2}$ are spatial relations, $\beta_{1,2}$ and $\beta_{2,2}$ are properties and $\beta_{1,0} = \beta_{2,0}$ (class) then $P(\{R_1, R_2\}|C_i) = P(\beta_{1,0} \cap \beta_{1,1} \cap \beta_{2,1} \cap \beta_{1,2} \cap \beta_{2,2} \,|\, C_i) =$

$$P(\beta_{1,0} \cap \beta_{1,1} \cap \beta_{2,1} \,|\, C_i) \cdot P(\beta_{1,2} \cap \beta_{2,2} \,|\, \beta_{1,0} \cap \beta_{1,1} \cap \beta_{2,1} \cap C_i)$$

The first term takes into account the relations of the rules while the second term refers to the conditional probability of satisfying the property predicates in the rules given the relations. By means of the naïve Bayes assumption, the probabilities can be factorized as follows:

$$P(\beta_{1,0} \cap \beta_{1,1} \cap \beta_{2,1} \,|\, C_i) = P(\beta_{1,1} \,|\, C_i) \cdot P(\beta_{2,1} \,|\, C_i)$$
$$P(\beta_{1,2} \cap \beta_{2,2} \,|\, \beta_{1,0} \cap \beta_{1,1} \cap \beta_{2,1} \cap C_i) = P(\beta_{1,2} \,|\, \beta_{1,1} \cap \beta_{2,1} \cap C_i) \cdot P(\beta_{2,2} \,|\, \beta_{1,1} \cap \beta_{2,1} \cap C_i)$$

Since $\beta_{1,2}$ and $\beta_{2,2}$ do not depend from $\beta_{2,1}$ and $\beta_{1,1}$ respectively, then:

$$P(\beta_{1,2} \cap \beta_{2,2} \,|\, \beta_{1,0} \cap \beta_{1,1} \cap \beta_{2,1} \cap C_i) = P(\beta_{1,2} \,|\, \beta_{1,1} \cap C_i) \cdot P(\beta_{2,2} \,|\, \beta_{2,1} \cap C_i)$$

By generalizing to a set of rules we have:

$$P(C_i)P(o|C_i) = P(C_i) \prod_{k \in |R|} \left(P(relations_k \,|\, C_i) \prod_j P(property_{k,j} | relations_k, C_i)\right) \tag{3}$$

where the term $relations_k$ represents the event that the set of spatial relations expressed in the k-th rule is satisfied, while the term $property_{k,j}$ represents the event that the j-th property of the k-th rule is satisfied.

If $relations_k = \{ relation(Set_1,Set_2) \mid Set_1,Set_2 \in \{S\} \cup \{R_k, 1 \le k \le m\}, Set_1 \ne Set_2 \}$ is a set of binary relations between spatial objects (either task relevant or reference) involved in the k-th rule, the probability $P(relations_k|C_i)$ is computed by means of the naïve Bayes assumption:

$$P(relations_k|C_i) = \prod_{l \in |relations_k|} P(relation(Set_{l_1}, Set_{l_2}) \mid C_i)$$

where:

$$P(relation(Set_{l_1}, Set_{l_2}) \mid C_i) = P(relation(Set'_{l_1}, Set'_{l_2})) = \frac{|\, relation(Set'_{l_1}, Set'_{l_2}) \,|}{|\, Set'_{l_1} \,| \cdot |\, Set'_{l_2} \,|} \quad (4)$$

where Set'_l is a subset of objects in Set_l that are related, by means of spatial relations, with objects in S of class C_i, while $|\, relation(Set'_{l_1}, Set'_{l_2}) \,|$ is the number of relations between objects of Set'_{l_1} and objects of Set'_{l_2}.

To compute the probability $P(property_{k,j}|relations_k, C_i)$ in (3), we use the Laplace estimation:

$$P(property_{k,j}|relations_k, C_i) = \frac{|\, relations_k \wedge property_{k,j} \wedge C_i \,| + 1}{|\, relations_k \wedge C_i \,| + F} \quad (5)$$

where F is the number of possible admissible values of the property. Laplace's estimate is used in order to avoid null probabilities in equation (2). In practice, the value at the nominator is the number of target objects of class C_i that are related to other spatial objects by means of spatial relations expressed in $relations_k$ and for which $property_{k,j}$ is satisfied. The value of the denominator is the number of target objects of class C_i that are related to other spatial objects by means of spatial relations expressed in $relations_k$ plus F.

In order to avoid the problem that the same relation or the same property is considered more than once in the computation of probabilities in formula (3), the values computed in formula (4) and (5) are effectively determined and included in formula (3) only if the values have not been computed before.

5 A Spatial Associative Classification Framework

The integration of multi-level spatial association rules discovery with naïve Bayesian classification is realized in a spatial associative classification system based on a client-server model (see Fig. 1). Both the spatial association rule miner SPADA and the multi-relational naïve Bayes classifier are on the server side, so that several data mining tasks can be run concurrently by multiple users. SPADA fully exploits the flexibility of ILP to specify the background knowledge *BK* (i.e hierarchies and domain specific knowledge) as well as the language bias *LB* (i.e. search constraints). Hierarchies are expressed by a collection of ground atoms and represent spatial objects at different granularity level while domain specific knowledge is expressed as sets of definite clauses and support a spatial qualitative reasoning. Conversely, search

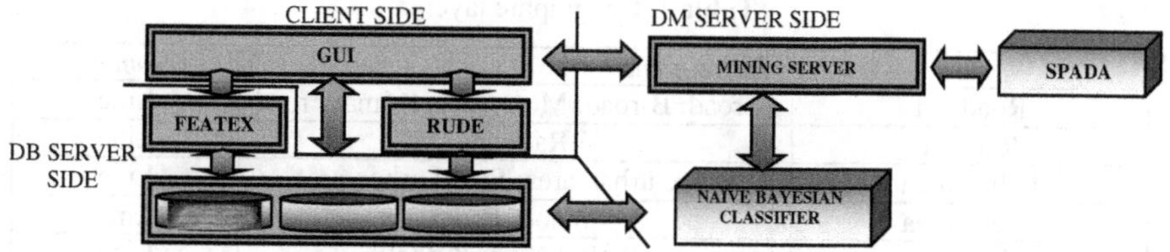

Fig. 1. Spatial associative classification system.

constraints are used to bias the search in order to fulfil user expectations. In this framework, constraints are also used to partially fix the structure of extracted rules in order to discover spatial association rules that contain only the class label in the head. For each granularity level, extracted rules concur in building the spatial classification model by exploiting a multi-relational naïve Bayesian classifier integrated with the SDB.

On the client side, the framework includes a Graphical User Interface (GUI), which provides users with facilities for controlling all parameters of the mining process.

SPADA, like many other association rule mining algorithms, cannot process numerical data properly, so it is necessary to perform a discretization of numerical features with a relatively large domain. For this purpose, the framework includes in the client side the module RUDE (relative unsupervised discretization algorithm) which discretizes a numerical attribute of a relational database in the context defined by other attributes [18].

The SDB (Oracle Spatial) can run on a third computation unit. Many spatial features (relations and attributes) can be extracted from spatial objects stored in the SDB. Feature extraction requires complex data transformation processes to make spatial relations explicit and representable as ground Prolog atoms. Therefore, a middle layer module, named FEATEX (Feature Extractor), is required to make possible a loose coupling between SPADA and the SDB by generating features of spatial objects (points, lines, or regions). The module is implemented as an Oracle package of procedures and functions, each of which computes a different feature [2]. Transformed data are also stored in SDB tables.

6 The Application: Mining North West England Census Data

In this section we present a real-world application concerning the mining of both spatial association rules and classification models for geo-referenced census data interpretation. We consider both census and digital map data provided in the context of the European project SPIN! (Spatial Mining for Data of Public Interest) [22]. They concern Greater Manchester, one of the five counties of North West England (NWE). Greater Manchester is divided into ten metropolitan districts, each of which is decomposed into censual sections or wards, for a total of two hundreds and fourteen wards. Spatial analysis is enabled by the availability of vectorized boundaries of the 1998 census wards as well as by other Ordnance Survey digital maps of NWE, where several interesting layers are found, namely road net, rail net, water net, urban area and green area (see Table 1).

Table 1. Geographic layers.

	Layer name	*Geometry*
Road net	A-road; B-road; Motorway; Primary road	Line
Rail net	Railway	Line
Urban area	Large urban area; Small urban area	Line
Green area	Wood; Park:	Line
Water net	Water; River; Canal	Line
Greater Manchester Ward	Ward	Region

Census data are available at ward level. They provide socio-economic statistics (e.g. mortality rate, that is, the percentage of deaths with respect to the number of inhabitants) as well as some measures describing the deprivation level. Indeed, the material deprivation of an area may be estimated according to information provided by Census combined into single index scores [1]. Over the years different indices have been developed for different applications: the Jarman Underprivileged Area Score was designed to measure the need for primary care, the indices developed by Townsend and Carstairs have been used in health-related analyses, while the Department of the Environment's Index (DoE) has been used in targeting urban regeneration funds. Thereby, we have considered the values of Jarman index, Townsend index, Carstairs index and DoE index. The higher the index value the more deprived a ward is. Both index values as well as mortality rate are all numeric and have been discretized by means of RUDE. More precisely, Jarman index, Townsend index, DoE index and Mortality rate have been automatically discretized in (*low, high*), while Carstairs index has been discretized in (*low, medium, high*).

For this application, we have considered Greater Manchester wards as reference (target) objects. In particular, three different experimental settings have been analysed by varying the target property among mortality rate, Jarman index and DoE index. We have chosen Jarman and DoE indices because they are defined on the basis of different social factors. For each setting, we have focused our attention on investigating dependencies between the target property and socio-economic factors represented in census data as well as geographical factors represented in linked topographic maps. These dependencies are detected in form of spatial association rules having only the target property in the head. Rules in this form may be employed for spatial subgroup mining, that is, discovery of interesting groups of spatial objects with respect to a certain property of interest [13] as well as for classification purpose.

For this analysis, we have formulated queries involving the FEATEX *relate* function to compute topological relationships between reference objects and task relevant objects. For instance, a relationship extracted by FEATEX is *crosses(ward_135, urbareaL_151)*, where *ward_#* denotes a specific Greater Manchester ward, while *urbanareaL#* refers to a large urban area crossing the interested ward. The topological relationship *crosses* is computed according to the 9-intersection model [9]. The number of computed relationships is 784,107.

To support a spatial qualitative reasoning, a domain specific knowledge (BK) has been expressed in form of a set of rules. Some of these rules are:

crossed_by_urbanarea(X,Y) :- connects(X,Y), is_a(Y, urban_area). ...
crossed_by_urbanarea(X,Y) :- inside(X,Y), is_a(Y, urban_area).

Here the use of the predicate *is_a* hides the fact that a hierarchy has been defined for spatial objects which belong to the urban area layer. In detail, five different hierarchies have been defined to describe the following layers: road net, rail net, water net, urban area and green area (see Fig. 2). The hierarchies have depth three and are straightforwardly mapped into three granularity levels. They are also part of the BK.

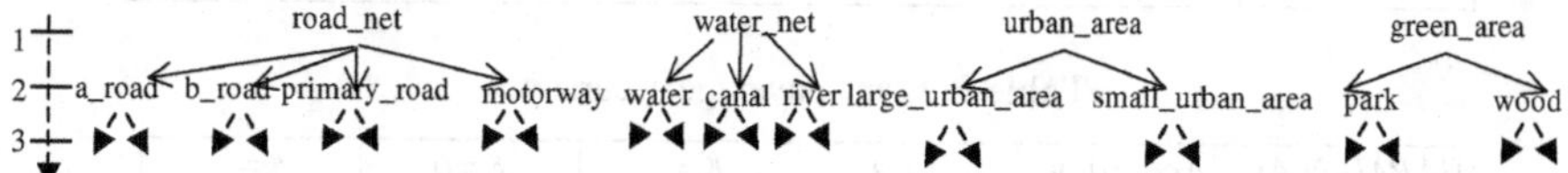

Fig. 2. Spatial hierarchies defined for road net, water net, urban area and green area.

Finally, we have specified a language bias (LB) both to constrain the search space and to filter out uninteresting spatial association rules. In particular, we have ruled out all spatial relations (e.g. crosses, inside, and so on) directly extracted by FEATEX and asked for rules containing topological predicates defined by means of BK. Moreover, by combining the rule filters *head_constraint([mortality_rate(_),1,1)* and *rule_head_length(1,1)* we have asked for rules containing only *mortality rate* in the head. Similar considerations apply to the classification tasks concerning the Jarman and the DoE indices. In addition, we have specified the maximum number *K* of refinement steps (i.e. number of literals in the body of rules).

For each setting, a ten-fold cross validation has been performed and results are evaluated. For instance, by analyzing spatial association rules extracted with parameters *minsup* = 0.1, *minconf* = 0.6 we discover the following rule:

mortality_rate(A, high) ← *is_a(A, ward), crossed_by_urbanarea(A, B),*
is_a(B, urban_area), townsendidx_rate(A, high) (40.72%, 72.47%)

which states that a high mortality rate is observed in a ward *A* that includes an urban area *B* and has a high value of Townsend index. The support (40.72%) and the high confidence (72.47%) confirm a meaningful association between a geographical factor, such as living in deprived urban areas, and a social factor, such as the mortality rate. It is noteworthy that SPADA generates the following rule:

mortality_rate(A, high) ← *is_a(A,ward), crossed_by_urbanarea(A,B),*
is_a(B, urban_area) (56.7%, 60.77%)

which has a greater support and a lower confidence. These two association rules show together an unexpected association between Townsend index and urban areas. Apparently, this means that this deprivation index is unsuitable for rural areas.

At a granularity level 2, SPADA specializes the task relevant object B by generating the following rule which preserves both support and confidence:

mortality_rate(A, high) ← *is_a(A, ward), crossed_by_urbanarea(A, B),*
is_a(B, urban_areaL), townsendidx_rate(A,high) (40.72%, 72.47%)

This rule clarifies that the urban area B is large.

The average predictive accuracy of mined multi-level spatial classification model is evaluated by varying *minsup*, *minconf* and *K* for each setting,. Results are reported in Table 2, 3 and 4. In the first setting, results show that, predictive accuracy of the Bayesian classifier is slightly better than the accuracy (0.567) of the trivial classifier that returns the most probable class. We explain this result with the inherent complex-

Table 2. Mortality Rate average accuracy.

MORTALITY Avg. Accuracy		*K=4*	*K=5*	*K=6*	*K=7*
minsup=0.1	*Level=1*	0.5932	0.5915	0.5932	0.628
minconf=0.6	*Level=2*	0.5932	0.596	0.5932	0.628
minsup=0.2	*Level=1*	0.5932	0.602	0.5932	0.623
minconf=0.65	*Level=2*	0.5932	0.602	0.5932	0.623

Table 3. Jarman average accuracy.

JARMAN Avg. Accuracy		*K=4*	*K=5*	*K=6*	*K=7*
minsup=0.1	*Level=1*	0.8176	0.8176	0.8176	0.8176
minconf=0.6	*Level=2*	0.8176	0.8176	0.8176	0.8176
minsup=0.2	*Level=1*	0.528	0.528	0.528	0.528
minconf=0.8	*Level=2*	0.528	0.528	0.6272	0.6705

Table 4. DoE average accuracy.

DoE Avg. Accuracy		*K=4*	*K=5*	*K=6*	*K=7*
minsup=0.1,	*Level=1*	0.912	0.912	0.912	0.912
minconf=0.6	*Level=2*	0.912	0.912	0.912	0.912
minsup=0.2,	*Level=1*	0.875	0.875	0.875	0.821
minconf=0.8	*Level=2*	0.875	0.9028	0.883	0.874

ity of the task. Different conclusions can be drawn from both Jarman and DoE results, where the Bayesian classifiers significantly improve the trivial classifiers (acc. 0.542 and 0.625, respectively). Another consideration is that the average predictive accuracies of classification models discovered at higher granularity levels (i.e. level=2) are always better or equal to the corresponding accuracies at lowest levels. This means that the classification model takes advantage of the use of the hierarchies defined on spatial objects. Furthermore, results show that by decreasing the number of extracted rules (higher support and confidence) we have lower accuracy. This means that there are several rules that strongly influence classification results and often such rules are not characterized by high values of support and confidence. Finally, we observe that, generally, the higher the number of refinement steps, the better the model.

7 Conclusions

In this paper we have presented a spatial associative classifier that combines spatial association rule discovery with naïve Bayes classification. Domain specific knowledge may be defined as a set of rules that makes possible the qualitative spatial reasoning. In addition, hierarchies on spatial objects are expressed by a collection of ground atoms and are exploited to mine classification models at different granularity levels. Search constraints are used to bias the spatial association rules discovery in order to fulfil user expectations. In particular, constraints are also used to partially fix the structure of extracted rules in order to discover spatial association rules that contain only the class label in the head. Finally, for each granularity level, extracted rules concur in building the spatial classification model by exploiting a multi-relational naïve Bayesian classifier integrated with the SDB.

Experiments on real-world spatial data show that the use of different levels of granularity generally increases the accuracy of the mined classification model. As future work, we intend to frame the work within the context of hierarchical Bayesian classifiers, in order to exploit the multi-level nature of extracted association rules.

Acknowledgments

We would like to thank Jim Petch, Keith Cole and Mohammed Islam (University of Manchester) for expert collection, collation, editing and delivery of the several data sets made available through Manchester Computing in the context of the IST European project SPIN! (Spatial Mining for Data of Public Interest).

References

1. Andrienko, G., Andrienko, N.: Exploration of heterogeneous spatial data using interactive geo-visualization tools: study of deprivation indices in North-West England. North-West England Report. IST European project SPIN!(Spatial Mining for Data of Public Interest).
2. Appice, A., Ceci, M., Lanza, A., Lisi, F.A. Malerba, D.: Discovery of Spatial Association Rules in Georeferenced Census Data: A Relational Mining Approach, Intelligent Data Analysis. Special issue of Mining Official Data. 7(6) (2003).
3. Baralis, E., Garza, P.; Majority Classification by Means of Association Rules., Knowledge Discovery in Databases PKDD'03, LNCS 2838, Springer-Verlag (2003), 35-46.
4. Clark, P., Matwin, S.: Using qualitative models to guide induction learning. Proceedings of International Conference of Machine Learning, Morgan Kaufmann, (1993), 49-56.
5. Domingos, P. Pazzani, M.: On the optimality of the simple Bayesian classifier under zeo-ones loss. Machine Learning, 29 (2-3), (1997), 103-130.
6. Dong, G., Zhang, X., Wong, L. Li, J.: Classification by aggregating emerging patterns. Proceedings of DS'99 (LNCS 1721), Japan, (1999).
7. Dzeroski, S., Lavrac, N. (eds.): Relational Data Mining. Springer-Verlag, Berlin, (2001).
8. Egenhofer, M.J.: Reasoning about Binary Topological Relations. Proceedings of the Second Symposium on Large Spatial Databases, Zurich, Switzerland, (1991) 143-160.
9. Egenhofer, M.J., Franzosa, R.: Point-Set Topological Spatial Relations, International Journal of Geographical Information Systems, 5(2), (1991), 61-174.
10. Ester, M. Kriegel, H.P., Sander J.: Spatial Data Mining: A Database Approach. Proceedings International Symposium on Large Databases, Berlin, (1997), 47-66.
11. Fürnkranz, J., Flach, P.A.: An analysis of rule evaluation metrics. Proceedings of International Conference of Machine Learning, Morgan Kaufmann, (2003).
12. Güting, R.H.: An introduction to spatial database systems. VLDB Journal, 4(3) (1994)
13. Klösgen, W., May, M.: Spatial Subgroup Mining. Proceedings of European Syposium of Principles of Knowledge Discovery in Database PKDD'02. Springer-Verlag, (2002).
14. Koperski, K., Han, J.: Discovery of Spatial Association Rules in Geographic Information Databases. Advances in Spatial Databases. LNCS 951, Springer-Verlag, (1995) 47-66.
15. Koperski, K.: Progressive Refinement Approach to Spatial Data Mining,Ph.D. thesis, Computing Science, Simon Fraser University, (1999).
16. Lisi, F.A., Malerba, D.: Inducing Multi-Level Association Rules from Multiple Relations. Machine Learning, (2004), to appear.
17. Liu, B., Hsu, W., Ma, Y.: Integratine classification and association rule mining. Proceedings of Knowledge Discovery in Databases KDD'98, New York, (1998).
18. Ludl, M.C., Widmer, G.: Relative Unsupervised Discretization for Association Rule Mining. PKDD'00, LNCS 1910, Springer-Verlag, (2000), 148-158.

19. Malerba D., A. Appice, & N. Vacca. SDMOQL: An OQL-based Data Mining Query Language for Map Interpretation Tasks. Proc. of the Workshop on Database Technologies for Data Mining (DTDM'02), Prague, Czech Republic, March 25-27, 2002
20. Malerba, D., Esposito, F., Lanza, A., Lisi, F.A., Appice, A.: Empowering a GIS with Inductive Learning Capabilities: The Case of INGENS. Journal of Computers, Environment and Urban Systems, Elsevier Science, 27 (2003). 265-281.
21. Malerba, D.: Learning Recursive Theories in the Normal ILP Setting, Fundamenta Informaticae, 57, 1, (2003), 39-77.
22. May, M.: Spatial Knowledge Discovery: The SPIN! System. In: Fullerton, K. (ed.): Proceedings of the 6th EC-GIS Workshop, Lyon, JRC, Ispra, (2000).
23. Preparata, F., Shamos, M.: Computational Geometry: An Introduction. Springer-Verlag, New York (1985).
24. Shekhar, S., Schrater, P.R., Vatsavai, R. R., Wu, W., Chawla, S.: Spatial Contextual Classification and Prediction Models for Mining Geospatial Data. IEEE Transaction on Multimedia, 4(2) (2002) 174-188.
25. Pazzani, M., Mani, S., Shankle, W.R.: Beyond concise and colorful: learning intelligible rules. In Proceedings of Knowledge Discovery in Databases KDD'97, (1997).
26. Wrobel, S.: Inductive logic programming for knowledge discovery in databases. In: Džeroski, S., N. Lavra (eds.): Relational Data Mining, Springer: Berlin, (2001) 74-101.

AutoPart: Parameter-Free Graph Partitioning and Outlier Detection*

Deepayan Chakrabarti

Carnegie Mellon University
deepay@cs.cmu.edu

Abstract. Graphs arise in numerous applications, such as the analysis of the Web, router networks, social networks, co-citation graphs, etc. Virtually all the popular methods for analyzing such graphs, for example, k-means clustering, METIS graph partitioning and SVD/PCA, require the user to specify various parameters such as the number of clusters, number of partitions and number of principal components. We propose a novel way to group nodes, using information-theoretic principles to choose both the number of such groups and the mapping from nodes to groups. Our algorithm is completely *parameter-free*, and also scales practically linearly with the problem size. Further, we propose novel algorithms which use this node group structure to get further insights into the data, by finding outliers and computing distances between groups. Finally, we present experiments on multiple synthetic and real-life datasets, where our methods give excellent, intuitive results.

1 Introduction – Motivation

Large, sparse graphs arise in many applications, under several guises. Consequently, because of their importance and prevalence, the problem of discovering structure in them has been widely studied in several domains, such as social networks, co-citation networks, ecological food webs, protein interaction graphs and many others. Such structure can be used for getting insights into the graph, for example, for detecting "communities".

Problem Description: A graph $G(V, E)$ has a set E of edges connecting any pair of nodes from a set V. Our definition includes both directed and undirected

* This material is based upon work supported by the National Science Foundation under Grants No. IIS-9817496, IIS-9988876, IIS-0083148, IIS-0113089, IIS-0209107 IIS-0205224 INT-0318547 SENSOR-0329549 EF-0331657 IIS-0326322 by the Pennsylvania Infrastructure Technology Alliance (PITA) Grant No. 22-901-0001, and by the Defense Advanced Research Projects Agency under Contract No. N66001-00-1-8936. Additional funding was provided by donations from Intel, and by a gift from Northrop-Grumman Corporation. Any opinions, findings, and conclusions or recommendations expressed in this material are those of the author(s) and do not necessarily reflect the views of the National Science Foundation, or other funding parties.

J.-F. Boulicaut et al. (Eds.): PKDD 2004, LNAI 3202, pp. 112–124, 2004.

graphs. We want algorithms that discover structure in such datasets, and provide insights into them. Specifically, our goals are:

(G1) Clusters: "Similar" nodes should be grouped into "natural" clusters.
(G2) Outliers: Edges deviating from the overall structure should be tagged as outliers.
(G3) Inter-cluster Distances: For any pair of clusters, a measure of the "distance" between them should be defined.

In addition, the algorithms should have the following main properties:

(P1) Automatic: We want a principled and intuitive problem formulation, such that the user does not need to set *any* parameters.
(P2) Scalable: They should scale up for large, possibly disk resident graphs.
(P3) Incremental: They should allow online recomputation of results when new nodes and edges are added; this will allow the method to adapt to new incoming data from, say, web crawls.

In this paper, we propose algorithms to accomplish these objectives. Intuitively, we seek to group nodes so that the adjacency matrix is divided into rectangular/square regions as "similar" or "homogeneous" as possible. These regions of varying density would succinctly summarize the underlying structure of associations between nodes. In short, our method will take as input a matrix like in Figure 3(a) and produce Figure 3(g) as the output, without any human intervention.

The layout of the paper is as follows. In Section 2, we survey the related work. Subsequently, in Section 3, we formulate our data description model starting from first principles. Based on this, in Section 3.3 we outline a two-level framework to find homogeneous blocks in the adjacency matrices of graphs and develop an efficient, *parameter-free* algorithm to discover them. In Section 3.5, we use this structure to find outlier edges in the graph, and to calculate distances between node groups. In Section 4, we evaluate our algorithms, demonstrating good results on several real and synthetic datasets. Finally, we conclude in Section 5.

2 Survey

There has been quite a bit of work on graph partitioning. The prevailing methods are METIS [1] and spectral partitioning [2]. Both approaches have attracted a lot of interest and attention; however, both need the user to specify k, that is, the number of pieces the graph should be broken into. Moreover, they typically also require a measure of imbalance between the two pieces of each split. The *Markov Clustering* [3] method uses random walks, but is slow. Girvan and Newman [4] iteratively remove edges with the highest "stress" to eventually find disjoint communities, but the algorithm is again slow. Flake et al. [5] use the max-flow min-cut formulation to find communities around a seed node; however, the selection of seed nodes is not fully automatic.

Table 1. Table of symbols.

Symbol	Definition
D	Square binary adjacency matrix of a given graph
$d_{i,j}$	Entry in cell (i,j) of D; $d_{i,j} := 0$ or 1
n	Length of each side of D
k	Number of node groups
k^*	Optimal number of groups
$\mathcal{G}$	Node $\rightarrow$ group map
$\mathcal{G}_x$	Group corresponding to node x
$D_{i,j}$	Submatrix of links from group i to j
a_i	Number of nodes in group i
a_i, a_j	Dimensions of $D_{i,j}$
$n(D_{i,j})$	Number of elements in $D_{i,j}$; $n(D_{i,j}) := a_i a_j$
$w(D_{i,j})$	Weight of $D_{i,j}$ = number of "1"s in $D_{i,j}$
$P_{i,j}$	Density of "1"s in $D_{i,j}$; $P_{i,j} := w(D_{i,j})/n(D_{i,j})$
$H(p)$	Binary Shannon entropy function
$C(D_{i,j})$	Code cost for $D_{i,j}$
$T(D; k, \mathcal{G})$	Total cost for D

Remotely related are clustering techniques. Every row in the adjacency matrix can be envisioned as a multi-dimensional point. Several methods have been developed to cluster a cloud of n points in m dimensions, for example, k-means, k-harmonic means, CURE, BIRCH, Chameleon, LSI and others [6–8]. However, most current techniques require a user-given parameter, such as k for k-means. One solution called X-means [9] uses BIC to determine k. However, several of the clustering methods suffer from the dimensionality curse (like the ones that require a covariance matrix); others may not scale up for large datasets. Also, in our case, the points and their corresponding vectors are semantically related (each node occurs as a point *and* as a component of each vector); most clustering methods do not consider this. Other related work includes information-theoretic co-clustering (ITCC) [10]. However, the focus there is on lossy compression, whereas we employ a lossless MDL-based compression scheme. No MDL-like principle is yet known for lossy encoding, and hence, the number of clusters in ITCC cannot (yet) be automatically determined. Besides these, there has been work on conjunctive clustering [11] and community detection [12].

In conclusion, the above methods miss one or more of our prerequisite properties, typically not being automatic (P1). Next, we present our solution.

3 Proposed Method

Our goal is to find patterns in a large graph, with no user intervention, as shown in Figure 3. How should we decide the number of node groups k along with the assignments of nodes to their "proper" groups?

Compression as a Guide: We introduce a novel approach and propose a general, intuitive model founded on compression, and more specifically, on the *MDL*

(Minimum Description Language) principle [13]. The idea is the following: the binary $n \times n$ matrix represents *associations* between the n nodes of the graph (corresponding to rows and columns in the adjacency matrix). If we mine this information properly, we could reorder the adjacency matrix so that "similar" nodes are grouped with each other. Then, the adjacency matrix would consist of homogeneous rectangular/square blocks of high(low) density, representing the fact that certain node groups have more(less) connections with other groups. To compress the matrix, we would prefer to have only a few blocks, each of them being very homogeneous. However, having more groups lets us create more homogeneous blocks (at the extreme, having n groups gives n^2 perfectly homogeneous blocks of size 1×1). Thus, the best compression scheme must achieve a tradeoff between these two factors, and this tradeoff point indicates the best number of node groups k. We accomplish this by a novel application of the overall MDL philosophy, where the compression costs are based on the number of bits required to transmit both the "summary" of the node groups, as well as each block given the groups. Thus, the user does not need to set any parameters; our algorithm chooses them so as to minimize these costs.

3.1 Compression Scheme for a Binary Matrix

Let $D = [d_{i,j}]$ denote an $n \times n$ adjacency matrix. Each graph node corresponds to one row and column in this matrix. We assume that $n \geq 1$. Let us index the rows and columns as $1, 2, \ldots, n$.

Let k denote the number of disjoint node groups. Let us index the groups as $1, 2, \ldots, k$. Let

$$\mathcal{G} : \{1, 2, \ldots, n\} \rightarrow \{1, 2, \ldots, k\}$$

denote the assignments of nodes to groups. We can rearrange the underlying data matrix D so that all nodes corresponding to group 1 are listed first, followed by nodes in group 2, and so on. Such a rearrangement, implicitly, sub-divides the matrix D into k^2 smaller two-dimensional rectangular/square blocks, denoted by $D_{i,j}, i, j = 1, \ldots, k$. The more homogeneous these blocks, the better compression we can get, and so, the better the choice of $\mathcal{G}$. Table 1 lists the symbols used later.

We now describe a two-part code for the matrix D. The first part will be a *description complexity* involved in describing the blocks formed by $\mathcal{G}$ and the second part will be the actual *code* for the matrix given information about the blocks. A good choice of $\mathcal{G}$ will compress the matrix well, which will lead to low total encoding cost.

Description Cost: The description complexity (ie., information about the rectangular/square blocks) consists of the following terms:

1. Send the number of nodes n using $\log^\star(n)$ bits, where $\log^\star(x) = \log_2(x) + \log_2 \log_2(x) + \ldots$ with only the positive terms being retained [14]. This term is independent of $\mathcal{G}$ and k, and, hence, while useful for actual transmission of the data, will not figure in our framework.

2. Send the node permutations using $n\lceil \log n \rceil$ bits, respectively. Again, this term is also independent of $\mathcal{G}$ and k.
3. Send the number of groups k in $\log^\star k$ bits.
4. Send the number of nodes in each node group. Let us suppose that $a_1 \geq a_2 \geq \ldots \geq a_k \geq 1$. Compute $\bar{a}_i = \left(\sum_{t=i}^{k} a_t\right) - k + i$ for all $i = 1, \ldots, k-1$. Now, the desired quantities can be sent using $\sum_{i=1}^{k-1} \lceil \log \bar{a}_i \rceil$ bits
5. For each block $D_{i,j}$ $(i, j = 1, \ldots, k)$, send $w(D_{i,j})$, namely, the number of "1"s in $D_{i,j}$ using $\lceil \log(a_i a_j + 1) \rceil$ bits.

Code Cost: Suppose that the entire preamble specified above (containing information about the square and rectangular blocks) has been sent. We now transmit the actual matrix given this information. We can calculate the density $P_{i,j}$ of "1"s in $D_{i,j}$ using the description code above. The number of bits required to transmit $D_{i,j}$ is

$$\begin{aligned} C(D_{i,j}) &= n(D_{i,j}) H(P_{i,j}) \\ &= -w(D_{i,j}) \log (P_{i,j}) - [n(D_{i,j}) - w(D_{i,j})] \log(1 - P_{i,j}) \end{aligned} \quad (1)$$

where H is the binary Shannon entropy function, $n(D_{i,j}) = a_i a_j$, and all logarithms are base 2. Summing over all the $D_{i,j}$ submatrices:

$$\text{Code cost} = \sum_{i=1}^{k} \sum_{j=1}^{k} C(D_{i,j}) \quad (2)$$

Total Encoding Cost: We can now write the total cost for the matrix D, with respect to a given k and $\mathcal{G}$ as:

$$T(D; k, \mathcal{G}) := \log^\star k + \sum_{i=1}^{k-1} \lceil \log \bar{a}_i \rceil + \sum_{i=1}^{k} \sum_{j=1}^{k} \lceil \log(a_i a_j + 1) \rceil + \sum_{i=1}^{k} \sum_{j=1}^{k} C(D_{i,j}) \quad (3)$$

ignoring the costs $\log^\star(n)$ and $n\lceil \log n \rceil$ since they are independent of $\mathcal{G}$ and k.

3.2 Problem Formulation

We want an algorithm that can optimally choose k^* and $\mathcal{G}^*$ so as to minimize $T(D; k^\star, \mathcal{G}^\star)$. Typically, such problems are computationally hard, and hence, in this paper, we shall pursue feasible practical strategies. We solve the problem by a two-step iterative process: First, find a good node grouping $\mathcal{G}$ for a given number of node groups k; and second, search for the number of node groups k. For the former, we describe an iterative minimization algorithm to find a $\mathcal{G}$ that effectively finds a minimum, given a fixed number of node groups k. Then, we outline an effective heuristic strategy that searches over k to minimize the total cost $T(D; k, \mathcal{G})$.

3.3 Algorithms

In the previous section we established our goal: Among all possible values for k, and all possible node groups $\mathcal{G}$, pick an arrangement which reduces the total compression cost as much as possible, as MDL suggests (model plus data). Although theoretically pleasing, Equation 3 does not tell us *how* to go about finding the best arrangement - it can only pinpoint the best one, among several candidates! The question is: *how can we generate good candidates?*

We answer this question in two steps:

1. [`InnerLoop`] For a given k, find a good arrangement $\mathcal{G}$.
2. [`OuterLoop`] Efficiently search for the best k ($k = 1, 2, \ldots$).

Algorithm InnerLoop (Finding $\mathcal{G}$ given k):

1. Let t denote the iteration index. Initially, set $t = 0$. If no $\mathcal{G}(0)$ is provided, start with an arbitrary $\mathcal{G}(0)$ mapping nodes into k node groups. For this initial partition, compute the submatrices $D_{i,j}(t)$, and the corresponding distributions $P_{i,j}(t)$.
2. For every node x, splice the corresponding row into k parts $x_{row,1}, \ldots, x_{row,k}$ according to $\mathcal{G}(t)$ (i.e., $x_{row,1} = \{d_{x,u} | \mathcal{G}_u(t) = 1\}$ and so on). Similarly, splice the column into k parts $x_{col,1}, \ldots, x_{col,k}$. Compute the number of "1"s $w(x_{row,j})$ and $w(x_{col,j})$ ($j = 1 \ldots k$) for all these parts. Now, assign node x to node group $\mathcal{G}_x(t+1)$ such that

$$\begin{aligned}
\mathcal{G}_x\,(t+1) = &\arg\min_{1 \le i \le k} \\
&\left\{ \sum_{j=1}^{k} - \left[w(x_{row,j}) \log P_{i,j}(t) + (n(x_{row,j}) - w(x_{row,j})) \log(1 - P_{i,j}(t)) \right. \right. \\
&+ w(x_{col,j}) \log P_{j,i}(t) + (n(x_{col,j}) - w(x_{col,j})) \log(1 - P_{j,i}(t))\big] \\
&+ d_{x,x} \left[\log P_{i,\mathcal{G}_x(t)}(t) + \log P_{\mathcal{G}_x(t),i}(t) - \log P_{i,i}(t) \right] \\
&\left. + (1 - d_{x,x}) \left[\log(1 - P_{i,\mathcal{G}_x(t)}(t)) + \log(1 - P_{\mathcal{G}_x(t),i}(t)) - \log(1 - P_{i,i}(t)) \right] \right\}
\end{aligned} \quad (4)$$

   where the first two lines denote the cost of shifting the row and column corresponding to node x to a new group, while the last two lines account for the "double-counting" of the cell $d_{x,x}$ in the adjacency matrix.
3. With respect to $\mathcal{G}(t+1)$, recompute the matrices $D_{i,j}^{t+1}$, and the corresponding distributions $P_{i,j}^{t+1}$.
4. If there is no decrease in total cost, stop; otherwise, set $t = t + 1$, go to step 2, and iterate.

Fig. 1. Algorithm `InnerLoop`.

The `InnerLoop` algorithm iterates over several possible settings of $\mathcal{G}$ for the same number of node groups k. Each iteration improves (or maintains) the code cost, as stated in the theorem below.

Theorem 1. *After ach iteration of* `InnerLoop`*, the code cost decreases or remains the same. The proof is omitted for lack of space.*

The loop finishes when the total cost stops improving. Note that it is possible for some groups to be empty, but this is not a problem. The complexity of `InnerLoop` is $O(w(D) \cdot k \cdot I)$ where I is the number of iterations.

Algorithm OuterLoop (Finding k):

1. Let T denote the search iteration index. Start with $T = 0$ and $k(0) = 1$.
2. At iteration T, try to increase k: $k(T+1) = k(T) + 1$. Split the node group r with maximum entropy per node, i.e.,

$$r := \arg\max_{1 \le i \le k} \sum_{1 \le j \le k} \frac{n(D_{i,j})H\big(P_{i,j}\big) + n(D_{j,i})H\big(P_{j,i}\big)}{a_i}$$

   Construct an initial label map $\mathcal{G}(T+1)$ as follows: For every node x that belongs to group r (i.e., for every $1 \le x \le n$ such that $\mathcal{G}_x(T) = r$), place it into the new group $k(T+1)$ (i.e., set $\mathcal{G}_x(T+1) = k(T+1)$) if and only if it decreases the per-node entropy of the group r, i.e., if and only if

$$\sum_{1 \le j \le k} \frac{n(D'_{r,j})H\big(P'_{r,j}\big) + n(D'_{j,r})H\big(P'_{j,r}\big)}{a_r - 1} < \sum_{1 \le j \le k} \frac{n(D_{r,j})H\big(P_{r,j}\big) + n(D_{j,r})H\big(P_{j,r}\big)}{a_r}$$

   where $D'_{r,j}$ is $D_{r,j}$ without node x. Otherwise let $\mathcal{G}_x(T+1) = r = \mathcal{G}_x(T)$. If we move node x to the new group, we also update $D_{r,j}$ and $D_{j,r}$ (for all $1 \le j \le k$) accordingly.
3. Run the `InnerLoop` algorithm with initial $\mathcal{G} = \mathcal{G}(T+1)$ and $k = k(T+1)$ to find a new node mapping $\mathcal{G}(T+1)$ and the corresponding total cost.
4. If there is no decrease in total cost, stop and return $k^* = k(T)$ and $\mathcal{G}^* = \mathcal{G}(T)$. Otherwise, set $T = T + 1$ and continue.

Fig. 2. Algorithm `OuterLoop`.

The `OuterLoop` algorithm tries to look for good values of k. It chooses the node group with the maximum entropy per node, and splits it into two groups. The nodes put into the new group are exactly the ones whose removal reduces the entropy per node in the original group. As shown below in Theorem 2, this split never decreases the code cost.

Theorem 2. *On splitting any node group, the code cost either decreases or remains the same. The proof is omitted due to lack of space.*

By Theorem 1, the same holds for `InnerLoop`. Therefore, the entire algorithm also decreases the code cost (Eq. 2). However, the description complexity evidently increases with k. We have found that, in practice, this search strategy performs very well. The `OuterLoop` algorithm is run k^* times, so the overall complexity of the search is $O(w(D)(k^*)^2 I)$. In practice, $I \le 20$ is always enough.

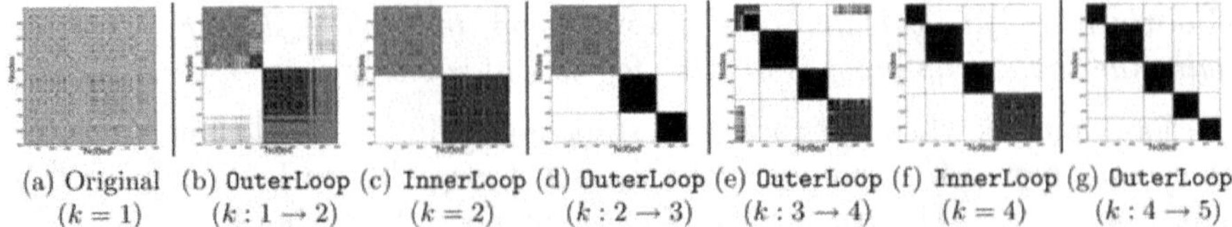

(a) Original ($k = 1$) (b) OuterLoop ($k : 1 \to 2$) (c) InnerLoop ($k = 2$) (d) OuterLoop ($k : 2 \to 3$) (e) OuterLoop ($k : 3 \to 4$) (f) InnerLoop ($k = 4$) (g) OuterLoop ($k : 4 \to 5$)

Fig. 3. *Algorithm execution snapshots:* Starting with a randomly permuted "caveman" matrix (a), the algorithm applies `OuterLoop` and `InnerLoop` till the final structure (g) is revealed. We omit the `InnerLoop` results when they produce no improvement. Iterations of `OuterLoop` are separated by vertical lines for clarity.

Figure 3 shows an execution snapshot of the full algorithm on a randomly permuted "caveman" matrix (ie., a block diagonal matrix [15]) with Zipfian cave-sizes. `OuterLoop` increases the number of node groups while `InnerLoop` rearranges nodes between groups. No plots are shown when the `InnerLoop` does not decrease the total cost. The correct final result is shown in plot (g).

3.4 Online Recomputations

When new nodes are obtained (such as from new crawls for a Web graph), we can put them into the node groups which minimize the increase in total encoding cost due to their addition. Based on the same principle, when new edges are found, the corresponding nodes can be reassigned to new node groups. The algorithm can then be run again with this initialization till it converges. Similar methods apply for node/edge deletions. Thus, new additions or deletions can be handled without full recomputations.

3.5 Using the Block Structure

Having found the underlying structure of a graph in the form of node groups, we can utilise this information to further mine the data. Again, we use our information-theoretic approach to answer several tough problems efficiently, using the node groupings found by the previous algorithms.

Outlier Edges: Which edges between nodes are abnormal/suspicious? Intuitively, an outlier shows some deviation from normality, and so it should hurt attempts to compress data. Thus, an edge whose removal significantly reduces the total encoding cost is an outlier. Our algorithm is: Find the block where removal of an edge leads to the maximum immediate reduction in cost (that is, no iterations of the `InnerLoop` and `OuterLoop` algorithms are performed). All edges within that block contribute equally to the cost, and so all of them are considered outliers.

$$\text{"Outlierness" of edge } (u,v) := T(D'; k, \mathcal{G}) - T(D; k, \mathcal{G}) \quad (5)$$

where $D' = D$ except that $d'_{u,v} = 0$. This can be used to *rank* the edges in terms of their "outlierness".

Table 2. Dataset characteristics.

Dataset	Num nodes	Num edges	Remarks
CAVE	900	170,800	Five "caves" with zipfian sizes
CAVE-Noisy	900	190,117	10% noise added to the above
NOISE	100	1,831	Pure white noise
EPINIONS	75,888	508,960	"Who-trusts-whom" data
DBLP	6,090	175,494	Coauthorship and cocitation data

"Distance" Between Node Groups: How "close" are two node groups to each other? Following our information theory footing, we propose the following criterion: If two groups are "close", then combining the two into one group should not lead to a big increase in encoding cost. Based on this, we define "distance" between two groups as the relative increase in encoding cost if the two were merged into one:

$$Dist(i,j) := \frac{Cost(merged) - Cost(i) - Cost(j)}{Cost(i) + Cost(j)} \tag{6}$$

where only the nodes in groups i and j are used in computing costs. We experimented with other measures (such as the absolute increase in cost) but Eq 6 gave the best results. The cost of computing both outliers and distances between groups is independent of the number of non-zeros $w(D)$, and so both can be performed for large graphs.

4 Experiments

We did experiments to answer the following questions: (i) how good is the quality of the node groups found, (ii) how well do our algorithms find outlier edges, (iii) do our measures of "distances" between node groups make sense, and (iv) how well does our method scale up. All our experiments require no input parameters, which rules out other methods like METIS or spectral partitioning.

We used several datasets (see Table 2), both real and synthetic. The synthetic ones were: **(1)** `CAVE`, representing a social network of "cavemen" [15], that is, a block-diagonal matrix of variable-size blocks (or "caves"; members of a cave form a clique, and know only those from their own cave), **(2)** `CAVE-Noisy`, created by adding noise (10% of the number of non-zeros) , and **(3)** `NOISE`, with pure white noise. The real-world datasets are: **(4)** `EPINIONS`, a "who-trusts-whom" social graph of `www.epinions.com` users [16], and **(5)** `DBLP`, a graph obtained from `www.informatik.uni-trier.de/~ley/db`, with the nodes being authors in `SIGMOD, ICDE, VLDB, PODS` or `ICDT` (database conferences); two nodes are linked by an edge if the two authors have co-authored a paper or one has cited a paper by the other (thus, this graph is undirected). We performed experiments on other datasets too; the results were similar, and are not reported to save space. Our implementation was done in MATLAB (version 6.5 on Linux) using sparse matrices. The experiments were performed on an Intel Xeon 2.8GHz machine with 1GB RAM.

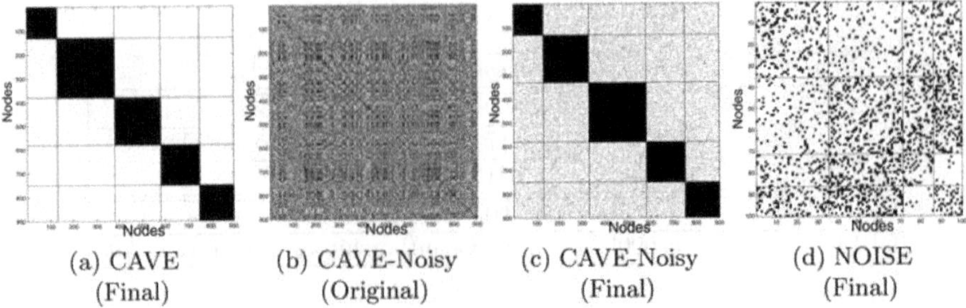

(a) CAVE (Final)
(b) CAVE-Noisy (Original)
(c) CAVE-Noisy (Final)
(d) NOISE (Final)

Fig. 4. *Synthetic datasets*: (a) Our method gives the intuitively correct groups for CAVE (Figure 3(a) shows the original graph). (b,c) The results remain the same in spite of noise in CAVE-Noisy, showing the robustness of the algorithm. (d) The NOISE dataset shows 4 groups, which are explained by the patterns emerging due to randomness, such as the "almost-empty" and "more-dense" blocks.

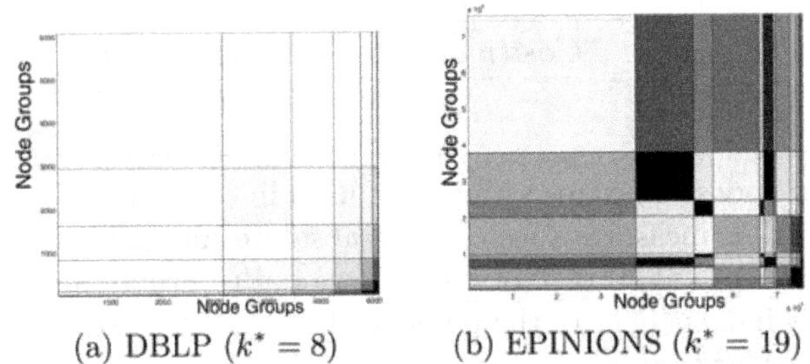

(a) DBLP ($k^* = 8$)
(b) EPINIONS ($k^* = 19$)

Fig. 5. *Real datasets*: Shaded blocks are shown instead of the actual points; darker shades correspond denser blocks. The plots show how the algorithm has separated the graph into large but extremely sparse, and small but very dense groups. Most well-known database researchers show up in the dense regions of plot (a), as expected.

4.1 Quality

Results – Synthetic Data: Figure 4 shows the groupings found by our method on several synthetic datasets. For the noise-free `CAVE` matrix, we get exactly the intuitively correct groups (plot a). When noise is present (plot b), we still get the correct groups (plot c), demonstrating the robustness of our algorithm. Plot (d) shows 4 groups for the `NOISE` graph. This is expected; it is well known that spurious patterns emerge even when we have pure noise, and our algorithm finds blocks of clearly lower or higher density.

Results – Real Data: Figure 5 shows the groupings found on several real-world datasets. For the `DBLP` dataset, eight groups were found. Group 8 is comprised of only Michael Stonebraker, David DeWitt and Michael Carey; these are well-known people who have a lot of papers and citations. The other groups show decreasing number of connections but increasing sizes, with group 1 being the largest but having the lowest connectivity. Similarly, for the `EPINIONS` graph, we find a small dense "core" group which has very high connectivity, and then

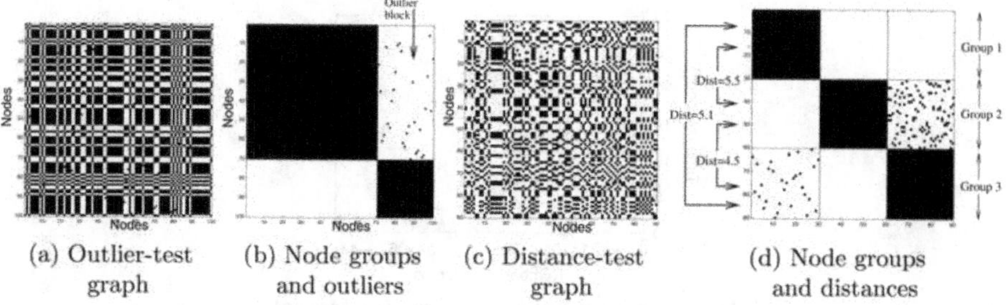

(a) Outlier-test graph | (b) Node groups and outliers | (c) Distance-test graph | (d) Node groups and distances

Fig. 6. *Outliers and group distances:* Plot (b) shows the node groups found for graph (a). Edges in the top-right block are correctly tagged as outliers. Plot (d) shows the node groups and group distances for graph (c). Groups 2 and 3 (having the most "bridges") are tagged as the closest groups. Similarly, groups 1 and 2 are the farthest.

larger and less heavily-connected groupings. Thus, our method gives intuitive results for real-world graphs too.

4.2 Outlier Edges

To test our algorithm for picking outliers, we use a synthetic dataset as in Figure 6(a). The node groups found are shown in 6(b). Our algorithm tags all edges whose removal would best compress the graph as outliers. Thus, all edges "across" the two groups are chosen as outliers under this principle (since all edges in a block contribute equally to the encoding cost), as shown in Figure 6(b). Thus, the intuitively correct outliers are found.

4.3 Distances Between Node Groups

To test for node-group distances, we use the graph in 6(c) with 6(d) showing the structure found. The three caves have equal sizes but the number of "bridge" edges between groups varies. This is correctly picked up by our algorithm, which ranks groups with more "bridges" as being closer to each other. Thus, groups 2 and 3 are tagged as the "closest" groups, while groups 1 and 2 are "farthest".

4.4 Scalability

Figure 7 shows wall-clock times (in seconds) of our MATLAB implementation. The dataset is a "caveman" graph with 3 caves; the size of the graph and the number of edges in it are varied for the experiment, with the relative proportions of cave sizes being kept fixed. The execution time increases linearly with respect to the number of non-zeros, as expected from our order-of-complexity computation. Thus, our proposed method can scale to large graphs.

5 Conclusions

We considered the problem of finding the underlying structure in a graph. We introduced a novel approach and proposed a general, intuitive model founded on

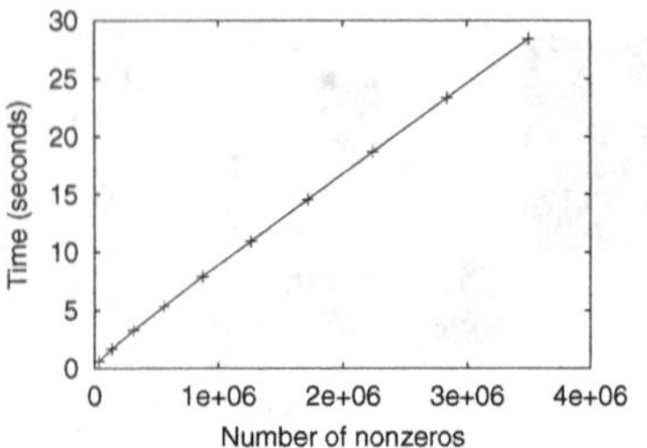

Time versus number of edges (nonzeros)

Fig. 7. *Scalability:* On a 3-cave graph, wall-clock execution time grows linearly with the number of edges. Thus, our method can scale to large graphs.

lossless compression and information-theoretic principles. Based on this model, we provided novel algorithms for finding node groups and outlier edges, as well as for computing distances between node groups, thus fulfilling all our goals (G1)-(G3) from Section 1. Our algorithms are fully automatic and parameter-free, scalable and allow online computations, achieving properties (P1)-(P3). Finally, we evaluated our method on several real and synthetic datasets, where it produced excellent and intuitive results.

Acknowledgements

We would like to thank Dr. Faloutsos at CMU and the reviewers for their insightful comments and suggestions.

References

1. Karypis, G., Kumar, V.: Multilevel algorithms for multi-constraint graph partitioning. In: Proc. SC98. (1998) 1–13
2. Andrew Y.Ñg, Michael I.J̃ordan, Y.W.: On spectral clustering: Analysis and an algorithm. In: Proc. NIPS. (2001) 849–856
3. van Dongen, S.M.: Graph clustering by flow simulation. PhD thesis, Univesity of Utrecht (2000)
4. Girvan, M., Newman, M.E.J.: Community structure in social and biological networks. In: Proc. Natl. Acad. Sci. USA. Volume 99. (2002)
5. Flake, G.W., Lawrence, S., Giles, C.L.: Efficient identification of Web communities. In: KDD. (2000)
6. Zhang, B., Hsu, M., Dayal, U.: K-harmonic means - a spatial clustering algorithm with boosting. In: Proc. 1st TSDM. (2000) 31–45
7. Han, J., Kamber, M.: Data Mining: Concepts and Techniques. Morgan Kaufmann (2000)
8. Deerwester, S., Dumais, S.T., Furnas, G.W., Landauer, T.K., Harshman, R.: Indexing by latent semantic analysis. JASI **41** (1990) 391–407
9. Pelleg, D., Moore, A.: X-means: Extending K-means with efficient estimation of the number of clusters. In: Proc. 17th ICML. (2000) 727–734

10. Dhillon, I.S., Mallela, S., Modha, D.S.: Information-theoretic co-clustering. In: Proc. 9th KDD. (2003) 89–98
11. Mishra, N., Ron, D., Swaminathan, R.: On finding large conjunctive clusters. In: Proc. 16th COLT. (2003) 448–462
12. Reddy, P.K., Kitsuregawa, M.: An approach to relate the web communities through bipartite graphs. In: Proc. 2nd WISE. (2001) 302–310
13. Rissanen, J.: Modeling by shortest data description. Automatica **14** (1978) 465–471
14. Rissanen, J.: Universal prior for integers and estimation by minimum description length. Annals of Statistics **11** (1983) 416–431
15. Watts, D.J.: Small Worlds: The Dynamics of Networks between Order and Randomness. Princeton Univ. Press (1999)
16. Richardson, M., Domingos, P.: Mining knowledge-sharing sites for viral marketing. In: KDD, Edmonton, Canada (2002) 61–70

Properties and Benefits of Calibrated Classifiers

Ira Cohen and Moises Goldszmidt

Hewlett-Packard Research Laboratories
1501 Page Mill Rd., Palo Alto, CA 94304
{ira.cohen,moises.goldszmidt}@hp.com

Abstract. A calibrated classifier provides reliable estimates of the true probability that each test sample is a member of the class of interest. This is crucial in decision making tasks. Procedures for calibration have already been studied in weather forecasting, game theory, and more recently in machine learning, with the latter showing empirically that calibration of classifiers helps not only in decision making, but also improves classification accuracy. In this paper we extend the theoretical foundation of these empirical observations. We prove that (1) a well calibrated classifier provides bounds on the Bayes error (2) calibrating a classifier is guaranteed not to decrease classification accuracy, and (3) the procedure of calibration provides the threshold or thresholds on the decision rule that minimize the classification error. We also draw the parallels and differences between methods that use receiver operating characteristic (ROC) curves and calibration based procedures that are aimed at finding a threshold of minimum error. In particular, calibration leads to improved performance when multiple thresholds exist.

1 Introduction

In a decision making task, in order to evaluate different courses of action, it is useful to obtain accurate likelihood estimates of the alternatives. Pattern classifiers can be used to provide automated mappings between situations (represented by features) and outcomes (represented by the class membership). Yet, to be applicable to decision making problems, we require a reliable estimate of the true probability of class membership of each test sample. We will use the term *calibrated* to refer to classifiers with reliable estimates of the class membership probabilities. A successful classifier in terms of classification accuracy is not necessarily calibrated, e.g., the Naive Bayes classifier.Procedures for calibrating classifiers have been proposed in different contexts: In weather prediction tasks [1], in game theory [2, 3], and more recently in the context of pattern classification [4, 5]. Zadrozny and Elkan were also the first to notice the need of calibrating classifiers when used as decision making aids.

Our own incentive to study calibration came from applying probabilistic based classifiers to the problem of characterizing and forecasting the I/O response time of large storage arrays given passive observations. As these forecasts are used for scheduling purposes, we also need to accompany each forecast with an accurate estimate of the probability of the forecast. We applied a variant

J.-F. Boulicaut et al. (Eds.): PKDD 2004, LNAI 3202, pp. 125–136, 2004.

of the calibration procedure suggested in [1, 4] and noticed that in addition to producing more accurate estimates, the classification accuracy of our induced classifiers increased. While these empirical results agree with those of Zadrozny and Elkan [4, 5], a theoretical guarantee that calibration cannot degrade classification performance was still missing. Our investigation of the calibration produced the following results which we prove in Sections 3 and 4. First, we can bound the Bayes error using the same parameters that result from the calculations needed for calibration. Second, we are *guaranteed* that the classification accuracy of the original classifier does not decrease as a consequence of calibration. Moreover, the classification accuracy can actually increase. Third, using the calibration process we can compute a threshold or thresholds in the decision rule of the classifier that minimize the classification error. We show that when a single threshold is derived from the calibration procedure, the result is equivalent to finding the point of minimum error in an ROC curve [6, 7]. However, when calibration produces multiple thresholds on the decision rule, the error achieved with those is lower than that of any single threshold derived from the ROC based methods. Thus, in addition to producing more accurate estimates of a-posteriori probabilities, calibration obviates the need for using ROC based methods for finding optimal thresholds.

The rest of the paper is organized as follows. Section 2 introduces formally the notions of calibration, refinement and Brier score. Sections 3 and 4 contain the proofs of our main results. Section 5 illustrates the effects of calibration on classifiers induced on real data, observing also the effect of the sample size on the process of calibration. Finally, Section 6 discusses and summarizes the main results.

2 Notation and Preliminary Definitions

A classifier takes an incoming vector of features $\mathbf{X}$ and maps it to a class label. We will use C to denote the *class variable* the values of which are called *classes*. Throughout this paper we assume a binary classification problem, i.e., one in which C takes one of two values. We will use $(1, 0)$, or $(c, \bar{c})$ to denote a specific instantiation of C. Each instantiation of $\mathbf{X}$, denoted by $\mathbf{x}$ is a *sample*. We assume that all samples are i.i.d.

Let $p(C|\mathbf{X})$ be the *true* a-posteriori distribution of the class given the features. The optimal classification rule, that is, the optimal function that maps a sample $\mathbf{x}$ to one of the values of C, under the 0-1 cost function, is the maximum a-posteriori (MAP) rule [8]:

$$g^*(\mathbf{x}) = argmax_{c'=(0,1)}[p(C = c'|\mathbf{x})], \tag{1}$$

The decision rule $g^*(\mathbf{x})$ is called the Bayes optimal decision and

$$\begin{aligned} \mathbf{e}_{\mathrm{B}} &= \sum_{\mathbf{x}} p(g^*(\mathbf{x}) \neq c|C = c)\, p(\mathbf{x}) \\ &= \sum_{\mathbf{x}} min(p(C = 1|\mathbf{x}), p(C = 0|\mathbf{x}))p(\mathbf{x}), \end{aligned} \tag{2}$$

is the associated probability of error. This error is known as the Bayes error (or Bayes risk), and it is the minimum probability of error achievable with the given set of features[1].

Given that $p(C|\mathbf{X})$ is unknown, one strategy for classification is to induce an estimate $\hat{p}(C|X)$ of the a-posteriori probability, and then use a decision rule $\hat{g}(X)$ such that the classification error, given by

$$CE = \sum_{\mathbf{x}} p(\hat{g}(\mathbf{x}) \neq c|C = c)\, p(\mathbf{x}) \tag{3}$$

is minimized. We note that plugging in $\hat{p}(C|X)$ into the decision rule in Eq. 1 may not be optimal [6], since given the errors and biases embedded in the estimate $\hat{p}(C|\mathbf{X})$ the threshold of 0.5, implicit in Eq. 1, may not minimize the error in Eq. 3. We return to this subject in Section 4, where we show the link between calibration and the decision rule that minimizes Eq. 3.

The classification error provides one way to evaluate classifiers. However, when using the classifier output as a basis for decision making, we need a score that takes into account not only the prediction accuracy of the classifier, but also the quality of the estimate $\hat{p}(C|\mathbf{X})$. One such score is the Brier score [9]. The Brier score is one of a class of so-called *proper* scores [1] which are used in evaluating the subjective probability assessment of forecasters. For the binary classification case, the Brier score is given as the average squared difference between the forecaster's probability of $C = 1$ and the true label:

$$BS = \frac{1}{n} \sum_{i=1}^{n} (\hat{p}(C = 1|\mathbf{x}_i) - c_i)^2, \tag{4}$$

where n is the number of samples. Among the various intuitive justifications of this score, the following one is based on decision theoretic considerations. Assume that the agent (classifier or forecaster), should pay a price proportional to the confidence with which it asserts its decision. The Brier score uses the probability of the estimate as providing the appropriate penalty. Note that in Eq. 4, if the agent predicts $C = 1$ with high probability but $c_i = 0$ the penalty will be higher than if he predicts $C = 1$ with low probability. Thus, the lower the Brier score, the lower is the penalty assessed to the agent.

The notion of calibration can be derived directly from the Brier score. We need some preliminary definitions. Let $t \in [0, 1]$ denote the a-posteriori probability assessment of a forecaster. Following [1], we assume that t takes on a finite number of values on the interval $[0, 1]$. We denote by R_t the set of feature values for which the classifier density, $\hat{p}(C = 1|\mathbf{x})$, yields a forecast probability t, namely:

$$R_t = \{\mathbf{x} \in \mathbf{X} : \hat{p}(C = 1|\mathbf{x}) = t\}. \tag{5}$$

[1] Note that the summation over $\mathbf{X}$ implies finite values for the features; for continuous features the summation is replaced by integration. Throughout the paper we maintain the summation over $\mathbf{X}$, but note that the analysis holds for continuous features as well.

Let $\pi(t)$ be the probability that the forecaster predicts $C = 1$ with probability t on a random instance. $\pi(t)$ can also be thought of as the frequency at which the forecaster predicted $C = 1$ with probability t on a set of N samples, with $N \to \infty$. As such, given the probability density of the features, $p(\mathbf{x})$, $\pi(t)$ can be expressed as:

$$\pi(t) = \sum_{\mathbf{x} \in R_t} p(\mathbf{x}) \,. \tag{6}$$

Let $p(C = 1|t)$ be the probability that $C = 1$ given that the forecaster predicts $C = 1$ with probability t. The Brier score can be rewritten as (see [1] for derivation):

$$BS = \sum_t \pi(t)(t - p(c|t))^2 + \sum_t \pi(t)p(c|t)\,(1 - p(c|t)). \tag{7}$$

The first term is a measure of the *calibration*, and the second term is a measure of the *refinement* of the forecaster, denoted as $\mathbf{R}$. Calibration indicates how close is the probability assessment of the forecaster on $C = 1$ to the frequency with which $C = 1$ occurs (in reality). Note that for calibration to be 0, t has to be $p(c|t)$ for every t. A *well-calibrated* forecaster is one with calibration equal to 0. The notion of calibration fits our purposes, since the probability assessments of a well-calibrated agent, can be used in decision making as an indication of its confidence of the classification label provided.

Refinement scores the *usefulness* of each forecast. As an illustration, assume that we live in a place that rains 50% of the time. Thus a forecaster that always announces rain with 50% confidence is calibrated, yet not very useful in helping to plan a picnic for the following day. Ideally we would like estimates that are close to certainty. The more concentrated $p(c|t)$ is towards 0 or 1, the more refined the classifier. To minimize the overall Brier score, a forecaster has to be both well-calibrated and refined. Thus, if two classifiers are well-calibrated, the one with the lower Brier score is also more refined. We describe the relationship between bias, Bayes error, calibration and refinement in the next section.

3 The Brier Score, Bias, and the Bayes Error

In the following we show that being well-calibrated is a weaker condition than being unbiased. Loosely speaking, a well-calibrated classifier is an "on average" unbiased classifier. We also show that we can use the notion of refinement (second term in Eq. 7) as a bound on the Bayes error. In particular, twice the refinement of a well-calibrated classifier is an upper bound on the Bayes error; and, in the case that the classifier is unbiased, then its refinement is a lower bound on the Bayes error. Section 5 illustrates the practical implications of the various approximations made when calibrating in practice.

3.1 The Bias/Calibration Relationship

Being well-calibrated requires that $t = p(c|t)$. Using Bayes rule we write $p(c|t)$ as $\frac{p(c,t)}{\pi(t)}$ which can be further rewritten as:

$$p(c|t) = \frac{\sum_{\mathbf{x} \in R_t} p(\mathbf{x})\, p(c|\mathbf{x})}{\sum_{\mathbf{x} \in R_t} p(\mathbf{x})}, \tag{8}$$

where $\pi(t)$ in the denominator is replaced using Eq. 6. The numerator states that the probability of the joint event that the class variable takes its c value and that the classifier states this with probability t, is the result of summing over these precise events in feature space (i.e, for $\mathbf{x} \in R_t$). Given our assumption regarding the i.i.d. nature of the samples, this holds. We can now state the following:

Proposition: An unbiased classifier is also well-calibrated.

Proof. For an unbiased classifier, $\lim_{n\to\infty} \hat{p}(c|\mathbf{x}) = p(c|\mathbf{x})$ for every $\mathbf{x}$, where n is the number of samples. Therefore, as $n \to \infty$, for every t: $\forall \mathbf{x} \in R_t$, $t = p(c|\mathbf{x})$. Replacing $p(c|\mathbf{x})$ with t in Eq.(8) yields $p(c|t) = t$, which is the condition for calibration to be 0.□

However, a well-calibrated classifier might not be unbiased. We see from Eq.(8) that for a well-calibrated classifier, its forecast, $\hat{p}(c|x) = t$ for $\mathbf{x} \in R_t$, is a normalized average of the true a-posteriori probability in the region defined by R_t. Clearly, one can construct cases where the classifier is biased, but $p(c|t) = t$ for all t: for example, suppose we have $X = \{1, 2\}$, $p(X = 1) = 0.5$ and $p(c|X = 1) = 0.2$ and $p(c|X = 2) = 0.6$. Suppose also that the classifier always predicts c with $\hat{p}(c|X) = 0.4$ for any X (hence on $t = 0.4$ has non-zero probability). Obviously, the classifier is biased. However, from Eq. 8 we have that $p(c|t) = 0.4$ and the classifier is well-calibrated.

3.2 The Bayes Error-Refinement Relationship

We start by defining a t dependent error measure:

$$e_t = \sum_t \pi(t) min(t, 1-t). \tag{9}$$

e_t essentially mirrors the Bayes error formula of Eq.2, but as we will see, e_t upper bounds the Bayes error. We are now ready to state the following result:

Theorem 1 *Given a well-calibrated classifier, whose forecasts are $\hat{p}(c|x)$, and given the true a-posteriori probability $p(c|x)$ with corresponding Bayes error rate $\mathbf{e}_\mathsf{B}$, the following holds: $\mathbf{e}_\mathsf{B} \le e_t \le 2\mathbf{R}$.*

Proof. Recall that for a well-calibrated classifier, $t = p(c|t)$. Making the appropriate substitution in the second term of Eq. 7, the refinement $\mathbf{R}$ can be written as:$\mathbf{R} = \sum_t \pi(t)t(1-t)$. It is easy to show that for $0 \le t \le 1$, $min(t, 1-t) \le 2 \cdot t(1-t)$, from which follows that, $e_t \le 2\mathbf{R}$.Now we have to show $\mathbf{e}_\mathsf{B} \le e_t$. We rewrite the expressions for the Bayes error in terms of t and R_t:

$$\mathbf{e}_\mathsf{B} = \sum_t \sum_{\mathbf{x} \in R_t} p(\mathbf{x})\, min(p(c|\mathbf{x})\,, 1 - p(c|\mathbf{x})). \tag{10}$$

We use Eq. 6 to substitute the $\pi(t)$ term in Eq. 9 and obtain:

$$e_t = \sum_t \sum_{\mathbf{x} \in R_t} p(\mathbf{x})\, min(t, 1-t). \tag{11}$$

With this reformulation, all we have to show is that for every $\mathbf{x}$ in every R_t, $p(\mathbf{x})\, min(p(c|\mathbf{x})\,, 1-p(c|\mathbf{x})) \leq p(\mathbf{x})\, min(t, 1-t)$. We have two cases, when $t \leq 0.5$ and when $t > 0.5$. We proceed with the proof for the first case. The proof for the second case is completely analogous. For the case where $t \leq 0.5$ we can write:

$$\sum_{\mathbf{x}\in R_t} p(\mathbf{x})\, min(t, 1-t) = t \cdot \sum_{\mathbf{x}\in R_t} p(\mathbf{x})$$

Using Eq. 8 and the fact that the classifier is well calibrated we replace t in the right hand side of the above equation to get:

$$\begin{aligned} t \sum_{\mathbf{x}\in R_t} p(\mathbf{x}) &= \frac{\sum_{\mathbf{x}\in R_t} p(\mathbf{x})\, p(c|\mathbf{x})}{\sum_{\mathbf{x}\in R_t} p(\mathbf{x})} \cdot \sum_{\mathbf{x}\in R_t} p(x) \\ &= \sum_{\mathbf{x}\in R_t} p(\mathbf{x})\, p(c|\mathbf{x})\,. \end{aligned} \tag{12}$$

In going over all $\mathbf{x} \in R_t$, we have two cases, depending on whether $p(c|\mathbf{x}) < 0.5$ or $p(c|\mathbf{x}) \geq 0.5$ [2]. Let $\mathbf{x}^-$ be such that $p(c|\mathbf{x}^-) < 0.5$. Thus we get that $min(p(c|\mathbf{x}^-)\,, 1 - p(c|\mathbf{x}^-)) = p(c|\mathbf{x}^-)$. It follows then that Eqs. 10 and Eq. 12 are equal for all such cases. Let now $\mathbf{x}' \in R_t$ be such that $p(c|\mathbf{x}') > 0.5$. For that $\mathbf{x}'$, $min(p(c|\mathbf{x}')\,, 1 - p(c|\mathbf{x}')) = 1 - p(c|\mathbf{x}')$. So, while e_t sums over $p(c|\mathbf{x}')$, as in Eq. 12, the Bayes error adds the smaller term, $1 - p(c|\mathbf{x}')$. It follows that $\mathbf{e}_{\mathrm{B}} \leq e_t$. □

From the proof, we see that 'looseness' in the upper bound on the Bayes error occurs whenever for certain $\mathbf{x} \in R_t$, $p(c|\mathbf{x})$ is on the other side of $1/2$ with respect to t. For t's that are close to 0 or 1, there is less of a chance for such $\mathbf{x}$'s to exist (see Eq. 8), while t close to $1/2$ has higher chances of occurrence for such cases. Therefore, a well-calibrated classifier with $\pi(t)$ that has mass close to 0 and 1 is not only more refined, but also provides a tighter bound on the Bayes error.

If the classifier is unbiased, we can provide a stronger result. In this case we know that asymptotically, $t = p(c|\mathbf{x})$ for every $\mathbf{x} \in R_t$ and we have that $R \leq \mathbf{e}_{\mathrm{B}}$. This follows from the fact that $\mathbf{e}_{\mathrm{B}} = e_t$ when the classifier is unbiased, and from the fact that for $0 \leq t \leq 1$, the relation $t(1-t) \leq min(t, 1-t)$ holds.

4 Calibration, Classification Error and ROC Curves

As discussed in Section 2, in order to minimize the classification error given by Eq. 3, we need to find the appropriate decision rule. This, in turn, translates to finding a probabilistic threshold α, so that we can classify a sample $\mathbf{x}$ as belonging to class c, when $\hat{p}(c|\mathbf{x}) \geq \alpha$. In this section we provide a procedure for finding α in terms of calibration. The intuition is as follows. If we had the real density $p(C|\mathbf{X})$, the optimal decision rule is given by Eq. 1, which in turn implies that $\alpha = 0.5$. Now, the process of calibrating may be seen as the process of bringing

[2] Recall that these $\mathbf{x}$ samples are placed in R_t according to the value of $\hat{p}(c|\mathbf{x})$.

$\hat{p}(C|X)$ closer to the real density. Calibrating a classifier is a mapping from $\hat{p}(c|x)$ to $p(c|t)$. In fact the procedures proposed in [4,5] essentially implement this mapping. Thus, under certain conditions we outline below the optimal threshold α^* of the original classifier is one where in the calibration mapping $p(c|\alpha^*) = 0.5$.

To formalize this intuitions we need to express the classification error in terms of the calibration mapping density. Suppose that our decision function on t is such that we say $C = 0$ if $t \leq \alpha$ and $C = 1$ if $t > \alpha$, where $0 \leq \alpha \leq 1$ (note that for the plug-in decision rule $\alpha = 0.5$). Given the density of $\pi(t)$ on t, the classification error is a function of α and is written as:

$$P_{error}(\alpha) = \int_0^\alpha p(C=1|t)\,p(t)\,dt + \int_\alpha^1 (1 - p(C=1|t))p(t)\,dt, \qquad (13)$$

where now t takes any value on the interval $[0,1]$, and is not limited to a discrete set as in the previous section. The first integral in Eq. 13 is the (weighted) area under the calibration map, $p(C=1|t)$, for which we predict class 0; this area is proportional to the probability that we missed instances that had label of 1. The second integral provides the proportion of the error for which we predict 1, but the actual class label was 0. Borrowing terms from signal detection theory, the first term is proportional to the probability of missed detection (detection of class 1), and the second integral is proportional to the probability of false detection (or false alarm). These areas are illustrated as the marked regions in Figures 1(a) and (b). We can now state the following:

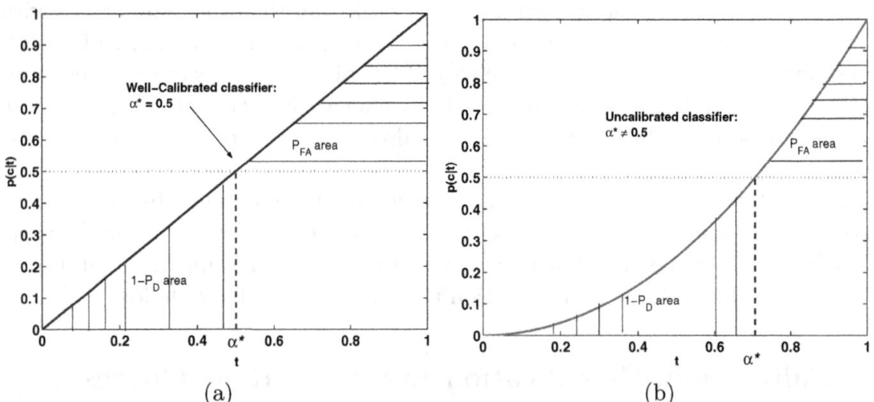

Fig. 1. (a) Illustration of the calibration map of a well-calibrated (diagonal line) and a non-calibrated classifier in (b).

Theorem 2 *Given a classifier with a-posteriori probabilities t, density $\pi(t)$ and a calibration map $p(C=1|t)$, where $p(C=1|t)$ does not cross 1/2 more than once, the threshold α on t which achieves minimum probability of error, i.e., $\alpha^* = \arg\min_\alpha P_{error}(\alpha)$ is given as α s.t. $p(C=1|t=\alpha) = 0.5$.*

Proof. Taking the derivative of $P_{error}(\alpha)$ with respect to α yields:

$$\frac{dPerror}{d\alpha} = 2 \cdot p(C = 1|t = \alpha) - 1. \tag{14}$$

Setting the derivative to 0 yields $p(C = 1|\alpha^*) = 1/2.\square$

The reason why the calibration map provides the optimal threshold on t for minimizing the probability of error is quite simple: the calibration map can be thought of as a new well-calibrated classifier, with a single feature t - thus the threshold of 1/2 on this new classifier is optimal. Because we require that the calibration map does not cross 1/2 more then once, there is a (single) threshold on our "feature" t that achieves the minimum error.

The function $p(C = 1|t)$ can also be used to create ROC curves. To see this, recall that an ROC curve plots the probability of detection, $P_D = P(Predict\ C = 1|Truth\ is\ C = 1)$, against the probability of false alarm, $P_{FA} = P(Predict\ C = 1|Truth\ is\ C = 0)$, created by varying a threshold (e.g., likelihood ratio). The threshold is varied so that we start from perfect detection, but maximum false alarm, to no false alarms, but minimum detection. We already stated that the two integrals composing P_{error} in Eq. 13 are directly related to P_D and P_{FA}, and to put it more accurately:

$$P_D(\alpha) = \frac{1 - \int_0^\alpha p(C = 1|t)\,p(dt)}{p(C = 1)}$$
$$P_{FA}(\alpha) = \frac{\int_\alpha^1 (1 - p(C = 1|t))p(dt)}{1 - p(C = 1)}, \tag{15}$$

thus by varying the threshold α, we can generate the entire ROC curve using the calibration map. At this point it is clear that methods that find the threshold of minimum error from ROC curves [6,7] produce the exact same result as the calibration procedure, when the calibration map does not cross 1/2 more than once.

However, the calibration procedure generalizes more than what can be achieved with the ROC method. Theorem 2 can be extended to the case where the calibration map crosses 1/2 more than once, requiring multiple thresholds on the original decision function for minimizing the error: given multiple thresholds on the decision function we can rewrite equation 13 (splitting the integral based on the number of needed thresholds) and find that minimizing the probability of error for any number of thresholds still occurs when the calibration map is 1/2. Such cases could occur with classifiers that output a-posteriori probabilities that are ranked incorrectly. For example, suppose that one class is split into several clusters in space, and a classifier (for example, a linear one) separates well some clusters, leaving other clusters far from the decision boundary. The resultant calibration map of such classifiers would cross 0.5 at several places, but the point of minimum error is still at $p(C = 1|t) = 0.5$. Thus, inverting the calibration map when at that point provides several thresholds on the decision rule. We illustrate the above with a two dimensional example, shown in Figure 2(a). The class marked with circles (class "1") consists of two clusters which are divided by the class marked with x's (class "0"). Learning a Logistic regression classifier on the data leads to a single linear boundary (shown in the figure), which does well at separating one cluster, but leaves the second one very far from the boundary. Thus, data from that cluster have higher probability of belonging to class "0" than data from class "0" itself. Figure 2(b) shows the calibration map of the Logistic regression classifier. The map crosses 0.5 at two values, thus leading to two decision boundaries (with the same slope of the original, but two different intersects). With these boundaries, both clusters of class "1" are well separated, and the resultant error is significantly lower, reducing from 10% with the original boundary to 5.5% with the new boundaries.

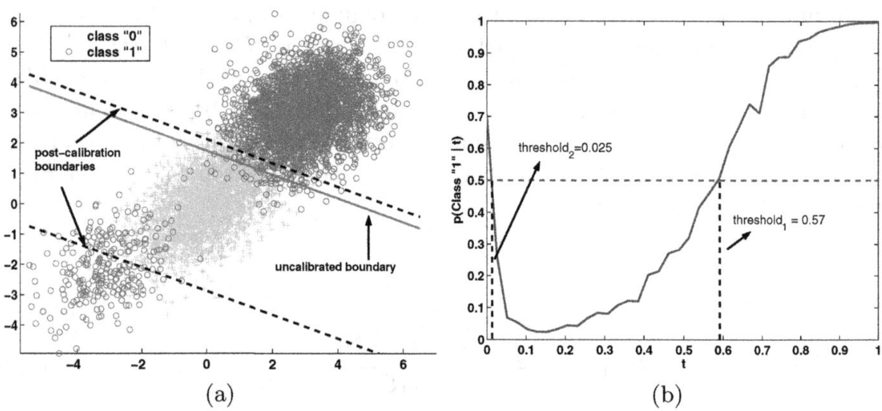

Fig. 2. Example of calibration finding multiple thresholds on the decision rule. (a) Decision boundaries before and after calibration superimposed on data. (b) The calibration map of the original linear classifier.

The results above can also be extended from the 0-1 loss to the general loss function, for which c_{01} is the cost of predicting class 0 when the true class is 1 and c_{10} is the cost of predicting class 1 and the true class is 0. The Bayes decision rule minimizing the *risk* under this loss function calls for classifying a sample $\mathbf{x}$ as 1 if $p(C = 1|\mathbf{x}) > \frac{c_{10}}{c_{10}+c_{01}}$ [10].As with the classification error under the 0-1 loss, applying the threshold $\frac{c_{10}}{c_{10}+c_{01}}$ on the *estimated* classifier, $\hat{p}(C = 1|\mathbf{x})$, may not minimize the risk under the generalized loss. However, using the same arguments given in Theorem 2, finding the thresholds which minimize the generalized loss function for a given classifier is the value on t for which $p(C = 1|t) = \frac{c_{10}}{c_{10}+c_{01}}$.

5 Calibration with Finite Data

With finite data sets, we want to estimate $p(C = 1|\hat{p}(C = 1|x))$ reliably. A procedure for this estimation was provided in [4, 5], where $\hat{p}(C = 1|x)$ is binned on the interval [0, 1] and the calibration map is estimated by counting the number of samples that fall into each bin. The procedure was originally suggested as a method for calibrating Naive Bayes classifiers, but is applicable to any classifier that outputs probabilities, or a distance measure that can be converted to probabilities (e.g., Tree-augmented Naive Bayes [11], Logistic regression, mixture models, and SVMs). The empirical success of calibration on various (typically large sized data sets) has been shown in previous works – in this section we aim at providing insight to the finite sample effects that can arise with calibration.

Estimating the calibration map involves learning a function from a scalar input ($\hat{p}(C = 1|\mathbf{x})$ to a scalar output. Thus, it is insensitive to the number of features in the classifier. The estimation is sensitive though to the sample size and to the number of bins used in the estimation procedure. We evaluate the effect of the sample size on the calibration procedure, thus we use learning curves, showing the various performance metrics before and after calibration.

We use the calibration procedure for prediction of I/O response time of individual requests to an enterprize storage array. Our data is based on an anonymized month-long trace of requests to an Hewlett Packard XP 512 storage array collected by the Storage Systems group at Hewlett-Packard Laboratories between 27 September and 27 October 2002. The raw traces are transformed to 10 features that describe queue lengths, locality and sequentiality, as measured by the server issuing the I/O request to the storage array. The problem is transformed to a binary classification problem by determining that any response time faster or equal to 1.5 msec is considered *fast*, while any response time slower than 1.5 msec is considered *slow*.

The data consists of 686091 training data and 343046 test data. We build two competing models to predict the correct class for the I/O request. The first is the Naive Bayes classifier, with Gaussian conditional distribution for the numerical feature and multinomial distribution for a locality feature. The second is a mixture of regression (MoR) classifier. The MoR model finds a mixture of regressors between the features and response time, which provides a distribution of response time for each value of the features, from which we can compute the a-posteriori probability of the response time being fast or slow. With the full training data, the Naive Bayes model achieves 82.18% accuracy before calibration and 85.60% after calibration, a significant improvement. The MoR model improves from 85.50% to 86.16%, a more modest improvement, to be expected from a model that is more naturally calibrated compared to Naive Bayes. The learning curves, both of accuracy and the Brier score, are shown in Figure 3.

For generating the learning curves we fix the number of bins used in the calibration procedure at 20, and average the results measured on the test set of 5 trials for each point on the curve. We see that for the Naive Bayes classifier, calibration improves accuracy and the Brier score early on the curve (already at 200 training samples), while for the already almost calibrated MoR, the calibration procedure does not produce a significant benefit to performance until fairly large training sets are available. Observing the changes in the Brier score, it appears that both models achieve near convergence to a calibrated classifier as early as after 1000 samples. It is also important to note that for the MoR and sample sizes smaller than 400, the calibration procedure slightly degrades performance because of overfitting. These experiments illustrate that models that are far from being calibrated benefit from calibration even with few data; for classification, any change in the decision boundary in the right direction has a large effect. However, models that are close to being calibrated are more sensitive to noise in the calibration map, and are more prone to overfitting with small data sets. We discuss possible ways to overcome these effects in the summary.

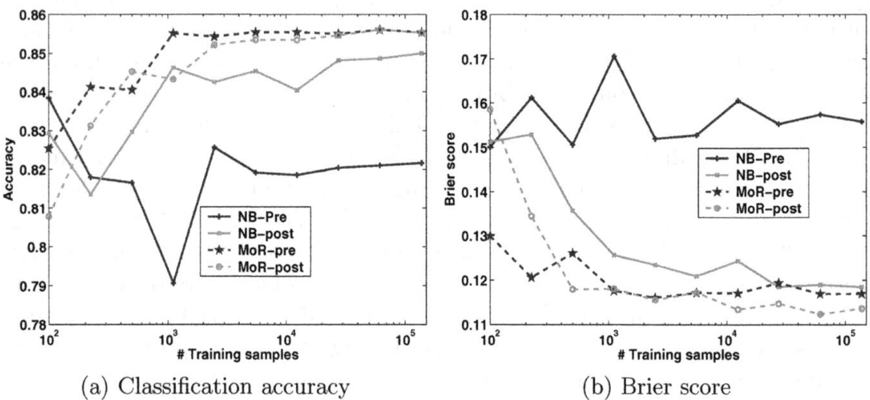

(a) Classification accuracy (b) Brier score

Fig. 3. Learning curves of Naive Bayes and MoR for the I/O prediction data.

6 Summary

In this paper we characterize the mathematical relation between calibration and bounds on the Bayes error and the use of calibration to find thresholds in the decision rule minimizing a classifier's error. These theoretical results, coupled with mounting empirical evidence in the literature, illustrate the importance and value of calibrating classifiers for classification and decision making.

The result relating calibration with the decision rule that minimizes the classification error, produces an effective procedure for finding optimal thresholds in this decision rule. This also establishes a direct relationship with ROC curves, a relationship which was informally alluded to in [6], and is formalized in this paper. As with any learning algorithm, finite sample effects have to be considered; our learning curve experiments show that a simple calibration procedure performs well with large training sets, but can cause overfitting with small training sets. Reducing the possibility of overfitting can be done by smoothing of the calibration map or estimating a smooth function (such as the sigmoid) as the calibration map [5]. As a note, the number of thresholds on the decision function would depend on the smoothing function used, e.g., with a sigmoid, only one threshold can be found, which might not be always desirable. We also observe that calibration is more beneficial, even at small sample sizes, for classifiers that are inherently not calibrated (such as Naive Bayes), compared to calibration of classifiers that are more naturally calibrated (such as logistic regression).

Future work includes providing bounds on how the estimation error of the calibration map affects the estimation of the optimal thresholds for classification and the payoff in terms of decision making.Such bounds could help avoid overfitting, especially with small sample sizes. Extending the method beyond binary classification problems is another research direction; similar to methods extending ROC curves beyond binary classification [7]. We are also exploring the use of calibration in semi-supervised learning, helping eliminate the possibility of performance degradation when using unlabeled data to learning classifiers, a phenomenon that occurs with biased models that output uncalibrated a-posteriori probabilities [12].

Acknowledgments

We thank Terence Kelly both for his help and suggestions and his work on the I/O response time prediction. We also thank Kim Keeton for providing the I/O data, Tom Fawcett for his comments on ROC curves, George Forman and Charles Elkan for providing feedback on the paper.

References

1. DeGroot, M., Fienberg, S.: The comparison and evaluation of forecasters. The statistician **32** (1983) 12–22
2. Fundenberg, D., Levine, D.: An easier way to calibrate. Games and economic behavior **29** (1999) 131–137
3. Foster, D., Vohra, R.V.: Asymptotic calibration. Biometrika **85** (1998) 379–390
4. Zadrozny, B., Elkan, C.: Obtaining calibrated probability estimates from decision trees and naive Bayesian classifiers. In: ICML. (2001)
5. Zadrozny, B., Elkan, C.: Transforming classifier scores into accurate multiclass probability estimates. In: Knowledge Discovery and Data Mining. (2002)
6. Fawcett, T.: ROC graphs: Notes and practical considerations for data mining representation. Technical Report HPL-2003-4, Hewlett-Packard Labs, Palo Alto, CA (2003)
7. Lachiche, N., Flach, P.: Improving accuracy and cost of two-class and multi-class probabilistic classifiers using ROC curves. In: ICML. (2003) 416–423
8. Devroye, L., Gyorfi, L., Lugosi, G.: A Probabilistic Theory of Pattern Recognition. Springer Verlag, New York (1996)
9. Brier, G.: Verification of forecasts expressed in terms of probability. Monthly weather review **78** (1950) 1–3
10. Duda, R.O., Hart, P.E., Stork, D.: Pattern Classification. John Wiley and Sons, New York (2001)
11. Friedman, N., Geiger, D., Goldszmidt, M.: Bayesian network classifiers. Machine Learning **29** (1997) 131–163
12. Cozman, F.G., Cohen, I., Cirelo, M.: Semi-supervised learning of mixture models. In: ICML. (2003) 99–106

A Tree-Based Approach to Clustering XML Documents by Structure

Gianni Costa[1], Giuseppe Manco[1], Riccardo Ortale[2], and Andrea Tagarelli[2]

[1] ICAR-CNR – Institute of Italian National Research Council
Via Pietro Bucci 41c, 87036 Rende (CS), Italy
{costa,manco}@icar.cnr.it
[2] DEIS, University of Calabria
Via Pietro Bucci 41c, 87036 Rende (CS), Italy
{ortale,tagarelli}@si.deis.unical.it

Abstract. We propose a novel methodology for clustering XML documents on the basis of their structural similarities. The idea is to equip each cluster with an *XML cluster representative*, i.e. an XML document subsuming the most typical structural specifics of a set of XML documents. Clustering is essentially accomplished by comparing cluster representatives, and updating the representatives as soon as new clusters are detected. We present an algorithm for the computation of an XML representative based on suitable techniques for identifying significant node matchings and for reliably merging and pruning XML trees. Experimental evaluation performed on both synthetic and real data shows the effectiveness of our approach.

1 Introduction

As the heterogeneity of XML sources increases, the need for organizing XML documents according to their structural features has become challenging. In such a context, we address the problem of inferring structural similarities among XML documents, with the adoption of clustering techniques. This problem has several interesting applications related to the management of Web data. For example, structural analysis of Web sites can benefit from the identification of similar documents, conforming to a particular schema, which can serve as the input for wrappers working on structurally similar Web pages. Also, query processing in semistructured data can take advantage from the re-organization of documents on the basis of their structure. Grouping semistructured documents according to their structural homogeneity can help in devising indexing techniques for such documents, thus improving the construction of query plans.

The problem of comparing semistructured documents has been recently investigated from different perspectives [5, 14, 4, 3, 8]. Recent studies have also proposed techniques for clustering XML documents. [7] describes a partitioning method that clusters documents, represented in a vector-space model, according to both textual contents and structural relations among tags. The approach in [13] proposes to measure structural similarity by means of an XML-aware

J.-F. Boulicaut et al. (Eds.): PKDD 2004, LNAI 3202, pp. 137–148, 2004.

edit distance, and applies a standard hierarchical clustering algorithm to evaluate how closely cluster documents correspond to their respective DTDs.

In our opinion, the main drawback of the above approaches is the lack of a notion of *cluster prototype*, i.e. a summarization of the relevant features of the documents belonging to a cluster. The notion of cluster prototype is crucial in most significant application domains, such as wrapper induction, similarity search, and query optimization. Indeed, in the context of wrapper induction, the efficiency and effectiveness of the extraction techniques strongly rely on the capability of rapidly detecting homogeneous subparts of the documents under consideration. Similarity search can substantially benefit from narrowing the search space. In particular, this can be achieved by discarding clusters whose prototypes exhibit features which do not comply with the target properties specified by a user-supplied query.

To the best of our knowledge, the only approach devising a notion of cluster prototype is [11]. Indeed, the authors propose to compare documents according to a structure graph, *s-graph*, summarizing the relations between elements within documents. Since the notion of s-graph can be easily generalized to sets of documents, the comparison of a document with respect to a cluster can be easily accomplished by means of their corresponding s-graphs. However, a main problem with the above approach relies on the loose-grained similarity which occurs. Indeed, two documents can share the same prototype s-graph, and still have significant structural differences, such as in the hierarchical relationship between elements. It is clear that the approach fails in dealing with application domains, such as wrapper generation, requiring finer structural dissimilarities.

In this paper we propose a novel methodology for clustering XML documents by structure, which is based on the notion of *XML cluster representative*. A cluster representative is a prototype XML document subsuming the most relevant structural features of the documents within a cluster. The intuition at the core of our approach is that a suitable cluster prototype can be obtained as the outcome of a proper overlapping among all the documents within a given cluster. Actually, the resulting tree has the main advantage of retaining the specifics of the enclosed documents, while guaranteeing a compact representation. This eventually makes the proposed notion of cluster representative extremely profitable in the envisaged applications: in particular, as a summary for the cluster, a representative highlights common subparts in the enclosed documents, and can avoid expensive comparisons with the individual documents in the cluster.

The proposed notion of cluster representative relies on the notions of XML tree *matching* and *merging*. Specifically, given a set of XML documents, our approach initially builds an *optimal matching tree*, i.e. an XML tree that is built from the structural resemblances that characterize the original documents. Then, in order to capture all such peculiarities within a cluster, a further tree, called a *merge tree*, is built to include those document substructures that are not recurring across the cluster documents. Both trees are exploited for suitably computing a cluster representative as will be later detailed. Finally, a hierarchical clustering algorithm exploits the devised notion of representative to partition XML documents into structurally homogeneous groups. Experimental evalua-

Input: A set $\mathcal{S} = \{t_1, \ldots, t_n\}$ of XML document trees;
Output: A cluster partition $\mathcal{P} = \{C_1, \ldots, C_k\}$ of $\mathcal{S}$.
Method:
let $\mathcal{P} := \{C_1, \ldots, C_n\}$, where initially $C_i = \{t_i\}$;
set $r_i := t_i$ as the representative for C_i;
compute a tree-distance matrix M_d, where $M_d(i, j) = d(t_i, t_j)$;
repeat
choose clusters C_i and C_j such that $d(rep(C_i), rep(C_j))$ is minimized;
compute the representative $r = rep(r_i, r_j)$ for cluster $C = C_i \cup C_j$;
set $\mathcal{P} := \mathcal{P} - \{C_i, C_j\} \cup \{C\}$, and update M_d;
until $\mathcal{P}$ **has maximal quality**;

Fig. 1. The *XRep* algorithm for clustering XML documents.

tion performed on both synthetic and real data states the effectiveness of our approach in identifying document partitions characterized by a high degree of homogeneity.

2 Problem Statement

Clustering is the task of organizing a collection of objects (whose classification is unknown) into meaningful or useful groups, namely *clusters*, based on the interesting relationships discovered in the data. The goal is grouping highly-similar objects into individual partitions, with the requirement that objects within distinct clusters are dissimilar from one another.

Several clustering algorithms [10] can be suitably adapted for clustering semistructured data. We concentrate on hierarchical approaches, which are widely known as providing clusters with a better quality, and can be exploited to generate cluster hierarchies. Fig.1 shows *XRep*, an adaptation of the agglomerative hierarchical algorithm to our problem. Each XML tree (derived by parsing the corresponding XML document) is initially placed in its own cluster, and a pair-wise tree distance matrix is computed. The algorithm then walks into an iterative step in which the least dissimilar clusters are merged. As a consequence, the distance matrix is updated to reflect this merge operation. The overall process is stopped when an optimal partition (i.e. a partition whose intra-distance within clusters is minimized and inter-distance between clusters is maximized) is reached. In this paper, we follow the approach devised in [9], and address the problem of clustering XML documents in a parametric way. More precisely, the general scheme of the *XRep* algorithm is parametric to the notions of *distance measure* and *cluster representative*.

Concerning the distance measure, we choose to adapt the *Jaccard coefficient* [10] to the context of XML trees. A first measure can be straightforwardly defined by considering the feature space representing the set of labels (i.e. tag names) associated with the nodes in a tree: if we denote with $tag(t)$ the set of tag names for a tree t, then we define as $d_J^{(1)}(t_1, t_2) = 1 - \frac{|tag(t_1) \cap tag(t_2)|}{|tag(t_1) \cup tag(t_2)|}$ the Jaccard distance between two trees t_1 and t_2. An alternative (and more refined) definition is given by taking into account the paths in the trees rather than only the node labels. More precisely, $d_J^{(2)}(t_1, t_2) = 1 - \frac{|path(t_1) \cap path(t_2)|}{\max\{|path(t_1)|, |path(t_2)|\}}$, where

$path(t_i)$ denotes the set of paths in t_i, and $path(t_1) \cap path(t_2)$ is the set of common paths between t_1 and t_2.

Intuitively, the representative of a cluster of XML documents is a document which effectively synthesizes the most relevant structural features of the documents in the cluster. The notion of representative can be formalized as follows.

Definition 1. *Given a set $\mathcal{U}$, equipped with a distance function $d : \mathcal{U} \times \mathcal{U} \mapsto \mathbb{R}$, and a set $\mathcal{S} = \{t_1, \ldots, t_n\} \subseteq \mathcal{U}$ of XML document trees, the representative of $\mathcal{S}$ (denoted by $rep(\mathcal{S})$) is the tree t^* that minimizes the sum of the distances:*

$$t^* = rep(\mathcal{S}) \in \mathcal{U} \iff t^* = argmin_{t \in \mathcal{U}} f(t)$$

where $f(t) = \sum_{i=1}^{n} d(t_i, t)$. □

The computation of the representative of a set turns out to be a hard problem if the above distance measures are adopted. Therefore we exploit a suitable heuristic for addressing the above minimization problem. Viewed in this respect, our goal is to find a *lower-bound-tree* and an *upper-bound-tree* for the optimal representative. The lower-bound-tree (resp. upper-bound-tree) is a tree on which any node deletion (resp. node insertion) leads to a worsening in function f. Thus, a representative can be heuristically computed by traversing the search space delimited by the above trees. Two alternative greedy strategies can be devised: either a growing approach, which iteratively adds nodes to the lower-bound, or a pruning approach, which iteratively removes nodes from the upper-bound. In the following, we will denote the lower-bound-tree and the upper-bound-tree as *optimal matching tree* and *merge tree*, respectively. Notice that the optimal matching tree represents a stopping condition for the pruning approach, whereas the merge tree is always a sub-optimal solution since it contains the optimal representative. Dually, the merge tree defines a stopping condition for the growing approach, whereas the optimal matching tree is a sub-optimal solution since it is contained in the optimal representative.

We develop a pruning approach in which the computation of an XML cluster representative consists of the following three main stages: the construction of an optimal matching tree, the computation of a merge tree, and the pruning of the merge tree. Fig.3 sketches an algorithm which has been developed according to the above three stages.

3 Mining Representatives from XML Trees

We give now some definitions which are at the basis of our approach. A tree t is a tuple $t = (r_t, V_t, E_t, \lambda_t)$ where $V_t \subseteq \mathbb{N}$ is the set of nodes, $E_t \subseteq V_t \times V_t$ is the set of edges, r_t is the root node of t, and $\lambda_t : V_t \mapsto \Sigma$ is a node labelling function where Σ is an alphabet of node labels. In particular, we say that an *XML tree* is a tree where Σ is an alphabet of *element tags*. Moreover, let $depth_t(v)$ denote the depth level of node v in t, with $depth_t(r_t) = 0$, and let $path_t(v) = \langle v_{i_1} = r_t, v_{i_2}, \ldots, v_{i_p} = v \rangle$ denote the list of p nodes that lead up to the node v from the root r_t.

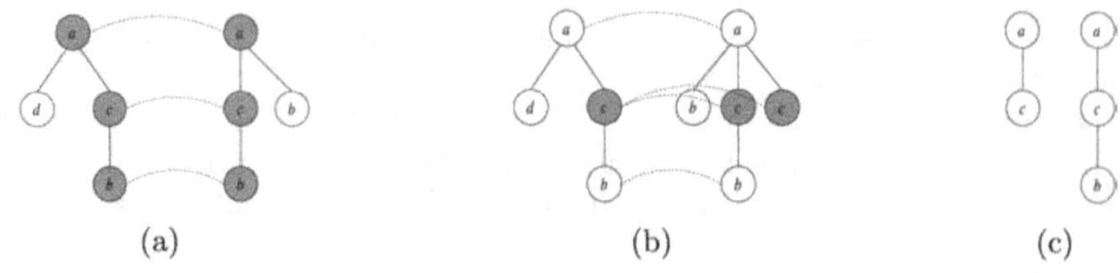

Fig. 2. (a) Strong and (b) multiple matching nodes, and (c) their trees.

Definition 2 (strong matching). *Given two trees t_1 and t_2, and two nodes $v \in V_{t_1}$, $w \in V_{t_2}$, a* strong matching *$match(v, w)$ between v and w exists if $\lambda_{t_1}(v_i) = \lambda_{t_2}(w_i)$ and $depth_{t_1}(v_i) = depth_{t_2}(w_i)$, for each pair of nodes (v_i, w_i) such that $v_i \in path_{t_1}(v)$ and $w_i \in path_{t_2}(w)$.* □

The above definition states that two nodes, v and w, have a strong matching if v and w together with their respective ancestors share both the same label (i.e. tag name) and depth level. Fig.2(a) displays an example of strong matching among the colored nodes.

The detection of matching nodes between two trees allows the construction of a new tree, called a *matching tree*, which resembles the *intersection* of the original trees.

Definition 3 (matching tree). *Given two trees t_1 and t_2, a tree $t = (r_m, V_m, E_m, \lambda_m)$ is a* matching tree, *denoted by $t = match(t_1, t_2)$, if the following conditions hold:*

1. *there exist two mappings $f_1 : t \mapsto t_1$ and $f_2 : t \mapsto t_2$ associating nodes and edges in t with a subtree in t_1 and t_2;*
2. *for each $u \in V_m$, there exists a strong matching between $v = f_1(u)$ and $w = f_2(u)$ (i.e. $match(v, w)$ holds); moreover, $\lambda_m(u) = \lambda_{t_1}(v) = \lambda_{t_2}(w)$;*
3. *$f_1(r_m) = r_{t_1}$, and $f_2(r_m) = r_{t_2}$; moreover, for each $e = (u, v) \in E_m$, $f_1(e) = (f_1(u), f_1(v))$ and $f_2(e) = (f_2(u), f_2(v))$.* □

Notice that, in general, multiple matchings may occur when a node in a tree has a matching with more than one node in a different tree. More formally, given two trees t_1 and t_2, a node $v \in V_{t_1}$ has a *multiple matching* if $\exists w', w'' \in V_{t_2}$ such that both $match(v, w')$ and $match(v, w'')$ hold. An example of multiple matching between nodes in two trees is shown in Fig.2(b). Multiple matchings trigger ambiguities in defining matching trees: Fig.2(c) represents two alternative matching trees for the documents in Fig.2(b).

3.1 XML Tree Matching

In order to capture as many structural affinities as possible, we are interested in finding matching trees with maximal size. Formally, a matching tree $t_m = match(t_1, t_2)$ is an *optimal matching tree* for two trees t_1 and t_2 if there not

Input:
An XML tree $r_1 = \langle r_{r_1}, V_{r_1}, E_{r_1}, \lambda_{r_1} \rangle$ as representative of cluster C_1, and
an XML tree $r_2 = \langle r_{r_2}, V_{r_2}, E_{r_2}, \lambda_{r_2} \rangle$ as representative of cluster C_2.
Output:
An XML tree *rep* as representative of cluster $C = C_1 \cup C_2$.
Method:
compute the matching matrix M_m, with size $(|V_{r_1}| \times |V_{r_2}|)$;
compute the marking vectors V_{m_1}, V_{m_2}, where $V_{m_1}.size = |V_{r_1}|$ and $V_{m_2}.size = |V_{r_2}|$;
set $m_1 := |\{v_i \in V_{r_1} | V_{m_1}[i] \neq -1\}|$, and $m_2 := |\{v_i \in V_{r_2} | V_{m_2}[i] \neq -1\}|$;
if $(m_1 > m_2)$
  $match :=$ `buildMatch`$(r_1, r_2, V_{m_1}, V_{m_2})$; $merge :=$ `buildMerge`$(r_1, r_2, V_{m_1}, V_{m_2})$;
else
  $match :=$ `buildMatch`$(r_2, r_1, V_{m_2}, V_{m_1})$; $merge :=$ `buildMerge`$(r_2, r_1, V_{m_2}, V_{m_1})$;
$rep :=$ `prune`$(C_1 \cup C_2, merge, match)$;
return *rep*;

Function `buildMatch`$(t_1, t_2, V_{m_1}, V_{m_2})$: t;
$t := t_1$;
for each $v_i \in V_{t_1}, V_{m_1}[i] = -1$ **do**
  $remove(t, v_i)$;   /* *removes the subtree rooted at v_i from t* */
let $I_j = \{v_{i_1}, \ldots, v_{i_h} \in V_{t_1} \mid V_{m_1}[i_p] = j,\ p \in [1..h]\}$;
for each I_j **do**
  $removeDuplicates(t, I_j)$;   /* *removes duplicated paths from t* */
return t;

Function `buildMerge`$(t_1, t_2, V_{m_1}, V_{m_2})$: t;
$t := t_1$;
for each $v_i \in V_{t_1}$ **do**
  let $J = \{w_{j_1}, \ldots, w_{j_h} \in V_{t_2} \mid V_{m_2}[j_p] = i,\ p \in [1..h]\}$;
  let $v \in V_{t_1}$ such that $(v, v_i) \in E_{t_1}$;
  $insert(t, v, v_i, |J| - 1)$;   /* *inserts node v_i as a child of v into t, $|J| - 1$ times* */
for each $w_i \in V_{t_2}, V_{m_2}[i] = -1$ **do**
  let $w_j \in V_{t_2}$ such that $(w_j, w_i) \in E_{t_2}$, and $v_h \in V_{t_1}$ such that $V_{m_2}[j] = h$;
  $insert(t, v_h, w_i)$;   /* *inserts node w_i as a child of v_h into t* */
return t;

Function `prune`(C, t, t') : r;
set $r := t$;
do
  let $\mathcal{L} \subseteq V_r$ be the set of leaf nodes in r;
  compute $d_0 := \sum_{t \in C} d(t, r)$;
  for each $v_l \in \mathcal{L}$ **do**
    compute $r^{(l)} := removeLeaf(r, v_l)$;
  $l^* = \arg\min_{v_l} [\sum_{t \in C} d(t, r^{(l)})]$;
  set $d^* := \sum_{t \in C} d(t, r^{(l^*)})$;
  if $(d^* < d_0)$
    $r := r^{(l^*)}$;
while $d^* < d_0$ and $V_r \subseteq V_{t'}$;
return r;

Fig. 3. The algorithm for the computation of an XML cluster representative.

exists another matching tree $t'_m = match(t_1, t_2) \neq t_m$ such that $|V_{t_m}| \geq |V_{t'_m}|$. We describe a dynamic-programming technique for building an optimal matching tree from two XML trees. The technique consists of three steps: *i*) detection of matching nodes, *ii*) selection of best matchings, and *iii*) optimal matching tree construction.

Matching detection. Given two trees t_1 and t_2, the detection of matching nodes is performed building a $(|V_{t_1}| \times |V_{t_2}|)$ matching matrix M_m. In this matrix, the generic (i, j)-th element corresponds to nodes $v_i \in V_{t_1}$ and $w_j \in V_{t_2}$, and contains a weight $\omega_m(v_i, w_j)$ to be associated with the matching between v_i and w_j. Initially, the weight is 1 if $match(v_i, w_j)$ holds, and 0 otherwise. In order to ease the construction of the matching matrix, nodes are enumerated by level, thus guaranteeing a particular block structure for M_m. Indeed, for each level

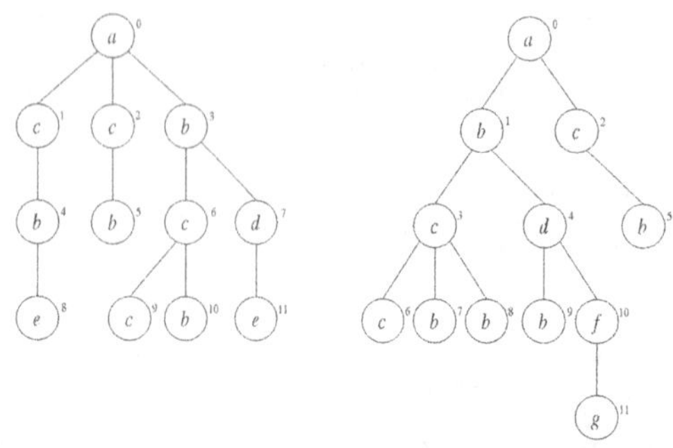

(a) Examples XML trees t_1 and t_2

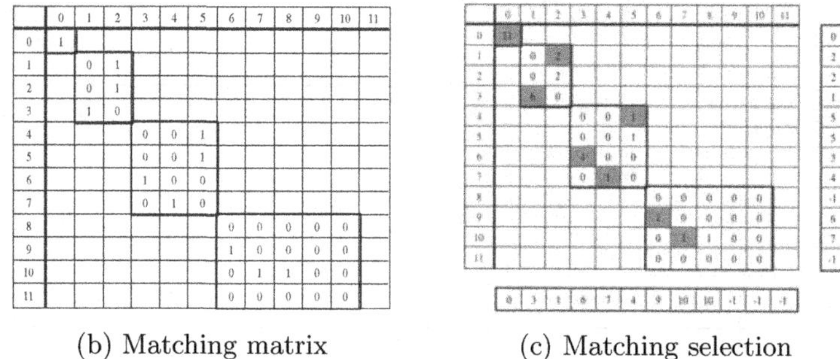

(b) Matching matrix (c) Matching selection

Fig. 4. Data structures for the construction of an optimal matching tree.

k, a sub-matrix $M_m(k)$ collects the matchings among the nodes in t_1 and t_2 with depth equal to k. Fig.4(a) displays two example XML trees with numbered nodes. The corresponding matching matrix is shown in Fig.4(b).

Selection of best matchings. The problem of multiple matchings can be addressed by discarding those matchings which are less relevant according to the weighting function ω_m. The weight $\omega_m(v, w)$, associated to two matching nodes $v \in V_{t_1}$ and $w \in V_{t_2}$, is computed by taking into account the matches between the children nodes of both v and w. Formally, $\omega_m(v, w) = 1 + \sum_{i,j} \omega_m(v_i, w_j)$, where nodes v_i, w_j are such that $(v, v_i) \in E_{t_1}$ and $(w, w_j) \in E_{t_2}$. Fig.4(c) shows the weights associated with each possible node pair.

Multiple matchings relative to any node of t_1 (resp. t_2) can be detected by checking multiple entries with non-zero values within the corresponding row (resp. column) of M_m. We now describe the process for detecting multiple matchings. In the following we focus on the identification of nodes within t_1 that have multiple matchings with those in t_2: the dual situation (i.e. identification of nodes in t_2 having multiple matching with nodes in t_1) has a similar treatment.

Let $v_i \in V_{t_1}$ denote the node corresponding to the i-th row in M_m, and let $J_{v_i} = \{j_1, \ldots, j_h\}$ be the set of column indexes, corresponding to the nodes $w_{j_1}, \ldots, w_{j_h}$ of t_2, such that $M_m(i, j_k) > 0$ (i.e. such that $\omega_m(v_i, w_{j_k}) > 0$), $k = [1..h]$. Thus, v_i exhibits multiple matchings if $|J_{v_i}| > 1$. For each node $v_i \in V_{t_1}$,

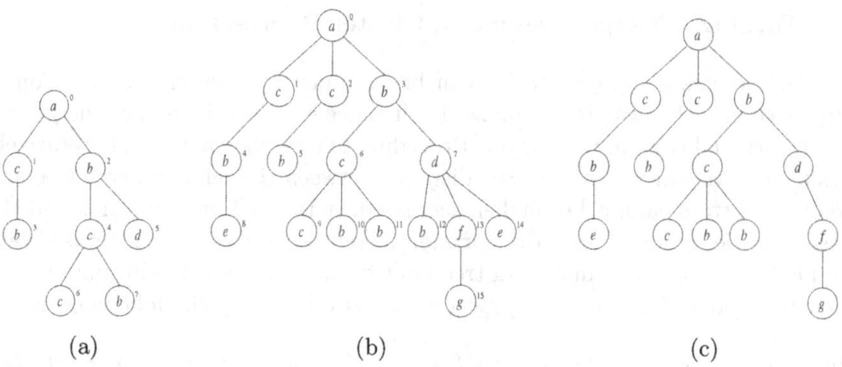

Fig. 5. (a) Lower-bound (optimal matching tree), (b) upper-bound (merge tree), and (c) optimal representative tree relative to the trees of Fig.4(a).

the *best matching* node corresponds to the column index $j^*_{v_i} = \arg\max_{j_1,\ldots,j_h} \{M_m(i,j_1),\ldots,M_m(i,j_h)\}$. If the maximum in $\{M_m(i,j_1),\ldots,M_m(i,j_h)\}$ is not unique we choose $j^*_{v_i}$ to be the minimum index. The overall best matchings for nodes of t_1 can be easily tracked by using a *marking vector* $V_{m_1} = \{j^*_{v_1},\ldots,j^*_{v_n}\}$, whose generic i-th entry indicates the node of t_2 with which $v_i \in V_{t_1}$ has the best matching. We set $V_{m_1}[i] = -1$ if the node $v_i \in V_{t_1}$ has no matching. Fig.4(c) shows the marking vectors V_{m_1} and V_{m_2} associated with t_1 and t_2, respectively.
Optimal matching tree construction. An optimal matching tree is effectively built by exploiting the above marking vectors: it suffices that all nodes with no matching are discarded. Fig.5(a) shows the optimal matching tree computed for t_1 and t_2 of Fig.4(a). As we can see in the figure, the optimal matching tree is obtained from t_1 by removing nodes $2, 5, 8, 11$.

3.2 Building a Merge Tree

The optimal matching tree of two documents represents an optimal intersection between the documents. The notion of *merge tree* resembles an optimized *union* of the original trees. Notice that, firstly an optimal matching tree has to be detected, in order to avoid redundant nodes to be added. Indeed, a trivial merge tree could be simply built as the union of the trees under investigation. Function `buildMerge` in Fig.3 details the construction of a merge tree, which takes into account nodes discarded while building the optimal matching tree. To this purpose, given two trees t_1 and t_2, we first consider nodes in t_1 having duplicate nodes, and insert such duplicates into the merge tree. Next, nodes in t_2 which do not match with any node in t_1 are added.

Fig.5(b) shows the merge tree associated to the trees of Fig.4(a). Nodes $8, 11$ from t_1 and $9, 10, 11$ from t_2 have no matching, whereas nodes $2, 5$ from t_1 and 8 from t_2 exhibit multiple matchings.

3.3 Turning a Merge Tree into a Cluster Representative

An effective cluster representative can be obtained by removing nodes from a merge tree in such a way to minimize the distance between the refined merge tree and the original trees in the cluster. Procedure `prune`, shown in Fig.3, iteratively tries to remove leaf nodes until the distance between the refined merge tree and the original trees cannot be further decreased. It is worth noticing that, on the basis of the definition of procedure `prune`, the representative of a cluster is always bounded by the optimal matching tree built from the documents in that cluster. The correctness of the pruning procedure is established by the following result.

Theorem 1. *Let* t_1, t_2 *be two XML trees. Moreover, let* $t_M = merge(t_1, t_2)$, $t_m = match(t_1, t_2)$ *and* $t^* = rep(\{t_1, t_2\})$. *Then,* $t_m \subseteq t^* \subseteq t_M$. □

Let us consider again the trees t_1 and t_2 of Fig.4(a) and their associated merge tree $merge(t_1, t_2)$ of Fig.5(b). Suppose that t_1 and t_2 belong to the same cluster C. In order to compute the representative tree for C, the pruning procedure is initially applied to the set of leaves $\mathcal{L} = \{5, 8, ..., 12, 14, 15\}$. If we choose to adopt the $d_J^{(2)}$ distance, the procedure computes an initial intra-cluster distance $d_0^C = 5/8$. This distance is reduced to $4/7$ as leaf node 14 is removed. Yet, d_0^C can be decreased by removing node 12. Since at this point no further node contributes to the minimization of d_0^C, the pruning process ends. Fig.5(c) shows the cluster representative resulting from pruning the merge tree in Fig.5(b), with the adoption of the $d_J^{(2)}$ distance.

4 Evaluation

We evaluated the effectiveness of *XRep* by performing experiments on both synthetic and real data. In the former case, we mainly aimed at assessing the effectiveness of our clustering scheme with respect to some prior knowledge about the structural similarities among the XML documents taken into account. Specifically, we exploited a synthetic data set that comprises seven distinct classes of XML documents, where each such class is a structurally homogeneous group of documents randomly generated from a previously chosen DTD. Tests were performed in order to investigate the ability of *XRep* in catching such groups.

To the purpose of assembling a valuable data set, we developed an automatic generator of synthetic XML documents, that allows the control of the degree of structural resemblance among the document classes under investigation. The generation process works as follows. Given a seed DTD $\texttt{DTD}_0$, a similarity threshold τ, and a number k of classes, the generator randomly yields a set S_τ^k of k different DTDs, hereinafter called class DTDs, that individually retain at most τ percent of the element definitions within $\texttt{DTD}_0$. The k class DTDs are eventually leveraged to generate as many collections of conforming XML documents, on the basis of suitable statistical models ruling the occurrences of the document elements [8].

The seed DTD was manually developed and exhibits a quite complex structure. For the sake of brevity, we only focus on its major features. $\mathtt{DTD}_0$ contains 30 distinct element declarations that adopt neither attributes nor recursion. Non empty elements contain at most 4 children. Yet, the occurrences of such elements are suitably defined by exploiting all kinds of operators, namely $+,*,?$, and $|$. Finally, the tree-based representation of any XML document conforming to $\mathtt{DTD}_0$ has a depth that is equal to 6.

Each test on synthetic data was performed on a distinct set of seven class DTDs, sampled from $\mathtt{DTD}_0$, at increasing values of the similarity threshold τ: we chose τ to be respectively equal to 0.3, 0.5, and 0.8.

Real XML documents were extracted from six different collections available on Internet:

- Astronomy, 217 documents extracted from an XML-based metadata repository, that describes an archive of publications owned by the *Astronomical Data Center* at NASA/GSFC.
- Forum, 264 documents concerning messages sent by users of a Web forum.
- News, 64 documents concerning press news from all over the world, daily collected by *PR Web*, a company that provides free online press release distribution.
- Sigmod, 51 documents concerning issues of SIGMOD Record. Such documents were obtained from the XML version of the ACM SIGMOD Web site produced within the *Araneus* project [6].
- Wrapper, 53 documents representing wrapper programs for Web sites, obtained by means of the *Lixto* system [2].
- Xyleme_Sample, a collection of 1000 documents chosen from the *Xyleme*'s repository, which is populated by a Web crawler using an efficient native XML storage system [12].

The distribution of tags within these documents is quite heterogeneous, due to the complexity of the DTDs associated with the classes, and to the semantic differences among the documents. In particular, wrapper programs may have substantially different forms, as a natural consequence of the structural differences existing among the various Web sites they have been built on: thus, the skewed nature of the documents in Wrapper should be taken into account. Also, documents sampled from Xyleme exhibit a more evident heterogeneity, since they have been crawled from very different Web sources.

Clustering results were evaluated by exploiting the standard *precision* and *recall* measures [1]. However, in the case of Xyleme_Sample, we had no knowledge of an a-priori classification. As a consequence, we resorted to an internal quality criterium that takes into account the compactness of the discovered clusters. More precisely, given a cluster partition $\mathcal{P} = \{\mathcal{C}_1, \ldots, \mathcal{C}_n\}$, where $\mathcal{C}_i = \{x^i_1, \ldots, x^i_{n_i}\}$, we defined an *intra-cluster distance* measure as $\mathcal{IC}(\mathcal{P}) = \frac{1}{n}\sum_{\mathcal{C}_i \in \mathcal{P}} \frac{1}{n_i} \sum_{x \in \mathcal{C}_i} d(x, rep(\mathcal{C}_i))$.

Table 1 summarizes the quality values obtained testing *XRep* on both synthetic and real data. All the experiments have been carried out by adopting the Jaccard distance $d_J^{(2)}$ introduced in Section 2. Tests on synthetic data evaluated

Table 1. Quality results.

type	docs	avg size	classes	clusters	τ	precision	recall	F-measure	$\mathcal{IC}$
synth	1400	0.13KB	7	7	0.3	0.979	0.978	0.978	0.219
synth	1400	0.81KB	7	7	0.5	0.802	0.909	0.852	0.304
synth	1400	3.19KB	7	7	0.8	0.689	0.773	0.728	0.369
real	649	5.74KB	5	5	-	1	1	1	0.208
real	500	8.56KB	-	7	-	-	-	-	0.376
real	1000	9.42KB	-	9	-	-	-	-	0.43

the performance of *XRep* on three collections of 1400 documents (200 documents for each class DTD). Experimental evidence highlights the overall accuracy of *XRep* in distinguishing among classes of XML documents characterized by different average sizes due to different choices for the threshold τ. As we can see, *XRep* exhibits an excellent behavior for $\tau = \{0.3, 0.5\}$, while the acceptable performance reported on row 3 (i.e. $\tau = 0.8$) is due to the intrinsic difficulty in catching minimal differences in the structure of the involved XML documents. Indeed, two clearly distinct class DTDs, say $\mathtt{DTD}_i$ and $\mathtt{DTD}_j$, may share a number of element definitions inducing similar paths within the conforming XML documents. If such definitions assign multiple occurrences to the elements of the common paths, the initial class separation between $\mathtt{DTD}_i$ and $\mathtt{DTD}_j$ may be potentially vanished by a strong degree of document similarity due to a large number of common paths in the corresponding XML trees.

Tests on real data considered separately the first five collections (649 XML documents with an average size that is equal to $5.74KB$), and the Xyleme_Sample collection. In the first case, *XRep* showed amazingly optimal accuracy in identifying even latent differences among the involved real documents. As far as Xyleme_Sample is concerned, we conducted two experiments (rows 5 and 6 in Table 1), where in the first one we considered only one and a half of the dataset. However, as we expected, in both cases intra-cluster distance provides fairly good values: this is mainly due to the high heterogeneity which characterizes documents in Xyleme_Sample.

5 Conclusions and Further Work

We presented a novel methodology for clustering XML documents, focusing on the notion of *XML cluster representative* which is capable of capturing the significant structural specifics within a collection of XML documents. By exploiting the tree nature of XML documents, we provided suitable strategies for tree matching, merging, and pruning. Tree matching allows the identification of structural similarities to build an initial substructure that is common to all the XML document trees in a cluster, whereas the phase of tree merging leads to an XML tree that even contains uncommon substructures. Moreover, we devised a suitable pruning strategy for minimizing the distance between the documents in a cluster and the document built as the cluster representative. The clustering framework was validated both on synthetic and real data, revealing high effectiveness.

We conclude by mentioning some directions for future research. The approach described in the paper has to be considered an initial approach to clustering tree-structured XML data. Further notions of cluster representative can be investigated, e.g. by relaxing the requirement that a prototype corresponds to a single XML document. Indeed, there are many cases in which a collection of XML documents is better summarized by a forest of subtrees, where each subtree represents a given peculiarity shared by some documents in the collection. A typical case raises, for instance, when the collection has an empty matching tree, and still exhibits significant homogeneities.

References

1. R. Baeza-Yates and B. Ribeiro-Neto. *Modern Information Retrieval.* ACM Press Books. Addison Wesley, 1999.
2. R. Baumgartner, S. Flesca, and G. Gottlob. Visual web information extraction with Lixto. In *Proc. VLDB'01 Conf.*, pages 119–128, 2001.
3. E. Bertino, G. Guerrini, and M. Mesiti. A matching algorithm for measuring the structural similarity between an XML document and a DTD and its applications. *Information Systems*, 29(1), 2004.
4. S. Chawathe et al. Change detection in hierarchically structured information. In *Proc. SIGMOD'96 Conf*, pages 493–504, 1996.
5. G. Cobena, S. Abiteboul, and A. Marian. Detecting changes in XML documents. In *Proc. ICDE'02 Conf.*, pages 41–52, 2002.
6. V. Crescenzi, G. Mecca, and P. Merialdo. Roadrunner: Towards automatic data extraction from large web sites. In *Proc. VLDB'01 Conf.*, pages 109–118, 2001.
7. A. Doucet and H. A. Myka. Naive clustering of a large XML document collection. In *Proc. INEX'02 Workshop*, 2002.
8. S. Flesca et al. Detecting structural similarities between XML documents. In *Proc. WebDB'02 Workshop*, 2002.
9. F. Giannotti, C. Gozzi, and G. Manco. Clustering transactional data. In *Proc. ECML-PKDD'02 Conf.*, pages 175–187, 2002.
10. A. K. Jain and R. C. Dubes. *Algorithms for Clustering Data.* Prentice-Hall, 1988.
11. W. Lian et al. An efficient and scalable algorithm for clustering XML documents by structure. *IEEE TKDE*, 16(1):82–96, 2004.
12. L. Mignet, D. Barbosa, and P. Veltri. The XML Web: a First Study. In *Proc. WWW'03 Conf.*, pages 500–510, 2003.
13. A. Nierman and H. V. Jagadish. Evaluating structural similarity in XML documents. In *Proc. WebDB'02 Workshop*, 2002.
14. Y. Wang, D.J. DeWitt, and J. Cai. X-Diff: A fast change detection algorithm for XML documents. In *Proc. ICDE'03 Conf.*, pages 519–530, 2003.

Discovery of Regulatory Connections in Microarray Data

Michael Egmont-Petersen, Wim de Jonge, and Arno Siebes

Institute of Information and Computing Sciences, Utrecht University,
Padualaan 14, De Uithof, Utrecht, The Netherlands
Michael@cs.uu.nl

Abstract. In this paper, we introduce a new approach for mining regulatory interactions between genes in microarray time series studies. A number of preprocessing steps transform the original continuous measurements into a discrete representation that captures salient regulatory events in the time series. The discrete representation is used to discover interactions between the genes. In particular, we introduce a new across-model sampling scheme for performing Markov Chain Monte Carlo sampling of probabilistic network classifiers. The results obtained from the microarray data are promising. Our approach can detect interactions caused both by co-regulation and by control-regulation.

1 Introduction

In bioinformatics, we are faced with an increasing amount of data that characterize the structure and function of different living organisms. Still more experimental data such as sequences (nucleotides, proteins) and gene activities (mRNA expression ratios) are generated either in the biology laboratory or in a clinical setting. The ever-expanding datasets fuel a growing demand for new datamining techniques that can help to discover possible relations between the biological entities under study and couple the different sources of data. Such datamining techniques should be able to cope with many variables that may exhibit complex dependency relations. We present a new cross-model sampling Markov Chain Monte Carlo algorithm, which we test by learning Bayesian network classifiers to predict regulatory relations between a set of predictor genes and a target gene.

Microarrays were introduced in the nineties as a means for studying in parallel the expression of all genes pertaining to a particular organism. One of the ultimate goals is to discover which genes are involved in the regulation of others, the so-called *regulatory pathways*. Microarrays measure the relative abundance of mRNA, corresponding to each known gene transcribed at a certain time t in a particular organism under study. So the prospect of microarrays is that of an aid that can help to identify functional roles of genes and eventually enrich the knowledge of the complex relations between the genotype and the phenotype of the organism under study.

Microarray time series experiments are conducted in order to study significant dynamic expression patterns. One goal of a time series experiment is to investigate which genes regulate others. It is to be expected that some genes that are controlled by the

J.-F. Boulicaut et al. (Eds.): PKDD 2004, LNAI 3202, pp. 149–160, 2004.

same transcription factor show a similar but lagged expression pattern over time, when the expression of the particular transcription factor varies. We make a distinction between *co-regulation* and *controlled regulation*. Two genes are said to be positively co-regulated when the change in relative abundance of the genes has the same first- order derivatives with respect to time. Two genes are said to be inversely co- regulated when the change in relative abundance of the genes has the opposite first-order derivatives with respect to time. Two perfectly co-regulated genes can have expression patterns with different amplitudes. One or more genes (the regulators) are said to control the expression of a particular gene (the target) when the expressions of the regulator genes directly influence the expression of the target gene.

Under conditions where particular genes are co-regulated or one or more regulator genes control the expression of a target gene, one would expect co-variation between the expressions of these genes over time. Our goal is to develop a datamining approach that can discover dynamic patterns of co-regulation and control regulation between sets of genes. Clustering techniques and correlation measures have been used extensively to identify groups of genes that are likely to be functionally related, see, e.g., Datta & Datta for an overview [1]. However, the standard clustering techniques do not take post-transcriptional and post-translational lag times into account. More importantly, in mining the vast amount of time series array data for putative control regulation relations, *lags* between expression levels of genes may contain indicative clues as to which genes code for proteins that act as regulators for others.

In this article, we present a novel datamining method for finding possible regulatory relations between small sets of genes, based on time-course microarray data, see, e.g., [2]. In the sequel, we regard the normalized (relative) expression levels of each gene as a time signal. We introduce preprocessing steps that transform such a time signal into "salient features", points in time that may disclose possible lagged interactions between genes. From this discrete representation, we train dynamic Bayesian networks to predict regulatory events of specific target genes using a novel MCMC-approach. Our new method is evaluated on microarray data obtained from the experiments by Spellman et al. [2]. The results are promising. Most of the regulatory relations found could be corroborated by literature.

2 Microarray Data

Our goal is to discover and interpret statistical relations between the relative expression of genes. For that purpose, we need to choose a suitable representation scheme for time series microarray data. Generally, each spot indicates the average relative (log) expression of mRNA corresponding to a particular gene R_i. The expression ratio of gene R_i can be seen as a continuous stochastic variable, characterized by the probability density function $p(R_i)$. Each variable R_i can either be a predictor or a target, relative to the other variables entering the model. We use t to denote the time step at which variable R_i is being measured and discretize the arraydata $R_i(t)$, $t \in \{t_0, \ldots, t_T\}$ into the following three categories: *change*, *local minimum* and *local maximum*. This differs from the approach by others [3, 4], who make a distinction between up-regulated, medium regulated and down-regulated gene expression. Our representation is different in the sense that it combines successive expression ratios $R_i(t_{v-1})$, $R_i(t_v)$ and $R(t_{v+1})$ into fea-

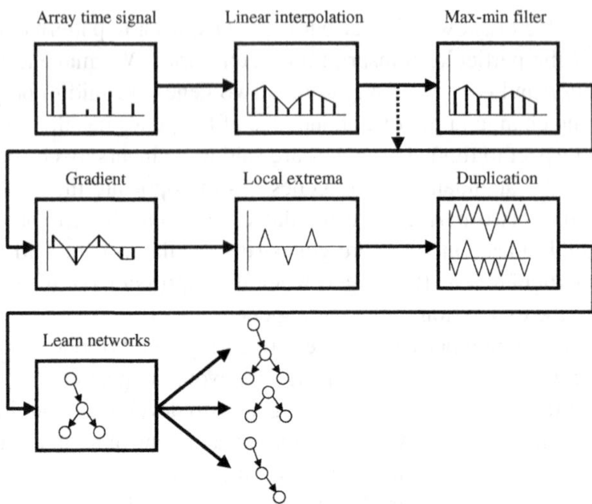

Fig. 1. In total six preprocessing steps are performed before our new datamining algorithm is applied to the dataset: 1) the log ratios of each gene are computed, 2) linear interpolation results in uniformly sampled log expression ratios, 3) the (optional) max-min filter removes transient extrema, 4) convolution with the first-order derivative of the Gaussian function results in derivatives of the expression ratios, 5) the local extrema are defined as time points at which the sign of the first-order derivative changes, 6) the selected target gene is coded as a binary variable by duplication, 7) MCMC-learning of local genetic networks.

tures that capture the local dynamics (local extrema) of the expression ratios. With our representation, relations are discovered between the most likely time points at which a gene (eventually its associated protein) is active (local maximum) and inactive (local minimum). Our approach makes it possible to establish a regulatory relation between a transcription factor with small absolute changes in expression ratio, and a target gene, because the amplitude is disregarded.

The preprocessing steps consist of 1) computation of the log-ratio per gene, 2) linear interpolation, 3) max-min filtering (optional) and 4) detection of local minima and local maxima using the derivative operator from the linear scale space. In the steps 5) and 6), the local extrema are identified and the number of observations doubled.

2.1 Interpolation

Computation of the derivatives over each gene entails the application of (linear) filters. Filtering requires that the signal be uniformly sampled over time. We use a linear nearest neighbor scheme to interpolate non-uniformly sampled time series because this scheme can never introduce new local minima or maxima. Interpolation results in a uniformly sampled time series $t,\ t \in \{1, \ldots, T\}$ of expression rations, $R_i(t)$ for gene i.

2.2 Max-Min Filter

To cope with transient changes in the first order derivative as a result of noise, we incorporate an extra (optional) preprocessing step consisting of the morphological max-min filter [5]. An advantage of the max-min filter is that non-transient extrema in the original signal are left unaffected. The max-min filter is defined as

$$K(t) = \frac{\max_{t_1 \in b(t)} \left(y = \min_{t_2 \in b(t_1)} (R(t_2)) \right) + \min_{t_1 \in b(t)} \left(y = \max_{t_2 \in b(t_1)} (R(t_2)) \right)}{2} \tag{1}$$

When the width of the window $b(t)$ exceeds zero, small inflections become saddle points, otherwise $K(t) = R(t)$.

2.3 Regularized Differentiation

Transformation of the continuous expression ratios $K_i(t)$ into the desired discrete representation: *change*, *local minimum* and *local maximum*, requires the computation of derivatives, $\partial K_i(t)/\partial t$. We use operators from the linear scale space [6, 7] to transform differentiation into a well-posed problem [8] by means of regularization. Regularized derivatives of a discrete time series are obtained by convolution with the first-order derivative of the Gaussian function

$$g\prime(t; \mu, \sigma) = \frac{-\sqrt{2}}{2\sqrt{\pi}\sigma^3} \cdot (t - \mu) \cdot \exp\left(-\frac{(t-\mu)^2}{2\sigma^2} \right) \tag{2}$$

Convolution with $g\prime$ results in

$$H(t) = g\prime(t; 0, \sigma) * K(t) = \int_{-\infty}^{\infty} g\prime(\tau; 0, \sigma) \cdot K(t - \tau)\, d\tau \tag{3}$$

When the sign of $H(t)$ changes between two consecutive time steps, $H(t-\delta) < 0$ but $H(t+\delta) > 0$, this indicates a *local minimum* whereas $H(t-\delta) > 0$ but $H(t+\delta) < 0$ indicates a *local maximum*. When there is no change in sign, the time step $H(t)$ gets the label *change*.

2.4 Data Representation and Modeling

Our goal is to identify possible co-regulatory and control-regulatory relations between sets of genes. With $\mathcal{C}(R_i, R_j)$ we indicate co-regulation between the genes R_i and R_j, whereas $\mathcal{T}(R_i \to R_j)$ indicates that gene R_i controls the regulation of gene R_j. An important difference between co-regulation and controlled regulation is that co-regulation is a commutative relation, whereas controlled regulation is assumed not to be commutative. Consequently, the inclusion of lags in the time series should, in theory, make it possible to discern putative control regulations from co-regulations. The continuous valued variable $H(t)$ is discretized by the function f. This results in a discrete time series per gene, $x_{i,t} = f(H_i(t-\delta), H_i(t+\delta))$, with $X_{i,t} = x_{i,t}$, $x_{i,t} \in \{\text{min}, \text{change}, \text{max}\}$.

We propose to model possible gene interactions using dynamic Bayesian network classifiers. In the remaining part of the paper, we use X to indicate the set of predictor genes and C the target gene. Figure 2 indicates which types of relations may be found by our approach. We use a lagged time series model in the following way. At each time

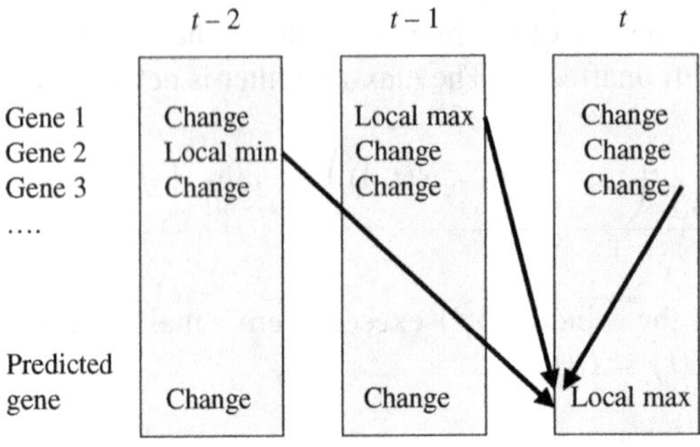

Fig. 2. Significant regulatory events of the expression of a target gene are being predicted by regulatory events pertaining to other genes earlier and at the same time as the target. Thereby, both control regulation and co-regulation with predictor genes can be modeled.

step t, the outcome of one target gene C_t, c_t, should be predicted by the outcomes of the predictor genes $x_{i,\tau}$, $\tau \in \{t-\lambda, .., t\}$ with λ, $\lambda \geq 0$, indicating the maximal lag that can be accounted for. This representation results in the following matrix of $n = r \times \lambda$ (potential) predictor variables

$$\boldsymbol{X}_t = \begin{bmatrix} x_{1,t-\lambda} & x_{1,t-\lambda+1} & \cdots & x_{1,t} \\ x_{2,t-\lambda} & \cdots & \cdots & x_{2,t} \\ \cdots & \cdots & \cdots & \cdots \\ x_{r,t-\lambda} & \cdots & \cdots & x_{r,t} \end{bmatrix} \tag{4}$$

from which the outcome of C_t is being predicted. The data $(\boldsymbol{X}_t, C_t)$, $t \in \{1, \ldots, T\}$, constitutes the basic training set. We use a datamining algorithm that performs concomitant feature and model selection in order to estimate the most likely lagged classifier model. Connections to features (possible regulator genes) that contribute to predicting the outcome of C_t are likely to be included in the model whereas genes that do not improve the predictive performance remain disconnected.

For most of the genes, a local extremum occurs much less frequently than a change. Consequently, it is likely that many correlations appear between genes of which the expression changes. To ensure that only local minima and local maxima of the target gene are being predicted, we choose to reduce the number of possible outcomes of the target variable to just two: *local minimum* and *local maximum*. To retain the three basic outcomes, we double the number of observations resulting in a final training set $\boldsymbol{D}$

- $(\boldsymbol{x}_t, c_t = \mathit{max})$, becomes $\boldsymbol{d}_s = (\boldsymbol{x}_t, c_t = \mathit{max})$ and $\boldsymbol{d}_{s+1} = (\boldsymbol{x}_t, c_t = \mathit{max})$
- $(\boldsymbol{x}_t, c_t = \mathit{min})$, becomes $\boldsymbol{d}_s = (\boldsymbol{x}_t, c_t = \mathit{min})$ and $\boldsymbol{d}_{s+1} = (\boldsymbol{x}_t, c_t = \mathit{min})$
- $(\boldsymbol{x}(t), c_t = \mathit{change})$, is doubled into an ambiguous prediction of c_t, $\boldsymbol{d}_s = (\boldsymbol{x}_t, c_t = \mathit{min})$ and $\boldsymbol{d}_{s+1} = (\boldsymbol{x}_t, c_t = \mathit{max})$

with $s = 1, \ldots, 2T-1$. Note that we do not add any information to $\boldsymbol{D}$ that is not included in the original data. Only the number of observations doubles, a fact that can easily be accounted for if one wants to estimate, e.g., the variance of the model outcome.

3 Dynamic Bayesian Network Classifiers

Before describing in detail how to build dynamic Bayesian classifiers, we briefly consider previous work. Friedman [9] pioneered with his Bayesian network approach to modelling gene interactions. A separate variable indicates the cell cycle phase, i.e., time. Husmeier [3] used a dynamic Bayesian network to model the lagged relations between genes using the likelihood of the graph as a scoring metric. Husmeier acknowledges the problem imposed by the limited size of available microarray time series. As earlier stated, we choose to predict the change in expression of individual genes by a classifier.

A probabilistic network classifier $M = (G, \boldsymbol{\theta})$ consists of a structural model specification, the directed graph G, and the parameters, $\boldsymbol{\theta}$, with the (un)conditional probability $\theta_{i,j,\pi(i)} = P(D_i = d_j \mid \pi(D_i) = \boldsymbol{d}_{\pi(D_i)})$. The notation $\pi(D_i) = \boldsymbol{d}_{\pi(D_i)}$ indicates the values of the parents of node D_i in the graph G (the parents constitute the nodes with arcs pointing directly to node D_i). Computation of the posterior probability distribution $P(C|\boldsymbol{X})$ is specified by the directed graph. It follows from the chain rule that the joint probability $P(\boldsymbol{d}) = P(c, \boldsymbol{x})$ is computed from

$$P(\boldsymbol{d}) = \prod_{i=1}^{k+1} P(D_i = d_j \mid \pi(D_i) = \boldsymbol{d}_{\pi(D_i)}) \tag{5}$$

A little manipulation of Bayes formula yields the posterior probability associated with class label c_j

$$P(c_j|\boldsymbol{x}) = \frac{P(c_j, \boldsymbol{x})}{\sum_m P(c_m, \boldsymbol{x})} \tag{6}$$

3.1 Learning Probabilistic Network Classifiers

Probabilistic network classifiers [10] have to be learned from a dataset $\boldsymbol{D}$. In the past, complete graphical models have successfully been learned using the approach introduced by Madigan & York [11]. However, their version of the MCMC-algorithm is not appropriate for learning probabilistic network classifiers, because it samples complete graphs drawn from the conditional distribution $P(G \mid \boldsymbol{D})$. Instead, we introduce a novel Markov Chain Monte Carlo technique based on the principles of Reversible Jump MCMC [12] to sample the posterior distribution probabilistic network classifiers. We make a simplification that leads to a less complex across-model sampling scheme than RJMCMC. Consequently, we can omit the Jacobian determinant term.

Let the variables in the learning database $\boldsymbol{D}$ be separated into a set of predictor variables X and a classification variable C, $D = (C, X)$. Our goal is to sample models from the following target distribution $P(L(C) \mid \boldsymbol{D})$, with $L(C)$ a score function (also called loss function [13])

$$P(L(C) \mid \boldsymbol{X}, G, \boldsymbol{\theta}^*, \boldsymbol{D}) = \prod_{\boldsymbol{d} \in \boldsymbol{D}} l(P(c \mid \boldsymbol{x}, G, \boldsymbol{\theta}^*, \boldsymbol{d}); v, \gamma) \tag{7}$$

with l the modified step function

$$l(y; v, \gamma) = \begin{cases} \gamma: & y \leq \frac{1}{2} - v \\ \frac{1}{2}: & |y - \frac{1}{2}| < v \\ 1 - \gamma: & y > \frac{1}{2} + v \end{cases} \tag{8}$$

for which it holds that $l(C = c) \in (0, 1)$, $\sum_c l(C = c) = 1$. The modified step function has two parameters, the *span of indeterminacy* v and the *bounding probability* γ. The parameter v determines the range of posterior probabilities regarded as ties, resulting in an intermediate score. The bounding probability $0.5 > \gamma > 0$ determines the gain or loss obtained by classifying a case correctly or wrongly, respectively. The score $L(C)$ can be considered a genuine probability, similar to the likelihood $P(\boldsymbol{D} \mid G)$ applied by Madigan & York. The distribution $P(L(C) \mid \boldsymbol{D})$ cannot be sampled directly, hence we perform the following factorization yielding a hierarchical Bayesian model:

$$\begin{aligned} &P(L(C) \mid \boldsymbol{D}) = \\ &\textstyle\sum_G P(L(C) \mid \boldsymbol{X}, G, \boldsymbol{D}) \\ &P(G \mid \boldsymbol{X}) P(\boldsymbol{X} \mid k, \boldsymbol{D}) P(k \mid \boldsymbol{D}) \end{aligned} \tag{9}$$

with G the directed graph, $\boldsymbol{X}$ the observations corresponding to the subset of selected predictor variables, and k the number of selected predictor variables. Computation of $P(L(C) \mid \boldsymbol{X}, G, \boldsymbol{D})$ requires a closed form solution to

$$P(L(C) \mid \boldsymbol{X}, G, \boldsymbol{D}) = \int_{\boldsymbol{\theta}} P(L(C) \mid \boldsymbol{X}, G, \boldsymbol{\theta}, \boldsymbol{D})\, P(\boldsymbol{\theta} \mid \boldsymbol{X}, G, \boldsymbol{D})\, d\boldsymbol{\theta} \tag{10}$$

in which $P(L(C) \mid \boldsymbol{X}, G, \boldsymbol{\theta}, \boldsymbol{D})$ is the probability of the score $L(C)$, given the parameter vector $\boldsymbol{\theta}$, the data associated with the predictor variables $\boldsymbol{X}$, the acyclic graph G and the database $\boldsymbol{D}$. As no closed form is presently available, we suggest to use instead $P(L(C) \mid \boldsymbol{X}, G, \boldsymbol{\theta}^*, \boldsymbol{D})$ with $\boldsymbol{\theta}^*$ the maximum-likelihood estimate of the parameter vector[1]. Note that the model G does not change as a function of $\boldsymbol{\theta}$. Since G does not depend on $\boldsymbol{\theta}$, conditioning on $\boldsymbol{\theta}^*$, the most likely parameter vector, will not strongly bias the estimate of $P(L(C) \mid \boldsymbol{X}, G, \boldsymbol{D})$. However, this approximation necessitates the use of a regularization prior. The following derivation is based on work presented elsewhere [14]. The variance of $\ln(L(C))$ equals the sum of the variances of $\ln(l(\boldsymbol{x}_i); v, \gamma)$, pertaining to the individual cases i

$$\sigma^2_{\ln(P(L(C)))} = \left(\ln\left(\frac{1}{2} + \gamma\right) - \ln\left(\frac{1}{2} - \gamma\right)\right)^2 \sum_i (p_i - p_i^2) \tag{11}$$

The probability p_i is in fact the probability per case that resampling the training set leads to the same winner resulting in $l(\boldsymbol{x}_i; v, \gamma) = 0.5 + \gamma$. Conversely, $1 - p_i$ is an error rate for a correctly classified case i. Consequently, we subtract $\sigma^2_{\ln(L(C))}$ from $\ln\{P(L(C) \mid \boldsymbol{X}, G, \boldsymbol{\theta}^*, \boldsymbol{D})\}$.

[1] This motivates our choice of score function in the first place.

The Markov Chain Monte Carlo algorithm should preferably not be biased towards a certain number of features or model complexity. Hence, we propose to use a uniform prior $P(k \mid \boldsymbol{D})$ on the size of the feature set k. For each feature set size k, each feature subset should be equally likely, so $P(\boldsymbol{X} \mid k, \boldsymbol{D})$ is also uniform. Finally, for a particular feature set, each possible model utilizing this feature subset should have the same prior, so $P(G \mid \boldsymbol{X})$ is uniform. We define the one-step look ahead neighbourhood of the graph G consisting of the directed acyclic graphs of classifiers that can be constructed by adding one arc to G or deleting one arc from G. The neighborhood $NB_C(G)$ is subdivided into four disjoint subsets

$$NB_C(G) = \{NB_C(G+1_F), NB_C(G-1_F), NB_C(G+1_M), NB_C(G-1_M)\} \quad (12)$$

The subset $NB_C(G+1_F)$ contains the graphical models in $NB_C(G)$ where the addition of an arc implies that $G\prime$ contains one feature variable more than G. The subset $NB_C(G-1_F)$ contains the models in $NB_C(G)$ where the deletion of an arc implies that $G\prime$ contains one feature variable less than G. The subset $NB_C(G+1_M)$ contains the models in $NB_C(G)$ where the addition of an arc increases the complexity of $G\prime$, but where G and $G\prime$ include the same feature variables. $NB_C(G-1_M)$ contains the models in $NB_C(G)$ where the deletion of an arc decreases the complexity of $G\prime$, but where G and $G\prime$ include the same feature variables. Define the appropriate proposal distribution q_C:

$$q_C(G \to G\prime) = \begin{cases} u < \frac{1}{4} & q_1(|NB_C(G+1_F)|^{-1}) \\ \frac{1}{4} \le u < \frac{1}{2} & q_2(|NB_C(G-1_F)|^{-1}) \\ \frac{1}{2} \le u < \frac{3}{4} & q_3(|NB_C(G+1_M)|^{-1}) \\ \frac{3}{4} \le u & q_4(|NB_C(G-1_M)|^{-1}) \end{cases} \quad (13)$$

with $u \sim \mathcal{U}(0,1)$. The proposals q_1, q_2, q_3 and q_4 result in a classifier pertaining to each of the four disjoint sub-neighborhoods, $NB_C(G+1_F)$, $NB_C(G-1_F)$, $NB_C(G+1_M)$ or $NB_C(G-1_M)$, respectively. The proposal distribution q_C implements the uniform priors, $P(G \mid \boldsymbol{X})$, $P(\boldsymbol{X} \mid k, \boldsymbol{D})$ and $P(k \mid \boldsymbol{D})$. So in each proposal, the MCMC-algorithm with the same probability chooses to add a feature, delete a feature, increase the model complexity or simplify the model (the two latter moves keep the same feature subset). The resulting Metropolis-Hastings ratio becomes

$$\frac{P(L(C) \mid \boldsymbol{X}_q, G_q, \boldsymbol{\theta}^*_q, \boldsymbol{D})\, P_q(\,(\boldsymbol{X}_q, k_q) \to (\boldsymbol{X}, k)\,)\, V}{P(L(C) \mid \boldsymbol{X}, G, \boldsymbol{\theta}^*, \boldsymbol{D})\, P_q(\,(\boldsymbol{X}, k) \to (\boldsymbol{X}_q, k_q)\,)\, V_q} \quad (14)$$

with q indicating the new proposal, the regularization terms

$$\ln(V_q) = -\alpha\, \sigma^2_{\ln(P(L(C)|X_q \cdots))} \quad \text{and} \quad \ln(V) = -\alpha\, \sigma^2_{\ln(P(L(C)|X \cdots))}.$$

The proposal probabilities, $P_q(\,(\boldsymbol{X}_q, k_q) \to (\boldsymbol{X}, k)\,)$ and $P_q(\boldsymbol{X}, k \to \boldsymbol{X}_q, k_q)$ correct for parts of the model space where one or more of the sub-neighborhoods are empty.

4 Experiments

To validate the applicability of our method on a true biological system, we used the yeast cell-cycle expression dataset from Spellman et al. [2]. The yeast cell cycle is a highly regulated process, with a central role for a class of genes named cyclins. Cyclins are transiently expressed in different phases of the cell-cycle, and team up with a cyclin-dependent kinase (CDK). Together, the cyclins and the kinases regulate the expression and/or activity of transcription factors, which in turn regulate the expression of genes that are directly involved in the diverse processes that prepare a yeast cell for division. We used an experiment where cells were initially synchronized, and subsequently followed in time as they progressed through the cell cycle.

The cyclins CLB2 and CLN3 are functional partners of the essential CDK CDC28. Clb2p/Cdc28p posttranscriptionally regulates transcription factors Mcm1p/Fkh2p through Ndd1p. We followed the expression of CDC28, CLB2, MCM1, and several target genes of MCM1/FKH2 to determine whether this genetic network could be identified using our method. Cln3p/Cdc28p are known to regulate the activity of the Swi5p transcription factor; their expression and the expression of target genes of Swi5p were analyzed. Ste12p, another transcription factor acting in concert with Mcm1p, was also analyzed in concert with some of its target genes.

To investigate the influence of our signal processing steps, parameter settings of the max-min filter and the scale-space transformation were varied, and co- and controlled regulatory events were compared to actual regulatory interactions described in the literature [15]. Finally, our co-regulatory relations were compared with the results obtained from hierarchical clustering (Euclidean distance measure). A summary of our results per target gene is presented in Table 1.

We varied the settings of the max-min filter and the scale-space transformation. In total 29 time points were sampled from the Spellman data. To obtain a data set with a uniformly sampled time, nearest neighbor linear interpolation was applied to a few time steps. This interpolation scheme was chosen because it can never introduce new extrema in the time series. Subsequently, dynamic predictor variables were extracted with lags ranging from 2, 1 and 0 time steps (with each time step corresponding to 10 minutes). As a complete time series with all three lags is required, only 27 time points were available. After preprocessing (Fig. 1), in total 54 (doubled) data points were available. The following genes were considered as targets: CLB1, BUD4, SWI4, CDC6, AGA1, ASH1, CDC45, CDC47, CTS1, FUS1 and MFA2. As predictive feature variables, the following variables were included: MCM1, STE12, CDC28, CLB2, CLN3 and SWI5. Corresponding to each target gene, the MCMC-algorithm was run 10.000 iterations. The most likely and second-most likely feature subsets occurring in the Markov chain were identified, see Table 1. We could find some co-regulations and controlled regulations with every setting applied. The max-min filter was important for the end result; when not applied, many spurious correlations were found, likely due to the relatively high noise in the signal corresponding to the lower expressed regulatory genes. The higher σ^2 was set, the more significant our results were. With σ^2 set at 4, only one false positive interaction $\mathcal{T}$(CLN3 $\rightarrow$ MFA2) was detected, yet some co-regulatory events were missed, that were apparent when σ^2 was set to 2. Since we were primarily interested in controlled regulation, we used the max-min filter set at

Table 1. Selected target genes listed in the most left column at different lag times are indicated by their names. Results consistent with co-regulation (also close in hierarchical clustering) or controlled regulation are indicated with a Y(es) in the 'valid' and 'close' columns, respectively. Spurious correlation is indicated with a N(o). The question marks indicate possible controlled regulations, where regulatory genes were co-regulated with their targets. The parentheses (...) indicate observations pertaining to the second-most likely model found by MCMC.

Target gene	Lag(0)	Valid	Close	Lag(-1, -2)	Valid
CLB1	CLB2	Y	Y		
BUD4	CLB2	Y	Y		
SWI4				CLB2	Y
CDC6				CLB2	Y
AGA1				CLB2	Y
ASH1				CLB2 (SWI5)	Y (Y)
CDC45				CLB2 (MCM1)	Y (Y)
CDC47	CDC28	N	N		
CTS1				SWI5	Y
FUS1	SWI5 (MCM1)	N (?)	N		
MFA2				CLN3 (CLB2)	N (Y)

$w = 3$, and chose σ^2 to be set at 2 in the scale space transformation. The regularization parameter α was set to 10.

5 Discussion

Our method relies solely on the timing of expression ratios of mRNAs, corresponding to the genes under investigation. It is possible to imagine that regulators, when altered in level, can change the level of their target genes at a given time in the near future. We expect the time course of regulatory events to be limited by diffusion of the molecules within the cell, and the rate of transcription of a target gene, and therefore we expect controlled regulations to occur within the time frame of minutes. Since one time point represents 7 minutes in the dataset under investigation, only the lag (0) co-regulation and lag (-1), and lag (-2) controlled regulatory events were taken into account.

Microarray data are inherently noisy, and only describe the expression of mRNA levels, ruling out the possibility to directly detect interactions due to cellular processes occurring after translation (e.g., mRNA decay, protein modifications, protein degradation). Despite these obstacles, our combination of preprocessing, coupled to selection of predictive features from a group of potential regulatory genes, allowed for robust detection of interactions.

The parameter settings of the max-min filter and the scale-space transformation had considerable influence on the genes detected in our method. Several factors can account for this. Firstly, transcription factors are expressed at a low level, resulting in a higher variation in expression due to the inherent noise of microarray experiments. The max-min filter and the scale-space transformation both smoothen these smaller variations, resulting in a trade-off between noise suppression and sensitivity. Secondly, when a higher value for σ^2 is used in the scale-space transformation, a bias is introduced which can alter the timing of regulatory events. Finally, controlled regulations occurring within a time interval of seven minutes (the sampling time in the experiment) will be detected as co-regulations. Within the limits of the experimental set-up, we cannot catch these regulatory events, a shortcoming that could be circumvented by sampling at shorter time intervals.

We identified ten controlled regulatory events in the set of genes we analyzed, of which only one correlation turned out to be spurious. Two additional co-regulated genes ($\mathcal{C}(\mathrm{CDC28}, \mathrm{CDC47})$, $\mathcal{C}(\mathrm{MCM1}, \mathrm{FUS1})$) may represent controlled regulations characterized by shorter lags than the sampling time. It is interesting to note that the expression of these genes was distant in cluster analysis, a consequence of the expression ratios being inversed in sign, yet co-regulated. An example of the latter was $\mathcal{C}(\mathrm{MCM1}, \mathrm{FUS1})$, the high frequency and regular spacing of extrema lead us to conclude that the detected correlation was due to co-linearity, because a controlled stimulatory interaction is expected to show a lagged co-regulation with extrema being of the same sign. In the future we will include prior knowledge, such as whether a regulatory gene stimulates or represses transcription of a target gene, to circumvent this problem.

In summary, we present a proof of concept for a new method to extract regulatory interactions from microarray time series data. Despite the noisy character of the data and other experimental limitations, ten out of thirteen detected control-regulatory events corresponded to published experimental data, whereas one of the three false positives can be corrected using prior knowledge. Future approaches, incorporating knowledge about biological systems, will be expected to yield an even higher predictive accuracy.

6 Conclusion

In this article, we introduced a completely new approach to discovering putative regulative relations between genes studied in time series microarray experiments. The preprocessing steps make it possible to capture dynamic relations between sets of genes. Using Markov Chain Monte Carlo sampling and a new hierarchical Bayesian model, we discover control regulations and co-regulations between sets of genes. The method works well as it results in small compact graphs that reflect experimentally verified regulatory relations between genes. The predictive variables included in the (second) most likely graphs often exert control upon the target gene. Among 15 regulatory relations found, only 2 were spurious. In the future, we will evaluate our approach further on simulation data to get more insight into the parameter settings and on other real data sets to validate the method's appropriateness.

References

1. Datta, S., Datta, S.: Comparisons and validation of statistical clustering techniques for microarray gene expression data. Bioinformatics **19** (2003) 459–466

2. Spellman, P., Sherlock, G., M.Q., Z., Iyer, V., Anders, K., Eisen, M., Brown, P., Botstein, D., Futcher, B.: Comprehensive identification of cell cycle-regulated genes of the yeast saccharomyces cerevisiae by microarray hybridization. Molecular Biology of the Cell **9** (1998) 3273–3297
3. Husmeier, D.: Sensitivity and specificity of inferring genetic regulatory interactions from microarray experiments with dynamic bayesian networks. Bioinformatics **19** (2003) 2271–2282
4. Smith, V., Jarvis, E., Hartemink, A.: Evaluating functional network inference using simulations of complex biological systems. Bioinformatics **18** (2002) S216–S224
5. Verbeek, P., Vrooman, H., van Vliet, L.: Low-level image-processing by max min filters. Signal Processing **15** (1988) 249–258
6. Florack, L., ter Haar Romeny, B., Koenderink, J., Viergever, M.: Scale and the differential structure of images. Image and Vision Computing **10** (1992) 376–388
7. Lindeberg, T.: Scale-space for discrete signals. IEEE Transactions on Pattern Analysis and Machine Intelligence **12** (1990) 234–254
8. Lindeberg, T., ter Haar Romeny, B. In: Linear scale-space II: Early visual operations. Kluwer Academic Publishers, Dordrecht (1994) 39–72
9. Friedman, N., Linial, M., Nachman, I., Pe'er, D.: Using bayesian networks to analyze expression data. Journal of Computational Biology **7** (2000) 601–620
10. Friedman, N., Geiger, D., Goldszmidt, M.: Bayesian network classifiers. Machine learning **29** (1997) 131–163
11. Madigan, D., York, J.: Bayesian graphical models for discrete-data. International statistical review **63** (1995) 215–232
12. Green, P.: Reversible jump markov chain monte carlo computation and bayesian model determination. Biometrika **82** (1995) 711–732
13. Duda, R., Hart, P.: Pattern classification and scene analysis. John Wiley & Sons, New York (1973)
14. Egmont-Petersen, M., Feelders, A., Baesens, B.: Confidence intervals for probabilistic network classifiers. To appear in Computational Statistics and Data Analysis (2004)
15. Mendenhall, M., Hodge, A.: Regulation of cdc28 cyclin-dependent protein kinase activity during the cell cycle of the yeast *saccharomyces cerevisiae*. Microbiology and Molecular Biology Reviews **62** (1998) 1191–1243

Learning from Little: Comparison of Classifiers Given Little Training

George Forman and Ira Cohen

Hewlett-Packard Research Laboratories
1501 Page Mill Rd., Palo Alto, CA 94304
{ghforman,icohen}@hpl.hp.com

Abstract. Many real-world machine learning tasks are faced with the problem of small training sets. Additionally, the class distribution of the training set often does not match the target distribution. In this paper we compare the performance of many learning models on a substantial benchmark of binary text classification tasks having small training sets. We vary the training size and class distribution to examine the *learning surface*, as opposed to the traditional *learning curve*. The models tested include various feature selection methods each coupled with four learning algorithms: Support Vector Machines (SVM), Logistic Regression, Naive Bayes, and Multinomial Naive Bayes. Different models excel in different regions of the learning surface, leading to meta-knowledge about which to apply in different situations. This helps guide the researcher and practitioner when facing choices of model and feature selection methods in, for example, information retrieval settings and others.

1 Motivation and Scope

Our goal is to advance the state of meta-knowledge about selecting which learning models to apply in which situations. Consider these four motivations:

1. Information Retrieval: Suppose you are building an advanced search interface. As the user sifts through the each page of ten search results, it trains a classifier on the fly to provide a ranking of the remaining results based on the user's positive or negative indication on each result shown thus far. *Which learning model should you implement to provide greatest precision under the conditions of little training data and a markedly skewed class distribution?*

2. Semi-supervised Learning: When learning from small training sets, it is natural to try to leverage the many unlabeled examples. A common first phase in such algorithms is to train an initial classifier with the little data available and apply it to select additional predicted-positive examples and predicted-negative examples from the unlabeled data to augment the training set before learning the final classifier (e.g. [1]). With a poor choice for the initial learning model, the augmented examples will pollute the training set. *Which learning model is most appropriate for the initial classifier?*

J.-F. Boulicaut et al. (Eds.): PKDD 2004, LNAI 3202, pp. 161–172, 2004.

Table 1. Summary of test conditions we vary.

P = 1..40	Positives in training set	Feature Selection Metrics:	
N = 1..200	Negatives in training set	IG	Information Gain
FX= 10..1000	Features selected	BNS	Bi-Normal Separation
Performance Metrics:		Learning Algorithms:	
TP10	True positives in top 10	NB	Naive Bayes
TN100	True negatives in bottom 100	Multi	Multinomial Naive Bayes
F-measure	2×precision×recall÷(precision+recall)	Log	Logistic Regression
	(harmonic avg. of precision & recall)	SVM	Support Vector Machine

3. Real-World Training Sets: In many real-world projects the training set must be built up from scratch over time. The period where there are only a few training examples is especially long if there are many classes, e.g. 30–500. Ideally, one would like to be able train the most effective classifiers at any point. *Which methods are most effective with little training?*

4. Meta-knowledge: Testing all learning models on each new classification task at hand is an agnostic and inefficient route to building high quality classifiers. The research literature must continue to strive to give guidance to the practitioner as to which (few) models are most appropriate in which situations. Furthermore, in the common situation where there is a shortage of training data, cross-validation for model selection can be inappropriate and will likely lead to over-fitting. Instead, one may follow the *a priori* guidance of studies demonstrating that some learning models are superior to others over large benchmarks.

In order to provide such guidance, we compare the performance of many learning models (4 induction algorithms × feature selection variants) on a benchmark of hundreds of binary text classification tasks drawn from various benchmark databases, e.g, Reuters, TREC, and OHSUMED.

To suit the real-world situations we have encountered in industrial practice, we focus on tasks with small training sets and a small proportion of positives in the test distribution. Note that in many situations, esp. information retrieval or fault detection, the ratio of positives and negatives provided in the training set is unlikely to match the target distribution. And so, rather than explore a learning *curve* with matching distributions, we explore the entire learning *surface*, varying the number of positives and negatives in the training set independently of each other (from 1 to 40 positives and 1 to 200 negatives). This contrasts with most machine learning research, which tests under conditions of (stratified) cross-validation or random test/train splits, preserving the distribution.

The learning models we evaluate are the cross product of four popular learning algorithms (Support Vector Machines, Logistic Regression, Naive Bayes, Multinomial Naive Bayes), two highly successful feature selection metrics (Information Gain, Bi-Normal Separation) and seven settings for the number of top-ranked features to select, varying from 10 to 1000.

We examine the results from several perspectives: precision in the top-ranked items, precision for the negative class in the bottom-ranked items, and F-measure, each being appropriate for different situations. For each perspective, we determine which models consistently perform well under varying amounts of training data. For example, Multinomial Naive Bayes coupled with feature selection via Bi-Normal Separation can be closely competitive to SVMs for precision, performing significantly better when there is a scarcity of positive training examples.

The rest of the paper is organized as follows. The remainder of this section puts this study in context to related work. Section 2 details the experiment protocol. Section 3 gives highlights of the results with discussion. Section 4 concludes with implications and future directions.

1.1 Related Work

There have been numerous controlled benchmark studies on the choice of feature selection (e.g. [2]) and learning algorithms (e.g. [3]). Here, we study the cross-product of the two together, and the results bear out that different algorithms call for different feature selection methods in different circumstances. Further, our study examines the results both for maximizing F-measure and for maximizing precision in the top (or bottom) ranked items – metrics used in information retrieval and recommenders.

A great deal of research is based on 5-fold or 10-fold cross-validation, which repeatedly trains on 80–90% of the benchmark dataset. In contrast, our work focuses on learning from *very small* training sets – a phase most training sets go through as they are being built up. Related work by [4] shows that Naive Bayes often surpasses Logistic Regression in this region for UCI data sets. We extend these results to the text domain and to other learning models.

We vary the number of positives and negatives in the training set as independent variables, and examine the learning surface of the performance for each model. This is most akin to the work by [5], in which they studied the effect of varying the training distribution and size (an equivalent parameterization to ours) for the C4.5 decision tree model on a benchmark of UCI data sets, which are not in the text domain and do not require feature selection. They measured performance via accuracy and area under the ROC curve. We measure the performance at the middle (F-measure) and both extreme ends of the ROC curve (precision in the top/bottom scoring items) – metrics better focused on practical application to information retrieval, routing, or semi-supervised learning. For example, when examining search engine results, one cares about precision in the top displayed items more than the ROC ranking of *all* items in the database.

The results of studies such as ours and [5] can be useful to guide research in developing methods for learning under greatly unbalanced class distributions, for which there is a great deal of work [6]. Common methods involve over-sampling the minority class or under-sampling the majority class, thereby manipulating the class distribution in the training set to maximize performance. Our study elucidates the effect this can have on the learning surface for many learning models.

Table 2. Description of benchmark datasets.

Dataset	Cases	Features	Classes	Dataset	Cases	Features	Classes
Whizbang/Cora	1800	5171	36	TREC/fbis	2463	2000	17
OHSUMED/Oh0	1003	3182	10	TREC/La1	3204	31472	6
OHSUMED/Oh5	918	3012	10	TREC/La2	3075	31472	6
OHSUMED/Oh10	1050	3238	10	TREC/tr11	414	6429	9
OHSUMED/Oh15	913	3100	10	TREC/tr12	313	5804	8
OHSUMED/ohscal	11162	11465	10	TREC/tr21	336	7902	6
Reuters/Re0	1504	2886	13	TREC/tr23	204	5832	6
Reuters/Re1	1657	3758	25	TREC/tr31	927	10128	7
WebACE/wap	1560	8460	20	TREC/tr41	878	7454	10
				TREC/tr45	690	8261	10

2 Experiment Protocol

Here we describe how the study was conducted, and as space allows, why certain parameter choices were made. Table 1 shows the parameter settings we varied and defines the abbreviations we use hereafter.

Learning Algorithms: We evaluate the learning algorithms listed in Table 1 using the implementation and default parameters of the WEKA machine learning library (version 3.4) [7].

Naive Bayes is a simple generative model in which the features are assumed to be independent of each other given the class variable. Despite its unrealistic independence assumptions, it has been shown to be very successful in a variety of applications and settings (e.g. [8, 9]). The multinomial variation of Naive Bayes was found by [10] to excel in text classification.

Regularized logistic regression is a commonly used and successful discriminative classifier in which a class a-posteriori probability is estimated from the training data using the logistic function [11]. The WEKA implementation is a multinomial logistic regression model with a ridge estimator, believed to be suitable for small training sets.

The Support Vector Machine, based on risk minimization principles, has proven well equipped in classifying high-dimensional data such as text [2, 12, 3]. We use a linear kernel and varied the complexity constant C; all of the results presented in this paper are for C=1 (WEKA's default value); a discussion of the results when we varied C is given in the discussion section. (The WEKA v3.4 implementation returns effectively boolean output for two-class problems, so we had to modify the code slightly to return an indication of the Euclidean distance from the separating hyperplane. This was essential to get reasonable TP10 and TN100 performance.)

Feature Selection: Feature selection is an important and often under estimated component of the learning model, as it accounts for large variations in performance. In this study we chose to use two feature selection metrics, namely Information Gain (IG) and Bi-Normal Separation (BNS). We chose these two based on a comparative study of a dozen features ranking metrics indicating that

Table 3. Experiment procedure.

```
For each of the 19 multi-class dataset files:
 For each of its classes C
      where there are >=50 positives (C) and >=250 negatives (not C):
  For each of 5 random split seeds:
   Randomly select 40 positives and 200 negatives for set MaxTrain,
        leaving the remaining cases in the testing set.
   For P = 1..40:
    For N = 1..200:
     Select as the training set the first P positives
         and the first N negatives from MaxTrain.
     // Task established.  Model parameters follow. //
     For each feature selection metric IG, BNS:
       Rank the features according to the feature selection metric
             applied to the training set only.
       For FX = 10,20,50,100,200,500,1000:
         Select the top FX features.
         For each algorithm SVM, Log, NB, Multi:
           Train on the training set of P positives and N negatives.
           Score all items in the testing set.
           For each performance measure TP10, F-measure, TN100:
             Measure performance.
```

these two are top performers for SVM; classifiers learned with features selected via IG tended to have better precision; whereas BNS proved better for recall, overall improving F-measure substantially [2]. We expected IG to be superior for the goal of precision in the top ten.

Benchmark Data Sets: We used the prepared benchmark datasets available from [2, 13], which stem originally from benchmarks such as Reuters, TREC, and OHSUMED. They comprise 19 text datasets, each case assigned to a single class. From these we generate many binary classification tasks identifying one class as positive vs. all other classes. Over all such binary tasks, the median percentage of positives is ~5%. We use binary features, representing whether the (stemmed) word appears at all in the document. The data are described briefly in Table 2. For more details, see Appendix A of [2].

Experiment Procedure: The experiment procedure is given in Table 3 as pseudo-code. Its execution consumed ~5 years of computation time, run on hundreds of CPUs in the HP Utility Data Center. Overall there are 153 2-class data sets which are randomly split five times yielding 765 binary tasks for each P and N. The condition on the second loop ensures that there are 40 positives available for training plus at least 10 others in the test set (likewise, 200 negatives for training and at least 50 for testing). Importantly, feature selection depends only on the training set, and does not leak information from the test set.

Because the order of items in MaxTrain is random, later selecting the *first* P or N cases amounts to random selection; this also means that the performance for (P=3,N=8) vs. (P=**4**,N=8) represents the same test set, and the same train-

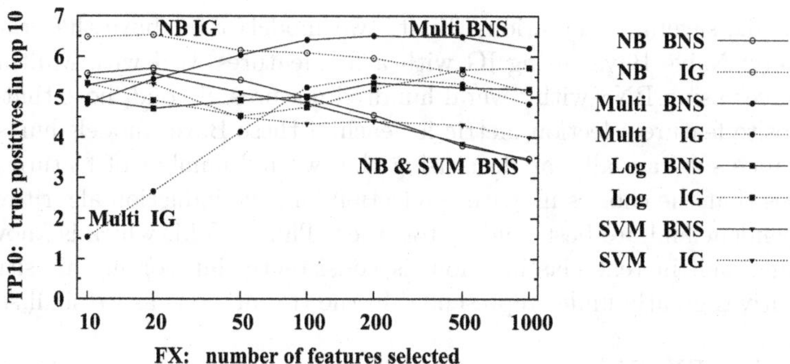

Fig. 1. Average TP10 performance for each learning model given P=5 positives and N=200 negatives, varying the number of features selected FX. *(For more readable color graphs, see http://www.hpl.hp.com/techreports/2004/HPL-2004-19R1.html)*

ing set with one additional positive added at random, i.e. they may be validly interpreted as steps on a learning curve.

Performance Metrics: We analyze the results independently for each of the following metrics:

1. The TP10 metric is the number of true positives found in the 10 test cases that are predicted most strongly by the classifier to be positive.

2. The TN100 metric is the number of true negatives found in the 100 test cases most predicted to be negative. Because of the rarity of positives in the benchmark tasks, scores in the upper 90's are common (TN10 is nearly always 10, hence the use of TN100).

3. F-measure is the harmonic average of precision and recall for the positive class. It is superior to grading classifiers based on accuracy (or error rate) when the class distribution is skewed.

Different performance metrics are appropriate in different circumstances. For recommendation systems and information retrieval settings, where results are displayed to users incrementally with the most relevant first, the metric TP10 is most appropriate. It represents the precision of the first page of results displayed. For information filtering or document routing, one cares about both the precision and the recall of the individual hard classification decisions taken by the classifier. F-measure is the metric of choice for considering both together. For semi-supervised learning settings where additional positive (negative) cases are sought in the unlabeled data to expand the training set, the appropriate metric to consider is TP10 (TN100). Maximum precision is called for in this situation, or else the heuristically extended training set will be polluted with noise labels.

3 Experiment Results

We begin by examining an example set of results for the average TP10 performance over all benchmark tasks, where the training set has P=5 positives and N=200 negatives. See Fig.1. We vary the number of features selected along the logarithmic x-axis.

We make several observations. First, two models rise above the others by ~15% here: Naive Bayes using IG with a few features, tied with Multinomial Naive Bayes using BNS with several hundred features. Second, note that using the opposite feature selection metric for each of these Bayes models hurts their performance substantially, as does using the 'wrong' number of features. This illustrates that the choices in feature selection and the induction algorithms are interdependent and are best studied together. Third, SVM, which is known for performing well in text classification, is consistently inferior in this situation with positives greatly under represented in the training set, as we shall see.

Condensing FX Dimension: These results are for only a single value of P and N. Given the high dimensionality of the results, we condense the FX dimension hereafter, presenting only the best performance obtained over any choice for FX. Because this maximum is chosen based on the *test* results, this represents an upper bound on what could be achieved by a method that attempts to select an optimal value of FX based on the training set alone. Condensing the FX dimension allows us to expose the differences in performance depending on the learning algorithm and feature selection metric. Furthermore, for the practitioner, it is typically easy to vary the FX parameter, but harder to change the implemented algorithm or feature selection metric.

Visualization: With this simplification, we can vary the number of positives P and negatives N in the training set to derive a *learning surface* for each of the 8 combinations of algorithm and feature selection metric. We begin by illustrating a 3-D perspective in Fig.2a showing the learning surfaces for just three learning models: Multinomial Naive Bayes, SVM and Logistic Regression, each with BNS feature selection. The performance measure here is the average number of true positives identified in the top 10 (TP10). From this visualization we see that Log-BNS is dominated over the entire region, and that between the remaining two models, there are consistent regions where each performs best. In particular, the SVM model substantially under performs the Mulinomial Naive Bayes when there are very few positives. The two are competitive with many positives and negatives.

This 3-D perspective visualization becomes difficult if we display the surfaces for each of the eight learning models. To resolve this, we plot all surfaces together and then view the plot from directly above, yielding the topo-map visualization shown in Fig.2b. This reveals only the best performing model in each region. The visualization also indicates the absolute performance (z-axis) via isoclines, like a geographical topo-map.

While a topo-map shows the model that performed *best* for each region, it does not make clear *by how much* it beat competing models. For this, we show two z-axis cross-sections of the map near its left and right edges. Figures 2c–d fix the number of negatives at N=10 and N=200, comparing the performance of all eight models as we vary the number of positives P. Recall that by design the test set for each benchmark task is fixed as we vary P and N, so that we may view these as learning curves as we add random positive training examples.

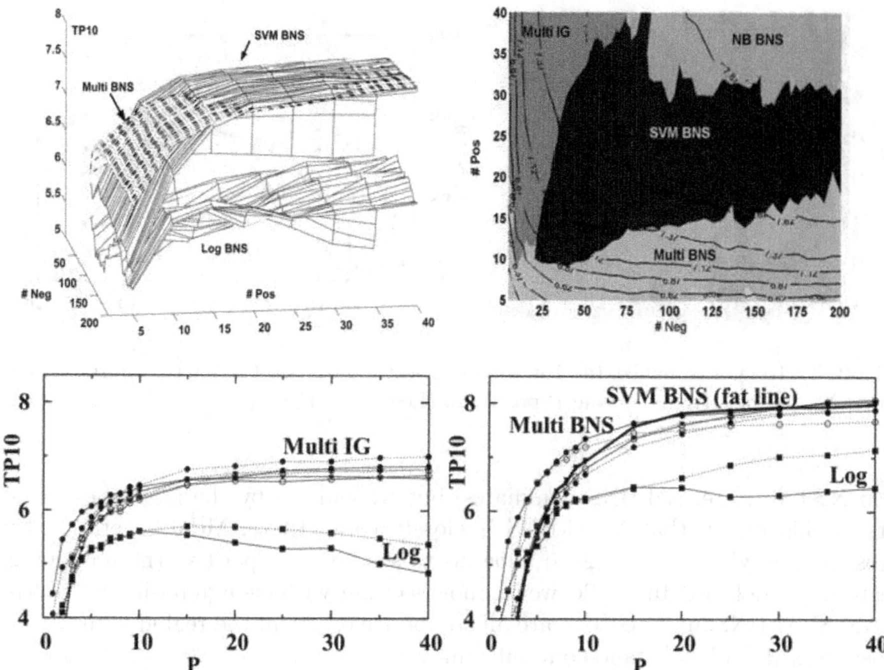

Fig. 2. TP10 performance. (a) Learning surfaces for three models (SVM-BNS, Multi-BNS and Log-BNS) as we vary the number of positives and negatives in the training set. (b) Topo-map of best models – 3D surfaces of all models viewed from above. Isoclines show identical TP10 performance. (c) Cross-section at N=10 negatives, varying P positives. (d) Cross-section at N=200 negatives, varying P positives. (Legend in Fig.1.)

TP10 Results: Our initial impetus for this study was to determine which learning models yield the best precision in the top 10 given little training data. We find that the answer varies as we vary the number of positives and negatives, but that there are consistent regions where certain models excel. This yields meta-knowledge about when to apply different classifiers. See Fig.2b. We observe that BNS is generally the stronger feature selection metric, except roughly where the number of positives exceeds the number of negatives in the training set along the y-axis, where Multi-IG dominates.

Recall that the test distribution has only a few percent positives, as is common in many tasks. So, a random sample would fall in the region near the x-axis having <10% positives where Multi-BNS dominates. Observe by the horizontal isoclines in most of this region that there is little to no performance improvement for increasing the number of negatives. In this region, the best action one can take to rapidly improve performance is to provide more *positive* training examples (and similarly near the y-axis).

The isoclines show that the best TP10 performance overall can be had by providing a training set that has an over representation of positives, say P>30

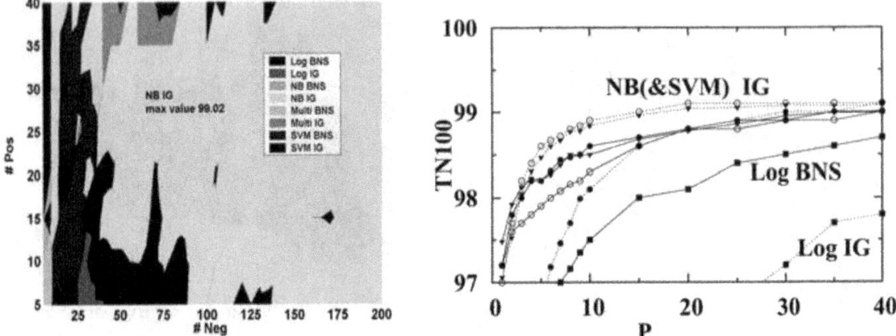

Fig. 3. TN100 performance. (a) Topo-map of best models for TN100. (b) Cross-section given N=200 negatives, varying P positives. (Legend in Fig.1.)

and N>100. Here, NB-BNS dominates, but we can see by the mottling of colors in this region that Multi-BNS is closely competitive. More generally, the cross-section views in Figs.2c–d allow us to see how competitive the remaining learning models are. In Fig.2c, we can now see that with enough positives, Multi-BNS, SVM-BNS and NB-BNS are all competitive, but in the region with <=15 positives, Multi-BNS stands out substantially.

TN100 Results: We also determined which learning model yields the most true negatives in the bottom-ranked list of 100 test items. Although no model is dominant everywhere, NB-IG is a consistent performer, as shown in Fig.3a, especially with many negatives. In the N=200 cross-section shown in Fig.3b, we see that it and SVM-IG substantially outperform the other models. Overall, performance is very high (over 98% precision) – a fact that is not surprising considering that there are many more negatives than positives in the test sets. Unfortunately, no further performance improvement is attained after ~20 positives (though one may speculate for P>40).

F-Measure Results: Next we compare the learning models by F-measure. Figure 4a shows the topo-map of the best performing models over various regions. Observing the isoclines, the greatest performance achieved is by SVM-BNS, with appropriate oversampling of positives. If random sampling from the test distribution, most labels found will be negative, and put us in the region of poor F-measure performance along the x-axis, where NB-IG dominates. We see in the cross-section in Fig.4b with P=5 fixed and varying the number of negatives N, that NB-IG dominates all other models by a wide margin and that its performance plateaus for N>=30 whereas most other models experience declining performance with increasing negatives. Likewise, in Fig.4c with N=10 fixed, the performance of all models declines as we increase the number of positives. Finally, in Fig.4d with N=200 fixed, we see that substantial gains are had by SVM with either feature selection metric as we obtain many positive training examples. With few positives, such as obtained by random sampling, SVM becomes greatly inferior to NB-IG or Multi-BNS.

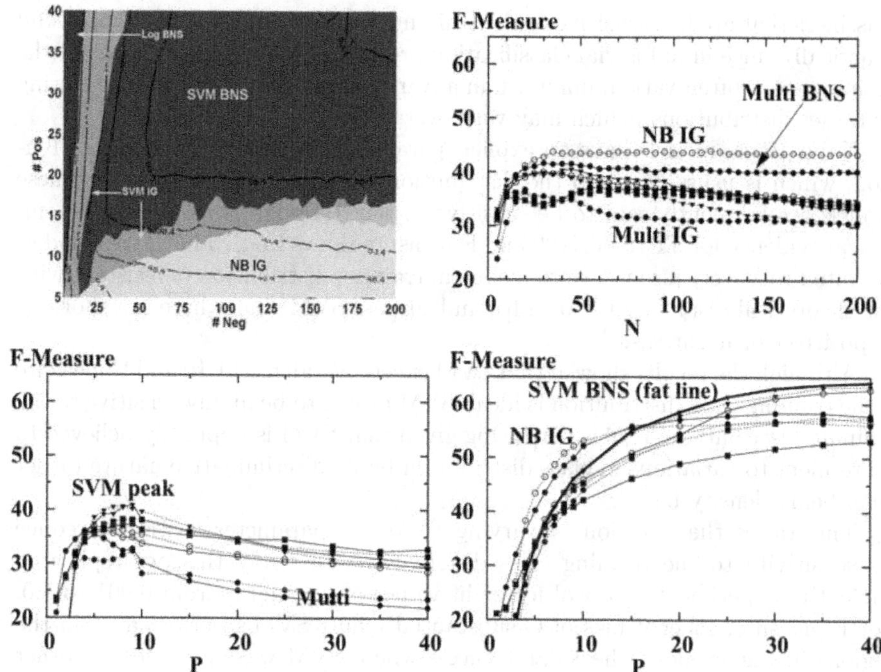

Fig. 4. F-measure performance. (a) Topo-map of best models for F-measure. (b) Cross-section at P=5 positives, varying N negatives. (c) Cross-section at N=10 negatives, varying P positives. (d) Cross-section at N=200 negatives, varying P positives. (Legend in Fig.1.)

Observing the isoclines in Fig.4a, the best approach for maximizing F-measure at all times while building a training set incrementally from scratch is to use SVM (-BNS or else -IG) and keep the class distribution of the training set at roughly 20% positives by some non-random sampling method.

3.1 Discussion

Generalizing from the notion of a learning curve to a learning *surface* proves to be a useful tool for gaining insightful meta-knowledge about regions of classifier performance. We saw that particular classifiers excel in different regions; this may be constructive advice to practitioners who know in which region they are operating. The learning surface results also highlight that performance can be greatly improved by non-random sampling that somewhat favors the minority class on tasks with skewed class distributions (however, balancing P=N is unfavorable). This is practical for many real-world situations, and may substantially reduce training costs to obtain satisfactory performance.

Because of the ubiquitous research practices of random sampling and cross-validation, however, researchers routinely work with a training set that matches the distribution of the test set. This can mask the full potential of classifiers under study. Furthermore, it hides from view the research opportunity to develop classifiers that are less sensitive to the training distribution. This would be useful practically since in industrial classification problems the class distribution of the training set is often varied, unknown in advance, and does not match the testing or target distributions, which may vary over time.

Naive Bayes models have an explicit parameter reflecting the class distribution, which is ususally set to the distribution of the training set. Hence, these models are frequently said to be sensitive to the training distribution. The empirical evidence for TP10, TN100 and F-measure shows that Naive Bayes models are often relatively insensitive to a shift in training distribution (consistent with the theoretical results by Elkan [14]), and surpass SVM when there is a shortage of positives or negatives.

Although the results showed that SVM excels overall for TP10 and F-measure if the training class distribution is ideal, SVM proves to be highly sensitive to the training distribution. This is surprising given that SVM is popularly believed to be resilient to variations in class distribution by its discriminative nature rather than being density-based.

This raises the question of varying SVM's C parameter to try to reduce its sensitivity to the training class distribution. To study this, we replicated the entire experiment protocol for eight values of C ranging from 0.001 to 5.0. For F-measure, other values of C substantially hurt SVM's performance in the regions in Fig.4a along the x- and y-axes where SVM was surpassed by other models. In the large region where SVM already dominates, however, some values of C increased performance – at the cost of making the learning surface *more* sensitive to the training distribution. For P=40 and N=200, F-measure can be increased by as much as ~5% when C=0.1, but then its performance declines drastically if the number of positives is reduced below 20.

The additional results were similar for TP10. Refer to the topo-map in Fig.2b. No value of C made SVM surpass the performance of Multi-BNS in the region along the x-axis. At the upper right, performance could be increased by fortunate choices for C (by as much as ~3% for P=40 and N=200, which exceeds the performance of NB-BNS slightly). This boost comes at the cost of very poor performance in the lower region along the x-axis. Finally, some values of C made SVM competitive with Multi-IG in the top left, but again made performance much worse as we decrease the number of positives.

Overall, varying C does not lead to fundamentally different conclusions about the regions of performance. It does not address the issue of making the choice of classifier insensitive to the operating region. Furthermore, in regions where performance can be improved, it remains to be seen whether the optimal C value can be determined automatically via cross-validation. With small training sets, such cross-validation may only lead to over fitting the training set, without practical improvement.

4 Summary

This paper compared the performance of different classifiers with settings often encountered in real situations: small training sets especially scarce in positive examples, different test and train class distributions, and skewed distributions of positive and negative examples. Visualizing the performance of the different classifiers using the learning surfaces provides meta-information about which models are consistent performers under which conditions. The results showed that feature selection should not be decoupled from the model selection task, as different combinations are best for different regions of the learning surface.

Future work potentially includes expanding the parameters, classifiers and datasets studied, and validating that the meta-knowledge successfully transfers to other text (or non-text) classification tasks.

References

1. Liu, B., Dai, Y., Li, X., Lee, W.S., Yu, P.S.: Building text classifiers using positive and unlabeled examples. In: Intl. Conf. on Data Mining. (2003) 179–186
2. Forman, G.: An extensive empirical study of feature selection metrics for text classification. Journal of Machine Learning Research **3** (2003) 1289–1305
3. Yang, Y., Liu, X.: A re-examination of text categorization methods. In: ACM SIGIR Conf. on Research and Development in Information Retrieval. (1999) 42–49
4. Ng, A.Y., Jordan, M.I.: On discriminative vs. generative classifiers: A comparison of logistic regression and naive Bayes. In: Neural Information Processing Systems: Natural and Synthetic. (2001) 841–848
5. Weiss, G.M., Provost, F.: Learning when training data are costly: The effect of class distribution on tree induction. Journal of Artificial Intelligence Research **19** (2003) 315–354
6. Japkowicz, N., Holte, R.C., Ling, C.X., Matwin, S., eds.: AAAI Workshop: Learning from Imbalanced Datasets, TR WS-00-05, AAAI Press (2000)
7. Witten, I.H., Frank, E.: Data Mining: Practical machine learning tools with Java implementations. Morgan Kaufmann, San Francisco (2000)
8. Duda, R.O., E.Hart, P.: Pattern Classification and Scene Analysis. John Wiley and Sons (1973)
9. Domingos, P., Pazzani, M.: Beyond independence: conditions for the optimality of the simple Bayesian classifier. In: Proc. 13th International Conference on Machine Learning. (1996) 105–112
10. McCallum, A., Nigam, K.: A comparison of event models for naive Bayes text classification. In: AAAI-98 Workshop on Learning for Text Categorization. (1998)
11. le Cessie, S., van Houwelingen, J.: Ridge estimators in logistic regression. Applied Statistics **41** (1992) 191–201
12. Joachims, T.: Text categorization with support vector machines: Learning with many relevant features. In: European Conf. on Machine Learning. (1998) 137–142
13. Han, E., Karypis, G.: Centroid-based document classification: Analysis & experimental results. In: Conference on Principles of Data Mining and Knowledge Discovery. (2000) 424–431
14. Elkan, C.: The foundations of cost-sensitive learning. In: International Joint Conference on Artificial Intelligence. (2001) 973–978

Geometric and Combinatorial Tiles in 0–1 Data

Aristides Gionis, Heikki Mannila, and Jouni K. Seppänen

Helsinki Institute for Information Technology,
University of Helsinki and Helsinki University of Technology, Finland

Abstract. In this paper we introduce a simple probabilistic model, hierarchical tiles, for 0–1 data. A basic tile (X, Y, p) specifies a subset X of the rows and a subset Y of the columns of the data, i.e., a rectangle, and gives a probability p for the occurrence of 1s in the cells of $X \times Y$. A hierarchical tile has additionally a set of exception tiles that specify the probabilities for subrectangles of the original rectangle. If the rows and columns are ordered and X and Y consist of consecutive elements in those orderings, then the tile is geometric; otherwise it is combinatorial. We give a simple randomized algorithm for finding good geometric tiles. Our main result shows that using spectral ordering techniques one can find good orderings that turn combinatorial tiles into geometric tiles. We give empirical results on the performance of the methods.

1 Introduction

The analysis of large 0–1 data sets is an important area in data mining. Several techniques have been developed for analysing and understanding binary data; association rules [3] and clustering [15] are among the most well-studied. Typical problems in association rules is that the correlation between items is defined with respect to arbitrarily chosen thresholds, and that the large size of the output makes the results difficult to interpret. On the other hand, clustering algorithms define distances between points with respect to all data dimensions, making it possible to ignore correlations among subsets of dimensions – an issue that has been addressed with subspace-clustering approaches [1, 2, 7, 8, 12].

One of the crucial issues in data analysis is finding good and understandable models for the data. In the analysis of 0–1 data sets, one of the key questions can be formulated simply as "Where are the ones?". That is, one would like to have a simple, understandable, and reasonably accurate description of where the ones (or zeros) in the data occur.

We introduce a simple probabilistic model, hierarchical tiles, for 0–1 data. Informally, the model is as follows. A *basic tile* $\tau = (X, Y, p)$ specifies a subset X of the rows and a subset Y of the columns of the data, i.e., a rectangle, and gives a probability p for the occurrence of 1 in the cells of $X \times Y$. A *hierarchical tile* τ consists of a basic tile plus a set of exception tiles, i.e., $\tau = (\tau_0, \{\tau_1, \ldots, \tau_k\})$, where each τ_i is a tile. The tiles $\tau_1, \ldots, \tau_k$ are assumed to be defined on disjoint subrectangles of τ_0. For an illustrative example, actually computed by our algorithm on one of our real data sets, see Figure 1. Given a point $(x, y) \in X \times Y$, the tile τ predicts the probability associated with τ_0, unless (x, y) belongs to the subset defined by an exception tile τ_i, for some $i \geq 1$; in this case the prediction is the prediction given by that particular τ_i. Thus a

J.-F. Boulicaut et al. (Eds.): PKDD 2004, LNAI 3202, pp. 173–184, 2004.

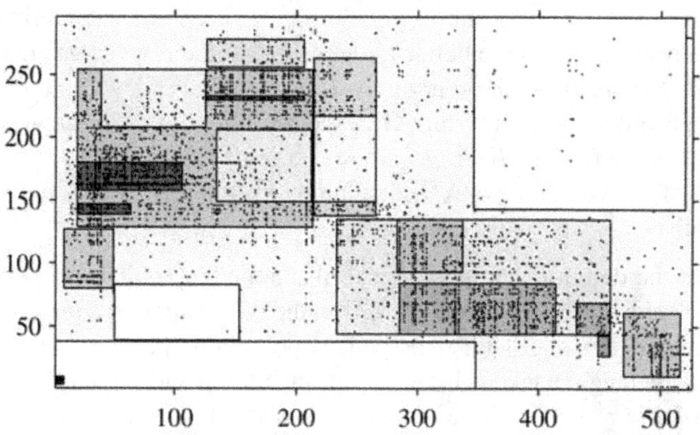

Fig. 1. Hierarchical tiling obtained for one of the data sets, `Paleo2`. The darkness of each rectangle depicts the associated probability.

hierarchical tile for which τ_0 covers the whole set $X \times Y$ defines a probability model for the set[1].

There are two types of tiles. If the rows and columns are ordered and X and Y are ranges on those orderings, then the tile is *geometric*; if X and Y are arbitrary subsets then the tile is *combinatorial*. Given a data set with n rows and m columns, there are $\Theta(n^2m^2)$ possible geometric basic tiles, but $\Theta(2^n2^m)$ possible combinatorial basic tiles. Thus combinatorial tiles are a much stronger concept, and finding the best combinatorial tiles is much harder than finding the best geometric tiles.

In this paper we first give a simple randomized algorithm for finding geometric tiles. We show that the algorithm finds with high probability the tiles in the data. We then move to the question of finding combinatorial tiles. Our main tool is spectral ordering, based on eigenvector techniques [9]. We prove that using spectral ordering methods one can find orderings on which good combinatorial tiles become geometric. We evaluate the algorithms on real data, and indicate how the tiling model gives accurate and interpretable results. The rest of the paper is organized as follows. In Section 2 we define formally the problem of hierarchical tiling, and in Section 3 we describe our algorithms. We present our experiments in Section 4, and in Section 5 we discuss the related work. Finally, Section 6 is a short conclusion.

2 Problem Description

The input to the problem consists of a 0–1 data matrix A with m rows R and n columns C. For row i and column j, the (i, j) entry of A is denoted by $A(i, j)$.

Rectangles. As we already mentioned, we distinguish between combinatorial and geometric rectangles. A *combinatorial rectangle* $r_c(A, X, Y)$ of the matrix A, defined for

[1] Our model can easily be extended to the case where each basic tile has a probability parameter for each column in Y; this leads the model to the direction of subspace clustering. For simplicity of exposition we use the formulation of one parameter per basic tile.

a subset of rows $X \subseteq R$ and a subset of columns $Y \subseteq C$, is a submatrix of A on X and Y. Geometric rectangles are defined assuming that the rows R and columns C of A are *ordered*. We denote such ordering by $R = \langle r_1, \ldots, r_m \rangle$ with $r_1 < \ldots < r_m$, where '$<$' is an ordering relationship. Given an ordering on R, a *range* X of R is a subset of *consecutive* rows of R. A *geometric rectangle* $r_g(A, X, Y)$ is now defined as the submatrix of A over the rows X and columns Y, where X and Y are ranges of R and C, respectively.

Tiles. To make the definition of hierarchical tiles noncircular we use a concept of the level, which tells how deep the nesting is. Given the data matrix A, a *basic tile*, or *level-0 tile* τ^0 is a rectangle r of A with an associated probability p, i.e., $\tau^0 = (r, p)$. Entries of A inside the rectangle r take value 1 with probability p and value 0 with probability $1-p$. A *level-k tile* τ^k consists of a basic tile and a set of exception tiles; the exception tiles are of level at most $k-1$. We write $\tau^k = (\tau^0, \{\tau_1, \ldots, \tau_m\})$, where $\tau^0 = (r, p)$ and each τ_i is a tile of level at most $k-1$. We require that the exception tiles $\tau_1, \ldots, \tau_m$ are *disjoint* and they are *contained* in τ^0. Finally, with each tile we associate a *domain*. The domain $\mathrm{Dom}(\tau^0)$ of a basic tile $\tau^0 = (r, p)$ is the rectangle r. The domain $\mathrm{Dom}(\tau^k)$ of a level-k tile $\tau^k = (\tau^0, \{\tau_1, \ldots, \tau_m\})$ is the domain of τ^0.

Prediction and likelihood. Given a position (i, j) in the data matrix A, the *prediction* $q(\tau^k, i, j)$ of a tile τ^k for (i, j) is defined recursively. For a basic tile $\tau^0 = (r, p)$ the prediction $q(\tau^0, i, j)$ is p (a basic tile predicts what it says). If $\tau^k = ((r, p), \{\tau_1, \ldots, \tau_m\})$, then $q(\tau^k, i, j) = p$, if $(i, j) \notin \bigcup_{l=1}^{m} \mathrm{Dom}(\tau_l)$ (if (i, j) is outside all exception tiles of τ^k). Otherwise, let t be the index such that $(i, j) \in \mathrm{Dom}(\tau_t)$; then $q(\tau^k, i, j) = q(\tau_t, i, j)$ (the prediction of the tile is the prediction of its exception that contains (i, j)). Let $A(r) = \{A(i, j) \mid (i, j) \in r\}$ be the restriction of data matrix A on the rectangle r. Given a tile $\tau^k = ((r, p), \{\tau_1, \ldots, \tau_m\})$ the likelihood of data $A(r)$ given τ^k is defined in the normal way:

$$L(A(r) \mid \tau^k) = \prod_{i,j} q(\tau^k, i, j)^{A(i,j)} (1 - q(\tau^k, i, j))^{1-A(i,j)}.$$

Hierarchical tiling problem. The problem of finding hierarchical tiles that explain the data matrix as well as possible can now be formulated as finding the tile $\tau = ((A, p), \{\tau_1, \ldots, \tau_m\})$ that maximizes the likelihood $L(A \mid \tau)$. However, in order to avoid overfitting the data (very complex tiles that fit the data perfectly, e.g., using tiles at the level of single matrix entries) one needs to penalize for solutions with high complexity. Using Minimum Description Length (MDL) arguments, we define the *score* of $A(r)$ with respect to τ as $s(A(r) \mid \tau) = cK - \log L(A(r) \mid \tau)$, where K is a measure of total complexity of τ, and c is a scaling constant between complexity and minus log-likelihood. The complexity measure K is a function of the total number of tiles in τ – counting τ itself, its exceptions, the exceptions of its exceptions, and so on. If we denote the total number of tiles of τ by $|\tau|$ then K is defined to be $K = |\tau| \log |A(r)|$. The factor $\log |A(r)|$ is due to the fact that as the size of the data grows we need more information bits to specify the tiles, accounting to more complex models. The problem of finding hierarchical tiles can now be defined as follows: Given data matrix A, find the tile τ with the lowest score $s(A \mid \tau)$. The tile τ can be of any level, but it is required that $\mathrm{Dom}(\tau) = A$, i.e., it should cover the whole data matrix.

3 Algorithms

3.1 Geometric Tiles

We start describing our algorithm for discovering geometric tiles by first considering very simple cases, and then we discuss how to extend the ideas for the more complex situations. The simplest case is when we consider finding only one tile. Given a specific geometric rectangle $r = (A, \langle a, b\rangle, \langle c, d\rangle)$ of A to be used as the domain of the tile, the only choice to be made is the tile probability. As one can see easily, the maximum likelihood estimate for the tile probability is the frequency of the ones $f(r)$ in r. The frequency $f(r)$ can be computed in constant time, assuming that accumulating sums have been computed for all entries of the matrix: if $Ac(i, j)$ denotes the sum of 1s inside the rectangle $(A, \langle 1, 1\rangle, \langle i, j\rangle)$ then

$$f(r) = \frac{Ac(b,d) - Ac(b,c-1) - Ac(a-1,d) + Ac(a-1,c-1)}{(b-a+1)(d-c+1)}.$$

When the domain of the tile is not given, in principle one can try all possible rectangles r, evaluate the likelihood $L(A(r) \mid \tau(f(r), r))$ for each r, and select the tile that maximizes the likelihood. However, considering all rectangles is prohibitively expensive, since there are $\Theta(m^2n^2)$ different choices.

Designing an efficient algorithm to find a tile whose likelihood is provably not much worse than the likelihood of the best tile is a very interesting problem. However, it appears quite challenging: the likelihood function is not monotone with respect to tile containment, so there are no obvious ways to prune away potential tiles. Here we suggest a local-search algorithm for finding good tiles. The idea is to start with a random rectangle, and try to expand it or shrink it in each of the four directions. Expanding a rectangle r_0 in one direction, say to the right, is done in a sequence of geometric steps: for $r_0 = (\langle a, b\rangle, \langle c, d\rangle)$, we try all rectangles $(\langle a, b\rangle, \langle c, d+1\rangle)$, $(\langle a, b\rangle, \langle c, d+2\rangle)$, $(\langle a, b\rangle, \langle c, d+4\rangle)$, and so on, until the right boundary of the matrix is reached. The same expansion technique performed for other directions, and shrinking is done in a similar way. Out of all rectangles tried, the one with the largest likelihood is selected, call it r_1. If the likelihood of r_1 is larger than the likelihood of r_0, then a new expansion/shrinking phase starts from r_1. The process continues until a rectangle is found whose likelihood does not increase in an expansion/shrinking phase. A total of T random trials with different starting rectangles r_0 is performed, and the rectangle with the largest likelihood over all trials is given as the result.

Lemma 1. *Assume that the data matrix A contains i.i.d. bits with probability q, with the exception of one geometric rectangle R, which contains i.i.d. bits with probability $p \neq q$ and whose number of rows and columns is a constant fraction of the number of rows and columns (resp.) of the matrix A. Then, the local search method with random restarts will find R with high probability, i.e., probability bounded away from zero in the limit of infinite data.*

Due to space limitations the proof of all our claims is deferred to the full version of the paper.

Next we discuss how to find a larger collection of tiles with large likelihood. Our method employs the algorithm for finding one tile in a greedy fashion: Find the $(k+1)$-st tile with the best likelihood, given the k tiles that have been found so far. When searching for the next best tile, tiles that overlap existing tiles are not considered. This is checked during the expansion phase. To decide the number of tiles to be selected, the MDL score function $s(A \mid \tau)$ is used. When $s(A \mid \tau)$ stops decreasing, no more tiles are selected. For constructing the tile hierarchies, we have implemented and experimented with four different strategies:

Top down: At each step, the next tile is selected to be only at the same level, or included in already existing tiles.

Bottom up: The next tile is selected to be at the same level, or to include already existing tiles.

Mixed: The next tile is allowed to be anywhere as long as it does not overlap with existing tiles.

Single level: Only tiles of level 0 are selected.

Notice that the search space of the mixed strategy is the union of the search spaces of the other strategies, thus, one would expect the mixed strategy to outperform the others. The single-level strategy finds non-hierarchical tilings.

3.2 Combinatorial Tiles

In many applications, the rows and columns of the data set are not ordered, so it is important to be able to find combinatorial tiles. In this section we discuss our approach for this task. The basic idea is to transform the problem of finding combinatorial tiles to the previous case of finding geometric tiles. The transformation is done by ordering the rows and the columns of the data set, so that the rows and columns that are involved in combinatorial tiles become consecutive in the ordering. In this way, it is sufficient to search for geometric tiles in the ordered data set.

As we will see, it is not always possible to find such an ordering, since a data set might contain too many combinatorial tiles, and no single ordering can simultaneously transform all of them into geometric tiles. However, we will show that if a good ordering exists, our method will find it. The ordering method is based on the *spectral* properties of the data set. We next give a brief overview of the spectral techniques [9].

Consider a set of objects W and a symmetric matrix $S = (s_{ij})$ that specifies the similarity s_{ij} for each pair of objects (i,j). The *Laplacian matrix* of S is defined as the symmetric and zero-sum matrix $L_S = D_S - S$, where D_S is the diagonal matrix whose (i,i)-th entry is the sum of the i-th row (or column) of S, that is, $d_i = \sum_j s_{ij}$. Let e be the vector having value 1 in all its entries. Since all rows (and columns) of L_S sum to zero, we have $L_S e = 0$, which means that e is an eigenvector of L_S, corresponding to the eigenvalue 0. Because L_S is a symmetric positive semidefinite matrix, all of its eigenvalues are real and nonnegative, and therefore 0 is the smallest eigenvalue. The second smallest eigenvalue of L_S is called the *Fiedler value*, and the corresponding vector is called *Fiedler vector* [11]. One can show that the Fiedler value is given by

$$\min_{\substack{x^T e=0 \\ x^T x=1}} x^T L_S x = \min_{\substack{x^T e=0 \\ x^T x=1}} \sum_{i,j} s_{ij}(x_i - x_j)^2 \tag{1}$$

and the Fiedler vector is a vector that achieves the minimum subject to the constraints $x^T e = 0$ and $x^T x = 1$. A vector x can be viewed as *mapping* from objects in W to real numbers. In particular, the object i in W is mapped to the i-th coordinate x_i of x. If we view Equation (1) as an energy function $F_S(x) = x^T L_S x$, then the Fiedler vector v has the property that it minimizes $F_S(x)$ over all vectors x that satisfy the constraints $x^T e = 0$ and $x^T x = 1$. Intuitively, because of the terms $s_{ij}(x_i - x_j)^2$, minimizing $F_S(x)$ results to mapping "similar" objects to "near-by" values. The two constraints have a simple interpretation: $x^T e = 0$ requires the vector x to be orthogonal to the trivial solution vector e, i.e., to have zero mean, and $x^T x = 1$ amounts to fixing the scale of the solution, i.e., the variance of x is 1.

Our method uses Fiedler vectors to order the rows and the columns of a data set A. The idea is to consider each row as an "object" and define the similarity matrix $S = (s_{ij})$ for pairs of rows. Two natural definitions of row similarity is the *Hamming similarity* and *dot-product similarity*. The Hamming similarity h_{ij} between rows i and j is the number of common values, while the dot-product similarity c_{ij} is defined to be the number of common 1s. Both similarity definitions are used in our experiments. The method computes the row-row similarity matrix S for the data matrix A, and then it computes the Fiedler vector of the Laplacian L_S. The rows are ordered on the basis of their Fiedler-vector coordinates. The columns are ordered with the same method, independently of the rows.

Next we will show that for a data set generated from a simple combinatorial tiling model, the spectral algorithm will discover the correct structure. We begin with some definitions. An $n \times n$ matrix S has (k, d, s_1, s_2)-*block structure* if the n indices of S can be partitioned in k blocks $B_1, \ldots, B_k$ as follows: (i) the size of each block is greater than d, (ii) for all $i, j \in B_l$ we have $s_{ij} = \tilde{s}_l \geq s_1$, i.e., the value s_{ij} for indices within a block is a constant greater than s_1, (iii) for all $i \in B_l$ and $j \in B_t$ with $l \neq t$ we have $s_{ij} = \tilde{s}_{lt} \leq s_2$, i.e., the value s_{ij} for indices in different blocks is a constant smaller than s_2. We say that a vector x *respects* the structure of a (k, d, s_1, s_2)-block matrix S if for every triple of indices (i, j, h) with $x_i \leq x_j \leq x_h$ it cannot be the case that $i \in B_l$, $j \in B_t$, $h \in B_g$, and $l = g \neq t$.

Lemma 2. *Let S be an $n \times n$ object-similarity matrix that has $(2, d, s_1, s_2)$-block structure. Consider $S' = S + E$, where E is a symmetric matrix. Then the Fiedler vector of $L_{S'}$ respects the structure of S, provided that $|L_E| = o(d(s_1 - s_2))$, where $|L_E|$ is the norm of the matrix L_E.*

The situation is more complex when the similarity matrix has more than 2 blocks, as the associated Fiedler vectors form a subspace of dimension greater than 1. For example, consider a matrix with 3 blocks. A Fiedler solution is to assign all objects at three distinct values: one value for all objects within the same block. A different Fiedler solution uses only two distinct values: one value for objects in the first and second blocks and one value for objects in the third block. Note that Lemma 2 does not hold for the second solution, because if we break ties arbitrarily, most likely the objects of the first two blocks will not respect the block structure of the matrix.

To address the problem that in a Fiedler solution more than one blocks might be mapped to the same value, we have used a *recursive* application of the spectral algorithm: we first divide the coordinates of the Fiedler vector in two groups so that the sum

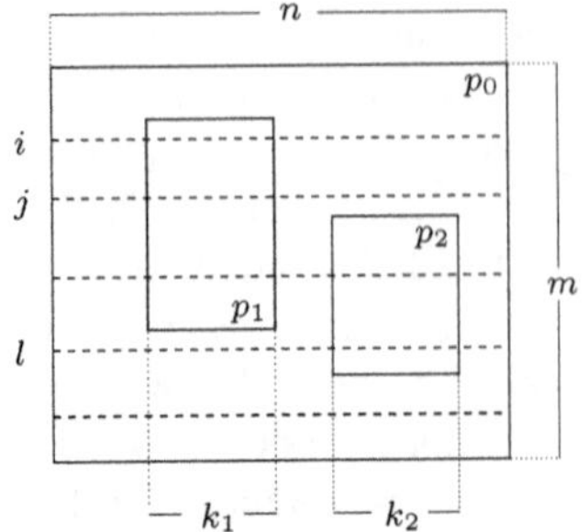

Fig. 2. Illustrative example of a data set generated from a "simple" tiling model.

1 0 0 1
1 0 1 0
0 1 1 0
0 1 0 1

Fig. 3. A case in which no ordering can turn all combinatorial rectangles into geometric rectangles.

of variances of the two groups is minimized, and then we apply recursively the spectral method in each of the two groups. A recursive step is performed only if the sum of variances in the two groups is less than half of the variance of all coordinates. Since the two groups that minimize the sum of variances cannot overlap, the groups can be determined optimally by sorting the coordinates and searching for the best breakpoint. We believe that using the recursive application of the spectral algorithm, Lemma 2 can be proven for matrices with more than two blocks, and this was verified in our experiments.

We complete our argument by showing that for a data set generated from a "simple" tiling model, the conditions of Lemma 2 hold. Consider, for example, the case of the simple data set shown in Figure 2: a 1 is generated at each entry of the data set with probability p_0, except in the two tiles, where a 1 is generated with probabilities p_1 and p_2, respectively. The tiles considered are combinatorial and the task is to reorder the rows and columns so that the geometric tiles shown in the figure emerge.

Consider the row-row similarity matrix S, where the Hamming similarity is used; a similar argument can be made for the dot-product similarity. A block is defined by the set of rows that intersect the same tiles, for example, rows i and j in Figure 2 belong to the same block. If the probability of the data at the entry A_{ih} is p_{ih}, then the similarity s_{ij} between rows i and j can be written as a sum of independent Bernoulli trials, i.e., $s_{ij} = \sum_{h=1}^{n} W_{ijh}$, where W_{ijh} is 1 with probability $p_{ijh} = p_{ih}p_{jh} + (1 - p_{ih})(1 - p_{jh})$ and 0 otherwise. Then, the *expected* similarity between rows i and j is $E[s_{ij}] = \sum_{h=1}^{n} p_{ijh}$. For example, the expected similarity between rows i and j in Figure 2 is $E[s_{ij}] = (n - k_1)(p_0^2 + (1 - p_0)^2) + k_1(p_1^2 + (1 - p_1)^2)$. We define the "simple" tiling model by making the following assumptions.

(i) Each block consists of a constant fraction of the total number of rows, i.e., $\Theta(m)$.

(ii) For each row i, the expected similarity $E[s_{ij}]$ is maximized for rows j in the same block. This assumption is reasonable in many case, for example, in Figure 2 it holds if, say, p_0 is less than 1/2, and p_1 and p_2 are greater than p_0. We assume that the expected similarity of two rows i and j in the same block is $E[s_{ij}] \geq s_1$, while the expected similarity of two rows i and j in different blocks is $E[s_{ij}] \leq s_2$. Furthermore, we assume that $s_2 - s_2 = \Theta(n)$. Again this assumption holds for the situation depicted in Figure 2 for, say, $p_0 = 0.2, p_1 = 0.8, p_2 = 0.7$.

Theorem 1. *For a simple tiling model as described above, the spectral algorithm will discover the correct ordering or rows and columns with high probability. The probability is taken over the generation of particular data instances from the model.*

In more complicated situations, it is possible that there is no ordering that turns all combinatorial tiles in the data into geometric tiles simultaneously. For example, the matrix in Figure 3 has four combinatorial tiles of 1s of size 2×1, but no reordering can bring these tiles together as geometric tiles. A possible solution to this problem is to first find the best tile in one ordering of the data, then reorder in a way that disregards the tiles already found, and then continue finding tiles. We feel that this approach would seriously detract from the interpretability of the results, so we restrict ourselves to finding tiles in one ordering only.

4 Experimental Evaluation

We used four real data sets to test our tiling algorithms. The first two sets contain information about fossil findings: the rows correspond to sites and the columns to genera. The first, `Paleo1`, contains 124 sites and 139 genera, and the second, `Paleo2`, 526 sites and 296 genera. The third data set, `Course`, contains information about Masters-level course registrations at the University of Helsinki Department of Computer Science. The set has 102 courses and 1739 students, with an average of 3.0 course registrations per student. For the fourth data set, `Movie`, we took the smaller of the two MovieLens movie-rating data sets[2], and turned it into a 0–1 matrix by mapping the high ratings 4 and 5 to 1, and the lower ratings (and non-ratings) to 0. The resulting data set contains ratings on 1682 movies by 943 users, with an average of 58.7 movies per user.

For each of the data sets, we first reordered both the rows and the columns by spectral ordering as outlined in Section 3.2, using both cosine and Hamming similarity as the similarity function. Then we ran the tiling algorithm described in Section 3 until it had found 50 tiles, using 100 random restarts per tile. The algorithm has four alternatives for the search strategy: top-down, bottom-up, mixed, and single-level.

Figure 4 shows how the log-likelihood behaves as a function of the number of tiles. Plots are shown for all the data sets and all strategies applied, but only for the cosine similarity. We see that mixed and top-down strategies outperform bottom-up and single-level. The reason that top-down is better than bottom-up and as good as mixed is probably the following: since we start selecting tiles greedily so that the total likelihood becomes as large as possible, we favor large tiles in the beginning, and it is thus more beneficial to recurse into those tiles than combining them to form even larger tiles.

Figure 5 shows some examples of tilings found with the different strategies. The top-down strategy found tilings very similar to those of the mixed strategy, and single-level was very similar to bottom-up. The probabilities of the tiles are shown in shades of grey, so that white corresponds to 0 and black to 1. The figure supports our hypothesis of why top-down and mixed outperform bottom-up and single-level: all strategies have found some large almost-empty tiles, but mixed and top-down can recurse into them to find the exceptions, whereas bottom-up and single-level only keep tiling the untiled

[2] http://www.grouplens.org

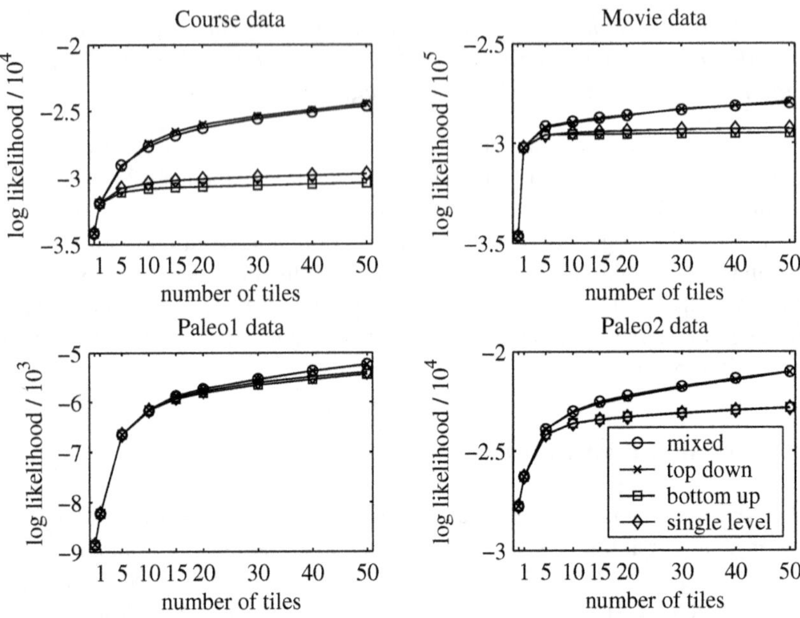

Fig. 4. Log-likelihood of model as a function of the number of tiles.

area, which has fewer opportunities. The tilings found on the Hamming-sorted data look somewhat more balanced than those of the cosine-sorted data, since the Hamming measure has grouped dense subsets in two corners of the matrix.

One parameter in our algorithm is the number T of the number of random restarts used when selecting each tile. To assess the effect of this parameter, we varied its value and computed the total log-likelihood of 10-tile models for two data sets, `Paleo2` and `Course`. The results are shown in Figure 6; each log-likelihood value shown is the average from 200 runs. As is to be expected, there is a diminishing-returns phenomenon, and after some point it helps very little to increase the number of restarts.

As an example of the interpretability of the results, we show one tile and its smaller exception tile in the `Course` data. The larger tile has 308 students and the following 11 courses, and its probability is 2.9% (which is relatively high compared to the background tile's 0.1%):

- User interface research
- Object architectures
- Simulation methods
- Implementation of the Linux system
- Object databases (*)
- Architectures of object systems (*)
- Research course in object languages
- Computer-aided co-operation
- Mobile workstations
- Algorithm technology
- Object technology

The exception tile consists of 115 students and the two courses marked with (*) above, and has probability 17.0%. Thus, there is a tight core of students interested in object-oriented technologies, and around that core there are more students who study object-oriented methodology, user interfaces, and some applications. (The probabilities

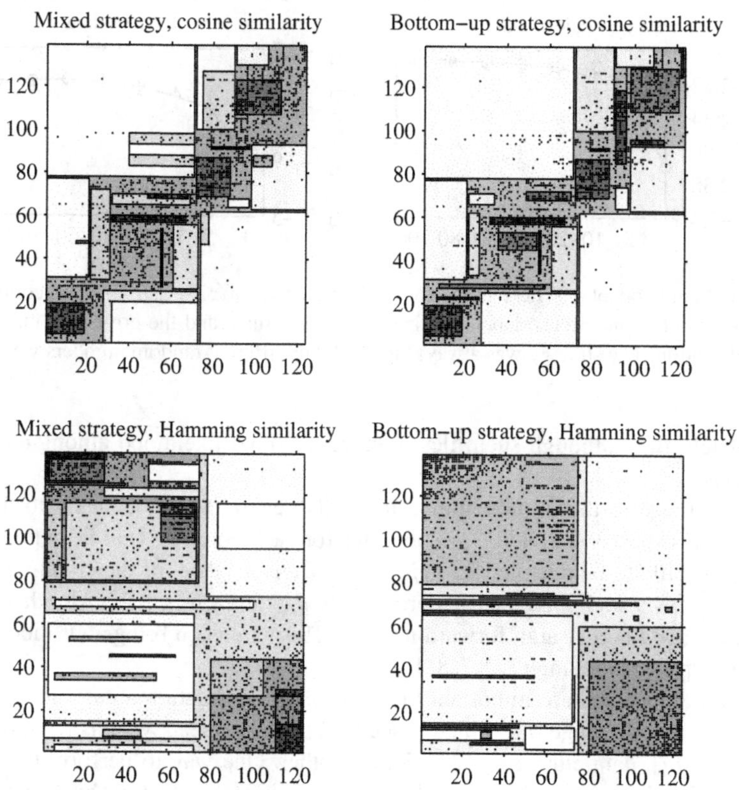

Fig. 5. 25-tile models of the `Paleo1` data found using two of the four different search strategies and two different orderings of the matrix.

may seem low, but many of these courses are in fact small seminars that have only been organized once.)

5 Related Work

Hierarchical tiles are partly motivated by the classical work of Rivest on finding decision lists [22]. Related is also the work on ripple-down rules [13]. A PAC-learning algorithm for finding hierarchical concepts was given in [16].

A lot of work on the analysis of 0–1 data has focused on finding frequent item sets and association rules [3, 6, 14]. A tile can be viewed as a frequent itemset: the tile's columns are the the items and the rows are the supporting transactions. A key difference with these approaches is that our method allows for errors, and also that a tile with many items might be selected even if its support is low. In addition, the greedy nature of our algorithm allows the user to look only at the first k tiles found, and interpret them as the tiles that best explain the data set among all tilings of size k.

A related problem is that of finding maximal empty rectangles in data [10, 18]. The crucial difference to our task is that we do not require tiles to be completely empty

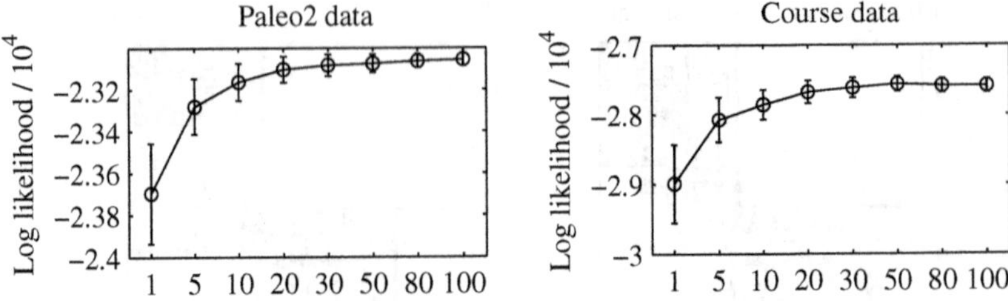

Fig. 6. Log-likelihood of 10-tile models for `Paleo2` and `Course` data as a function of the number of restarts. The circles denote the average of 200 runs, and the error bars indicate the standard deviation. The strategy was always "mixed", but different random numbers were used.

or completely full, although such tiles do have maximal likelihood among otherwise similar tiles.

As mentioned in the introduction, our results generalize immediately to the case where the tiles specify individual probabilities for each column. That is, define a generalized basic tile to be a triple $(X, Y, \bar{p})$, where X is a subset of the columns, Y is a subset of the rows, and $\bar{p}$ associates a probability p_A for each $A \in Y$. All other definitions are changed in a straightforward manner. This extension brings our model quite close to subspace clustering [1, 2, 7, 8, 12, 19].

Spectral algorithms are important tools for many application areas and they have been used in a wide range of problems, such as, solving linear systems [21], ordering problems [4, 17], data clustering [20, 23], and other. One way to perform the sorting more efficiently is to apply the spectral technique only to a subset of the data and then to refine the ordering of the whole data. [5]

6 Concluding Remarks

We have defined the concept of hierarchical tiles, and shown how they give a natural probabilistic model for 0–1 data. We gave a simple algorithm for finding geometric tiles, and showed some of its properties. We discussed the use of spectral ordering methods for finding good orderings. Our main theoretical result is that under certain assumptions the orderings produced by spectral techniques are such that strong combinatorial tiles become actually geometric tiles in the ordering. We demonstrated the applicability of the notion of hierarchical tiles by giving example results on real data.

References

1. C. Aggarwal and P. Yu. Finding generalized projected clusters in high dimensional spaces. In *SIGMOD*, 2000.
2. R. Agrawal, J. Gehrke, D. Gunopulos, and P. Raghavan. Automatic subspace clustering of high dimensional data for data mining applications. In *SIGMOD*, 1998.
3. R. Agrawal, T. Imielinski, and A. Swami. Mining associations between sets of items in large databases. In *SIGMOD*, 1993.

4. J. Atkins, E. Boman, and B. Hendrickson. A spectral algorithm for seriation and the consecutive ones problem. *SIAM Journal on Computing*, 28(1), 1999.
5. A. Beygelzimer, C.-S. Perng, and S. Ma. Fast ordering of large categorical datasets for better visualization. In *SIGKDD*, 2001.
6. T. Calders and B. Goethals. Mining all non-derivable frequent itemsets. In *PKDD*, 2002.
7. C. Cheng, A. Fu, and Y. Zhang. Entropy-based subspace clustering for mining numerical data. In *SIGKDD*, 1999.
8. Y. Cheng and G. Church. Biclustering of expression data. In *ISMB*, 2000.
9. F. Chung. *Spectral graph theory*. American Mathematical Society, 1997.
10. J. Edmonds, J. Gryz, D. Liang, and R. J. Miller. Mining for empty spaces in large data sets. *Theor. Comput. Sci.*, 296(3):435–452, 2003.
11. M. Fiedler. Algebraic connectivity of graphs. *Czech. Math. J.*, 23, 1973.
12. J. Friedman and J. Meulman. Clustering objects on subsets of attributes. *JRSS B*, 2004.
13. B. Gaines and P. Compton. Induction of ripple-down rules applied to modeling large databases. *JIIS*, 5(3), 1993.
14. J. Han, J. Wang, Y. Lu, and P. Tzvetkov. Mining top-k frequent closed patterns without minimum support. In *ICDM*, 2002.
15. A. Jain, M. Murty, and P. Flynn. Data clustering: A review. *ACM Computing Surveys*, 1999.
16. J. Kivinen, H. Mannila, and E. Ukkonen. Learning hierarchical rule sets. In *COLT*, 1992.
17. Y. Koren and D. Harel. Multi-scale algorithm for the linear arrangement problem. Technical Report MCS02-04, The Weizmann Institute of Science, 2002.
18. B. Liu, L.-P. Ku, and W. Hsu. Discovering interesting holes in data. In *IJCAI*, 1997.
19. T. Murali and S. Kasif. Extracting conserved gene expression motifs from gene expression data. In *Pac. Symp. Biocomp.*, volume 8, 2003.
20. A. Ng, M. Jordan, and Y. Weiss. On spectral clustering: Analysis and an algorithm. In *NIPS*, 2001.
21. A. Pothen, H. Simon, and L. Wang. Spectral nested dissection. Technical Report CS-92-01, Pennsylvania State University, Department of Computer Science, 1992.
22. R. Rivest. Learning decision lists. *Machine Learning*, 2(3), 1987.
23. H. Zha, X. He, C. Ding, M. Gu, and H. Simon. Bipartite graph partitioning and data clustering. In *CIKM*, 2001.

Document Classification Through Interactive Supervision of Document and Term Labels

Shantanu Godbole, Abhay Harpale, Sunita Sarawagi, and Soumen Chakrabarti

IIT Bombay
Powai, Mumbai, 400076, India
shantanu@it.iitb.ac.in

Abstract. Effective incorporation of human expertise, while exerting a low cognitive load, is a critical aspect of real-life text classification applications that is not adequately addressed by batch-supervised high-accuracy learners. Standard text classifiers are supervised in only one way: assigning labels to whole documents. They are thus deprived of the enormous wisdom that humans carry about the significance of words and phrases in context. We present HIClass, an interactive and exploratory labeling package that actively collects user opinion on feature representations and choices, as well as whole-document labels, while minimizing redundancy in the input sought. Preliminary experience suggests that, starting with essentially an unlabeled corpus, very little cognitive labor suffices to set up a labeled collection on which standard classifiers perform well.

1 Introduction

Motivated by applications like spam filtering, e-mail routing, Web directory maintenance, and news filtering, text classification has been researched extensively in recent years [1–3]. State-of-the-art classifiers now achieve up to 90% accuracy on well-known benchmarks. Almost all machine learning text classification research assumes some fixed, simple class of feature representation (such as bag-of-words), and at least a partially labeled corpus. Statistical learners also depend on the deployment scenario to be reasonably related to the training population.

Many of these assumptions do not hold in real-life applications. Discrimination between labels can be difficult unless features are engineered and selected with extensive human knowledge. Often, there is no labeled collection to start with. In fact, even the label set may not be specified up front, and must evolve with the user's understanding of the application. Several projects reported at the annual Operational Text Classification workshops [4] describe applications spanning law, journalism, libraries and scholarly publications in which automated, batch-mode techniques were not satisfactory; substantial human involvement was required before a suitable feature set, label system, labeled corpus, rule base, and resulting system accuracy were attained. However, not all the techniques used in commercial systems are publicly known, and few general principles can be derived from these systems.

J.-F. Boulicaut et al. (Eds.): PKDD 2004, LNAI 3202, pp. 185–196, 2004.

There is much scope for building machine learning tools which engage the user in an active dialog to acquire human knowledge about features and document labels. When such supervision is available only as label assignments, *active learning* has provided some clear principles [5–7] and strategies for maximum payoffs from the dialog. We wish to extend the active learning paradigm significantly to include both feature engineering and document labeling conversations, exploiting rapidly increasing computing power to give the user immediate feedback on her choices.

Our contributions: In this paper we present the design of a system HIClass (Hyper Interactive text Classification) for providing this tight interaction loop. We extend SVMs to naturally absorb human inputs in the form of feature engineering, term inclusion/exclusion and term and document labels. In the past, such actions were performed through ad hoc means and as a distinct processing step before classification construction. We make these more effective by (1) providing the user easy access to a rich variety of summaries about the learnt model, the input data and aggregate performance measures, (2) drawing the user's attention to terms, classes or documents in greatest need of inspection, and (3) helping the user assess the effect of every choice on the performance of the system on test data.

Outline: We describe the HIClass workbench in Section 2 and review the design choices and various modes of user interaction. Section 3 describes our method of active learning on documents with modifications to handle multi-labeled data and methods to reduce the cognitive load on the user. Section 4 introduces the idea of active learning on terms treating them as first class entities. We report our experiences with the workbench and experimental results in Section 5. We review related work in Section 6 and conclude in Section 7.

2 The HIClass Workbench for Text Classification

We present an overview of HIClass in Fig. 1. The lower layer shows the main data entities and the main processing units. There is a small pool of labeled documents (partitioned by sampling into a training and test set) and a large unlabeled pool. The feature extractor turns documents into feature vectors. Features are usually words, but the user can interactively refine features to be more complex; this is described next. The system can store and access by name multiple classifiers with their fitted parameters at any given time, assisting comparative analysis of parameters, performance on held-out data, and drill-down error diagnostics. The upper layer shows the prominent menus/modes in which a user can interact with the system. Next we describe the important building blocks shown in Fig. 1.

2.1 Document and Classification Models

The first step of the design of HIClass is to choose a flexible classification model that (1) suits state-of-the-art automated learners and (2) can be easily interpreted and tuned by the user.

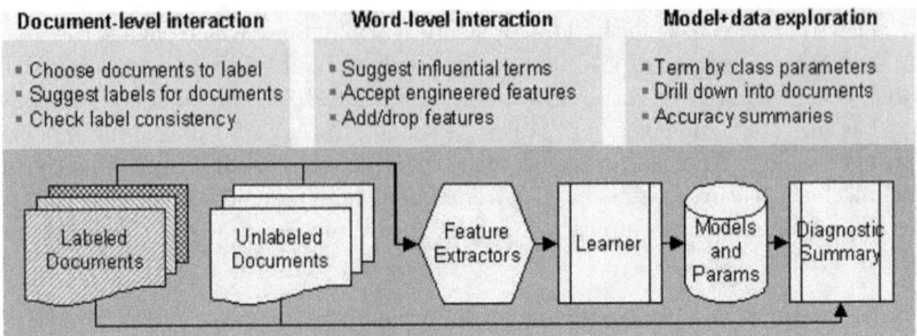

Fig. 1. The architecture of HIClass.

A document is a bag of features. Usually, features are words after minor processing like stemming and case-normalization. But the user can also (dynamically) define features to reflect domain knowledge. E.g., month names or currency names may be conflated into synthetic features. On the other hand, the user may notice harmful conflation between "blood bank" and "bank", and define "blood bank" as a single compound feature. We will continue to use *term*, *word* and *feature* interchangeably where no confusion can result. Documents are represented as unit vectors.

Labeled documents can be associated with more than one class in general. HIClass supports **linear additive** classifier models, where each class c is associated with a set of weights $w_1^c, \dots w_T^c$ corresponding to the T terms in a vocabulary. Each document is represented by a vector of non-negative weights $\boldsymbol{x} = (x_1, \dots, x_T)$, each component corresponding to a feature. The classifier assigns a document all class labels c for which $\mathbf{w}^c \cdot \boldsymbol{x} + b_c \geq 0$ where b_c is a scalar per-class bias parameter. As documents vectors have only non-negative components, both magnitude and sign of components of w^c give natural interpretations of salience of terms.

The linear additive model generalizes a number of widely-used classifiers, including naive Bayes (NB), maximum entropy, logistic regression, and support vector machines (SVMs). Here we focus on SVMs. Given documents d_i with labels $y_i \in \{-1, +1\}$, a two-class linear SVM finds a vector $\mathbf{w}$ and a scalar constant b, such that for all documents $y_i(\mathbf{w}_c \cdot d_i + b) \geq 1$, and $||\mathbf{w}_c||$ is minimized.

When the application demands more than two classes, one can (1) rewrite the above optimization slightly, with one $\mathbf{w}$ vector per class, so that the discriminant $\mathbf{w}_{c_j} \cdot d_i + b_j$ is largest for the correct class c_j; or (2) build an ensemble of SVMs, each playing off one class against another ("one-vs-one"), and assigning the document to the class that wins the largest number of matches; or (3) build an ensemble of SVMs, as many as there are classes, each predicting an yes/no label for its corresponding class ("one-vs-rest" or "one-vs-others"). In practice, all these approaches are comparable in accuracy [8]. We use one-vs-others as it is easily extended to make multi-labeled prediction and is efficient.

2.2 Exploration of Data, Model, Performance Summaries

HIClass provides support for viewing the trained classifier scores, aggregate as well as drill-down statistics about terms, documents and classes, and standard accuracy measures. Aggregate statistics like per-class population, similarity distribution, and uncertainty distribution can be viewed.

After building an initial classifier using the starting labeled set L, the user can view the learnt model as a matrix of class-by-term scores. Simple inspection of term-class scores in an OLAP-like tool enables the expert to propose changes to per-class classification models like including and excluding certain terms. The user-interface allows easy movement from a term-centric to a class-centric analysis where the user can see documents with terms indicative of belongingness to classes.

With every proposed change, the user can study the impact of the change by observing its performance on the test data. The user can inspect graphs for the accuracy and $F1$ of the whole system or for individual classes across iterations. The user can identify classes which hamper the overall performance of the system and can concentrate on them further by adding more labeled documents or fine tuning important relevant terms. The user can also inspect a confusion matrix between any two classes that reveals their overlap, allowing the user to inspect and tune discriminating terms by inspecting the results of a binary SVM on the classes.

2.3 Feature Engineering

Fast evaluation over a variety of test data enables a user to easily identify limitations of a trained model and perhaps the associated feature set. Most users, on inspection of this set of scores, will be able to propose a number of modifications to the classifiers. Some of these modifications may not impact performance on the available test set but could be beneficial in improving the robustness and performance of the classifier in the long term. For the Reuters dataset, close inspection of some of the terms shown to have a high positive weight for the class *crude* reveals:

- "Reagan" is found to be a positive indicator of the class *crude* though proper names should be identified and treated differently.
- "Ecuador" and "Ecuadorean" reveal insufficient stemming.
- "World bank" and "Buenos Aires" should always occur together as a bi-gram; "Union", a high weight term for *crude*, should be associated with "Pacific Union" in *crude*, but as "Soviet Union" in other classes.
- Month names, currencies, date formats, proper nouns should be recognized and grouped into appropriate aggregate features.

2.4 Document Labeling Assistant

When unlabeled data is abundant and labeled data is limited, a user can choose to add labels to some of the unlabeled documents. Active learning has proven to be highly effective in interactively choosing documents for labeling so that the total number of documents to be labeled is minimized. HIClass provides a number of mechanisms to lighten the user's cognitive load in the document labeling process. Details of appear in Section 3.

2.5 Term-Level Active Learning

The high accuracy of linear SVMs at text classification [1] suggests that class membership decision depends on the combination of "soft" evidence from a class-conditional subset of the corpus vocabulary. E.g., high rate of occurrence of one or more of the words *wicket*, *run*, *stump*, and *ball* leads us to believe a document is about (the game) cricket. Given enough training documents, good classifiers can learn the importance of these terms. However, in the initial stages of bootstrapping a labeled corpus, it is far more natural for the user to directly specify these important features as being positively associated with the class "cricket", rather than scan and label long documents containing these words.

HIClass allows users to label terms with classes just like documents. We expect the cognitive load of labeling terms to be lower because the user does not have to waste time reading long documents. We help the user in spotting such terms by doing active learning on terms. This is elaborated in Section 4.

3 Active Labeling of Documents

The system starts with a small training pool of labeled documents L and a large pool of unlabeled documents U. Assume that the number of class labels is k and each document can be assigned multiple labels. We train k one-vs-others SVMs on L. Our goal during active learning is to pick some unlabeled documents about whose predictions the classifier is *most uncertain*. Various measures are used for calculating uncertainty with SVMs [6]. However, these assume binary, single-labeled documents. We extend these to the multi-class, multi-labeled setting as described next.

3.1 Uncertainty

Each unlabeled document gets k discriminant values, one from each SVM in the one-vs-others ensemble. We arrange these values on the number line, and find the largest gap between adjacent values. A reasonable policy for multilabel classification using one-vs-others SVMs is that discriminant values to the right of the gap (larger values) correspond to SVMs that should be assigned a positive label to the document and the rest should be negative.

We need this policy because, in our experience with one-vs-others ensembles, as many as 30% of documents may be labeled negative by all members of the ensemble. For single label classification, it is common to pick the maximum discriminant even if it is negative. Our policy may be regarded as an extension of this heuristic to predict multiple labels.

With this policy, we declare that document to be most uncertain whose this largest gap is the smallest among all documents. When documents are restricted to have one label, this reduces to defining certainty (confidence) in terms of the gap between the highest scoring and the second highest scoring class.

3.2 Bulk-Labeling

The user could label these uncertain documents one by one. But experience suggests that we can do better: often, many of these document are quite similar, and if we could present tight clusters that the user can label all at once, we can reduce the cognitive load on the user and speed up the interaction.

We pick the u most uncertain documents and compute pairwise vector-space similarity between documents in the uncertain set, and prepare for the user a cluster/subset of fixed size (set by the parameter s) that has the largest sum of pairwise similarities.

When showing these uncertain clusters to the user, we also provide an ordered list of suggested labels. The ordering is created by taking the centroid of each uncertain cluster and finding its similarity to the k centroids of positive training data of the k classes. Fig. 2 summarizes the active bulk-labeling process for documents.

Start with a labeled pool L and an unlabeled pool U.
while user wants to continue with active labeling **do**
  Train a A-vs-not$A$ SVM ensemble on T
  Calculate uncertainty on all documents in U:
  for all documents $d \in U$ **do**
    Get k scores by applying the k SVMs to d. Find the largest gap in score values.
  end for
  Sort the $|U|$ gaps in ascending order and add top u to the uncertain set.
  Select the s most similar documents from top u
  Suggesting ranked list of labels for the group s:
  for all k classes **do**
    Find similarity between centroid of s and centroid of positive training data of class k
  end for
  Sort these distances in a suggested list of classes
  Present s and the ranked list of k suggestions to the user for active labeling
  Accept multi-labeled suggestions for all documents in s. Check for conflicts
  Add these s documents to L with user provided labels and remove from U
end while

Fig. 2. The algorithm for active learning on documents.

(An alternative is to use the existing classifier itself to propose suggestions based on the confidence with which the documents in the uncertain cluster are classified into various classes. However, we feel keeping the same suggestion list for all documents in each uncertain cluster reduces the cognitive load on the user. Also, empirically we found in the initial stages this provides better suggestion than the SVMs.)

The user provides feedback to the system by labeling all documents in an uncertain cluster in one shot. The labeled documents are inspected by a conflict check module for consistency. We defer discussion of this topic due to lack of space. Once the user confirms the labels, the newly labeled documents are removed from U and added to L. The system then iterates back to re-training the SVM ensemble.

4 Active Learning Involving Terms

As mentioned in Section 2.5, users generally find it easier to bootstrap the labeled set using trigger terms (that they already know) rather than tediously scrutinize lengthy documents for known triggers. We demonstrate this with an example from the Reuters dataset. We trained two SVMs using the *interest* class in Reuters; the first trained with a single document per class and the second trained with 50 documents per class. For each SVM, we report some terms corresponding to the maximum positive weights in the table. The SVM using more data contains terms like "rate" "fe" (foreign exchange), "pct" (percent), and "interest": that a user can readily recognize as being positively associated with the label *interest* that are missing from the first SVM.

Num labeled=1		Num labeled=50	
Term	w	Term	w
forecast	0.40	rate	2.08
bank	0.29	fe	1.97
noon	0.20	pct	1.65
account	0.20	market	1.26
oper	0.14	custom	1.01
market	0.14	interest	0.92
england	0.09	forecast	0.92
		stg	0.87
		bank	0.83

We allow a direct process of proposing trigger terms within the additive linear framework. We believe such manual addition of terms will be most useful in the initial phases to bootstrap a starting classifier which is subsequently strengthened using document-level active learning. We propose a mechanism analogous to active learning on documents to help a user spot such terms. SVMs treat labeled terms as mini-documents whose vector representation has a 1 at the term's position and 0 everywhere else, resulting in standard unit length document vectors.

We develop a criterion for term active learning that is based on the theoretically optimum criterion of minimizing uncertainty on the unlabeled set but avoids the exhaustive approach required to implement it [5–7] by exploiting the special nature of single-term documents.

Consider adding a term t whose current weight is w_t in the trained SVM. For terms not in any of the labeled documents $w_t = 0$. Suppose we add t as a "mini-document" with the user-assigned label y_t. Let the new SVM weight vector be w'. Since the term t is a mini-document whose vector has $x_t = 1$ and $\forall t' \neq t, x_{t'} = 0$, we can assume that in the new w' only w_t is changed to a new w'_t and no other $w_{t'}$ is affected significantly. This is particularly true for terms that do not already appear in the labeled set. From the formulation of SVMs, $y_t(w'_t + b) \geq 1$.

If the current w_t is such that $|w_t + b| \geq 1$ then adding t will probably not have any affect. So we consider only those ts where $|w_t + b| < 1$. Adding t with a label +1 will enforce $w'_t + b = 1$ i.e., $w'_t = 1 - b$ and with a label of -1 will make it $w'_t = -1 - b$. For each possible value of $y_t = c$, we get a new value of $w'_t(c)$. Thus we can directly compute the new uncertainty of each unlabeled document x by computing the *change* in the distance from separator value as $(w'_t(c) - w_t)x_t$, since uncertainty is inversely proportional to distance from the separator. Let $Pr(c, t)$ be the probability that the term t will be assigned to class c, as our weighing factor. We estimate $Pr(c, t)$ by the fraction of documents containing term t which have been predicted to belong to class c. We then compute the weighted uncertainty $WU(t)$ for a term t as $WU(t) = \sum_c U(c, t)Pr(c, t)$ and then select the term with the smallest $WU(t)$ for labeling. Other details and approximate variants can be found in [9]. This gives us a way to compute the total uncertainty over the unlabeled set without retraining a SVM for each candidate term.

5 Experimental Study

We have experimented with several text classification tasks ranging from well-established benchmarks like Reuters and 20-newsgroups to more noisy classification tasks, like the Outdoors dataset, chosen from Web directories [11]. It is difficult to quantify the many ways in which HIClass is useful. Therefore we pick a few measures like the benefits of active learning with terms and document to report as performance numbers. We also present some results which quantify the cognitive load on the user and try to show how HIClass eases the user's interaction and labeling process.

HIClass consists of roughly 5000 lines of C++ code for the backend and 1000 lines of PHP scripts to manage frontend user interactions. The frontend is a web browser, readily available on any user's desktop. XML is used to pass messages between the frontend and the server backend. LibSVM [12] is used as the underlying SVM classifier.

All our development and experiments were done on a dual-processor P3 server running Debian Linux and with 2GB RAM. Due to space limitations we report numbers for fixed settings of some of our system parameters. Further experiments can be found in [9]. Unless otherwise stated, the number of initial documents per class is set to 1, the number of documents selected for bulk labeling is 5 and the number of uncertain documents over which we pick similar clusters (the parameter u of Section 3) is set to 75.

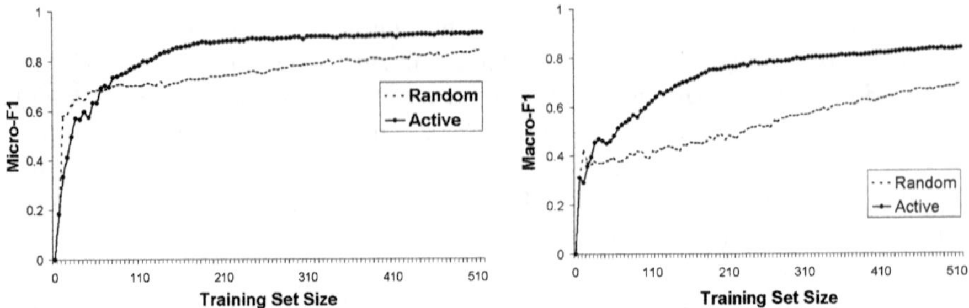

Fig. 3. Reuters - Micro and Macro-averaged F1 on held-out test data while increasing training set size, randomly versus using document level active learning.

5.1 Document-Level Active Learning

We now show how active learning on documents can reduce the number of documents for which the user needs to provide labels in a multiclass, multi-labeled settting. We started with one document in each class and added 5 documents in each round. All graphs are averaged over 30 random runs. Fig. 3 compares the micro and macro averaged $F1$, of selecting 5 documents per round using active learning and using random selection for Reuters (similar results with other datasets omitted due to lack of space). We see that active learning outperforms randomly adding documents to L and reaches its peak $F1$ levels faster.

5.2 Reducing Labeling Effort

We next show the effectiveness of the two techniques that we proposed in Section 3 for reducing the effort spent for labeling a document. For lack of space we only show results with Reuters in this sub-section.

Quality of Suggestions. We quantify the quality of suggestions provided to the user by the average rank of the true labels in the suggested list. We see in Figure 4 that even in the initial stages of active learning the true classes on an average are within rank 4 whereas the total number of possible classes is 20 for this dataset. We also see that the suggestions with u fixed at 75 are better than at 10 as expected.

Bulk-Labeling. We quantify the benefit of bulk-labeling by measuring **inverse similarity**, defined as the number of true distinct labels in a batch of s documents as a fraction of the total number of document-label pairs in the batch. So, if $s = 5$ and each document in a batch has one label and all of them are the same, then the inverse similarity is $\frac{1}{5}$.

It is reasonable to assume that the cognitive load of labeling is proportional to the number of distinct labels that the user has to assign. Thus, Fig. 5 establishes that our chosen set of similar documents reduce cognitive load by a

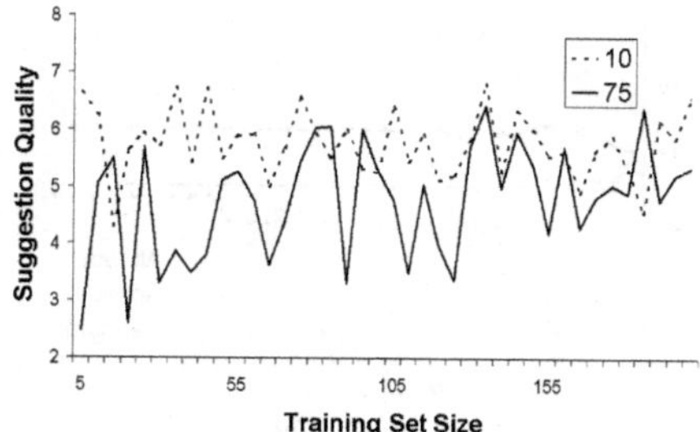

Fig. 4. Reuters – Quality of suggestion measured as the rank at which correct labels are found in the suggested labels.

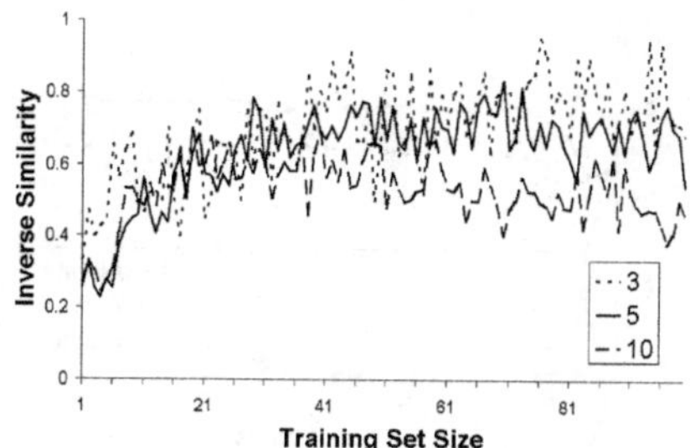

Fig. 5. Reuters – Benefits of bulk-labeling measured as inverse similarity defined in section 5.2.

factor of 2. The benefits are higher in the initial stages because then there are several documents with high uncertainty to choose from. With higher number of documents per batch, the benefits get larger.

We cannot set s to be very high because there is a tradeoff between *reducing effort per label* by bulk labeling similar documents and *increasing number of labels* by possibly including redundant documents per batch. If we calculated labeling cost in terms of *number of documents* to be labeled, the optimum strategy is to label the most uncertain single document per batch. But the effort the user has to spend in deciding on the right label for rapidly changing document contexts will be high. The right tradeoff can only be obtained through experience and will vary with different classification tasks and also the user's experience and familarity with the data.

5.3 Term-Level Active Learning

Our goal here is to evaluate the efficacy of training with labeled terms. We take all available labeled documents for a class and train a one-vs-rest SVM for that class. All single-term documents that are predicted as positive or negative with very large margins (above $b/3$ here) are labeled with the predicted class and the rest are not labeled. We then start with a SVM trained initially with a single labeled document on each side and keep adding these collected labeled terms in order of the magnitude of their weights (the AllData method). We also evaluate the performance of our term level active learning described in 4. However, we use an approximation algorithm which is less time-intensive and computationally efficient. We select terms with higher values of $f(t)$ where $f(t) = (\sum_{i \in pos} x_{i_t} - \sum_{i \in neg} x_{i_t}) * (N - (pos - neg)(b + \dot{w}_t))$ where N is the total number of unlabeled documents, and pos and neg refer to number of positive and negative unlabeled documents.

In Fig. 6 we show the resulting accuracy on 8 classes of Reuters and 3 classes of the 20-newsgroups dataset. Active learning on terms clearly works as expected though the gains are small. This is to be expected since SVMs are trained with very few terms instead of entire documents. Random selection performs much

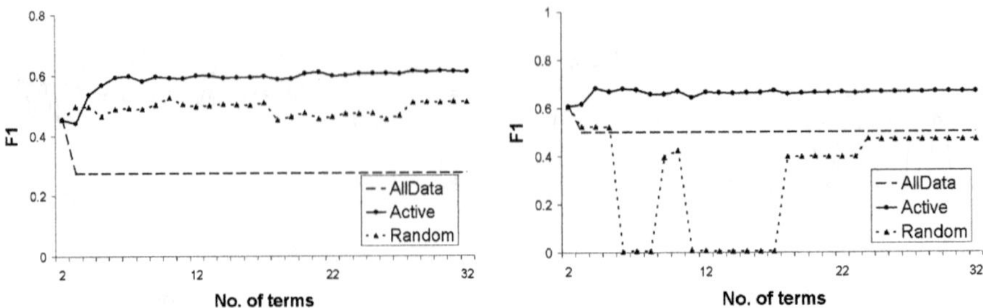

Fig. 6. Adding labeled terms in score order Reuters (left) and 20-newsgroups (right).

worse in both the datasets. This confirmed our intuition that term-level active learning is best viewed as a bootstrapping technique followed by document-level active learning.

6 Related Work

Most earlier work on applying active learning to text categorization [6, 10] assume a single binary SVM whereas our proposed scheme is for multiple one-vs-othersSVMs and for multi-labeled classification. Active learning has also recently been applied to the problem of selecting missing attributes of labeled instances whose values should be filled in by the user [13]. This is different from our setting of term active learning because our goal is to add terms as additional labeled instances. The notion of labeling terms is used in [14] for building lexicons of terms related to a concept. So the goal there is not to assign documents to categories but to exploit the co-occurance patterns of terms in documents to categorize terms.

7 Conclusion

We have described HIClass, an interactive workbench for text classification which combines the cognitive power of humans with the power of automated learners to make statistically sound decisions. The system is based on active learning, starting with a small pool of labeled documents and a large pool of unlabeled documents. We introduce the novel concept of active learning on terms for text classification. We describe our OLAP-like interface for browsing the term-class matrix of the classifier cast as a linear additive model. The user can tune weights of terms in classes leading to better, more understandable classifiers. HIClass provides user continuous feedback on the state of the system, drawing her attention to classes, documents, and terms which would benefit by manual tuning.

References

1. T. Joachims. Text categorization with support vector machines: learning with many relevant features. In *Proceedings of ECML-98.*
2. K. Nigam, J. Lafferty, and A. McCallum. Using maximum entropy for text classification. In *IJCAI'99 Workshop on Information Filtering.*

3. J. Zhang and Y. Yang. Robustness of regularized linear classification methods in text categorization. In *SIGIR*, 2003.
4. Third workshop on Operational Text Classification OTC 2003. In conjunction with *SIGKDD 2003.*
5. D. A. Cohn, Z. Ghahramani, and M. I. Jordan. Active learning with statistical models. In *Advances in Neural Information Processing Systems*, 1995.
6. S. Tong and D. Koller. Support vector machine active learning with applications to text classification. *Journal of Machine Learning Research*, 2:45–66, Nov. 2001.
7. Y. Freund, H. S. Seung, E. Shamir, and N. Tishby. Selective sampling using the query by committee algorithm. *Machine Learning*, 28(2-3):133–168, 1997.
8. C. Hsu and C. Lin. A comparison of methods for multi-class support vector machines. In *IEEE Transactions on Neural Networks*, 13(2002), 415-425, 2001.
9. A. Harpale. Practical alternatives for active learning with applications to text classification. Master's Thesis, IIT Bombay, 2004.
10. A. K. McCallum and K. Nigam. Employing EM in pool-based active learning for text classification. In *Proceedings of ICML-98.*
11. S. Sarawagi, S. Chakrabarti, and S. Godbole. Cross-training: Learning probabilistic mappings between topics. In *SIGKDD*, 2003.
12. C.C. Chang, and C.J. Lin. LIBSVM: a library for support vector machines, 2001. `http://www.csie.ntu.edu.tw/~cjlin/libsvm/`
13. D. Lizotte, O. Madani, and R. Greiner. Budgeted learning of naive-bayes classifiers. In *UAI*, 2003.
14. H. Avancini, A. Lavelli, B. Magnini, F. Sebastiani, and R. Zanoli. Expanding domain-specific lexicons by term categorization. In *SAC*, 2003.

Classifying Protein Fingerprints

Melanie Hilario[1], Alex Mitchell[2], Jee-Hyub Kim[1],
Paul Bradley[2], and Terri Attwood[3]

[1] Artificial Intelligence Laboratory, University of Geneva, Switzerland
{Melanie.Hilario,Jee.Kim}@cui.unige.ch
[2] European Bioinformatics Institute, Hinxton, Cambridge CB10 1SD, UK
{mitchell,pbradley}@ebi.ac.uk
[3] School of Biological Sciences, University of Manchester, UK
attwood@bioinf.man.ac.uk

Abstract. Protein fingerprints are groups of conserved motifs which can be used as diagnostic signatures to identify and characterize collections of protein sequences. These fingerprints are stored in the PRINTS database after time-consuming annotation by domain experts who must first of all determine the fingerprint type, i.e., whether a fingerprint depicts a protein family, superfamily or domain. To alleviate the annotation bottleneck, a system called PRECIS has been developed which automatically generates PRINTS records, provisionally stored in a supplement called prePRINTS. One limitation of PRECIS is that its classification heuristics, handcoded by proteomics experts, often misclassify fingerprint type; their error rate has been estimated at 40%. This paper reports on an attempt to build more accurate classifiers based on information drawn from the fingerprints themselves and from the SWISS-PROT database. Extensive experimentation using 10-fold cross-validation led to the selection of a model combining the ReliefF feature selector with an SVM-RBF learner. The final model's error rate was estimated at 14.1% on a blind test set, representing a 26% accuracy gain over PRECIS' handcrafted rules.

1 Motivation and Background

Protein fingerprints are groups of conserved amino acid motifs drawn from multiple sequence alignments that are used to characterise protein families. The PRINTS database [1] is a compendium of more than 1800 diagnostic fingerprints for protein families, superfamilies and domains. It provides large amounts of handcrafted annotation, aiming to document the constituent protein families and to rationalise the conserved regions in functional and structural terms. The annotation procedure is exhaustive and time consuming, and consequently PRINTS remains relatively small by comparison with other, largely automatically-derived signature databases.

To address this issue, automation of fingerprint production and annotation has been investigated. The PRINTS group has previously developed PRECIS [9], an annotation tool which generates protein reports from related SWISS-PROT entries. Though this approach has worked well overall, PRECIS has areas of

J.-F. Boulicaut et al. (Eds.): PKDD 2004, LNAI 3202, pp. 197–208, 2004.

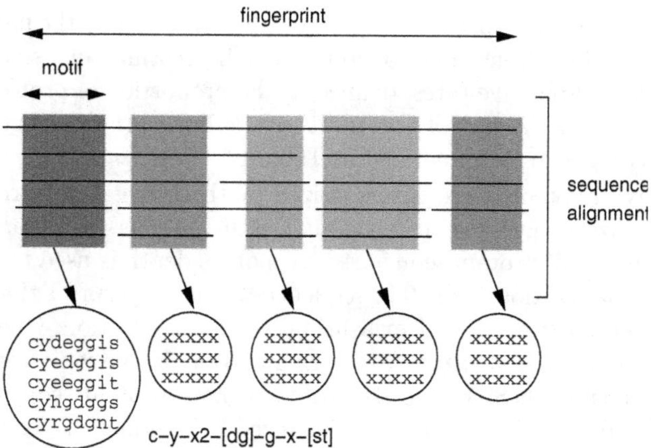

Fig. 1. Schema of a protein fingerprint. Each row is a protein sequence and each column an amino acid. Solid rectangles depict conserved regions or motifs.

limitation. The annotation generated by the tool inevitably lags that which could be derived from current literature. This is due to the fact it is almost entirely dependent on information stored in SWISS-PROT, which, despite the valiant efforts of a team of annotators, cannot be kept up to date. Another limitation of PRECIS is that its relatively simple heuristics often misclassify fingerprints. Broadly speaking, fingerprints may be diagnostic for a gene family or superfamily (united by a common function), or a domain family (united by a common structural motif). The type of fingerprint has implications on the kind of information and the level of detail to be reported within the annotation. Increased accuracy of fingerprint classification would therefore help ensure that the correct information was processed to generate appropriate annotation.

2 Task and Data Representation

The goal of the work reported in this paper was to replace PRECIS' handcrafted heuristics with classification models extracted from data. These heuristics determined fingerprint type through an analysis of SWISS-PROT database records concerning protein sequences within the fingerprint. Before turning to SWISS-PROT, we decided to investigate whether a fingerprint's physical parameters could be used as discriminators to improve classification. As shown in Fig. 1, a fingerprint is basically a multiple sequence alignment with a number of conserved regions or motifs. A fingerprint can be characterized in terms of three distinct entities: the fingerprint itself and its component motifs and proteins. We are therefore confronted with a multirelational learning problem which can be addressed most naturally using a relational approach. This paper focuses on an alternative approach which propositionalizes the task representation by aggregating protein and motif characteristics over the fingerprint.

Fingerprint. The fingerprint as a whole can be described by the number of motifs and proteins it contains. The coherence of a fingerprint can also be expressed by true and partial positive rates, defined as the proportion of protein sequences that match all or only a part of the motifs in the fingerprint respectively. These fingerprint statistics are summarized in Table 1-I.

Motif. Individual motifs are characterized by their size and degree of conservation. Motif size is assessed in terms of length (number of amino acids) and depth (the number of protein sequences). A motif's depth is used to compute its coverage, i.e., the fraction of protein sequences in the fingerprint that match the motif. We explored two ways of measuring motif conservation. One alternative was to estimate a motif's entropy by averaging over the entropies of its individual columns (residues). Since motifs involving more protein sequences tend to have higher entropies, the result was normalized by dividing the average entropy over the number of sequences. The main objection of proteomics experts to the entropy-based approach was that it takes no account of domain knowledge concerning differential distances between amino acids; entropy computations assume a zero-or-one distance between residues, whereas it is a known biological fact that certain residues are more closely related than others. This knowledge has been codified in substitution matrices, among which Blosum matrices have been shown to achieve better overall performance [7]. On the basis of the Blosum-62 matrix, we computed a motif's blosum score by averaging over the blosum scores of its individual residues. In the absence of strong prior arguments in favor of

Table 1. Predictive information for fingerprint classification.

Description		Variables
I. Fingerprint		
Number of motifs		nmt
Number of proteins		npr
True positive rate		tpr
Partial positive rate		ppr
II. Motif		
Motif length (average, std, median, min, max)		mlen-A|S|D|N|X
Motif coverage (average, stdev, median, min, max)		mcov-A|S|D|N|X
Motif entropy (average, stdev, median, min, max)		ment-A|S|D|N|X
Motif blosum score (average, stdev, median, min, max)		msco-A|S|D|N|X
Intermotif distance (average, stdev, median, min, max)		mdis-A|S|D|N|X
III. Protein sequence		
SWISS-PROT ID: fraction of proteins with an ID		pSP
- LHS	frac of proteins whose LHS length ⩾ 3|4 chars	pN3, pN4
	frac of proteins with common first 1|2|3|4 chars in LHS	mj1, mj2, mj3, mj4
	entropy of LHS averaged over first 1|2|3|4 chars	e1, e2, e3, e4
- RHS	frac of proteins with a common RHS (species)	mjr
	entropy of RHS taken as a unit	er
CC	similarity: sequence belongs to family	cc-belongs
CC	similarity: sequence contains domain	cc-contains

either entropy or blosum scores, we decided to retain both, leaving it up to the feature selection process (Sec. 3.2) to sort out their relative effectiveness. A final characteristic concerns the distance between a motif and its nearest neighbor in the fingerprint; intuitively, large lengths and small intermotif gaps suggest closely related protein sequences. These motif characteristics are summarized in Table 1-II. For propositional learning, where the training unit is the fingerprint, we summarized each characteristic by computing its average, standard deviation, median, minimum and maximum over all fingerprint motifs.

Sequence. Each protein sequence is uniquely identified by its SWISS-PROT/TrEMBL ID or accession number. In an approach similar to that taken by PRECIS, we use these codes to retrieve the SWISS-PROT entry for the protein and examine this for information concerning the individual protein or the family to which it belongs. The SWISS-PROT ID field itself is particularly informative by virtue of its structure. It is composed of two parts separated by an underscore; the left hand side (LHS) denotes the protein type and the right hand side (RHS) the species. PRECIS' classification heuristics focus on the LHS, which tends to be homogeneous among members of a protein family. PRECIS searches for a common root of at least 2 characters in a set of sequence IDs; if such a root is found in at least 75% of these, the fingerprint is assumed to represent a family.

Rather than imposing fixed thresholds as PRECIS does, we simply isolated features that might correlate with fingerprint type and expressed them in terms of relative frequencies – e.g., the relative frequency of SWISS-PROT IDs in a set of proteins, or the proportion of IDs whose LHS is at least 3 characters long. We used two features to simulate PRECIS' homogeneity heuristic: (1) the majority score, defined as the proportion of LHSs sharing the most frequent common root of 1-4 characters, and (2) entropy as averaged over the first 1-4 characters of the LHS. For the right hand side, homogeneity was also quantified by the majority score and entropy, but computed this time over the RHS as a whole. This asymmetric processing of the 2 ID components aims to mimic unwritten conventions that appear to govern assignment of protein names in SWISS-PROT. In the LHS, biological homogeneity is suggested by the length of the leftmost common substring in a set of protein names; for instance the perfect uniformity of the LHS in JAK1_HUMAN, JAK1_MOUSE, JAK1_BRARE and JAK1_CYPCA suggests a tightly knit baselevel family while the 4th-letter variations in BAXA_HUMAN, BAXB_HUMAN and BAXD_HUMAN reflect interfamily differences within a superfamily. However, these conventions are implicit and short of perfectly consistent, hence the need for adaptive induction from examples rather than formulation as hard and fast rules.

Finally, we follow PRECIS' reliance on SWISS-PROT's CC similarity field, which often contains information about the family membership of a protein. This field's value can take the form belongs to <family-or-superfamily-name> or contains <domain-name>. However the information is not always consistent for all proteins in a fingerprint; rather than a boolean indicating the presence or absence of the flag words 'belongs to' or 'contains', we compute the proportion of proteins containing one or the other (whichever is more frequent).

3 Data Preprocessing and Mining Methods

3.1 Missing Value Imputation

The SWISS-PROT ID field is a valuable source of hints concerning the class of proteins in a fingerprint. Unfortunately, many proteins have no associated SWISS-PROT IDs. As a result for 7.5% of the training examples, all 12 features concerning the LHS and RHS of SWISS-PROT IDs had missing values. These values are clearly not 'missing completely at random' as defined in [8], since their presence or absence is contingent on the value of another feature, the fraction of SWISS-PROT identified proteins. This precludes the use of simple data completion methods such as replacement by means, which have the added drawback of underestimating variance. In addition, the distribution of incomplete features was diverse and far from normal; we thus imputed missing values using a non parametric technique based on K-nearest neighbors [11].

3.2 Feature Selection

The initial data representation described in Sec. 2 contained a total of 45 features or predictive variables, 30 based on the initial fingerprint and 15 on information culled from SWISS-PROT. It was not obvious which of these features were discriminating or redundant or even harmful. To obtain the minimal feature set needed to obtain reasonable performance, several feature selection methods were investigated and their impact on classification accuracy evaluated. We compared two variable ranking methods based on information gain or mutual information ($I(X,Y) = H(X) - H(X|Y) = H(Y) - H(Y|X)$), and symmetrical uncertainty ($U(X,Y) = 2\left[\frac{H(X)+H(Y)-H(X,Y)}{H(X)+H(Y)}\right]$). To account for feature interaction, we included ReliefF and correlation-based feature selection (CFS). CFS selects feature sets rather than individual features [6]; while ReliefF scores individual features, its nearest-neighbor based approach evaluates each feature in the context of all others and integrates the impact of irrelevant, noisy or redundant features [10].

3.3 Algorithm and Model Selection

To ensure coverage of the space of possible hypotheses, we investigated learning algorithms with clearly distinct biases. Among the basic algorithms we used were logic-based learning algorithms that build decision trees and rules (J48 and Part [12], variants of C5.0 tree and C5.0 rules respectively; Ltree, which builds oblique hyperplanes contrary to the orthogonal decision borders built by C5.0 [5]); density-estimation based learners like Naïve Bayes (NBayes) and instance-based learning (IBL, a variant of K-nearest-neighbors); linear discriminants (Lindiscr) and their diverse extensions such as multilayer perceptrons (MLPs), and support vector machines (SVMs). Details on each of these learning approaches can be found in, e.g., [4].

These methods represent different points along the bias-variance spectrum: NBayes and LinDiscr are extremely high-bias algorithms; at the other extreme, decision trees and rules, IBL, MLPs and SVMs are high-variance algorithms which can yield simple or very complex models depending on user-tuned complexity parameters. From the viewpoint of feature evaluation, sequential methods like orthogonal decision trees and rules consider predictive features successively while so-called parallel learners like NBayes, IBL, LinDiscr, MLPs, and SVMs evaluate all features simultaneously. Ltree is a hybrid sequential-parallel algorithm as it builds decision tree nodes sequentially but can create linear combinations of features at each node.

This set of basic algorithms was completed by two ensemble methods, boosted decision trees (C5.0boost) and RandomForest [2]. Ensemble methods build multiple models and classify new instances by combining (usually via some form of weighted voting) the decisions of the different models. Both methods use decision trees as base learners but differ in the way they diversify the training set. Boosting produces new variants of the training set by increasing the weights of instances misclassified by the model built in the previous training cycle, effectively obliging the learner to focus on the more difficult examples. RandomForest produces data variants by selecting features instead of examples. At each node, RandomForest randomly draws a subset of K (a user-defined parameter) features and then selects the test feature from this typically much smaller subset.

In order to find a reasonably good hypothesis, learning algorithms should be assessed in a variety of parameter settings. From the set of candidates described above, only the high-bias algorithms, NBayes and Lindiscr, have no complexity parameters; however NBayes has variants based on whether continuous variables are discretized (D) or not, in which case probabilities are computed either by assuming normality (N) or via kernel-based density estimation (K). For all the others, we tested a number of parameter settings and used only the best settings for inter-algorithm comparison. The main complexity parameter of recursive partitioning algorithms (J48, Part, and Ltree) is the C parameter, which governs the amount of postpruning performed on decision trees and rules. Its default value is 0.25 for J48 and Part and 0.10 for Ltree. We tried values of C from 1 to 50 in increments of 5. In IBL, the parameter K (the number of nearest neighbors to explore) produces a maximal-variance model when assigned the default value of 1. At the other extreme, with K = the number of training instances, IBL degenerates to the default majority rule. We explored the behavior of IBL with K=1, 10, 25, 40, 55, and 70. The topology of MLPs is governed by H, the number of hidden units. We tested H=1, 10, 23, 50, 75, and 100. RandomForest is governed by 2 main parameters – the number of trees which form the commitee of experts (I) and the number of features (K) to select for each tree. We explored combinations of I = 10, 25, 50, and 100 and K = 3, 6, 12, 18, 24. Finally, SVM complexity parameters depend on the type of kernel used. We tried polynomial kernels with degree E = 1 and 2 as well as radial basis function (Gaussian) kernels, with gamma (G) or width = 0.01 (default), 0.05, 0.1, 0.15, and 0.2. In addition, the regularization parameter C governs the trade-off between the empirical risk or training error and the complexity of the hypothesis. We explored values of C from 1 (default) to 100 in increments of 10.

4 Experimentation and Results

4.1 Experimental Strategy

The dataset contained 1842 fingerprint records from version 37 of the PRINTS database. 1487 cases were used as the design set, i.e., as training and validation sets for algorithm, feature, and model selection. The rest (355 cases) was held out for blind testing of the trained models. All experiments were conducted using stratified 10-fold cross-validation. Prior to training, missing values of incomplete examples were imputed as described in Sec. 3.1. Contrary to KNN-based missing value imputation, which is unsupervised, all feature selection techniques used (Sec. 3.2) rely on the training class labels and therefore had to be nested within the cross-validation loop.

4.2 Results on the Initial Feature Set

Table 2, column 3, summarizes the cross validated error rates of the learning algorithms. To provide a basis for comparison, the top give two baseline errors. The first is the traditional default error obtained by assigning all examples to the majority class. The class distribution of the 1842-instance training set is as follows: domain = 0.05, family = 0.54, superfamily=0.41. The majority rule thus yields a baseline error of 45.6% on the training set. A second yardstick, specific to the given task, is the error rate obtained by applying PRECIS' handcrafted classification heuristics. A simulation run of these heuristics on both the design set and the blind test set revealed an error rate of around 40%.

The obvious result is that the error rates of all learning algorithms are significantly better than both the default error of ~46% and the PRECIS error of ~40%. The advantage gained from data mining leaves no room for doubt. Note that the lowest errors in this application are obtained by either ensemble

Table 2. Error rates on the full 45-feature set. Each row gives the optimal parameter setting found for the given method, its cross-validation (CV) error on the design dataset, and its final test error on the holdout (HO) set.

Method	Parameters	CV error	HO error
Default		45.60	46.19
PRECIS		39.55	40.28
SVM-RBF	G=0.05, C=50	14.06	14.65
RandomForest	I=100, K=6	14.59	17.46
C5.0boost	B=10, C=0.1	15.13	18.59
MLP	H=10	15.13	16.62
IBL	K=10	15.47	19.44
Lindiscr	-	15.80	17.18
LTree	C=0.05	16.27	17.46
J48	C=0.01	16.48	19.15
Part	C=0.05	19.97	21.69
NBayes	K	23.20	27.07

methods (RandomForest and C5.0boost) or parallel learning algorithms which examine all features simultaneously (SVM-RBF, MLP, IBL). On the contrary, learning algorithms which test individual features sequentially (Ltree, J48, Part) are gathered together at the lower end of the performance scale. NBayes turned out to be the least accurate in all our comparative experiments, with kernel-based density estimation obtaining slightly better results than the variants that discretize continuous features or assume a normal distribution. NBayes aside, this clear performance dichotomy between parallel and sequential algorithms will be observed constantly in this study under different experimental settings.

The statistical significance of the differences in error rate is not clearcut. Without adjustment for multiple comparisons, the difference between the five lowest error rates is not statistically significant at the 1% level. However, after applying the Bonferroni adjustment for a total of around 500 pairwise tests, all statistically significant differences vanish among the first 8 models. Nevertheless, we selected the model with the lowest nominal error – SVM-RBF with a kernel width of 0.05 and a complexity parameter C of 50.

The result of algorithm and model selection was then validated on the blind test set. Since cross-validation produces a different model at each iteration, the selected SVM-RBF parameterization was rerun on the full training set and applied to the blind test set of 355 examples. The error rate obtained was 14.65%, confirming that the observed cross-validation error of 14.06% resulted not from overfitting but from effective generalization. As a countercheck, the other candidate models were also run on the holdout set; the results are shown in the last column of Table 2. The difference between the cross-validation and the blind test error is less than 0.6% for SVM-RBF but varies between 1.2% and 4% for all other algorithms, the highest blind test error (NBayes) exceeding 27%. This remarkable stability of SVM-RBF, added to its predictive accuracy, parsimony, and reasonable computational speed, confirms and magnifies the advantage of SVM-RBF over the other learning methods on this specific classification task.

4.3 Results of Feature Selection

To find the minimal feature set needed for accurate prediction and see which features were truly discriminating, we applied the feature selection methods described in Section 3.2. The number of features to retain was automatically determined by the subset selector CFS in backward search mode but had to be supplied by the user for the three feature rankers ReliefF, InfoGain, and SymmU (we tried 32, 36, and 40 features).

Results are summarized in Table 3. Each row shows the specific combination of model parameters, feature selection method, and number of selected features that produced the lowest cross-validation error (col. 5) for a given learning algorithm (col. 1). The first obvious finding is that feature selection improves performance for all learning algorithms except SVM-RBF, which achieves the same error rate with 36 features as with the initial set of 45 features. Nevertheless, SVM-RBF conserves its top rank; in fact, the most remarkable result is that the overall ranking of learning algorithms remains the same before and after feature selec-

Table 3. Cross-validation and holdout error rates after feature selection.

Method	Parameters	Feature selector	# features	CV error	HO error
SVM-RBF	G=0.05, C=90	ReliefF	36	14.09	14.08
RandomForest	I=25, K=12	InfoGain	40	14.19	16.61
C5.0boost	C=0.3	ReliefF	32	14.79	16.90
MLP	H=10	ReliefF	40	14.86	16.90
IBL	K=10	SymmU	32	14.93	18.31
Lindiscr	-	ReliefF	40	15.40	17.46
LTree	C=0.05	SymmU	32	15.53	18.59
J48	C=0.10	SymmU	32	15.53	19.72
Part	C=0.10	CFS	7-10	17.35	18.03
NBayes	K	CFS	7-10	18.02	23.66

tion. To validate the observed results, we reran these ten learning configurations on the full training set and applied the resulting models to the blind test set. Here again, we observe the same phenomenon as on the initial feature set: the holdout error of SVM-RBF is 14.08%, practically identical to its cross-validation error of 14.06%; for all other algorithms the holdout error was higher than the cross-validation error by an average of 2.84%.

5 Discussion

This section addresses two issues related to the findings described above. First, what is the source of SVM-RBF's generalization power on this particular task? Second, how can we assess the relative impact of domain-specific (e.g. Blosum scores) and domain-independent features (e.g. entropy) on the discriminatory ability of the trained model?

One hypothesis that might explain the performance of SVM-RBF on this task is its approach to multiclass learning. Rather than solve a C-class problem directly like most of the other algorithms studied, it builds a decision frontier by building and combining the responses of $\binom{C}{2}$ pairwise binary classifiers. In this sense SVM-RBF could be viewed as an ensemble method and the rankings given in Tables 2 and 3 would simply confirm the widely observed efficacy of model combination versus individual models in many classification tasks. This hypothesis is however weakened by the fact that Lindiscr follows the same pairwise binary approach to multiclass problems and yet displays worse performance. To see more clearly into the issue, we reran J48, Part, IBL, NBayes and MLP with the same parameters as in Table 2, but this time adopting SVM's pairwise binary classification strategy. Holdout error increased slightly for MLP and J48, and improved somewhat for the others. However, no error improvement led to a performance level comparable to SVM-RBF's 14.65% holdout error. While the binary pairwise approach may have favorably affected accuracy, it cannot be considered the main source of SVM-RBF's generalization performance.

A complementary explanation can be found by comparing the performance of parallel and sequential learners, as noted in Sec. 4.2. Recursive partitioning methods generally fared badly on this problem, as shown clearly in Tables 2 and 3. Exceptions are the two ensemble methods where the myopia of sequential feature testing is compensated by iterative resampling, whether instance-wise (boosting) or feature-wise (RandomForest). These cases aside, parallel algorithms take the top six performance ranks, with or without feature selection. The clear performance dichotomy between parallel and sequential algorithms suggests a strong interaction among the 45 features distilled from fingerprints. This conjecture is further supported by results of sensitivity analyses described below.

The second issue concerns the relative contributions of domain-specific (e.g., Blosum scores, PRECIS heuristics) and domain-independent features (e.g. entropy measures) to the discriminatory power of the final classifier. We examined separately features describing motif conservation and those describing the set of collected proteins. Motif conservation in a fingerprint is depicted by 2 groups of features, one based on domain-specific Blosum scores and another on generic entropy measures (Sec. 2). To compare their relative effectiveness, we removed each feature set at a time and trained the selected SVM-RBF learner on the remaining features. Each time, the resulting increase in error was taken to quantify the impact of the excised feature set on classification performance. Finally, we removed both feature sets simultaneously to estimate their combined impact.

The results are shown in Table 4(a). Error increase was slight for both feature sets and neither appeared to have a convincingly higher impact on accuracy than the other. Even the combined impact on performance differed little from the individual contribution of one or the other. Blosum scores and entropy not only seem to have roughly equivalent and redundant predictive impact; their combined contribution is scarcely greater. We see two possible explanations: either motif conservation has a minor role in discriminating fingerprint types, or an adequate representation of motif conservation remains to be found.

Knowledge-based and knowledge-poor features concerning fingerprint proteins displayed quite different trends. One set of features embodied expert knowledge underlying the PRECIS heuristics while another set comprised less informed

Table 4. Impact of knowledge-based and knowledge-poor features. Error" and "Perf Impact" indicate respectively the error and error increase (wrt to the baseline) entailed by removal of a given feature set.

	CV		HO	
Baseline: Full feature set (SVM-RBF)	14.06		14.65	
(a) Features describing motif conservation				
	Error	Perf Impact	Error	Perf Impact
Uninformed (entropy)	15.33	1.28	16.34	2.28
Knowledge-based (Blosum scores)	14.85	0.81	16.62	2.56
Both	15.06	1.01	16.62	2.56
(b) Protein-related features				
	Error	Perf Impact	Error	Perf Impact
Uninformed (see Section 2)	24.88	10.83	29.86	15.80
Knowledge-based (PRECIS rules)	14.53	0.47	15.49	1.44
Both	27.10	13.05	33.52	19.47

features such as simple statistical and entropy measures on the left and right components of SWISS-PROT protein IDs. We followed the same procedure as above to quantify their respective contributions to generalization power; these are summarized in Table 4(b). When PRECIS-based features were removed, hold-out error increased by 1.44% whereas removal of uninformed features incurred a degradataion of 15.8% (cross-validation errors display the same behavior). It is clear that uninformed features are contributing much more to classification performance (note however that the 3 knowledge-based features are heavily outnumbered by the 12 uninformed features). Remarkably, when both feature sets were deleted, error climbed to 27.10%, much more than the sum of their individual contributions.

To summarize, these motif- and protein-centered views of fingerprints reveal two distinct feature interaction scenarios. In one case, the domain-specific and domain-independent feature sets have roughly comparable contributions to predictive accuracy; their combination seems to add nothing to either alone, but it is unclear which should be kept. In the second case one feature set is clearly more effective than the other, but their combined contribution to generalization performance suggests a synergy that individual feature rankers or sequential learners are at pains to capture. This could explain the observed superiority of parallel learners like SVMs and MLPs on this particular problem.

6 Conclusion and Future Work

Since this classification task is a preliminary step in the time-consuming process of annotating protein fingerprints, it is important to achieve high accuracy in order to avoid even more tedious backtracking and database entry revision. The approach described in this paper achieved a 26% accuracy increase relative to the performance of expert rules; the goal of ongoing and future work is to decrease further the residual error of 14.1%. There is a diversity of ways to achieve this; we explored 2 alternatives with negative results.

The first unfruitful track is the relational learning approach. As seen in Section 2, protein fingerprints have a natural multirelational flavor since they gather information on diverse object types – the fingerprints themselves and their component motifs and protein sequences. Relational learning thus seemed to be a way of gaining accuracy via increased expressive power. However, our experiments in relational instance-based learning were inconclusive; they incurred considerably higher computational costs but did not yield better performance than the propositional approach reported above.

The second track explored was the combination of multiple learned models. We investigated the efficacy of combining learned models with uncorrelated errors to obtain a more accurate classifier. We measured the pairwise error correlation of the 10 classifiers in Table 2. SVM-RBF, NBayes and Part were among those that had the lowest pairwise error correlations. We built an ensemble model which combined the predictions of these three learners by a simple majority vote. The error rate of the combined model was 15.87% on the training set and 18.03%

on the holdout test set – in both cases higher than that of SVM-RBF alone. Several other model combinations based on low error correlation were explored; they all yielded higher error rates than at least one of their component base learners.

Other hopefully more promising paths remain to be explored. Ongoing work is focused on correcting data imbalance to increase accuracy. Protein domain families represent less than 5% of PRINTS records, and we are adapting to this task a set of class rebalancing techniques that have proved effective in another application domain [3]. Perhaps the biggest remaining challenge is that of bringing more discriminatory information to bear on the classification task. Integrating information from databases other than SWISS-PROT is a feasible solution in the short term. But given the time lag between the production of new data and their availability in structured databases, we may ultimately have to mine the biological literature to gather fresh insights on the 14% of protein fingerprints that currently defy classification.

Acknowledgements

The work reported above was partially funded by the European Commission and the Swiss Federal Office for Education and Science in the framework of the BioMinT project.

References

1. T. K. Attwood, P. Bradley, D. R. Flower, A. Gaulton, N. Maudling, and A. L. Mitchell et al. PRINTS and its automatic supplement, prePRINTS. *Nucleic Acids Research*, 31(1):400–402, 2003.
2. L. Breiman. Random forests. *Machine Learning*, 45:5–32, 2001.
3. G. Cohen, M. Hilario, H. Sax, and S. Hugonnet. Data imbalance in surveillance of nosocomial infections. In *Proc. International Symposium on Medical Data Analysis*, Berlin, 2003. Springer-Verlag.
4. R. Duda, P. Hart, and D. Stork. *Pattern Classification*. Wiley, 2000.
5. J. Gama and P. Brazdil. Linear tree. *Intelligent Data Analysis*, 3:1–22, 1999.
6. M. Hall. Correlation-based feature selection for discrete and numeric class machine learning. In *Proc. 17th International Conference on Machine Learning*, 2000.
7. S. Henikoff and J. G. Henikoff. Amino acid substitution matrices from protein blocks. *Proc. National Academy of Sciences USA*, 89:10915–10919, November 1992.
8. R. J. Little and D. B. Rubin. *Statistical Analysis with Missing Data*. Wiley, 1987.
9. A. L. Mitchell, J. R. Reich, and T. K. Attwood. PRECIS–protein reports engineered from concise information in SWISS-PROT. *Bioinformatics*, 19:1664–1671, 2003.
10. M. R. Sikonja and I. Kononenko. Theoretical and empirical analysis of relieff and rrelieff. *Machine Learning*, 53:23–69, 2003.
11. O. Troyanskaya, M. Cantor, G. Sherlock, P. Brown, T. Hastie, R. Tibshirani, D. Botstein, and R. Altman. Missing value estimation methods for DNA microarrays. *Bioinformatics*, 17(6):520–525, 2001.
12. I. Witten and E. Frank. *Data Mining. Practical Machine Learning Tools and Techniques with Java Implementations*. Morgan Kaufmann, 2000.

Finding Interesting Pass Patterns from Soccer Game Records

Shoji Hirano and Shusaku Tsumoto

Department of Medical Informatics, Shimane University, School of Medicine
89-1 Enya-cho, Izumo, Shimane 693-8501, Japan
hirano@ieee.org, tsumoto@computer.org

Abstract. This paper presents a novel method for finding interesting pass patterns from soccer game records. Taking two features of the pass sequence – temporal irregularity and requirements for multiscale observation – into account, we have developed a comparison method of the sequences based on multiscale matching. The method can be used with hierarchical clustering, that brings us a new style of data mining in sports data. Experimental results on 64 game records of FIFA world cup 2002 demonstrated that the method could discover some interesting pass patterns that may be associated with successful goals.

1 Introduction

Game records of sports such as soccer and baseball provide important information that supports the inquest of each game. Good inquest based on the detailed analysis of game records makes the team possible to clearly realize their weak points to be strengthened, or, superior points to be enriched. However, the major use of the game records is limited to the induction of basic statistics, such as the shoot success ratio, batting average and stealing success ratio. Although video information may provide useful information, its analysis is still based on the manual interpretation of the scenes by experts or players.

This paper presents a new scheme of sports data mining from soccer game records. Especially, we focus on discovering the features of pass transactions, which resulted in successful goals, and representing the difference of strategies of a team by the pass strategies. Because a pass transaction is represented as a temporal sequence of the position of a ball, we used clustering of the sequences. There are two points that should be technically solved. First, the length of a sequence, number of data points constituting a sequence, and intervals between data points in a sequence are all irregular. A pass sequence is formed by concatenating contiguous pass events; since the distance of each pass, the number of players translating the contiguous passes are by nature difference, the data should be treated as irregular sampled time-series data. Second, multiscale observation and comparison of pass sequences are required. This is because a pass sequence represents both global and local strategies of a team. For example, as a global strategy, a team may frequently use side-attacks than counter-attacks.

J.-F. Boulicaut et al. (Eds.): PKDD 2004, LNAI 3202, pp. 209–218, 2004.

As a local strategy, the team may frequently use one-two pass. Both levels of strategies can be found even in one pass sequence; one can naturally recognize it from the fact that a video camera does zoom-up and zoom-out of a game scene. In order to solve these problems, we employed multiscale matching [1], [2], a pattern recognition based contour comparison method.

The rest of this paper is organized as follows. Section 2 describes the data structure and preprocessing. Section 3 describes multiscale matching. Section 4 shows experimental results on the FIFA world cup 2002 data and Section 5 concludes the results.

2 Data Structure and Preprocessing

2.1 Data Structure

We used the high-quality, value-added commercial game records of soccer games provided by Data Stadium Inc., Japan. The current states of pattern recognition technique may enable us to automatically recognize the positions of ball and players [3], [4], [5], however, we did not use automatic scene analysis techniques because it is still hard to correctly recognize each action of the players.

The data consisted of the records of all 64 games of the FIFA world cup 2002, including both heats and finals, held during May-June, 2002. For each action in a game, the following information was recorded: time, location, names(number) of the player, the type of event (pass, trap, shoot etc.), etc. All the information was generated from the real-time manual interpretation of video images by a well-trained soccer player, and manually stored in the database. Table 1 shows an example of the data. In Table 1, 'Ser' denotes the series number, where a series denotes a set of contiguous events marked manually by expert. The remaining fields respectively represent the time of event occurrence ('Time'), the type of event ('Action'), the team ID ('T_1') and player ID (P_1) of one acting player 1, the team ID ('T_2') and player ID (P_2) of another acting player 2, spatial position of player 1 (X_1, Y_1), and spatial position of player 2 (X_1, Y_1), Player 1 represents the player who mainly performed the action. As for pass action, player 1 represents the sender of a pass, and player 2 represents the receiver of the pass. Axis X corresponds to the long side of the soccer field, and axis Y corresponds

Table 1. An example of the soccer data record.

Ser	Time	Action	T_1	P_1	T_2	P_2	X_1	Y_1	X_2	Y_2
1	20:28:12	KICK OFF	Senegal	10			0	-33		
1	20:28:12	PASS	Senegal	10	Senegal	19	0	-50	-175	50
1	20:28:12	TRAP	Senegal	19			-175	50		
1	20:28:12	PASS	Senegal	19	Senegal	14	-122	117	3004	451
1	20:28:14	TRAP	Senegal	14			3004	451		
⋮				⋮						
169	22:18:42	P END	France	15			1440	-685		

to the short side. The origin is the center of the soccer field. For example, the second line in Table 2 can be interpreted as: Player no. 10 of Senegal, locating at (0,-50), sent a pass to Player 19, locating at (-175,50).

2.2 Preprocessing

We selected the series that contains important 'PASS' actions that resulted in goals as follows.

1. Select a series containing an 'IN GOAL' action.
2. Select a contiguous 'PASS' event. In order not to divide the sequence into too many subsequences, we regarded some other events as contiguous events to the PASS event; for example, TRAP, DRIBBLE, CENTERING, CLEAR, BLOCK. Intercept is represented as a PASS event in which the sender's team and receiver's team are different. However, we included an intercept into the contiguous PASS events for simplicity.
3. From the Selected contiguous PASS event, we extract the locations of Player 1, X_1 and Y_1, and make a time series of locations $p(t) = \{(X_1(t), Y_1(t))|1 \leq t \leq T\}$ by concatenating them. For simplicity, we denote $X_1(t)$ and $Y_1(t)$ by x(t) and y(t) respectively.

Figure 1 shows an example of spatial representation of a PASS sequence generated by the above process. Table 2 provides an additional information, the raw data that correspond to Figure 1. In Figure 1 the vertical line represents the axis connecting the goals of both teams. Near the upper end (+5500) is the goal of France, and near the lower end is the goal of Senegal. This example PASS sequence represents the following scene: Player no. 16 of France, locating at (-333,3877), send a pass to player 18. Senegal cuts the pass at near the center of the field, and started attack from the left side. Finally, Player no. 11 of Senegal made a CENTERING, and after several block actions of France, Player no. 19 of Senegal made a goal.

By applying the above preprocess to all the IN GOAL series, we obtained N sequences of passes $P = \{p_i|1 \leq i \leq N\}$ that correspond to N goals, where i of p_i denote the i-th goal.

3 Multiscale Comparison and Grouping of the Sequences

For every pair of PASS sequences $\{(p_i, p_j) \in P|1 \leq i < N, i < j \leq N\}$, we apply multiscale matching to compare their dissimilarity. Based on the resultant dissimilarity matrix, we perform grouping of the sequences using conventional hierarchical clustering [6].

Multiscale Matching is a method to compare two planar curves by partly changing observation scales. We here briefly explain the basic of multiscale matching. Details of matching procedure are available in [2].

Let us denote two input sequences to be compared, p_i and p_j, by A and B. First, let us consider a sequence $x(t)$ containing X_1 values of A. Multiscale

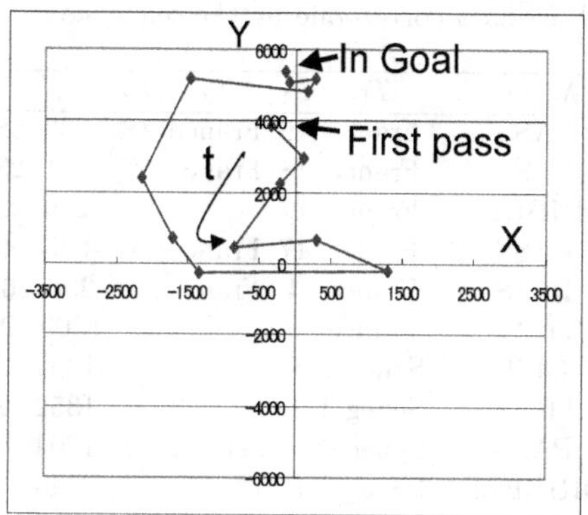

Fig. 1. Spatial representation of a PASS sequences.

representation of $x(t)$ at scale σ, $X(t,\sigma)$ can be obtained as a convolution of $x(t)$ and a Gaussian function with scale factor σ as follows.

$$X(t,\sigma) = \int_{-\infty}^{+\infty} x(u) \frac{1}{\sigma\sqrt{2\pi}} e^{-(t-u)^2/2\sigma^2} du \tag{1}$$

where the gauss function represents the distribution of weights for adding the neighbors. It is obvious that a small σ means high weights for close neighbors, while a large σ means rather flat weights for both close and far neighbors. A sequence will become more flat as σ increases, namely, the number of inflection points decreases. Multiscale representation of $y(t)$, $Y(t,\sigma)$ is obtained similarly.

The curvature of point t at scale σ is obtained as follows.

$$K(t,\sigma) = \frac{X'Y'' - X''Y'}{(X'^2 + Y'^2)^{3/2}}, \tag{2}$$

where X', X'', Y' and Y'' denote the first- and second-order derivatives of $X(t,\sigma)$ and $Y(t,\sigma)$ by t, respectively. The m-th order derivative of $X(t,\sigma)$, $X^{(m)}(t,\sigma)$, is derived as follows.

$$X^{(m)}(t,\sigma) = \frac{\partial^m X(t,\sigma)}{\partial t^m} = x(t) \otimes g^{(m)}(t,\sigma). \tag{3}$$

Figure 2 provides an illustrative example of multiscale description of A.

Next, we divide the sequence $K(t,\sigma)$ into a set of convex/concave subsequences called segments based on the place of inflection points. A segment is a subsequence whose ends correspond to the two adjacent inflection points, and can be regarded as a unit representing substructure of a sequence.

Let us assume that a pass sequence $A^{(k)}$ at scale k is composed of R segments. Then $A^{(k)}$ is represented by

$$A^{(k)} = \left\{ a_i^{(k)} \mid i = 1, 2, \cdots, R^{(k)} \right\}, \tag{4}$$

Table 2. Raw data corresponding the sequence in Figure 1.

Ser	Time	Action	T_1	P_1	T_2	P_2	X_1	Y_1	X_2	Y_2
47	20:57:07	PASS	France	16	France	18	-333	3877	122	-2958
47	20:57:08	PASS	France	18	France	17	122	2958	-210	-2223
47	20:57:10	DRIBBLE	France	17			-210	2223	-843	-434
47	20:57:14	PASS	France	17	France	4	-843	434	298	-685
47	20:57:16	PASS	France	4	France	6	298	685	1300	217
47	20:57:17	TRAP	France	6			1300	217		
47	20:57:19	CUT	Senegal	6			-1352	-267		
47	20:57:19	TRAP	Senegal	6			-1352	-267		
47	20:57:20	PASS	Senegal	6	Senegal	11	-1704	702	-2143	2390
47	20:57:21	DRIBBLE	Senegal	11			-2143	2390	-1475	5164
47	20:57:26	CENTERING	Senegal	11			-1475	5164		
47	20:57:27	CLEAR	France	17			175	4830		
47	20:57:27	BLOCK	France	16			281	5181		
47	20:57:27	CLEAR	France	16			281	5181		
47	20:57:28	SHOT	Senegal	19			-87	5081		
47	20:57:28	IN GOAL	Senegal	19			-140	5365		

where $a_i^{(k)}$ denotes the i-th segment of $A^{(k)}$ at scale $\sigma^{(k)}$. By applying the same process to another input sequence B, we obtain the segment-based representation of B as follows.

$$B^{(h)} = \left\{ b_j^{(h)} \mid j = 1, 2, \cdots, S^{(h)} \right\} \tag{5}$$

where $\sigma^{(h)}$ denote the observation scale of B and $S^{(h)}$ denote the number of segments at scale $\sigma^{(h)}$.

The main procedure of multiscale matching is to find the best set of segment pairs that minimizes the total segment difference. Figure 3 illustrates the process. For example, five contiguous segments A_1 - A_5 at the lowest scale of Sequence A are integrated into one segment A_6 at the middle scale, and the integrated segment A_6 well matches to one segment B_1 in Sequence B at the lowest scale. Thus the set of the five segments in Sequence A and the one segment in Sequence B will be considered as a candidate for corresponding subsequences. While, another pair of segments A_0 and B_0 will be matched at the lowest scale. In this way, matching is performed throughout all scales.

There is one restriction in determining the best set of the segments. The resultant set of the matched segment pairs must not be redundant or insufficient to represent the original sequences. Namely, by concatenating all the segments in the set, the original sequence must be completely reconstructed without any partial intervals or overlaps. The matching process can be fasten by implementing dynamic programming scheme [2].

Dissimilarity $d(a_i^{(k)}, b_j^{(h)})$ of two segments $a_i^{(k)}$ and $b_i^{(h)}$ is defined as follows.

$$d(a_i^{(k)}, b_j^{(h)}) = \frac{\mid \theta_{a_i}^{(k)} - \theta_{b_j}^{(h)} \mid}{\theta_{a_i}^{(k)} + \theta_{b_j}^{(h)}} \left| \frac{l_{a_i}^{(k)}}{L_A^{(k)}} - \frac{l_{b_j}^{(h)}}{L_B^{(h)}} \right| \tag{6}$$

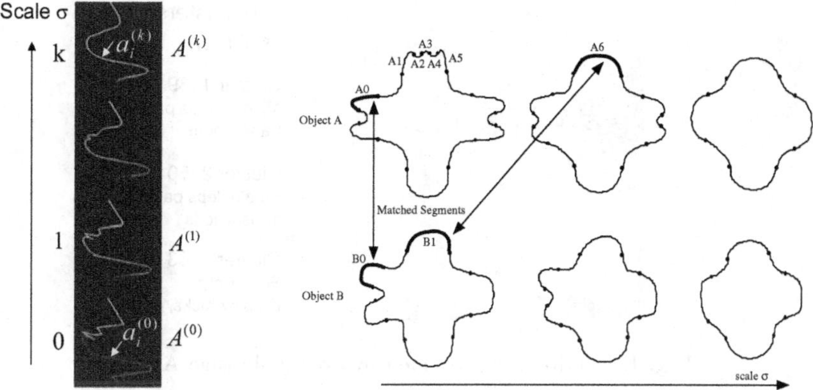

Fig. 2. An example of multiscale description.

Fig. 3. An illustrative example of multiscale matching.

where $\theta_{a_i}^{(k)}$ and $\theta_{b_j}^{(h)}$ denote rotation angles of tangent vectors along segments ${a_i}^{(k)}$ and ${b_j}^{(h)}$, $l_{a_i}^{(k)}$ and $l_{b_j}^{(h)}$ denote the length of segments, $L_A^{(k)}$ and $L_B^{(h)}$ denote the total length of sequences$A$ and B at scales $\sigma^{(k)}$ and $\sigma^{(h)}$, respectively.

The total difference between sequences A and B is defied as a sum of the dissimilarities of all the matched segment pairs as

$$D(A, B) = \sum_{p=1}^{P} d(a_p^{(0)}, b_p^{(0)}), \tag{7}$$

where P denotes the number of matched segment pairs.

4 Experimental Results

We applied the proposed method to the action records of 64 games in the FIFA world cup 2002 described in Section 2. First let us summarize the procedure of experiments.

1. Select all IN GOAL series from original data.
2. For each IN GOAL series, generate a time-series sequence containing contiguous PASS events. In our data, there was in total 168 IN GOAL series excluding own goals. Therefore, we had 168 time-series sequences, each of which contains the sequence of spatial positions $(x(t), y(t))$.
3. For each pair of the 168 sequences, compute dissimilarity of the sequence pair by multiscale matching. Then construct a 168×168 dissimilarity matrix.
4. Perform cluster analysis using the induced dissimilarity matrix and conventional agglomerative hierarchical clustering (AHC) method.

The following parameters were used in multiscale matching: the number of scales = 30, scale interval = 1.0, start scale = 1.0. In order to elude the problem of

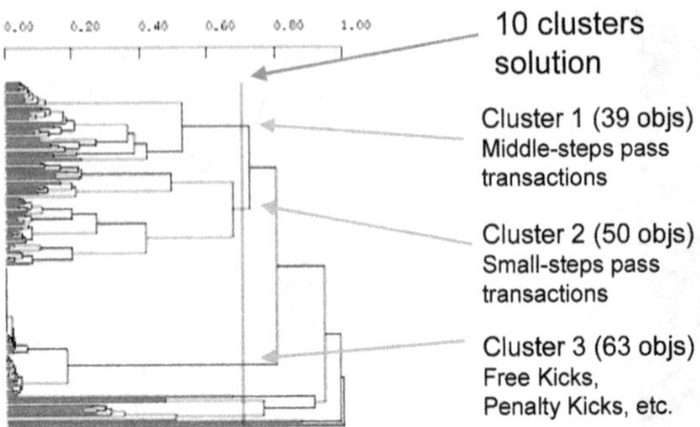

Fig. 4. Dendrogram obtained by average-linkage AHC.

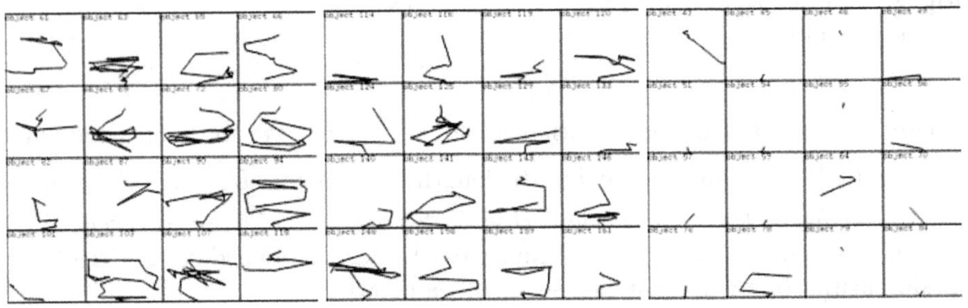

Fig. 5. Example sequences in cluster 1. **Fig. 6.** Example sequences in cluster 2. **Fig. 7.** Example sequences in cluster 3.

shrinkage at high scales, the shrinkage correction method proposed by Lowe et al. [7] was applied.

Figure 4 provides a dendrogram generated obtained using average-linkage AHC. Thirteen clusters solution seemed to be reasonable according to the step width of dissimilarity. However, from visual inspection, it seemed better to select 10 clusters solution, because the feature of clusters was more clearly observed. Figure 5 - 7 provide some examples of sequences clustered into the three major clusters 1, 2, 3 in Figure 4, respectively. Cluster 1 contained complex sequences, each of which contained many segments and often included loops. These sequences represented that the goals were succeeded after long, many steps of pass actions, including some changes of the ball-owner team. On the contrary, cluster 2 contained rather simple sequences, most of which contained only several segments. These sequences represented that the goals were obtained after interaction of a few players. Besides, the existence of many long line segment implied the goals might be obtained by fast break. Cluster 3 contained remarkably short sequences. They represented special events such as free kicks, penalty kicks and corner kicks, that made goals after one or a few touches. These observations demonstrated that the sequences were clustered according to the steps/complexity of the pass pass routes.

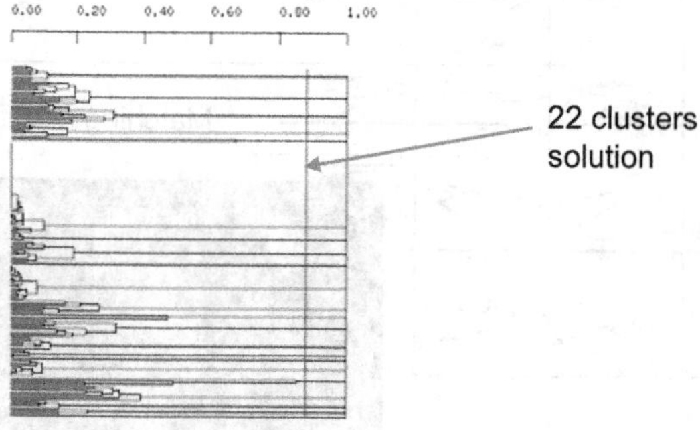

Fig. 8. Dendrogram obtained by complete-linkage AHC.

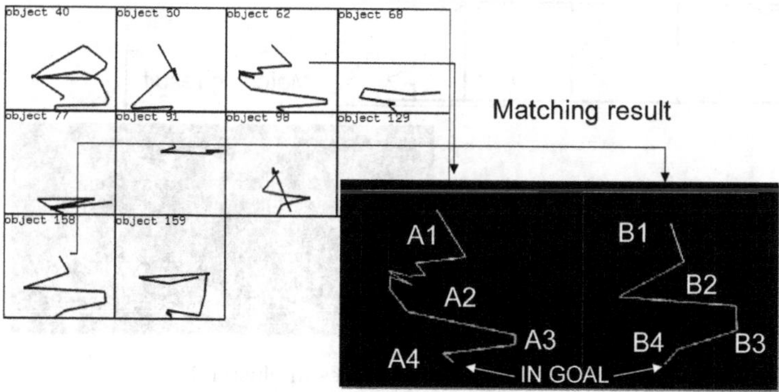

Fig. 9. Example sequences in cluster 15.

Figure 4 provides a dendrogram generated obtained using complete-linkage AHC. Because we implemented multiscale matching so that it produces a pseudo, maximum dissimilarity if the sequences are too different to find appropriate matching result, some pairs of sequences were merged at the last step of the agglomerative linkage. This gave the dendrogram in Figure 4 a little unfamiliar shape. However, complete-linkage AHC produced more distinct clusters than average-linkage AHC.

Figure 9 - 11 provide examples of sequences clustered into the major clusters 15 (10 cases), 16 (11 cases) and 19 (4 cases), for 22 clusters solution. Most of the sequences in cluster 15 contained sequences that include cross-side passes. Figure 9 right represents a matching result of two sequences in this cluster. A matched segment pair is represented in the same color, with notation of segment number A1-B1, A2-B2 etc. The result demonstrates that the similarity of pass patterns - right (A1-B1), cross (A2-B2), centering (A3-B3), shoot (A4-B4) were successfully captured. Sequences in cluster 16 contained loops. Figure 10 right shows a matching result of two sequences in this cluster. Although the directions of goals were different in these two sequences, correspondence between the loops, cross-side passes, centerings and shoots are correctly captured. This is because

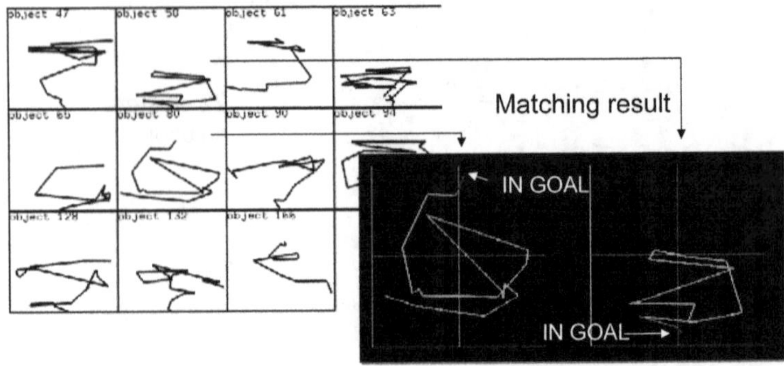

Fig. 10. Example sequences in cluster 16.

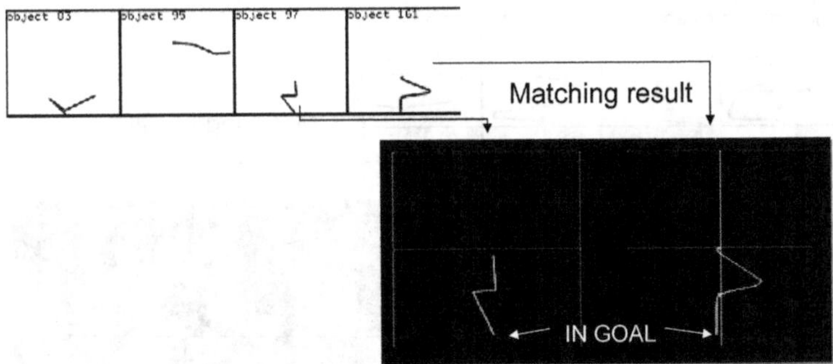

Fig. 11. Example sequences in cluster 19.

multiscale matching is invariant for affine transformations. Cluster 19 contained short step sequences. The correspondence of the segments were also successfully captured as shown in Figure 11.

5 Conclusions

In this paper, we have presented a new method for finding interesting pass patterns from time-series soccer record data. Taking two characteristics of the pass sequence – irregularity of data and requirements of multiscale observation – into account, we developed a cluster analysis method based on multiscale matching, which may build a new scheme of sports data mining. Although the experiments are in the preliminary stage and subject to further quantitative evaluation, the proposed method demonstrated its potential for finding interesting patterns in real soccer data.

Acknowledgments

The authors would like to express their sincere appreciation to Data Stadium Inc. for providing valuable dataset. This research was supported in part by the Ministry of Education, Science, Sports and Culture, Grant-in-Aid for Young Scientists (B), #16700144, 2004.

References

1. F. Mokhtarian and A. K. Mackworth (1986): Scale-based Description and Recognition of planar Curves and Two Dimensional Shapes. IEEE Transactions on Pattern Analysis and Machine Intelligence, PAMI-8(1): 24-43
2. N. Ueda and S. Suzuki (1990): A Matching Algorithm of Deformed Planar Curves Using Multiscale Convex/Concave Structures. IEICE Transactions on Information and Systems, J73-D-II(7): 992–1000.
3. A. Yamada, Y. Shirai, and J. Miura (2002): Tracking Players and a Ball in Video Image Sequence and Estimating Camera Parameters for 3D Interpretation of Soccer Games. Proceedings of the 16th International Conference on Pattern Recognition (ICPR-2002), 1:303–306.
4. Y. Gong, L. T. Sin, C. H. Chuan, H. Zhang, and M. Sakauchi (1995): Automatic Parsing of TV Soccer Programs. Proceedings of the International Conference on Multimedia Computing and Systems (ICMCS'95), 167–174.
5. T. Taki and J. Hasegawa (2000): Visualization of Dominant Region in Team Games and Its Application to Teamwork Analysis. Computer Graphics International (CGI'00), 227-238.
6. B. S. Everitt, S. Landau, and M. Leese (2001): Cluster Analysis Fourth Edition. Arnold Publishers.
7. Lowe, D.G (1989): Organization of Smooth Image Curves at Multiple Scales. International Journal of Computer Vision, 3:119–130.

Discovering Unexpected Information for Technology Watch

François Jacquenet and Christine Largeron

Université Jean Monnet de Saint-Etienne
EURISE
23 rue du docteur Paul Michelon
42023 Saint-Etienne Cedex 2
{Francois.Jacquenet,Christine.Largeron}@univ-st-etienne.fr

Abstract. The purpose of technology watch is to gather, process and integrate the scientific and technical information that is useful to economic players. In this article, we propose to use text mining techniques to automate processing of data found in scientific text databases. The watch activity introduces an unusual difficulty compared with conventional areas of application for text mining techniques since, instead of searching for frequent knowledge hidden in the texts, the target is unexpected knowledge. As a result, the usual measures used for knowledge discovery have to be revised. For that purpose, we have developed the UnexpectedMiner system using new measures for to estimate the unexpectedness of a document. Our system is evaluated using a base that contains articles relating to the field of machine learning.

1 Introduction

In recent years, business sectors have become more and more aware of the importance of mastering strategic information. Businesses are nevertheless increasingly submerged with information. They find it very difficult to draw out the strategic data needed to anticipate markets, take decisions and interact with their social and economic environment. This has led to the emergence of business intelligence [5, 6, 14] that can be defined as the set of actions involved in retrieving, processing, disseminating and protecting legally obtained information that is useful to economic players. When the data analyzed is scientific and technical, the more specific term used is technology watch, meaning the monitoring of patents and scientific literature (articles, theses, etc.).

A watch process can be broken down into four main phases: a needs audit, data collection, processing of the data collected and integration and dissemination of the results. Our focus in this article is mainly on the third phase. For the purpose of automatically processing the data collected, data mining techniques are attractive and seems to be particularly suitable considering that most of the data is available in digital format.

Data mining has grown rapidly since the mid 90's with the development of powerful new algorithms that enable large volumes of business data to be pro-

J.-F. Boulicaut et al. (Eds.): PKDD 2004, LNAI 3202, pp. 219–230, 2004.

cessed [3]. When the data considered comes in the form of texts, whether structured or not, the term used is text mining [8]. By analogy with data mining, text mining, introduced in 1995 by Ronan Feldman [4], is defined by Sebastiani [19] as the set of tasks designed to extract the potentially useful information, by analysis of large quantities of texts and detection of frequent patterns. In fact text mining is already a wide area of research that provides useful techniques that can be used in the context of technology watch. Losiewicz et al. [12] for example show that clustering techniques, automatic summaries, information extraction can be of great help for business leaders. Zhu and Porter [22, 21] show how bibliometrics can be used to detect technology opportunities from competitors information found in electronic documents. Another use of text mining techniques for technology watch was the works of B. Lent et al. in [10] that tried to find new trends from an IBM patent database using sequential pattern mining algorithms [1]. The idea was to observe along the time, sequences of words that were not frequent in patents at a particular period and that became frequent later. In a similar way, but in the the Topic Detection Tracking[1] framework, Rajaraman et al. [17] proposed to discover trends from a stream of text documents using neural networks. In these cases, text mining techniques are mainly used to help managers dealing with large amount of data in order to find out frequent useful information or discover some related works linked with their main concerns. Nevertheless, one important goal of technology watch and more generally business intelligence is to detect new, unexpected and hence generally infrequent information. Thus, the algorithms for extracting frequent patterns that are commonly used for data mining purposes are inappropriate to this area. Indeed, as the name implies, these tools are tailored for information that occurs frequently in a database. This is no doubt one of the main reasons why the software packages marketed so far fail to fulfill knowledge managers needs adequately.

From that assessment, some researchers have tried to focus on what they called, rare events, unexpected information, or emerging topics, etc, depending on the papers. For example, Bun et al. [2] proposed a system to detect emerging topics. From a set of Web sites, their system is able to detect changes in the sites and then scans the words that appear in the changes in order to find emerging topics. Thus, the system is not able to detect unexpected information from Web sites they visit for the first time. Matsumura et al. [13] designed a system to discover emerging topics between Web communities. Based on the KeyGraph algorithm [15], their system is able to analyze and visualize co-citations between Web pages. Communities, each having members with common interests are obtained as graph-based clusters, and an emerging topic is detected as a Web page relevant to multiple communities. However this system is not relevant for corpora of documents that are not necessarily organized as communities.

[1] http://www.nist.gov/speech/tests/tdt

More recently many researches have focused on novelty detection, more specifically in the context of a particular track of the TREC challenges[2]. Many systems have been design in the context of that challenge, nevertheless, they only deal with corpora of sentences and not corpora of texts. Moreover, for most systems, the users need to provide examples of novel sentences. Then the system uses some similarity functions in order to compare new sentences with known novel sentences.

WebCompare, developed by Liu et al. [11] proposed the users to find unexpected information from competitors' Web sites. Hence, a user of the system has to give a set of URLs of Web pages and the URL of his Web site. Then WebCompare is able to find the pages that contain unexpected information with respect to the user's Web site. The unexpectedness of a Web page is evaluated given a measure based on the TF.IDF paradigm. In fact, WebCompare is probably the most related work to our concerns.

Next section presents the global architecture of the system we designed, and called UnexpectedMiner, to automate the Technology Watch process. Section 3 presents the various measures we proposed to mine unexpected information in texts. Section 4 presents some experiments we made to show the efficiency of each measure with respect to each other before we conclude with some future works.

2 The UnexpectedMiner System

In the technology watch area, we have developed the UnexpectedMiner system aimed at extracting documents that are relevant to the knowledge manager from a corpus of documents inasmuch as they deal with topics that were unexpected and previously unknown to the manager. In addition, the system must specifically treat the knowledge manager's request without relying to any large extent on her or his participation. Finally, an important feature we wanted to build into our system is versatility, i.e. a system that is not dedicated to a particular field or topic.

Keeping in mind those objectives, we propose a system made up of several modules as illustrated in Figure 1.

2.1 Pre-processing of Data

In the first phase, the technology watch manager specifies the needs by producing a number of reference documents. In the remainder of this article, this set of documents shall be designated by R while $|R|$ refers to their number. In practice, between ten and twenty documents should be enough to target the scope of interest for the technology watch. The system must then review new documents

[2] The track "novelty detection" at TREC conferences has appear for the first time at the TREC 2002 conference. Papers presented to this conference and next may be found at http://trec.nist.gov

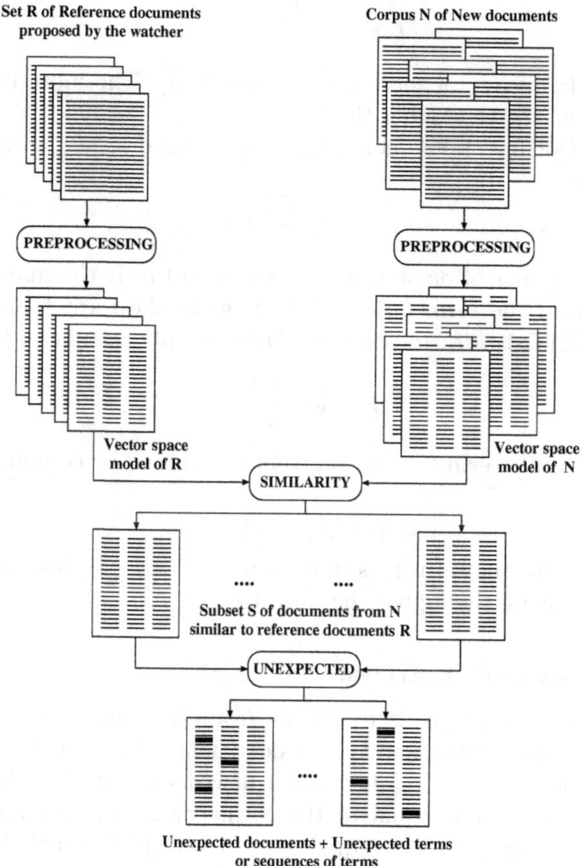

Fig. 1. UnexpectedMiner system architecture

in various corpora made available and retrieve innovative information. This set of new documents shall be referred to below as N and $|N|$ is its cardinal.

Sets R and N then undergo the pre-processing phase. The module designed for that purpose includes a number of conventional processing steps such as removing irrelevant elements and stop words from the documents (logos, url addresses, tags, etc.) and carrying out a morphological analysis of the words in the phrases extracted. Finally, each document is classically represented in vector form. The document d_j is thus considered to be a set of indexed terms t_i where each indexed term is in fact a word in document d_j. An index written out as $T = \{t_1, t_2, ..., t_m\}$ lists all the terms encountered in the documents. Each document is thus represented by a weights vector $\boldsymbol{d_j} = (w_{1,j}, w_{2,j}, ..., w_{m,j})$ where $w_{i,j}$ represents the weight of term t_i in document d_j. If the term t_i does not appear in the document d_j , then $w_{i,j} = 0$. To compute the frequency of a term in a document, we use the TF.IDF formula [18]. TF (Term Frequency) is the relative frequency of term t_i in a document d_j defined by:

$$tf_{i,j} = \frac{f_{i,j}}{max_l f_{l,j}}$$

where $f_{i,j}$ is the frequency of term t_i in document d_j. The more frequent the term t_i in document d_j, the higher the $tf_{i,j}$.

IDF (Inverse Document Frequency) measures the discriminatory power of term t_i defined by:

$$idf_i = \log_2 \frac{Nd}{n_i} + 1$$

where Nd is the number of documents processed and n_i is the number of documents that contain the term t_i. The less frequent the term t_i in the set of documents, the higher is idf_i. In practice, IDF is simply calculated by:

$$idf_i = \log \frac{Nd}{n_i}$$

The weight $w_{i,j}$ of a term t_i in a document d_j is found by combining the two previous criteria:

$$w_{i,j} = tf_{i,j} \times idf_i$$

The more frequent the term t_i is in document d_j and the less frequent it is in the other documents, the higher the weight $w_{i,j}$.

2.2 Similar Document Retrieval

The goal of the second module is to extract from database N new documents which are the most similar to the reference documents R provided by the knowledge manager. The similarity s_{jk} between a new document $d_j \in N$ and a reference document $d_k \in R$ is equal to the cosine measure, commonly used in information retrieval systems. It is equal to the cosine of the angle between the vectors that represent these documents:

$$s_{jk} = \frac{\boldsymbol{d_j} \bullet \boldsymbol{d_k}}{|\boldsymbol{j}| \times |\boldsymbol{k}|}$$

where

$$\boldsymbol{d_j} \bullet \boldsymbol{d_k} = \sum_i w_{i,j} \times w_{i,k}$$

$$|\boldsymbol{j}| = \sqrt{\sum_{i=1,m} w_{i,j}^2}$$

The mean similarity s_j of a new document $d_j \in N$ with the set of reference documents R is equal to:

$$s_j = \frac{1}{|R|} \sum_{k=1}^{|R|} s_{jk}$$

After classifying the mean similarity of new documents in the descending order, a subset S is extracted from N. This is the set of the new documents that are the most similar to those supplied as reference documents by the knowledge manager.

2.3 Unexpected Information Retrieval

The core of the UnexpectedMiner system is the unexpected information retrieval module. The purpose of this module is to find the documents d_j of S that contain unexpected information with respect to $R \cup S - \{d_j\}$. Indeed, a document d_j is highly unexpected if, while similar to documents of $R \cup S - \{d_j\}$, it contains information that is found neither in any other document of S nor in any document of R. This module is described in detail in the section below.

3 Measures of the Unexpectedness of a Document

Five measures are proposed for assessing the unexpectedness of a document.

3.1 Measure 1

The first measure is derived directly from the criterion proposed by Liu et al. [11] for discovering unexpected pages on a Web site. It is defined by:

$$M1(d_j) = \frac{\sum_{i=1}^{m} U^1_{i,j,c}}{m}$$

with

$$U^1_{i,j,c} = \begin{cases} 1 - \frac{tf_{i,c}}{tf_{i,j}} & if \ tf_{i,c}/tf_{i,j} \leq 1 \\ 0 & else \end{cases}$$

where d_j is a document in S and $D_c = R \cup S - \{d_j\}$ is the document obtained by combining all the reference documents in R with the similar documents except d_j.

The main drawback with this measure is that it gives the same value for both the terms t_i and $t_{i'}$ that occur with different frequencies in a new document $d_j \in S$ once these terms do not occur in D_c (in other words, in the other documents in $R \cup S - \{d_j\}$). Now it would be desirable to get an unexpectedness value $U^1_{i,j,c}$ for t_i greater than the value $U^1_{i',j,c}$ found for $t_{i'}$ when t_i is more frequent than $t_{i'}$ in d_j. This is particularly the case when t_i pertains to a word that has never been encountered before whereas $t_{i'}$ is a misspelled word. This consideration led us to propose and experiment other measures for assessing the unexpectedness of a document.

3.2 Measure 2

With the second measure, the unexpectedness of a term t_i in a document $d_j \in S$ in relation to all of the other documents D_c is defined by:

$$U^2_{i,j,c} = \begin{cases} tf_{i,j} - tf_{i,c} \ if \ tf_{i,j} - tf_{i,c} \geq 0 \\ 0 \qquad\qquad else \end{cases}$$

Just as in with $M1$, the unexpectedness of a document d_j is equal to the mean of the unexpectedness values associated with the terms representing d_j:

$$M2(d_j) = \frac{\sum_{i=1}^{m} U^2_{i,j,c}}{m}$$

This second measure gets rid of the drawback in the first. Indeed, if we go back to the previous example, if the term t_i occurs more frequently than $t_{i'}$ in document d_j and that neither appear in D_c, then:

$$U^2_{i,j,c} > U^2_{i',j,c}$$

3.3 Measure 3

With the previous measures, only the terms were considered. In the technology watch area and information retrieval likewise, it is often the association of several terms, e.g. "data mining", which is operative. On that basis, we decided to represent each document by terms and sequences of consecutive terms.

We then used an algorithm for sequential pattern mining based on the algorithm from Agrawal [1] for extracting sets of frequent sequences of consecutive terms in the textual data.

For terms and sequences of consecutive terms, we defined a third measure that is an adaptation of M2 in which:

$$tf_{i,j} = \frac{f_{i,j}}{max_l f'_{l,j}}$$

where $max_l f'_{l,j}$ is the maximum frequency observed in the terms and sequences of consecutive terms.

However, neither of these three measures takes into account the discriminatory power of a term as expressed by IDF. This inadequacy can partially be overcome by combining all of the documents. Nonetheless, it seemed to us valuable to design unexpectedness measures that make direct use of this information, as with the two methods described below.

3.4 Measure 4

The fourth measure makes direct use of the discriminatory power idf_i, of the term t_i by evaluating the unexpectedness of a document d_j through the sum of the weights $w_{i,j}$ of the terms t_i that represent it (remember $w_{i,j} = tf_{i,j} \times idf_i$):

$$M4(d_j) = \sum_{i=1}^{m} w_{i,j}$$

With this measure, two documents d_j and d'_j may nonetheless have same unexpectedness value in spite of the fact that the weights of the terms representing the first document are equal while those for the second document are very different.

3.5 Measure 5

To overcome the limitation of $M4$, the fifth measure proposed assigns the highest weight in a document's vector of representation as that document's unexpectedness value:

$$M5(d_j) = \max_l w_{l,j}$$

Tests were performed to evaluate this system and compare the various measures. These are described in the next section.

4 Experiments

4.1 Corpus and Evaluation Criteria Used

The reference set R comprises 18 scientific articles in English dealing with Machine Learning none of which deal with some particular topics such as "Support Vector Machines, Affective Computing, Reinforcement Learning,... etc". The N base comprises 57 new documents, 17 of which are considered by the knowledge manager to be similar to the reference documents. Among these 17, 14 deal with topics that the latter considers unexpected.

For the purposes of evaluating UnexpectedMiner, we used the *precision* and *recall* criteria defined by J.A. Swets [20]. In our system, *precision* measures the percentage of documents extracted by the system that are truly unexpected. *Recall* measures the percentage of unexpected documents found in the N document corpus by the system. These are conventional criteria in the area of information retrieval and we shall not consider them in any further detail here.

4.2 Evaluation of the Five Measures

Because the main contribution of this work is to define new means for measuring the unexpectedness of a document, the module that implements those measures was first evaluated independently from the module for the extraction of similar documents, and then in combination with all the other modules.

We therefore first restricted the S base to the 17 new documents considered by the knowledge manager to be similar to the reference documents R. The results obtained in terms of recall and precision using the five measures defined above are provided in figures 2 to 6 where the number of documents extracted by the system is given on the x-axis

Whereas the N base comprises by far a majority of documents that deal with unexpected topics (14 documents out of 17), only the $M1$ measure is unable to return them first since the precision value is 0% considering just the two first documents extracted (figure 2) while this value is 100% for the other measures (figures 3 to 6).

The results achieved with the $M2$ (figure 3) and $M3$ measures (figure 4) are more satisfactory. But it is measures $M4$ and $M5$ that first return the most documents that deal with unexpected topics. Indeed, precision continues to be equal to 100% even considering up to six documents for $M4$ (figure 5) and up to seven for $M5$ (figure 6).

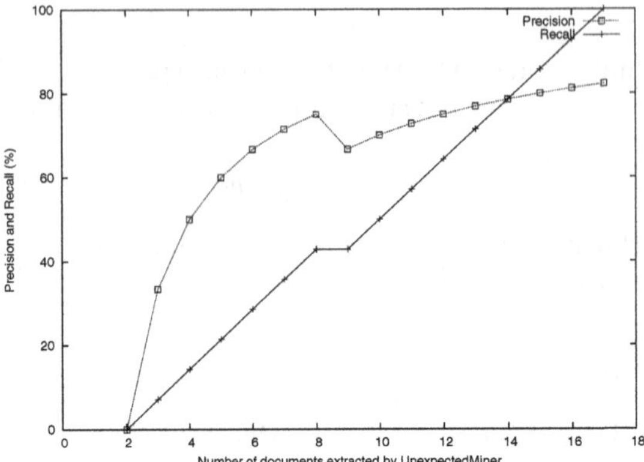

Fig. 2. Precision and recall: measure 1

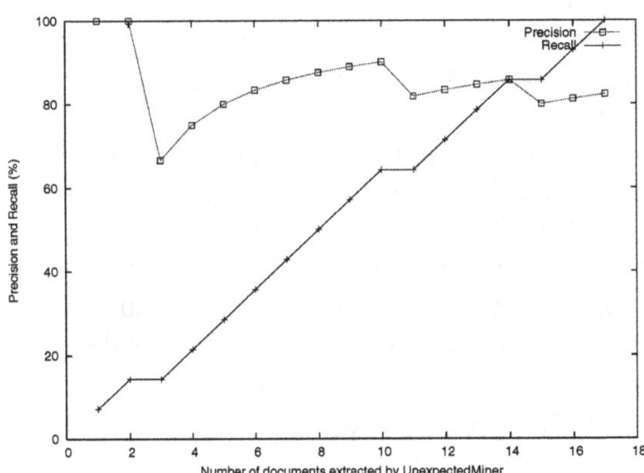

Fig. 3. Precision and recall: measure 2

4.3 Evaluation of the Whole System

We finally evaluated the complete system in respect of the N base that comprised 57 new documents. Among the 15 first documents considered similar to the reference documents by the system, only 9 actually were, i.e., a precision rate of 60 % and a recall rate of 52.9 %. Among those 9 documents, 7 dealt with unexpected topics. Under this second experiment, only measure $M1$ is unable to first extract a document that deals with an unexpected topic: precision is 0% whereas it is 100% for $M2$, $M3$, $M4$ and $M5$ that also correctly identify the same unexpected document. It is noteworthy that unexpected documents are not detected as well by the $M1$ measure since the recall is 100% only when the number of documents extracted is equal to the number of documents supplied to the system. The performance of $M2$ and $M4$ are more or less comparable but once again it is measure $M5$ which first extracts the documents relating to unexpected topics. However, this measure rather often assigns the same value

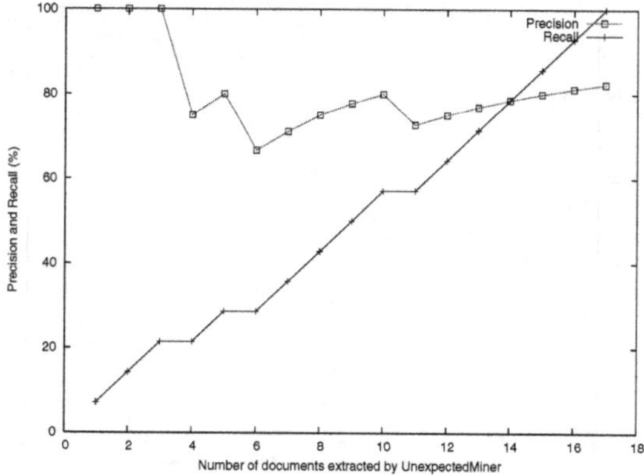

Fig. 4. Precision and recall: measure 3

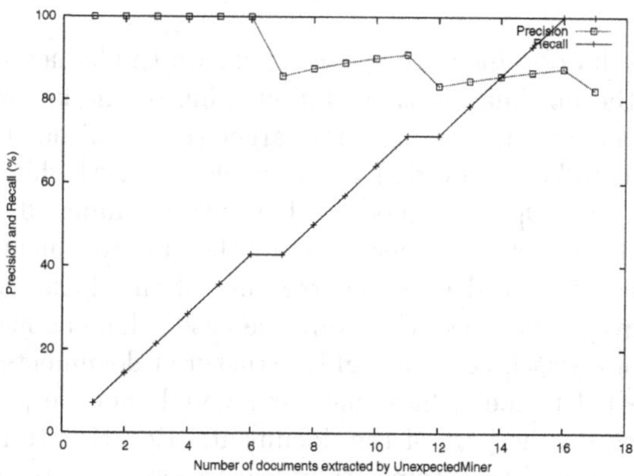

Fig. 5. Precision and recall: measure 4

to several documents. Finally, although the results achieved with $M3$ are somewhat less satisfactory, there is a match between the sequences of unexpected words that it returns and those being sought, i.e. "support vector machine" or "reinforcement learning". It is worth noting that a useful feature of the UnexpectedMiner system in this respect is that it indicates the words or sequences of words that most contributed to making a documents submitted to the system an unexpected document.

5 Conclusion

We have developed a watch system that is designed to extract relevant documents from a text corpus. Documents are relevant when they deal with topics that were unexpected and previously unknown to the knowledge manager. Several measures of the unexpectedness of a document were proposed and compared.

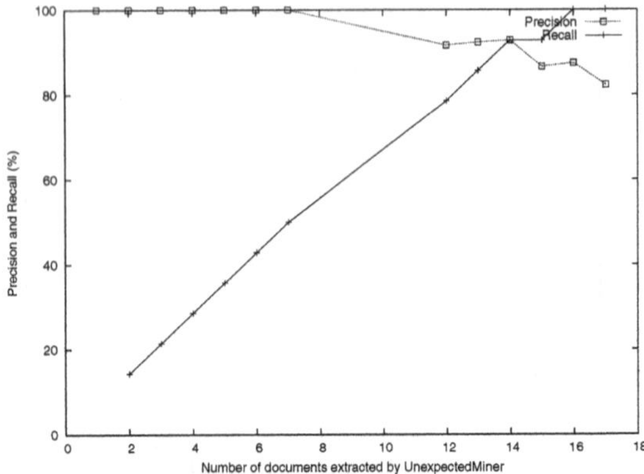

Fig. 6. Precision et recall: measure 5

Although the results obtained are encouraging, we think they can still be improved. To find the information that is interesting to the knowledge manager, it therefore appears essential to properly target the set of documents in which this information is to be retrieved. Thus, it would be worthwhile studying other measures of similarity [9] in the module that extract similar documents to the set of reference documents. Furthermore, another improvement to the system might be achieved by considering the structure of the documents [16]. In the area of competitive intelligence, this could be easily done as most of the bases used for technology watch contain highly structured documents (such as XML files for examples). Unexpectedness measures could then be parameterized by weights depending on the part of the documents. Boosting techniques such as in [7] could then be used to automatically learn those weights and discover unexpected information.

References

1. R. Agrawal and R. Srikant. Mining sequential patterns. In *Eleventh International Conference on Data Engineering*, pages 3–14, Taipei, Taiwan, 1995. IEEE.
2. K. K. Bun and M. Ishizuka. Emerging topic tracking system. In *Proceedings of the International Conference on Web Intelligence*, LNAI 2198, pages 125–130, 2001.
3. U.M Fayyad, G. Piatetsky, P. Smyth, and R. Uthurusamy. *Advances in Knowledge Discovery and Data Mining.* AAAI/MIT Press, 1996.
4. R. Feldman and I. Dagan. Knowledge discovery from textual databases. In *Proceedings of KDD95*, pages 112–117, 1995.
5. B. Gilad and J. Herring. *The Art and Science of Business Intelligence Analysis.* JAI Press, 1996.
6. C. Halliman. *Business Intelligence Using Smart Techniques : Environmental Scanning Using Text Mining and Competitor Analysis Using Scenarios and Manual Simulation.* Information Uncover, 2001.

7. M.V. Joshi, R. Agarwal, and V. Kumar. Predicting rare classes: Can boosting make any weak learner strong? In *Proceedings of the 7th ACM SIGKDD International Conference on Knowledge Discovery and Data Mining*, pages 297–306. ACM, 2002.
8. Y. Kodratoff. Knowledge discovey in texts: A definition and applications. In *Proceedings of the International Symposium on Methodologies for Intelligent Systems*, LNAI 1609, pages 16–29, 1999.
9. L. Lebart and M. Rajman. Computing similarity. In *Handbook of Natural Language Processing*, pages 477–505. Dekker, 2000.
10. B. Lent, R. Agrawal, and R. Srikant. Discovering trends in text databases. In *Proceedings of KDD'97*, pages 227–230. AAAI Press, 14–17 1997.
11. B. Liu, Y. Ma, and P. S. Yu. Discovering unexpected information from your competitors' web sites. In *Proceedings of KDD'2001*, pages 144–153, 2001.
12. P. Losiewicz, D.W. Oard, and R. Kostoff. Textual data mining to support science and technology management. *Journal of Int. Inf. Systems*, 15:99–119, 2000.
13. N. Matsumura, Y. Ohsawa, and M. Ishizuka. Discovery of emerging topics between communities on www. In *Proceedings Web Intelligence'2001*, pages 473–482, Maebashi, Japan, 2001. LNCS 2198.
14. L. T. Moss and S. Atre. *Business Intelligence Roadmap: The Complete Project Lifecycle for Decision-Support Applications.* Addison-Wesley, 2003.
15. Y. Ohsawa, N. E. Benson, and M. Yachida. Keygraph: Automatic indexing by co-occurrence graph based on building construction metaphor. In *Proceedings of the Advances in Digital Libraries Conference*, pages 12–18, 1998.
16. B. Piwowarski and P. Gallinari. A machine learning model for information retrieval with structured documents. In *Proceedings of the International Conference on Machine Learning and Data Mining in Pattern Recognition*, LNCS 2734, pages 425–438, July 2003.
17. K. Rajaraman and A.H. Tan. Topic detection, tracking and trend analysis using self-organizing neural networks. In *Proceedings of PAKDD'2001*, pages 102–107, Hong-Kong, 2001.
18. G. Salton and M. J. McGill. *Introduction to Modern Information Retrieval.* McGraw-Hill, 1983.
19. F. Sebastiani. Machine learning in automated text categorization. *ACM Computing Surveys*, 34(1):1–47, March 2002.
20. J.A. Swets. Information retrieval systems. *Science*, 141:245–250, 1963.
21. D. Zhu and A.L. Porter. Automated extraction and visualization of information for technological intelligence and forecasting. *Technological Forecasting and Social Change*, 69:495–506, 2002.
22. D. Zhu, A.L. Porter, S. Cunningham, J. Carlisie, and A. Nayak. A process for mining science and technology documents databases, illustrated for the case of "knowledge discovery and data mining". *Ciencia da Informação*, 28(1):7–14, 1999.

Scalable Density-Based Distributed Clustering

Eshref Januzaj[1], Hans-Peter Kriegel[2], and Martin Pfeifle[2]

[1] Braunschweig University of Technology, Software Systems Engineering
januzaj@sse.cs.tu-bs.de
http://www.sse.cs.tu-bs.de
[2] University of Munich, Institute for Computer Science
{kriegel,pfeifle}@dbs.ifi.lmu.de
http://www.dbs.ifi.lmu.de

Abstract. Clustering has become an increasingly important task in analysing huge amounts of data. Traditional applications require that all data has to be located at the site where it is scrutinized. Nowadays, large amounts of heterogeneous, complex data reside on different, independently working computers which are connected to each other via local or wide area networks. In this paper, we propose a scalable density-based distributed clustering algorithm which allows a user-defined trade-off between clustering quality and the number of transmitted objects from the different local sites to a global server site. Our approach consists of the following steps: First, we order all objects located at a local site according to a quality criterion reflecting their suitability to serve as local representatives. Then we send the best of these representatives to a server site where they are clustered with a slightly enhanced density-based clustering algorithm. This approach is very efficient, because the local determination of suitable representatives can be carried out quickly and independently from each other. Furthermore, based on the scalable number of the most suitable local representatives, the global clustering can be done very effectively and efficiently. In our experimental evaluation, we will show that our new scalable density-based distributed clustering approach results in high quality clusterings with scalable transmission cost.

1 Introduction

Density-based clustering has proven to be very effective for analyzing large amounts of heterogeneous, complex data, e.g. for clustering of complex objects [1][4], for clustering of multi-represented objects [9], and for visually mining through cluster hierarchies [2]. All these approaches require full access to the data which is going to be analyzed, i.e. the data has to be located at one single site. Nowadays, large amounts of heterogeneous, complex data reside on different, independently working computers which are connected to each other via local or wide area networks (LANs or WANs). Examples comprise distributed mobile networks, sensor networks or supermarket chains where check-out scanners, located at different stores, gather data unremittingly. Furthermore, international companies such as DaimlerChrysler have some data which is located in Europe and some data in the US. Those companies have various reasons why the data cannot be transmitted to a central site, e.g. limited bandwidth or security aspects. Another example is WAL-MART featuring the largest civil database in the world, consisting of more than 200 terabytes of data [11]. Every night all data is transmitted to

J.-F. Boulicaut et al. (Eds.): PKDD 2004, LNAI 3202, pp. 231–244, 2004.

Betonville from the different stores via the largest privately hold satellite system. Such a company would greatly benefit, if it were possible to cluster the data locally at the stores, and then determine and transmit suitable local representatives which allow to reconstruct the complete clustering at the central in Betonville. The transmission of huge amounts of data from one site to another central site is in some application areas almost impossible. In astronomy, for instance, there exist several highly sophisticated space telescopes spread all over the world. These telescopes gather data unceasingly. Each of them is able to collect 1GB of data per hour [5] which can only, with great difficulty, be transmitted to a global site to be analyzed centrally there. On the other hand, it is possible to analyze the data locally where it has been generated and stored. Aggregated information of this locally analyzed data can then be sent to a central site where the information of different local sites are combined and analyzed. The result of the central analysis may be returned to the local sites, so that the local sites are able to put their data into a global context.

In this paper, we introduce a scalable density-based distributed clustering algorithm which efficiently and effectively detects information spread over several local sites. In our approach, we first compute the density around each locally located object reflecting its suitability to serve as a representative of the local site. After ordering the objects according to their density, we send the most suitable local representatives to a server site, where we cluster the objects by means of an enhanced DBSCAN [4] algorithm. The result is sent back to the local sites. The local sites update their clustering based on the global model, e.g. merge two local clusters to one or assign local noise to global clusters.

This paper is organized as follows: In Section 2, we review the related work in the area of density-based distributed clustering. In Section 3, we discuss a general framework for distributed clustering. In Section 4, we present our quality driven approach for generating local representatives. In Section 5, we show how these representatives can be used for creating a global clustering based on the information transmitted from the local sites. In Section 6, we present the experimental evaluation of our *SDBDC* (Scalable Density-Based Distributed Clustering) approach showing that we achieve high quality clusterings with relative little information. We conclude the paper in Section 7 with a short summary and a few remarks on future work.

2 Related Work on Density-Based Distributed Clustering

Distributed Data Mining (DDM) is a dynamically growing area within the broader field of Knowledge Discovery in Databases (KDD). Generally, many algorithms for distributed data mining are based on algorithms which were originally developed for parallel data mining. In [8] some state-of-the-art research results related to DDM are resumed.

One of the main data mining tasks is clustering. There exist many different clustering algorithms based on different paradigms, e.g. density-based versus distance-based algorithms, and hierarchical versus partitioning algorithms. For more details we refer the reader to [7].

To the best of our knowledge, the only density-based distributed clustering algorithm was presented in [6]. The approach presented in [6] is based on the density-based partitioning clustering algorithm DBSCAN. It consists of the following steps. First, a DBSCAN algorithm is carried out on each local site. Based on these local clusterings, cluster representatives are determined. Thereby, the number and type of local representatives is fixed. Only so called special core-points are used as representatives. Based on these local representatives, a standard DBSCAN algorithm is carried out on the global site to reconstruct the distributed clustering. The strong point of [6] is that it tackles the complex and important problem of distributed clustering. Furthermore, it was shown that a global clustering carried out on about 20% of all data points, yields a clustering quality of more than 90% according to the introduced quality measure.

Nevertheless, the approach presented in [6] suffers from three drawbacks which are illustrated in Figure 1 depicting data objects located at 3 different local sites.

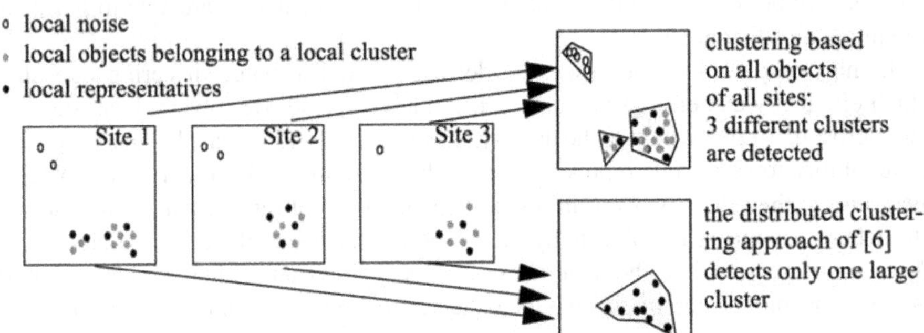

Fig. 1. Local noise on different local sites.

- First, local noise is ignored. The clustering carried out on the local sites ignores the local noise located in the upper left corner of each site. Thus the distributed clustering algorithm of [6] does not detect the global cluster in the upper left corner.
- Second, the number of representatives is not tuneable. Representatives are always special core-points of local clusters (cf. black points in Figure 1). The number of these special core-points is determined by the fact that each core point is within the ε-range of a special core-point. Dependent on how the DBSCAN algorithm walks through a cluster, the special core-points are computed.
- Third, these special core points might be located at the border of the clusters (cf. Figure 1). Of course, it is much better if the representatives are not located at the trailing end of a cluster, but are central points of a cluster. Representatives located at the border of local clusters might lead to a false merging of locally detected clusters when carrying out a central clustering. This is especially true, if we use high ε-values for the clustering on the server site, e.g. in [6] a high and static value of $\varepsilon_{rmglobal} = 2\varepsilon_{\text{local}}$ was used. The bottom right corner of Figure 1 shows that badly located representatives along with high $\varepsilon_{\text{global}}$-values might lead to wrongly merged clusters.

To sum up, in the example of Figure 1, for instance, the approach of [6] would only detect one cluster instead of three clusters, because it cannot deal with local noise and tend to merge clusters close to each other. Our new SDBDC approach enhances the approach presented in [6] as follows:

- We deal effectively and efficiently with the problem of local noise.
- We do not produce a fixed number of local representatives, but allow the user to find an individual trade-off between cluster quality and runtime.
- Our representatives reflect dense areas tending to be in the middle of clusters.
- Furthermore, we propose a more effective way to detect the global clustering based on the local representatives. We do not apply a DBSCAN algorithm with a fixed ε-value. Instead we propose to use an enhanced DBSCAN algorithm which uses different ε-values for each local representative r depending on the distribution of the objects represented by r.

3 Scalable Density-Based Distributed Clustering

Distributed Clustering assumes that the objects to be clustered reside on different sites. Instead of transmitting all objects to a central site (also denoted as server) where we can apply standard clustering algorithms to analyze the data, the data is analyzed independently on the different local sites (also denoted as clients). In a subsequent step, the central site tries to establish a global clustering based on the local models, i.e. the local representatives. In contrast to a central clustering based on the complete dataset, the central clustering based on the local representatives can be carried out much faster.

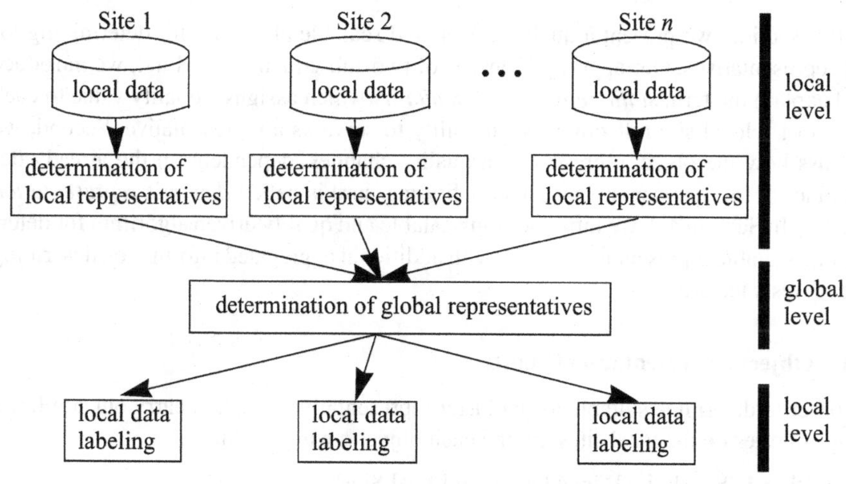

Fig. 2. Distributed clustering.

Distributed Clustering is carried out on two different levels, i.e. the local level and the global level (cf. Figure 2). On the local level, all sites analyse the data independently from each other resulting in a local model which should reflect an optimum trade-off between complexity and accuracy. Our proposed local models consist of a set of representatives. Each representative is a concrete object from the objects located at the local site. Furthermore, we augment each representative r with a suitable covering radius indicating the area represented by r. Thus, r is a good approximation for all objects residing on the corresponding local sites and are contained in the covering area of r.

Next, the local model is transferred to a central site, where the local models are merged in order to form a global model. The global model is created by analysing the local representatives. This analysis is similar to a new clustering of the representatives with suitable global clustering parameters. To each local representative a global cluster identifier is assigned. The resulting global clustering is sent to all local sites.

If a local object is located in the covering area of a global representative, the cluster-identifier from this representative is assigned to the local object. Thus, we can achieve that each site has the same information as if their data were clustered on a global site, together with the data of all the other sites. To sum up, distributed clustering consists of three different steps (cf. Figure 2):

- Determination of a local model
- Determination of a global model which is based on all local models
- Updating of all local models

In this paper, we will present effective and efficient algorithms for carrying out step 1 and step 2. For more details about step 3, the relabeling on the local sites, we refer the interested reader to [6].

4 Quality Driven Determination of Local Representatives

In this section, we present a quality driven and scalable algorithm for determining local representatives. Our approach consists of two subsequent steps. First, we introduce and explain the term *static representation quality* which assigns a quality value to each object of a local site reflecting its suitability to serve as a representative. Second, we discuss how the object representation quality changes, dependent on the already determined local representatives. This quality measure is called *dynamic representation quality*. In Section 4.2, we introduce our scalable and quality driven algorithm for determining suitable representatives along with additional aggregated information describing the represented area.

4.1 Object Representation Quality

In order to determine suitable local cluster representatives, we first carry out similarity range queries on the local sites around each object o with a radius ε.

Definition 1 (Similarity Range Query on Local Sites)
Let O be the set of objects to be clustered and $d : O \times O \rightarrow IR_0^+$ the underlying distance function reflecting the similarity between two objects. Furthermore, let $O_i \subseteq O$ be the set of objects located at site i. For each object $o \in O$ and a query range $\varepsilon \in IR_0^+$, the similarity range query $sim_{range} : O_i \times IR_0^+ \rightarrow 2^{O_i}$ returns the set.

$$sim_{range}(o, \varepsilon) = \{o_i \in O_i | d(o_i, o) \leq \varepsilon\}$$

After having carried out the range queries on the local sites, we assign a static representation quality $StatRepQ(o, \varepsilon)$ to each object o w.r.t. a certain ε-value.

Definition 2 (Static Representation Quality $StatRepQ$)
Let $O_i \subseteq O$ be the set of objects located at site i. For each object $o \in O_i$ and a query range $\varepsilon \in IR_0^+$, $StatRepQ : O_i \times IR_0^+ \rightarrow IR_0^+$ is defined as follows:

$$StatRepQ(o, \varepsilon) = \sum_{o_i \in sim_{range}(o, \varepsilon)} \varepsilon - d(o_i, o)$$

For each object o_i contained in the ε-range of a query object o, we determine the distance to the border of the ε-range query, i.e. we weight each object o_i in the ε-range of o by $\varepsilon - d(o_i, o)$. This value is the higher, the closer o_i is to o. Then the quality measure $StatRepQ(o, \varepsilon)$ sums up all the values $\varepsilon - d(o_i, o)$ for all objects located in the ε-range of our query object. Obviously, $StatRepQ(o, \varepsilon)$ is the higher, the more objects are located in the ε-range around o and the closer these objects are to o. Figure 3 illustrates that the highest $StatRepQ(o, \varepsilon)$ value is assigned to those objects which intuitively seem to be the most suitable representatives of a local site. The figure shows that the value $StatRepQ(A, \varepsilon)$ is much higher than the value $StatRepQ(A', \varepsilon)$, reflecting the more central role of object A compared to object A'.

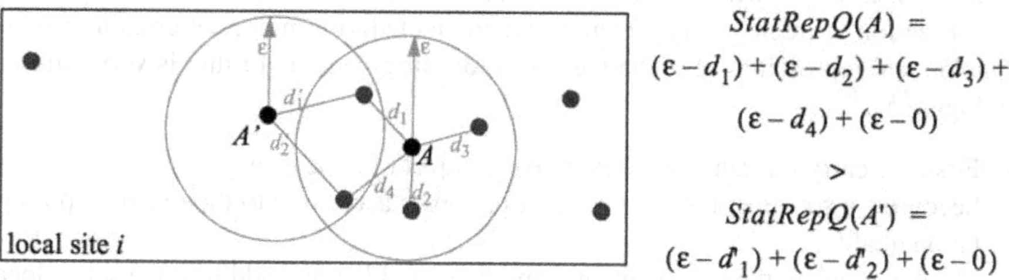

Fig. 3. Static representation quality.

Next we define a dynamic representation quality $DynRepQ(o, \varepsilon, Rep_i)$ for each local object o. This quality measure depends on the already determined set of local representatives Rep_i of a site i and the radius of our ε-range query.

Definition 3 (Dynamic Representation Quality $DynRepQ$)
Let $O_i \subseteq O$ be the set of objects located at site i and $Rep_i \subseteq O_i$ the set of the already determined local representatives of site i. Then, $DynRepQ : O_i \times IR_0^+ \times 2^{O_i} \rightarrow IR_0^+$ is defined as follows:

$$DynRepQ(o, \varepsilon, Rep_i) \quad \sum_{\substack{o_i \in sim_{range}(o, \varepsilon) \\ \forall r \in Rep_i : o_i \notin sim_{range}(r, \varepsilon)}} \varepsilon - d(o_i, o)$$

$DynRepQ(o, \varepsilon, Rep_i)$ depends on the number and distances of the elements found in the ε-range of an object o, which are not yet contained in the ε-range of a former local representative. For each object o which has not yet been selected as a representative, the value $DynRepQ(o, \varepsilon, Rep_i)$ gradually decreases with an increasing set of local representatives, i.e. an increasing set Rep_i. Figure 4 illustrates the different values of $DynRepQ(B, \varepsilon, Rep_i)$ for two values of the set Rep_i. If $Rep_i = \{\}$, the value $DynRepQ(B, \varepsilon, Rep_i)$ is much higher than if the element A is included in Rep_i.

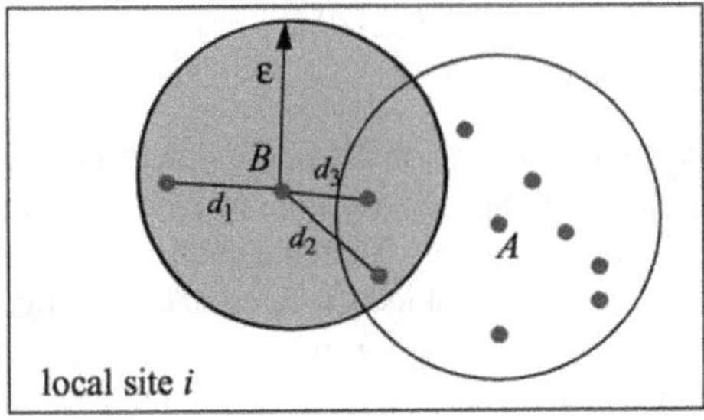

$$DynRepQ(B, \varepsilon, \{\ \}) = (\varepsilon - d_1) + (\varepsilon - d_2) + (\varepsilon - d_3) + (\varepsilon - 0)$$

$$>$$

$$DynRepQ(B, \varepsilon, \{A\}) = (\varepsilon - d_1) + (\varepsilon - 0)$$

Fig. 4. Dynamic representation quality.

4.2 Scalable Calculation of Local Representatives

In this subsection, we will show how we can use the quality measures introduced in the last subsection to create a very effective and efficient algorithm for determining a set of suitable local representatives. The basic idea of our greedy algorithm is very intuitive (cf. Figure 5).

- First, we carry out range queries for each object of a local site.
- Second, we sort the objects in descending order according to their static representation quality.
- Third, we delete the first element from the sorted list and add it to the set of local representatives.
- Fourth, we compute the dynamic representation quality for each local object which has not yet been used as a local representative and sort these objects in descending order according to their dynamic representation quality.
- If we have not yet determined enough representatives, we continue our algorithm with step 3. Otherwise, the algorithm stops.

Obviously, the algorithm delivers the most suitable local representatives at a very early stage of the algorithm. After having determined a new local representative, it can be sent to a global site without waiting for the computation of the next found representative. As we decided to apply a greedy algorithm for the computation of our local representatives, we will not revoke a representative at a later stage of the algorithm. So the algorithm works quite similar to ranking similarity queries known from database systems allowing to apply the cursor principle on the server site. If the server decides that it has received enough representatives from a local site, it can close the cursor, i.e. we do not have to determine more local representatives. The termination of the algorithm can either be determined by a size-bound or an error-bound stop criterion [10].

```
Oi          set of objects located at site i;
ε           ε-range value;
ALGORITHM DeterminationOfLocalRepresentatives;
BEGIN
   Rep_i :={};                                  // set of local representatives;
   FOR EACH o ∈ O_i DO
      compute StatRepQ(o,ε);
   END FOR;
   SortRepList := <(o1,StatRepQ(o1,ε)), ..., (o|Oi|,StatRepQ(o|Oi|,ε))| i≤j =>StatRepQ(oi,ε) ≥StatRepQ(oj,ε)>;
   WHILE NOT stop_criterion (Rep_i) DO
      Rep_i := Rep_i + SortRepList[1];
      FOR EACH o ∈ O_i − REP_i DO
         compute DynRepQ(o,ε,Rep_i);
      END FOR;
      SortRepList := <(o1,DynRepQ(o1,ε,Rep_i)), ..., (o|Oi-REPi|,DynRepQ(o|Oi-REPi|,ε,Rep_i))|
                 i ≤ j =>DynRepQ(oi,ε,Rep_i) ≥ DynRepQ(oj,ε,Rep_i)>;
   END WHILE;
END.
```

Fig. 5. Scalable calculation of local representatives.

This approach is especially useful if we apply a clustering algorithm on the server site which efficiently supports incremental clustering as, for instance, DBSCAN [3].

For all representatives included in a sequence of local representatives, we also compute their Covering Radius $CovRad$, indicating the element which has the maximum distance from the representative, and the number $CovCnt$ of objects covered by the representative.

Definition 4 (Covering Radius and Covering Number of Local Representatives)
Let $O_i \subseteq O$ be the set of objects located at site i and $Rep_i = \{r_{i_1}, \ldots, r_{i_n}\}$ the sequence of the first n local representatives where $\{r_{i_1}, \ldots, r_{i_n}\} \subseteq O_i$. Then the covering radius $CovRad : O_i \times IR_0^+ \times 2^{O_i} \rightarrow IR_0^+$ and the covering number $CovCnt : O_i \times IR_0^+ \times 2^{O_i} \rightarrow IR_0^+$ of the i_{n+1}th representative are defined as follows:

$$
\begin{aligned}
&CovRad(r_{i_{n+1}}, \varepsilon, Rep_{i_n}) = \\
&\qquad max\Big\{\varepsilon - d(o, r_{i_{n+1}}) | \forall o \in O_i \forall r \in Rep_{i_n} \\
&\qquad : o \in sim_{range}(r_{i_{n+1}}, \varepsilon) \land o \notin sim_{range}(r, \varepsilon)\Big\} \\
&CovCnt(r_{i_{n+1}}, \varepsilon, Rep_{i_n}) = \\
&\qquad \big|\{o | \forall o \in O_i \forall r \in Rep_{i_n} : o \in sim_{range}(r_{i_{n+1}}, \varepsilon) \land o \notin sim_{range}(r, \varepsilon)\}\big|
\end{aligned}
$$

Figure 6 depicts the $CovRad$ and $CovCnt$ values for two different representatives of site i. Note that the computation of $CovRad$ and $CovCnt$ can easily be integrated into the computation of the representatives as illustrated in Figure 5. The local representatives along with the corresponding values $CovRad$ and $CovCnt$ are sent to the global site in order to reconstruct the global clustering, i.e. we transmit the following sequence consisting of n local representatives from site i:

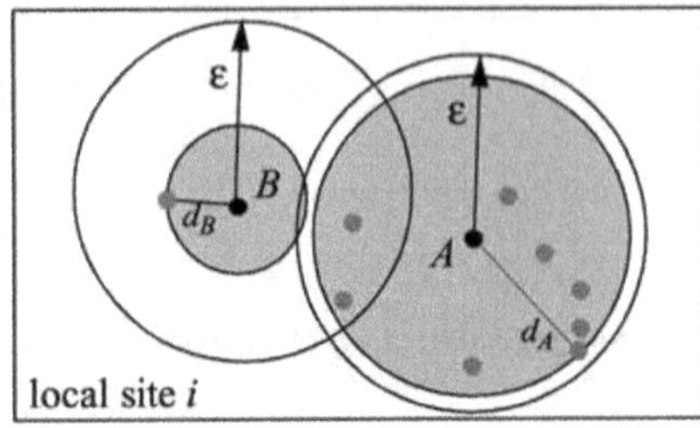

$CovRad(A, \langle\,\rangle) = d_A$

$CovCnt(A, \langle\,\rangle) = 9$

$CovRad(B, \langle A \rangle) = d_B$

$CovCnt(B, \langle A \rangle) = 2$

sent to global site: < $(A, d_A, 9)$, $(B, d_B, 2)$>

Fig. 6. Covering radius *CovRad* and covering number *CovCnt*.

$$\Big\langle (r_{i_1}, CovRad(r_{i_1}, \varepsilon, \{\}), CovCnt(r_{i_1}, \varepsilon,))$$

$$\vdots$$

$$(r_{i_n}, CovRad(r_{i_n}, \varepsilon, \{r_{i_1}, \ldots, r_{i_{n-1}}\}), CovCnt(r_{i_n}, \varepsilon, \{r_{i_1}, \ldots, r_{i_{n-1}}\}))\Big\rangle.$$

For simplicity, we will write $CovRad(r_{i_j})$ instead of $CovRad(r_{i_j}, \varepsilon, \{r_{i_1}, \ldots, r_{i_{j-1}}\})$ and $CovCnt(r_{i_j})$ instead of $CovCnt(r_{i_j}, \varepsilon, \{r_{i_1}, \ldots, r_{i_{j-1}}\})$, if the set of already transmitted representatives and the used ε-values are clear from the context.

5 Global Clustering

On the global site, we apply an enhanced version of DBSCAN adapted to clustering effectively local representatives. We carry out ε-range queries around each representative. Thereby, we use a specific ε-value $\varepsilon(r_i)$ for each representative r_i (cf. Figure 7). The ε-value $\varepsilon(r_i)$ is equal to the sum of the following two components. The first component consists of the basic ε-value which would be used by the original DBSCAN algorithm and which was used for the range queries on the local sites. The second component consists of the specific $CovRad(r_i)$ value of the representative r_i, i.e we set $\varepsilon(r_j) = \varepsilon + CovRad(r_i)$.

The idea of this approach is as follows (cf. Figure 8). The original DBSCAN algorithm based on all data of all local sites would carry out an ε-range query around each point of the data set. As we perform the distributed clustering only on a small fraction of these weighted points, i.e. we cluster on the set of the local representatives transmitted from the different sites, we have to enlarge the ε-value by the $CovRad(r_i)$ value of the actual representative r_i. This approach guarantees that we can find all objects in the enlarged ε-range query which would have been found by any object represented by the actual local representative r_i. For instance in Figure 7, the representative r_{j_1} is within the ε-range of the local object o_i represented by r_{i_1}. Only because we use the enlarged ε-range $\varepsilon(r_{i_1})$, we detect that the two representatives r_{i_1} and r_{j_1} belong to the same cluster.

Furthermore, we weight each local representative r_i by its $CovCnt(r_i)$ value, i.e. by the number of local objects which are represented by r_i. By taking these weights into

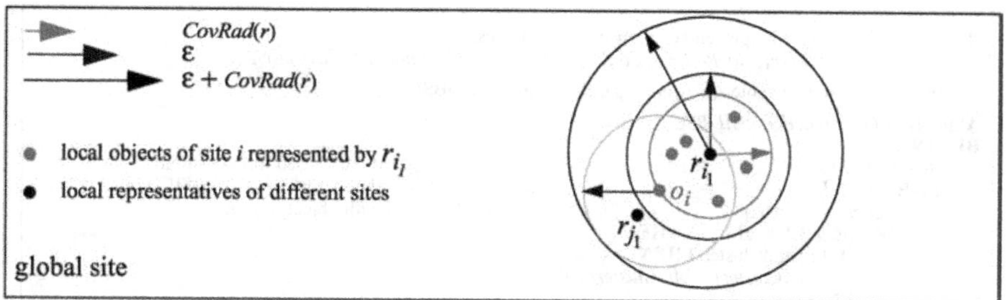

Fig. 7. Global clustering on varying $\varepsilon(r_i)$-parameters for the different representatives r_i.

account, we can detect whether local representatives are core-points, i.e. points which have more than $MinPts$ other objects in their ε-range. For each core-point r_j contained in a cluster C, we carry out an enlarged ε-range query with radius $\varepsilon(r_j)$ trying to expand C. In our case, a local representative r_j might be a core-point although less than $MinPts$ other local representatives are contained in its $\varepsilon(r_j)$-range. For deciding whether r_j is a core-point, we have to add up the number of objects $CovCnt(r_i)$ represented by the local representatives r_i contained in the $\varepsilon(r_j)$-range of r_j (cf. Figure 8).

6 Experimental Evaluation

We evaluated our SDBDC approach based on three different 2-dimensional point sets where we varied both the number of points and the characteristics of the point sets. Figure 9 depicts the three used test data sets A (8700 objects, randomly generated data/clusters), B (4000 objects, very noisy data) and C (1021 objects, 3 clusters) on the central site.

In order to evaluate our SDBDC approach, we equally distributed the data set onto the different client sites and then compared SDBDC to a single run of DBSCAN on all data points. We carried out all local clusterings sequentially. Then, we collected all representatives of all local runs, and applied a global clustering on these representatives. For all these steps, we used a Pentium III/700 machine. In all experiments, we measured the overall number of transmitted local representatives, which primarily influences the overall runtime. Furthermore, we measured the cpu-time needed for the distributed clustering consisting of the maximum time needed for the local clusterings and the time needed for the global clustering based on the transmitted local representatives.

We measured the quality of our SDBDC approach by the quality measure introduced in [6]. Furthermore we compared our approach to the approach presented in [6] where for the three test data sets about 17% of all local objects were used as representatives. Note that this number is fixed and does not adapt to the requirements of different users, i.e high clustering quality or low runtime.

Figure 10 shows the trade-off between the clustering quality and the time needed for carrying out the distributed clustering based on 4 different local sites.

Figure 10a shows clearly that with an increasing number of local representatives the overall clustering quality increases. For the two rather noisy test data sets A and

```
R               set of all representatives from all local sites;
                // R ={(r1, CovRad(r1), CovCnt(r1)), .., (r|R|, CovRad(r|R|), CovCnt(r|R|));
ε, MinPts       ε-range value and MinPts parameter used by DBSCAN
Algorithm DistributedGlobalDBSCAN
BEGIN
    ActClusterId := 1;                                  // ClusterId = 0 is used for NOISE and
    FOR i =1 .. |R| DO                                  // ClusterId = -1 for UNCLASSIFIED objects
        ActObj := R.get(i);                             // select the ith object from R
        IF ActObj.ClusterId = -1 THEN
            IF ExpandCluster THEN
                ActClusterId:=ActClusterId +1;
            END IF;
        END IF;
    END FOR;
END.

ExpandCluster: Boolean;
BEGIN
    seeds := RangeQuery(ActObj, ε+CovRad(ActObj));      // range query with enlarged radius around ActObj
    CntObjects := 0;
    FOR i = 1 .. |seeds| DO
        CntObjects := CntObjects + seeds[i].CovCnt;     // all objects represented by representatives
    END FOR;
    IF CntObjects < MinPts THEN                         // Object ActObj is not a core object
        ActObj.ClusterID := 0                           // ClusterID 0 is used for NOISE
        RETURN FALSE;
    ELSE                                                // Object o is a core object
        FOR i = 1 .. |seeds| DO
            IF seeds[i].ClusterId = {-1, 0} THEN
                seeds[i].ClusterId := ActClusterId;
            END IF;
        END FOR;
        delete ActObj from seeds;
        WHILE seeds NOT EMPTY DO
            ActObj := seeds[1];
            neighborhood := RangeQuery(ActObj, ε+CovRad(ActObj));   // range query with enlarged radius
            CntObjects := 0
            FOR i = 1 .. |neighborhood| DO
                CntObjects := CntObjects + neighborhood[i].CovCnt;
            END FOR;
            IF CntObjects >= MinPts THEN                // ActObj is a core object
                FOR i = 1 .. |neighborhood| DO
                    p := neighborhood[i];
                    IF p.ClusterId = {-1, 0} THEN       // object p is UNCLASSIFIED or NOISE
                        IF p.ClusterId = -1 THEN        // object p is UNCLASSIFIED
                            add p to seeds;
                        END IF;
                        p.ClusterId := ActClusterID;
                    END IF;
                END FOR;
            END IF;
            delete ActObj from seeds;
        END WHILE;
        RETURN TRUE;
    END IF;
END;
```

Fig. 8. Distributed global DBSCAN algorithm.

B reflecting real-world application ranges, we only have to use about 5% of all local objects as representatives in order to achieve the same clustering quality as the one achieved by the approach presented in [6].

Figure $10b_1$ shows the speed up w.r.t. the transmission cost we achieve when transmitting only the representatives determined by our SDBDC approach compared to the transmission of all data from the local sites to a global site. We assume that a local ob-

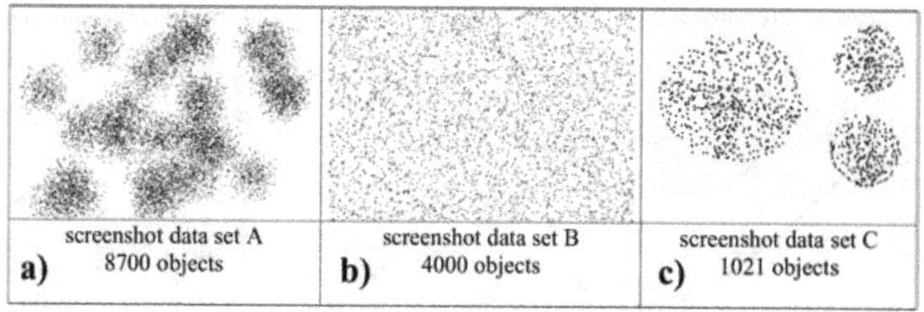

Fig. 9. Used test data sets. **a)** test data set A, **b)** test data set B, **c)** test data set C.

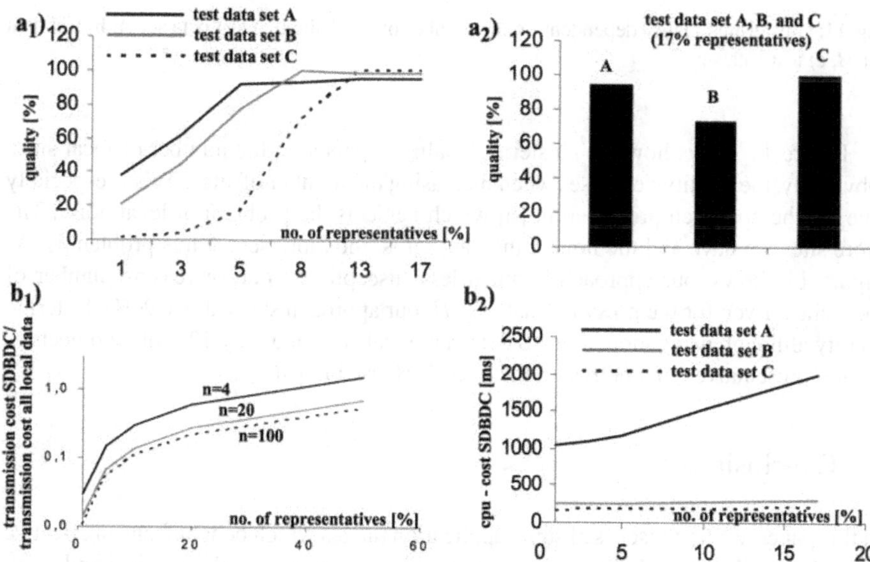

Fig. 10. Trade-off between runtime and clustering quality (4 sites). **a)** clustering quality (**a$_1$**) SDBDC **a$_2$**) approach presented in [6]), **b)** runtime (**b$_1$**) transmission cost **b$_2$**) cpu-cost for local and global clustering).

ject is represented by n bytes and that both $CovRad(r_i)$ and $CovCnt(r_i)$ need about 4 bytes each. For realistic values of n, e.g. $n = 100$, a more than three times lower representative number, e.g. 5% used by the SDBDC approach compared to 17% used by the approach presented in [6], results in a 300% speed up w.r.t. the overall transmission cost which dominate the overall runtime cost (cf. Figure 10b$_1$).

Figure 10b$_2$ depicts the sum of the maximum cpu-time needed for the clustering on the local site and the cpu-time needed for the clustering on the global site. A small number of local representatives terminates the generation of the local representatives at an early stage leading to a short runtime for computing the required local representatives. Furthermore, the global clustering can be carried out the more efficiently, the smaller the overall number of local representatives is.

To sum up, a small number of local representatives accelerates the SDBDC approach considerably. If we use about 5% of all objects as representatives, we can achieve a high quality and, nevertheless, efficient distributed clustering.

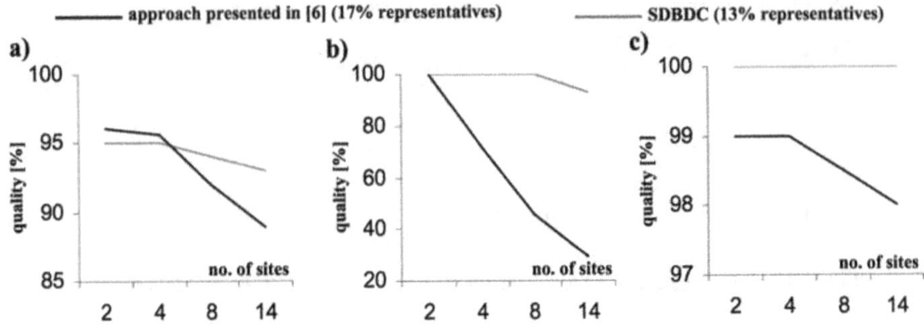

Fig. 11. Clustering Quality dependent on the number of local sites. **a)** test data set A, **b)** test data set B, **c)** test data set C.

Figure 11 shows how the clustering quality depends on the number of local sites. Obviously, the quality decreases when increasing the number of sites. This is especially true for the approach presented in [6] which neglects the problem of local noise. The more sites we have and the noisier the data set is, the more severe this problem is. As Figure 11 shows, our approach is much less susceptible to an increasing number of local sites. Even for the noisy test data set B, our approach stays above 90% clustering quality although using more than 10 defferent local sites and only 13% of all objects as local representaives (in contrast to the fixed 17% used in [6]).

7 Conclusions

In this paper, we first discussed some application ranges which benefit from an effective and efficient distributed clustering algorithm. Due to economical, technical and security reasons, it is often not possible to transmit all data from different local sites to one central server site where the data can be analysed by means of clustering. Therefore, we introduced an algorithm which allows the user to find an individual trade-off between clustering-quality and runtime. Our approach first analyses the data on the local sites and orders all objects o according to a quality criterion $DynRepQ(o)$ reflecting whether the actual object is a suitable representative. Note that this quality measure depends on the already determined representatives of a local site. After having transmitted a user dependent number of representatives to the server, we apply a slightly enhanced DBSCAN clustering algorithm which takes the covering radius and the number of objects covered by each representative r_i into account, i.e. the server site clustering is based on the aggregated information $CovRad(r_i)$ and $CovCnt(r_i)$ describing the area on a local site around a representative r_i. As we produce the local representatives in a give-me-more manner and apply a global clustering algorithm which supports efficient incremental clustering, our approach allows to start with the global clustering algorithm as soon as the first representatives are transmitted from the various local sites. Our experimental evaluation showed that the presented scalable density-based distributed clustering algorithm allows effective clustering based on relatively little information, i.e. without sacrificing efficiency and security.

In our future work, we plan to develop hierarchical distributed clustering algorithms which are suitable for handling nested data.

References

1. Ankerst M., Breunig M. M., Kriegel H.-P., Sander J.: "OPTICS: Ordering Points To Identify the Clustering Structure", Proc. ACM SIGMOD, Philadelphia, PA, 1999, 49-60.
2. Brecheisen S., Kriegel H.-P., Kröger P., Pfeifle M.: "Visually Mining Through Cluster Hierarchies", Proc. SIAM Int. Conf. on Data Mining, Orlando, FL, 2004.
3. Ester M., Kriegel H.-P., Sander J., Wimmer M., Xu X.: "Incremental Clustering for Mining in a Data Warehousing Environment", Proc. 24th Int. Conf. on Very Large Databases (VLDB), New York City, NY, 1998, 323-333.
4. Ester M., Kriegel H.-P., Sander J., Xu X.: "A Density-Based Algorithm for Discovering Clusters in Large Spatial Databases with Noise", Proc. 2nd Int. Conf. on Knowledge Discovery and Data Mining (KDD), Portland, OR, AAAI Press, 1996, 226-231.
5. Hanisch R. J.: "Distributed Data Systems and Services for Astronomy and the Space Sciences", in ASP Conf. Ser., Vol. 216, Astronomical Data Analysis Software and Systems IX, eds. N. Manset, C. Veillet, D. Crabtree (San Francisco: ASP) 2000.
6. Januzaj E., Kriegel H.-P., Pfeifle M.: "DBDC: Density-Based Distributed Clusteringö, Proc. 9th Int. Conf. on Extending Database Technology (EDBT), Heraklion, Greece, 2004, 88-105.
7. Jain A. K., Murty M. N., Flynn P. J.: "Data Clustering: A Review", ACM Computing Surveys, Vol. 31, No. 3, Sep. 1999, 265-323.
8. Kargupta H., Chan P. (editors) : "Advances in Distributed and Parallel Knowledge Discovery", AAAI/ MIT Press, 2000.
9. Kailing K., Kriegel H.-P., Pryakhin A., Schubert M.: "Clustering Multi-Represented Objects with Noiseö, Proc. 8th Pacific-Asia Conf. on Knowledge Discovery and Data Mining, Sydney, Australia, 2004.
10. Orenstein J. A.: "Redundancy in Spatial Databasesö, Proc. ACM SIGMOD Int. Conf. on Management of Data, 1989, 294-305.
11. http://www.walmart.com.

Summarization of Dynamic Content in Web Collections

Adam Jatowt and Mitsuru Ishizuka

University of Tokyo, 7-3-1 Hongo, Bunkyo-ku, 113-8656 Tokyo, Japan
{jatowt,ishizuka}@miv.t.u-tokyo.ac.jp

Abstract. This paper describes a new research proposal of multi-document summarization of dynamic content in web pages. Much information is lost in the Web due to the temporal character of web documents. Therefore adapting summarization techniques to the web genre is a promising task. The aim of our research is to provide methods for summarizing volatile content retrieved from collections of topically related web pages over defined time periods. The resulting summary ideally would reflect the most popular topics and concepts found in retrospective web collections. Because of the content and time diversities of web changes, it is necessary to apply different techniques than standard methods used for static documents. In this paper we propose an initial solution to this summarization problem. Our approach exploits temporal similarities between web pages by utilizing sliding window concept over dynamic parts of the collection.

1 Introduction

In document summarization research summaries are usually built from newspaper articles or some static documents. However in the age of the growing importance of the Web, it is becoming necessary to focus more on the summarization of web pages. Until now, few methods have been proposed that are especially designed for summarization in web genre (e.g., [3], [4]). The Web is a dynamic and heterogeneous environment. These characteristics cause difficulties for adapting traditional text analysis techniques into the web space. One of the most important differences between web pages and other document formats is the capability of the latter ones to change their content and structure in time. Many popular web pages continuously change, evolve and provide new information. Thus one should regard a web document as a dynamic object or as a kind of slot assigned to the URL address. This approach enables to consider volatile content, which is inserted or deleted from web documents for summary creation. Such summarization task differs from the standard multi-document summarization in the sense that it focuses on the changed contents of web pages (Figure 1).

There are several cases where summarization of changes in web documents could be beneficial. For example a user may be interested in knowing what was popular in his favorite web collection during given period of time. It can be too difficult for him to manually access each web document for discovering important changes. By carefully choosing information sources, one can construct a web collection which is informative about a particular topic. Such collection would be considered as a single, complex information source about the user's area of interest. Then main events and popular changes concerning user-defined topic could be acquired to the extent, which depends on the quality and characteristics of the input collection.

J.-F. Boulicaut et al. (Eds.): PKDD 2004, LNAI 3202, pp. 245–254, 2004.

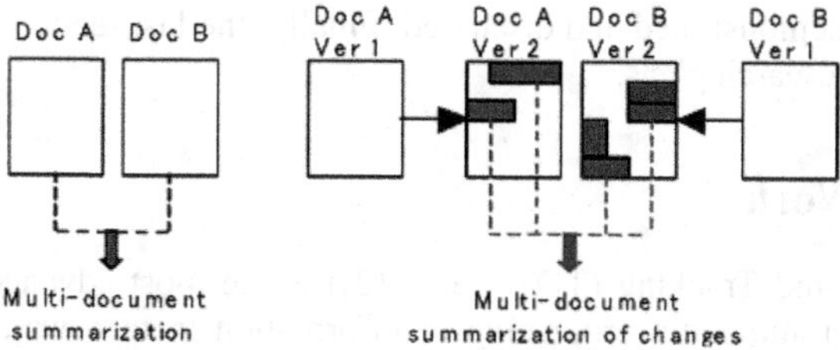

Fig. 1. Difference between traditional and new summarization. In the new one temporal versions of two or more documents are compared to reveal their changes, which are later summarized.

Another motivation for our research comes from the observation that current search engines cannot retrieve all changing data of web pages. Thus much information is lost because the web content changes too fast for any system to crawl and store every modified version of documents.

The method presented in this paper can be generally applied to any types of web pages. However, perhaps some modified approach could be more efficient for particular kinds of web documents like for example newswires, company web pages or mailing lists. Anyway, due to the large number of different page types and the difficulty of their classification we have attempted to provide generic summarization solutions, which are not tailored for any specific kinds of documents. Another concern is that different types of web pages have different frequencies and sizes of changes. Our approach works well for dynamic web pages, which have enough changing content so that meaningful summaries can be created. Therefore for rather static web documents the output may not be satisfactory enough and, in such a case, some existing document summarization methods (e.g., [8]) could work better. The speed and the size of changes of a web page can be approximated as the average change frequency and the average size of changes over the whole summarization period. Additionally to obtain a meaningful summary there should be a continuity of topics in temporal versions of a web page. Therefore we make an assumption here that the topical domain and the main characteristics of a document do not change rapidly so that a short-term summarization could be feasible. In other words, we assume semantical and structural continuity of different versions of the same web page.

In our approach we have focused generally on singular web pages. Thus any links or neighboring pages have been neglected. However the algorithm can be extended to work with the collection of web sites or any groups of linked web documents. In these cases a given depth of penetration can be defined for tracking such networks of web pages. For example, we can examine any pages, which are linked from the company home page that is pages such as: company products, staff, vacancies etc. An intuitive solution is to combine all these pages together into one single document representing the selected part of the web site. The content of each joined page could have lower scores assigned depending on the distance from the starting web page. In this way all web sites or other sub-groups of pages in the collection would be treated as single web documents where the connectivity-based weighting scheme is applied to the content of every single page.

The rest of the paper is organized as follows. In the next section we discuss related research work. Sections 3 and 4 present dynamic characteristics of web collections and our methodology for summarization of changes. In Section 5 the results of the experiments are demonstrated and discussed. Finally, the last section contains conclusions and future research plans.

2 Related Work

Topic Detection and Tracking (TDT) (e.g., [2]) is the most advanced research area which focuses on automatic processing of information from news articles. TDT attempts to recognize and classify events from online news streams or from retrospective news corpora. In our case we want to use collections of arbitrary kinds of web pages rather then news articles only. Thus we aim at detecting not only events reported by newswires but any popular concepts in a given topical domain representing user's interest.

Additionally, TDT or other automatic news mining applications like for example Google News [5] concentrate more on tracking and detecting particular events than on generating their topical summaries. The part of research, which centers on temporal summarization of news articles is represented by: [1], [9], [11], [14]. In [1] novelty and usefulness measures are applied for sentences extracted from newswire resources in order to generate temporal summaries of news topics. Newsblaster [9] or NewsInEssence [11] are other examples of applications developed for producing automatic summaries of popular events. The authors use some pre-selected resources of the newswire type for input data. Finally TimeMines [14] is a system for finding and grouping significant features in documents based on chi-square test.

There is a need for an application that could summarize new information from any, decided by users, kinds of resources. WebInEssence [12] is a web-based multi-document summarization and recommendation system that meets the above requirement. However, our approach is different in the sense that we attempt to do temporal summarization of documents, that is, summarization of their "changes" or dynamic content, instead of considering web pages as static objects. Temporal single-document summarization of web documents has been recently proposed in [7]. Multi-document summarization of common changes in online web collections has been shown in ChangeSummarizer system [6], which uses web page ranking and static contexts of dynamic document parts. Nevertheless, despite of the popularity of Web, there is still a lack of applications for retrospective summarization of changes in web documents.

3 Changes in Web Collections

There are two simple methods for obtaining topical collections of web pages. In the first case one may use any available web directory like for example ODP [10]. However, there is quite a limited number of topical choices in existing web directories, which additionally may have outdated contents. It means that a user cannot choose any arbitrary topic that he or she requires but is rather restricted to the pre-defined, general hierarchy of domains. Another straightforward way to obtain a web collection is to use search engine. In this case any combination of terms can be issued providing more freedom of choice. However the responding set of web pages may not always be completely relevant to the user's interest. Therefore an additional examination of search results is often necessary. Additionally, one should also filter collected documents to reject any duplicate web pages since they could considerably degenerate the final summary output.

In the next step, web page versions are automatically downloaded with some defined frequency. The interval t between the retrieval of each next version of a single web page should be chosen depending on the temporal characteristics of the collection. The longer period t, the lower the recall of changes is due to the existence of short-life content as it often happens in the case of newswire or popular pages. Some parts of web pages may change more than one time during interval t what poses a risk that the information can be lost. On the other hand, high frequency of page sampling should result in the increased recall but naturally also in the higher usage of network resources. Let $C_a = \{C_1, C_2,.., C_n\}$ be a set of all changes occurring in a single web page a during given interval and $F_a = \{F_1, F_2,..., F_u\}$ a set of discovered changes. If we assume that the page changes with a constant frequency t_a then the recall of changes can be approximated as:

$$R_a = \frac{|F_a|}{|C_a|} \approx \frac{t_a}{t} \quad \text{if} \quad t_a \leq t . \qquad (1)$$
$$R_a = \frac{|F_a|}{|C_a|} = 1 \quad \text{if} \quad t_a > t .$$

Let T denote the whole time interval for which a summary will be created. Assuming short period T, which embraces only a few intervals t, we obtain a small number of pages that have any changes. In this case the influence of these web pages on the final summary will be relatively high. Thus probably the final summary could have lower quality with regards to the real changes in the topic of collection, since only few web pages are determining the summary. In case of a choice of long T containing many intervals t, we expect more changes to be detected, which cause a low influence of a single web page on the final summary.

For two similar web pages the delay of the reaction to an event occurring at a particular point of time can be different. We assume that these web pages always report the most important and popular events concerning user's area of interest. It is usually expected that a newswire source would mention the particular information in a matter of hours or days. However it may take longer time in the case of other type of web pages which are more static. We call this difference a "time diversity" of web pages in order to distinguish it from the "content diversity". The choice of too short T may result in poor precision of the final summary because the reactions to a particular event could be spread in time in different web pages. However, on the other hand, longer T increases the probability that many unrelated and off-topic changes from the collection are taken into consideration what may cause the reduced quality of an output.

4 Methodology

To extract the changing content, two consecutive versions of every web page are compared with each other. The comparison is done on the sentence level. Sentences from proximate versions of a document are compared so that inserted and deleted ones can be detected. We have decided to focus only on a textual content of web documents. Thus pictures and other multimedia are discarded. There are two types of textual changes that can occur in a page: an insertion and a deletion. If a particular sentence appears only in the later version of a web page then it is regarded as an insertion. In case it can be found only in the previous version we define such sentence as a deletion.

Next, standard text preprocessing steps are conducted such as stemming and stop-words removal. We consider words and bi-grams extracted from the changes in the collection as a selected pool of features. Each such a term is scored depending on its distribution in the dynamic parts of collection during interval *T*. The term scoring method assumes that popular concepts are found in the same type of changes in high number of web page versions which are in close proximity to each other. Therefore terms appearing in changed parts of many documents will have higher scores assigned than the ones that are found in changed sections of only a few web pages (Equation 2). Moreover, a term that appears frequently inside the changes of many versions of a single web page should also have its score increased. However, in the concept of the "popularity", document frequency of the term is more important than its term frequency therefore the equation part concerning document frequency has an exponential character. Document frequency *DF* is the number of document versions that contain given term. Term frequency TF_j is the frequency of the term inside the dynamic part of a single document version *j*. In Equation 2 term frequency is divided by the number of all terms inside each change, that is the size of a change S_j, and averaged over all web page samples *N*n* where *N* is the number of different web documents and *n* the number of versions of each web page. In general, the basic scoring scheme is similar to the well-known *TFIDF* weighting [13].

$$S_{term} = \frac{\sum_{j=1}^{N*n} \frac{TF_j}{S_j}}{N*n} * \exp(DF) \tag{2}$$

As it has been mentioned before there are two possible types of changes: the deleted and the inserted change. Intuitively, deletions should be considered as a kind of out-dated content, which is no longer important and thus can be replaced by a new text. However if many web documents have deleted similar content in the same time then one may expect that some important event, expressed by this content, has been completed. In this case terms occurring in such deletions should have high value of importance assigned. On the other hand terms which are found in many proximate insertions will also have high scores. Finally the overall score of a term will be a combination of both partial scores calculated over deleted and inserted textual contents.

Let d_x be a deletion and i_x an insertion of a single web page, where *x* is the number of the web document version (Figure 2). Specifically d_x, i_x indicate the content that was deleted from the *x-1* version and the content that was added to the *x* version of the web page.

The total amount of deletions *D* of a single web page over period *T*, called a "negative change" of the page, is the union of the deleted text for all page versions:

$$D = \bigcup_{x=1}^{x=n} d_x . \tag{3}$$

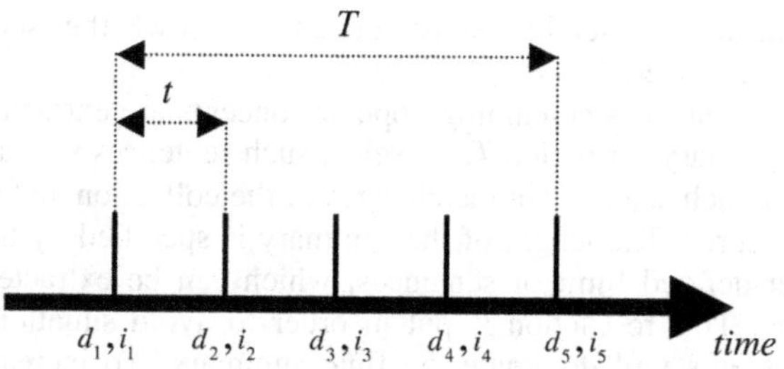

Fig. 2. Temporal representation of changes in a web page.

On the other hand, the whole pool of insertions expressed by I is described as a "positive change" of the document (Equation 4).

$$I = \bigcup_{x=1}^{x=n} i_x \,. \tag{4}$$

We want to assign maximum weights to terms, which are inserted or deleted from high number of documents in proximate time. In this way we use temporal similarity of terms in a similar fashion as TDT considers bursts of event-type data. The range of this similarity is determined by the user in the form of a sliding window of length L. The window moves through the sequentially ordered collection so that only L/t versions of each web document are considered in the same time (Figure 3). Terms are extracted from the positive and negative types of changes inside every window and are scored according to the weighting scheme from Equation 1. However, now the differences between document and term frequencies in both kinds of changes are considered. The score of a term in each window position is denoted by S_{term}^{win} and expressed as:

$$S_{term}^{win} = \frac{\sum_{j=1}^{N*n} \left| \frac{TF_j^I}{S_j^I} - \frac{TF_j^D}{S_j^D} \right| * \exp\left| DF^I - DF^D \right|}{N * n * L} \,. \tag{5}$$

In this equation the superscripts I and D denote the respecting types of changes inside one window position. Thus the term and document frequencies of each term are calculated only for the area restricted by the window. The overall term score (Equation 6) is the average distribution of the term in changes inside all window positions Nw.

$$S_{term}^{overall} = \frac{\sum_{win=1}^{Nw} S_{term}^{win}}{Nw} \,. \tag{6}$$

If inside many window positions a term was occurring generally in one type of changes then its overall score will be quite high. On the other hand, terms, which exhibit almost equal distributions in positive and negative types of changes inside majority of window positions, will have assigned low scores. In other words we favor terms that occur in bursts of deletions or bursts of insertions in the substantial number of window positions. This is implemented by considering the absolute values of differences in term and document frequencies of both types of changes (Equation 5). The length of the window is chosen by the user depending on whether short- or long-term concepts are to be discovered.

In the last step, sentences containing popular concepts are extracted and presented to the user as a summary for period *T*. To select such sentences we calculate the average term score for each sentence in the changes of the collection and retrieve the ones with the highest scores. The length of the summary is specified by the user. We also implement a user-defined limit of sentences, which can be extracted from a single document version. This restriction is put in order to avoid situations where one or only a few documents will dominate the final summary. To increase the summary understandability we add preceding and following sentences surrounding selected, top-scored sentences. Additionally to minimize summary redundancy we calculate cosine similarities between all sentences and reject the redundant ones. Lastly, sentences are arranged in the temporal order and are provided with links to their original web documents to enable users the access to the remaining parts of pages.

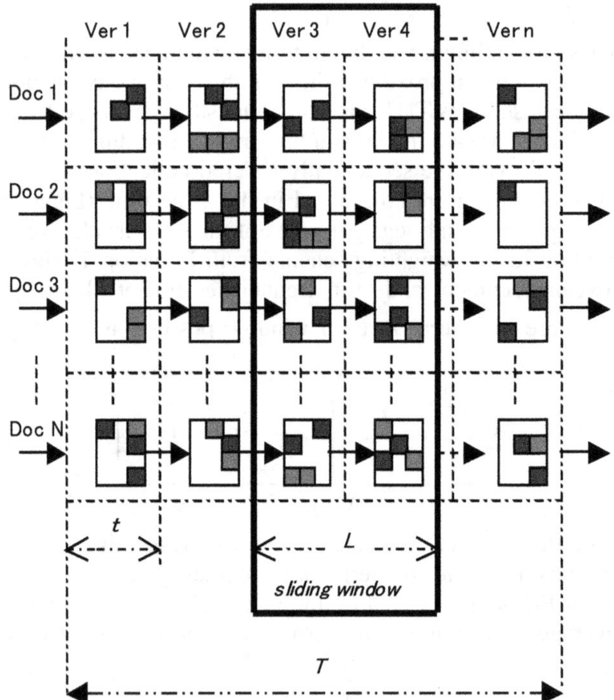

Fig. 3. Sliding window in the web collection of *N* documents with *n* versions. Dark areas symbolize insertions and deletions in web page versions.

5 Experiment

The results of our experiment are presented for the web collection which was built after issuing the query "EU enlargement" to the search engine and downloading 200 top-ranked web pages. We have manually filtered duplicate pages and documents which were topically distant from the meaning of the query. Change tracking was performed with the time delay t of 3 days during interval from 12th March to 12th May 2004. Table 1 displays the top scored sentences for this query.

Unfortunately, to the best of our knowledge there is no annotated corpus available, which could be used for the evaluation of change summarization in web collections. Since the creation of such corpus is not a straightforward task we have restricted the evaluation here to presenting an example of a summary and to discussing a simplified, experimental scenario, which shows the influence of different window lengths on the term score. Given the diversity of the types of web pages and discussed topics for this quite general query, it should not be surprising that results may not constitute coherent and high quality summary. Intuitively, it is very important to construct the appropriate web collection with closely related pages. We have noticed that results are better for narrow, topics, where documents tend to be more related topically with each other.

Table 1. Summary for "EU enlargement" query.

Europe reunited means a stronger, democratic and more stable continent, with a single market providing economic benefits for all its 450 million citizens. The European Union has come a long way since the original six member states joined forces to create the European Coal and Steel Community in 1951 and the European Economic Community in 1957, calling upon the peoples of Europe "who share their ideas to join their efforts." The six became nine in 1973, and had grown to 15 by 1995.
An enlarged EU: an opportunity for health? 1st May 2004 represents a both historical and symbolic landmark in the process of European integration. Initiated more than 50 years ago, the concept of the European Union now includes some of the countries that once belonged to the former Communist block. While undoubtedly a milestone in the "ever wider and closer" this new enlargement also raises a whole range of serious challenges particularly in terms of health and health policies.
Bulgaria has made economic progress but still has a long way to go and must close down nuclear power plants. The state of the Romanian economy is far from where it needs to be for EU accession. It must make more progress and improve its child-care institutions.
Never before has the European Union invited such a large group of countries which has had such a remarkably different social and economic system. This challenge will be especially large in the field of Structural Funds, the mechanism of the European Union that aims to achieve economic and social cohesion across the European territory.

Let us imagine a situation when only two instances of the same term are present in the changes of a collection. The term occurs once as a deletion and once as an insertion of a single document during period T. The rest of the changes of the collection can be empty. Different relative locations of both instances of the term should result in adequate term scores. In Figure 4 we plot the score of the term against the relative positions of its both instances. One instance of the term, for example an insertion, is fixed in the middle of the horizontal axis (value 9 in Figure 4) representing sequential versions of documents. In other words the term is occurring in the inserted part of 9th version of the document. The other instance (deletion) can be placed in any position in the collection. We will put it in every consecutive version of the document and compute term scores for all such positions. Thus starting from the beginning of the horizontal axis until the end of T we move the deletion instance and calculate the score of the term. In result the term score is plotted against all possible distances between both term instances. The score depends also on the window length. In Figure 4 three different lengths of the sliding window are used for the score calculation. Thus in general, the graph shows the influence of the size of the sliding window and the temporal closeness of both changes (temporal distance between the term instances) on the final term score.

From Figure 4 we see that in the middle of the horizontal axis the overall score is equal to zero since both instances of the term nullify each other. The relative term score is decreasing when the distance between both term instances becomes smaller. This decline starts later for windows with smaller length *L*, which embrace only few web page versions. Thus for a short length of the window the algorithm works in a short-term mode and assigns high scores to terms which can change often in close time throughout the summarization interval. On the other hand, for the high value of *L* long-term changes are favored, so in this case, the temporal granularity of an event discovery is diminished. Therefore for wider windows some short-life events can be overlooked. However, unlike in the former case there is the advantage of the reduced effect of temporal diversity between different web pages.

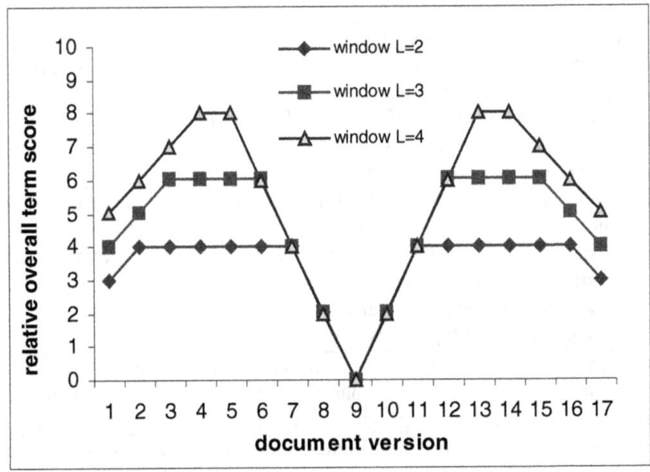

Fig. 4. Score for two opposite instances of the same term for different window lengths where the first instance occurs only in the middle of period *T* (the center of the horizontal axis) while the second one is placed in any document version.

6 Conclusions

We have introduced a new research area of summarization of dynamic content in retrospective web collections and have proposed an initial method, which employs sliding window over insertion and deletion types of changes. Our approach focuses on the temporal aspects of summarization in web genre. We have proposed to treat deletions as a part of dynamic data of web documents and invented a combined scoring approach for both types of changes. The advantages and challenges of summarization in dynamic web collections have been also discussed.

Currently we investigate evaluation methods which would enable us to compare different approaches to this summarization task. In the future we also would like to focus on the summarization of changes of the whole web sites and to make experiments with diverse kinds of web pages. Apart from that, we would like to take into consideration more attributes of web documents. Besides textual content there are some other changeable elements of web pages that can be exploited for summarization purposes.

References

1. Allan, J., Gupta, R., Khandelwal, V.: Temporal Summaries of News Topics. Proceedings of the 24th Annual ACM SIGIR Conference on Research and Development in Information Retrieval. New Orleans, USA (2001) 10-18
2. Allan, J. (ed.): Topic Detection and Tracking: Event-based Information Organization. Kluwer Academic Publishers. Norwell MA, USA (2002)
3. Berger, A. L., Mittal, V. O.: OCELOT: a System for Summarizing Web Pages. Proceedings of the 23rd ACM SIGIR Conference on Research and Development in Information Retrieval. Athens, Greece (2000) 144-151
4. Buyukkokten, O., Garcia-Molina, H., Paepcke, A.: Seeing the Whole in Parts: Text Summarization for Web Browsing on Handheld Devices. Proceedings of the 10th International WWW Conference. Hong Kong, Hong Kong (2001) 652-662
5. Google News: http://news.google.com
6. Jatowt, A., Khoo, K. B., Ishizuka, M.: Change Summarization in Web Collections. Proceedings of the 17th International Conference on Industrial and Engineering Applications of Artificial Intelligence and Expert Systems. Ottawa, Canada (2004) 653-662
7. Jatowt, A., Ishizuka, M.: Web Page Summarization Using Dynamic Content. Proceedings of the 13th International World Wide Web Conference. New York, USA (2004) 344-345
8. Mani, I., Maybury, M.T. (eds.): Advances in Automatic Text Summarization. MIT Press, Cambridge MA, USA (1999)
9. McKeown, K., Barzilay, R., Evans, D., Hatzivassiloglou, V., Klavans, J.L., Nenkova, A., Sable, C., Schiffman, B., Sigelman, S.: Tracking and Summarizing News on a Daily Basis with Columbia's Newsblaster. Proceedings of Human Language Technology Conference. San Diego, USA (2002)
10. Open Directory Project (ODP): http://dmoz.org
11. Radev, D., Blair-Goldensohn, S., Zhang, Z., Raghavan, S.R.: NewsInEssence: A System for Domain-Independent, Real-Time News Clustering and Multi-Document Summarization. In Human Language Technology Conference. San Diego, USA (2001)
12. Radev, D., Fan, W., Zhang, Z.: WebInEssence: A Personalized Web-Based Multi-Document Summarization and Recommendation System. In NAACL 2001 Workshop on Automatic Summarization. Pittsburgh, USA (2001) 79-88
13. Salton, G., Buckley, C.: Term Weighting Approaches in Automatic Text Retrieval. Information Processing and Management, Vol. 24, No 5, (1988) 513-523
14. Swan, R., Jensen, D.: TimeMines: Constructing Timelines with Statistical Models of Word Usage. In ACM SIGKDD 2000 Workshop on Text Mining, Boston MA, USA (2000) 73-80

Mining Thick Skylines over Large Databases

Wen Jin[1], Jiawei Han[2], and Martin Ester[1]

[1] School of Computing Science, Simon Fraser University
{wjin,ester}@cs.sfu.ca
[2] Department of Computer Science, Univ. of Illinois at Urbana-Champaign
hanj@cs.uiuc.edu

Abstract. People recently are interested in a new operator, called *skyline* [3], which returns the objects that are not dominated by any other objects with regard to certain measures in a multi-dimensional space. Recent work on the skyline operator [3, 15, 8, 13, 2] focuses on efficient computation of skylines in large databases. However, such work gives users only *thin skylines*, i.e., single objects, which may not be desirable in some real applications. In this paper, we propose a novel concept, called *thick skyline*, which recommends not only skyline objects but also their nearby neighbors within ε-distance. Efficient computation methods are developed including (1) two efficient algorithms, *Sampling-and-Pruning* and *Indexing-and-Estimating*, to find such thick skyline with the help of statistics or indexes in large databases, and (2) a highly efficient *Microcluster-based algorithm* for mining thick skyline. The *Microcluster-based method* not only leads to substantial savings in computation but also provides a concise representation of the thick skyline in the case of high cardinalities. Our experimental performance study shows that the proposed methods are both efficient and effective.

1 Introduction

In query-answering, people recently are interested in a new operator, called *skyline operator*[3]. *Given a set of n objects, the* ***skyline*** *refers to those that are not dominated by any other object. An object p* ***dominates*** *another object q, noted as $p \succ q$, if p is as good or better in all dimensions and better in at least one dimension.* A typical example is illustrated in Figure 1, showing the skyline of hotels with dimensions of the Distance (to the beach) and the Price. The hotels (a, b, c, d, e, f) are the skylines ranked as the *best* or *most satisfying* hotels.

The skyline operator can be represented by an (extended) SQL statement. An example Skyline Query of New York hotels corresponding to Figure 1 in SQL would be: *SELECT * FROM Hotels WHERE city='New York' SKYLINE OF Price min, Distance min*, where *min* indicates that the Price and the Distance attributes should be minimized. For simplicity, we assume that skylines are computed with respect to *min* conditions on all the dimensions, though it can be a combination with other condition such as *max*[3].

Most existing work on skyline queries has been focused on efficient computation of skyline objects in large databases [3, 15, 8, 13, 2]. However, the results

J.-F. Boulicaut et al. (Eds.): PKDD 2004, LNAI 3202, pp. 255–266, 2004.

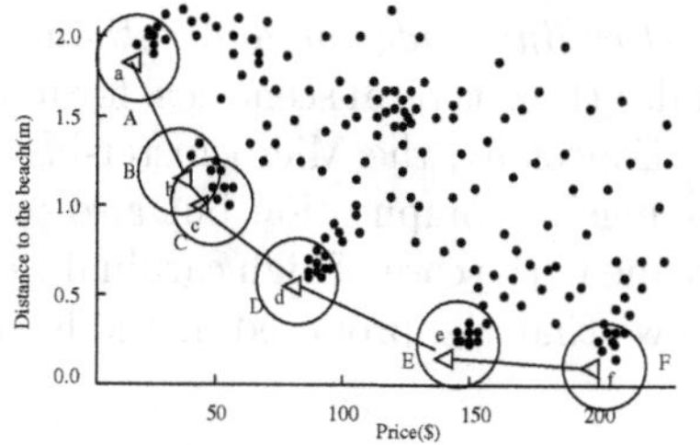

Fig. 1. Thick Skyline of N.Y. hotels.

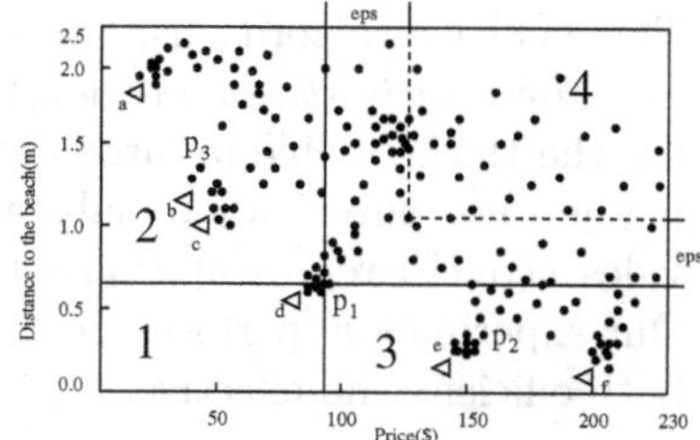

Fig. 2. Sampling objects to pruning.

obtained by the skyline operator may not always contain satisfiable information for users. Let's examine an example: Given the hotels in Figure 1, a conference organizer needs to decide the conference location. He usually will be interested in the following questions: 1. Can we find a bunch of *skyline* hotels which are nearby so as to provide good choices for the attendees? 2. If a skyline hotel is occupied, is there any nearby hotel which, though not ranked as high as skyline hotel, can still be a good candidate?

Apparently, the above questions cannot be answered directly by pure skyline computation as candidates which have similar attribute to skylines are not provided. Another problem in most of the existing studies is that they are based on the assumption of small skyline cardinality [13, 3, 8]. However, skyline objects could be many, making it inconvenient for users to browse and manually choose interesting ones. To address the problem of either two few or too many skyline objects, it seems to be natural to consider a compact and meaningful structure to represent the skyline and its neighborhood.

In this paper, we propose an extended definition of *skyline*, develop a novel data mining technique to skyline computation, and study the interesting patterns related to the skyline. The concept of skyline is extended to *generalized skyline* by pushing a user-specific constraint into skyline search space. For simplicity, the user-specific constraint is defined as the ε-neighbor of any skyline object. *Thick skyline* composed of the generalized skyline objects is the focus of the paper.

Mining the thick skyline is computationally expensive since it has to handle skyline detection and nearest neighbor search, which both require multiple database scans and heavy computation. Can we design algorithms that remove the computational redundancy in skyline detection and nearest neighbor search as much as possible? Furthermore, even in the same database system, different configurations may be required for different applications. For example, some may only allow the dataset to exist as a single file, others may have additional support with different types of index, such as B-tree, R-tree or CF-tree. How can we develop nice approaches to cope with these situations respectively?

Our contributions in this paper are as follows:

- A novel model of *thick skyline* is proposed that extends the existing skyline operator based on the distance constraint of skyline objects and their nearest neighbors.

- Three efficient algorithms, *Sampling-and-Pruning*, *Indexing-and-Estimating* and *Microcluster-based*, are developed under three typical scenarios, for mining the thick skyline in large databases. Especially, the Microcluster-based method not only leads to substantial savings in computation but also provides a concise representation of thick skyline in the case of high cardinalities.
- Our experimental performance study shows that the proposed methods are both efficient and effective.

The remaining of the paper is organized as follows. Section 2 overviews related work on the skyline. Sections 3 gives the definition of thick skyline and describes our proposed three algorithms. The results of our performance study are analyzed in Section 4. Section 5 concludes the paper with the discussion of future research directions.

2 Related Work

The skyline computation originates from the maximal vector problem in computational geometry, proposed by Kung et al. [7]. The algorithms developed [9, 14] usually suits for a small dataset with computation done in main memory. One variant of maximal vector problem, which is related to but different from the notion of thick skyline, is the *maximal layers* problem[11, 4] which aims at identifying different layers of maximal objects.

Borzsonyi et al. first introduce the skyline operator over large databases [3] and also propose a divide-and-conquer method. The method based on [7, 12] partitions the database into memory-fit partitions. The partial skyline objects in each partition is computed using a main-memory-based algorithm [14, 9], and the final skyline is obtained by merging the partial results. In [15], the authors proposed two progressive skyline computing methods. The first employs a bitmap to map each object and then identifies skyline through bitmap operations. Though the bit-wise operation is fast, the huge length of the bitmap is a major performance concern. The second method introduces a specialized B-tree which is built for each combination list of dimensions that a user might be interested in. Data in each list is divided into batches. The algorithm processes each batch with the ascending index value to find skylines.

Kossmann et al. present an online algorithm, NN, based on the nearest neighbor search. It gives a big picture of the skyline very quickly in all situations. However, it has raw performance when large amount of skyline needs to be computed. The current most efficient method is BBS (branch and bound skyline), proposed by Papadias et al., which is a progressive algorithm to find skyline with optimal times of node accesses [13]. Balke et al. [2] in their paper show how to efficiently perform distributed skyline queries and thus essentially extend the expressiveness of querying current Web information systems. They also propose a sampling scheme that allows to get an early impression of the skyline for subsequent query refinement.

3 The Thick Skyline and Mining Algorithms

Definition 1. (Generalized Skyline) *Given a d-dimensional database X and the skyline objects set $\{s_1, s_2, \ldots, s_m\}$, the generalized skyline is the set of all the following objects:*

- *the skyline objects, and*
- *the non-skyline objects which are ε-neighbors of a skyline object.*

We categorize the generalized skyline object into three classes: (a) a single skyline object, called *outlying skyline object*, (b) a *dense skyline object*, which is in a set of nearby skyline objects, and (c) a *hybrid skyline object*, which is in a set consisting of a mixture of nearby skyline objects and non-skyline objects.

From the data mining point of view, we are particularly interested in identifying the patterns of skyline information represented by clusters of types (b) and (c), which leads to the definition of the thick skyline.

Definition 2. (Thick Skyline) *Given a d-dimensional database X, the thick skyline is composed of all dense skyline and hybrid skyline objects.*

In this section, we explore different approaches to mining thick skyline in a d-dimensional database X with size $|X|$ under three typical situations. The first approach applies sampling and pruning technique to the relational files, and exploits the statistics of the database. The second approach estimates and identifies the range of thick skyline based on general index structures in relational database, which is not only suitable for thick skyline computation, but also composable with other relational operators. The third approach exploits the special summarization structure of micro-clusters which is widely used in data mining applications, and finds the thick skyline using bounding and pruning technique.

3.1 Sampling-and-Pruning Method

Sampling-and-Pruning method runs with the support of a database system where statistics, such as order and quantile in each dimension, can be obtained from the system catalog. The identification of thick skyline relies on the comparisons between objects. Clearly, the access order of objects crucially determines the number of comparisons. Hence we wish to pick some objects with high dominating capacity at the beginning to prune many dominated objects. As nearest neighbor search method [8] is expensive [13], a sampling method is developed.

The basic idea is as follows. We first randomly sample k ($k \ll |X|$)objects S with high dominating capacity as initial "seeds". Several criteria are required during the sampling step: (1) It prefers to choose objects with smaller values in dimensions which appear to be more dominating, and (2) the k objects are not dominated by each other. Each object of S can be taken temporarily as "skyline" objects to compare with other objects.

If the values in each dimension are distributed independently, an alternative but more aggressive sampling method [1] can also be applied to construct each of the k sampling objects by choosing d (smaller) values in each dimension (i.e., such k objects may not necessarily be in the dataset). Figure 2 shows a 2-dimensional hotel dataset partitioned into regions 1, 2, 3 and 4 by a randomly sampled object p_1. The pruning capacity of this sampling can be analyzed probabilistically. Assuming there are n objects and the largest values in Price and Distance axis are s and t respectively. Obviously, if p_1 is chosen properly, region 1 should not be empty, which will lead to the pruning of region 4. Otherwise it is a poor sampling object. Suppose the coordinates of p_1 is (u, v), the probability of region 1 being empty is $(\frac{s \cdot t - u \cdot v}{s \cdot t})^n = (1 - \frac{u}{s} \cdot \frac{v}{t})^n$. If u, v are chosen as the $\left\lceil \sqrt{n \ln n} \right\rceil$th smallest value in Price and Distance respectively, i.e. u and v are relatively small according to criteria 1, then the probability is $(1 - \frac{\sqrt{n \ln n}}{n} \cdot \frac{\sqrt{n \ln n}}{n})^n = (1 - \frac{\ln n}{n})^n \leq e^{-\ln n}[10] = \frac{1}{n}$, which is very small.

In the thick skyline computation process, all those non-skyline objects need to be investigated during the comparison step. In order to avoid unnecessary comparisons, we introduce a strongly dominating relationship and a lemma is deduced for pruning many non ε-neighbors of any skyline.

Definition 3. (Strongly Dominating Relationship) *An object $p \in X$ strongly dominates another object $q \in X$, noted as $p \rhd q$, if $p + \varepsilon$ dominates q, i.e. $\forall i$, $1 \leq i \leq d$, $p_i + \varepsilon \leq q_i$, and $p_i + \varepsilon < q_i$ in at least one dimension. On the other hand, q is a strongly dominated object.*

Lemma 1. *Given a dataset X, objects p and q, if $p \rhd q$, then q cannot be a thick skyline object.*

The strongly dominating relationship is illustrated by Figure 2, where objects strongly dominated by p_1 are in the dashed-lines rectangle.

Due to the space limitation, we briefly describe the algorithm as follows: First, sampling data S are generated and temporarily added to thick skyline list. In the pruning process, if an object x is strongly dominated by an object s in S, it is removed. If it is not only a dominated object but also an ε-neighbor of s, it is added to the neighbor list of s. If x dominates s, remove s and its strongly dominated neighbors by x and add x into the list. Finally after the pruning process, the thick skyline of a small amount of remaining object can be computed using any method such as the Indexing-and-Estimating Method.

3.2 Indexing-and-Estimating Method

Based on database index such as B-tree, by combining range estimate of the batches in the "minimum dimension" index [15] with an elaborate search technique, we can find the thick skyline within one scan of the database.

Assume that X is partitioned into d lists such that an object $p = (p_1, p_2, \ldots, p_d)$ is assigned to the i-th list ($1 \leq i \leq d$) if and only if p_i is the minimum among all dimensions. Table 1 shows an example. Objects in each list are sorted in ascending order of their minimum coordinate (*minC*, for short). A *batch* in the i-th list consists of objects having the same *minC*. In computing of skylines, initially the

Table 1. The index approach.

List1		List2	
$a(1,9)$	$minC = 1$	$k(9,1)$	$minC = 1$
$b(2,10)$	$minC = 2$	$i(3,2), m(6,2)$	$minC = 2$
$c(4,8)$	$minC = 4$	$h(4,3), n(8,3)$	$minC = 3$
$g(5,6)$	$minC = 5$	$l(10,4)$	$minC = 4$
$d(6,7)$	$minC = 6$	$f(7,5)$	$minC = 5$
$e(9,10)$	$minC = 9$		

first batches of all the lists are accessed and the one with the minimum $minC$ is processed. We assume the current minimum batches in the two lists of Table 1 are $minC_1$ and $minC_2$ respectively. Since $\{a\}$ and $\{k\}$ have identical $minC$, the algorithm picks $\{a\}$ and adds it to the skyline list. As the next batch $\{b\}$ has $minC_1 = 2$, $\{k\}$ in list 2 with $minC_2 = 1$ is processed and inserted into the list as it is not dominated by a. Then, the next batch handled is $\{b\}$ in list 1, where b is dominated by a in the list. Similarly, batch $\{i, m\}$ is processed and i is added to the skyline. At this step, no further batches need to be processed as the remaining objects in both lists are dominated by i and the skyline are $\{a, i, k\}$. During the search of a skyline, the range where its ε-neighbors exist is given in the following lemma.

Lemma 2. *Given d index lists of X, and a skyline object $p = (p_1, p_2, \ldots, p_d)$ is in the batch $minC = p_i$ of the ith list, then:*
(a) the ε-neighbors of p can only possibly exist in the batch range $[p_i - \varepsilon, p_i + \varepsilon]$ of the i-th list; and the batch range $\left[p_j - \varepsilon, p_j + \frac{\varepsilon}{\sqrt{2}}\right]$ of the j-th list ($j \neq i$);
(b) p does not have any ε-neighbor in jth list ($j \neq i$) if $(p_j - p_i) > \sqrt{2} \cdot \varepsilon$.

Proof. *(a) The proof of bounds in the i-th list and the lower bound in the j-th list are straightforward. Assume a ε-neighbor of p in the j-th list is $p' = (p'_1, p'_2, \ldots, p'_d)$ and p' exists in a batch with $minC = p'_j > p_j + \frac{\varepsilon}{\sqrt{2}}$, then $p'_i > p'_j > p_j + \frac{\varepsilon}{\sqrt{2}} > p_i + \frac{\varepsilon}{\sqrt{2}}$, we have $|p'_i - p_i| > \frac{\varepsilon}{\sqrt{2}}$ and $\left|p'_j - p_j\right| > \frac{\varepsilon}{\sqrt{2}}$, so $(\sum_{i=1}^{d} |p'_i - p_i|^2)^{\frac{1}{2}} > \varepsilon$, contradicting the definition. (b) As shown in Figure 3 (Eps is ε), all the objects in the j-th list can only appear in area I, while i-th in II. $dist(p, q) = p_j - p_i$, and $dist(p, o)$, is the shortest distance from p to any object in I. If $dist(p, o) > \varepsilon$, which means $dist(p, q) > \sqrt{2} \cdot \varepsilon$, then no ε-neighbor exist in I.*

In the dynamic scanning process of each list, if a skyline object p in the i-th list is found, how far shall we go back to find some of its ε-neighbors in the j-th list if the lower range bound is smaller than the $minC_j$ of the current batch? For those neighbors residing in batches greater than $minC_j$, we can certainly leave them to the remaining sequential scan. We show that only a ε length sliding window around the current batch $minC_j$ (denoted as SW_{minC_j}) needs to be maintained for each list, hence avoiding repeated backwards scans of the list and incur more overhead. The batch number $minC$ within the slide window SW_{minC_j} satisfies $minC_j - \varepsilon < minC < minC_j$.

Lemma 3. *Given d index lists of X, a skyline object $p = (p_1, p_2, \ldots, p_d)$ is found in the i-th list, while the current batch in the j-th list is $minC_j$ ($1 \leq j \leq d$ and $i \neq j$), if there are ε-neighbors of p existing in the batches with $minC \leq minC_j$ in the j-th list, then they can only exist in the slide window SW_{minC_j}.*

Proof. *Since batch $minC_i$ in the i-th list is the one the skyline searching algorithm is now handling, $minC_i \geq minC_j$. Also we have $p_j > p_i$ and $p_i = minC_i$. The lower bound of the batch range $p_j - \varepsilon > p_i - \varepsilon = minC_i - \varepsilon \geq minC_j - \varepsilon$ which is covered by the slide window SW_{minC_j}.*

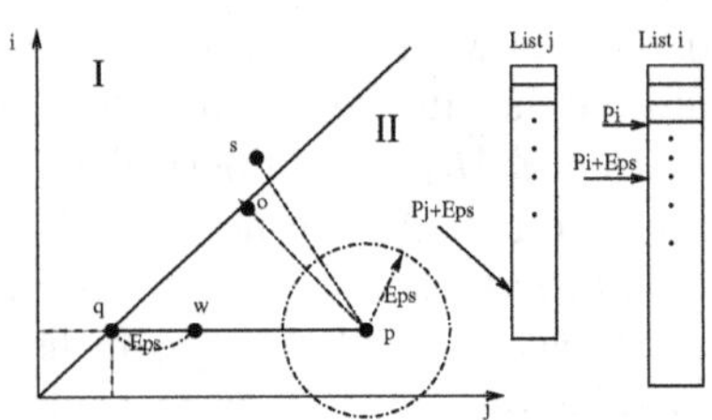

Fig. 3. Evaluate Neighborhood Scope.

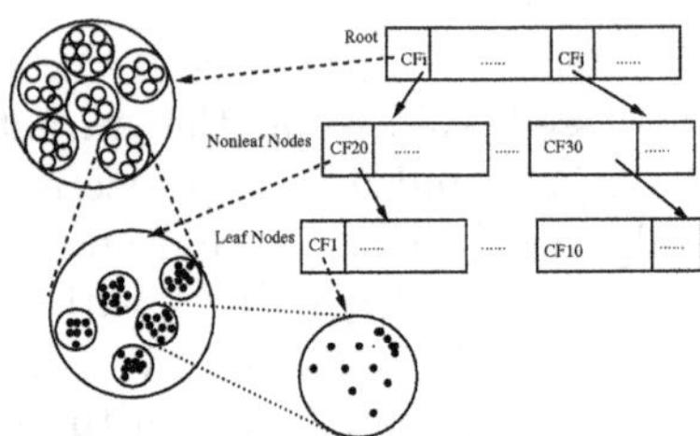

Fig. 4. Microclusters.

Lemma 3 also ensures an important property: Whenever an object is out of the slide window, it will never become an ε-neighbor for any skyline which enables us to find thick skyline within one scan. The algorithm pseudocode is as below.

Algorithm 1 An Indexing-and-Estimating Method.
Input: B-tree of d lists index and distance threshold ε.
Output: The thick skyline T.
Method:

$S = \emptyset; T = \emptyset$;
FOR $i = 1$ to d DO;
    $SW_i = \emptyset; upper_i = |list_i|; minC_i = \min list_i$;
WHILE $(Thin - Skyline - Search - Unfinished)$ DO;
    Choose the batch with $\min minC_1, \ldots, minC_d$, say $minC_k$;
    Check each object p in this batch;
    IF p is a skyline object THEN
        $S = S \cup \{p\}$;
        IF $(p_j - p_i) < \sqrt{2} \cdot \varepsilon$ THEN
            update $upper_j$ to $p_j + \frac{\varepsilon}{\sqrt{2}}$; check SW_j for ε neighbor ;
            IF any q is a ε neighbor THEN
                $T = T \cup \{q\}$;
    ELSE IF p is an ε-neighbor THEN
            $T = T \cup \{p\}$;
    Move $list_k$ to next batch and update SW_k;
WHILE $list_1 < upper_1 \vee \ldots \vee list_d < upper_d$ DO;
    scan objects to find ε neighbors and add to T;
$T = T \cup S$; Output thick skyline T;

The algorithm initiates skyline list and ε-neighbors list (Step 1), current batches, slide windows and the upper bound batch to scan in each list (Step 2-3). Each object p in the minimum $minC_i$ is compared with the skyline list (Step 6). If p is a skyline object, the corresponding range is updated, and part of p's ε-neighbors may be found in the slide windows (Step 8-12), while others are left to the remaining access of the lists(Step 13-14). Step 16-17 calculates the possible ε-neighbors in the range when thin skyline search finishes. Finally, it will output both skyline objects and ε-neighbors (Step 18).

3.3 Microcluster-Based Method

In order to scale-up data mining methods to large databases, a general strategy is to apply data compression or summarization. For example, we can partition the database into microclusters based on CF-tree [16, 6].

Definition 4. (Microcluster) *A microcluster for a set of d-dimensional objects* $X_1 \ldots X_n$, $X_i = (x_i^1 \ldots x_i^d)$, *is defined as a* $(4 \cdot d + 1)$*-tuple* $(\overline{CF1^x}, \overline{CF2^x}, \overline{CF3^x}, \overline{CF4^x}, n)$, *where* $\overline{CF1^x}$, $\overline{CF2^x}$, $\overline{CF3^x}$, *and* $\overline{CF4^x}$ *each represents a vector of d entries. The definition of each of these entries is as follows:*

- *The p-th entry of* $\overline{CF1^x}$ *is equal to* $\sum_{j=1}^{n} x_j^p$.
- *The p-th entry of* $\overline{CF2^x}$ *is equal to* $\sum_{j=1}^{n} (x_j^p)^2$.
- *The p-th entry of* $\overline{CF3^x}$ *is equal to* $\min_{j=1}^{n}(x_j^p)$.
- *For each dimension, the data with the minimum distance to the origin is maintained in* $\overline{CF4^x}$.
- *The number of data points is maintained in* n.

The centroid x_a and radius r_a of a microcluster mc_a can be represented as: $x_a = \frac{\overline{CF1^x}}{n}$, and $r_a = (\frac{\sum_{j=1}^{n}(x_j - x_a)^2}{n})^{\frac{1}{2}} = (\frac{\overline{CF2^x} + n \cdot x_a - 2 \cdot x_a \cdot \overline{CF1^x}}{n})^{\frac{1}{2}}$. The minimum distance $mdist_o$ from a microcluster to the origin is determined by $\overline{CF4^x}$.

For mining thick skyline, the database is partitioned into a set of microclusters with radius r_i (r_i can be around ε) in the leaf nodes of an extended CF-tree. Each non-leaf node represents a larger microcluster consisting of all its sub-microclusters as shown in Figure 4. There may exist overlap between microclusters and some methods [5, 6] can be used to remedy this problem.

The dominating relationship can be applied to the microclusters. For any two microclusters mc_a and mc_b, if $x_a \succ \overline{CF3_b^x}$, then $mc_a \succ mc_b$, that is, the objects in mc_b must be dominated by some objects in mc_a. As the number of microclusters is much less than that of objects, the computation cost is very low.

Let us now examine the neighborhood relationship between microclusters. Supposed object p is in a microcluster mc_a, the distance between mc_a and any microcluster mc_b is represented as: $dist_m(mc_a, mc_b) = dist(x_a, x_b) - r_a - r_b$. If $dist_m(mc_a, mc_b) < \varepsilon$, then mc_b and mc_a are *ε-neighboring microclusters* for p.

The basic idea of mining thick skyline is as follows. Instead of accessing every object in the dataset, we only need to identify the microclusters that contain skyline objects (called *skylining microclusters*), then find which microclusters are their ε-neighbors. The thick skyline objects can finally be determined from those microclusters. Skylining microclusters is an appropriate summarization of thick skyline in the case of large number of skylines or dynamic dataset.

The algorithm starts at the root node of the CF-tree and searches the microcluster in the ascending order of distance $mdist_o$. Initially, the algorithm selects the minimum one. Since the CF-tree is a hierarchical structure, the corresponding microcluster mc_i in the leaf node can be quickly located. mc_i is a skylining microcluster and is added to a heap h_1 sorted by the distance $mdist_o$. Then the algorithm selects the microcluster with the next minimum distance to the origin. If $\overline{CF3^x}$ of the selected microcluster is dominated by the centroid of any microcluster in h_1, it cannot contain skyline objects and we simply skip it. If it is strongly dominated by any microcluster in h_1, it can be pruned. Otherwise, the selected microcluster will be added to h_1. The algorithm continues until all of them are visited. As only the statistics of the microcluster is accessed, the cost is low.

Afterwards, the algorithm visits heap h_1, and extracts the microcluster mc_i' at the top of the heap. Within mc_i', object $\overline{CF4^x}$ is the skyline object, and the remaining objects are examined for skylineness in the order of $mdist_o$ (property guaranteed by [8]). Then a group ε-neighbors search for all the skyline objects in mc_i' is launched by searching *ε-neighboring microclusters*. Using the extended CF-tree, we simply check whether mc_i' intersects with larger microclusters in the root node, then with the non-leaf nodes, and finally locate the desired microclusters in the leaf nodes. The search complexity is bounded by the tree height and the intersected number of microclusters in the tree. The objects in these neighboring microclusters are examined whether they are ε-neighbors of skyline objects in mc_i'. The microcluster mc_i' is then removed from h_1, and the algorithm terminates until h_1 is empty. Based on the above description, The pseudocode for the Microcluster-based algorithm is as follows.

Algorithm 2 A Microcluster-based Method.
Input: m microclusters, and the distance threshold ε.
Output: The thick skyline.
Method:

$S = \emptyset$; $T = \emptyset$; $heap_1 = \emptyset$;
WHILE any mc_i with $\min mdist_o$ not visited DO
    IF $\neg(mc_j \in heap_1 \succ mc_i)$ THEN
        Add mc_i to $heap_1$;
WHILE $heap_1$ is not empty DO
    Select mc_i' at the top of $heap_1$;
    Select object p in mc_i' with $\min mdist_o$;
    IF $\neg(p \in S \succ p)$ THEN;
        Add p to S;
    Find neighboring microclusters of mc_i';
    Add ε-neighbors of skyline in mc_i' to T;
Output thick skyline $S \cup T$;

4 Experiments

In this section, we report the results of our experimental evaluation in terms of efficiency and effectiveness. We compare the runtime performance and evaluate several factors such as the choice of ε. We focus on the cost in the computing stage instead of pre-processing stage such as index or CF-tree building. Following similar data generation methods in [3], we employ two types of datasets: independent databases where the attribute values of tuples are generated using uniform distributions and anti-correlated datasets where the attribute values are good in one dimension but are bad in one or all other dimensions. The dimensionality of datasets d is between 2 and 5, the value of each dimension is in the integer range [1, 1000] and the number of data (cardinality) N is between 100k and 10M. We have implemented the three proposed methods in C++. All the experiments are conducted on Intel 1GHZ processor with 512M RAM.

• Runtime Performance To investigate the runtime versus different dimensionalities, We use dataset with cardinality 1M. Figures 5 and 6 depict the result in independent and anti-correlated distribution respectively. In both cases, Indexing-and-Estimating method achieves best performance in small dimensionality(d=2), due to its list structure being most suitable for relative small dataset and skyline size. Microcluster-based method is best towards large dimensionality ($d > 2$) and large skyline size, and the Sampling-and-Pruning method ranks the third.

Figures 7 and 8 show the runtime w.r.t. varied cardinality in independent and anti-correlated distributed 3-d datasets respectively($\varepsilon = 1$). In both cases, Microcluster-based method starts to over compete Indexing-and-Estimating when $N > 600K$ due to its region pruning and good scalability of hierarchical structure. As there is no index to facilitate computation, Sampling-and-Indexing still ranks the third, but the run time is not bad even when cardinality N=10M.

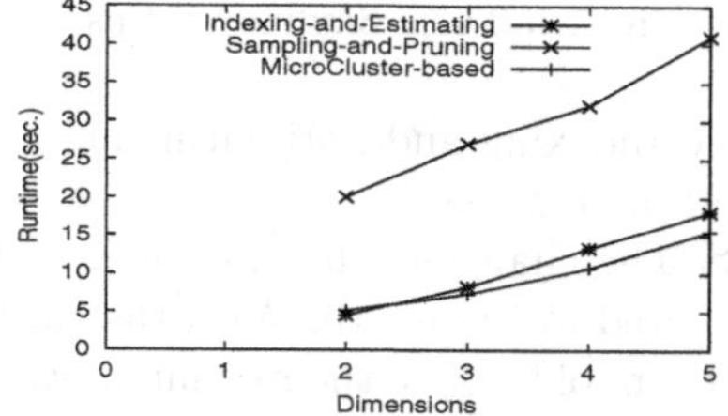

Fig. 5. Runtime vs. Dimensionality(I).

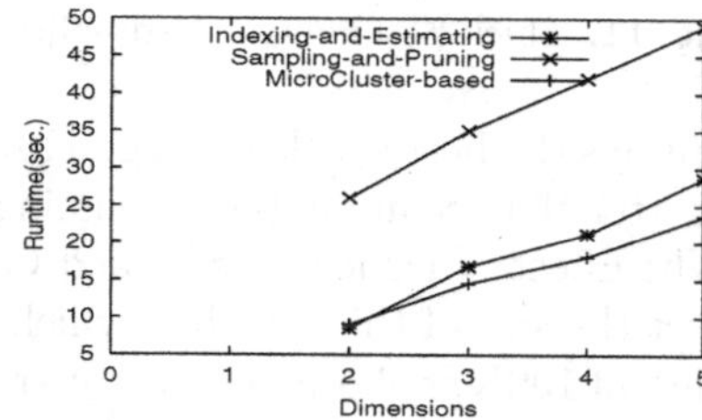

Fig. 6. Runtime vs. Dimensionality(II).

• The Effect of ε Obviously, the choice of ε values will affect the size of thick skyline. ε is usually small w.r.t. the domain bound, reflecting the "local neighborhood", and can be recommended by the system as an initial parameter for the future interaction. When we increase ε value from 1 to 30 in 1M independent 3-d dataset, Figure 12 and Figure 9 show that both the number of thick skyline and the run time of all algorithms increase. In particular, Microcluster-based method

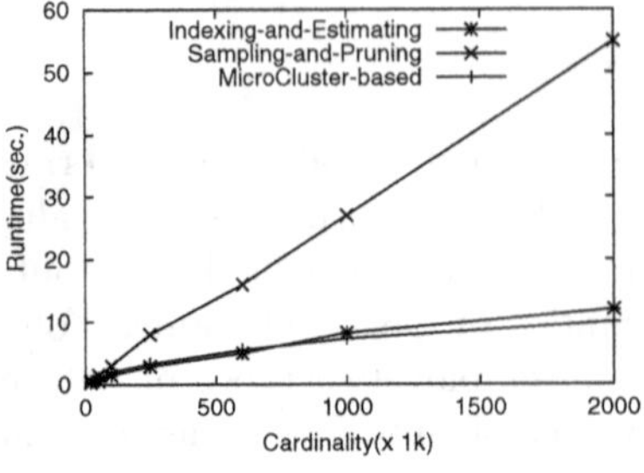

Fig. 7. Runtime vs. Cardinality(I).

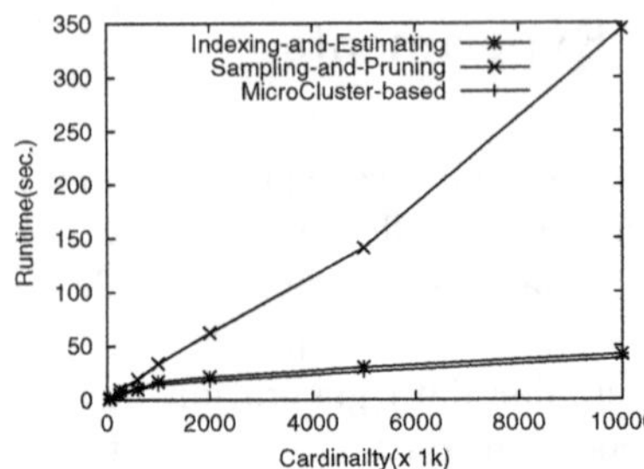

Fig. 8. Runtime vs. Cardinality(II).

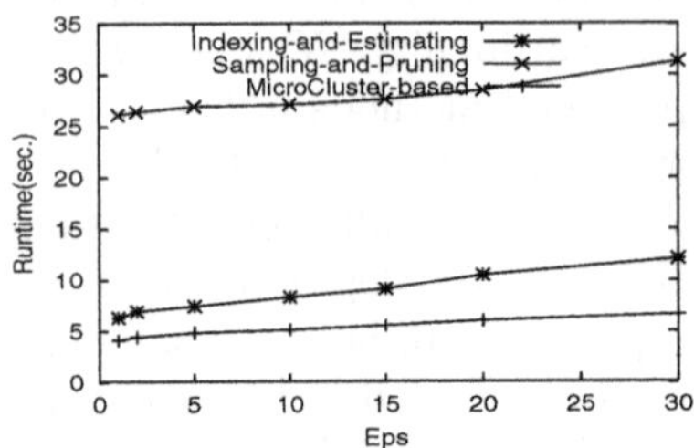

Fig. 9. Runtime vs. Eps.

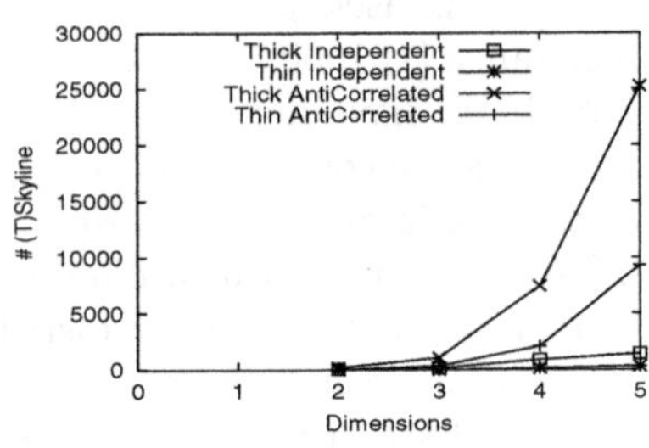

Fig. 10. Dimensionality vs. (T)Skyline.

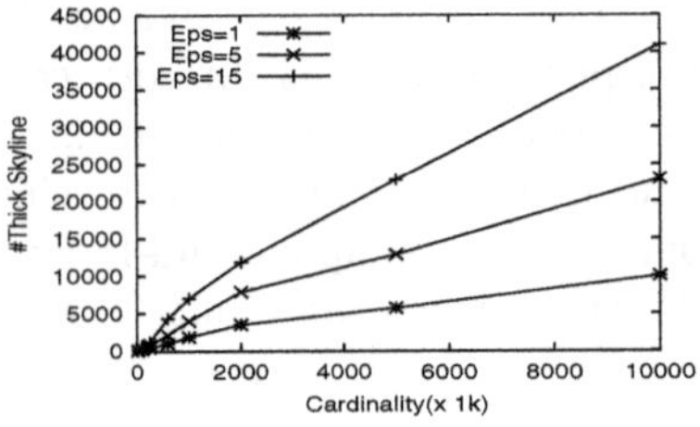

Fig. 11. Thick Skyline vs. Cardinality.

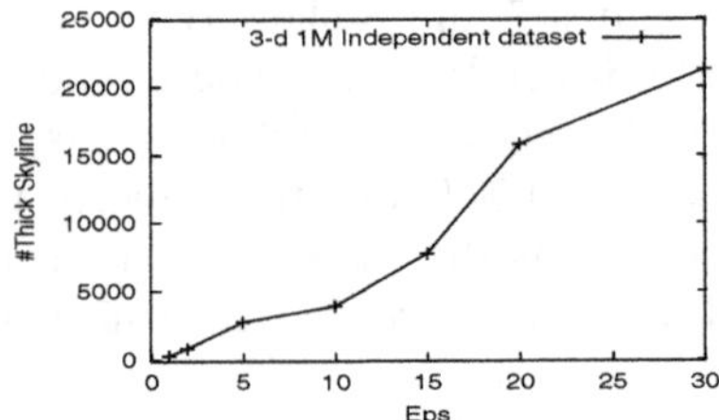

Fig. 12. Thick Skyline vs. Eps.

is always the best and keeps a good scalability. Indexing-and-Estimating method is better than Sampling-and-Pruning method in runtime.

• The Effect of Dimensionality and Cardinality The change of dimensionality will affect the size of thick skyline which is illustrated in Figure 10. With the cardinality of 100K, ε chosen as the square root of sum of 0.1% of the maximum value in each dimension, Figure 10 shows the size of both thin and thick skylines in different dimensions. We notice that if dimensionality increases, the number of the thick skyline objects increases dramatically for both independent and anti-correlated distributed dataset, with the latter having higher speed. The affect of cardinality is shown in Figure 11.

5 Conclusions

The paradigm of rank-aware query processing, in particular, the new *skyline operator*, has recently received a lot of attention in database community. In this paper, we propose a novel notion of *thick skyline* based on the distance constraint of a skyline object from its neighbors. The task of mining thick skyline is to recommend skyline objects as well as their ε-distance neighbors. We develop three algorithms, Sampling-and-Pruning, Indexing-and-Estimating and Microcluster-based, to find such thick skylines in large databases. Our experiments demonstrate the efficiency and effectiveness of the algorithms. We believe the notion of thick skyline and mining methods not only extends the skyline operator in database query, but also provides interesting patterns for data mining tasks. Future work includes mining thick skylines in data streams and combining the task with other regular data mining process.

References

1. J. L. Bentley, K. L. Clarkson, D. B. Levine. Fast linear expected-time alogorithms for computing maxima and convex hulls. In *SODA 1990.*
2. W.-T. Balke, U. Güntzer, J. X. Zheng. Efficient Distributed Skylining for Web Information Systems In *EBDT 2004.*
3. S. Borzsonyi, D. Kossmann, and K. Stocker, The Skyline Operator. In *ICDE 2001.*
4. T. Cormen, C. E. Leiserson, R. L. Rivest and C. Stein. Introduction to Algorithms, second edition. The MIT Press, 2001.
5. Alexander Hinneburg, Daniel A. Keim Optimal Grid-Clustering: Towards Breaking the Curse of Dimensionality in High-Dimensional Clustering. VLDB 1999: 506-517
6. W. Jin, K. H. Tung, and J. Han. Mining Top-n Local Outliers in Very Large Databases. In *KDD 2001.*
7. H. T. Kung et. al. On finding the maxima of a set of vectors. *JACM*, 22(4), 1975.
8. D. Kossmann, F. Ramsak, and S. Rost. Shooting Stars in the Sky: An Online Algorithm for Skyline Queries. In *VLDB 2002.*
9. J. Matousek. Computing dominances in E^n. *Inf. Process. Lett.*, 38(5), 1991.
10. R. Motwani, P. Raghavan. Randomized Algorithms. *Cambridge Univ. Press*, 1995.
11. F. Nielsen. Output-sensitive peeling of convex and maximal layers. Thesis, 1996
12. F. P. Preparata, and M. I. Shamos. Computational Geometry: An Introduction. Springer-Verlag, 1985.
13. D. Papadias, Y. Tao, G. Fu, and B. Seeger. An Optimal and Progressive Algoroithm for Skyline Queries. In *SIGMOD'03.*
14. I. Stojmenovic and M. Miyakawa. An Optimal Parallel Algorithm for Solving the Maximal Elements Problem in the Plane. In *Parallel Computing, 7(2)*, 1988.
15. K. Tan et al. Efficient Progressive Skyline Computation. In *VLDB 2001.*
16. T. Zhang, R. Ramakrishnan, and M. Livny. BIRCH: an efficient data clustering method for very large databases. In *SIGMOD'96.*

Ensemble Feature Ranking

Kees Jong[1], Jérémie Mary[2], Antoine Cornuéjols[2],
Elena Marchiori[1], and Michèle Sebag[2]

[1] Department of Mathematics and Computer Science
Vrije Universiteit Amsterdam, The Netherlands
{cjong,elena}@cs.vu.nl
[2] Laboratoire de Recherche en Informatique, CNRS-INRIA
Université Paris-Sud Orsay, France
{mary,antoine,sebag}@lri.fr

Abstract. A crucial issue for Machine Learning and Data Mining is Feature Selection, selecting the relevant features in order to focus the learning search. A relaxed setting for Feature Selection is known as Feature Ranking, ranking the features with respect to their relevance.
This paper proposes an ensemble approach for Feature Ranking, aggregating feature rankings extracted along independent runs of an evolutionary learning algorithm named *ROGER*. The convergence of ensemble feature ranking is studied in a theoretical perspective, and a statistical model is devised for the empirical validation, inspired from the complexity framework proposed in the Constraint Satisfaction domain. Comparative experiments demonstrate the robustness of the approach for learning (a limited kind of) non-linear concepts, specifically when the features significantly outnumber the examples.

1 Introduction

Feature Selection (FS) is viewed as a major bottleneck of Supervised Machine Learning and Data Mining [13,10]. For the sake of the learning performance, it is highly desirable to discard irrelevant features prior to learning, especially when the number of available features significantly outnumbers the number of examples, as is the case in Bio Informatics. FS can be formalized as a combinatorial optimization problem, finding the feature set maximizing the quality of the hypothesis learned from these features. Global approaches to this optimization problem, referred to as wrapping methods, actually evaluate a feature set by running a learning algorithm [20,13]; for this reason, the wrapping approaches hardly scale up to large size problems. Other approaches combine GA-based feature selection with ensemble learning [9].

A relaxed formalization of FS, concerned with feature ranking (FR) [10], is presented in section 2. In the FR setting, one selects the top ranked features, where the number of features to select is specified by the user [11] or analytically determined [19].

A new approach, inspired from bagging and ensemble learning [4] and referred to as ensemble feature ranking (EFR) is introduced in this paper. EFR

J.-F. Boulicaut et al. (Eds.): PKDD 2004, LNAI 3202, pp. 267–278, 2004.

aggregates several feature rankings independently extracted from the same training set; along the same lines as [4, 5], it is shown that the robustness of ensemble feature ranking increases with the ensemble size (section 3).

In this paper, EFR is implemented using GA-based learning. Practically, we used the *ROGER* algorithm (*ROC-based Genetic Learner*) first presented in [17], that optimizes the so-called AUC criterion. The AUC, the area under the Receiver Operating Characteristics (ROC) curve has been intensively studied in the ML literature since the late 90's [3, 6, 14, 16]. The ensemble feature ranking aggregates the feature rankings extracted from hypotheses learned along independent *ROGER* runs.

The approach is validated using a statistical model inspired from the now standard Constraint Satisfaction framework known as Phase Transition paradigm [12]; this framework was first transported to Machine Learning by [8]. Seven order parameters are defined for FS (section 4); the main originality of the model compared to previous ones [10] is to account for (a limited kind of) non-linear target concepts. A principled and extensive experimental validation along this model demonstrates the good performance of Ensemble Feature Ranking when dealing with non linear concepts (section 5). The paper ends with a discussion and perspectives for further research.

2 State of the Art

Without aiming at an exhaustive presentation (see [10] for a comprehensive introduction), this section introduces some Feature Ranking algorithms. *ROGER* is then described for the sake of completeness.

Notations used throughout the paper are first introduced. Only binary concept learning is considered in the following. The training set $\mathcal{E}$ includes n examples, $\mathcal{E} = \{(\mathbf{x}_i, y_i),\ \mathbf{x}_i \in \mathbb{R}^d,\ y_i \in \{-1, 1\},\ i = 1 \dots n\}$. The i-th example is described from d continuous feature values; label y_i indicates whether the example pertains to the target concept (positive example) or not (negative example).

2.1 Univariate Feature Ranking

In univariate approaches, a score is associated to each feature independently from the others. In counterpart for this simplicity, univariate approaches are hindered by feature redundancy; indeed, features correlated to the target concept will be ranked first, no matter whether they offer little additional information.

The feature score is computed after a statistical test, quantifying how well this feature discriminates positive and negative examples. For instance the Mann-Whitney test, reported to support the identification of differentially relevant features [15], associates to the k-th feature the score defined as $Pr(x_{i,k} > x_{j,k} \mid y_i > y_j)$, i.e. the fraction of pairs of (positive, negative) examples such that feature k ranks the positive example higher than the negative one. This criterion coincides with the Wilcoxon rank sum test, which is equivalent to the AUC criterion [21].

2.2 Univariate FR + Gram Schmidt Orthogonalization

A sophisticated extension of univariate approaches, based on an iterative selection process, is presented in [19]. The score associated to each feature is proportional to its cosine with the target concept:

$$score(k) = \frac{\sum_{i=1}^{n} x_{i,k} \cdot y_i}{\sqrt{\sum_{i=1}^{n} x_{i,k}^2}}$$

The two-step iterative process i) determines the current feature k maximizing the above score; ii) projects all remaining features and the target concept on the hyperplane perpendicular to feature k. The stopping criterion is based on an analytic study of the random variable defined as the cosine of the target concept with a random uniform feature.

Though this approach addresses the limitations of univariate approaches with respect to redundant features, it still suffers from the myopia of greedy search strategies (with no backtrack).

2.3 ML-Based Approaches

As mentioned earlier on, an alternative to univariate approaches is to exploit the output of a machine learning algorithm, which assumedly takes into account every feature one by one in relation with the other ones [13].

When learning a linear hypothesis ($h(\mathbf{x}) = \sum_{i=1}^{d} w_i x_i$ $[+b]$), one associates a score to each feature k, namely the square of the weight w_k; the higher the score, the more relevant the feature is *in combination with the other features*.

A two-step iterative process, termed SVM-Recursive Feature Elimination, is proposed by [11]. In each step, i) a linear SVM is learned, the features are ranked by decreasing absolute weight; ii) the worst features are removed.

Another approach, based on linear regression [1], uses a randomized approach for better robustness. Specifically, a set of linear hypotheses are extracted from independent subsamples of the training set, and the score of the k-th feature averages the feature weight over all hypotheses learned from these subsamples. However, as subsamples must be significantly smaller than the training set in order to provide diverse hypotheses, this approach might be limited in application to domains with few available examples, e.g. DNA array mining.

Another work, more loosely related, is concerned with learning an ensemble of GA-based hypotheses extracted along independent runs [9], where: i) the underlying GA-inducer looks for good feature subsets; and ii) the quality of a feature subset is measured from the accuracy of a k-nearest neighbor or euclidean decision table classification process, based on these features.

2.4 *ROGER* (ROC-Based Genetic learneR)

ROGER is an evolutionary learning algorithm first presented in [18, 17]. Using elitist evolution strategies ($(\mu + \lambda)$-ES), it determines hypotheses maximizing the Area Under the ROC curve (AUC) [3, 14]. As already mentioned, the AUC criterion was shown equivalent to the Wilcoxon statistics [21].

ROGER allows for constructing a limited kind of non linear hypotheses. More precisely, a hypothesis h measures the weighted L_1 distance to some point $\mathbf{c}$ in the instance space $\mathbb{R}^d$. Formally, to each genetic individual $Z = (w_1, \ldots, w_d, c_1, \ldots, c_d)$ is associated the hypothesis h_Z defined as:

$$h_Z : \mathbf{x} = (x_1, \ldots, x_d) \in \mathbb{R}^d \mapsto \mathbb{R}, \qquad h_Z(\mathbf{x}) = \sum_{i=1}^{d} w_i \times |x_i - c_i|$$

This way, *ROGER* explores search space $\mathbb{R}^{2d}$, with size linear in the number of features while possibly detecting some non-linearities in the data. *ROGER* maximizes the fitness function $\mathcal{F}$, where $\mathcal{F}(Z)$ is computed as the Wilcoxon statistics associated to h_Z ($\mathcal{F}(Z) = Pr(h_Z(\mathbf{x}_i) > h_Z(\mathbf{x}_j) | y_i > y_j)$).

3 Ensemble Feature Ranking

This section describes an ensemble approach to feature ranking which will be implemented using the *ROGER* algorithm above. The properties of ensemble feature ranking are first examined from a theoretical perspective.

3.1 Notations

Inspired from ensemble learning [4] and randomized algorithms [5], the idea is to combine independent feature rankings into a hopefully more robust feature ranking. Formally, let O_t denote a feature ranking (permutation on $\{1, ..d\}$). With no loss of generality, we assume that features are enumerated with decreasing relevance (e.g. feature i is more relevant than feature j iff $i < j$).

Let $O_1, \ldots, O_T$ be T independent, identically distributed feature rankings. For each feature pair (i, j) let $N_{i,j}$ denote the number of O_t that rank feature i before feature j and let $Y_{i,j}$ be true iff $N_{i,j} > \frac{T}{2}$. We start by showing that a feature ranking can be constructed from variables $Y_{i,j}$, referred to as ensemble feature ranking (EFR); the EFR quality is then studied.

3.2 Consistent Ensemble Feature Ranking

In order to construct an ensemble feature ranking, variables $Y_{i,j}$ must define a transitive relation, i.e. $Y_{i,k}$ is true if $Y_{i,j}$ and $Y_{j,k}$ are true; when this holds for all i, j, k, feature rankings $O_1, \ldots O_T$ are said *consistent*.

The swapping of feature pairs (i, j) is observed from the boolean random variables $X_{i,j}$ ($X_{i,j}(O_t) = ((O_t(i) < O_t(j)) \neq (i < j))$. Inspired from [16], it is assumed that variables X_{ij} are independent Bernoulli random variables with same probability p. Although the independence assumption is certainly not valid (see discussion in [16]), it allows for an analytical study of EFR, while rigorously combining permutations raises more complex mathematical issues.

Lemma. *Let* $p = Pr((O_t(i) < O_t(j)) \neq (i < j))$ *denote the swapping rate of feature rankings* O_t*, and assume that* $p = \frac{1}{2} - \varepsilon$, $\varepsilon > 0$, *(that is, each ranking does a little better than random guessing wrt every pair of attributes). Then*

$$Pr(Y_{i,j} \text{ false } \mid i < j) \leq e^{-2\varepsilon^2 T}$$

Proof. Follows from Hoeffding's inequality.

Proposition 1. *Under the same assumption,* $O_1, \ldots O_T$ *are consistent with probability 1 as* T *goes to infinity.*
Proof. It must be noted first that from Bayes rule, $Pr(i < j \mid Y_{i,j} \text{ true }) = Pr(Y_{i,j} \text{ true } \mid i < j)$ (as $Pr(i < j) = Pr(Y_{i,j} \text{ true }) = \frac{1}{2}$).
Assume that $Y_{i,j}$ and $Y_{j,k}$ are true. After the working assumption that the $X_{i,j}$ are independent,

$$Pr(i < j, j < k \mid Y_{i,j} \wedge Y_{j,k} \text{ true }) = Pr(i < j \mid Y_{i,j} \text{ true}) \cdot Pr(j < k \mid Y_{j,k} \text{ true})$$

Therefore after the lemma, $Y_{i,j}$ and $Y_{j,k}$ true imply that $i < k$ and hence that $Y_{i,k}$ is true, with probability going exponentially fast to 1 as T goes to infinity.

3.3 Convergence

Assuming the consistency of the feature rankings $O_1, \ldots, O_T$, the ensemble feature ranking O^* is naturally defined, counting for each feature i the number of features j that are ranked before i by over half the O_t ($O^*(i) = \#\{Y_{j,i} \text{ true}, j = 1..d\}$, where $\#A$ is meant for the size of set A).

The convergence of ensemble feature ranking is studied with respect to the probability of misranking a feature i by at most τ indices ($Pr(|O^*(i) - i| \geq \tau)$). Again, for the simplicity of this preliminary analytical study, it is assumed that the misranking probability does not depend on the "true" rank of feature[1] i.

Proposition 2. *Let* p^* *denote the probability for the ensemble feature ranking to to swap two features,* $p^* = Pr(Y_{ij} \neq (i < j))$*, and let* $\tau = (d-1)p^* + \varepsilon$, $\varepsilon > 0$.

$$\textit{Then} \qquad Pr(|O^*(i) - i| \geq \tau) \leq e^{-\frac{2\varepsilon^2}{d-1}}$$

Proof. Feature i is misranked by at least τ indices if there exists at least τ features j in the remaining $d-1$ features, such that $Y_{ij} \neq (i < j)$.
Let $\mathcal{B}(d-1, p^*)$ denote the binomial distribution of parameters $d-1$ and p^*, then $Pr(|O^*(i) - i| \geq \tau) < Pr(\mathcal{B}(d-1, p^*) \geq \tau)$, where after Hoeffding's inequality,

$$Pr(\mathcal{B}(d-1, p^*) - (d-1)p^* > \varepsilon) \leq e^{-\frac{2\varepsilon^2}{d-1}}$$

[1] Clearly, this assumption does not hold, as the probability of misranking top or bottom features is biased compared to other features. However, this preliminary study focuses on the probability to largely misrank features, e.g. the probability of missing a top 10 feature when discarding the 50% features ranked at the bottom.

The good asymptotic behavior of ensemble feature ranking then follows from the fact that: i) the swapping rate p^* of the EFR decreases with the size T of the ensemble, *exponentially amplifying the advantage of the elementary* feature ranking over the random decision [5]; ii) the distribution of the EFR misranking error is centered on $p^* \times (d-1)$, d being the total number of features.

4 Statistical Validation Model

Before proceeding to experimental validation, it must be noted that the performance of a feature selection algorithm is commonly computed from the performance of a learning algorithm based on the selected features, which makes it difficult to compare standalone FS algorithms.

To sidestep this difficulty, a statistical model is devised, enabling the direct evaluation of the proposed FR approach. This model is inspired from the statistical complexity analysis paradigm developed in the Constraint Satisfaction community [12], and first imported in the Machine Learning community by Giordana and Saitta [8]. This model is then discussed wrt [10].

4.1 Principle

In the statistical analysis paradigm, the problem space is defined by a set of order parameters (e.g. the constraint density and tightness in CSPs [12]). The performance of a given algorithm is viewed as a random variable, observed in the problem space. To each point in the problem space (values of the order parameters), one associates the average behavior of the algorithm over all problem instances with same value of the order parameters.

This paradigm has proved insightful in studying the scalability of prominent learning algorithms, and detecting unexpected "failure regions" where the performance abruptly drops to that of random guessing [2].

4.2 Order Parameters

Seven order parameters are defined for Feature Selection:

- The number n of examples.
- The total number d of features.
- The number r of relevant features. A feature is said to be relevant iff it is involved in the definition of the target concept, see below.
- The type l of target concept, linear ($l=1$) or non-linear ($l=2$), with

$$l = 1: \quad y(\mathbf{x}) = 1 \quad \text{iff} \quad (\textstyle\sum_{i=1}^{r} x_i > s) \tag{1.1}$$
$$l = 2: \quad y(\mathbf{x}) = 1 \quad \text{iff} \quad (\textstyle\sum_{i=1}^{r} (x_i - .5)^2 < s) \tag{1.2}$$

- The redundancy ($k = 0$ or 1) of the relevant features. Practically, redundancy ($k = 1$) is implemented by replacing r of the irrelevant features, by linear random combinations of the r relevant ones[2].
- The noise rate e in the class labels: the class associated to each example is flipped with probability e.
- The noise rate σ in the dataset feature values: each feature value is perturbed by adding a Gaussian noise drawn after $\mathcal{N}(0, \sigma)$.

4.3 Artificial Problem Generator

For each point $(n, d, r, l, k, e, \sigma)$ in the problem space, independent instances of learning problems are generated after the following distribution.

All d features of all n examples are drawn uniformly in $[0, 1]$. The label of each example is computed as in equation (1.1) (for $l = 1$) or equation (1.2) (for $l = 2$) [3]. In case of redundancy ($k = 1$), r irrelevant features are selected and replaced by linear combinations of the r relevant ones. Last, the example labels are randomly flipped with probability e, and the features are perturbed by addition of a Gaussian noise with variance σ.

The above generator differs from the generator proposed in [10] in several respects. [10] only considers linear target concepts, defined from a linear combination of the relevant features; this way, the target concept differentially depends on relevant features, while all features have the same relevance in our model. In contrast, the proposed model investigates linear as well as a (limited kind of) non-linear concepts.

4.4 Format of the Results

Feature rankings are evaluated and compared using a ROC-inspired setting. To each index $i \in \{1, d\}$ is associated the fraction of true relevant features (respectively, the fraction of irrelevant, or falsely relevant, features) with rank higher than i, denoted $TR(i)$ (resp. $FR(i)$). The curve $\{(FR(i), TR(i)), i = 1, \ldots, d\}$ is referred to as ROC-FS curve associated to $\mathcal{O}$.

The ROC-FS curve shows the trade-off achieved by the algorithm between the two objectives of setting high ranks (resp. low ranks) to relevant (resp. irrelevant) features. The ROC-FS curve associated to a perfect ranking (ranking all relevant features before irrelevant ones), reaches the global optimum $(0, 1)$ (no irrelevant feature is selected, $FR = 0$, while all relevant features are selected, $TR = 1$).

The inspection of the ROC-FS curves shows whether a Feature Ranking algorithm consistently dominates over another one. The curve also gives a precise

[2] Since any subset of r features selected among the r relevant ones plus the r redundant ones is sufficient to explain the target concept, the true relevance rate is set to 1. when at least r features have been selected among the true $2r$ ones.

[3] The threshold s referred to in the target concept definition is set to $r/2$ in equation (1.1) (respectively $r/12$ in equation (1.2)), guaranteeing a balanced distribution of positive and negative examples. The additional difficulties due to skewed example distributions are not considered in this study.

picture of the algorithm performance; the beginning of the curve shows whether the top ranked features are actually relevant, suggesting an iterative selection approach as in [19]; the end of the curve shows whether the low ranked features are actually irrelevant, suggesting a recursive elimination procedure as in [11].

Finally, three indicators of performance are defined on a feature ranking algorithm. The first indicator measures the probability for the best (top) ranked feature to be relevant, noted p_b, reflecting the FR potential for a selection procedure. The second indicator measures the worst rank of a relevant feature, divided by d, noted p_w, reflecting the FR potential for an elimination procedure.
A third indicator is the area under the ROC-FS curve (AUC), taken as global indicator of performance (the optimal value 1 being obtained for a perfect ranking).

5 Experimental Analysis

This section reports on the experimental validation of the EFR algorithm described in section 3. The results obtained are compared to the state of the art [19] using the cosine criterion. Both algorithms are compared using the ROC-FS curve and the performance measures introduced in section 4.4.

5.1 Experimental Setting

A principled experimental validation has been conducted along the formal model defined in the previous section. The number d of features is set to 100, 200 and 500. The number r of relevant features is set to $d/20, d/10$ and $d/5$. The number n of examples is set to $d/2$, d and $2d$. Linear and non-linear target concepts are considered ($l = 1$ or 2), with redundant ($k = 1$) and non-redundant ($k = 0$) feature sets. Last, the label noise e is set to 0, 5 and 10%, and the variance σ of the feature Gaussian noise is set to 0., .05 and .10.

In total 972 points $(d, r, m, l, k, e, \sigma)$ of the problem space are considered. For each point, 20 datasets are independently generated. For each dataset, 15 independent *ROGER* runs are executed to construct an ensemble feature ranking $\mathcal{O}$; the associated indicators p_b, p_w and the *AUC* are computed, and their median over all datasets with same order parameters is reported.

The reference results are obtained similarly from the cosine criterion [19]: for each point of the problem space, 30 datasets are independently generated, the cosine-based feature ranking is evaluated from indicators p_b, p_w and the *AUC*, and the indicator median over all 30 datasets is reported.

Computational runtimes are measured on PC Pentium-IV; the algorithms are written in C++. *ROGER* is parameterized as a (20+200)-ES with self adaptive mutation, uniform crossover with rate .6, uniform initialization in [0, 1], and a maximum number of 50,000 fitness evaluations[4].

[4] All datasets and *ROGER* results are available at `http://www.lri.fr/~sebag/EFRDatasets` and `http://www.lri.fr/~sebag/EFResults`

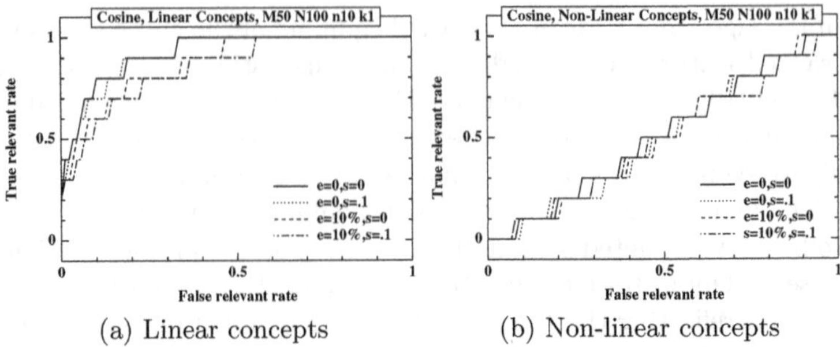

(a) Linear concepts (b) Non-linear concepts

Fig. 1. Cosine criterion: Median ROC-FS curves over 30 training sets on Linear and Non-Linear concepts, with $d = 100$, $n = d/2$, $r = d/10$, Non redundant features.

Table 1. The cosine ranking criterion: Probability p_b of top ranking a relevant feature, Median relative rank p_w of the worst ranked relevant feature, Area under the ROC-FS curve.

					Linear concepts, No redundancy			Linear concepts, Redundancy			Non Linear concepts, No redundancy			Non Linear concepts, Redundancy		
n	d	r	e	σ	p_b	p_w	AUC	p_b	p_w	AUC	p_b	p_w	AUC	p_b	p_w	AUC
50	100	10	0	0	0.87	.33	0.920	0.97	.10	0.97	0.03	.93	0.49	0.17	.42	0.74
50	100	10	0	0.1	0.9	.33	0.916	0.9	.10	0.97	0.03	.94	0.49	0.1	.45	0.74
50	100	10	10%	0	0.87	.47	0.87	0.77	.16	0.95	0.1	.93	0.49	0.1	.44	0.73
50	100	10	10%	0.1	0.8	.56	0.848	0.83	.17	0.95	0.03	.93	0.51	0.1	.44	0.75
100	100	10	0	0	1	.18	0.97	1	.10	0.99	0	.91	0.53	0.23	.42	0.76
100	100	10	0	0.1	1	.22	0.966	1	.10	0.99	0.03	.90	0.52	0.17	.41	0.75
100	100	10	10%	0	0.93	.29	0.944	0.97	.10	0.98	0.17	.92	0.52	0.27	.47	0.76
100	100	10	10%	0.1	0.93	.36	0.934	0.97	.10	0.97	0.1	.92	0.52	0.3	.46	0.74

5.2 Reference Results

The performance of the cosine criterion for linear and non-linear concepts is illustrated on Fig. 1, where the number d of features is 100, the number n of examples is 50, and the number of relevant features r is 10. The performance indicators are summarized in Table 1; complementary results, omitted due to space limitations, show similar trends for higher values of d.

An outstanding performance is obtained for linear concepts. With twice as many features as examples, the probability p_b of top ranking a relevant feature is around 90%. A graceful degradation of p_b is observed as the noise rate increases, more sensitive to the label noise than to the feature noise. The relevant features are in the top p_w features, where p_w varies from 1/3 to roughly 1/2.

The performance steadily improves when the number of examples increases, p_b reaching 100% and p_w ranging from 1/5 to 1/3 for $n = d$.

In contrast, the cosine criterion behaves no better than random ranking for non-linear concepts; this is visible as the ROC-FS curve is close to the diagonal, and the situation does not improve by doubling the number of examples. The seemingly better performances for redundant features is explained as the true

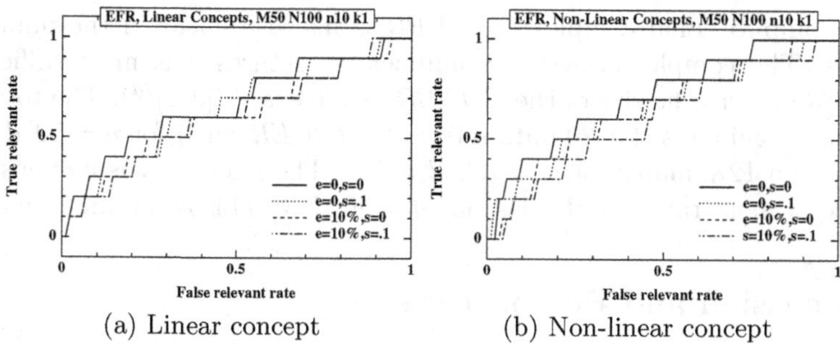

(a) Linear concept (b) Non-linear concept

Fig. 2. EFR performance: Median ROC-FS curves over 20 training sets on Linear and Non-Linear concepts, with $d = 100$, $n = d/2$, $r = d/10$, Non redundant features.

Table 2. Ensemble Feature Ranking with ROGER: Probability p_b of top ranking a relevant feature, Median relative rank p_w of the worst ranked relevant feature, Area under the ROC-FS curve.

					Linear concepts						Non Linear concepts					
					No redundancy			Redundancy			No redundancy			Redundancy		
n	d	r	e	σ	p_b	p_w	AUC	p_b	p_w	AUC	p_b	p_w	AUC	p_b	p_w	AUC
50	100	10	0	0	0.5	.92	0.67	0.65	.29	0.86	0.20	.75	0.71	0.50	.26	0.88
50	100	10	0	0.1	0.5	.80	0.63	0.75	.30	0.85	0.45	.82	0.68	0.25	.33	0.84
50	100	10	10%	0	0.35	.94	0.61	0.45	.31	0.85	0.25	.81	0.68	0.30	.32	0.83
50	100	10	10%	0.1	0.35	.89	0.62	0.60	.40	0.82	0.25	.88	0.61	0.35	.28	0.83
100	100	10	0	0	0.85	.79	0.79	0.90	.23	0.92	0.55	.63	0.81	0.80	.20	0.92
100	100	10	0	0.1	0.50	.74	0.77	0.95	.21	0.92	0.60	.72	0.78	0.50	.22	0.90
100	100	10	10%	0	0.55	.77	0.72	0.65	.27	0.89	0.65	.78	0.77	0.50	.20	0.91
100	100	10	10%	0.1	0.65	.82	0.75	0.45	.28	0.88	0.40	.72	0.75	0.40	.26	0.87

relevance rate models the probability of extracting at most r features among $2r$ ones.

5.3 Evolutionary Feature Ranking

The performance of EFR is measured under the same conditions (Fig. 2, Table 2). EFR is clearly outperformed by the cosine criterion in the linear case. With twice as many features as examples, the probability p_b of top ranking a relevant feature ranges between 35 and 50% (non redundant features), against 80 and 90% for the reference results. When the number of examples increases, p_b increases as expected; but p_b reaches 55 to 85% against 93 to 100% for the reference results.

In contrast, EFR does significantly better than the reference criterion in the non-linear case. Probability p_b ranges around 30%, compared to 3% and 10% for the reference results with $n = 50$ and p_b increases up to circa 55% when n increases up to 100.

With respect to computational cost, the cosine criterion is linear in the number of examples and in $d \log d$ wrt the number of features; the runtime is negligible in the experiment range.

The computational complexity of EFR is likewise linear in the number of examples. The complexity wrt the number of features d is more difficult to assess as d governs the size of the *ROGER* search space ($[0,1]^{2d}$). The total cost is less than 6 minutes (for 20 data sets $\times$ 15 *ROGER* runs) for $n = 50, d = 100$ and less than 12 minutes for $n = 100, d = 100$. The scalability is demonstrated in the experiment range as the cost for $n = 50, d = 500$ is less than 23 minutes.

6 Discussion and Perspectives

The contribution of this paper is based on the exploitation of the diverse hypotheses extracted along independent runs of evolutionary learning algorithms, here *ROGER*. This collection of hypotheses is exploited for ensemble-based feature ranking, extending the ensemble learning approach [4] to Feature Selection and Ranking [10].

As should have been expected, the performances of the Evolutionary Feature Ranker presented are not competitive with the state of the art for linear concepts. However, the flexibility of the hypothesis search space explored by *ROGER* allows for a breakthrough in (a limited case of) non-linear concepts, even when the number of examples is a fraction of the number of features.

These results are based on experimental validation over 9,000 datasets, conducted after a statistical model of Feature Ranking problems. Experiments on real-world data are underway to better investigate the EFR performance, and the limitations of the simple model of non-linear concepts proposed.

Further research will take advantage of multi-modal evolutionary optimization heuristics to extract diverse hypotheses from each *ROGER* run, hopefully reducing the overall computational cost of the approach and addressing more complex learning concepts (e.g. disjunctive concepts).

Acknowledgments

We would like to thank Aad van der Vaart. The second, third and last authors are partially supported by the PASCAL Network of Excellence, IST-2002-506778.

References

1. J. Bi, K.P. Bennett, M. Embrechts, C.M. Breneman, and M. Song. Dimensionality reduction via sparse support vector machines. *J. of Machine Learning Research*, 3:1229–1243, 2003.
2. M. Botta, A. Giordana, L. Saitta, and M. Sebag. Relational learning as search in a critical region. *J. of Machine Learning Research*, 4:431–463, 2003.
3. A.P. Bradley. The use of the area under the ROC curve in the evaluation of machine learning algorithms. *Pattern Recognition*, 1997.
4. L. Breiman. Arcing classifiers. *Annals of Statistics*, 26(3):801–845, 1998.
5. R. Esposito and L. Saitta. Monte Carlo theory as an explanation of bagging and boosting. In *Proc. of IJCAI'03*, pages 499–504. 2003.

6. C. Ferri, P. A. Flach, and J. Hernández-Orallo. Learning decision trees using the area under the ROC curve. In *Proc. ICML'02* , pages 179–186. Morgan Kaufmann, 2002.
7. Y. Freund and R.E. Shapire. Experiments with a new boosting algorithm. In L. Saitta, editor, *Proc. ICML'96* , pages 148–156. Morgan Kaufmann, 1996.
8. A. Giordana and L. Saitta. Phase transitions in relational learning. *Machine Learning*, 41:217–251, 2000.
9. C. Guerra-Salcedo and D. Whitley. Genetic approach to feature selection for ensemble creation. In *Proc. GECCO'99* , pages 236–243, 1999.
10. I. Guyon and A. Elisseeff. An introduction to variable and feature selection. *J. of Machine Learning Research*, 3:1157–1182, 2003.
11. I. Guyon, J. Weston, S. Barnhill, and V. Vapnik. Gene selection for cancer classification using support vector machines. *Machine Learning*, 46:389–422, 2002.
12. T. Hogg, B.A. Huberman, and C.P. Williams (Eds). *Artificial Intelligence: Special Issue on Frontiers in Problem Solving: Phase Transitions and Complexity*, volume 81(1-2). Elsevier, 1996.
13. G.H. John, R. Kohavi, and K. Pfleger. Irrelevant features and the subset selection problem. In *Proc. ICML'94* , pages 121–129. Morgan Kaufmann, 1994.
14. C.X. Ling, J. Hunag, and H. Zhang. AUC: a better measure than accuracy in comparing learning algorithms. In *Proc. of IJCAI'03*, 2003.
15. M. S. Pepe, G. Longton, G. L. Anderson, and M. Schummer. Selecting differentially expressed genes from microarray experiments. *Biometrics*, 59:133–142, 2003.
16. S. Rosset. Model selection via the AUC. In *Proc. ICML'04* . Morgan Kaufmann, 2004, to appear.
17. M. Sebag, J. Azé, and N. Lucas. Impact studies and sensitivity analysis in medical data mining with ROC-based genetic learning. In *IEEE-ICDM03* , pages 637–640, 2003.
18. M. Sebag, J. Azé, and N. Lucas. ROC-based evolutionary learning: Application to medical data mining. In *Artificial Evolution VI*, pages 384–396. Springer Verlag LNCS 2936, 2004.
19. H. Stoppiglia, G. Dreyfus, R. Dubois, and Y. Oussar. Ranking a random feature for variable and feature selection. *J. of Machine Learning Research*, 3:1399–1414, 2003.
20. H. Vafaie and K. De Jong. Genetic algorithms as a tool for feature selection in machine learning. In *Proc. ICTAI'92* , pages 200–204, 1992.
21. L. Yan, R. H. Dodier, M. Mozer, and R. H. Wolniewicz. Optimizing classifier performance via an approximation to the Wilcoxon-Mann-Whitney statistic. In *Proc. of ICML'03* , pages 848–855. Morgan Kaufmann, 2003.

Privately Computing a Distributed k-nn Classifier*

Murat Kantarcıoğlu and Chris Clifton

Purdue University, Department of Computer Sciences
250 N University St
West Lafayette, IN 47907-2066 USA
+1-765-494-6408, Fax: +1-765-494-0739
{kanmurat,clifton}@cs.purdue.edu

Abstract. The ability of databases to organize and share data often raises privacy concerns. Data warehousing combined with data mining, bringing data from multiple sources under a single authority, increases the risk of privacy violations. Privacy preserving data mining provides a means of addressing this issue, particularly if data mining is done in a way that doesn't disclose information beyond the result. This paper presents a method for privately computing $k - nn$ classification from distributed sources without revealing any information about the sources or their data, other than that revealed by the final classification result.

1 Introduction

The growing amount of information stored in different databases has lead to an increase in privacy concerns. Bringing data from multiple sources under one roof may improve processing and mining of the data, but it also increases the potential for misuse. Privacy is important, and concerns over privacy can prevent even the legitimate use of data. For example, the *Data-Mining Moratorium Act* introduced in the U.S. Senate would have forbid *any* "data-mining program" within the U.S. Department of Defense. Privacy concerns must be addressed, or such over-reaction may prevent beneficial uses of data mining.

Consider the case of a physician who wants to learn the most likely diagnosis for a patient by looking at diagnoses of similar symptoms at other hospitals. Specifically, the physician wants to use a *k-nearest neighbor* (k-nn) classifier to predict the disease of the patient. Revealing the patients particular test results may not be a threat to privacy (if only the physician knows the identity of the patient) but privacy of the different hospitals may be at risk. If this procedure is implemented naïvely, the researcher may learn that two patients with the same medical test results are diagnosed with different diseases in different hospitals. This could damage the reputations of the hospitals. The possibility of such incidents may prevent hospitals from participating in such a diagnostic tool. The

* This material is based upon work supported by the National Science Foundation under Grant No. 0312357.

J.-F. Boulicaut et al. (Eds.): PKDD 2004, LNAI 3202, pp. 279–290, 2004.

obvious question is, can this be done without revealing anything other than the final classification? The answer is yes: This paper presents an efficient method with provable privacy properties for k-nn classification.

This work assumes data is horizontally partitioned, i.e., each database is able to construct its own k-nearest neighbors independently. The *distributed* problems are determining which of the local results are the closest globally, and finding the majority class of the global k-nearest neighbors. We assume that attributes of the instance that needs to be classified are not private (i.e., we do not try to protect the privacy of the query issuer); we want to protect the privacy of the data sources. The approach makes use of an untrusted, non-colluding party: a party that is not allowed to learn anything about any of the data, but is trusted not to collude with other parties to reveal information about the data.

The basic idea is that each site finds its own k-nearest neighbors, and encrypts the class with the public key of the site that sent the instance for classification (querying site). The parties compare their k-nearest neighbors with those of all other sites – except that the comparison gives each site a random share of the result, so no party learns the result of the comparison. The results from all sites are combined, scrambled, and given to the untrusted, non-colluding site. This site combines the random shares to get a comparison result for each pair, enabling it to sort and select the global k-nearest neighbors (but without learning the source or values of the items). The querying site and the untrusted, non-colluding site then engage in a protocol to find the class value. Each site learns nothing about other sites (the comparison results appears to be randomly chosen bits.) The untrusted site sees $k * n$ encrypted results. It is able to totally order the results, but since it knows nothing about what each means or where it comes from, it learns nothing. The querying site only sees the final result.

Details of the algorithm are given in Section 3, along with a discussion of the privacy of the method. Computation and communication costs are given in Section 4. First, we discuss related work and relevant background.

2 Related Work

Finding the k-nearest neighbors of a multidimensional data point q [1] and building k-nn classifiers [2] have been well studied, but not in the context of security.

Interest has arisen in privacy-preserving data mining. One approach is to add "noise" to the data before the data mining process, and use techniques that mitigate the impact of the noise on the data mining results. The other approach is based on protecting the privacy of distributed sources. This was first addressed for the construction of decision trees. This work closely followed the secure multiparty computation approach discussed below, achieving "perfect" privacy, i.e., nothing is learned that could not be deduced from one's own data and the resulting tree. The key insight was to trade computation and communication cost for accuracy, improving efficiency over the generic secure multiparty computation method. Methods have since been developed for association rules, K-means and EM clustering, and generalized approaches to reducing the number of "on-line" parties required for computation. For a survey of this area see [3]. This paper falls in the latter class: privacy preserving distributed data mining work. The goal is to provably prevent disclosure of the "training data" as much as possible, disclosing only what is inherent in the classification result.

To better explain the concept of provable privacy of distributed sources, we give some background on Secure Multiparty Computation. Yao introduced a solution with his millionaire's problem: Two millionaires want to know who is richer, without disclosing their net worth[4]. Goldreich proved there is a secure solution for *any* functionality[5]. The idea is that the function to be computed is represented as a combinatorial circuit. The idea is based on computing random shares of each wire in the circuit: the exclusive-or of the shares is the correct value, but from its own share a party learns nothing. Each party sends a random bit to the other party for each input, and makes its own share the exclusive-or (xor) of its input and the random bit. The parties then run a cryptographic protocol to learn shares of each gate. At the end, the parties combine their shares to obtain the final result. This protocol has been proven to produce the desired result, and to do so without disclosing anything except that result.

The cost of circuit evaluation for large inputs has resulted in several algorithms for more efficiently computing specific functionality. We do make use of circuit evaluation as a subroutine for privately determining if $a \geq b$. We also use the definitions and proof techniques of Secure Multiparty Computation to verify the privacy and security properties of our algorithm; these will be introduced as needed. First, we give the details of the algorithm itself.

3 Secure k-nn Classification

We first formally define the problem. Let R be the domain of the attributes and C be the domain of the class values. Let D_i denote the database of instances at site S_i. Let (x, d, k) be the query originated by site O, where $x \in R$ is the instance to be classified, and $d : R \times R \rightarrow [0, 1]$ is a distance function used to determine which k items are closest to x (e.g., Euclidean distance, although any metric could be used provided each site can compute $d(x, x_j)$ for every x_j in its database.) Given the data instance x, our goal is to find the k nearest neighbors of x in the union of the databases and return the class of the majority of those neighbors as the predicted class of x:

$$C_x = \text{Maj}\left(\prod_c \left(\underset{(x_i,c_i)\in D_1\cup D_2 \ldots \cup D_n}{\text{argmin}_k} (d(x_i, x))\right)\right)$$

where $\prod$ is the projection function and Maj is the majority function.

The security/privacy goal is to find C_x while meeting the following criteria:

- No site except O will be able to predict C_x better than looking at (x, d, k) and its own database D_i (E.g., if Site S_i has k points x_i such that $d(x, x_i) = 0$, it is likely that the majority class of the x_i will be the result); and
- No site learns anything about the source of x_i except its own.

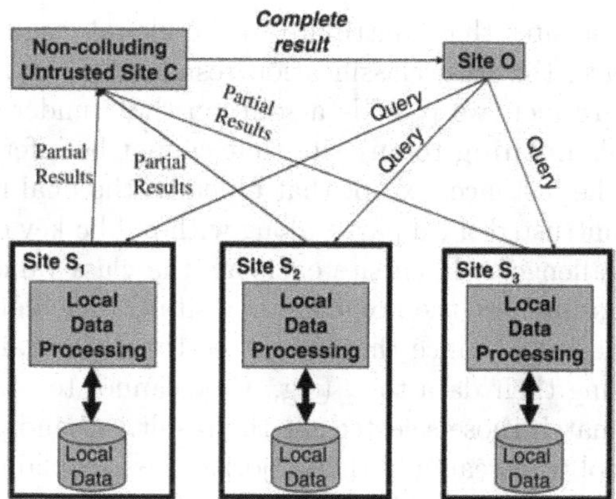

Fig. 1. Information flow in secure k-nn classification

Theorem 1 shows this by proving what *is* disclosed. Assuming the number of sites n and the query (x, d, k) are public, site O learns only the result C_x. The other sites learn only what they can infer from their own data and the query.

To achieve a secure and efficient solution, we make use of an untrusted, non-colluding site, an approach first suggested in [6]. Such a site learns nothing in isolation, however by colluding with other sites it could obtain information that should not be revealed. Therefore, the only trust placed in the site is that it not collude with any other sites to violate privacy. Although this seems like a strong assumption, it often occurs in real life. For example, bidders or sellers on e-bay assume that e-bay is not colluding with other bidders or sellers against them. We emphasize that the untrusted site learns nothing except the public values k and n. A diagram showing the information flow is given in Figure 1.

3.1 The Algorithm

Given the query, each site can find its own closest k items without exchanging information. These $n * k$ candidate items must contain the k nearest neighbors of x; what remains is to find the closest k among the candidates and return the majority class of those k instances to site O. This poses two security challenges:

1. Determining which of the $n * k$ items are the k nearest neighbors without revealing anything about the items, where they come from, or their distance to the instance that needs to be classified; and
2. Learning the class value while disclosing it to only the originating site.

At first glance, this appears simple – have each site send their local k-nn distances and classes to a third party C. C learns the result, but also learns distances and the sites that contributes to the global k-nn result. A slightly more sophisticated approach is for C to publish a candidate distance value, and have local sites return the number of items within that distance, refining the value until k items are within the distance. Then the class value can be computed.

This still reveals the sites that contribute to the global result, the distances to the queried instance, the final classification result to C, and more. To ensure privacy concerns are met, we provide a solution that (under the non-collusion assumption) reveals **nothing** to any site that cannot be inferred by looking at its own data and the instance, except that O learns the final result.

The use of an untrusted third party, along with public-key encryption, makes it easy to solve challenge 2. Each site encrypts the class value with the public key of O before passing it to the non-colluding site C. The data source sites are then left out of the loop – since they never see the data again, they can learn nothing after passing their data to C (e.g., they cannot test to see if their own encrypted values match those selected for the result.) C and O will participate in a special protocol to reveal only the majority class (explained later.)

Meeting challenge 1 is more difficult. Sending the distance $d(x_i, x) = d_i$ to C, even with encrypted results, reveals information about the location of the points. Instead site C is sent an $n * k - 1$ length vector for each point x_i, containing the results of comparing x_i with all other points. This enables C to order the points and select the k nearest neighbors. Since the existence of a distance metric implies a total ordering exists, and the number of points is fixed, C learns nothing.

Two problems remain. The first is that we must prevent C from learning which point comes from which site, or it can learn the source of the k nearest neighbors (among other things.) This is easily addressed – all points and their associated comparison vectors are sent to one of the data sites S_s, which combines them and scrambles the order before passing them on to C. Public key encryption, using C's public key, prevents S_s from learning anything.

The second issue is more challenging: How do we build the comparison vectors at site S_i without S_i learning about values at other sites? If two sites just compare items, they both learn something about the data at the other site (e.g., if S_i's items are all closer to x than S_j's items, both learn that S_j does not have an item that contributes to the result.) For this we use the share-splitting idea from secure multiparty computation. Instead of containing the results of comparing x_i with other items, the vector contains a *random share* of the comparison result. E.g., if d_i for x_i is smaller than d_j for x_j, then the comparison $d_i < d_j$ should return 0. Either the element of the d_i vector corresponding to d_j and the element of the d_j vector corresponding to d_i both contain 0, or they both contain 1 – $0 \oplus 0 = 1 \oplus 1 = 0$. However, knowing only one share tells nothing: a share 0 could mean either $0 \oplus 0 = 0$ or $0 \oplus 1 = 1$. From d_i's view, the share has equal probability of being 0 or 1 (a random choice), so it learns nothing.

To generate random shares of the comparison, we return to secure multiparty computation. We stop the generic circuit comparison method before combining shares to learn the final result. In other words, given two integers a and b, secure comparison of a, b ($f : \{0,1\}^* \times \{0,1\}^* \longmapsto \{0,1\} \times \{0,1\}$) is defined as follows:

$$f(a,b) = \begin{cases} (1 \oplus r, r) \text{ if } a > b \\ (0 \oplus r, r) \text{ if } a < b \end{cases}$$

where each site sees only one component of the function output. This states that if $a > b$ the xor of the shares of the participating sites will be 1, otherwise the xor of the shares will be 0. Using this function, each site can compare its elements with all other elements and learn nothing about the result of the comparisons.

Two additional details. First, the identifiers used to track the d_j in the comparison share vector for d_i must not disclose anything. One option would be for the combining site S_s to assign identifiers, but this would require independent encryption of the single bits of the comparison shares, and single bit encryption is problematic. Instead, each site generates pseudo-random unique identifiers site C cannot distinguish from random. One simple and secure way to handle this problem is using a pseudo-random permutation function. With a fixed key, DES is assumed to be such a function. The sites agree on a key K, and each site S_i generates k identifiers by evaluating $E_K(ik), E_K(ik+1), \ldots, E_K(ik+k-1)$. Since encryption is assumed to be secure and a permutation, each site will get non-intersecting identifiers that appear random to any (polynomial time) adversary not knowing K (in particular, C). The identifiers are also used to determine which of a comparison pair will be the left-hand side in a comparison: The item with the smaller identifier corresponds to x in the comparison function $f(x, y)$.

The second detail is equality. Identifying that two items are at equal distances reveals information. We must disambiguate consistently, without giving even a probabilistic estimate on the likelihood of equality. The solution is to add extra low-order bits to each distance, based on a unique mapping from the identifier that appears random to C - the same trick used to generate identifiers. The distance used for comparison is actually $d||E_u(ik+j)$, where the encryption function E is as above, but with a different key. This ensures that distances are unique, guaranteeing a total ordering of equal distances.

Protocol 1 gives the algorithm details. Note that at the end of the k-nearest neighbor selection phase, C has the class of the k-nearest neighbors encrypted with E_o. Assuming the usage of Blum-Goldwasser encryption, each class value $class_i$ will have ciphertext of the form $(\acute{r}, class_i \oplus r)$, where O has enough information to determine r given $\acute{r}$, enabling decryption to get $class_i$. Instead of sending these values to O, C will xor each of these values with a random value r_i. C then sends $(\acute{r}, class_i \oplus r \oplus r_i)$ to O. O decrypts to get $class'_i = class_i \oplus r_i$, indistinguishable (to O) from a random value. O and C now use the generic secure circuit evaluation approach to evaluate the majority function:

$$\mathrm{Maj}(class'_1 \oplus r_1, \ldots, class'_k \oplus r_k).$$

This is a simple circuit with size complexity dependent on k and number of distinct classes. The cost is dominated by the k-nearest neighbor selection phase.

To clarify, we give an example for $k = 1$ and $n = 3$.

Example 1. Given $(x, d, 1)$, each site finds its 1-nearest neighbor. Assume that site S_1 has $(d(x, x_1), \prod_c(x_1, c_1) = (0.1, c_1)$, site S_2 has (x_2, c_2) at distance 0.2, and site S_3 has (x_3, c_3) at 0.15. After generating random identifiers, S_1 has $(3, 0.1, c_1)$, S_2 has $(1, 0.2, c_2)$, and S_3 has $(2, 0.15, c_3)$. (For simplicity we omit the low-order bit disambiguation.) In generating the comparison vector for c_1 and c_2, S_1 notes that $c_1.id = 3 > 1$, so it has the right argument of $f(a, b)$ and generates a random bit (say 0) as its share of the output. Since $0.2 \geq 0.1$, S_2

Protocol 1 Privacy-preserving k-nn classification algorithm

Require: n sites S_i, $1 \le i \le n$, each with a database D_i; permuting site S_s (where s may be in $1, \ldots, n$); and untrusted non-colluding site C (distinct from S_i or S_s). Query (x, d, k) generated by originating site O. Public encryption keys E_c for site C and E_o for site O, key generation function E_K and E_u known only to the S_i.

for all sites S_i, in parallel **do**
  {Build vector of random key, distance, and result for local closest k}
  Select k items $(d(x_i, x), \prod_{c_i}(x_i, c_i))$ with smallest $d(x_i, x)$ from D_i into N_i
  $R_i = \emptyset$
  for $j = 0..k-1$ {Compute identifiers and "extended" local distances} **do**
    $R_i = R_i \cup \{(E_K(ik+j), N_i[j].d || E_u(ik+j),\ E_o(N_i[j].result))\}$ {$||$ is string concatenation}
  end for
  $ER_i = \emptyset$
  for each $(id, d, E_o(c)) \in R_i$ {Comparison phase} **do**
    $v = \emptyset$
    for each site $h = i \ldots n$ {If $i = h$, just generate values locally.} **do**
      for $j = 0 \ldots k-1$ **do**
        if $id < R_h[j].id$ **then**
          $v = v \cup\ \{(R_h[j].id, S_i\text{'s share of } f(d, R_h[j].d)\}$
          $v_{hj} = v_{hj} \cup \{(id, S_j\text{'s share of } f(d, R_h[j].d)\}$
        else if $id > R_h[j].id$ **then**
          $v = v \cup\ \{(R_h[j].id, S_i\text{'s share of } f(R_h[j].d, d)\}$
          $v_{hj} = v_{hj} \cup \{(id, S_j\text{'s share of } f(R_h[j].d, d)\}$
        end if
      end for
    end for
    $ER_i = ER_i \cup (id, E_c(v), E_o(c))$
  end for
  send ER_i to S_s
end for
{At site S_s: Permutation phase}
set $ER = \cup_{i=1}^{n}(ER_i)$
permute ER and send it to C
{At site C: k nearest neighbor selection phase}
Decrypt the encrypted shares of the comparison results
Use the pairwise comparisons to find the global k nearest neighbors
Let R be the set of encrypted class values($E_o(c)$) of the global k nearest neighbor
for all $R.E_o(c_i)$ {encrypted as $(\acute{r}, c_i \oplus r)$} **do**
  $NR[i] = R.E_o(c_i) \oplus$ random r_i
end for
Site C sends NR to O
{At site O:}
for all $NR[i]$ {$= (\acute{r}, c_i \oplus r \oplus r_i)$} **do**
  Find r using $\acute{r}$ and the private key
  $NR_d[i] = c_i \oplus r \oplus r_i \oplus r$
end for
Find Maj from the random shares of C and NR_d using secure circuit evaluation.

learns that its share should be $1 \oplus 0 = 1$. (Neither S_2 or S_1 learns the other's share, or the comparison result.) Secure comparisons with the other sites are performed, giving each site tuples containing its share of the comparison with all other items. These are encrypted with C's public key to give:

$$S_1 : (3, E_c((1,1),(2,0)), E_o(c_1))$$
$$S_2 : (1, E_c((2,1),(3,0)), E_o(c_2))$$
$$S_3 : (2, E_c((1,1),(3,1)), E_o(c_3))$$

The above are sent to S_3, which permutes the set, strips source information, and sends it to C. Site C decrypts the comparison share vectors to get:

$$(2, ((1,1),(3,1)), E_o(c_3))$$
$$(3, ((1,1),(2,0)), E_o(c_1))$$
$$(1, ((2,1),(3,0)), E_o(c_2))$$

C now compares the items to find the nearest neighbor. As an example, to compare items 2 and 3, we take the pair $(3,1)$ from the first (2) row and the pair $(2,0)$ from the second (3) row. Combining the share portions of these pairs gives $1 \oplus 0 = 1$, so $d(x, x_2) \geq d(x, x_3)$. Likewise, comparing 1 and 3 gives $0 \oplus 1 = 1$, so $d(x, x_1) \geq d(x, x_3)$. Therefore, x_1 is closest to x. C sends $E_o(c_1)$ to O, which decrypts to get c_1, the correct result. (With $k > 1$, C and O would engage in a protocol to determine which c_i was in the majority, and send $E_o(c_i)$ to O.)

3.2 Security of the Protocol

We now prove that Protocol 1 is secure. We assume that sites O and C are not among the S_i. We have discussed the need for C being a separate site. O cannot be a data source, as it would be able to recognize its own $E_o(x)$ among the results, thus knowing if it was the source of some of the k nearest neighbors.

To define security we use definitions from the Secure Multiparty Computation community, specifically security in the semi-honest model. Loosely speaking, a semi-honest party follows the rules of the protocol, but is free to try to learn additional information from what it sees during the execution of the protocol.

The formal definition is given in [7]. Basically, it states that the view of each party during the execution of the protocol can be effectively simulated knowing only the input and the output of that party. Extending this definition to multiple parties is straightforward. The key idea is that of *simulation*: If we can simulate what is seen during execution of the protocol knowing only our own input and our portion of the final output, then we haven't learned anything from the information exchanged in a real execution of the protocol. Under certain assumptions, we can extend our protocols to malicious parties (those that need not follow the protocol). Due to space limitations we omit the discussion here.

We need one additional tool to prove the security of the protocol. The encrypted items seen by S_s and C during execution of the protocol may disclose some information. The problem is that two items corresponding to the same plaintext map to the same ciphertext. If multiple items are of the same class (as would be expected in k-nn classification), the permuting site S_s would learn the class entropy in the k-nn of each site as well as the number of identical results between sites. The comparison site C would learn this for the data as a whole. Neither learns the result, but something of the distribution is revealed.

Fortunately, the cryptography community has a solution: *probabilistic* public-key encryption. The idea is that the same plaintext may map to different ciphertexts, but these will all map back to the same plaintext when decrypted. Using probabilistic public-key encryption for E_o allows us to show Protocol 1 is secure. (Deterministic public-key encryption is acceptable for E_c, as the set of $nk-1$ identifiers in the set v are different for every item, so no two plaintexts are the same.) The Blum-Goldwasser probabilistic encryption scheme[8], with a cipher text of the form $(\acute{r}, M \oplus r)$ for message M, is one example. In this, given $\acute{r}$ and the private key, it is possible to compute r to recover the original message.

Theorem 1. *Protocol 1 privately computes the k-nn classification in the semi-honest model where there is no collusion; only site O learns the result.*

PROOF. To show that Protocol 1 is secure under the semi-honest model, we must demonstrate that what each site sees during the execution of the protocol can be simulated in polynomial time using only its own input and output. Specifically, the output of the simulation and the view seen during the execution must be computationally indistinguishable. We also use the general composition theorem for semi-honest computation: if g securely reduces to f and there is a way to compute f securely, then there is a way to compute g securely. In our context, f is the secure comparison of distances, and g is Protocol 1. We show that our protocol uses comparison in a way that reveals nothing.

We first define the simulator for the view of site S_i. Before the comparison phase, S_i can compute its view from its own input. The comparison phase involves communication, so simulation is more difficult. If we look at a single comparison, S_i sees several things. First, it sees the identifier $R_h[j].id$. Since S_i knows E_K, h, and j; the simulator can generate the exact identifier, so the simulator view is identical to the actual view. It also sees a share of the comparison result. If $i = h$, S_i generates the values locally, and the simulator does the same. If not local, there are two possibilities. If $id > R_h[j].id$, it holds the second argument, and generates a random bit as its share of the comparison result. The simulator does the same. Otherwise, the secure comparison will generate S_i's share of the comparison. Assume $d < R_h[j].d$: S_i's share is $0 \oplus r$, where r is S_h's randomly chosen share. Assuming S_h is equally likely to generate a 1 or 0, the probability that S_i's share is 1 is 0.5. This is independent of the input – thus, a simulator that generates a random bit has the same likelihood of generating a 1 as S_i's view in the real protocol. The composition theorem (and prior work on secure comparison) shows the algorithm so far is privacy preserving.

We can extend this argument to the entire set of comparisons seen by S_i during execution of the protocol. The probability that the simulator will output a particular binary string x for a given sequence of comparisons is $\frac{1}{2^{nk-1}}$. Since actual shares of the comparison result are chosen randomly from a uniform distribution, the same probability holds for seeing x during actual execution:

$$Pr\left[VIEW_{S_i}^{v_j} = x\right] = \frac{1}{2^{nk-1}}$$
$$= Pr\left[Simulator_i = x\right]$$

Therefore, the distribution of the simulator and the view is the same for the entire result vectors. Everything else is simulated exactly, so the views are computationally indistinguishable. Nothing is learned during the comparison phase.

The sites S_i now encrypt the result vectors; again the simulator mimics the actual protocol. Since the sources were indistinguishable, the results are as well.

The next step is to show that S_s learns nothing from receiving ER_i. Site S_s can generate the identifiers $ER_i[j].id$ it will receive, as in simulating the comparison phase. By the security definitions of encryption, the $E_c(v)$ must be computationally indistinguishable from randomly generated strings of the same length as the encrypted values, provided no two v are equal (which they cannot be, as discussed above.) Likewise, the definition of probabilistic encryption ensures that the $E_o(x)$ are computationally indistinguishable from randomly generated strings of the same length. Since E_c and E_o are public, S_s knows the length of the generated strings. The simulator chooses a random string from the domain of E_c and E_o; the result is computationally indistinguishable from the view seen during execution of the protocol. (If S_s is one of the S_i, the simulator must reuse the ER_s generated during the comparison simulation instead of generating a new one.)

C receives $n * k$ tuples consisting of an identifier, an encrypted comparison set v, and encrypted class value $E_o(c)$. Since the identifiers are created with an encryption key unknown to C, the values are computationally indistinguishable from random values. The simulator for C randomly selects $k * n$ identifiers from a uniform distribution on the domain of ER_i. The outcomes $E_o(c)$ are simulated the same as by S_s above. The hardest part to simulate is the comparison set. Since the comparison produces a total ordering, C cannot simply generate random comparison results. Instead, the simulator for C picks an identifier i_1 to be the closest, and generates a comparison set consisting of all the other identifiers and randomly chosen bits corresponding to the result shares. It then inserts into the comparison set for each other identifier i_k the tuple consisting of i_1 and the appropriate bit so that the comparison of i_1 with i_k will show i_1 as closest to q. For example, if $i_1 \geq i_k$, then $f(i_k, i_1)$ should be 1. If the bit for i_1's share is chosen to be 0, the tuple $(i_1, 1)$ is placed in i_k's comparison set. By the same argument used in the comparison phase, this simulator generates comparison values that are computationally indistinguishable from the view seen by C. Since the actual identifiers are computationally indistinguishable, and the value of the comparison is independent of the identifier value, the order of identifiers generated by C is computationally indistinguishable from the order in the real execution. The simulator encrypts these sets with E_c to simulate the data received.

In the final stage, O sees the $NR[i]$. The simulator for O starts with $NR_d[i] = c_i \oplus r_i$. The one-time pad r_i (unknown to O) ensures NR_d can be simulated by random strings of the length of $NR_d[i]$. Xor-ing the $NR_d[i]$ with r simulates NR_d. The final step reveals $E_o(c)$ to O, where c is the majority class. Since O knows the result c, the simulator generates $E_o(c)$ directly. Applying the composition theorem shows that the combination of the above simulation with the secure circuit evaluation is secure.

We have shown that there is a simulator for each site whose output is computationally indistinguishable from the view seen by that site during execution of the protocol. Therefore, the protocol is secure in the semi-honest model. □

The algorithm actually protects privacy in the presence of malicious parties, providing O and C do not collude. The proof is omitted due to space restrictions.

4 Communication and Computation Cost Analysis

Privacy is not free. Assume m is the size required to represent the distance, and q bits are required to represent the result. A simple insecure distributed k-nn protocol would have the S_i send their k nearest neighbor distances/results to O, for $O(nk(m+q))$ bit communication cost. The computation by O could easily be done in $O(nk\log(k))$ comparisons. (One pass through the data, inserting each item into the appropriate place in the running list of the k nearest neighbors.) Although we do not claim this is optimal, it makes an interesting reference point.

In Protocol 1, each site performs k^2 comparisons with every other site. There are $\binom{n}{2}$ site combinations, giving $O(n^2k^2)$ comparisons. An m bit secure comparison has communication cost $O(mt)$, where t is based on the key size used for encryption. Thus, the total communication cost of the comparison phase of Protocol 1 is $O(n^2k^2mt)$ bits. Assuming Blum-Goldwasser encryption, each site then sends $O(nk+t)$ bits of encrypted comparison shares for each item, plus the $O(q+t)$ result, to S_s and on to C. This gives $O(n^2k^2+nkq+nkt)$ bits. C sends $O(k(q+t))$ bits of encrypted result to O. The dominating factor is the secure comparisons, $O(n^2k^2mt)$.

The computation cost is dominated by encryption, both direct and in the oblivious transfers (the dominating cost for secure circuit evaluation). There are $O(nk)$ encryptions of the query results, each of size q, and $O(nk)$ encryptions of comparison sets of size $O(nk)$. The dominating factor is again the $O(n^2k^2)$ secure comparisons. Each requires $O(m)$ 1 out of 2 oblivious transfers. An oblivious transfer requires a constant number of encryptions, giving $O(n^2k^2m)$ encryptions as the dominating computation cost. Assuming RSA public-key encryption for the oblivious transfer, the bitwise computation cost is $O(n^2k^2mt^3)$.

The parallelism inherent in a distributed system has a strong impact on the execution time. Since the secure comparisons may proceed in parallel, the time complexity $O(nk^2mt^3)$. Batching the comparisons between each pair of sites allows all comparisons to be done in a constant number of rounds. Thus, the dominating *time* factor would appear to be decryption of the nk comparison sets, each of size $O(nk)$. Note that m must be greater than $\log(nk)$ to ensure no equality in distances, so unless n is large relative to the other values the comparisons are still likely to dominate. Once decrypted, efficient indexing of the comparison vectors allows the same $O(nk\log(k))$ cost to determine the k nearest neighbor as in the simple insecure protocol described above.

A more interesting comparison is with a fully secure k-nn algorithm based directly on secure circuit evaluation. For n parties and a circuit of size C, the generic method requires $O(n^2C)$ 1 out of 2 oblivious transfers: a communication complexity $O(n^2Ct)$. To compare with the generic method, we need a lower bound on the size of a circuit for k-nn classification on nk $(m+q)$-bit inputs. An obvious lower bound is $\Omega(nk(m+q))$: the circuit must (at least) be able to process all data. This gives a bit complexity of $O(n^2nk(m+q)t)$. Our method clearly wins if $n > k$ and is asymptotically superior for fixed k; for $n \leq k$ the question rests on the complexity of an optimal circuit for k-nn classification.

5 Conclusions

We have presented a provably secure algorithm for computing k-nn classification from distributed sources. The method we have presented is not cheap – $O(n^2k^2)$ where n is the number of sites – but when the alternative is not performing the task at all due to privacy concerns, this cost is probably acceptable. This leads us to ask about lower bounds: can we *prove* that privacy is not free. Privacy advocates often view privacy versus obtaining knowledge from data as an either/or situation. Demonstrating that knowledge can be obtained while maintaining privacy, and quantifying the associated costs, enables more reasoned debate.

References

1. Korn, F., Sidiropoulos, N., Faloutsos, C., Siegel, E., Protopapas, Z.: Fast nearest neighbor search in medical image databases. In Vijayaraman, T.M., Buchmann, A.P., Mohan, C., Sarda, N.L., eds.: Proceedings of 22th International Conference on Very Large Data Bases, Mumbai (Bombay), India, VLDB, Morgan Kaufmann (1996) 215–226
2. Fukunaga, K.: Introduction to statistical pattern recognition (2nd ed.). Academic Press Professional, Inc. (1990)
3. : Special section on privacy and security. SIGKDD Explorations **4** (2003) i–48
4. Yao, A.C.: How to generate and exchange secrets. In: Proceedings of the 27th IEEE Symposium on Foundations of Computer Science, IEEE (1986) 162–167
5. Goldreich, O., Micali, S., Wigderson, A.: How to play any mental game - a completeness theorem for protocols with honest majority. In: 19th ACM Symposium on the Theory of Computing. (1987) 218–229
6. Feige, U., Kilian, J., Naor, M.: A minimal model for secure computation. In: 26th ACM Symposium on the Theory of Computing (STOC). (1994) 554–563
7. Goldreich, O.: General Cryptographic Protocols. In: The Foundations of Cryptography. Volume 2. Cambridge University Press (2004)
8. Blum, M., Goldwasser, S.: An efficient probabilistic public-key encryption that hides all partial information. In Blakely, R., ed.: Advances in Cryptology – Crypto 84 Proceedings, Springer-Verlag (1984)

Incremental Nonlinear PCA for Classification*

Byung Joo Kim[1] and Il Kon Kim[2]

[1] Youngsan University School of Network and Information Engineering, Korea
bjkim@ysu.ac.kr
[2] Kyungpook National University Department of Computer Science, Korea
ikkim@knu.ac.kr

Abstract. The purpose of this study is to propose a new online and nonlinear PCA(OL-NPCA) method for feature extraction from the incremental data. Kernel PCA(KPCA) is widely used for nonlinear feature extraction, however, it has been pointed out that KPCA has the following problems. First, applying KPCA to patterns requires storing and finding the eigenvectors of a kernel matrix, which is infeasible for a large number of data N. Second problem is that in order to update the eigenvectors with an another data, the whole eigenspace should be recomputed. OL-NPCA overcomes these problems by incremental eigenspace update method with a feature mapping function. According to the experimental results, which comes from applying OL-NPCA to a toy and a large data problem, OL-NPCA shows following advantages. First, OL-NPCA is more efficient in memory requirement than KPCA. Second advantage is that OL-NPCA is comparable in performance to KPCA. Furthermore, performance of OL-NPCA can be easily improved by relearning the data. For classification extracted features are used as input for least squares support vector machine. In our experiments we show that proposed feature extraction method is comparable in performance to a Kernel PCA and proposed classification system shows a high classification performance on UCI benchmarking data and NIST handwritten data set.

Keywords: Incremental nonlinear PCA, Kernel PCA, Feature mapping function, LS-SVM

1 Introduction

In many pattern recognition problem it relies critically on efficient data representation. It is therefore desirable to extract measurements that are invariant or insensitive to the variations within each class. The process of extracting such measurements is called *feature extraction.* Principal Component Analysis(PCA)[1] is a powerful technique for extracting features from possibly high-dimensional data sets. For reviews of the existing literature is described in [2][3][4]. Traditional PCA, however, has several problems. First PCA requires a batch computation

* This study was supported by a grant of the Korea Health 21 R&D Project, Ministry of Health & Welfare, Republic of Korea (02-PJ1-PG6-HI03-0004)

J.-F. Boulicaut et al. (Eds.): PKDD 2004, LNAI 3202, pp. 291–300, 2004.

step and it causes a serious problem when the data set is large i.e., the PCA computation becomes very expensive. Second problem is that, in order to update the subspace of eigenvectors with another data, we have to recompute the whole eigenspace. Finial problem is that PCA only defines a linear projection of the data, the scope of its application is necessarily somewhat limited. It has been shown that most of the data in the real world are inherently non-symmetric and therefore contain higher-order correlation information that could be useful[5]. PCA is incapable of representing such data. For such cases, nonlinear transforms is necessary. Recently kernel trick has been applied to PCA and is based on a formulation of PCA in terms of the dot product matrix instead of the covariance matrix[8]. Kernel PCA(KPCA), however, requires storing and finding the eigenvectors of a $N \times N$ kernel matrix where N is a number of patterns. It is infeasible method when N is large. This fact has motivated the development of incremental way of KPCA method which does not store the kernel matrix. It is hoped that the distribution of the extracted features in the feature space has a simple distribution so that a classifier could do a proper task. But it is point out that extracted features by KPCA are global features for all input data and thus may not be optimal for discriminating one class from others[6]. This has naturally motivated to combine the feature extraction method with classifier for classification purpose. In this paper we propose a new classifier for on-line and nonlinear data. Proposed classifier is composed of two parts. First part is used for feature extraction. To extract nonlinear features, we propose a new feature extraction method which overcomes the problem of memory requirement of KPCA by incremental eigenspace update method incorporating with an adaptation of feature mapping function. Second part is used for classification. Extracted features are used as input for classification. We take Least Squares Support Vector Machines(LS-SVM)[7] as a classifier. LS-SVM is reformulations to the standard Support Vector Machines(SVM)[8]. SVM typically solving problems by quadratic programming(QP). Solving QP problem requires complicated computational effort and needs more memory requirement. LS-SVM overcomes this problem by solving a set of linear equations in the problem formulation. Paper is composed of as follows. In Section 2 we will briefly explain the incremental eigenspace update method. In Section 3 nonlinear PCA is introduced and to make nonlinear PCA incrementally feature mapping function is explained. Proposed classifier combining LS-SVM with proposed feature extraction method is described in Section 4. Experimental results to evaluate the performance of proposed classifier is shown in Section 5. Discussion of proposed classifier and future work is described in Section 6.

2 Incremental Eigenspace Update Method

In this section, we will give a brief introduction to the method of incremental PCA algorithm which overcomes the computational complexity and memory requirement of standard PCA. Before continuing, a note on notation is in order. Vectors are columns, and the size of a vector, or matrix, where it is important, is

denoted with subscripts. Particular column vectors within a matrix are denoted with a superscript, while a superscript on a vector denotes a particular observation from a set of observations, so we treat observations as column vectors of a matrix. As an example, A^i_{mn} is the ith column vector in an $m \times n$ matrix. We denote a column extension to a matrix using square brackets. Thus $[A_{mn}b]$ is an $(m \times (n+1))$ matrix, with vector b appended to A_{mn} as a last column.
To explain the incremental PCA, we assume that we have already built a set of eigenvectors $U = [u_j], j = 1, \cdots, k$ after having trained the input data $\mathbf{x}_i, i = 1, \cdots, N$. The corresponding eigenvalues are Λ and $\bar{\mathbf{x}}$ is the mean of input vector. Incremental building of eigenspace requires to update these eigenspace to take into account of a new input data. Here we give a brief summarization of the method which is described in [9]. First, we update the mean:

$$\overline{x}' = \frac{1}{N+1}(N\overline{x} + x_{N+1}) \tag{1}$$

We then update the set of Eigenvectors to reflect the new input vector and to apply a rotational transformation to U. For doing this, it is necessary to compute the orthogonal residual vector $\hat{h} = (Ua_{N+1} + \overline{x}) - x_{N+1}$ and normalize it to obtain $h_{N+1} = \frac{h_{N+1}}{\|h_{N+1}\|_2}$ for $\| h_{N+1} \|_2 > 0$ and $h_{N+1} = 0$ otherwise. We obtain the new matrix of Eigenvectors $U^{'}$ by appending h_{N+1} to the eigenvectors U and rotating them :

$$U' = [U, h_{N+1}]R \tag{2}$$

where $R \in \mathbf{R}_{(\mathbf{k+1})\times(\mathbf{k+1})}$ is a rotation matrix. R is the solution of the eigenproblem of the following form:

$$DR = R\Lambda' \tag{3}$$

where Λ' is a diagonal matrix of new Eigenvalues. We compose $D \in \mathbf{R}_{(\mathbf{k+1})\times(\mathbf{k+1})}$ as:

$$D = \frac{N}{N+1}\begin{bmatrix} \Lambda & 0 \\ 0^T & 0 \end{bmatrix} + \frac{N}{(N+1)^2}\begin{bmatrix} aa^T & \gamma a \\ \gamma a^T & \gamma^2 \end{bmatrix} \tag{4}$$

where $\gamma = h^T_{N+1}(x_{N+1} - \bar{x})$ and $a = U^T(x_{N+1} - \bar{x})$. Though there are other ways to construct matrix D[10,11], the only method ,however, described in [9] allows for the updating of mean.

2.1 Eigenspace Updating Criterion

The incremental PCA represents the input data with principal components $a_{i(N)}$ and it can be approximated as follows:

$$\widehat{x}_{i(N)} = Ua_{i(N)} + \bar{x} \tag{5}$$

To update the principal components $a_{i(N)}$ for a new input x_{N+1} , computing an auxiliary vector η is necessary. η is calculated as follows:

$$\eta = \left[U\widehat{h}_{N+1}\right]^T (\overline{x} - \overline{x}') \tag{6}$$

then the computation of all principal components is

$$a_{i(N+1)} = (R')^T \begin{bmatrix} a_{i(N)} \\ 0 \end{bmatrix} + \eta, \qquad i = 1, \cdots, N+1 \tag{7}$$

The above transformation produces a representation with $k+1$ dimensions. Due to the increase of the dimensionality by one, however, more storage is required to represent the data. If we try to keep a k-dimensional eigenspace, we lose a certain amount of information. It is needed for us to set the criterion on retaining the number of eigenvectors. There is no explicit guideline for retaining a number of eigenvectors. Here we introduce some general criteria to deal with the model's dimensionality:

- Adding a new vector whenever the size of the residual vector exceeds an absolute threshold;
- Adding a new vector when the percentage of energy carried by the last Eigenvalue in the total energy of the system exceeds an absolute threshold, or equivalently, defining a percentage of the total energy of the system that will be kept in each update;
- Discarding Eigenvectors whose Eigenvalues are smaller than a percentage of the first Eigenvalue;
- Keeping the dimensionality constant.

In this paper we take a rule described in b). We set our criterion on adding an Eigenvector as $\lambda'_{k+1} > 0.7\bar{\lambda}$ where $\bar{\lambda}$ is a mean of the λ. Based on this rule, we decide whether adding u'_{k+1} or not.

3 Online and Nonlinear PCA

A prerequisite of the incremental eigenspace update method is that it has to be applied on the data set. Furthermore incremental PCA builds the subspace of eigenvectors incrementally, it is restricted to apply the linear data. But in the case of KPCA this data set $\Phi(x^N)$ is high dimensional and can most of the time not even be calculated explicitly. For the case of nonlinear data set, applying feature mapping function method to incremental PCA may be one of the solutions. This is performed by so-called *kernel-trick*, which means an implicit embedding to an infinite dimensional Hilbert space[8](i.e. feature space) F.

$$K(x, y) = \Phi(x) \cdot \Phi(y) \tag{8}$$

Where K is a given kernel function in an input space. When K is semi positive definite, the existence of Φ is proven[8]. Most of the case, however, the mapping Φ is high-dimensional and cannot be obtained explicitly except polynomial kernel function. We can easily derive polynomial feature mapping function as following procedure. Let d =2, $x = (x_1, x_2), y = (y_1, y_2))$ then $(x \cdot y)^2 = (x_1^2, \sqrt{2}x_1x_2, x_2^2)(y_1^2, \sqrt{2}y_1y_2, y_2^2)^T = (\phi(x) \cdot \phi(y))$. Now it is then easy to see that

$(\phi(x)) = (x_1^2, \sqrt{2}x_1x_2, x_2^2)$. In case of polynomial feature mapping function there is no difference in performance according to degree d[8]. By this result, we only need to apply the polynomial feature mapping function to one data point at a time and do not need to store the $N \times N$ kernel matrix.

4 Proposed Classification System

In earlier Section 3 we proposed an incremental nonlinear PCA method for nonlinear feature extraction. Feature extraction by incremental nonlinear PCA effectively acts a nonlinear mapping from the input space to an implicit high dimensional feature space. It is hoped that the distribution of the mapped data in the feature space has a simple distribution so that a classifier can classify them properly. But it is point out that extracted features by nonlinear PCA are global features for all input data and thus may not be optimal for discriminating one class from others. For classification purpose, after global features are extracted using they must be used as input data for classification. There are many famous classifier in machine learning field. Among them neural network is popular method for classification and prediction purpose. Traditional neural network approaches, however have suffered difficulties with generalization, producing models that can overfit the data. To overcome the problem of classical neural network technique, support vector machines(SVM) have been introduced. The foundations of SVM have been developed by Vapnik and it is a powerful methodology for solving problems in nonlinear classification. Originally, it has been introduced within the context of statistical learning theory and structural risk minimization. In the methods one solves convex optimization problems, typically by quadratic programming(QP). Solving QP problem requires complicated computational effort and need more memory requirement. LS-SVM overcomes this problem by solving a set of linear equations in the problem formulation. LS-SVM method is computationally attractive and easier to extend than SVM.

5 Experiment

To evaluate the performance of proposed classification system, experiment is performed by following step. First we evaluate the feature extraction ability of online and nonlinear PCA(OL-NPCA). The disadvantage of incremental method is their accuracy compared to batch method even though it has the advantage of memory efficiency. So we shall apply proposed method to a simple toy data and image data set which will show the accuracy and memory efficiency of incremental nonlinear PCA compared to APEX model proposed by Kung[15] and batch KPCA. Next we will evaluate the training and generalization ability of proposed classifier on UCI benchmarking data and NIST handwritten data set. To do this, extracted features by OL-NPCA will be used as input for LS-SVM.

5.1 Toy Data

To evaluate the feature extraction accuracy and memory efficiency of OL-NPCA compared to APEX and KPCA we take nonlinear data used by Scholkoff[5]. Totally 41 training data set is generated by:

$$y = x^2 + 0.2\varepsilon : \ \varepsilon \ from \ N(0,1), x = [-1,1] \tag{9}$$

First we compare feature extraction ability of OL-NPCA to APEX model. APEX model is famous principal component extractor based on Hebbian learning rule. Applying toy data to OL-NPCA we finally obtain 2 eigenvectors. To evaluate the performance of two methods on same condition, we set 2 output nodes to standard APEX model.

Table 1. Performance evaluation of OL-NPCA and APEX

Method	Iteration	Learning Rate	$\parallel w_1 \parallel$	$\parallel w_2 \parallel$	$\cos\theta_1$	$\cos\theta_2$	MSE
APEX	50	0.01	0.6827	1.4346	0.9993	0.7084	14.8589
APEX	50	0.05				do not converge	
APEX	500	0.01	1.0068	1.0014	0.9995	0.9970	4.4403
APEX	500	0.05	1.0152	1.0470	0.9861	0.9432	4.6340
APEX	1000	0.01	1.0068	1.0014	0.9995	0.9970	4.4403
APEX	1000	0.05	1.0152	1.0470	0.9861	0.9432	4.6340
OL-NPCA	100		1	1	1	1	0.0223

In table 1 we experimented APEX method on various conditions. Generally neural network based learning model has difficulty in determining the parameters; for example learning rate, initial weight value and optimal hidden layer node. This makes us to conduct experiments on various conditions. $\parallel w \parallel$ is norm of weight vector in APEX and $\parallel w \parallel = 1$ means that it converges stable minimum. $cos\theta$ is angle between Eigenvector of KPCA and APEX, OL-NPCA respectively. $cos\theta$ of Eigenvector can be a factor of evaluating accuracy how much OL-NPCA and APEX is close to accuracy of KPCA. Table 1 nicely shows the two advantages of OL-NPCA compared to APEX: first, performance of OL-NPCA is better than APEX; second, the performance of OL-NPCA is easily improved by re-learning. Another factor of evaluating accuracy is reconstruction error. Reconstruction error is defined as the squared distance between the image of x_N and reconstruction when projected onto the first i principal components.

$$\delta = |\Psi(x_N) - P_l\Psi(x_N)|^2 \tag{10}$$

In here P_l is the first i principal component. The MSE(Mean Square Error) value of reconstruction error in APEX is 4.4403 whereas OL-NPCA is 0.0223. This means that the accuracy of OL-NPCA is superior to standard APEX and similar to that of batch KPCA. Above results of simple toy problem indicate that OL-NPCA is comparable to the batch way KPCA and superior in terms of accuracy.

Next we will compare the memory efficiency of OL-NPCA compared to KPCA. To extract nonlinear features, OL-NPCA only needs D matrix and R matrix whereas KPCA needs kernel matrix. Table 2 shows the memory requirement of each method. Memory requirement of standard KPCA is 93 times more than OL-NPCA. We can see that OL-NPCA is more efficient in memory requirement than KPCA and has similar ability in extracting nonlinear features. By this simple toy problem we can show that OL-NPCA has similar ability in extracting nonlinear features compare to KPCA and more efficient in memory requirement than KPCA.

Table 2. Memory efficiency of OL-NPCA compared to KPCA on toy data

	KPCA	OL-NPCA
Kernel matrix	41 X 41	none
R matrix	none	3 X 3
D matrix	none	3 X 3
Efficiency ratio	93.3889	1

5.2 Reconstruction Ability

To compare the reconstruction ability of incremental eigenspace update method proposed by Hall to APEX model we conducted experiment on US National Institute of Standards and Technology(NIST) handwritten data set. Data has been size-normalized and 16 X 16 images with their values scaled to the interval [0,1]. Applying this data to incremental eigenspace update method we finally obtain 6 Eigenvectors. As earlier experiment we set 6 output nodes to standard APEX method. Figure 1 shows the original data and their reconstructed images by incremental eigenspace update method and APEX respectively. We can see that reconstructed features by incremental eigenspace update method is more clear and similar to original image compared to APEX method.

5.3 UCI Machine Learning Repository

To test the performance of proposed classifier for real world data, we enlarge our experiment to the Cleveland heart disease data and wine data obtained from the UCI Machine Learning Repository. Detailed description of data is available from web site(http://www.ics.uci.edu/ mlearn/MLSummary.html). In this problem we randomly split training data as 80% and remaining as test data. A RBF kernel has been taken with and obtained by 10-fold cross-validation procedure to select the optimal hyperparameter. Table 3 shows the learning and generalization ability by proposed classifier.

By this result we can see that proposed classification system classifies well on specific data.

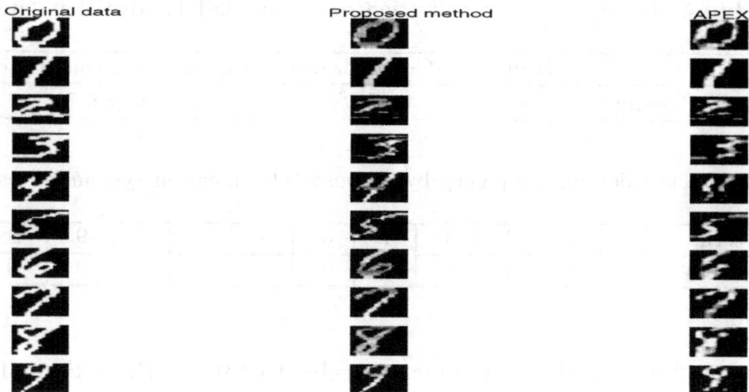

Fig. 1. Reconstructed image by OL-NPCA and APEX

Table 3. Training and generalization result by proposed classifier on UCI Machine Learning Repository

	Training	Generalization	Eigenvalue update criterion
Cleveland heart-disease	100%	97.35%	$\lambda' > 0.7\overline{\lambda}$
Wine data	100%	98.04%	$\lambda' > 0.7\overline{\lambda}$

5.4 NIST Handwritten Data Set

To validate the above results on a widely used pattern recognition benchmark database, we conducted classification experiment on the NIST data set. This database originally contains 15,025 digit images. For computational reasons, we decided to use a subset of 2000 data set, 1000 for training and 1000 for testing. In this problem we use multiclass LS-SVM classifier proposed by Suykens[16]. An important issue for SVM is model selection. In [17] it is shown that the use of 10-fold cross-validation for hyperparameter selection of LS-SVMs consistently leads to very good results. In this problem RBF kernel has been taken and hyperparameter $\gamma_1 = 1.5198$, $\gamma_2 = 179.731$, $\gamma_3 = 10.51$, $\gamma_4 = 12.81$ and $\sigma_1 = 67.416$, $\sigma_2 = 656.351$, $\sigma_3 = 54.349$, $\sigma_4 = 57.909$ are obtained by 10-fold cross-validation technique. The results on the NIST data are given in Table 4 and 5. For this widely used pattern recognition problem, we can see that proposed classification system classifies well on given data.

6 Conclusion and Remarks

This paper is devoted to the exposition of a new technique on extracting nonlinear features and classification system from the incremental data. To develop this technique, we apply an incremental eigenspace update method to KPCA with an polynomial feature mapping function approach. Proposed OL-NPCA has following advantages. Firstly, OL-NPCA has similar feature extracting performance

Table 4. Training and generalization result on NIST handwritten data

	Training	Generalization	Eigenvalue update criterion
Proposed Classifier	100%	98.7%	$\lambda' > 0.7\bar{\lambda}$

Table 5. Misclassification frequency by proposed classification system on test data

Pattern	0	1	2	3	4	5	6	7	8	9	Total
Frequency	0	0	0	3	4	0	0	6	0	0	13

for incremental and nonlinear data comparable to batch KPCA. Secondly, OL-NPCA is more efficient in memory requirement than batch KPCA. In batch KPCA the $N \times N$ kernel matrix has to be stored, while for OL-NPCA requirements are $O((k+1)^2)$. Here $k(1 \leq k \leq N)$ is the number of eigenvectors stored in each eigenspace updating step, which usually takes a number much smaller than N. Thirdly, OL-NPCA allows for complete incremental learning using the eigenspace approach, whereas batch KPCA recomputes whole decomposition for updating the subspace of eigenvectors with another data. Finally, experimental results show that extracted features from OL-NPCA lead to good performance when used as a pre-preprocess data for a LS-SVM.

References

1. Tipping, M.E. and Bishop, C.M. :Mixtures of probabilistic principal component analysers. Neural Computation 11(2), (1998) 443-482
2. Kramer, M.A.:Nonlinear principal component analysis using autoassociative neural networks. AICHE Journal 37(2),(1991) 233-243
3. Diamantaras, K.I. and Kung, S.Y.:Principal Component Neural Networks: Theory and Applications. New York John Wiley & Sons, Inc.(1996)
4. Kim, Byung Joo. Shim, Joo Yong. Hwang, Chang Ha. Kim, Il Kon, "Incremental Feature Extraction Based on Emperical Feature Map," Foundations of Intelligent Systems, volume 2871 of Lecture Notes in Artificial Intelligence, pp 440-444, 2003
5. Softky, W.S and Kammen, D.M, "Correlation in high dimensional or asymmetric data set: Hebbian neuronal processing," Neural Networks vol. 4, pp.337-348, Nov. 1991.
6. Gupta, H., Agrawal, A.K., Pruthi, T., Shekhar, C., and Chellappa., R., "An Experimental Evaluation of Linear and Kernel-Based Methods for Face Recognition," accessible at http://citeseer.nj.nec.com.
7. Suykens, J.A.K. and Vandewalle, J.:Least squares support vector machine classifiers. Neural Processing Letters, vol.9, (1999) 293-300
8. Vapnik, V. N.:Statistical learning theory. John Wiley & Sons, New York (1998)
9. Hall, P. Marshall, D. and Martin, R.: Incremental eigenalysis for classification. In British Machine Vision Conference, volume 1, September (1998)286-295
10. Winkeler, J. Manjunath, B.S. and Chandrasekaran, S.:Subset selection for active object recognition. In CVPR, volume 2, IEEE Computer Society Press, June (1999) 511-516

11. Murakami, H. Kumar.,B.V.K.V.:Efficient calculation of primary images from a set of images. IEEE PAMI, 4(5), (1982) 511-515
12. Scholkopf, B. Smola, A. and Muller, K.R.:Nonlinear component analysis as a kernel eigenvalue problem. Neural Computation 10(5), (1998) 1299-1319
13. Tsuda, K., "Support vector classifier based on asymmetric kernel function," Proc. ESANN, 1999.
14. Mika, S.:Kernel algorithms for nonlinear signal processing in feature spaces. Master's thesis, Technical University of Berlin, November (1998)
15. Diamantaras, K.I. and Kung, S.Y, Principal Component Neural Networks: Theory and Applications, New York John Wiley&Sons, Inc. 1996.
16. Suykens, J.A.K. and Vandewalle, J.: Multiclass Least Squares Support Vector Machines, In: Proc. International Joint Conference on Neural Networks (IJCNN'99), Washington DC (1999)
17. Gestel, V. Suykens, T. J.A.K. Lanckriet, G. Lambrechts, De Moor, A. B. and Vandewalle, J., "A Bayesian Framework for Least Squares Support Vector Machine Classifiers," Internal Report 00-65, ESAT-SISTA, K.U. Leuven.

A Spectroscopy* of Texts for Effective Clustering

Wenyuan Li[1], Wee-Keong Ng[1], Kok-Leong Ong[2], and Ee-Peng Lim[1]

[1] Nanyang Technological University, Centre for Advanced Information Systems
Nanyang Avenue, N4-B3C-14, Singapore 639798
liwy@pmail.ntu.edu.sg, {awkng,aseplim}@ntu.edu.sg
[2] School of Information Technology, Deakin University
Waurn Ponds, Victoria 3217, Australia
leong@deakin.edu.au

Abstract. For many clustering algorithms, such as k-means, EM, and CLOPE, there is usually a requirement to set some parameters. Often, these parameters directly or indirectly control the number of clusters to return. In the presence of different data characteristics and analysis contexts, it is often difficult for the user to estimate the number of clusters in the data set. This is especially true in *text* collections such as Web documents, images or biological data. The fundamental question this paper addresses is: "How can we effectively estimate the natural number of clusters in a given *text* collection?". We propose to use spectral analysis, which analyzes the eigenvalues (not eigenvectors) of the collection, as the solution to the above. We first present the relationship between a *text* collection and its underlying spectra. We then show how the answer to this question enhances the clustering process. Finally, we conclude with empirical results and related work.

1 Introduction

The bulk of data mining research is devoted to the development of techniques that solve a particular problem. Often, the focus is on the design of algorithms that outperform previous techniques either in terms of speed or accuracy. While such effort is a valuable endeavor, the overall success of knowledge discovery (i.e., the larger context of data mining) requires more than just algorithms for the data. With an exponential increase of data in recent years, an important and crucial factor to the success of knowledge discovery is to close the gap between the algorithms and the user.

A good example to argue a case for the above is clustering. In clustering, there is usually a requirement to set some parameters. Often, these parameters directly or indirectly control the number of clusters to return. In the presence of different data characteristics and analysis contexts, it is often difficult for the user to determine the correct number of clusters in the data set [1–3]. Therefore, setting these parameters require either detailed pre-existing knowledge of the data, or time-consuming trial and error. In the latter case, the user also needs

* **spectroscopy** *n.* the study of spectra or spectral analysis.

J.-F. Boulicaut et al. (Eds.): PKDD 2004, LNAI 3202, pp. 301–312, 2004.

sufficient knowledge to know what is a good clustering. Worse, if the data set is very large or has a high dimensionality, the trial and error process becomes very inefficient for the user.

To strengthen the case further, certain algorithms require a good estimate of the input parameters. For example, the EM [4] algorithm is known to perform well in image segmentation [5] when the number of clusters and the initialization parameters are close to their true values. Yet, one reason that limits its application is the poor estimate on the number of clusters. Likewise, a poor parameter setting in CLOPE [6] can dramatically increase its runtime. In all cases above, the user is likely to devote more time in parameter tuning rather than knowledge discovery. Clearly, this is undesirable.

In this paper, we provide a concrete instance of the above problem by studying the issue in the context of *text* collections, i.e., Web documents, images, biological data, etc. Such data sets are inherently large in size and have dimensionality in magnitude of hundreds to several thousands. And considering the domain specificity of the data, getting the user to set a value for k, i.e., the number of clusters, becomes a challenging task. In this case, a good starting point is to initialize k to the natural number of clusters.

This gives rise to the fundamental question that this paper addresses: "How can we effectively estimate the natural number of clusters for a given *text* collection?". Our solution is to perform a spectral analysis on the similarity space of the *text* collection by analyzing the eigenvalues (not eigenvectors) that encode the answer to the above question. Using this observation, we next provide concrete examples of how the clustering process is enhanced in a user-centered fashion. Specifically, we argue that spectral analysis addresses two key issues in clustering: it provides a means to quickly assess the cluster quality; and it bootstraps the analysis by suggesting a value for k.

The outline of this paper is as follows. In the next section, we begin with some preliminaries of spectral analysis and its basic properties. Section 3 presents our contribution on the use of normalized eigenvalues to answer the question we posed in this paper. Section 4 discusses a concrete example of applying the observation to enhance the clustering process. Section 5 presents the empirical results as evidence to the viability of our proposal. Section 6 discusses the related work, and Section 7 concludes this paper.

2 Preliminaries

Most algorithms perform clustering by embedding the data in some similarity space [7], which is determined by some widely-used similarity measures, e.g., cosine similarity [8]. Let $\mathbf{S} = (s_{ij})_{n\times n}$ be the similarity space matrix, where $0 \leqslant s_{ij} \leqslant 1$, $s_{ii} = 1$ and $s_{ij} = s_{ji}$, i.e., $\mathbf{S}$ is symmetric. Further, let $\mathcal{G}(\mathbf{S}) = \langle V, E, \mathbf{S}\rangle$ be the graph of $\mathbf{S}$, where V is the set of n vertices and $\mathbf{E}$ is the set of weighted edges. Each vertex v_i of $\mathcal{G}(\mathbf{S})$ corresponds to the i-th column (or row) of $\mathbf{S}$, and the weight of each edge $\widehat{v_i v_j}$ corresponds to the non-diagonal entry s_{ij}. For any two vertices (v_i, v_j), a larger value of s_{ij} indicates a higher connectivity between them, and vice versa.

Once we obtained $\mathcal{G}(\mathbf{S})$, we can analyze its spectra as we will illustrate in the next section. However, for ease of discussion, we establish the following basic facts of spectral graph theory below. Among them, the last fact about $\mathcal{G}(\mathbf{S})$ is an important property that we exploit: it depicts the relationship between the spectra of the disjoint subgraphs $\mathcal{G}(\mathbf{S}_i)$ and the spectra of $\mathcal{G}(\mathbf{S})$.

Theorem 1. *Let $\lambda_1 \geqslant \lambda_2 \geqslant \ldots \geqslant \lambda_n$ be the eigenvalues of $\mathcal{G}(\mathbf{S})$ such that $-1 \leqslant \lambda_i \leqslant 1$, $i = 1, 2, \cdots, n$. Then, the following holds: (i) $\sum \lambda_i = 0$, and $\lambda_1 = 1$; (ii) if $\mathcal{G}(\mathbf{S})$ is connected, then $\lambda_2 < 1$; (iii) the spectra of $\mathcal{G}(\mathbf{S})$ is the union of the spectra of its disjoint subgraphs $\mathcal{G}(\mathbf{S}_i)$.*

Proof. As shown in [9, 10].

In reality, the different similarity matrices are not normalized making it difficult to analyze them directly. In other words, the eigenvalues do not usually fall within $-1 \leqslant \lambda_i \leqslant 1$. Therefore, we need to perform an additional step to get Theorem 1: we transform $\mathbf{S}$ to a weighted Laplacian $\mathbf{L} = (\ell_{ij})$, where $\ell_{ij} \in [0; 2)$ is a normalized eigenvalue obtainable by the following:

$$\ell_{ij} = \begin{cases} 1 - \frac{s_{ij}}{d_i}, & i = j \\ -\frac{s_{ij}}{\sqrt{d_i, d_j}}, & s_{ij} \neq 0 \\ 0, & otherwise \end{cases} \tag{1}$$

where $d_i = \sum_j s_{ij}$ is the degree of vertex v_i in $\mathcal{G}(\mathbf{S})$. We then derive a variant of $\mathbf{L}$ defined as follows:

$$\mathsf{L} = \mathbf{D}^{-1/2}(\mathbf{S} - \mathbf{I})\mathbf{D}^{-1/2} \tag{2}$$

where $\mathbf{D}$ is the diagonal matrix, diag(d_i). From Equations 1 and 2, we can deduce eig$(\mathsf{L}) = \{1 - \lambda \,|\, \lambda \in \text{eig}(\mathbf{L})\}$, where eig$(\cdot)$ is the set of eigenvalues of $\mathbf{S}$. Notably, the eigenvalues in L maintains the same conclusions and properties of those found in $\mathbf{L}$. Thus, we now have a set of eigenvalues that can be easily analyzed. Above all, this approach does not require any clustering algorithm to find k. This is very attractive in terms of runtime and simplicity. Henceforth, the answer to our question is now mapped to a matter of knowing how to analyze eig(L). We will describe this in the next section.

3 Clustering and Spectral Properties

For ease of exposition, we first discuss the spectra properties of a conceptually disjoint data set, whose chosen similarity measure achieves a perfect clustering. From this simple case, we extend our observations to real-world data sets, and show how the value of k can be obtained.

3.1 A Simple Case

Assume that we have a conceptually disjoint data set, whose chosen similarity measure achieves a perfect clustering. In this case, the similarity matrix $\mathbf{A}$ will have the following structure:

$$\mathbf{A} = \begin{bmatrix} \mathbf{A}_{11} & \cdots & \mathbf{A}_{1k} \\ \vdots & \ddots & \vdots \\ \mathbf{A}_{k1} & \cdots & \mathbf{A}_{kk} \end{bmatrix} \begin{matrix} n_1 \\ \vdots \\ n_k \end{matrix} \qquad (3)$$
$$n_1 \cdots n_k$$

with the properties: all entries in each diagonal block matrix $\mathbf{A}_{ii}$ of $\mathbf{A}$ are 1; and all entries in each non-diagonal block matrix $\mathbf{A}_{ij}$ in $\mathbf{A}$ are 0. From this similarity matrix, we can obtain its eigenvalues in decreasing order [9], i.e.,

$$\lambda_i(\mathbf{A}) = \begin{cases} 1, & 1 \leqslant i \leqslant k \\ 0, & k < i \leqslant n \end{cases} \qquad (4)$$

Lemma 1. *Given a similarity matrix $\mathbf{S}$ as defined in Equation (3), where $n_1 + \cdots + n_k = n$; where each diagonal entry $\mathbf{S}_{ii}$ satisfies $0 < n_i - \|\mathbf{S}_{ii}\|_F < \delta(\delta \to 0)$; and where each non-diagonal entry $\mathbf{S}_{ij}$ satisfies $\|\mathbf{S}_{ij}\|_F \to 0$ ($\|\cdot\|_F$ is the Frobenius norm), then $\mathbf{S}$ achieves a perfect clustering of n clusters. At the same time, the spectra of $\mathcal{G}(\mathbf{S})$ exhibit the following properties:*

$$\begin{array}{ll} \lambda_i \to 1 & (i = 1, \cdots, k \text{ and } 0 < \lambda_i \leqslant 1) \\ |\lambda_i| \to 0 & (i = k+1, \cdots, n) \end{array} \qquad (5)$$

Proof. Let $\mathbf{E} = \mathbf{S} - \mathbf{A}$, where $\mathbf{A}$ is as defined in Equation (3). From definitions of $\mathbf{A}$ and $\mathbf{S}$, we obtain the following:

$$\left.\begin{array}{rr} 0 < n_i - \|\mathbf{S}_{ii}\|_F < \delta(\delta \to 0), & \|\mathbf{A}_{ii}\|_F = n_i \\ \|\mathbf{S}_{ij}\|_F \to 0, & \|\mathbf{A}_{ij}\|_F = 0 \end{array}\right\} \Rightarrow \|\mathbf{E}\|_F \to 0 \qquad (6)$$

where by the well-known property of the Frobenius norm, and the p matrix norm (where $p = 2$ [10]), we have:

$$\|\mathbf{E}\|_2 \leqslant \|\mathbf{E}\|_F \qquad (7)$$

and

$$|\lambda_i(\mathbf{A} + \mathbf{E}) - \lambda_i(\mathbf{A})| \leqslant \|\mathbf{E}\|_2, \quad (i = 1, \cdots, n) \qquad (8)$$

where $\|\cdot\|_2$ is the $p = 2$ matrix norm. Equation (7) states that the Frobenius norm of a matrix is always greater than or equal to the p matrix norm at $p = 2$, and Equation (8) defines the distance between the eigenvalues in $\mathbf{A}$ and its perturbation matrix $\mathbf{S}$. In addition, the sensitivity of the eigenvalues in $\mathbf{A}$ to its perturbation is given by $\|\mathbf{E}\|_2$. Hence, from Equations (6), (7), and (8), we can conclude that:

$$\lambda_i(\mathbf{S}) \to \lambda_i(\mathbf{A}), \quad (i = 1, \cdots, n) \qquad (9)$$

which when we combine with Equation (4), we arrive at Lemma 1.

Table 1. A small text collection taken and modified from [11]. It contains the titles of 12 technical memoranda: 5 about human-computer interaction; 4 about mathematical graph theory; and 3 about clustering. The topics are conceptually disjoint with two assumptions: (i) the italicized terms are the selected feature set; and (ii) the cosine similarity measure is used to compute **S**.

c1	*Human* machine *interface* for ABC *computer* applications
c2	A *survey* of *user* opinion of *computer system response time*
c3	The *EPS user interface* management *system*
c4	*System* and *human system* engineering testing of *EPS*
c5	Relation of *user* perceived *response time* to error measurement
m1	The generation of random, binary, ordered *trees*
m2	The intersection *graph* of paths in *trees*
m3	*Graph minors* IV: Widths of *trees* and well-quasi-ordering
m4	*Graph minors*: A *survey*
d1	Linguistic features and *clustering* algorithms for topical *document clustering*
d2	A comparison of *document clustering techniques*
d3	*Survey* of *clustering* Data Mining *Techniques*

Simply put, when the spectra distribution satisfies Equation (5), then **S** shows a good clustering, i.e., the intra-similarity approaches 1, and the inter-similarity approaches 0. As an example, suppose we have a collection with 3 clusters as depicted in Table 1. The 3 topics are setup to be conceptually disjoint, and the similarity measure as well as the feature set are selected such that the outcome produces 3 distinct clusters. In this ideal condition, the spectra distribution (as shown in Figure 1) behaves as per Equation (5).

Of course, real-world data sets that exhibit perfect clustering are extremely rare. This is especially the case for *text* collections, where its dimensionality is large but the data itself is sparse. In this case, most similarity measures do not rate two documents as distinctively similar, or different. If we perform a spectral analysis on the collection, we will end up with a spectra of $\mathcal{G}(\mathbf{S})$ that is very different from our example in Figure 1. As we will see next, this spectra distribution is much more complex.

3.2 Spectra Distribution in Large Data Sets

Point (iii) of Theorem 1 offers a strong conclusion between $\mathcal{G}(\mathbf{S})$ and its subgraphs. However, real-world data sets often exhibit a different characteristic. If we examine their corresponding $\mathcal{G}(\mathbf{S})$, we will see that the connections between $\mathcal{G}(\mathbf{S})$ and its subgraphs are weak, i.e., Lemma 1 no longer holds.

Fortunately, we can still judge the cluster quality and estimate the number of natural clusters with spectral analysis. In this section, we present the proofs that leads to the conclusion about cluster quality and k. But first, we need introduce the Cheeger constant. Let $SV \subset V$ of $\mathcal{G}(\mathbf{S})$. We define the volume of SV as:

$$\mathrm{vol}(SV) = \sum_{v \in SV} d_v \tag{10}$$

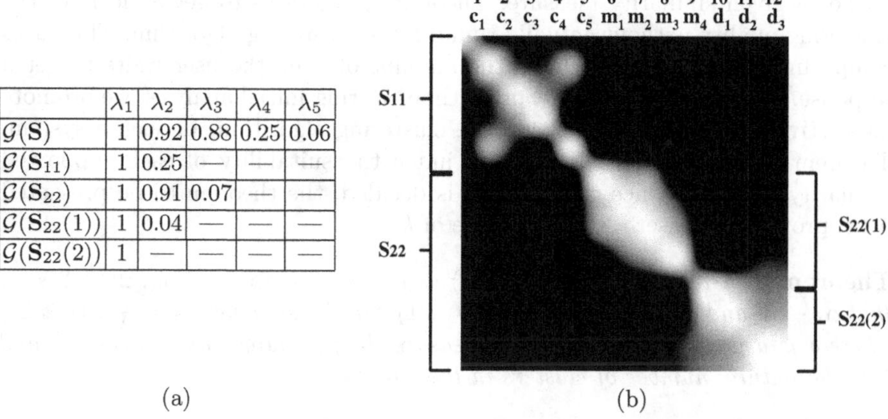

	λ_1	λ_2	λ_3	λ_4	λ_5
$\mathcal{G}(\mathbf{S})$	1	0.92	0.88	0.25	0.06
$\mathcal{G}(\mathbf{S}_{11})$	1	0.25	—	—	—
$\mathcal{G}(\mathbf{S}_{22})$	1	0.91	0.07	—	—
$\mathcal{G}(\mathbf{S}_{22}(1))$	1	0.04	—	—	—
$\mathcal{G}(\mathbf{S}_{22}(2))$	1	—	—	—	—

Fig. 1. The spectra distribution of the collection in Table 1: (a) Spectrum (> 0) of $\mathcal{G}(\mathbf{S})$ and its subgraphs; (b) a graphical representation of **S**. Note that all grey images in this paper are not "plots" of the spectra. Rather, they are a graphical way of summarizing the results of clustering for comparison/discussion purposes.

where d_v is the sum of all weighted edges containing vertex v. Further, let $E(\delta SV)$ be the set of edges, where each edge has one of its vertices in SV but not the other, i.e., $\overline{SV}$. Then, its volume is given by:

$$|E(\delta SV)| = \sum_{v_i \in SV, v_j \notin SV} \text{weight}(v_i, v_j) \tag{11}$$

and by Equations (10) and (11), we derive the Cheeger constant:

$$h(\mathcal{G}) = \min_{SV \subset V} \frac{|E(\delta SV)|}{\min(\text{vol}(SV), \text{vol}(\overline{SV}))} \tag{12}$$

which measures the optimality of the bipartition in a graph. The magnitude $|E(\delta SV)|$ measures the connectivity between SV and $\overline{SV}$ while vol(SV) measures the density of SV against V.

Since SV enumerates all subsets of V, $h(\mathcal{G})$ is a good measure that finds the best bipartition, i.e., $\langle SV, \overline{SV} \rangle$. Perhaps, more interesting is the observation that no other bipartition gives a better clustering than the bipartition determined by $h(\mathcal{G})$. Therefore, $h(\mathcal{G})$ can be used as an indicator of cluster quality, i.e., the lower its value, the better the clustering.

Theorem 2. *Given the spectra of $\mathcal{G}(\mathbf{S})$ as $1 = \lambda_1 \geqslant \lambda_2 \geqslant \cdots \geqslant \lambda_n$, if $\lambda_2 \to 1$, then there exists a good bipartition for $\mathcal{G}(\mathbf{S})$, i.e., a good cluster quality.*

Proof. From [9], we have the Cheeger inequality: $\frac{(1-\lambda_2)}{2} \leqslant h(\mathcal{G}) < \sqrt{2(1-\lambda_2)}$ that gives the bound of $h(\mathcal{G})$. By this inequality, if $\lambda_2 \to 1$, then $h(\mathcal{G}) \to 0$. And since $h(\mathcal{G}) \to 0$ implies a good clustering, we have the above.

For a given similarity measure, Theorem 2 allows us to get a "feel" of the clustering quality without actually running the clustering algorithm. This saves computing resources and reduces the amount of time the user waits to get a response. By minimizing this "waiting time" during initial analysis, we promote interactivity between the user and the clustering algorithm. In such a system, Theorem 2 can also be used to help judge the suitability of each supported similarity measure. Once the measure is decided, the theorem to be presented next, provides the user a starting value of k.

Theorem 3. *Given the spectra of* $\mathcal{G}(\mathbf{S})$ *as* $1 = \lambda_1 \geqslant \lambda_2 \geqslant \cdots \geqslant \lambda_n$, $\exists k \geqslant 2$ *such that* $\alpha_i \to 1$ *and* $\alpha_i - \alpha_{i+1} > \delta$ $(0 < \delta < 1)$ *for the sequence* $\alpha_i = \frac{\lambda_i}{\lambda_2}$, $(i \geqslant 2)$, *where* δ *is a predefined threshold to measure the first large gap between* α_i; *and* k *is the natural number of clusters in the data set.*

Proof. Since Theorem 2 applies to both $\mathcal{G}(\mathbf{S})$ and its subgraphs $\mathcal{G}(\mathbf{S}_{ii})$, then we can estimate the cluster quality of the bipartition in $\mathcal{G}(\mathbf{S}_{ii})$ (as well as its subgraphs). Combine with Point (iii) of Theorem 1, we can conclude that the number of eigenvalues in $\mathcal{G}(\mathbf{S})$ (that approach 1 and have large eigengaps) give the value of k, i.e., the number of clusters.

To cite an example for the above, we revisit Table 1 and Figure 1. By the Cheeger constant of $\mathcal{G}(\mathbf{S})$, $SV = \{c_1, c_2, c_3, c_4, c_5\}$ and $\overline{SV} = \{m_1, m_2, m_3, m_4, d_1, d_2, d_3\}$ produces the best bipartition. Thus, $\mathbf{S}_{11}$ represents the inter-similarities in SV and $\mathbf{S}_{22}$ represents inter-similarities in $\overline{SV}$. From Theorem 2, we can assess the cluster quality of $\mathcal{G}(\mathbf{S})$'s bipartition by λ_2. Also, we can recursively consider the bipartitions of the bipartitions of $\mathcal{G}(\mathbf{S})$, i.e., $\mathcal{G}(\mathbf{S}_{11})$ and $\mathcal{G}(\mathbf{S}_{22})$. Again, the Cheeger constant of $\mathcal{G}(\mathbf{S}_{22})$ shows that $\mathcal{G}(\mathbf{S}_{22}(1))$ and $\mathcal{G}(\mathbf{S}_{22}(2))$ are the best bipartition in the subgraph $\mathcal{G}(\mathbf{S}_{22})$. Likewise, the λ_2 of $\mathcal{G}(\mathbf{S}_{11})$, $\mathcal{G}(\mathbf{S}_{22})$, $\mathcal{G}(\mathbf{S}_{22}(1))$, and $\mathcal{G}(\mathbf{S}_{22})$ all satisfy this observation.

In fact, this recursive bisection of $\mathcal{G}(\mathbf{S})$ is a form of clustering using the Cheeger constant – the spectra of $\mathcal{G}(\mathbf{S}_{22})$ contains the eigenvalues of $\mathcal{G}(\mathbf{S}_{22}(1))$ and $\mathcal{G}(\mathbf{S}_{22}(2))$, and $\mathcal{G}(\mathbf{S})$ contains the eigenvalues of $\mathcal{G}(\mathbf{S}_{11})$ and $\mathcal{G}(\mathbf{S}_{22})$ respectively (despite with some small "fluctuations"). As shown in Figure 1(a), λ_2 of $\mathcal{G}(\mathbf{S})$ gives the cluster quality of the bipartition $\mathcal{G}(\mathbf{S}_{11})$ and $\mathcal{G}(\mathbf{S}_{22})$ in $\mathcal{G}(\mathbf{S})$; and λ_3 of $\mathcal{G}(\mathbf{S})$, which corresponds to λ_2 of $\mathcal{G}(\mathbf{S}_{22})$, gives the cluster quality indicator for the bipartition $\mathcal{G}(\mathbf{S}_{22}(1))$ and $\mathcal{G}(\mathbf{S}_{22}(2))$ in $\mathcal{G}(\mathbf{S}_{22})$, and so on.

Therefore, if there exist k distinct and dense diagonal squares (i.e., $\mathbf{S}_{ii}$ where $1 \leqslant i \leqslant k$) in the matrix, then λ_i of $\mathcal{G}(\mathbf{S})$ will be the cluster quality indicator for the i-th bipartition $(2 \leqslant i \leqslant k)$, and the largest k eigenvalues of $\mathcal{G}(\mathbf{S})$ give the estimated number of clusters in the data.

4 A Motivating Example

In this section, we discuss an example of how the theoretical observations discussed earlier work to close the gap between the algorithm and the user. For illustration, we assume that the user is given some unknown collection.

Table 2. The *text* collections used in our experiments to estimate k: we selected 4 classes of `classic` with each class containing 1,000 documents; 5 `newsgroups` with each newsgroup containing 500 documents; 2 categories of the `webset` with each category containing 600 documents.

Collections	Source	# Classes	# Documents
`classic`	ADI/CACM/CISI/CRAN/MED	5	5559
`newsgroup`	UseNet news postings	17	7473
`webset`	Categories in Yahoo [12]	10	6607

If the user does not have pre-existing knowledge of the data, there is a likelihood of not knowing where to start. In particular, all clustering algorithms directly or indirectly require the parameter k. Without spectral analysis, the user is either left guessing what value of k to start with; or expend time and effort to find k using one of the existing estimation algorithm. In the case of the latter, the user has to be careful in setting k_{max} (see Section 5.2) – if it's set too high, the estimation algorithm takes a long time to complete; if it's set too low, the user risks missing the actual value of k.

In contrast, our proposal allows the user to obtain an accurate value of k without setting k_{max}. Performance wise, this process is almost instantaneous in comparison to other methods that require a clustering algorithm. We believe this is important if the user's role is to *analyze* the data instead of *waiting* for the algorithms. Once an initial value of k is known, the user can commence clustering. Unfortunately, this isn't the end of cluster analysis.

Upon obtaining the outcome, the user usually faces another question: *what is the quality of this clustering?* In our opinion, there is no knowledge discovery when there is no means to judge the outcome. As a result, it is also at this stage where interactivity becomes important. On this issue, some works propose the use of constraints. However, it is difficult to formulate an effective constraint if the answer to the above is unknown. This is where spectral analysis plays a part. By Theorem 2, the user is given feedback about the cluster quality. At the same time, grey images (e.g., Figure 1(b)) can also be constructed to help the user gauge the outcome.

Depending on the feedback, the user may then wish to adjust k, or use another similarity measure. In either case, the user is likely to make a better decision with this assistance. Once the new parameters are decided, another run of the clustering algorithm begins. Our proposal would then kick in at the end of each run to provide the feedback to the user via Theorem 2. This interaction exists because different clustering objectives can be formulated on the same data set. At some point, the user may group overlapping concepts in one class. Other times, the user may prefer to separate them. In this aspect, our approach is non-intrusive and works in tandem with the user's intentions.

5 Empirical Results

The objective of our experiments is to provide the empirical evidence on the viability of our proposal. Due to space limitation, we only report a summary of our results here. The full details can be obtained from [13].

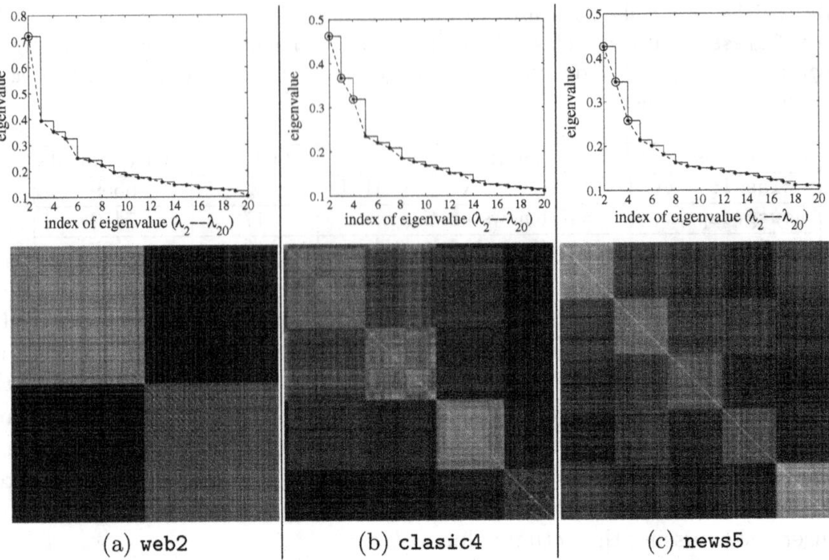

(a) `web2` (b) `clasic4` (c) `news5`

Fig. 2. The spectrum graphs (since λ_1 is always 1, our analysis begins from λ_2) and the graphical representation of their clustering for all 3 collections: the first two data sets are conceptually disjoint, and the last has overlapping concepts.

Figure 2(a) shows the `web2` collection with 2 class labels, and their topic being completely disjoint: *finance* and *sport*. In this case, we observe that λ_2 has a higher value than the others. Therefore, we conclude $k = 2$ and meanwhile, the high value of λ_2 (by Theorem 2) indicates that this is a good clustering. In the second case, `classic4` has 4 conceptually disjoint topics as observed in Figure 2(b). From its spectra graph, we see that λ_2, λ_3 and λ_4 show higher values and wider gaps than other eigenvalues. Again by Theorem 3, our method obtains the correct number of clusters, i.e., $k = 4$.

The third test is the most challenging. There are 5 topics: *atheism*, *comp.sys*, *comp.windows*, *misc.forsale* and *rec.sport*. They are not conceptually disjoint because *comp.sys* and *comp.windows* belong to the broader topic of *comp* in the newsgroup. This is also graphically echoed in Figure 2(c). When we apply our analysis, only λ_2, λ_3, and λ_4 have a higher value and a wider gap than the others. So by our theorem, $k = 4$. This conclusion is actually reasonable, since *comp* is more different than the other topics. If we observe the grey image in Figure 2(c), we see that the second and third squares appear to "meshed" together – an indication of similarity between *comp.sys* and *comp.windows*. Furthermore, *comp.sys*, *comp.windows* and *misc.forsale* can also be viewed as one topic. This is because *misc.forsale* has many postings on buying and selling of computer parts. Again, this can be observed in the grey image. On the spectra graph, λ_4 is much lower than λ_2, λ_3, and is closer to the remaining eigenvalues. Therefore, it is possible to conclude that λ_4 may not contribute to k, and hence $k = 3$ by Theorem 3. Strictly speaking, this is also an acceptable estimation. In other

Table 3. Comparison with 3 well-known indexes: the Calinski and Harabasz (CH) index; Krzanowski and Lai (KL) index; and Hartigan (Hart) index with 3 well-known clustering algorithms: bisecting k-means, graph-based, and hierarchical – a (√) indicates a correct estimation.

	Bisecting k-means			Graph-based			Hierarchical		
	`web2`	`classic4`	`news5`	`web2`	`classic4`	`news5`	`web2`	`classic4`	`news5`
CH	5	2	3	3	3	3	7	2	5 (√)
KL	29	17	22	27	21	22	22	21	9
Hart	6	13	4 (√)	6	10	4 (√)	1	2	1

words, the onus is on the user to judge the actual value of k, which is really problem-specific as illustrated in the Section 4.

Next, we compare the efficiency and effectiveness of our method. To date, most existing techniques require choosing an appropriate clustering algorithm, where it is iteratively ran with a predefined cluster number from 2 to k_{max}. The optimum k is then obtained by an internal indexed based on the clustering outcome. In this experiment, we compared our results against 3 widely used statistical methods [14] using 3 well-known clustering algorithms (see Table 3). From the table, our proposal remarkably outperforms all 3 methods in terms of accuracy. More exciting, our proposal is independent of any clustering algorithm, is well-suited to high dimensional data sets, has low complexity, and can be easily implemented with existing packages as shown in the following.

On the performance of our method, the complexity of transforming $\mathbf{S}$ to L is $O(h)$, where h is the number of non-zero entries in $\mathbf{S}$. Therefore, the eigenvalues can be efficiently obtained by the `Lanczos` method [10]. Since the complexity of each iteration in `Lanczos` is $O(h+n)$, the complexity of our method is therefore $O(k(h+n))$. If we ignore the very small k in real computations, the complexity of our method becomes $O(h+n)$. There are fast `Lanczos` packages (e.g., LANSO [15], PLANSO [16]) for such computation.

6 Related Works

Underlined by the fact that there is no clear definition of what is a good clustering [2, 17], the problem of estimating the number of clusters in a data set is arguably a difficult one. Over the years, several approaches to this problem have been suggested. Among them, the more well-known ones include cross-validation [18], penalized likelihood estimation [19, 20], resampling [21], and finding the 'knee' of an error curve [2, 22]. However, these techniques either make a strong parametric assumption, or are computationally expensive.

Spectral analysis has a long history of wide applications in the scientific domain. In the database community, eigenvectors have applications in areas such as information retrieval (e.g., singular value decomposition [23]), collaborative filtering (e.g., reconstruction of missing data items [24]), and Web searching [25]. Meanwhile, application areas of eigenvalues include the understanding of communication networks [9] and Internet topologies [26]. To the best of our knowledge, we have yet to come across works that use eigenvalues to assist cluster analysis. Most proposals that use spectral techniques for clustering focused on the use of eigenvectors, not eigenvalues.

7 Conclusions

In this paper, we demonstrate a concrete case of our argument on the need to close the gap between data mining algorithms and the user. We exemplified our argument by studying a well-known problem in clustering that every user faces when starting the analysis: "What value of k should we select so that the analysis converges quickly to the desired outcome?".

We answered this question, in the context of *text* collections, with spectral analysis. We show (both argumentatively and empirically) that if we are able to provide a good guess to the value of k, then we have a good starting point for analysis. Once the "ground" is known, data mining can proceed by changing the value of k incrementally from the starting point. This is often better than the trial and error approach. In addition, we also show that our proposal can be used to estimate the quality of clustering. This process, as part of cluster analysis, is equally important to the success of knowledge discovery. Our proposal contributes in part to this insight.

In the general context, the results shown here also demonstrate the feasibility to study techniques that bridge the algorithms and the user. We believe that this endeavor will play a pivotal role to the advancement of knowledge discovery. In particular, as data takes a paradigm shift into continuous and unbounded form, the user will no longer be able to devote time in tuning parameters. Rather, their time should be spent on interacting with the algorithms, such as what we have demonstrated in this paper.

References

1. Salvador, S., Chan, P.: Determining the number of clusters/segments in hierarchical clustering/segmentation algorithms. Technical report 2003-18, Florida Institute of Technology (2003)
2. Tibshirani, R., Walther, G., Hastie, T.: Estimating the number of clusters in a dataset via the gap statistic. Technical Report 208, Dept. of Statistics, Stanford University (2000)
3. Sugar, C., James, G.: Finding the number of clusters in a data set : An information theoretic approach. Journal of the American Statistical Association **98** (2003)
4. Dempster, A., Laird, N., Rubin, D.: Maximum likelihood from incomplete data via the em algorithm. Journal of Royal Statistical Society **39** (1977) 1–38
5. Evans, F., Alder, M., deSilva, C.: Determining the number of clusters in a mixture by iterative model space refinement with application to free-swimming fish detection. In: Proc. of Digital Imaging Computing: Techniques and Applications, Sydney, Australia (2003)

6. Yang, Y., Guan, X., You, J.: CLOPE: A fast and effective clustering algorithm for transactional data. In: Proc. of KDD, Edmonton, Canada (2002) 682–687
7. Strehl, A., Ghosh, J., Mooney, R.: Impact of similarity measures on web-page clustering. In: Proc. of AAAI Workshop on AI for Web Search. (2000) 58–64
8. Jain, A.K., Murty, M.N., Flynn, P.J.: Data clustering: A review. ACM Computing Surveys **31** (1999) 264–323
9. Chung, F.R.K.: Spectral Graph Theory. Number 92 in CBMS Regional Conference Series in Mathematics. American Mathematical Society (1997)
10. Golub, G., Loan, C.V.: Matrix Computations (Johns Hopkins Series in the Mathematical Sciences). 3rd edn. The Johns Hopkins University Press (1996)
11. Landauer, T., Foltz, P., Laham, D.: Introduction to latent semantic analysis. Discourse Processes **25** (1998) 259–284
12. Sinka, M.P., Corne, D.W.: A Large Benchmark Dataset for Web Document Clustering. In: Soft Computing Systems: Design, Management and Applications. IOS Press (2002) 881–890
13. Li, W., Ng, W.K., Ong, K.L., Lim, E.P.: A spectroscopy of texts for effective clustering. Technical Report TRC04/03 (`http://www.deakin.edu.au/~leong/papers/tr2`), Deakin University (2004)
14. Gordon, A.: Classification. 2nd edn. Chapman and Hall/CRC (1999)
15. LANSO: (Dept. of Computer Science and the Industrial Liason Office, Univ. of Calif., Berkeley)
16. Wu, K., Simon, H.: A parallel lanczos method for symmetric generalized eigenvalue problems. Technical Report 41284, LBNL (1997)
17. Kannan, R., Vetta, A.: On clusterings: good, bad and spectral. In: Proc. of FOCS, Redondo Beach (2000) 367–377
18. Smyth, P.: Clustering using monte carlo cross-validation. In: Proc. of KDD, Portland, Oregon, USA (1996) 126–133
19. Baxter, R., Oliver, J.: The kindest cut: minimum message length segmentation. In: Proc. Int. Workshop on Algorithmic Learning Theory. (1996) 83–90
20. Hansen, M., Yu, B.: Model selection and the principle of minimum description length. Journal of the American Statistical Association **96** (2001) 746–774
21. Roth, V., Lange, T., Braun, M., Buhmann, J.: A resampling approach to cluster validation. In: Proc. of COMPSTAT, Berlin, Germany (2002)
22. Tibshirani, R., Walther, G., Botstein, D., Brown, P.: Cluster validation by prediction strength. Technical report, Stanford University (2001)
23. Deerwester, S., Dumais, S., Landauer, T., Furnas, G., Harshman, R.: Indexing by latent semantic analysis. JASIS **41** (1990) 391–407
24. Azar, Y., Fiat, A., Karlin, A., McSherry, F., Saia, J.: Spectral analysis of data. In: ACM Symposium on Theory of Computing, Greece (2001) 619–626
25. Kleinberg, J.M.: Authoritative sources in a hyperlinked environment. Journal of the ACM **46** (1999) 604–632
26. Vukadinovic, D., Huan, P., Erlebach, T.: A spectral analysis of the internet topology. Technical Report 118, ETH TIK-NR (2001)

Constraint-Based Mining of Episode Rules and Optimal Window Sizes*

Nicolas Méger and Christophe Rigotti

INSA-LIRIS FRE CNRS 2672
69621 Villeurbanne Cedex, France
{nmeger,crigotti}@liris.cnrs.fr

Abstract. Episode rules are patterns that can be extracted from a large event sequence, to suggest to experts possible dependencies among occurrences of event types. The corresponding mining approaches have been designed to find rules under a temporal constraint that specifies the maximum elapsed time between the first and the last event of the occurrences of the patterns (i.e., a window size constraint). In some applications the appropriate window size is not known, and furthermore, this size is not the same for different rules. To cope with this class of applications, it has been recently proposed in [2] to specifying the maximal elapsed time between two events (i.e., a maximum gap constraint) instead of a window size constraint. Unfortunately, we show that the algorithm proposed to handle the maximum gap constraint is not complete. In this paper we present a sound and complete algorithm to mine episode rules under the maximum gap constraint, and propose to find, for each rule, the window size corresponding to a local maximum of confidence. We show that the extraction can be efficiently performed in practice on real and synthetic datasets. Finally the experiments show that the notion of local maximum of confidence is significant in practice, since no local maximum are found in random datasets, while they can be found in real ones.

1 Introduction

Many datasets are composed of a large sequence of events, where each event is described by a date of occurrence and an event type. Commonly mined descriptive patterns in these datasets are the so-called *episode rules.* Informally, an episode rule reflects how often a particular group G_1 of event types tends to appear close to another group G_2. A rule is associated with two measures, its frequency and its confidence, that have an intuitive reading similar to the one of frequency and confidence used for *association rules* [1]. These two measures respectively represent how often the two groups occur together (i.e., is the rule supported by many examples?) and how *strong* is this rule (i.e., when G_2 occurs, is G_1 appearing in many cases close to G_2?).

* This research is partially funded by the European Commission IST Programme - Accompanying Measures, AEGIS project (IST-2000-26450).

J.-F. Boulicaut et al. (Eds.): PKDD 2004, LNAI 3202, pp. 313–324, 2004.

Finding episode rules may provide interesting insight to experts in various domains. In particular, it has been shown to be very useful for alarm log analysis in the context of the TASA project [3]. More generally, it can also be applied, after an appropriated discretization to time series, and to spatial data (the temporal dimension is replaced by a spatial dimension) like for example in DNA sequences.

The standard *episode rule mining problem* is to find all episode rules satisfying given frequency and confidence constraints. There are two main approaches to find such rules. The first one, proposed and used by [7, 6] in the *Winepi* algorithm, is based on the occurrences of the patterns in a sliding window along the sequence. The second one, introduced in [5, 6] and supported by the *Minepi* algorithm, relies on the notion of *minimal occurrences* of patterns. Both techniques have been designed to be run using a maximum window size constraint that specifies the maximum elapsed time between the first and the last event of the occurrences of the patterns. More precisely, in the case of *Winepi*, the algorithm used a single window size constraint and must be executed again if the user wants to perform an extraction with a different window size. The other algorithm, *Minepi*, needs a maximal window size constraint to restrict reasonably the search space in practice, but can derive rules for several window sizes that are lesser than this maximal window size.

To our knowledge, no existing complete algorithm is able to extract episode rules without at least a maximum window size constraint (in addition to a frequency and a confidence constraint) on non-trivial datasets. In some applications the window size is not known beforehand, and moreover, the interesting window size may be different for each episode rule. To cope with this class of applications, it has been recently proposed in [2] to use a maximum gap constraint that imposes the maximal elapsed time between two consecutive events in the occurrences of an episode. This constraint allows the occurrences of larger patterns to spread over larger intervals of time not directly bounded by a maximum window size. This constraint is similar to the maximum gap constraint handled by algorithms proposed to find frequent sequential patterns in a base of sequences (e.g., [10, 11, 4]). A base of sequences is a large collection of sequences where each sequence is rather small, and the algorithms developed to mine such bases cannot be reused to extract episodes in a single large event sequence, because the notion of frequency of a pattern is very different in these two contexts. In a base of sequences the frequency of a pattern corresponds to the number of sequences in which the pattern occurs at least one time, and several occurrences of the pattern in the same sequence have no impact on its frequency. While in the case of an episode in an event sequence, the frequency represents the number of occurrences of the pattern in this sequence.

Thus [2] has proposed a new algorithm, but as we will show in Section 2.3 this algorithm is not complete. However, the contribution of [2] remains interesting because this work suggests that mining episode rules in practice could be done using a maximum gap constraint.

In this paper our contribution is twofold. Firstly, we present a sound and complete algorithm to extract episode rules satisfying frequency, confidence and maximum gap constraints. And secondly, we propose a way to find if it exists, for each rule, the smallest window size that corresponds to a local maximum of confidence for the rule (i.e., confidence is locally lower, for smaller and larger windows).

From a quantitative point of view, we present experiments showing that mining episode rules under the maximum gap constraint and finding the local maximums of confidence can be done in practice at reasonable extraction thresholds. From a qualitative point of view, these experiments advocated the fact that local maximums of confidence can be interesting suggestions of possible dependencies to the expert, because no local maximum have been found on synthetic random datasets, while they exist in real data. Finally, the experiments show that the number of rules satisfying the frequency and confidence constraints and having a local maximum of confidence is orders of magnitude lesser than the number of rules satisfying the frequency and confidence constraints only. So, in practice the expert has to browse only a very limited collection of extracted patterns.

This paper is organized as follows. The next section gives preliminary definitions and shows that the algorithm presented in [2] is incomplete. Section 3 introduces the algorithm *WinMiner* that handles the maximum gap constraints and finds the local maximums of confidence of episode rules. Section 4 presents experiments performed, and we conclude with a summary in Section 5.

2 Episode Rules and Local Maximum of Confidence

2.1 Preliminary Definitions

In this section we follow the standard notions of event sequence, episode, minimal occurrences and support used in [6] or give equivalent definition, when more appropriated to our presentation. The only noticeable difference is that our notion of occurrence incorporates the necessity for the occurrences to satisfy a maximum gap constraint.

Definition 1. ***(event, ordered sequence of events)*** Let E be a set of *event types*. An *event* is defined by the pair (e, t) where $e \in E$ and $t \in \mathbb{N}$. The value t denotes the time at which the *event* occurs. An *ordered sequence of events* s is a tuple $s = \langle(e_1, t_1), (e_2, t_2), ..., (e_n, t_n)\rangle$ such that $\forall i \in \{1, ..., n\}, e_i \in E \wedge t_i \in \mathbb{N}$ and $\forall i \in \{1, ..., n-1\}, t_i \leq t_{i+1}$.

Definition 2. ***(operator $\sqsubseteq$)*** Let α and β be two ordered sequences of events, then α is a subsequence of β, denoted $\alpha \sqsubseteq \beta$ iff α can be obtained by removing some elements of β or $\alpha = \beta$.

Definition 3. ***(event sequence)*** An *event sequence* S is a triple (s, T_s, T_e), where s is an ordered sequence of events of the form $\langle(e_1, t_1), (e_2, t_2), ..., (e_n, t_n)\rangle$ and T_s, T_e are natural numbers such that $T_s \leq t_1 \leq t_n \leq T_e$.

T_s and T_e respectively represent the starting time and the ending time of the event sequence. Notice that t_1 may differ from T_s and that t_n may differ from T_e.

Definition 4. ***(episode)*** An *episode* is a tuple α of the form $\alpha = \langle e_1, e_2, \ldots, e_k\rangle$ with $e_i \in E$ for all $i \in \{1, \ldots, k\}$. In this paper, we will use the notation $e_1 \rightarrow e_2 \rightarrow ... \rightarrow e_k$ to denote the episode $\langle e_1, e_2, ..., e_k\rangle$ where '$\rightarrow$' may be read as 'followed by'. We denote the empty episode by $\emptyset$.

Definition 5. ***(size, suffix and prefix of an episode)*** Let $\alpha = \langle e_1, e_2, ..., e_k\rangle$ be an episode. The *size* of α is denoted $|\alpha|$ and is equal to the number of elements of the tuple α, i.e., $|\alpha| = k$. The *suffix* of α is defined as an episode composed only by the last element of the tuple α, i.e., $suffix(\alpha) = \langle e_k\rangle$. The *prefix* of α is the episode $\langle e_1, e_2, ..., e_{k-1}\rangle$. We denote it as $prefix(\alpha)$.

Definition 6. ***(occurrence)*** An episode $\alpha = \langle e_1, e_2, ..., e_k\rangle$ *occurs* in an event sequence $S = (s, T_s, T_e)$ if there exists at least one *ordered sequence of events* $s' = \langle (e_1, t_1), (e_2, t_2), ..., (e_k, t_k)\rangle$ such that $s' \sqsubseteq s$ and $\forall i \in \{1, ..., k-1\}, 0 < t_{i+1} - t_i \leq gapmax$ with *gapmax* a user-defined threshold that represents the maximum time gap allowed between two consecutive events.

The interval $[t_1, t_k]$ is called an *occurrence* of α in S. The set of all the occurrences of α in S is denoted by $occ(\alpha, S)$.

These episodes and their occurrences correspond to the *serial* episodes of [6], up to the following restriction: the event types of an episode must occur at different time stamps in the event sequence. This restriction is imposed here for the sake of simplicity, and the definitions and algorithms can be extended to allow several event types to appear at the same time stamp. However, it should be noticed that this constraint applies on occurrences of the patterns, and not on the dataset (i.e., several events can occur at the same time stamp in the event sequence).

Definition 7. ***(minimal occurrence)*** Let $[t_s, t_e]$ be an occurrence of an episode α in the event sequence S. If there is no other occurrence $[t'_s, t'_e]$ such that $(t_s < t'_s \wedge t'_e \leq t_e) \vee (t_s \leq t'_s \wedge t'_e < t_e)$ (i.e., $[t'_s, t'_e] \subset [t_s, t_e]$), then the interval $[t_s, t_e]$ is called a *minimal occurrence* of α. The set of all minimal occurrences of α in S is denoted by $mo(\alpha, S)$.

Intuitively, a minimal occurrence is simply an occurrence that does not contain another occurrence of the same episode.

Definition 8. ***(width of an occurrence)*** Let $o = [t_s, t_e]$ be an occurrence. The time span $t_e - t_s$ is called the *width* of the occurrence o . We denote it as $width(o)$. The set of all occurrences (resp. minimal occurrences) of an episode α in an event sequence S having a width equal to w is denoted $occ(\alpha, S, w)$ (resp. $mo(\alpha, S, w)$).

Definition 9. ***(support of an episode)*** The *support* of an episode α in an event sequence S for a width w is defined as $Support(\alpha, S, w) = \sum_{0 \leq i \leq w} \mid mo(\alpha, S, i) \mid$. We also define the *general support* of α in S as $GSupport(\alpha, S) = \mid mo(\alpha, S) \mid$

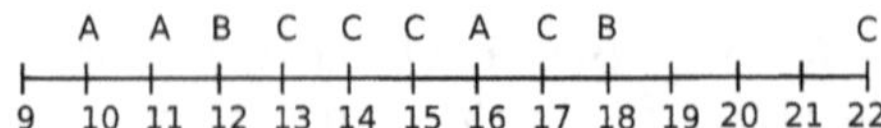

Fig. 1. Example of event sequence.

The notions of occurrence and support of an episode incorporate the satisfaction of the maximum gap constraint. However, for the sake of simplicity, the *gapmax* parameter does not appear explicitely in the notational conventions $mo(\alpha, S, i), Support(\alpha, S, w), GSupport(\alpha, S)$ and $mo(\alpha, S)$.

To illustrate some of the previous definitions, we consider the event sequence $S = (w, T_s, T_e)$ of Figure 1.

In this example, $T_s = 9$, $T_e = 22$, $w = \langle (A, 10), (A, 11), (B, 12), (C, 13), (C, 14), (C, 15), (A, 16), (C, 17), (B, 18), (C, 22)\rangle$. If we consider the episode $\alpha = A \rightarrow B$, and the constraint $gapmax = 3$, then $occ(\alpha, S) = \{[10, 12], [11, 12], [16, 18]\}$. It should be noticed that $[10, 18]$ does not belong to $occ(\alpha, S)$ since the constraint *gapmax* is not satisfied. The minimal occurrences of α are $mo(\alpha, S) = \{[11, 12], [16, 18]\}$. The occurrence $[10, 12]$ does not belong to $mo(\alpha, S)$ because it contains the occurrence $[11, 12]$. In the same way, in the case of episode $\beta = A \rightarrow B \rightarrow C$, we have $occ(\beta, S) = \{[10, 13], [10, 14], [10, 15], [11, 13], [11, 14], [11, 15]\}$ and $mo(\beta, S) = \{[11, 13]\}$. Some examples of support values are $GSupport(\alpha, S) = 2$, $Support(\alpha, S, 1) = 1$, $Support(\alpha, S, 2) = 2$ and $Support(\beta, S, 2) = 1$. It should be noticed that the support increases in a monotonic way, with respect to the width value.

2.2 Episode Rule and Local Maximum of Confidence

Definition 10. ***(episode rule)*** Let α and β be episodes such that $prefix(\beta) = \alpha$. An *episode rule* built on α and β is the expression $\alpha \Rightarrow suffix(\beta)$.

For example, if $\alpha = e_1 \rightarrow e_2$ and $\beta = e_1 \rightarrow e_2 \rightarrow e_3$, the corresponding episode rule is denoted $e_1 \rightarrow e_2 \Rightarrow e_3$. It should be noticed that the episode rules used in this paper are restricted to rules having a single event type in their right hand sides, but that the definitions and algorithms proposed can be extended in the case of right hand sides containing several event types.

Definition 11. ***(support and confidence of an episode rule)*** The *support* of an episode rule is defined by $Support(\alpha \Rightarrow suffix(\beta), S, w) = Support(\beta, S, w)$. The *confidence* of an episode rule is defined as follows:

$$Confidence(\alpha \Rightarrow suffix(\beta), S, w) = \frac{Support(\beta,S,w)}{Support(\alpha,S,w)}$$

Let γ be a user defined confidence threshold such that $0 \leq \gamma \leq 1$, and let σ be a user defined support threshold such that $0 < \sigma \leq 1$. Then, if $Confidence(r, S, w) \geq \gamma$ (resp. $Support(r, S, w) \geq \sigma$) the rule is said to be *confident* (resp. *frequent*) for the width w.

It should be noticed that, as for episodes, the support and confidence are defined with respect to a given width. The definition of confidence can be illustrated by the previous example (Figure 1). Knowing that $mo(A \rightarrow B, S) =$

$\{[11,12],[16,18]\}$ and $mo(A \rightarrow B \rightarrow C, S) = \{[11,13]\}$, we have $Confidence(A \rightarrow B \Rightarrow C, S, 2) = 1/2$.

Definition 12. ***(LM and FLM)*** A rule r is said to have a LM (*Local Maximum*) for a given width i on event sequence S iff the three following properties are satisfied:

- $Confidence(r,S,i) \geq \gamma \wedge Support(r,S,i) \geq \sigma$
- $\forall j,\ j < i \ \wedge \ Support(r,S,j) \geq \sigma \ \Rightarrow \ Confidence(r,S,i) > Confidence(r,S,j)$
- $\exists j,\ i < j \ \wedge \ Confidence(r,S,j) \leq Confidence(r,S,i) - (decRate{*}Confidence(r,S,i))$ with *decRate* a decrease threshold defined by the user, and $\forall\ k,\ i < k < j \ \Rightarrow \ Confidence(r,S,k) \leq Confidence(r,S,i)$

The rule r has a FLM(*First Local Maximum*) for width i iff r has a LM for width i, and r has no LM for width strictly lesser than i. A rule having at least one LM, and thus also a (single) FLM, is called a $FLM-rule$.

Intuitively, a LM for a rule r is a width w such that (1) r is frequent and confident for this width, (2) all lower width values such that r is frequent correspond to a strictly lower confidence, and (3) the next consecutive greater width values correspond also to a lower confidence until the confidence becomes lower than a given percentage (*decRate*) of confidence obtained for w. The figure 2 illustrates, for a given rule, possible variations of confidence with respect to the width values and the corresponding FLM are represented by a dot. The vertical axis represents the confidence of the rule and the horizontal dashed line indicates the confidence threshold γ. The horizontal axis represents the width values, and the vertical dashed line corresponds to a specific width, denoted w_σ, that is the width at which the rule turns out to be frequent. Two particular situations should be pointed out. Firstly, the bottom-left graphic where there is no FLM. And secondly, the bottom-right graphic, where the three first local maximums are not valid LM because there are not followed by a sufficient decrease of confidence.

2.3 Incompleteness of [2]

In [2] an algorithm has been proposed to extract frequent and confident episode rules under a maximum gap constraint, but the formal correctness of the algorithm has not been established. Indeed, we have remarked that the algorithm is not complete. To show this, we need to recall the definition of *window* of [6] : *"A window on an event sequence $S = (s, T_s, T_e)$ is an event sequence $W = (w, t_s, t_e)$ where $t_s < T_e$ and $t_e > T_s$, and w consists of those pairs (A,t) from s where $t_s \leq t < t_e$".*

Let $\mathcal{W}(S, win)$ be the set of all windows (w, t_s, t_e) on S such that $t_e - t_s = win$. Then [2] defines by $fr(\alpha) = \frac{|\{W \in \mathcal{W}(S,win) | \alpha \text{ occurs in } W\}|}{|\mathcal{W}(S,win)|}$ where $win = (|\alpha| - 1) \times gapmax$, the frequency of an episode α under a maximum gap constraint. In [2] the episode $A \rightarrow B \rightarrow C$ is considered only if the episodes $A \rightarrow B$

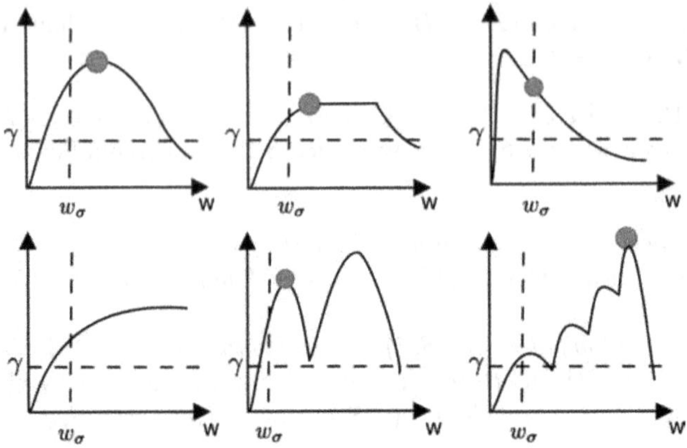

Fig. 2. Confidence vs width.

and $B \to C$ are frequent. Let S be the event sequence $(\langle(A,5),(B,8),(C,9)\rangle,1,10)$ and $gapmax = 4$. Then $fr(A \to B) = 1/12$ and $fr(A \to B \to C) = 4/16 = 1/4$. If the frequency threshold is $1/4$, then $A \to B$ is not frequent and thus $A \to B \to C$ is not considered by the algorithm even though it is frequent.

3 Extraction of All FLM-Rules

The proofs of the theorems and lemmas of this section are given in the extended version of this paper [9].

To extract the FLM-rules we propose an algorithm, called *WinMiner*, based on the so-called *occurrence list* approach, used in [6] and also in the context of mining sequential patterns in base of sequences (e.g., [11, 4]). The key idea of such methods is to store and use occurrences of patterns to compute the occurrences of longer patterns by means of a kind of temporal join operation. In our case, if we only keep the information about the location of all the minimal occurrences of the episodes of size k, it is not possible to produce all the minimal occurrences of the episodes of size $k+1$. Looking back to the example given Figure 1, we can see that there exists a minimal occurrence $[11, 18]$ of $A \to B \to C \to B$. If we only know $mo(A \to B \to C, S) = \{[11, 13]\}$ and $mo(B, S) = \{[12, 12], [18, 18]\}$, we can not determine the minimal occurrence of $A \to B \to C \to B$. This is because the first B occurs too early (before time stamp 18) and the second one occurs too late (the maximum gap constraint of 3 is not satisfied between time stamps 13 and 18). To overcome this difficulty, *WinMiner* is based on a notion of *minimal prefix occurrence* introduced in the next section.

3.1 Minimal Prefix Occurrence

Definition 13. ***(minimal prefix occurrence)*** Let $o = [t_s, t_e]$ be an occurrence of an episode α in the event sequence S, then o is a *minimal prefix occur-*

rence (*mpo*) of α iff $\forall$ $[t_1, t_2] \in mo(prefix(\alpha), S)$, if $t_s < t_1$ then $t_e \leq t_2$. We denote by $mpo(\alpha, S)$ the set of all *mpo* of α in S.

It is important to notice that $mpo(\alpha, S)$ is defined with respect to $mo(prefix(\alpha), S)$ and not with respect to $mo(\alpha, S)$. In the example depicted Figure 1, we have $mpo(A \rightarrow B \rightarrow C, S) = \{[11,13],[11,14],[11,15]\}$ and $mpo(B, S) = \{[12,12],[18,18]\}$. Then, using these sets it is possible to built $mpo(A \rightarrow B \rightarrow C \rightarrow B, S) = \{[11,18]\}$. As it can be intuitively noticed, the minimal occurrences are particular *mpo* and the minimal occurrences can be determined using the set of *mpo*. This is formally stated by the two following lemmas:

Lemma 1. *If* $[t_s, t_e] \in mo(\alpha, S)$ *then* $[t_s, t_e] \in mpo(\alpha, S)$ *and there is no* $[t_s, t'_e] \in mpo(\alpha, S)$ *such that* $t'_e < t_e$.

Lemma 2. *If* $[t_s, t_e] \in mpo(\alpha, S)$ *and there is no* $[t_s, t'_e] \in mpo(\alpha, S)$ *such that* $t'_e < t_e$, *then* $[t_s, t_e] \in mo(\alpha, S)$.

E/O-pair. The algorithm *WinMiner* handles an episode α and its *mpo* in a pair of the form (*episode, occurrences*) called E/O-pair. For a E/O-pair x, we denote respectively *x.Pattern* and *x.Occ*, the first and second element of the pair. The *x.Pattern* part represents the episode itself and *x.Occ* contains its *mpo* in a compact way. The *x.Occ* part is a set of pairs of the form (*Tbeg, TendSet*) where *Tbeg* represents a *mpo* starting time and *TendSet* is the set of the ending times of all *mpo* of *x.Pattern* starting at *Tbeg*. Intuitively, the interest of this representation is that, according to lemma 2, if we consider a pair (*Tbeg, TendSet*) then the interval [*Tbeg, min*(*TendSet*)] represents a minimal occurrence of *x.Pattern*.

3.2 Algorithm *WinMiner*

This algorithm extract all $FLM-rules$ in a event sequence S, according to the following user-specified parameters: a support threshold σ, a confidence threshold γ, a maximum time gap constraint *gapmax* and a decrease threshold *decRate*.

The algorithm is presented as Algorithm 1. First, it computes the *mpo* of all frequent episodes of size 1, using a function named *scan*, that is not detailed in this paper, but that simply determines the *mpo* of an episode by scanning S. The algorithm then calls the function *exploreLevelN* (Algorithm 2) to find in a depth-first way all episodes (of size greater than 1) such that their *GSupport* are greater or equal to the threshold σ. For a given episode *x.Pattern* this function extends the pattern on the right side with a frequent episode *y.Pattern* of size 1. This is performed by a call to the *join* function (Algorithm 3), that computes the new pattern *z.Pattern* = *x.Pattern* $\rightarrow$ *y.Pattern*, and also the corresponding set of *mpo* in *z.Occ* using the *mpo* of *x.Pattern* and *y.Pattern* stored respectively in *x.Occ* and *y.Occ*. The core part of this join operation is the line 6, where it checks that the new interval generated ($[t_s, t'_s]$) is an occurrence of *z.Pattern* satisfying the maximum gap constraint (condition $t'_s > t_s \wedge t'_s - t \leq gapmax$) and that this interval is a *mpo* of *z.Pattern* (condition $\forall (t_1, T'') \in x.Occ\ t_s < t_1 \Rightarrow \forall t_2 \in T'', t'_s \leq t_2$). Then, after the call to *join*, if the *GSupport* of $z.Pattern$ is greater or equal to σ, the Algorithm 2 determines, if it exists, the FLM of the rule built from the prefix and suffix of $z.Pattern$, by means of the function $findFLM$. Finally, the algorithm 2 considers, in a recursive way, the episodes that can be obtained by adding frequent episodes of size 1 to $z.Pattern$ itself.

Algorithm 1 (*WinMiner*)
Input: *S an event sequence S and E the set of event types.*

let $L_1 := \emptyset$
for all $e \in E$ **do**
    let $x.Pattern := e$
    let $x.Occ := scan(S, e)$
    if $\mid x.Occ \mid >= \sigma$
        let $L_1 := L_1 \cup \{x\}$
    fi
od
for all $x \in L_1$ **do**
    $exploreLevelN(x, L_1)$
fi

Algorithm 2 ($exploreLevelN$)
Input: *x a E/O-pair, and L_1 the set of E/O-pairs of frequent episodes of size 1.*

for all $y \in L_1$ **do**
    let $z := join(x, y)$
    if $|z.Occ| >= \sigma$
        $findFLM(x.Pattern \Rightarrow suffix(z.Pattern), x.Occ, z.Occ)$
        $exploreLevelN(z, L_1)$
    fi
od

Algorithm 3 ($join$) Input: *x and y, two E/O-pairs, containing an episode and its set of mpo, and where y corresponds to an episode of size 1.*

Output: *z, a E/O-pair containing the episode $x.Pattern \rightarrow y.Pattern$ and its set of mpo.*

let $z.Pattern := x.Pattern \rightarrow y.Pattern$
let $z.Occ := \emptyset$
for all $(t_s, T) \in x.Occ$ **do**
  let $L := \emptyset$
    for all $t \in T$ **do**
        let $EndingTimes := \{t'_s \mid \exists(t'_s, T') \in y.Occ \text{ such that } t'_s > t_s \wedge t'_s - t \leq gapmax \wedge \forall(t_1, T'') \in x.Occ, t_s < t_1 \Rightarrow \forall t_2 \in T'', t'_s \leq t_2\}$
        let $L := L \cup EndingTimes$
    od
    if $L \neq \emptyset$
        let $z.Occ := z.Occ \cup \{(t_s, L)\}$
    fi
od

Let us now consider the correctness of the approach.

Definition 14. Let S be an event sequence, then

- a E/O-pair x is sound iff $\forall(t_s, T) \in x.Occ, \forall t \in T$, $[t_s, t] \in mpo(x.Pattern, S)$.

- a E/O-pair x is complete iff $\forall [t_s, t_e] \in mpo(x.Pattern, S)$, $\exists (t_s, T) \in x.Occ$ s.t. $t_e \in T$.
- a E/O-pair x is non-redundant iff $\forall (t_s, T) \in x.Occ, \nexists (t_s, T') \in x.Occ$ s.t. $T \neq T'$.

The following theorem states the correctness of the function *join* (Algorithm 3):

Theorem 1 (correctness of *join*). *If x.Occ and y.Occ in the input of join are sound, complete and non-redundant, then z.Occ in the output is sound, complete and non-redundant.*

Theorem 2 (correctness of support counting). *Let S be an event sequence and z be a E/O-pair outputted by Algorithm 3, for sound, complete and non-redundant x.Occ and y.Occ in the input, then:*

- *the number of minimal occurrences of z.Pattern of width w is* $|mo(z.Pattern, S, w)| = |\{(t_s, T) \in z.Occ | min(T) - t_s = w\}|$
- $Support(z.Pattern, S, w) = \sum_{0 \leq i \leq w} |\{(t_s, T) \in z.Occ | min(T) - t_s = i\}|$
- $GSupport(\alpha, S) = |\{(t_s, T) \in z.Occ\}|$

Since $GSupport(\alpha, S) \geq \sigma \Rightarrow GSupport(prefix(\alpha), S) \geq \sigma$, then by the theorems 1 and 2, the depth-first enumeration of *WinMiner* is correct to find all episodes such that *GSupport* is greater or equal to σ. Furthermore, if an episode rule $prefix(\alpha) \Rightarrow suffix(\alpha)$ is frequent for a given width w then we also have $GSupport(\alpha, S) \geq \sigma$. Thus, by the theorems 1 and 2, we can also correctly find the support and confidence of a rule for a given width w, when the rule is frequent for this w.

So, we have at hand all the information necessary to determine if a rule such that $GSupport \geq \sigma$ is a FLM-rule and the width corresponding to its FLM. Due to space limitation, the corresponding algorithm (function *findFLM*) is not presented here, but can be found in the extended version of this paper [9].

4 Experiments

In this section we present experiments on a real dataset, using an implementation of *WinMiner* in C++, and performed on an Intel Pentium IV 2 GHz under a 2.4 Linux kernel (all the experiments were run using between 0.5 and 300 MB of RAM). Efficient implementation hints are given in [9]. Experiments on large random datasets are also presented in [9], and are not described here because of space limitation. These experiments show that the extractions can be done in practice in non-trivial cases and that no FLM-rule was found in these random datasets. Other experiments on atherosclerosis risk factors (atherosclerosis is the main cause of cardio-vascular diseases) are described in [8].

The experiments reported here were performed within the European Project AEGIS (IST-2000-26450) in collaboration with geophysicists to help them to

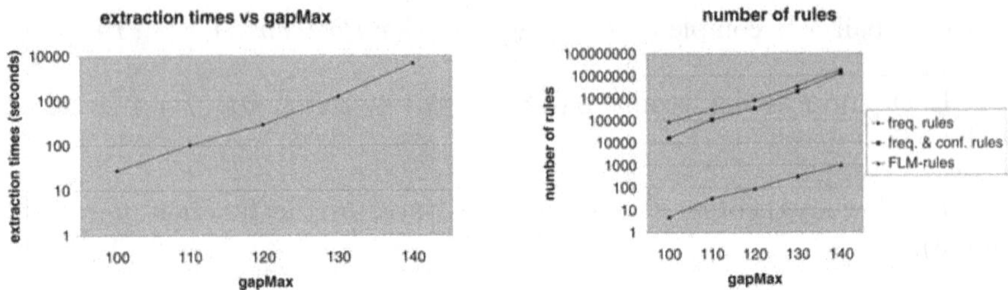

Fig. 3. Experiments on a seismic dataset.

find dependencies between earthquakes. In this paper, we only present experiments on a subset of the ANSS Composite Earthquake Catalog[1], that contains a series of earthquakes described by their locations, occurrence times and magnitudes. As the FLM-rules obtained in these experiments have suggested to the geophysicists some possible dependencies that are not at that time published in the geophysics literature, we cannot give the precise magnitudes and locations of the earthquakes considered in this subset of the catalog. After an appropriated discretization[2] the resulting dataset was built on 368 event types and contained 3509 events spread over 14504 time units of one day (about 40 years).

For the sake of conciseness, we only present, in Figure 3, extractions performed with $\sigma = 10$, $\gamma = 0.9$, $decRate = 30\%$ and $gapmax$ ranging from 100 to 140. The left graphic indicates that the running time is reasonable in practice (from less than 100 seconds to about 6700 seconds for the largest one). The right graphic presents, for each of the experiments, (1) the number of rules considered by *WinMiner* during the extraction (rules such that $GSupport \geq \sigma$), (2) the number of rules for which there exists a width w such that the rule is frequent and confident for w, and finally, (3) the number of FLM-rules. This graphic shows in particular that the collection of rules that are frequent and confident (for some width) is too huge to be handled by an expert, while the size of the collection of FLM-rules is several orders of magnitude smaller.

5 Conclusion

In this paper we presented a sound and complete algorithm to extract episode rules satisfying a maximum gap constraint. We also proposed to determine the window sizes corresponding to a local maximum of confidence. The experiments showed that extracting episode rules under the maximum gap constraint and finding the window sizes leading to a local maximum of confidence, can be efficiently performed in practice. Furthermore, no local maximum has been found

[1] Public world-wide earthquake catalog available at http://quake.geo.berkeley.edu/cnss/

[2] This preprocessing has been performed by Francesco Pacchiani of Laboratoire de Geologie at Ecole Normale Supérieure of Paris in the context of the AEGIS project.

on random datasets, while meaningful dependencies corresponding to local maximum of confidence have been found in a real seismic dataset.

References

1. R. Agrawal, T. Imielinski, and A. Swami. Mining association rules between sets of items in large databases. In P. Buneman and S. Jajodia, editors, *Proc. of the Int. Conf. SIGMOD'93*, pages 207–216, Washington D.C., USA, May 1993.
2. G. Casas-Garriga. Discovering unbounded episodes in sequential data. In *Proc. of the Int. Conf. PKDD'03*, pages 83–94, Croatia, September 2003. LNCS 2838.
3. K. Hatonen, M. Klemettinen, H. Mannila, P. Ronkainen, and H. Toivonen. Tasa: Telecomunications alarm sequence analyzer or: How to enjoy faults in your network. In *Int. Symp. NOMS'96*, pages 520–529, Kyoto, Japan, April 1996.
4. M. Leleu, C. Rigotti, J.-F. Boulicaut, and G. Euvrard. Constrained-based mining of sequential patterns over datasets with consecutive repetitions. In *Proc. of the Int. Conf. PKDD'03*, pages 303–314, Croatia, September 2003. LNCS 2838.
5. H. Mannila and H. Toivonen. Discovery of generalized episodes using minimal occurrences. In *Proc. of the 2nd Int. Conf. KDD'96*, pages 146–151, Portland, Oregon, August 1996.
6. H. Mannila, H. Toivonen, and A. Verkamo. Discovery of frequent episodes in event sequences. *Data Mining and Knowledge Discovery*, 1(3):259–298, November 1997.
7. H. Mannila, H. Toivonen, and I. Verkamo. Discovering frequent episodes in sequences. In *Proc. of the 1st Int. Conf. KDD'95*, pages 210–215, Canada, August 1995.
8. N. Méger, C. Leschi, N. Lucas, and C. Rigotti. Mining episode rules in STULONG dataset. Technical report, LIRIS Lab, Lyon, France, June 2004.
9. N. Méger and C. Rigotti. Constraint-based mining of episode rules and optimal window sizes. Technical report, LIRIS Lab, Lyon, France, June 2004.
10. R. Srikant and R. Agrawal. Mining sequential patterns: Generalizations and performance improvements. In *Proc. of the 5th Int. Conf. EDBT'96*, pages 3–17, Avignon, France, September 1996.
11. M. Zaki. Sequence mining in categorical domains: incorporating constraints. In *Proc. of the 9th Int. Conf. on CIKM'00*, pages 422–429, USA, November 2000.

Analysing Customer Churn in Insurance Data – A Case Study

Katharina Morik and Hanna Köpcke

Univ. Dortmund, Computer Science Department, LS VIII
morik@ls8.informatik.uni-dortmund.de
http://www-ai.cs.uni-dortmund.de

Abstract. Designing a new application of knowledge discovery is a very tedious task. The success is determined to a great extent by an adequate example representation. The transformation of given data to the example representation is a matter of feature generation and selection. The search for an appropriate approach is difficult. In particular, if time data are involved, there exist a large variety of how to handle them. Reports on successful cases can provide case designers with a guideline for the design of new, similar cases. In this paper we present a complete knowledge discovery process applied to insurance data. We use the TF/IDF representation from information retrieval for compiling time-related features of the data set. Experimental reasults show that these new features lead to superior results in terms of accuracy, precision and recall. A heuristic is given which calculates how much the feature space is enlarged or shrinked by the transformation to TF/IDF.

Keywords: preprocessing for KDD, insurance data analysis, customer relationship management, time-stamped data

1 Introduction

Insurance companies collect very large data sets, storing very many attributes for each policy they underwrite. The statistical analysis of their data has a long tradition ranging from the analysis of catastrophe insurance [1] and property and casuality insurance [2] to health insurance [3, 4]. The statistical methods are regularly used in order to model the credibility of customers, the amount of a single payment claimed, and the number of payments within a time span [5]. More recently, insurance companies also ask for data analysis in the context of their customer relationship management [6]. Direct marketing for cross- and up-selling is one of the current goals for data analysis. A special case is churn prediction, i.e. the prediction of policy termination before the end date. If those groups of customers or policies can be detected where the risk of churn is high, particular marketing actions can be pursued in order to keep the customers. Given this goal, it is not sufficient to model the distribution of churn or its overall likelihood, but policies or customers at risk should be actually identified. Then, the insurance salesmen can contact them. Along with the recent trend

J.-F. Boulicaut et al. (Eds.): PKDD 2004, LNAI 3202, pp. 325–336, 2004.

of customer relationship management, insurance companies move beyond statistical analysis. Knowledge discovery in insurance databases now builds upon datawarehousing projects in insurance companies (see, e.g., [7]).

In this paper, we want to present a knowledge discovery case whose solution was hard to find. We were provided with time-stamped data from the Swiss Life insurance company. There are many ways to handle time-related data, e.g., [8–13]. It is hard to select the appropriate approach [14]. The first question is, whether we actually need to handle the time information. There are cases, where the particular dates do not offer any additional information for the data analysis. A snapshot of the current state is sufficient for the prediction of the next state. In our case, ignoring the history of contracts did not succeed. Hence, we needed to take into account the time information. Since time was not given by equidistant events as in time series but by time-stamped changes to a contract, a promising approach is to learn event sequences. Another approach which has shown advantages in some knowledge discovery cases is the compilation of the time information into features. There are several compilation methods (e.g. windowing, single value decomposition [15]). We used the TF/IDF representation [16] from Information Retrieval to compile the time information into features. By the presentation of this case study we hope to offer a guideline for similar cases. Insurance is an important sector and, hence, many similar cases should exist. Moreover, we have turned the analysis into an efficient heuristic which applies to the raw data and estimates the size of the feature space after the transformation into TF/IDF attributes. The paper is organized as follows. First, we describe the insurance application in Section 2. Second, in Section 3 we describe the first experiments focusing on handling the history of contracts. Third, we describe the successful case in Section 4.1 and explain the effect of TF/IDF features to the size of the data set. We conclude by a proposal to gather successful cases of knowledge discovery at an abstract level and discuss related work in Section 5.

2 The Insurance Application

In the course of enhanced customer relationship management, the Swiss Life insurance company investigated opportunities for direct marketing [17]. A more difficult task was to predict churn in terms of a customer buying back his life insurance. Internal studies at the insurance company found that for some attributes the likelihood of churn differed significantly from the overall likelihood. However, these shifts of probabilities cannot be used for classification. Hence, we worked on knowledge discovery for the classification into early termination or continuation of policies. The discovered knowledge then selects customers at risk for further marketing actions and provides a basis to calculate the financial deposits needed in order to re-buy policies.

2.1 The Data

For the knowledge discovery task of churn prediction we received an anonymised excerpt of the data-warehouse of Swiss Life. The database excerpt consists of

12 tables with 15 relations between them. The tables contain information about 217,586 policies and 163,745 customers. The contracts belong to five kinds of insurances: life insurance, pension insurance, health insurance, incapacitation insurance, and funds bounded insurance. Every policy consists of average of 2 components. The table of policies has 23 columns and 1,469,978 rows. The table of components has 31 columns and 2,194,825 rows. Three additional tables store details of the components. The policies and components tables are linked indirectly by an additional table. If all records referring to the same policy and component (but at a different status at different times) are counted as one, there are 533,175 components described in the database. Concerning our prediction task, we can characterize the data in the following way:

Skewed data: Churn is only observed in 7.7% of the policies. Hence, we have the problem of skewed data [18].

High-dimensional data: Overall there are 118 attributes. If for the nominal attributes, their values would be transformed into additional Boolean attributes, we would have 2,181,401 attributes. In such a high-dimensional space visual data inspection or regular statistical methods fail. Even the MYSVM which is well suited for high-dimensional data cannot handle this number of attributes.

Sparse data: If the attribute values would be transformed into Boolean attributes, many of the attribute values would be zero.

Homogeneous accross classes: Those attribute values occurring frequently do so in the churn class as well as in the regular class.

All these characteristics contribute to the hardness of the knowledge discovery task. It becomes clear that the primary challenge lies in the mapping of the raw data into a feature space which allows an algorithm to learn. The feature space should be smaller than the original space, but should still offer the distinctions between the two classes, early termination of the contract and continuation. Finding the appropriate representation becomes even harder because of the time-stamps in the database. Each policy and each component may be changed throughout the period of a policy. For every change of a policy or a component, there is an entry in the policy table with the new values of the features, a unique mutation number, a code representing the reason of change, and the date of change. This means, that several rows of the policy table describe the history of the same policy. Each policy is on average changed 6 times, each component on average 4 times. Figure 1 shows the columns of the policy table which represent the time-stamps. The attribute VVID is the key identifying a contract. The attribute VVAENDNR is the unique number for a change. The attribute VVAENDART represents the reason of change. The attribute VVAENDDAT gives the date of change. As mentioned above, there are three alternative approaches to handling time-stamped data. The first choice that is successful quite often, is to ignore the time information. In our case, this means to select for each contract the row with the latest date (Section 3.1). The second choice is to explicitly model the sequential structures. In our case, this means that for each attribute of a contract, the begin date and the end date of a particular attribute

VVID	VVAENDNR	VVWIVON	VVWIBIS	VVAENDDAT	VVAENDART	...
16423	1	1946	1998	1946	1000	
16423	2	1998	1998	1998	27	
16423	3	1998	1998	1998	4	
16423	4	1998	1998	1998	54	
16423	5	1998	1998	1998	4	
16423	6	1998	9999	1998	61	
5016	1	1997	1999	1997	33	
5016	2	1999	2001	1999	33	
5016	3	2001	2001	2001	33	
5016	4	2001	2001	2001	33	
5016	5	2001	2002	2001	81	
5016	6	2002	9999	2002	94	
...						

history of a contract (rows with VVID 16423)

history of another contract (rows with VVID 5016)

Fig. 1. Extract of the policy table

value form a time interval (Section 3.2). The third choice is to compile the time information into the representation. Here, we counted for each attribute how often its value changed within one contract (Section 4).

3 First Experiments

3.1 Predicting Churn Without Time Information

Feature selection from the given database attributes is hardly obtained without reasoning about the application domain. Our first hypothesis was, that data about the customers could indicate the probability of contract termination. In this case, the changes of the contract components can be ignored. Therefore, we applied decision tree learning and MYSVM to customer data, sampling equally many customers who continued their policy and those who re-bought it[1]. 10 attributes from the customer tables were selected and a Boolean attribute *churn* was generated from the raw data. The resulting set of eleven attributes was transformed into the input formats of the learning algorithms. Decision tree learning achieved a precision of 57% and a recall of 80%. MYSVM obtained a precision of 11% and a recall of 57% with its best parameter setting (radial kernel) [21]. Trying association rule learning with the conclusion fixed to churn, did deliver correlated attributes. However, these were the same correlations that could be found for all customers [21]. The description of customers in the database does not entail the relevant information for predicting early contract termination. Changes in a customer's situation, e.g., buying a house, marriage, or child birth

[1] For decision tree and association rule learning we used WEKA [19], for support vector machine we used MYSVM [20].

is not stored. These events can only indirectly be observed by changes of the policy or its components.

In a second experiment we focused on the components table. From the 31 database attributes of the components table, 7 attributes which are just foreign keys, or code numbers, or the start and end date of a component state were ignored. 24 attributes of the components table were combined with a year (1960 - 2002), stating when this attribute was changed. Some of the 1008 combinations did not occur. The resulting table of 990 columns plus the target attribute *churn* shows in each row the complete set of changes of a component and whether the corresponding policy was re-bought, or not. Learning association rules with the conclusion fixed to churn clearly showed the peak of changes at 1998 where the Swiss law changed and many customers re-bought their contracts. Other changes were just the procedure of contract termination, such as finishing the component and the payment. Using mySVM , 44% precision and 87% recall were achieved using a linear kernel. These results show either that the data do not entail relevant information for churn prediction, or the representation prepared for learning was not well chosen. The first experiments were sufficient, however, to select only attributes that describe policies and no longer search within the customer data for reasons of early policy termination.

3.2 Predicting Churn on the Basis of Time Intervals

Taking into account the time aspect of the contract changes was considered an opportunity to overcome the rather disappointing results from previous experiments. There, time was just part of an attribute name. Now, we represented time explicitly. The time stamps of changes were used to create time intervals during which a particular version of the policy was valid. Relations between the time intervals were formulated using Allen's temporal relations [22]. Following the approach of Höppner [23] who applies the Apriori algorithm to one windowed time series [24], a modification to sets of time series has been implemented [25]. For each kind of an insurance, association rules about time intervals and their relations were learned according to two versions of time. The first version handles the actual dates, finding (again) that according to a change of Swiss law many customers bought back their contracts around 1998. The second version normalises the time according to the start of the contract such that time intervals of changes refer to the contract's duration. By filtering out the policies around 1998 we intended to prevent the analysis from misleading effects of the law change. Typical patterns of policy changes were found. For example, one rule states the following: *If a component is signed and sometimes after this the bonus is adjusted, then it is very likely that directly after the adjustment a profit payment is prolonged.* The prediction of churn was tried on the basis of both, component data and a combination of component and policy data, applying biased sampling such that churn and continuation became equally distributed. The rules learned from the set of histories leading to churn and the rules learned from the set of continued policy/component histories did not differ. The same interval relations were valid for both sets. Hence, the features representing a sequence of changes did not deliver a characterisation of churn. We are again left with the question whether there is nothing within the data that could be discovered, or whether we have just represented the data wrong for our task.

4 Using TF/IDF Features

In the first experiments we have taken into account the customer attributes, the policy attributes, the component attributes and the time intervals for the states of components and/or policy attributes. However, each change of a component was handled as a singleton event. Either there was a column for an attribute changing in a particular year, or there was a time interval for a state of an attribute. Given the yearly time granularity of the database, a component attribute can only change once in a year. Similarly, the time intervals were unique for one policy. Hence, we had the representation boiled down to Boolean attributes. Frequencies were only counted for all policy histories during the data mining step. Then, the values of a column were summed up, or the frequency of a relation between time intervals in all policies was determined. Whether the same component was changed several times within the same policy was not captured by the representation. Counting the frequencies of changes within one policy could offer the relevant information. It would be plausible that very frequent changes of a policy are an effect of the customer not being satisfied with the contract. If we transform the chosen excerpt from the raw data (about policies) into a frequency representation, we possibly condense the data space in an appropriate way. However, we must exclude the frequencies of those changes that are common to all contracts, e.g. because of a change of law. A measure from information retrieval formulates exactly this: term frequency and inverse document frequency (TF/IDF) [16]. *Term frequency* here describes how often a particular attribute a_i of c_j, the contract or one of its components, has been changed within a policy.

$$tf(a_i, c_j) = \| \{x \in time\,points \mid a_i\,of\,c_j\,changed\} \| \tag{1}$$

Document frequency here corresponds to the number of policies in which a_i has been changed. The set of all policies is written C.

$$df(a_i) = \| \{c_j \in C \mid a_i\,of\,c_j\,changed\} \| \tag{2}$$

Hence the adaptation of the TF/IDF measure to policy data becomes for each policy c_j:

$$tfidf(a_i) = tf(a_i, c_j) log \frac{\| C \|}{df(a_i)} \tag{3}$$

This representation still shrinks the data set, but not as much as does the Boolean representation.

4.1 Preprocessing

The first experiments have shown that the tables describing the customers can be ignored. We focused on the policy changes. We selected all attributes which describe the state of the policy, ways, number, and amount of payments, the type of the insurance, unemployment, and disablement insurance. We ignored attributes which link to other tables or cannot be changed (such as, e.g. the currency). As a result, 13 attributes plus the identifier from the original attributes of the policy table were selected. One of them is the reason entered for a change of a policy. There are 121 different reasons. We transformed these attribute values into Boolean attributes. Thus we obtained 134 features describing changes of a policy. The TF/IDF values for the 13 original attributes we calculated from the history of each contract. We ordered all rows concerning the same policy by its time-stamps and compared in each column the successive values in order to detect changes. The term frequency of an attribute is simply the number of its value changes. For the 121 newly created features we counted how often they occurred within the changes. It was now easy to calculate the document frequencies for each attribute as the number of policies with a term frequency greater than 0.

4.2 Results

Since churn is only observed in 7.7% of the contracts MYSVM learned from a very large sample. We performed a 10-fold cross-validation on 10,000 examples[2]. The test criteria were accuracy, precision, recall, and the F-measure. If we denote positives which are correctly classified as positive by A, positives which are classified as negative by C, negatives that are correctly classified as negative as D, and negatives that are classified as positive by B, we write the definitions:

$$Accuracy = \frac{A+D}{A+B+C+D} \quad Precision = \frac{A}{A+B} \quad Recall = \frac{A}{A+C} \tag{4}$$

In order to balance precision and recall, we used the F-measure:

$$F_\beta = \frac{(\beta^2+1)Prec(h)Rec(h)}{\beta^2 Prec(h) + Rec(h)} \tag{5}$$

where β indicates the relative weight between precision and recall. We have set $\beta = 1$, weighting precision and recall equally. This led to an average precision of 94.9%, an accuracy of 99.4%, and a recall of 98.2% on the test sets. We wondered, whether these excellent results were caused by the chosen feature space, or due to the advantages of MYSVM . Therefore, we have performed 10-fold crossvalidation also using the algorithms Apriori, J48, and Naive Bayes. Table 1 shows the results. All algorithms were trained on the original attributes and then on the TF/IDF attributes. J4.8 clearly seleted the 'reason of change' attributes by its decision trees, in both representations. For MYSVM , the advantage of the TF/IDF feature space compared with the original attributes is striking. However, for all other algorithms, the TF/IDF feature space is also superior to the original one. The representation of input features contributes to the success of learning

[2] We also tried biased sampling but this delivered no enhanced results.

Table 1. Results comparing different learning algorithms and feature spaces

	Apriori		J4.8	
	TF/IDF attr	Original attr	TF/IDF attr	Original attr
Accuracy	93.48%	94.3%	99.88%	97.82%
Precision	56.07%	84.97%	98.64%	96.53%
Recall	72.8%	18.39%	99.8%	70.08%
F-Measure	63.35%	30.24%	99.22%	81.21%
	mySVM		Naive Bayes	
	TF/IDF attr	Original attr	TF/IDF attr	Original attr
Accuracy	99.71%	26.65%	88.62%	87.44%
Precision	97.06%	8.73%	38.55%	32.08%
Recall	98.86%	100%	78.92%	77.72%
F-Measure	97.95%	16.06%	51.8%	45.41%

at least as much as the algorithm. How can we describe more formally, whether the transformation into TF/IDF attributes expands or shrinks the data set? Is this transformation very expensive in terms of the resulting size of the data set? The Euclidian length of the vectors can be used as a measure of compression. A vector $\boldsymbol{x}$ of n attributes has the Euclidian length R

$$R = \sqrt{\sum_{i=1}^{n} x_i^2} \tag{6}$$

The data space with the original 13 attributes could be such that each attribute is changed m times giving us $\sqrt{13} \cdot m$ – the worst case. We found in our data that $m = 15$, i.e. no attribute was changed more than 15 times. If all attributes had that many changes, the feature space for the TF/IDF attributes would be $R = \sqrt{13} \cdot 15 = 54.08$. In this case, the feature space were not sparse and TF/IDF features would only be used if learning could not distinguish the classes otherwise. Note, that we investigate here the characterization of feature spaces, not the learnability. That is, we want to see, what the data transformation does to our raw data. We now estimate the average case for our data set. We assume a ranking r of attributes with respect to their frequency, e.g., $r = 1$ for the most frequent attribute. Experimental data suggests that Mandelbrot distributions [26]

$$tf_r = \frac{c}{(k+r)^{\phi}} \tag{7}$$

with parameters c, k and ϕ provide a good fit. The average Euclidian length using the distribution is:

$$R^2 = \sum_{r=1}^{d} \left(\frac{c}{(r+k)^2} \right)^2 \tag{8}$$

In our case $d = 4$ which means that only four distinct attributes have a value greater than zero. We bound $R^2 \leq 37$ according to the Mandelbrot distribution and see that the average Euclidian length is significantly shorter than the worst case.

In order to ease the design process of new knowledge discovery cases, we want to transfer our analysis to new cases. In other words, we should know before the transformation whether the data space will be condensed, or not. A heuristic provides us with a fast estimate. We illustrate the heuristic using our case. We order the original table with time-stamps such that the states of the same contract are in succeeding rows. We consider each policy c_j a vector and calculate the frequency of changes for each of its n attributes $a_1...a_n$ in parallel "on the fly". We can then determine in one database scan the contract c_j with a maximum Euclidian length:

$$\hat{R} = \max_{c_j} \left(\sqrt{\sum_{i=1}^{n} tf(a_i, c_j)^2} \right) \tag{9}$$

If $\hat{R} \leq \sqrt{n}m$ where m is the maximum frequency, the transformation into TF/IDF features is worth a try, otherwise only strict learnability results could force us to perform the transformation. In our case $n = 13$ and $m = 15$ so that $\hat{R} = 22,91$ which is in fact less than $\sqrt{13} \cdot 15 = 54.08$.

5 Related Work and Conclusion

Time-related data include time series (i.e. equidistant measurements of one process), episodes made of events from one or several processes, and time intervals which are related (e.g., an interval overlaps, precedes, or covers another interval). Time-stamped data refer to a calendar with its regularities. They can easily be transformed into a collection of events, can most often be transformed into time intervals, and sometimes into time series. Time series are most often analysed with respect to a prediction task, but also trend and cycle recognition belong to the statistical standard (see for an overview [27, 28]). Following the interest in very large databases, indexing of time series according to similarity has come into focus [29, 30]. Clustering of time series is a related topic (cf. e.g., [31]) as is time series classification (cf. e.g., [32, 33]). The abstraction of time series into sequences of events or time intervals approximates the time series piecewise by functions (so do [12, 34, 35]). Event sequences are investigated in order to predict events or to determine correlations of events [15, 9, 11, 13, 36]. The approach of Frank Höppner abstracts time series to time intervals and uses the time relations of James Allen in order to learn episodes [23, 22]. Also inductive logic programming can be applied. Episodes are then written as a chain logic program, which expresses direct precedence by chaining unified variables and other time relations by additional predicates [37, 38]. Time-stamped data have been investigated in-depth in [10].

In this paper, we have presented a knowledge discovery case on time-stamped data, together with its design process. The design process can be viewed as the search for a data space which is small enough for learning but does not remove the crucial information. For insurance data, the design process is particularly difficult, because the data are skewed, extremely high-dimensional, homogeneous accross classes, and time-stamped. The particular challenge was the handling of time-stamped data. Neither the snapshot approach nor the intervals approach were effective in our insurance application. The key idea to solving the discovery task was to generate TF/IDF features for changes of policies. The compilation of time features into TF/IDF has been analysed. Moreover, we have turned the analysis into an efficient heuristic which applies to the raw data and estimates the size of the feature space after the transformation into TF/IDF attributes. Our report on the insurance case might be used as a blueprint for similar cases. The heuristic and the generation of frequency features for time-stamped data with respect to the aspect of state change has been implemented in the MiningMart system [39][3].

Acknowledgment

We thank Jörg-Uwe Kietz and Regina Zücker for their information about the insurance practice and the anonymised database. For stimulating discussions on the support vector machine, mathematics, and representation languages we thank Stefan Rüping wholeheartedly.

References

1. Goldie, C., Klüppelberg, C.: Subexponential distributions. In Adler, R., Feldman, R., Taqqu, M., eds.: A practical guide to heavy tails: Statistical techniques for analysing heavy tails. Birkhauser (1997)
2. C.Apte, Pednault, E., Weiss, S.: Data mining with extended symbolic methods. In: Procs. Joint Statistical Meeting. (1998) IBM insurance mining.
3. Pairceir, R., McClean, S., Scotney, B.: Using hierarchies, aggregates, and statisticalk models to discover knowledge from distributed databases. In: Procs. AAAI WOrkshop on Learning Statistical Models from Relational Data, Morgan Kaufmann (2000) 52 – 58
4. Lang, S., Kragler, P., Haybach, G., Fahrmeir, L.: Bayesian space-time analysis of health insurance data. In Schwaiger, M., O.Opitz, eds.: Exploratory Data Analysis in Empirical Research. Springer (2002)
5. Klugmann, S., Panjer, H., Wilmot, G.: Loss Models – Fram Doata to Decisions. Wiley (1998)
6. Staudt, M., Kietz, J.U., Reimer, U.: A data mining support environment and its application to insurance data. In: Procs. KDD. (1998) insurance mining.
7. Kietz, J.U., Vaduva, A., Zücker, R.: MiningMart: Metadata-driven preprocessing. In: Proceedings of the ECML/PKDD Workshop on Database Support for KDD. (2001)

[3] For more details you might visit *www.mmart.cs.uni-dortmund.de*. The described insurance case will also be published in terms of its meta-data. Of course, the business data are strictly confidential.

8. Agrawal, R., Psaila, G., Wimmers, E.L., Zaït, M.: Querying shapes of histories. In: Proceedings of 21st International Conference on Very Large Data Bases, Morgan Kaufmann (1995) 502–514
9. Baron, S., Spiliopoulou, M.: Monitoring change in mining results. In: Proceedings of the 3rd International Conference on Data Warehou-sing and Knowledge Discovery, Springer (2001) 51–60
10. Bettini, C., Jajodia, S., Wang, S.: Time Granularities in Databases, Data Mining, and Temporal Reasoning. Springer (2000)
11. Blockeel, H., Fürnkranz, J., Prskawetz, A., Billari, F.: Detecting temporal change in event sequences: An application to demographic data. In De Raedt, L., Siebes, A., eds.: Proceedings of the 5th European Conference on the Principles of Data Mining and Knowledge Discovery. Volume 2168 of Lecture Notes in Computer Science., Springer (2001) 29–41
12. Das, G., Lin, K.I., Mannila, H., Renganathan, G., Smyth, P.: Rule Discovery from Time Series. In Agrawal, R., Stolorz, P.E., Piatetsky-Shapiro, G., eds.: Proceedings of the Fourth International Conference on Knowledge Discovery and Data Mining (KDD-98), New York City, AAAI Press (1998) 16 – 22
13. Mannila, H., Toivonen, H., Verkamo, A.: Discovering frequent episode in sequences. In: Procs. of the 1st Int. Conf. on Knowledge Discovery in Databases and Data Mining, AAAI Press (1995)
14. Morik, K.: The representation race - preprocessing for handling time phenomena. In de Mántaras, R.L., Plaza, E., eds.: Proceedings of the European Conference on Machine Learning 2000 (ECML 2000). Volume 1810 of Lecture Notes in Artificial Intelligence., Berlin, Heidelberg, New York, Springer Verlag Berlin (2000)
15. Domeniconi, C., shing Perng, C., Vilalta, R., Ma, S.: A classification approach for prediction of target events in temporal sequences. In Elomaa, T., Mannoila, H., Toivonen, H., eds.: Principles of Data Mining and Knowledge Discovery. Lecture Notes in Artificial Intelligence, Springer (2002)
16. Salton, G., Buckley, C.: Term weighting approaches in automatic text retrieval. Information Processing and Management **24** (1988) 513–523
17. Kietz, J.U., Vaduva, A., Zücker, R.: Mining Mart: Combining Case-Based-Reasoning and Multi-Strategy Learning into a Framework to reuse KDD-Application. In Michalki, R., Brazdil, P., eds.: Proceedings of the fifth International Workshop on Multistrategy Learning (MSL2000), Guimares, Portugal (2000)
18. Bi, Z., Faloutsos, C., Korn, F.: The DGX distribution for mining massive, skewed data. In: 7th International ACM SIGKDD Conference on Knowledge Discovery and Data Mining, ACM (2001)
19. Witten, I., Frank, E.: Data Mining – Practical Machine Learning Tools and Techniques with Java Implementations. Morgan Kaufmann (2000)
20. Rüping, S.: mySVM-Manual. Universität Dortmund, Lehrstuhl Informatik VIII. (2000) http://www-ai.cs.uni-dortmund.de/SOFTWARE/MYSVM/.
21. Bauschulte, F., Beckmann, I., Haustein, S., Hueppe, C., El Jerroudi, Z., Koepcke, H., Look, P., Morik, K., Shulimovich, B., Unterstein, K., Wiese, D.: PG-402 Endbericht Wissensmanagement. Technical report, Fachbereich Informatik, Universität Dortmund (2002)
22. Allen, J.F.: Towards a general theory of action and time. Artificial Intelligence **23** (1984) 123–154
23. Höppner, F.: Discovery of Core Episodes from Sequences. In Hand, D.J., Adams, N.M., Bolton, R.J., eds.: Pattern Detection and Discovery. Volume 2447 of Lecture notes in computer science., London, UK, ESF Exploratory Workshop, Springer (2002) 1–12

24. Agrawal, R., Imielinski, T., Swami, A.: Mining association rules between sets of items in large databases. In: Proceedings of the ACM SIGMOD Conference on Management of Data, Washington, D. C. (1993) 207–216
25. Fisseler, J.: Anwendung eines Data Mining-Verfahrens auf Versicherungsdaten. Master's thesis, Fachbereich Informatik, Universität Dortmund (2003)
26. Mandelbrot, B.: A note on a class of skew distribution functions: Analysis and critique of a paper by H.A.Simon. Informationi and Control **2** (1959) 90 – 99
27. Box, G.E.P., Jenkins, G.M., Reinsel, G.C.: Time Series Analysis. Forecasting and Control. Third edn. Prentice Hall, Englewood Cliffs (1994)
28. Schlittgen, R., Streitberg, B.H.J.: Zeitreihenanalyse. 9. edn. Oldenburg (2001)
29. Keogh, E., Pazzani, M.: Scaling up dynamic time warping for datamining applications. In: Proceedings of the 6th ACM SIGKDD International Conference on Knowledge Discovery and Data Mining, ACM Press (2000) 285–289
30. Agrawal, R., Faloutsos, C., Swami, A.: Efficient similarity search in sequence databases. In: Proceedings of the 4th International Conference on Foundations of Data Organization and Algorithms. Volume 730., Springer (1993) 69–84
31. Oates, T., Firoiu, L., Cohen, P.R.: Using dynamic time warping to bootstrap hmm-based clustering of time series. In: Sequence Learning ? Paradigms, Algorithms, and Applications. Volume 1828 of Lecture Notes in Computer Science. Springer Verlag (2001) 35?–52
32. Geurts, P.: Pattern extraction for time series classification. In: Pro-ceedings of the 5th European Conference on the Principles of Data Mining and Knowledge Discovery. Volume 2168 of Lecture Notes in Computer Science., Springer (2001) 115–127
33. Lausen, G., Savnik, I., Dougarjapov, A.: Msts: A system for mining sets of time series. In: Proceedings of the 4th European Conference on the Principles of Data Mining and Knowledge Discovery. Volume 1910 of Lecture Notes in Computer Science., Springer Verlag (2000) 289–298
34. Guralnik, V., Srivastava, J.: Event detection from time series data. In: Proceedings of the fifth ACM SIGKDD international conference on Knowledge discovery and data mining, San Diego, USA (1999) 33 – 42
35. Morik, K., Wessel, S.: Incremental signal to symbol processing. In Morik, K., Kaiser, M., Klingspor, V., eds.: Making Robots Smarter – Combining Sensing and Action through Robot Learning. Kluwer Academic Publ. (1999) 185 –198
36. Mannila, H., Toivonen, H., Verkamo, A.: Discovery of frequent episodes in event sequences. Data Mining and Knowledge Discovery **1** (1997) 259–290
37. Klingspor, V., Morik, K.: Learning understandable concepts for robot navigation. In Morik, K., Klingspor, V., Kaiser, M., eds.: Making Robots Smarter – Combining Sensing and Action through Robot Learning. Kluwer (1999)
38. Rieger, A.D.: Program Optimization for Temporal Reasoning within a Logic Programming Framework. PhD thesis, Universität Dortmund, Dortmund, Germany (1998)
39. Morik, K., Scholz, M.: The MiningMart Approach to Knowledge Discovery in Databases. In Zhong, N., Liu, J., eds.: Intelligent Technologies for Information Analysis. Springer (2003) to appear.

Nomograms for Visualization of Naive Bayesian Classifier

Martin Možina[1], Janez Demšar[1], Michael Kattan[2], and Blaž Zupan[1,3]

[1] Faculty of Computer and Information Science, University of Ljubljana, Slovenia
{martin.mozina,janez.demsar}@fri.uni-lj.si

[2] Memorial Sloan Kettering Cancer Center, New York, NY, USA
kattanm@mskcc.org

[3] Dept. Mol. and Human Genetics, Baylor College of Medicine, Houston, TX, USA
blaz.zupan@fri.uni-lj.si

Abstract. Besides good predictive performance, the naive Bayesian classifier can also offer a valuable insight into the structure of the training data and effects of the attributes on the class probabilities. This structure may be effectively revealed through visualization of the classifier. We propose a new way to visualize the naive Bayesian model in the form of a nomogram. The advantages of the proposed method are simplicity of presentation, clear display of the effects of individual attribute values, and visualization of confidence intervals. Nomograms are intuitive and when used for decision support can provide a visual explanation of predicted probabilities. And finally, with a nomogram, a naive Bayesian model can be printed out and used for probability prediction without the use of computer or calculator.

1 Introduction

Compared to other supervised machine learning methods, naive Bayesian classifier (NBC) is perhaps one of the simplest yet surprisingly powerful technique to construct predictive models from labelled training sets. Its predictive properties have often been a subject of theoretical and practical studies (*e.g.* [1, 2]), and it has been shown that despite NBC's assumption of conditional independence of attributes given the class, the resulting models are often robust to a degree where they match or even outperform other more complex machine learning methods.

Besides good predictive accuracy, NBC can also provide a valuable insight to the training data by exposing the relations between attribute values and classes. The easiest and the most effective way to present these relations is through visualization. But while the predictive aspect of NBC has been much studied, only a few reports deal with visualization and explanation capabilities of NBC. In this, a notable exception is the work of Kononenko [2] and Becker *et al.* [3]. Kononenko introduced the concept of information that is gained by knowing the value of a particular attribute. When using the NBC for classification, Kononenko's information gains can offer an explanation on how the values of the attributes influenced the predicted probability of the class. Becker and coauthors

J.-F. Boulicaut et al. (Eds.): PKDD 2004, LNAI 3202, pp. 337–348, 2004.

proposed an alternative approach to visualization of NBC that is also available as Evidence Visualizer in the commercial data mining suite MineSet. Evidence Visualizer uses pie and bar charts to represent conditional probabilities, and besides visualization of the model offers interactive support to prediction of class probabilities.

In the paper, we propose an alternative method to visualization of a NBC that clearly exposes the quantitative information on the effect of attribute values to class probabilities and uses simple graphical objects (points, rulers and lines) that are easier to visualize and comprehend. The method can be used both to reveal the structure of the NBC model, and to support the prediction.

The particular visualization technique we rely on are nomograms. In general, a *nomogram* is any graphical representation of a numerical relationships. Invented by French mathematician Maurice d'Ocagne in 1891, the primary means of a nomogram was to enable the user to graphically compute the outcome of an equation without doing any calculus. Much later, Lubsen and coauthors [4] extended the nomograms to visualize a logistic regression model. They show the utility of such a device on a case for prediction of probability of diagnosis of acute myocardial infarction. Their nomogram was designed so that it can be printed on the paper and easily used by physicians to obtain the probability of diagnosis without resorting to a calculator or a computer. With an excellent implementation of logistic regression nomograms in a Design and hmisc modules for S-Plus and R statistical packages by Harrell [5], the idea has recently been picked up; especially in the field of oncology, there are now a number of nomograms used in daily clinical practice for prognosis of outcomes of different treatments that have been published for variety of cancer types (*e.g.* [6]; see also http://www.baylorcme.org/nomogram/modules.cfm).

Our NBC nomograms use a similar visualization approach to that of Harrell for logistic regression, and well conform to the NBC visualization design principles as stated by Becker [3]. In the paper, we first show how to adapt the NBC to be suitable for visualization with a nomogram. We also propose the means to compute confidence intervals for the contributions of attribute values and for the class probability, and include these in visualization. We discuss on the differences between our visualization approach and that of Evidence Visualizer, and compare NBC nomograms to those for logistic regression. We show that a simple adjustment of an NBC nomogram can support a visual comparison between NBC and logistic regression models. The particular benefits of the approach and ideas for the further work are summarized in the conclusion.

2 Naive Bayesian Nomogram

Let us start with an example. Fig. 1 shows a nomogram for a NBC that models the probability for a passenger to survive the disaster of the HMS Titanic. The nomogram, built from the well-known Titanic data (http://hesweb1.med.virginia.edu/biostat/s/data/), includes three attributes that report on the travelling class (first, second, and third class, or a crew member), age (adult or child), and gender of the passenger.

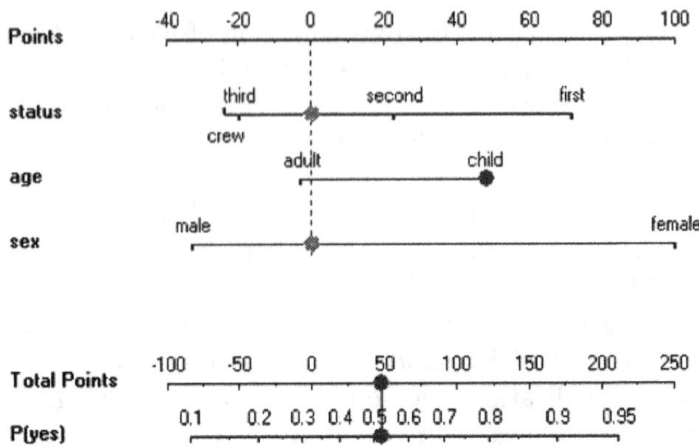

Fig. 1. A nomogram for prediction of survival probability of a passenger on HMS Titanic.

Of 2201 passengers on Titanic, 711 (32.3%) survived. To make a prediction, the contribution of each attribute is measured as a point score (topmost axis in the nomogram), and the individual point scores are summed to determine the probability of survival (bottom two axes of the nomogram). When the value of the attribute is unknown, its contribution is 0 points. Therefore, not knowing anything about the passenger, the total point score is 0, and the corresponding probability equals to the unconditional prior. The nomogram in Fig. 1 shows the case when we know that the passenger is a child; this score is slightly less than 50 points, and increases the posterior probability to about 52%. If we further know that the child travelled in the first class (about 70 points), the points would sum to about 120, with a corresponding probability of survival of about 80%.

Besides enabling the prediction, the naive Bayesian nomogram nicely reveals the structure of the model and the relative influences of the attribute values to the class probability. For the Titanic data set, gender is an attribute with the biggest potential influence on the probability of passenger's survival: being female increases the chances of survival the most (100 points), while being male decreases it (about -30 points). The corresponding line in the nomogram for this attribute is the longest. Of the three attributes age is apparently the least influential, where being a child increases the probability of survival. Most lucky were also the passengers of the first class for which – considering the status only – the probability of survival was much higher than the prior.

In the following we show how to represent the NBC in the way to be applicable for visualization with a nomogram. We further introduce confidence intervals, and discuss several other details on our particular implementation.

2.1 Derivation of Naive Bayesian Nomogram

Naive Bayesian rule to assess the probability of class c given an instance X with a set of attribute values $X = \langle a_i, a_2 \ldots a_n \rangle$ is:

$$P(c|X) = \frac{P(a_i, a_2 \ldots a_n|c)P(c)}{P(X)} = \frac{P(c)\prod_i P(a_i|c)}{P(X)} \tag{1}$$

We call class c a target class, since it will be the one represented in the nomogram. The probability of the alternative class (or alternative classes) $\bar{c}$ is $P(\bar{c}|X)$, and dividing the two we obtain:

$$\text{Odds} = \frac{P(c|X)}{P(\bar{c}|X)} = \frac{P(c)\prod_i P(a_i|c)}{P(\bar{c})\prod_i P(a_i|\bar{c})} \tag{2}$$

In terms of the log odds (logit $P = \log \frac{P}{1-P}$), this equation translates to:

$$\text{logit } P(c|X) = \text{logit } P(c) + \sum_i \log \frac{P(a_i|c)}{P(a_i|\bar{c})} \tag{3}$$

The terms in summation can be expressed as odds ratios (OR):

$$\frac{P(a_i|c)}{P(a_i|\bar{c})} = \frac{\frac{P(c|a_i)}{P(\bar{c}|a_i)}}{\frac{P(c)}{P(\bar{c})}} = \text{OR}(a_i) \tag{4}$$

and estimate the ratio of posterior to prior probability given the attribute value a_i [1]. We now take the right term in (3) and call it $F(c|X)$:

$$F(c|X) = \sum_i \log \frac{P(a_i|c)}{P(a_i|\bar{c})} = \sum_i \log \text{OR}(a_i) \tag{5}$$

and use it for the construction of the central part of the nomogram relating attribute values to point scores. The individual contribution (point score) of each known attribute value in the nomogram is equal to $\log \text{OR}(a_i)$, and to what we have referred as the sum of point scores corresponds to $F(c|X)$.

Using Eq. 5, we can now derive the central part of the nomogram from Fig. 1. As a target class, we will use the survival of the passengers. For each attribute value, we compute the individual contributions (*i.e.* point scores) from the instance counts in Table 1. For example, the log odds ratio for the passenger in the first class is 1.25, as the odds for surviving in the first class are $203/122 = 1.67$, unconditional odds for surviving are $711/1490 = 0.48$, and their log ratio is $\log(1.67/0.48) = 1.25$. Similarly, the log odds ratio for the second-class passengers is $\log \frac{118/167}{0.48} = 0.393$, and for the female passengers is $\log \frac{344/126}{0.48} = 1.744$. Notice that instead of relative frequencies used here, probabilities could also be estimated by other methods, like Laplace or m-estimate [8].

These and all other log odds ratios for different attribute values form the central part of the nomogram shown in Fig. 2. This is equal to the corresponding part of the nomogram from Fig. 1, except that for the latter we have re-scaled the units so that log odds ratio of 1.744 – a maximal absolute log odds ratio in the nomogram – represents 100 points. Log odds ratio is a concept that experts, like

[1] Our use of odds ratios is a bit different from that in logistic regression. Instead of relating posterior and prior probability, odd ratios in logistic regression relate the odds at the two different values of a binary attribute [7].

Table 1. Number of instances in Titanic data set with a particular value of an attribute and class.

ATTRIBUTE	VALUE	CLASS=YES	CLASS=NO
STATUS	FIRST	203	122
	SECOND	118	167
	THIRD	178	528
	CREW	212	673
AGE	ADULT	654	1438
	CHILD	57	52
SEX	MALE	367	1364
	FEMALE	344	126

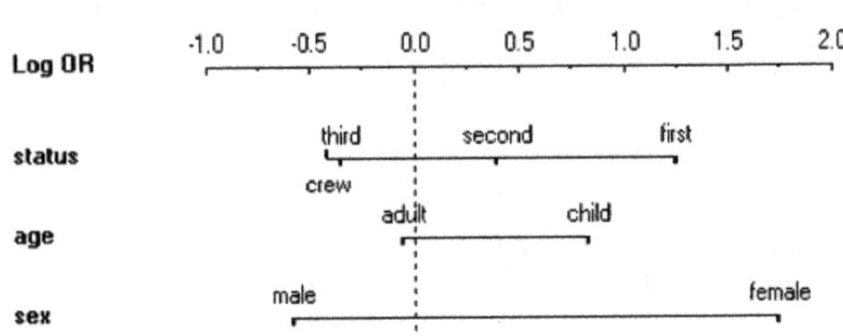

Fig. 2. A central part of the Titanic nomogram showing the log odds ratios for survival given different values of the attributes.

those from biomedical statistics, do understand and can interpret. For others, a scale with points from -100 to 100 may provide more comfort, and summing-up the integers may be easier than using real numbers. Also, it may be easier to compare the contributions of different attribute values in 100 points scale.

For the part of the nomogram that relates the sum of individual point scores to the class probability, we start from Eqs. 3 and 5 and obtain:

$$\log \frac{P(c|X)}{1 - P(c|X)} = \log \frac{P(c)}{1 - P(c)} + F(c|X) \tag{6}$$

From this, we compute the probability $P(c|X)$ as:

$$P(c|X) = [1 + e^{-\log P(c)/(1-P(c)) - F(c|X)}]^{-1} \tag{7}$$

The lower part of the nomogram, which relates the sum of points as contributed by the known attributes to the class probability, is then a tabulation of a function $P(c|X) = f[F(c|X)]$. For our Titanic example, this part of the nomogram is shown in Fig. 3.

2.2 Confidence Intervals

The point scores in the nomogram, *e.g.* the odds ratios $\mathrm{OR}(a_i)$, are estimated from the training data. It may be therefore important for the user of the nomo-

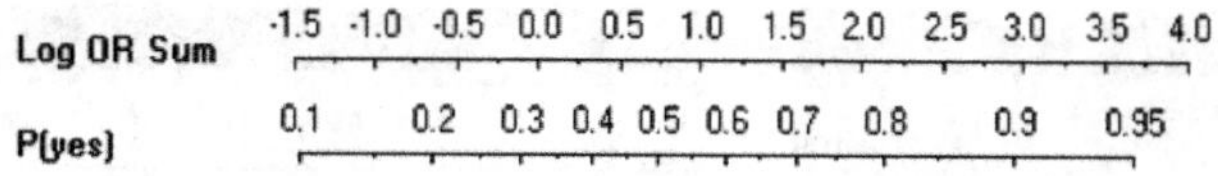

Log OR Sum -1.5 -1.0 -0.5 0.0 0.5 1.0 1.5 2.0 2.5 3.0 3.5 4.0

P(yes) 0.1 0.2 0.3 0.4 0.5 0.6 0.7 0.8 0.9 0.95

Fig. 3. The part of Titanic nomogram to determine the the probability of survival from the sum of log odds ratios as contributed from the known attribute values.

gram to know how much to trust these estimates. We can provide this information through confidence intervals.

The $1-\alpha$ confidence intervals of $\widehat{\mathrm{OR}}(a_i)$ are estimated as (see [7])

$$\widehat{\mathrm{OR}}(a_i) \pm z_{1-\alpha/2}\sqrt{\widehat{\mathrm{Var}}(\widehat{\mathrm{OR}}(a_i))} \tag{8}$$

where $\widehat{\mathrm{Var}}(\cdot)$ is computed as[2]:

$$\widehat{\mathrm{Var}}(\mathrm{logit}\ \hat{P}(c)) = [N\hat{P}(c)\hat{P}(\bar{c})]^{-1} \tag{9}$$

$$\widehat{\mathrm{Var}}(\widehat{\mathrm{OR}}(a_i)) = [N_{a_i}\hat{P}(c|a_i)\hat{P}(\bar{c}|a_i)]^{-1} - \widehat{\mathrm{Var}}(\mathrm{logit}\ \hat{P}(c)) \tag{10}$$

where N is a number of training examples, and N_{a_i} is a number of training examples that include attribute value a_i.

Fig. 4 shows a Titanic survival nomogram that includes the confidence intervals for $\alpha = 0.95$.

2.3 Implementation

We have implemented a NBC nomogram as a widget within a machine learning suite Orange [9]. The widget (see Fig. 4 for a snapshot) supports visualization of a nomogram, use of confidence intervals, and can, for each attribute value, plot a bar with height proportional to the number of particular instances.

Our implementation supports the classification. Attribute values (dots on attribute axis) can be moved across the axis, where we can also select values between two value marks (weighted distributions). Class probabilities and associated confidence intervals are updated instantaneously with any change in the data or corresponding naive Bayesian model.

3 Discussion and Related Work

Naive Bayesian nomograms, as presented in this paper, are a visualization technique that we adapted and extended from logistic regression [4, 5]. In this, we have considered the design requirements for visualization of NBC as proposed by

[2] Although we regard the computation of the confidence intervals for NBC as an important new addition for this method, our paper focuses on visualization of the confidence intervals and we omit the proof and derivation of confidence intervals due to space considerations.

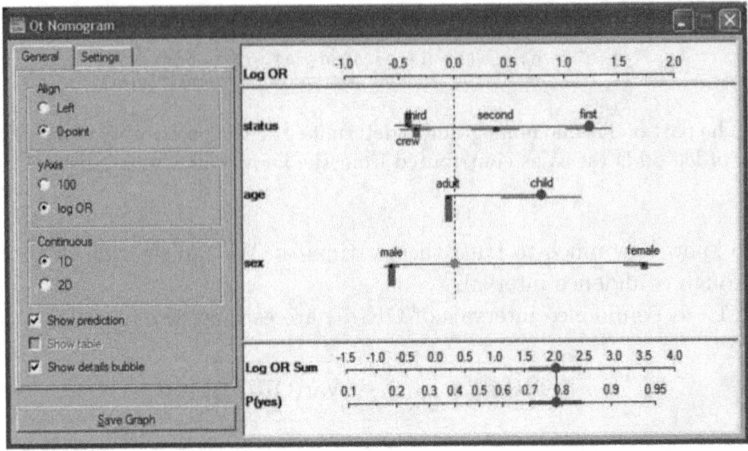

Fig. 4. Orange widget with the Titanic nomogram that includes confidence intervals for contributions of attribute values and class probabilities. For a woman travelling in the first class, the probability of survival is with 95% confidence between 0.87 and 0.92.

Becker *et al.* Below, we briefly review them and point out how we have addressed them. In the review of related work, we make a note on the explanation technique for NBC as proposed by Kononenko [2]. We then discuss the differences of our method with that of Evidence Visualizer [3]. Finally, we consider the related work on logistic regression nomograms, discuss the differences and highlight the advantages of naive Bayesian nomograms.

3.1 Design Principles

Becker *et al.* [3] list a number of design requirements for a visualization of a naive Bayesian classifier. With naive Bayesian nomograms, we address all of them. In particular, because of their simplicity, visualized NBC models should be easily understood by non-experts. In the nomogram, comparing the span of an attribute axis easily identifies the important attributes. The effects of each attribute value are also clearly represented in a nomogram, making it easy to spot a direction and magnitude of the influence. Attribute axis are aligned to zero-point influence (prior probability), which allows for a straightforward comparison of contributions across different values and attributes. Confidence intervals, in addition to histograms with record counts, inform the user about reliability of the estimates presented in the nomograms. Nomogram may also be a practical visualization method when dealing with larger number of attributes. For example, sorting attributes by their impact (*e.g.* highest overall/positive/negative influence) offers a means to study and gain a valuable insight into a large naive Bayesian models.

Nomogram-based class-characterization is particularly straightforward for domains with binary class: the zero-point influence line vertically splits the nomogram to the right (positive) and left (negative) part. The visualized class is characterized with the attribute values on the right, whereas the other class is characterized with values presented on the left side of the nomogram. Accordingly, the values farthest from the center are the most influential class indicators.

In our implementation, nomograms can be used for interactive, click-and-drag classification and what-if analysis. Alternatively, they can be printed out and used for probability prediction without the use of computer or calculator.

3.2 Related Work on Nomograms and Visualization of Naive Bayesian Classifier

In the early nineties, Kononenko [2] realized that besides good predictive performance NBCs can be used to *explain* to what degree and how attribute values influence the class probability when classifying an example. He showed that NBCs can be written as a sum of information contributed by each attribute, where a contribution of an attribute value a_i is $\log_2 P(c|a_i) - \log_2 P(c)$.

Zupan *et al* [10] plotted these contributions in a form similar to nomograms proposed in this paper. The main disadvantage of their approach, however, is that the partial attribute scores (influences) have to be summed for each class separately, and then normalized.

Despite the popularity of NBCs within machine learning community, there have not been many reports on methods to visualize it, and even fewer are practical implementations of these. A notable exception is Evidence Visualizer implemented within a data mining suite MineSet [3]. Evidence Visualizer offers two different views of NBCs, and we present an example using the Titanic NBC model in Fig. 5. In the pie chart display, there is a pie for each attribute-value combination, with slices corresponding to distinct classes and where the size of each slice is proportional to $P(a_i|c_j)$. The height of the pie is proportional to number of instances with an attribute value a_i. In the bar-chart representation, each bar signifies the evidence for a particular class c_j as proportional to $-\log(1 - P(a_i|c_j))$. Probability confidence intervals are displayed through color saturation. Evidence Visualizer can be used either in an exploratory way or for making predictions.

While models in Evidence Visualizer may be visually attractive, we believe that NBC nomograms as presented in this paper have several advantages. Both methods aim at visualizing the effects the attribute values have on class probability, where Evidence Visualizer uses bar/pie heights, size of pie slices and color saturations while NBC nomogram uses a simpler position-based representation. Visualization theory [11] gives a clear advantage to positional, line-based, visualization of quantitative information as opposed to that using more complex 2-D and 3-D objects. When comparing relative evidences contributed by different attribute values, a positional visualization should be clearer to that of comparing the size of the pie slices or heights of the bars. In particular, pie-charts have been criticized by Tufte [11] for their failure to order numbers along a visual dimension, and for reasons of poor interpretability of multiple pie charts. Ware [12] also reports that comprehension of absolute quantities when visualized through color saturations may be at least hard, while interpreting the visualization of

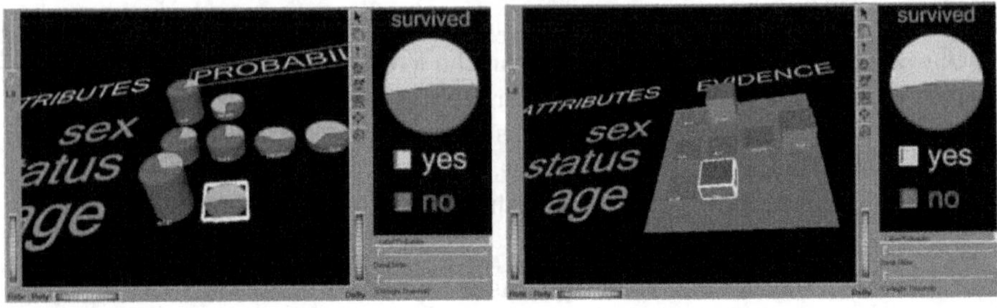

Fig. 5. Evidence Visualizer, a visualization of NBC from Titanic data set in MineSet.

confidence intervals through lines of variable length in the nomogram should be more straightforward.

A distinct advantage of a nomogram is that it uses simpler graphical objects and can thus be used in, for instance, systems for decision support on smaller-resolution devices such as handhelds or other mobile devices. The shortcoming of nomograms as compared to the Evidence Visualizer is in visualization in case of multiple-classes: the nomograms visualize one class at a time.

Our NBC nomograms stem from the work of Lubsen [4] and Harrell [5] on visualization of logistic regression. Logistic regression models the probability of target class value as:

$$P(c|X) = \frac{1}{1 + e^{-\beta_0 - \sum_i \beta_i x_i}} \tag{11}$$

where x_i is the value of i-th attribute being 0 or 1 in case of a binary attribute. Notice that m-valued nominal attributes can be encoded with $m - 1$ binary dummy variables. Log odds for the above probability is:

$$\text{logit } P(c|X) = \beta_0 + \sum_i \beta_i x_i \tag{12}$$

The position of a particular attribute value x_i in the logistic regression nomogram is determined through the product $\beta_i x_i$. In such nomogram, one of the values of the attributes will always be displayed at point 0, whereas others will span to the right or left depending on the sign of β. Adhering to the particular (and well-known) implementation by Harrell [5], all attributes are presented so that the leftmost value is drawn at the point 0, and the effects of aligning the attribute axis in this way are compensated through the appropriate change in β_0. The lower part of the nomogram (determination of class probability from the sum of attribute value points) is derived in a similar way as for NBC nomograms (see Fig. 6.a for a nomogram of logistic regression model for the Titanic data set).

There are several important differences between logistic and NBC nomograms, all stemming from the differences between the two modelling methods. NBC nomograms depict both negative and positive influences of the values of attributes, and differently from logistic regression nomograms, do not anchor one of the values to the 0-point. In this sense, logistic regression nomograms are

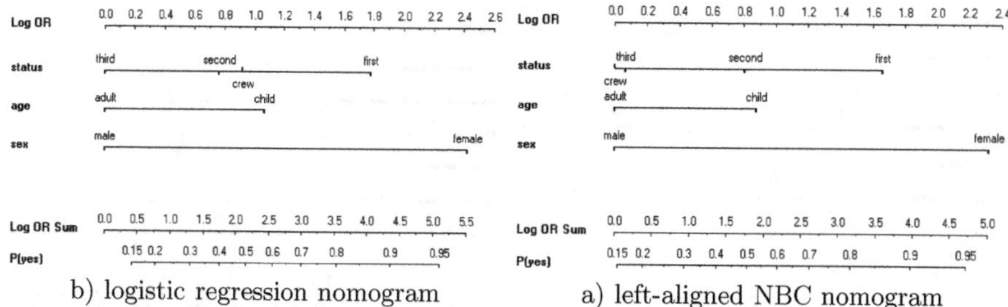

Fig. 6. Comparison of the nomograms for the Titanic data set.

less appropriate for class characterization. To illustrate this point, values that appear on logistic regression nomogram at 0-point are shown as having the same effect on the class probability, while they may be found at completely different positions in corresponding NBC nomograms. In that, we believe an NBC nomogram offers a better insight into the modelled domain. Another advantage is handling of unknown values. Namely, one needs to specify all attributes when reasoning with a logistic regression nomogram, while NBC nomograms offer a nice interactive one-value-at-a-time update of class probabilities.

To compare NBC and logistic regression nomograms, we can alter the presentation of the NBC nomogram so that the leftmost values for each of the attribute are aligned and their log odds ratio set to 0. We call this a *left-aligned NBC nomogram*. Alignment changes, *e.g.* offsets of the axis for each of the attributes, are reflected in an appropriate update of the lower part of the nomogram.

Fig. 6.a shows a left-aligned nomogram for the Titanic data set, and compares it to the logistic regression nomogram. The two nomograms look very alike, and the effect of the attributes is comparable even on the absolute, log odds ratio scale. The only noticeable difference is a position of the crew value of the attribute status. To analyze this, we used interaction analysis [13] which showed that this attribute strongly interacts with age (crew members are adults) and sex (most of them are male). It seems that the conditional attribute independence assumption of the NBC is most violated for this attribute and value, and hence the difference with logistic regression which is known to be able to compensate for the effects of attribute dependencies [7].

Fig. 7 shows a comparison of the two nomograms for another data set called voting. This is a data set from a UCI Machine Learning Repository [14] on sixteen key votes of each of the U.S. House of Representatives Congressmen of which we have selected six most informative votes for the figure). Visual comparison of the two nomograms reveals some similarities (the first three attributes) and quite some differences (the last three). As for Titanic, we have also noticed that attribute interaction analysis can help explain the differences between the two nomograms.

It is beyond this paper to compare NBC and logistic regression, which has otherwise received quite some attention recently [15]. With the above examples, however, we wanted to point out that nomograms may be the right tool for

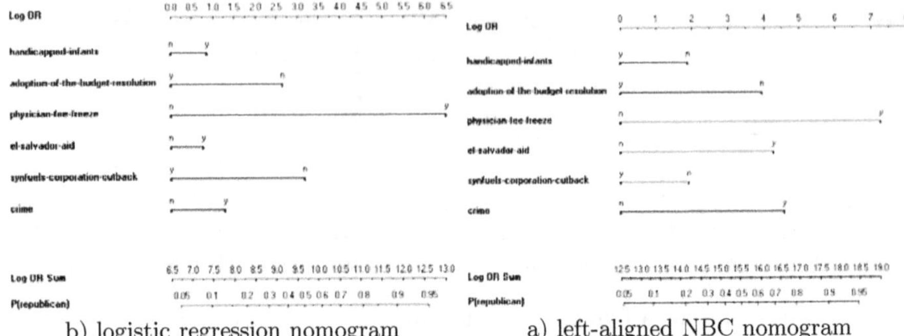

b) logistic regression nomogram a) left-aligned NBC nomogram

Fig. 7. Comparison of the nomograms for the voting data set.

experimental comparison of different models and modelling techniques, as it allows to easily spot the similarities and differences in the structure of the model.

4 Conclusion

In words of Colin Ware, "one of the greatest benefits of data visualization is the sheer quantity of information that can be rapidly interpreted if it is presented well" [12]. As the naive Bayesian classifier can be regarded as a simple yet powerful model to summarize the effects of attributes for some classification problems, it is also important to be able to clearly present the model to comprehend these effects and gain insight to the data.

In this paper, we show how we can adapt naive Bayesian classifiers and present them with a well established visualization technique called nomograms [5, 6]. The main benefit of this approach is simple and clear visualization of the complete model and the quantitative information it contains. The visualization can be used for exploratory analysis and decision making (classification), and we also show that it can be used effectively to compare different models, including those coming from logistic regression.

There are several aspects of naive Bayesian nomograms that deserve further attention and investigation. In the paper, our examples are all binary classification problems. Nomograms are intended to visualize the probability of one class against all others, and in principle for non-binary classification problems one would need to analyze several nomograms. Also, we have limited the scope of this paper to the analysis and presentation of data sets that include only nominal attributes. In principle, and as with logistic regression nomograms [5], one can easily present continuous relations, and we are now extending naive Bayesian nomograms in this way.

Acknowledgement

This work was supported, in part, by the program and project grants from Slovene Ministry of Science and Technology and Slovene Ministry of Information Society, and American Cancer Society project grant RPG-00-202-01-CCE.

References

1. Domingos, P., Pazzani, M.: Beyond independence: conditions for the optimality of the simple Bayesian classifier. In: Proceedings of the Thirteenth International Conference on Machine Learning, Bari, Italy, Morgan Kaufmann (1996) 105–112
2. Kononenko, I.: Inductive and bayesian learning in medical diagnosis. Applied Artificial Intelligence **7** (1993) 317–337
3. Becker, B., Kohavi, R., Sommerfield, D.: Visualizing the simple Bayesian classifier. In Fayyad, U., Grinstein, G., Wierse, A., eds.: Information Visualization in Data Mining and Knowledge Discovery. Morgan Kaufmann Publishers, San Francisco (2001) 237–249
4. Lubsen, J., Pool, J., van der Does, E.: A practical device for the application of a diagnostic or prognostic function. Methods of Information in Medicine **17** (1978) 127–129
5. Harrell, F.E.: Regression modeling strategies: with applications to linear models, logistic regression, and survival analysis. Springer, New York (2001)
6. Kattan, M.W., Eastham, J.A., Stapleton, A.M., Wheeler, T.M., Scardino, P.T.: A preoperative nomogram for disease recurrence following radical prostatectomy for prostate cancer. J Natl Cancer Inst **90** (1998) 766–71
7. Hosmer, D.W., Lemeshow, S.: Applied Logistic Regression. John Wiley & Sons, New York (2000)
8. Cestnik, B.: Estimating probabilities: A crucial task in machine learning. In: Proceedings of the Ninth European Conference on Artificial Intelligence. (1990) 147–149
9. Demšar, J., Zupan, B.: Orange: From experimental machine learning to interactive data mining. White Paper [`http://www.ailab.si/orange`], Faculty of Computer and Information Science, University of Ljubljana (2004)
10. Zupan, B., Demšar, J., Kattan, M.W., Beck, J.R., Bratko, I.: Machine learning for survival analysis: A case study on recurrence of prostate cancer. Artificial Intelligence in Medicine **20** (2000) 59–75
11. Tufte, E.R.: The visual display of quantitative information. Graphics Press, Cheshire, Connecticut (1983)
12. Ware, C.: Information Visualization: Perception for Design. Morgan Kaufmann Publishers (2000)
13. Jakulin, A., Bratko, I.: Analyzing attribute dependencies. In: Proc. of the 7th European Conference on Principles and Practice of Knowledge Discovery in Databases, Dubrovnik (2003) 229–240
14. Murphy, P.M., Aha, D.W.: UCI Repository of machine learning databases [`http://www.ics.uci.edu/~mlearn/mlrepository.html`]. Irvine, CA: University of California, Department of Information and Computer Science (1994)
15. Ng, A., Jordan, M.: On discriminative vs. generative classifiers: A comparison of logistic regression and naive bayes. In: Proc. of Neural Information Processing Systems). Volume 15. (2003)

Using a Hash-Based Method for Apriori-Based Graph Mining

Phu Chien Nguyen, Takashi Washio, Kouzou Ohara, and Hiroshi Motoda

The Institute of Scientific and Industrial Research, Osaka University
8-1 Mihogaoka, Ibaraki, Osaka, 567-0047, Japan
Phone: +81-6-6879-8541, Fax: +81-6-6879-8544

Abstract. The problem of discovering frequent subgraphs of graph data can be solved by constructing a candidate set of subgraphs first, and then, identifying within this candidate set those subgraphs that meet the frequent subgraph requirement. In Apriori-based graph mining, to determine candidate subgraphs from a huge number of generated adjacency matrices is usually the dominating factor for the overall graph mining performance since it requires to perform many graph isomorphism tests. To address this issue, we develop an effective algorithm for the candidate set generation. It is a hash-based algorithm and was confirmed effective through experiments on both real-world and synthetic graph data.

1 Introduction

Discovering frequent patterns of graph data, *i.e.*, frequent subgraph mining or simply graph mining, has attracted much research interest in recent years because of its broad application areas such as cheminformatics and bioinformatics [2, 7, 9, 11]. The kernel of frequent subgraph mining is subgraph isomorphism test, which is known to be NP-complete. The earliest studies to find subgraph patterns characterized by some measures from massive graph data, SUBDUE [3] and GBI [13], use greedy search to avoid high complexity of the subgraph isomorphism problem, resulting in an incomplete set of characteristic subgraphs.

Recent graph mining algorithms can mine a complete set of frequent subgraphs given a minimum support threshold. They can be roughly classified into two categories. Algorithms in the first category employ the same level-by-level expansion adopted in Apriori [1] to enumerate frequent subgraphs. Representative algorithms in this category include AGM [7] and FSG [9]. AGM finds all frequent induced subgraphs, which will be explained later, with a vertex-growth strategy. FSG, on the other hand, finds all frequent connected subgraphs based on an edge-growth strategy and was claimed to run faster than AGM. Algorithms in the second category use a depth-first search for finding candidate frequent subgraphs. A typical algorithm in this category is gSpan [12], which was reported to outperform both AGM and FSG in terms of computation time. However, a variant of AGM focusing on mining frequent connected subgraphs, AcGM [6], was claimed to be superior to FSG and comparable with gSpan.

J.-F. Boulicaut et al. (Eds.): PKDD 2004, LNAI 3202, pp. 349–361, 2004.

In the context of mining frequent itemsets by Apriori, the heuristic to construct the candidate set of frequent itemsets is crucial to the performance of frequent pattern mining algorithms [10]. Clearly, in order to be efficient, the heuristic should generate candidates with high likelihood of being frequent itemsets because for each candidate, we need to count its occurrences in all transactions. The smaller the candidate set, the less the processing cost required to enumerate frequent itemsets. This is even more decisive in the case of graph mining since to count support as well as to check the downward closure property for each candidate, we need to perform many graph and subgraph isomorphism tests.

To address this issue in mining frequent itemsets, an effective hash-based method called DHP [10] was introduced to Apriori. This algorithm uses hash counts to filter the candidates. Also, the generation of smaller candidate sets enables DHP to effectively trim the transaction database. In this paper, we propose using the hash-based approach combined with the transaction identifier (TID) list [1] for the AGM algorithm to filter the generated adjacency matrices before checking the downward closure property. The reason for applying this approach to AGM is that AGM is the only method so far which can mine general frequent subgraphs, including unconnected ones, where the complexity of candidate generation is huge comparing with the other Apriori-like graph mining algorithms. Under this condition, our hash-based approach is expected to be very effective. This approach is not limited to AGM, rather it can be applied to any Apriori-like graph mining algorithms such as AcGM or FSG. It is also expected to be applied to the Apriori-like algorithms for mining itemsets. As this paper aims at providing precise information about the way these modifications are implemented, we will start with presenting a rather detailed summary of the original AGM method as far as needed for describing the modifications.

2 Graph and Problem Definitions

A graph in which all of its vertices and edges have labels is mathematically defined as follows.

Definition 1 (Labeled Graph) *A labeled graph*

$$G = (V(G), E(G), L_V(V(G)), L_E(E(G)))$$

is made of four sets, the set of vertices $V(G) = \{v_1, v_2, ..., v_k\}$, *the set of edges connecting some vertex pairs in* $V(G)$: $E(G) = \{e_h = (v_i, v_j) | v_i, v_j \in V(G)\}$, *the set of vertex labels* $L_V(V(G)) = \{lb(v_i) | \forall v_i \in V(G)\}$, *and the set of edge labels* $L_E(E(G)) = \{lb(e_h) | \forall e_h \in E(G)\}$. *If* e_h *is undirected, both* (v_i, v_j) *and* (v_j, v_i) *belong to* $E(G)$. *The number of vertices,* $|V(G)|$, *is called the size of graph* G.

If all the vertices and edges of a graph have the same vertex and edge label associated with them, we will call it an *unlabeled* graph.

Definition 2 (Subgraph and Induced Subgraph) *Given a graph* $G = (V(G), E(G), L_V(V(G)), L_E(E(G)))$, *a subgraph of* G, $G_s = (V(G_s), E(G_s), L_V(V(G_s)), L_E(E(G_s)))$, *is a graph which fulfills the following conditions.*

$$V(G_s) \subset V(G), E(G_s) \subset E(G).$$

A subgraph whose size is k is called a k-subgraph.

If $V(G_s) \subset V(G)$ and $E(G_s)$ contains all the edges of $E(G)$ that connect vertices in $V(G_s)$, we call G_s an induced subgraph of G. In this case, G_s is said to be included in G and denoted as $G_s \subset G$.

The aforementioned graph G is represented by an adjacency matrix X which is a very well-known representation in mathematical graph theory [4, 5]. The transformation from G to X does not require much computational effort [6, 7].

Definition 3 (Adjacency Matrix) *Given a graph $G = (V(G), E(G), L_V(V(G)), L_E(E(G)))$, the adjacency matrix X is a square matrix which has the following (i, j)-element, x_{ij},*

$$x_{ij} = \begin{cases} num(lb(e_h)) & if\ e_h = (v_i, v_j) \in E(G) \\ 0 & if\ (v_i, v_j) \notin E(G) \end{cases},$$

where $i, j \in \{1, \ldots, |V(G)|\}$ and $num(lb(e_h))$ is a natural number assigned to an edge label $lb(e_h)$.

Each element of an adjacency matrix in the standard definition is either '0' or '1', whereas each element in Definition 3 can be the number of an edge label.

The vertex corresponding to the i-th row (i-th column) of an adjacency matrix is called the i-th vertex. The adjacency matrix of a graph of size k and the corresponding graph are denoted as X_k and $G(X_k)$, respectively. An identical graph structure can be represented by multiple adjacency matrices depending on the assignment of its rows and columns to its vertices. Such adjacency matrices are mutually convertible using the so-called transformation matrix [6, 7].

In order to reduce the number of candidates of frequent induced subgraphs, a coding method of the adjacency matrices needs to be introduced.

Definition 4 (Code of Adjacency Matrix) *In case of an undirected graph, the code of a vertex-sorted adjacency matrix X_k is defined as*

$$code(X_k) = x_{1,1}x_{1,2}x_{2,2}x_{1,3}x_{2,3}x_{3,3}x_{1,4} \cdots x_{k-1,k}x_{k,k},$$

where $x_{i,j}$ is the (i, j)-element. In case of a directed graph, it is defined as

$$code(X_k) = c_1 c_2 \ldots c_{\frac{k(k-1)}{2}},$$

where $c_{\frac{j(j-1)}{2}-(j-i-1)} = (|L_E(E(G))| + 1)x_{j,i} + x_{i,j}$ $(i < j)$.

A canonical form of adjacency matrices representing an identical graph is introduced to remove this ambiguity and to handle the matrices efficiently.

Definition 5 (Canonical Form) *The canonical form of adjacency matrices representing a graph is the unique matrix having a minimum (maximum) code. This minimum (maximum) code is referred to as the canonical label of the graph.*

The canonical label is the unique code that is invariant on the ordering of the vertices and edges in the graph.

Definition 6 (Graph and Subgraph Isomorphism) *Two graphs* $G_1 = (V(G_1), E(G_1), L_V(V(G_1)), L_E(E(G_1)))$ *and* $G_2 = (V(G_2), E(G_2), L_V(V(G_2)), L_E(E(G_2)))$ *are isomorphic if there is a bijection mapping between* $V(G_1)$ *and* $V(G_2)$ *such that* G_1 *and* G_2 *are identical, i.e., each edge of* $E(G_1)$ *is mapped to a single edge in* $E(G_2)$ *and vice versa. This mapping must also preserve the labels of vertices and edges.*

The problem of graph isomorphism is to determine an isomorphism between G_1 *and* G_2*. Meanwhile, the problem of subgraph isomorphism is to find an isomorphism between* G_2 *and a subgraph of* G_1*.*

Two graphs are isomorphic if and only if they have the same canonical label. Canonical labels therefore are extremely useful as they allow us to quickly compare two graphs regardless of the original vertex and edge ordering. However, the problem of determining the canonical label of a graph is equivalent to determining the isomorphism between two graphs [9]. Graph isomorphism is not known to be in P or NP-complete and subgraph isomorphism is known to be NP-complete [5].

Definition 7 (Graph Transaction and Graph Database) *A graph G is referred to as a graph transaction or simply a transaction, and a set of graph transactions GD, where* $GD = \{G_1, G_2, ..., G_n\}$*, is referred to as a graph database.*

Definition 8 (Support) *Given a graph database GD and a graph* G_s*, the support of* G_s *is defined as*

$$sup(G_s) = \frac{number\ of\ graph\ transactions\ G\ where\ G_s \subset G \in GD}{total\ number\ of\ graph\ transactions\ G \in GD}.$$

The (induced) subgraph having the support more than the minimum support specified by the user is called a frequent (induced) subgraph. A frequent (induced) subgraph whose size is k is called a k-frequent (induced) subgraph.

Definition 9 (Frequent Induced Subgraph Mining Problem) *Given a graph database GD and a minimum support (minsup), the problem is to derive all frequent induced subgraphs in GD.*

3 Algorithm of AGM-Hash

3.1 Candidate Generation

The generation of candidate frequent induced subgraphs in the AGM method is made by the levelwise search in terms of the size of the subgraphs [7]. Let X_k and Y_k be vertex-sorted adjacency matrices of two frequent induced subgraphs $G(X_k)$ and $G(Y_k)$ of size k. They can be joined if and only if the following conditions are met.

1. X_k and Y_k are identical except the k-th row and the k-th column, and the vertex labels of the first $(k-1)$ vertices are the same between X_k and Y_k.
2. In case that labels of the k-th vertex of $G(X_k)$ and the k-th vertex of $G(Y_k)$ are identical, $code(X_k) \leq code(Y_k)$. In case that they are not identical, $num(lb(v_k \in V(G(X_k)) < num(lb(v_k \in V(G(Y_k))))$.

If X_k and Y_k are joinable, their joining operation is defined as follows.

$$X_k = \begin{pmatrix} X_{k-1} & \boldsymbol{x}_1 \\ \boldsymbol{x}_2^T & 0 \end{pmatrix}, Y_k = \begin{pmatrix} X_{k-1} & \boldsymbol{y}_1 \\ \boldsymbol{y}_2^T & 0 \end{pmatrix},$$

$$Z_{k+1} = \begin{pmatrix} X_{k-1} & \boldsymbol{x}_1 & \boldsymbol{y}_1 \\ \boldsymbol{x}_2^T & 0 & z_{k,k+1} \\ \boldsymbol{y}_2^T & z_{k+1,k} & 0 \end{pmatrix} = \left(\begin{array}{cc|c} \multicolumn{2}{c|}{X_k} & \boldsymbol{y}_1 \\ & & z_{k,k+1} \\ \hline \boldsymbol{y}_2^T & z_{k+1,k} & 0 \end{array}\right),$$

$$lb(v_i \in V(G(Z_{k+1}))) = lb(v_i \in V(G(X_k))) = lb(v_i \in V(G(Y_k)))(i = 1, \ldots, k-1),$$
$$lb(v_k \in V(G(Z_{k+1}))) = lb(v_k \in V(G(X_k))),$$
$$lb(v_{k+1} \in V(G(Z_{k+1}))) = lb(v_{k+1} \in V(G(Y_k))).$$

X_k and Y_k are called the first and second generator matrix, respectively. The aforementioned condition 1 ensures that $G(X_k)$ and $G(Y_k)$ share the common structure except the vertex corresponding to the last row and column of each adjacency matrix. Meanwhile, condition 2 is required to avoid producing redundant adjacency matrices by exchanging X_k and Y_k, *i.e.*, taking Y_k as the first generator matrix and X_k as the second generator matrix. Unlike the joining of itemsets in which two frequent k-itemsets lead to a unique $(k+1)$-itemset [1, 10], the joining of two adjacency matrices of two frequent induced subgraphs of size k can lead to multiple distinct adjacency matrices of size $k+1$. Two elements $z_{k,k+1}$ and $z_{k+1,k}$ of Z_{k+1} are not determined by X_k and Y_k. In the case of mining undirected graph data, the possible graph structures for $G(Z_{k+1})$ are those wherein there is a labeled edge and those wherein there is no edge between the k-th vertex and the $(k+1)$-th vertex. Therefore, $(|L_E|+1)$ adjacency matrices satisfying $z_{k,k+1} = z_{k+1,k}$ are generated, where $|L_E|$ is the number of edge labels. In the case of mining directed graph data, $(|L_E|+1)^2$ adjacency matrices are generated since $z_{k,k+1} = z_{k+1,k}$ does not apply. The vertex-sorted adjacency matrix generated under the above conditions is called a *normal form*.

In the standard basket analysis, a $(k+1)$-itemset becomes a candidate frequent itemset only when all of its k-sub-itemsets are confirmed to be frequent itemsets [1], *i.e.*, the downward closure property is fulfilled. Similarly, the graph G of size $k+1$ is a candidate of frequent induced subgraphs only when all adjacency matrices generated by removing from the graph G the i-th vertex v_i $(1 \leq i \leq k+1)$ and all of its connected links are confirmed to be adjacency matrices of frequent induced subgraphs of size k. As this algorithm generates only normal-form adjacency matrices in the earlier k-levels, if the adjacency matrix of the graph generated by removing the i-th vertex v_i is a non-normal form, it must be transformed to a normal form to check if it matches one of the normal form matrices found earlier. An adjacency matrix X_k of a non-normal form is transformed into a normal form X'_k by reconstructing the matrix structure in a bottom up manner. The reader is referred to [7] for a detailed mathematical treatment of normalization.

Though this heuristic helps prune redundant candidates efficiently, it consumes much computational effort due to the large number of adjacency matrices produced. Moreover, the task of checking if a graph of size $(k+1)$ can be a candidate, *i.e.*, checking its downward closure property, is very expensive since we need to perform many graph isomorphism tests. In Section 3.3, we will introduce a hash-based method for AGM, called AGM-Hash, to reduce the number of adjacency matrices of one size larger before checking the downward closure property. This technique helps prune a significant number of redundant candidates right after the joining operation has been performed.

3.2 Frequency Counting

Once the candidate frequent induced subgraphs have been generated, we need to count their frequency. The naive strategy to achieve this end is for each candidate to scan each one of the graph transactions in the graph database and determine if it is contained or not using subgraph isomorphism. However, having to compute these isomorphisms is particularly expensive and this approach is not feasible for large datasets [9].

Frequency of each candidate frequent induced subgraph is counted by scanning the database after generating all the candidates of frequent induced subgraphs and obtaining their canonical forms. Every subgraph of each graph transaction G in the graph database can be represented by an adjacency matrix X_k, but it may not be a normal form in most cases. Since the candidates of frequent induced subgraphs in the AGM method are normal forms, it is required to derive the normal forms of every induced subgraph of G at each level. The frequency of each candidate is counted based on all normal forms of the induced subgraphs of G of same size and the canonical label of that candidate. When the value of the count exceeds the minimum support threshold, the subgraph is a frequent induced subgraph. Since some normal forms may represent the same graph and thus, the number of normal forms of induced subgraphs present in each transaction can be extremely large, we need to reduce that number of normal forms before counting support. Note that a normal form of size k is generated from two normal forms of size $k-1$. Therefore, if one of the two normal forms of size $k-1$ does not represent a frequent induced subgraph, then the normal form of size k cannot represent any frequent induced subgraph, and thus should not be considered for support counting. For this reason, AGM does not use normal forms of infrequent induced subgraphs to produce normal forms of any size larger.

In the AprioriTid algorithm [1], the transaction database is not used for counting support after the first pass. Rather, the time needed for the support counting procedure is reduced by replacing every transaction in the database by the set of candidate itemsets that occur in that transaction. This is done repeatedly at every iteration k. If a transaction does not contain any candidate

k-itemsets, then the adapted transaction database will not have an entry for this transaction. We employ this idea of ApriotiTid for counting support. The adapted graph database at level k is denoted as $\overline{C}^k$. Each member of the set $\overline{C}^k$ is of the form $< TID, \{X_k\} >$, where each X_k is a normal form of a potentially k-frequent induced subgraph present in the transaction with identifier TID. The member of $\overline{C}^k$ corresponding to transaction t is $< t.TID, C_t^k >$ with $C_t^k = \{c \in C^k \mid c \text{ contained in } t\}$, where C^k is the set of normal forms of candidate k-frequent induced subgraphs. If transaction t does not contain any candidate k-frequent induced subgraph, *i.e.*, $C_t^k = \emptyset$, then $\overline{C}^k$ will not have an entry for this transaction. As $\overline{C}^k$ will be used to generate normal forms of one size larger in order to count support for candidate $(k+1)$-frequent induced subgraphs, if $c \in C^k$ and c is not a normal form of any frequent induced subgraph, then c is removed from C_t^k.

3.3 Details of the AGM-Hash Algorithm

Fig. 1 gives the algorithmic form of AGM-Hash which is divided into two parts. Part 1 detects the set of 1-frequent induced subgraphs, *i.e.*, individual vertices that frequently occur in the graph database GD, and the set of their normal forms (*i.e.*, N^1). In addition, this part computes $\overline{C}^1$ which contains normal forms of 1-frequent induced subgraphs present in each graph transaction. Finally, AGM-Hash makes a hash table for 2-subgraphs (*i.e.*, H^2) using $\overline{C}^1$.

Part 2 consists of two phases. The first phase is to generate candidates of k-frequent induced subgraph C^k based on the normal forms of $(k-1)$-frequent induced subgraphs (*i.e.*, N^{k-1}) made in the previous pass and the hash table (*i.e.*, H^k), and calculates the canonical labels for each candidate. In the second phase, AGM-Hash generates the normal forms of induced subgraphs contained in each graph transaction $t \in \overline{C}^{k-1}$. Based on those normal forms, AGM-Hash counts the support for the candidates according to their canonical labels. If a candidate appears in transaction t, it will be added to C_t^k which is the set of candidates contained in t. These steps are repeated for every $t \in \overline{C}^{k-1}$. If $C_t^k \neq \emptyset$, it will be used for making the hash table for $(k+1)$-subgraphs (*i.e.*, H^{k+1}) and added to $\overline{C}^k$. In the following, the normal forms of k-frequent induced subgraphs (*i.e.*, N^k) and the set of k-frequent induced subgraphs are realized. Finally, AGM-Hash removes from C_t^k all normal forms of infrequent induced subgraphs, and from $\overline{C}^k$ all transactions t which do not contain any normal forms (*i.e.*, $C_t^k = \emptyset$). $\overline{C}^k$ is cached on the hard disk due to its huge size.

The subprocedures for the AGM-Hash algorithm are shown in Fig. 2. Same as AGM [7], AGM-Hash also generates adjacency matrices of size k based on N^{k-1}. However, similar to DHP [10], AGM-Hash employs the hash table (can be replaced by a bit vector) to test the validity of each adjacency matrix. A generated adjacency matrix of size k will be checked the downward closure property to be a candidate of k-frequent induced subgraphs only if that matrix is hashed into a hash entry whose value is larger than or equal to the minimum support

Part 1

```
$s = minsup|GD|$; //a minimum support
$F^1$ = { 1-frequent induced subgraphs };
$N^1$ = { normal forms of 1-frequent induced subgraphs };
$\overline{C}^1 = \{< t.TID, C_t^1 >| C_t^1$ are normal forms of 1-frequent induced subgraphs present in $t\}$;
set all the buckets of $H^2$ to zero;
forall entries $t \in \overline{C}^1$ do
    if $C_t^1 \neq \emptyset$ then do begin
        $N_t^2$ = gen-norm $(C_t^1)$; //generate normal forms of induced subgraphs in $t$
        forall normal forms of size 2, $X \in N_t^2$, do
            $H^2[h_2(X)] + +$;
    end
```

Part 2

```
$k = 2$;
while $C^k$ = gen-candidate$(N^{k-1}, H^k) \neq \emptyset$ do begin
    forall entries $t \in \overline{C}^{k-1}$ do begin
        $N_t^k$=gen-norm $(C_t^{k-1})$; //generate normal forms of induced subgraphs in $t$
        forall candidates $c \in C^k$ do
            if $c \in N_t^k$ then do begin
                $c.canonical\_label.count + +$; //increase count of corresponding can. label
                $C_t^k$ += $\{c\}$
            end
        if $C_t^k \neq \emptyset$ then do begin
            $\overline{C}^k$ += $< t.TID, C_t^k >$;
            makehasht$(C_t^k, H^{k+1})$;
        end
    end
    $N^k = \{c \in C^k \mid c.canonical\_label.count \geq s\}$;
    $F^k$ ={k-frequent induced subgraphs extracted from $N^k$};
    forall entries $t \in \overline{C}^k$ do begin
        forall $c \in C_t^k$ do
            if $c \notin N^k$ then $C_t^k$ -= $\{c\}$;
        if $C_t^k = \emptyset$ then $\overline{C}^k$-=$< t.TID, C_t^k >$;
    end
    k++;
end
```

Fig. 1. Main program of the AGM-Hash algorithm.

$s = minsup|GD|$. AGM-Hash is unique in that it employs graph invariants to make the hash function. In this paper, the hash function value of an induced subgraph is calculated based on its number of edges having the edge label of each type as well as that of vertices, with different scales between edge labels and vertex labels due to the need of diversity of hash values reflecting the graph topology. For example, given an induced subgraph X of size 4 which has two and four edges having the labels numbered 1 and 2, respectively, three and one vertices having the labels numbered 1 and 2, respectively, and a hash table whose size is 80, then $h_4(X) = (2 \times 10^1 + 4 \times 10^2) + (3 \times 1 + 1 \times 2)$ mod 80=25. Note that the number of k-candidate subgraphs depends on the number of k-subsets of vertices which has the largest value at $k = [N/2]$, where N is the number of vertices. It also depends on the number of edge labels. Therefore, the hash table size should be chosen reflecting this observation. In this paper the hash

Procedure C^k = gen-candidate(N^{k-1}, H^k)

$C^k = \emptyset$
foreach pair of normal forms, N_i^{k-1} and N_j^{k-1}, do
    if N_i^{k-1},N_j^{k-1} can be joined then
        C^k += { normal forms of size k, X, generated from N_i^{k-1}, N_j^{k-1} | $H^k[h_k(X)] \geq s$};
forall $c \in C^k$ do
    if c holds the downward closure property then
        calculate *c.canonical_label*;//calculate canonical label for c
    else C^k-={c};

Procedure makehasht(C_t^k, H^{k+1})

N_t^{k+1} = gen-norm(C_t^k);
forall normal forms of size $(k+1)$, $X \in N_t^{k+1}$, do
    $H^{k+1}[h_{k+1}(X)] + +$;

Procedure N_t^{k+1} = gen-norm(C_t^k)

$N_t^{k+1} = \emptyset$;
foreach pairs of normal forms of size k, C_i^k and $C_j^k \in C_t^k$, do
    if C_i^k and C_j^k can be joined then do begin
        construct X, a normal form of the $(k+1)$-induced subgraph in t whose vertices are $k+1$ vertices of C_i^k and C_j^k;
        $N_t^{k+1} + = \{X\}$;
    end

Fig. 2. Subprocedures for the AGM-Hash algorithm.

table size was empirically chosen as $2^{|L_E|} \times \binom{|T|}{k}$, where $|L_E|$ is the number of edge labels, $|T|$ is the average graph size in the graph database and $k = [|T|/2]$. Moreover, unlike the DHP method, AGM-Hash tests the validity of adjacency matrices before checking if they fulfill the downward closure property since graph isomorphism tests are particularly expensive.

4 Experiments

To assess the performance of the AGM-Hash algorithm, we conducted several experiments on both synthetic and real-world graph databases. The AGM-Hash method proposed in this paper was implemented by C++. All experiments were done on a Pentium 2.8 GHz machine with 2 GB main memory, running the Windows XP operating system. All the times reported are in seconds.

4.1 Simulation Results

The basic performance of the proposing method was examined using the graph-structured transactions that were artificially generated in a random manner. The number of vertices in a graph, is determined by the gaussian distribution having the average of $|T|$ and the standard deviation of 1. The edges are attached randomly with the probability of p. The vertex labels and edge labels are randomly

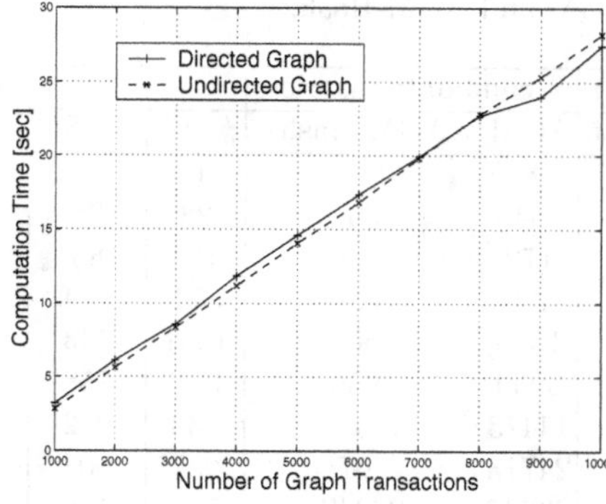

Fig. 3. Computation time v.s. number of transactions.

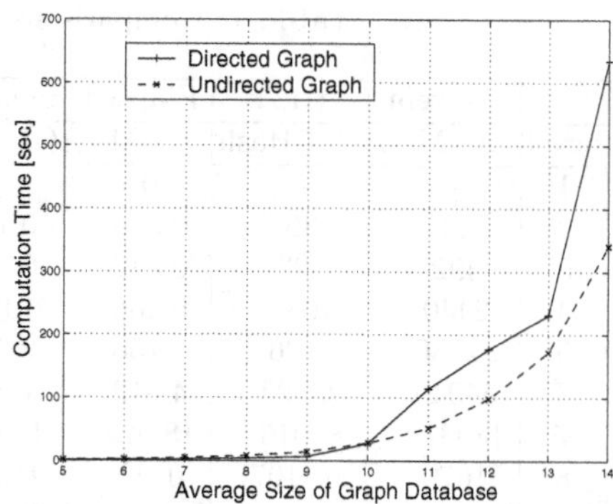

Fig. 4. Computation time v.s. average transaction size.

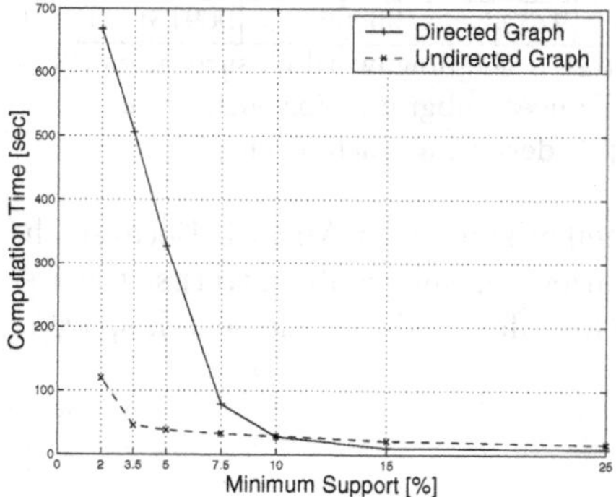

Fig. 5. Computation time v.s. minimum support.

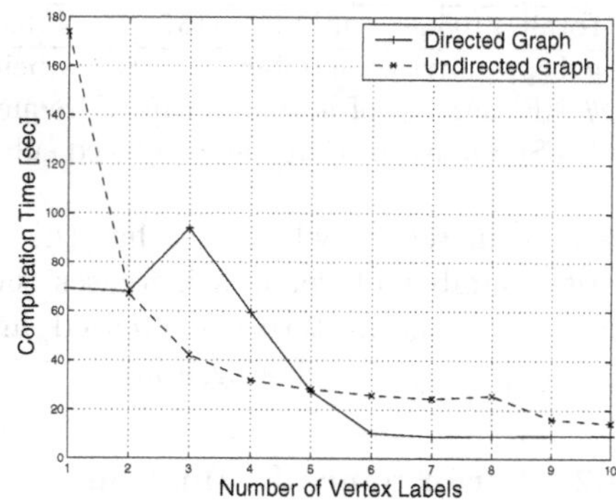

Fig. 6. Computation time v.s. number of vertex labels.

determined with equal probability. The number of vertex labels and the number of edge labels are denoted as $|L_V|$ and $|L_E|$, respectively. Similarly, L basic patterns of induced subgraphs having the average size of $|I|$ are generated. One of them is chosen by equal probability, *i.e.*, $1/L$, and overlaid on each transaction. The test data are generated for both cases of directed and undirected graphs. The default parameter values are $|GD| = 10000$, $|T| = 10$, $L = 10$, $|I| = 7$, $|L_V| = 5$, $|L_E| = 5$, $p = 50\%$ and $minsup$ (minimum support) $= 10\%$.

Figs. 3, 4, 5 and 6 show the computation time variations when $|GD|$, $|T|$, $minsup$ and $|L_V|$ are changed respectively in the frequent induced subgraph problem. The default parameter values were used in each figure except the parameter that was changed. The computation time is observed to increase almost linearly with $|GD|$ in Fig. 3 and to be exponential to $|T|$ in Fig. 4. Fig. 5 indicates that computation time decreases rapidly as $minsup$ increases. This is trivial since less $minsup$ results in larger number of frequent patterns. The observed tendencies in Figs. 3, 4 and 5 are almost identical with the properties of the Apriori algorithm [1]. Fig. 6 shows that computation time mostly decreases

Table 1. Comparisons of AGM and AGM-Hash.

	# kept matrices		Elapsed time (sec)		# candidate matrices			
Level	AGM	AGM-Hash	AGM	AGM-Hash	AGM	AGM-Hash	# NF	# FSG
1	-	-	0	0	8	8	6	6
2	84	36	0.016	0.187	84	36	28	28
3	492	223	0.062	0.422	172	129	103	86
4	2440	1084	0.266	1.047	436	406	371	214
5	12380	5826	1.156	2.781	1516	1507	1351	443
6	54040	23923	4.812	7.203	5050	4956	4595	758
7	180872	89916	18.922	19.031	14473	14384	12334	992
8	316796	157107	51.922	41.531	24713	24673	22377	904
9	310668	130337	94.969	67.359	26947	26947	25833	522
10	212740	47434	121.297	73.609	17576	17562	17562	174
11	94604	10732	133.984	75.328	5832	5832	5832	30
12	25432	526	137.375	75.516	526	526	526	2
Total	1210548	467144	137.375	75.516	97353	96966	90918	4159

kept matrices: number of matrices being tested the downward closure property.
NF: number of normal forms of frequent induced subgraphs for each level.
FSG: number of frequent induced subgraphs derived at each level.

as $|L_V|$ increases, which also has the tendency similar to Apriori. Because the larger number of vertices increases the variety of subgraph patterns, the frequency of each pattern is decreased, and this effect reduces the search space of the frequent induced subgraphs.

4.2 Performance Evaluation

The performance of the proposed AGM-Hash algorithm was compared with that of the original AGM method using a real-world graph database. This graph database consists of 82 mutagenesis compounds [8] whose average size is about 12.5. There are 7 graph transactions having the maximum size of 15 and 1 graph transaction having the minimum size of 9. The parameters evaluated include the number of adjacency matrices being tested the downward closure property, the number of candidates and computation time. It should be noted that unlike recently proposed methods, AGM and AGM-Hash find frequent induced subgraphs, *i.e.*, they can be either connected or unconnected. It is also worth noting that AcGM, which is a variant of AGM to mine frequent connected subgraphs, is fast enough [6] and it is AGM that needs improvement.

Table 1 shows the experimental results obtained from both AGM-Hash and AGM given a minimum support of 10%. The number of frequent induced subgraphs and the number of their normal forms are also shown together for reference. The maximum size of frequent induced subgraphs discovered by AGM and AGM-Hash is 12. As can be seen from Table 1, AGM-Hash efficiently reduces the number of adjacency matrices generated by the joining operation before passing them to the step of checking the downward closure property. This task, as shown earlier, is particularly expensive since it requires many graph isomorphism tests. As a result, computation time required by AGM-Hash is reduced significantly compared with that required by AGM. Meanwhile, the number of candidate matrices in the case of AGM-Hash is reduced slightly compared with that in AGM, which confirms the efficiency of the pruning method based on the Apriori-like principle and employed in AGM.

It can also be seen from Table 1 that computation time is less in the case of AGM-Hash since level 8 though the number of adjacency matrices being checked the downward closure property has been reduced significantly in earlier levels. This is mainly attributed to the additional cost required for making the hash table as well as for testing the validity of generated adjacency matrices.

5 Conclusion and Future Work

A hash-based approach to the AGM algorithm has been introduced to improve the method in terms of efficiency. The use of graph invariants for calculating the hash values and the combination of the hashing technique and the TID list can be highlighted as the main features of the proposed AGM-Hash method. Its performance has been evaluated on both the synthetic graph data and the real-world chemical mutagenesis data. The powerful performance of this approach under some practical conditions has been confirmed through these evaluations. Future work includes a study of the graph invariants to be used for making the hash table to further enhance the proposed method. Also, a design of the hash table for the purpose of deriving characterized frequent subgraphs is being considered. The AGM-Hash approach can be extended to any Apriori-like graph mining algorithm.

References

1. Agrawal, R. and Srikant, R. 1994. Fast algorithms for mining association rules. In Proceedings of the 20th VLDB Conference, pp.487–499.
2. Borgelt, C. and Berthold, M.R. 2002. Mining molecular fragments: Finding relevant substructures of molecules. In Proceedings of the 2nd IEEE International Conference on Data Mining (ICDM'02), pp.51–58.
3. Cook, D.J. and Holder, L.B. 1994. Substructure discovery using minimum description length and background knowledge, Journal of Artificial Intelligence Research, Vol.1, pp.231–255.
4. Diestel, R. 2000. Graph Theory. Springer-Verlag, New York.
5. Fortin, S. 1996. The graph isomorphism problem. Technical Report 96-20, University of Alberta, Edmonton, Alberta, Canada.
6. Inokuchi, A., Washio, T., Nishimura, K. and Motoda, H. 2002. A fast algorithm for mining frequent connected subgraphs. IBM Research Report RT0448, Tokyo Research Laboratory, IBM Japan.
7. Inokuchi, A., Washio, T. and Motoda, H. 2003. Complete mining of frequent patterns from graphs: Mining graph data. Machine Learning, Vol.50, pp.321–354.
8. Debnath, A.K., *et al.* 1991. Structure-activity relationship of mutagenic aromatic and heteroaromatic nitro compounds. Correlation with molecular orbital energies and hydrophobicity. Journal of Medical Chemistry, Vol. 34, pp.786–797.

9. Kuramochi, M. and Karypis, G. 2001. Frequent subgraph discovery. In Proceedings of the 1st IEEE International Conference on Data Mining (ICDM'01), pp.313–320.
10. Park, J.S., Chen, M-.S. and Yu, P.S. 1997. Using a hash-based method with transaction trimming for mining association rules. IEEE Trans. Knowledge and Data Engineering, Vol.9, No.5, pp.813–825.
11. Washio, T. and Motoda, H. 2003. State of the art of graph-based data mining. SIGKDD Explorations, Vol.5, No.1, pp.59–68.
12. Yan, X. and Han, J. 2002. gSpan: Graph-based structure pattern mining. In Proceedings of the 2nd IEEE International Conference on Data Mining (ICDM'02), pp.721–724.
13. Yoshida, K. and Motoda, H. CLIP: Concept learning from inference patterns. Artificial Intelligence, Vol.75, No.1, pp.63–92.

Evaluation of Rule Interestingness Measures with a Clinical Dataset on Hepatitis

Miho Ohsaki[1], Shinya Kitaguchi[2], Kazuya Okamoto[2], Hideto Yokoi[3], and Takahira Yamaguchi[4]

[1] Department of Information Systems Design, Doshisha Univerisity
1-3, Tataramiyakodani, Kyotanabe-shi, Kyoto 610-0321, Japan
mohsaki@mail.doshisha.ac.jp
[2] Department of Computer Science, Shizuoka University
3-5-1, Johoku, Hamamatsu-shi, Shizuoka 432-8011, Japan
{cs8037,cs9026}@cs.inf.shizuoka.ac.jp
[3] Department of Administration Engineering, Keio University
3-14-1, Hiyoshi, Kohoku-ku, Yokohama-shi, Kanagawa 223-8522, Japan
yamaguti@ae.keio.ac.jp
[4] Division of Medical Informatics, Chiba University Hospital
1-8-1, Inohana, Chuo-ku, Chiba-shi, Chiba 260-0856, Japan
yokoih@telemed.ho.chiba-u.ac.jp

Abstract. This research empirically investigates the performance of conventional rule interestingness measures and discusses their practicality for supporting KDD through human-system interaction in medical domain. We compared the evaluation results by a medical expert and those by selected measures for the rules discovered from a dataset on hepatitis. Recall, Jaccard, Kappa, CST, χ^2-M, and Peculiarity demonstrated the highest performance, and many measures showed a complementary trend under our experimental conditions. These results indicate that some measures can predict really interesting rules at a certain level and that their combinational use will be useful.

1 Introduction

Rule interestingness is one of active fields in Knowledge Discovery in Databases (KDD), and there have been many studies that formulated interestingness measures and evaluated rules with them instead of humans. However, many of them were individually proposed and not fully evaluated from the viewpoint of theoretical and practical validity. Although some latest studies made a survey on conventional interestingness measures and tried to categorize and analyze them theoretically [1–3], little attention has been given to their practical validity – whether they can contribute to find out really interesting rules.

Therefore, this research aims to (1) systematically grasp the conventional interestingness measures, (2) compare them with real human interest through an experiment, and (3) discuss their performance to estimate real human interest and their utilization to support human-system interaction-based KDD. The

J.-F. Boulicaut et al. (Eds.): PKDD 2004, LNAI 3202, pp. 362–373, 2004.

experiment required the actual rules from a real dataset and their evaluation results by a human expert. We determined to set the domain of this research as medical data mining and to use the outcome of our previous research on hepatitis [4] because medical data mining is scientifically and socially important and especially needs human-system interaction support for enhancing rule quality.

In this paper, Section 2 introduces conventional interestingness measures and selects dozens of measures suitable to our purpose. Section 3 shows the experiment that evaluated the rules on hepatitis with the measures and compared the evaluation results by them with those by a medical expert. It also discusses their performance to estimate real human interest, practicality to support KDD based on human-system interaction, and advanced utilization by combining them. Section 4 concludes the paper and comments on the future work.

2 Conventional Rule Interestingness Measures

The results of our and other researchers' surveys [1–3, 5] show that interestingness measures can be categorized with the several factors in Table 1. The subject to evaluate rules, a computer or human user, is the most important categorization factor. Interestingness measures by a computer and human user are called objective and subjective ones, respectively. There are more than forty objective measures at least. They estimate how a rule is mathematically meaningful based on the distribution structure of the instances related to the rule. They are mainly used to remove meaningless rules rather than to discover really interesting ones for a human user, since they do not include domain knowledge [6–17]. In contrast, there are only a dozen of subjective measures. They estimate how a rule fits with a belief, a bias, or a rule template formulated beforehand by a human user. Although they are useful to discover really interesting rules to some extent due to their built-in domain knowledge, they depend on the precondition that a human user can clearly formulate his/her own interest and do not discover absolutely unexpected knowledge. Few subjective measures adaptively learn real human interest through human-system interaction.

The conventional interestingness measures, not only objective but also subjective, do not directly reflect the interest that a human user really has. To avoid the confusion of real human interest, objective measure, and subjective measure, we clearly differentiate them. **Objective Measure**: The feature such as the

Table 1. The factors to categorize interestingness measures.

Factors	Meaning	Sub-factors
Subject	Who evaluates?	Computer / Human user
Object	What is evaluated?	Association rule / Classification rule
Unit	By how many objects?	A rule / A set of rules
Criterion	Based on what criterion?	Absolute criterion / Relative criterion
Theory	Based on what theory?	Number of instances / Probability / Statistics / Information / Distance of rules or attributes / Complexity of a rule

correctness, uniqueness, and strength of a rule, calculated by the mathematical analysis. It does not include human evaluation criteria. **Subjective Measure**: The similarity or difference between the information on interestingness given beforehand by a human user and those obtained from a rule. Although it includes human evaluation criteria in its initial state, the calculation of similarity or difference is mainly based on the mathematical analysis. **Real Human Interest**: The interest which a human user really feels for a rule in his/her mind. It is formed by the synthesis of cognition, domain knowledge, individual experiences, and the influences of the rules that he/she evaluated before.

This research specifically focuses on objective measures and investigates the relation between them and real human interest. We then explain the details of objective measures here. They can be categorized into some groups with the criterion and theory for evaluation. Although the criterion is absolute or relative as shown in Table 1, the majority of present objective measures are based on an absolute criterion. There are several kinds of criterion based on the following factors: Correctness – How many instances the antecedent and/or consequent of a rule support, or how strong their dependence is [6, 7, 13, 16], Generality – How similar the trend of a rule is to that of all data [11] or the other rules, Uniqueness – How different the trend of a rule is from that of all data [10, 14, 17] or the other rules [11, 13], and Information Richness – How much information a rule possesses [8]. These factors naturally prescribe the theory for evaluation and the interestingness calculation method based on the theory. The theory includes the number of instances [6], probability [12, 14], statistics [13, 16], information [7, 16], the distance of rules or attributes [10, 11, 17], and the complexity of a rule [8] (See Table 1). We selected the objective measures in Table 2 as many and various as possible for the experiment in Section 3. Note that many of them do not have the reference numbers of their original papers but those of survey papers in Table 2 to avoid too many literatures. We call GOI with the dependency coefficient value at the double of the generality one GOI-D, and vice versa for GOI-G and adopt the default value, 0.5, for the constant α of Peculiarity.

Now, we explain the motivation of this research in detail. Objective measures are useful to automatically remove obviously meaningless rules. However, some factors of evaluation criterion have contradiction to each other such as generality and uniqueness and may not match with or contradict to real human interest. In a sense, it may be proper not to investigate the relation between objective measures and real human interest, since their evaluation criterion does not include the knowledge on rule semantics and are obviously not the same of real human interest. However, our idea is that they may be useful to support KDD through human-system interaction if they possess a certain level of performance to detect really interesting rules. In addition, they may offer a human user unexpected new viewpoints. Although the validity of objective measures has been theoretically proven and/or experimentally discussed using some benchmark data [1–3], very few attempts have been made to investigate their comparative performance and the relation between them and real human interest for a real application [5]. Our investigation will be novel in this light.

Table 2. The objective measures of rule interestingness used in this research. **N:** Number of instances included in the antecedent and/or consequent of a rule. **P:** Probability of the antecedent and/or consequent of a rule. **S:** Statistical variable based on P. **I:** Information of the antecedent and/or consequent of a rule. **D:** Distance of a rule from the others based on rule attributes.

Measure Name (**Abbreviation**) [Reference Number of Literature]	**Theory**
Mathematical Definition	
Coverage [5]	**P**
$P(A)$, $P(A)$: Probability of antecedent.	
Prevalence [5]	**P**
$P(C)$, $P(C)$: Probability of consequent.	
Precision [3, 5]	**P**
$P(C\|A)$, $P(C\|A)$: Conditional probability of consequent for antecedent.	
Recall [5]	**P**
$P(A\|C)$, $P(A\|C)$: Conditional probability of antecedent for consequent.	
Support [1, 3, 5]	**P**
$P(C\|A) * P(A)$	
Specificity [5]	**P**
$P(\neg C\|\neg A)$, $\neg X$: Negation of X.	
Accuracy [5]	**P**
$P(C\|A) * P(A) + P(\neg C\|\neg A) * P(\neg A)$	
Lift [5]	**P**
$P(C\|A)/P(C)$	
Leverage [5]	**P**
$P(C\|A) - P(A) * P(C)$	
Added Value (**AV**) [3]	**P**
$P(C\|A) - P(C)$	
Relative Risk (**RR**) [1]	**P**
$P(C\|A)/P(C\|\neg A)$	
Jaccard [3]	**P**
$P(A \cap C)/\{P(A) + P(C) - P(A \cap C)\}$ $P(A \cap C)$: Probability of antecedent and consequent.	
Certainty Factor (**CF**) [3]	**P**
$\{P(C\|A) - P(C)\}/\{1 - P(C)\}$	
Odds Ratio (**OR**) [3]	**P**
$\{P(A \cap C) * P(\neg A \cap \neg C)\}/\{P(A \cap \neg C) * P(\neg A \cap C)\}$	
Yule's Q [3]	**P**
$(OR - 1)/(OR + 1)$	
Yule's Y [3]	**P**
$(\sqrt{OR} - 1)/(\sqrt{OR} + 1)$	
Kappa [3]	**P**
$\frac{P(A \cap C) + P(\neg A \cap \neg C) - P(A) * P(C) - P(\neg A) * P(\neg C)}{1 - P(A) * P(C) - P(\neg A) * P(\neg C)}$	
Klosgen's Interestingness (**KI**) [1, 3]	**P**
$\sqrt{P(A \cap C)} * \{P(C\|A) - P(C)\}$	
Brin's Interest (**BI**) [3]	**P**
$P(A \cap C)/\{P(A) * P(C)\}$	
Brin's Conviction (**BC**) [3]	**P**
$\{P(A) * P(\neg C)\}/P(A\|\neg C)\}$	

Gray and Orlowska's Interestingness weighting Dependency (**GOI-D**) [1, 5, 12]	**P**
$((\frac{P(C\mid A)}{P(A)*P(C)})^k - 1) * ((P(A) * P(C))^m$, k, m: Coefficients of dependency and generality.	
GOI weighting Generality (**GOI-G**) [1, 5, 12]	**P**
Definition is the same of GOI-D.	
Collective Strength (**CST**) [3]	**P**
$\frac{P(A\cap C)+P(\neg C\mid\neg A)}{P(A)*P(C)+P(\neg A)*P(\neg C)} * \frac{1-P(A)*P(C)-P(\neg A)*P(\neg C)}{1-P(A\cap C)-P(\neg C\mid\neg A)}$	
Credibility [5, 9]	**P, N**
$\beta_i * P(C) * \lvert P(R_i\mid C) - P(R_i)\rvert * T(R_i)$, β_i: Coefficient of normalization. $P(R_i)$: Probability of the rule R_i. $T(R_i)$: Number of instances in R_i.	
Laplace Correction (**LC**) [3]	**N**
$\{N(A\cap C)+1\}/\{N(A)+2\}$, $N(X)$: Number of instances in X.	
χ^2 Measure (χ^2-**M**) [5, 13]	**S**
$\sum_{event} \frac{(T_{event}-O_{event})^2}{T_{event}}$, $event$: $A \to C, A \to \neg C, \neg A \to C, \neg A \to \neg C$ T_{event}: Theoretical number of instances in $event$, O_{event}: Observed one.	
Gini Index (**Gini**) [3]	**S**
$P(A) * \{P(C\mid A)^2 + P(\neg C\mid A)^2\} + P(\neg A) * \{P(C\mid\neg A)^2 + P(\neg C\mid\neg A)^2\}$ $-P(C)^2 - P(\neg C)^2$	
Goodman and Kruskal's Interestingess (**GKI**) [3]	**S**
$\frac{\sum_i max_j P(A_i\cap C_j)+\sum_j max_i P(A_i\cap C_j)-max_i P(A_i)-max_j P(C_j)}{2-max_i P(A_i)-max_j P(C_j)}$	
Normalized Mutual Information (**NMI**) [3]	**I**
$\sum_i\sum_j P(A_i\cap C_j) * log_2 \frac{P(A_i\cap C_j)}{P(A_i)*P(C_j)} / \{-\sum_i P(A_i) * log_2 P(A_i)\}$	
J-Measure (**J-M**) [1, 3, 5, 7]	**I**
$P(C) * (KLD(C\mid A; C) + KLD(\neg C\mid\neg A; \neg C)$, KLD: Kullback-Leibler Distance	
Yao and Liu's Interestingness 1 based on one-way support (**YLI1**) [1]	**I**
$P(C\mid A) * log_2 \frac{P(A\cap C)}{P(A)*P(C)}$	
Yao and Liu's Interestingness 2 based on two-way support (**YLI2**) [1]	**I**
$P(A\cap C) * log_2 \frac{P(A\cap C)}{P(A)*P(C)}$	
Yao and Liu's Interestingness 3, the sum of possible YLI2 variations (**YLI3**) [1]	**I**
$P(A\cap C) * log_2 \frac{P(A\cap C)}{P(A)*P(C)} + P(A\cap \overline{C}) * log_2 \frac{P(A\cap\overline{C})}{P(A)*P(\overline{C})} +$ $P(\overline{A}\cap C) * log_2 \frac{P(\overline{A}\cap C)}{P(\overline{A})*P(C)} + P(\overline{A}\cap\overline{C}) * log_2 \frac{P(\overline{A}\cap\overline{C})}{P(\overline{A})*P(\overline{C})}$	
K-Measure (**K-M**) [5]	**I**
$KLD(C\mid A; C) + KLD(\neg C\mid\neg A; \neg C) - KLD(C\mid A; \neg C) + KLD(\neg C\mid\neg A; C)$	
ϕ Coefficient (ϕ) [3]	**N**
$\{P(A\cap C) - P(A) * P(C)\}/\sqrt{P(A) * P(C) * P(\neg A) * P(\neg C)}$	
Piatetsky-Shapiro's Interestingness (**PSI**) [3, 5, 6]	**N**
$N(A\cap C) - \frac{N(A)+N(C)}{N(U)}$, $N(U)$: Number of instances in universe. N_C: That in consequent. N_U: That in rule.	
Cosine Similarity (**CSI**) [3]	**N**
$P(A\cap C)/\sqrt{(P(A) * P(C))}$	
Gago and Bento's Interestingness (**GBI**) [11]	**D**
$\sum_{j=1}^{N_R} D(R_i, R_j)/N_R$, R_i : i-th rule. N_R: Number of rules. $D(R_i, R_j)$: Distance based on attribute overlap degree between i-th and j-th rules.	
Peculiarity [17]	**D**
$\sum_{i=1}^{N_a}\sum_{k=1}^{N_i} \lvert x_{ij} - x_{ik}\rvert^{\alpha}/N_a$, x_{ij}: j-th value of i-th attribute. N_a: Number of attributes. N_i: Number of values of i-th attribute. α: Constant.	

3 Evaluation Experiment of Objective Measures

3.1 Experimental Conditions

The experiment examined the performance of objective measures to estimate real human interest by comparing the evaluation by them and a human user. Concretely speaking, the selected objective measures and a medical expert evaluated the same medical rules, and their evaluation values were qualitatively and quantitatively compared. We used the objective measures in Table 2 and the rules and their evaluation results in our previous research [4].

Here, we note the outline of our previous research. We tried to discover new medical knowledge from a clinical dataset on hepatitis. The KDD process was designed to twice repeat a set of the rule generation by our mining system and the rule evaluation by a medical expert for polishing up the obtained rules. Our mining system was based on the typical framework of time-series data mining, a combination of the pattern extraction by clustering and the classification by a decision tree. It generated prognosis-prediction rules and visualized them as graphs. The medical expert conducted the following evaluation tasks: After each mining, he gave each rule the comment on its medical interpretation and one of the rule quality labels, which were Especially-Interesting (**EI**), Interesting (**I**), Not-Understandable (**NU**), and Not-Interesting (**NI**). **EI** means that the rule was a key to generate or confirm a hypothesis.

A few rules in the first mining inspired the medical expert to make a hypothesis, a seed of new medical knowledge: Contradict to medical common sense, GPT, which is an important medical test result to grasp hepatitis symptom, may change with three years cycle (See the left side in Fig. 1). A few rules in the second mining supported him to confirm the hypothesis and enhanced its reliability (See the right side in Fig. 1). As a consequence, we obtained a set of rules and their evaluation results by the medical expert in the first mining and that in the second mining. Three and nine rules received **EI** and **I** in the first mining, respectively. Similarly, two and six rules did in the second mining.

In our current research, the evaluation procedure by the objective measures was designed as follows: For each objective measure, the same rules as in our previous research were evaluated by the objective measure, sorted in the descending

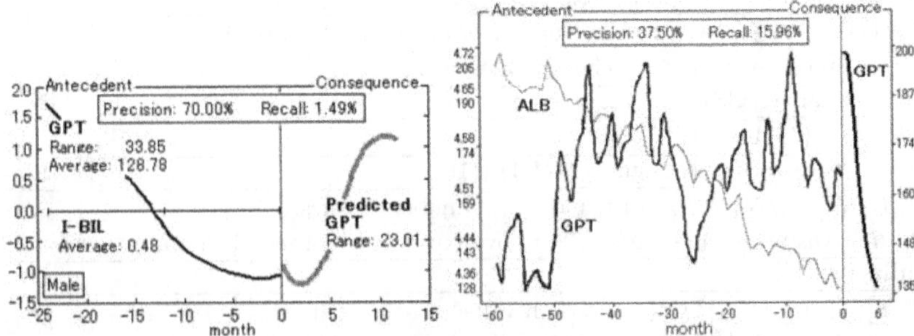

Fig. 1. The examples of highly valued rules in first (left) and second mining (right).

order of evaluation values, and assigned the rule quality labels. The rules from the top to the m-th were assigned **EI**, where m was the number of **EI** rules in the evaluation by the medical expert. Next, the rules from the $(m+1)$-th to the $(m+n)$-th were assigned **I**, where n was the number of **I** rules in the evaluation by the medical expert. The assignment of **NU** and **NI** followed the same procedure. We dared not to do evaluation value thresholding for the labeling. The first reason was that it is quite difficult to find the optimal thresholds for the all combinations of labels and objective measures. The second reason was that although our labeling procedure may not be precise, it can set the conditions of objective measures at least equal through simple processing. The last reason was that our number-based labeling is more realistic than threshold-based labeling. The number of rules labeled with **EI** or **I** by a human user inevitably stabilizes at around a dozen in a practical situation, since the number of evaluation by him/her has a severe limitation caused by his/her fatigue.

3.2 Results and Discussion

Fig. 2 and 3 show the experimental results in the first and second mining, respectively. We analyzed the relation between the evaluation results by the medical expert and the objective measures qualitatively and quantitatively. As the qualitative analysis, we visualized their degree of agreement to easily grasp its trend. We colored the rules with perfect agreement white, probabilistic agreement gray, and disagreement black. A few objective measures output same evaluation values for too many rules. For example, although eight rules were especially interesting (**EI**) or interesting (**I**) for the medical expert in second mining, the objective measure OR estimated 14 rules as **EI** or **I** ones (See Fig. 3). In that case, we colored such rules gray. The pattern of white (possibly also gray) and black cells for an objective measure describes how its evaluation matched with those by the medical expert. The more the number of white cells in the left-hand side, the better its performance to estimate real human interest.

For the quantitative analysis, we defined four comprehensive criteria to evaluate the performance of an objective measure. #1: Performance on **I** (the number of rules labeled with **I** by the objective measure over that by the medical expert. Note that **I** includes **EI**). #2: Performance on **EI** (the number of rules labeled with **EI** by the objective measure over that by the medical expert). #3: Number-based performance on all evaluation (the number of rules with the same evaluation results by the objective measure and the medical expert over that of all rules). #4: Correlation-based performance on all evaluation (the correlation coefficient between the evaluation results by the objective measure and those by the medical expert). The values of these criteria are shown in the right side of Fig. 2 and 3. The symbol '+' besides a value means that the value is greater than that in case rules are randomly selected as **EI** or **I**. Therefore, an objective measure with '+' has higher performance than random selection does at least. To know the total performance, we defined the weighted average of the four criteria as a meta criterion; we assigned 0.4, 0.1, 0.4, and 0.1 to #1, #2, #3, and

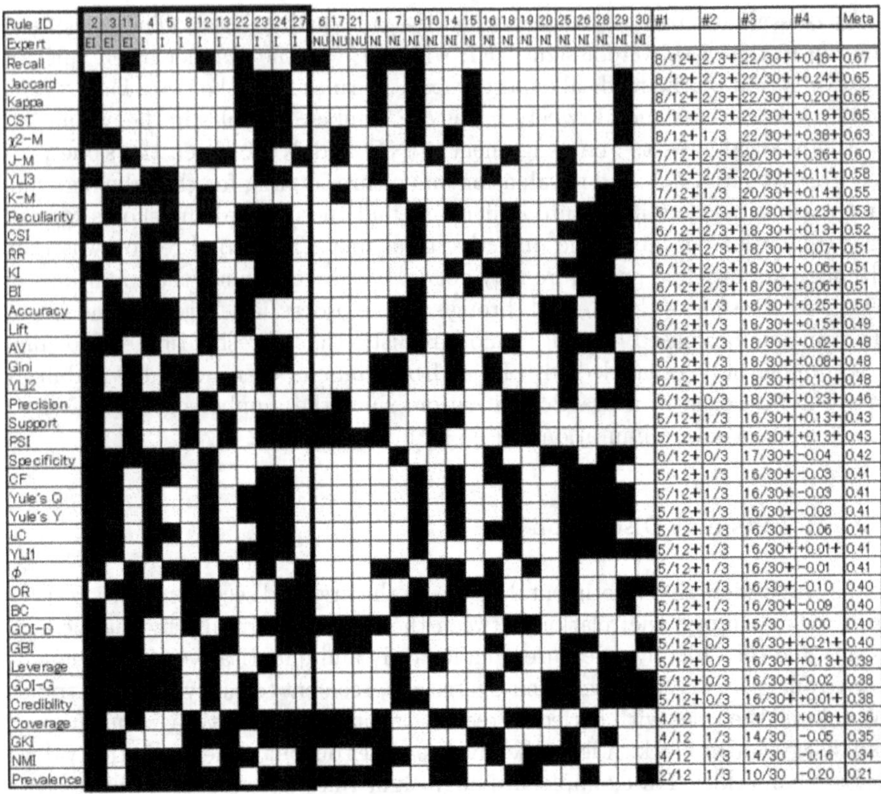

Fig. 2. The evaluation results by a medical expert and objective measures for the rules in first mining. Each column represents a rule, and each row represents the set of evaluation results by an objective measure. The rules are sorted in the descending order of the evaluation values given by the medical expert. The objective measures are sorted in the descending order of the meta criterion values. A square in the left-hand side surrounds the rules labeled with **EI** or **I** by the medical expert. White, gray, and black cells mean that the evaluation by an objective measure was perfectly, was probabilistically, and was not the same by the medical expert, respectively. The five columns in the right side show the performance on the four comprehensive criteria and the meta one. '+' means the value is greater than that of random selection.

#4, respectively, according to their importance. The objective measures were sorted in the descending order of the values of meta criterion.

The results in the first mining in Fig. 2 show that Recall demonstrated the highest performance, Jaccard, Kappa, and CST did the second highest, and χ^2-M did the third highest. Prevalence demonstrated the lowest performance, NMI did the second lowest, and GKI did the third lowest. The results in the second mining in Fig. 3 show that Credibility demonstrated the highest performance, Peculiarity did the second highest, and Accuracy, RR, and BI did the third highest. Prevalence demonstrated the lowest performance, Specificity did the

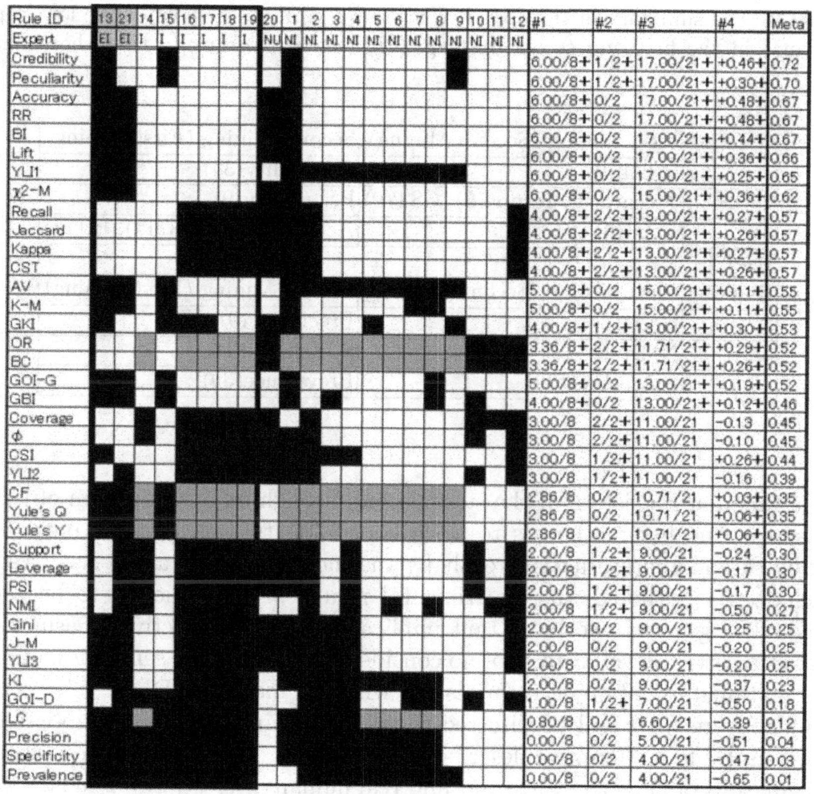

Rule ID	13	21	14	15	16	17	18	19	20	1	2	3	4	5	6	7	8	9	10	11	12	#1	#2	#3	#4	Meta
Expert	EI	EI	I	I	I	I	I	I	NU	NI	NI	NI	NI	NI	NI	NI	NI	NI	NI	NI	NI					
Credibility																						6.00/8+	1/2+	17.00/21+	+0.46+	0.72
Peculiarity																						6.00/8+	1/2+	17.00/21+	+0.30+	0.70
Accuracy																						6.00/8+	0/2	17.00/21+	+0.48+	0.67
RR																						6.00/8+	0/2	17.00/21+	+0.48+	0.67
BI																						6.00/8+	0/2	17.00/21+	+0.44+	0.67
Lift																						6.00/8+	0/2	17.00/21+	+0.36+	0.66
YLI1																						6.00/8+	0/2	17.00/21+	+0.25+	0.65
χ^2-M																						6.00/8+	0/2	15.00/21+	+0.36+	0.62
Recall																						4.00/8+	2/2+	13.00/21+	+0.27+	0.57
Jaccard																						4.00/8+	2/2+	13.00/21+	+0.26+	0.57
Kappa																						4.00/8+	2/2+	13.00/21+	+0.27+	0.57
CST																						4.00/8+	2/2+	13.00/21+	+0.26+	0.57
AV																						5.00/8+	0/2	15.00/21+	+0.11+	0.55
K-M																						5.00/8+	0/2	15.00/21+	+0.11+	0.55
GKI																						4.00/8+	1/2+	13.00/21+	+0.30+	0.53
OR																						3.36/8+	2/2+	11.71/21+	+0.29+	0.52
BC																						3.36/8+	2/2+	11.71/21+	+0.26+	0.52
GOI-G																						5.00/8+	0/2	13.00/21+	+0.19+	0.52
GBI																						4.00/8+	0/2	13.00/21+	+0.12+	0.46
Coverage																						3.00/8	2/2+	11.00/21	-0.13	0.45
ϕ																						3.00/8	2/2+	11.00/21	-0.10	0.45
OSI																						3.00/8	1/2+	11.00/21	+0.26+	0.44
YLI2																						3.00/8	1/2+	11.00/21	-0.16	0.39
CF																						2.86/8	0/2	10.71/21	+0.03+	0.35
Yule's Q																						2.86/8	0/2	10.71/21	+0.06+	0.35
Yule's Y																						2.86/8	0/2	10.71/21	+0.06+	0.35
Support																						2.00/8	1/2+	9.00/21	-0.24	0.30
Leverage																						2.00/8	1/2+	9.00/21	-0.17	0.30
PSI																						2.00/8	1/2+	9.00/21	-0.17	0.30
NMI																						2.00/8	1/2+	9.00/21	-0.50	0.27
Gini																						2.00/8	0/2	9.00/21	-0.25	0.25
J-M																						2.00/8	0/2	9.00/21	-0.20	0.25
YLI3																						2.00/8	0/2	9.00/21	-0.20	0.25
KI																						2.00/8	0/2	9.00/21	-0.37	0.23
GOI-D																						1.00/8	1/2+	7.00/21	-0.50	0.18
LC																						0.80/8	0/2	6.60/21	-0.39	0.12
Precision																						0.00/8	0/2	5.00/21	-0.51	0.04
Specificity																						0.00/8	0/2	4.00/21	-0.47	0.03
Prevalence																						0.00/8	0/2	4.00/21	-0.65	0.01

Fig. 3. The evaluation results by a medical expert and objective measures for the rules in second mining. See the caption of Fig. 2 for the details.

second lowest, and Precision did the third lowest. We summarized these objective measures in Table 3. As a whole, the following objective measures maintained their high performance through the first and second mining: Recall, Jaccard, Kappa, CST, χ^2-M, and Peculiarity. NMI and Prevalence maintained their low performance. Only Credibility changed its performance dramatically, and the other objective measures slightly changed their middle performance.

More than expected, some objective measures – Recall, Jaccard, Kappa, CST, χ^2-M, and Peculiarity – showed constantly high performance. They had comparatively many white cells and '+' for all comprehensive criteria. In addition, the mosaic-like patterns of white and black cells in Fig. 2 and 3 showed that the objective measures had almost complementary relationship for each other. The results and the medical expert's comments on them imply that his interest consisted of not only the medical semantics but also the statistical characteristics of rules. The combinational use of objective measures will be useful to reductively analyze such human interest and to recommend interesting rule candidates from various viewpoints through human-system interaction in medical KDD. One method to obtain the combination of objective measures is to formu-

Table 3. The summary of the objective measures with the highest or the lowest performance in the first and second mining. (N) means the rank in the other mining.

Top 3

Ranking	First Mining (Second Mining)	Second Mining (First Mining)
1	Recall(9)	Credibility(34)
2	Jaccard(9), Kappa(9), CST(9)	Peculiarity(9)
3	χ^2-M(8)	Accuracy(14), RR(11), BI(11)

Last 3

Ranking	First Mining (Second Mining)	Second Mining (First Mining)
37	GKI(15)	Precision(19)
38	NMI(30)	Specificity(22)
39	Prevalence(39)	Prevalence(39)

late a function consisting of the summation of weighted outputs from objective measures. Another method is to learn a decision tree using these outputs as attributes and the evaluation result by the medical expert as a class. We can conclude that although the experimental results are not enough to be generalized, they gave us two important implications: some objective measures will work at a certain level in spite of no consideration of domain semantics, and the combinational use of objective measures will help medical KDD.

Our future work will be directed to two issues including some sub-issues as shown in Table 4. Here, we describe their outlines. On Issue (i) the investigation/analysis of objective measures and real human interest, Sub-issue (i)-1 and (i)-2 are needed to generalize the current experimental results and to grasp the theoretical possibility and limitation of objective measures, respectively. Sub-issue (i)-3 is needed to establish the method to predict real human interest. We have already finished an experiment on Sub-issue (i)-1 and (i)-3, and will show their results soon. Sub-issue (i)-2 is now under discussion. The outcome of those empirical and theoretical researches will contribute to solving Issue (ii). Issue (ii) the utilization of objective measures for KDD support based on human-system interaction, assumes that the smooth interaction between a human user and a mining system is a key to obtain really interesting rules for the human user. Our previous research in Section 3.1 [4] and others' researches using the same dataset of ours [18] led us to this assumption. We think that smooth human-system interaction stimulates the hypothesis generation and confirmation of a human user, and actually it did in our previous research. Sub-issue (ii)-1 is needed to support such a thinking process in the post-processing phase of data mining. Our current idea is to develop a post-processing user interface in which a human user can select one among various objective measures and see the rules sorted with its evaluation values. We expect that the user interface will enhance the thinking from unexpected new viewpoints. Sub-issue (ii)-2 is the extension of Sub-issue (ii)-1; It comprehensively focuses on the spiral sequence of mining algorithm organization and post-processing. As the one of Sub-issue (ii)-2 solutions, now we are implementing an evaluation module, which uses the predicted real

Table 4. The outlines of issues in our future work.

Issue (i) Investigation/analysis of objective measures and real human interest.
Sub-issue (i)-1 Experiments with different datasets and medical experts. Sub-issue (i)-2 Mathematical analysis of objective measures. Sub-issue (i)-3 Reductive analysis of real human interest using the combination of objective measures.
Issue (ii) Utilization of objective measures for KDD support based on human-system interaction.
Sub-issue (ii)-1 Development of a post-processing user interface. Sub-issue (ii)-2 Development of a comprehensive KDD environment.

human interest with objective measures in Sub-issue (i)-3, into a constructive meta-learning system called CAMLET [19].

4 Conclusions and Future Work

This paper discussed how objective measures can contribute to detect interesting rules for a medical expert through an experiment using the rules on hepatitis. Recall, Jaccard, Kappa, CST, χ^2-M, and Peculiarity demonstrated good performance, and the objective measures used here had complementary relationship for each other. It was indicated that their combination will be useful to support human-system interaction. Our near-future work is to obtain the generic trend of objective measures in medical KDD. As an empirical approach, we have already finished another experiment with a clinical dataset on meningoencephalitis and a different medical expert and are comparing the experimental results on hepatitis and meningoencephalitis. As a theoretical approach, we are conducting the mathematical analysis of objective measure features. We will utilize these outcomes for supporting medical KDD based on system-human interaction.

References

1. Yao, Y. Y. Zhong, N.: An Analysis of Quantitative Measures Associated with Rules. Proceedings of Pacific-Asia Conference on Knowledge Discovery and Data Mining PAKDD-1999 (1999) 479–488
2. Hilderman, R. J., Hamilton, H. J.: Knowledge Discovery and Measure of Interest. Kluwer Academic Publishers (2001)
3. Tan, P. N., Kumar V., Srivastava, J.: Selecting the Right Interestingness Measure for Association Patterns. Proceedings of International Conference on Knowledge Discovery and Data Mining KDD-2002 (2002) 32–41
4. Ohsaki, M., Sato, Y., Yokoi, H., Yamaguchi, T.: A Rule Discovery Support System for Sequential Medical Data, – In the Case Study of a Chronic Hepatitis Dataset –. Proceedings of International Workshop on Active Mining AM-2002 in IEEE International Conference on Data Mining ICDM-2002 (2002) 97–102
5. Ohsaki, M., Sato, Y., Yokoi, H., Yamaguchi, T.: Investigation of Rule Interestingness in Medical Data Mining. Lecture Notes in Computer Science, Springer-Verlag (2004) will appear.

6. Piatetsky-Shapiro, G.: Discovery, Analysis and Presentation of Strong Rules. in Piatetsky-Shapiro, G., Frawley, W. J. (eds.): Knowledge Discovery in Databases. AAAI/MIT Press (1991) 229–248
7. Smyth, P., Goodman, R. M.: Rule Induction using Information Theory. in Piatetsky-Shapiro, G., Frawley, W. J. (eds.): Knowledge Discovery in Databases. AAAI/MIT Press (1991) 159–176
8. Hamilton, H. J., Fudger, D. F.: Estimating DBLearn's Potential for Knowledge Discovery in Databases. Computational Intelligence, 11, 2 (1995) 280–296
9. Hamilton, H. J., Shan, N., Ziarko, W.: Machine Learning of Credible Classifications. Proceedings of Australian Conference on Artificial Intelligence AI-1997 (1997) 330–339
10. Dong, G., Li, J.: Interestingness of Discovered Association Rules in Terms of Neighborhood-Based Unexpectedness. Proceedings of Pacific-Asia Conference on Knowledge Discovery and Data Mining PAKDD-1998 (1998) 72–86
11. Gago, P., Bento, C.: A Metric for Selection of the Most Promising Rules. Proceedings of European Conference on the Principles of Data Mining and Knowledge Discovery PKDD-1998 (1998) 19–27
12. Gray, B., Orlowska, M. E.: CCAIIA: Clustering Categorical Attributes into Interesting Association Rules. Proceedings of Pacific-Asia Conference on Knowledge Discovery and Data Mining PAKDD-1998 (1998) 132–143
13. Morimoto, Y., Fukuda, T., Matsuzawa, H., Tokuyama, T., Yoda, K.: Algorithms for Mining Association Rules for Binary Segmentations of Huge Categorical Databases. Proceedings of International Conference on Very Large Databases VLDB-1998 (1998) 380–391
14. Freitas, A. A.: On Rule Interestingness Measures. Knowledge-Based Systems, 12, 5–6 (1999) 309–315
15. Liu, H., Lu, H., Feng, L., Hussain, F.: Efficient Search of Reliable Exceptions. Proceedings of Pacific-Asia Conference on Knowledge Discovery and Data Mining PAKDD-1999 (1999) 194–203
16. Jaroszewicz, S., Simovici, D. A.: A General Measure of Rule Interestingness. Proceedings of European Conference on Principles of Data Mining and Knowledge Discovery PKDD-2001 (2001) 253–265
17. Zhong, N., Yao, Y. Y., Ohshima, M.: Peculiarity Oriented Multi-Database Mining. IEEE Transaction on Knowledge and Data Engineering, 15, 4 (2003) 952–960
18. Motoda, H. (eds.): Active Mining, IOS Press, Amsterdam, Holland (2002).
19. Abe, H. and Yamaguchi T.: Constructive Meta-Learning with Machine Learning Method Repository, IEA/AIE2004, LNAI3029 (2004) pp.502–511.

Classification in Geographical Information Systems

Salvatore Rinzivillo and Franco Turini

Dipartimento di Informatica,
v. Buonarroti, 2 - 56125 Pisa
University of Pisa, Italy
{rinziv,turini}@di.unipi.it

Abstract. The paper deals with the problem of knowledge discovery in spatial databases. In particular, we explore the application of decision tree learning methods to the classification of spatial datasets. Spatial datasets, according to the Geographic Information System approach, are represented as stack of layers, where each layer is associated with an attribute. We propose an ID3-like algorithm based on an entropy measure, weighted on a specific spatial relation (i.e. overlap). We describe an application of the algorithm to the classification of geographical areas for agricultural purposes.

1 Introduction

Spatial data are usually handled by means of a Geographical Information System, that is a system able to represent a territory along with its characteristics. In general a territory is represented as a stack of layers. Each of the layers is associated to a specific attribute, and the layer represents a partitioning of the territory according to the values of the attribute itself. In other words each of the partitions of the layer corresponds to an area of the territory associated to a specific value of the attribute.

Geographical Information Systems provide the user with the possibility of querying a territory for extracting areas that exhibit certain properties, i.e. given combinations of values of the attributes. Just as it is intuitive to extend standard database query language to embody inductive queries, we believe that an analogous approach can be explored for Geographical Information Systems, and, in general, for spatial databases. For example, finding a classification of areas for a given exploitation, agriculture say, or finding associations among areas, or even clustering areas seem to be sensible questions to answer. All the inductive algorithms for knowledge discovery in databases starts from considering a collection of "transactions", usually represented as a table of tuples. In the case of a spatial database, the notion of transaction naturally corresponds to the tuple of attribute values that characterize an area. That leaves us with two problems:

1. the selection of the transaction: i.e. which area with which values;
2. the exploitation of the area in the inductive algorithms.

J.-F. Boulicaut et al. (Eds.): PKDD 2004, LNAI 3202, pp. 374–385, 2004.

In this paper we address the problem of constructing a classification tree for classifying areas with respect to their properties. Given spatial databases that store a collection of layers for one or more territories, choose one of the layers as the output class and try to construct a classification tree capable of placing an area in the right class given the values of the other attributes with respect to the area. As a running example, we consider an agricultural territory, and we choose the kind of crop as the classification task, whereas the other attributes concern properties of the ground (amount of water etc.). We show how a suitably adapted ID3 algorithm can do the job. The crucial point is the introduction of a suitable entropy measure, that takes into account specific spatial relations.

In the rest of the section we briefly overview the work related to our research. In Section 2 we formalize the problem and in Section 3, the kernel of the paper, we discuss the algorithm both formally and informally by using the agriculture related running example. The Conclusions section is devoted to discussing our current and future work.

Knowledge Discovery in Spatial Databases. In the analysis of geographically referenced data it is important to take into account the spatial component of the data (i.e. position, proximity, orientation, etc.). Many methods have been proposed in the literature to deal with this kind of information.

Some techniques consider spatial data independently from non-spatial data during the learning step, and they relate them during pattern analysis. For example, in [2] an integrated environment for spatial analysis is presented, where methods for knowledge discovery are complemented by the visual exploration of the data presented on the map. The aim of map visualization is to prepare (or select) data for KDD procedures and to interpret the results. In [1], this environment is used for analyzing thematic maps by means of C4.5 [14] (for pattern extraction from non-spatial data). However, the iterative methods presented there keep spatial and non-spatial data separated, and geo-references are used as a means for visualization of the extracted patterns.

A tighter correlation between spatial and non-spatial data is given in [7], by observing that, in a spatial dataset, each object has an implicit neighborhood relation with the other objects in the set. The authors build a *neighborhood graph* where each node represents an object in the dataset and an edge between two nodes, say n_1 and n_2, represents a spatial relation between the objects associated with n_1 and n_2. The distance between two nodes in the graph (i.e. the number of edges in the path from the starting node to the target one) is used to weigh the properties associated to each node. This consideration gives the idea of *influence* between nodes, i.e. two nodes connected by a short path are likely to influence each other in the real world.

Some other papers explore the extension of knowledge discovery techniques to spatial domain. For example, in [11] an algorithm based on CLARANS for clustering spatial data is proposed.

In [9] and [10], two methods for extracting spatial association rules from a spatial dataset are presented.

Classification and ID3 Algorithm. A classification assigns a category to transactions according to the values of (some of) their attributes. Usually, the input is a table of tuples, where each row is a *transaction* and each column is an *attribute*. One of the attributes is chosen as the *class attribute*. The task is to build a model for predicting the value of the *class* attribute, knowing only the values of the others.

Many methods have been proposed for classification, such as neural networks, Bayesian classifiers and decision trees. We take into consideration decision tree models, since they proved to be very robust to noisy data, and in our context this property is crucial. The decision-tree classifiers proposed in the literature can be distinguished according to the statistical criterion used for splitting. For example CART [4] uses the *Gini index*, whereas *ID3* [13] and *C4.5* [14] use the *entropy* to measure (im)purity of samples. We focus our attention on ID3 algorithm and we show how it has been adapted to our purposes.

2 Spatial Data Model

Spatial data store the geometric description of geographical features, along with a state associated with these features. The state of a feature is a set of attributes. Many GIS tools organize spatial data as *vector* or *raster* maps. The correspondent attributes are stored in a set of tables related geographically to the geometric features [3, 6]. Digital maps are organized in *layers* (sometimes called *coverages* or *themes*). A layer represents a distinct set of geographic features in a particular domain. For example, a map of a country may have one layer representing cities, one representing roads, one representing rivers and lakes, and so on. Usually, a layer has several properties to control how it is rendered during visualization (i.e. the spatial extent, the range of scales at which the layers will be drawn on the screen, the colors for rendering features).

In [12], the Open GIS Consortium provides the specification for representing vector data to guarantee the interoperability and the portability of spatial data. We took advantage of the large availability of tools that support this specification (e.g. PostGIS [15], MySQL, GDAL libraries, JTS [8], GRASS GIS).

Given this organization of digital maps, it seems natural to maintain the same structure also during the knowledge discovery (and classification, in this case) task. The process of knowledge extraction aims at extracting implicit information from raw data. In a spatial dataset, the search for this implicit information is driven by the spatial relations among objects. For example in [7], a neighborhood graph is built to express these relationships. This graph is explored to extract useful information by finding, for example, similarities among paths.

Our point is to exploit the layer structure to select useful relations. For example, we may have interest to search “inter-layer” relations among objects (i.e. spatial relations where at least two members of the relation belong to distinct layers), rather than “intra-layer” relations. Thus, the heuristics proposed in [7] (i.e. following a path that tends to go away from its source) may be enriched by including only edges that represent “inter-layer” relations in the neighborhood graph. This assumption seems to be reasonable, since almost all papers on knowledge discovery present methods to extract patterns of this kind, i.e. patterns where the classes of objects involved belong to different layers. although the entity "layer" is not explicitly present.

We consider an input composed of n layers. Each layer L_i has only one geometric type: all the geometric features are polygons, or lines, or points. For the clarity of the presentation we consider for the moment only polygonal layers. Each layer has a set of attributes that describe the state of each object in the layer. We choose one of this attribute as the representative of the layer. Thus, each layer L_i represents a variable x_i over a domain X_i, where x_i is the chosen attribute. For example, a layer of the agricultural census data has several attributes associated with each polygon (e.g. the number of farms, the area of crop land, and other properties). If the attribute *number of farms* is chosen then each polygon in the layer can be considered as an instance of this variable.

Spatial Transactions. One of the layers is selected as the *class label* layer: we exploit these objects for selecting *spatial transactions*. Each polygon represents an "area of interest". These areas are related to areas in the other layers in order to extract a set of tuples where each value in a tuple corresponds to the value in each of the layers, with respect to the intersection with the area of interest. So, like for relational databases, we now have a set of transactions. While relational transactions are "measured" by counting tuple occurrences in a table, we use here, as it is intuitive, the area extension of each spatial transaction.

Example We introduce here a simple example that will be useful for the whole presentation. Consider an agricultural environment, where information about crops is organized in layers as described above. For the sake of readability, in Figure 1 layers are presented side by side, even if they should be considered stacked one on top of the other. In Figure 1(f) the areas of samples are reported. Layers describe measures of qualities of the soil, like the availability of water (fig.1(a)) or potassium (fig.1(b)), or other parameters of the environment, like climate conditions (fig.1(c)). The training set, namely the *class label* layer (fig.1(d)), provides information about classified sample regions. In this example, we have a training set that represents crop types for some areas in the region.

3 Spatial Classification by Decision Tree Learning

Our goal is to build a decision tree capable of assigning an area to a class, given the values of the other layers with respect to the area. Like in transaction classification, we follow two steps: first, we build a model from a set of samples, namely a *training set*; then, we use the model to classify new (unseen) areas.

The training set is determined by the spatial transactions extracted from the dataset. We focus our attention on spatial relations derivable from *9-intersection model* [5] according to the adopted spatial model [12].

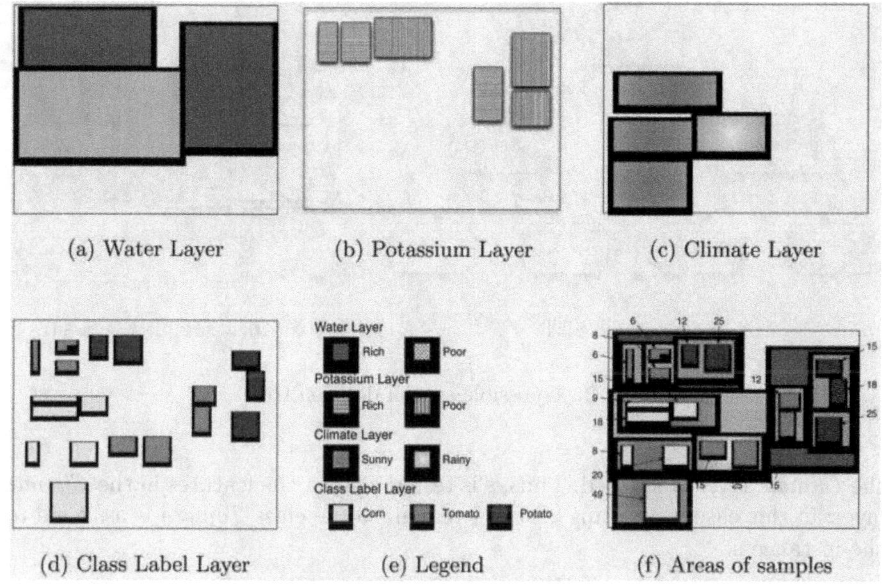

(a) Water Layer (b) Potassium Layer (c) Climate Layer

(d) Class Label Layer (e) Legend (f) Areas of samples

Fig. 1. An agricultural map

Spatial Decision Trees. A *Spatial Decision Tree* (SDT) is a rooted tree where *(i)* each internal node is a decision node over a layer, *(ii)* each branch denotes an outcome of the test and *(iii)* each leaf represents one of the class values.
A decision node n_i is associated with a layer L_i and with the attribute X_i of the layer. The outcoming edges are labeled with the possible values of X_i.

SDT Classification. An area A is classified by starting at the root node, testing the layer associated with this node and following the branch corresponding to the test result. Let $x_1, x_2, \ldots, x_m$ be the labels of the m edges of the root node. If A intersects an object of type x_j in the layer associated with the root node, then the edge labeled with x_j is followed. This testing process is repeated recursively starting from the selected child node until a leaf node is reached. The area A is classified according to the value in the leaf. When the query region A intersects several areas with distinct values, then all the corresponding branches are followed. The area A is split according to the layer values and each portion is classified independently.

Example In Figure 2(a) a spatial decision tree for the example in Figure 1 is presented. This decision tree classifies areas according to whether they are suitable for a type of crop rather than another. In particular, in this example we have three kind of crops: *Corn*, *Tomato* and *Potato*. Given a new instance **s** (marked with "??" in Figure 2(b)), we test **s** starting from the layer associated with the root node, i.e. the *Water* layer. Since **s** overlaps a water region whose value is *Poor*, the corresponding branch is followed and the node associated with

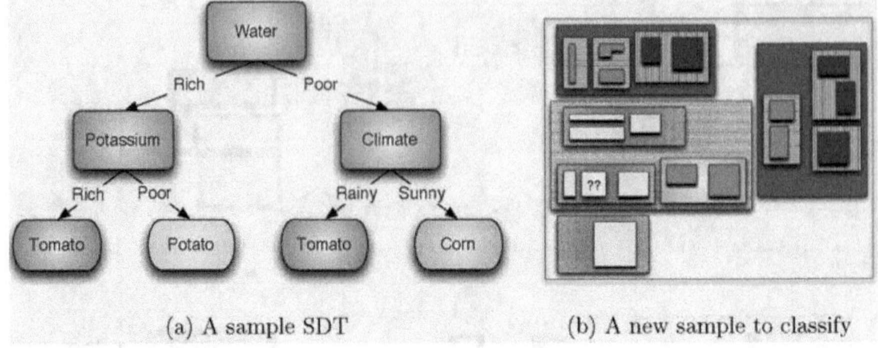

(a) A sample SDT (b) A new sample to classify

Fig. 2. A possible spatial decision tree

the *Climate* layer is selected. Thus, **s** is tested against the features in the *Climate* layer: in this case it overlaps a *Sunny* region, so the class *Tomato* is assigned to the instance **s**.

3.1 SDT Learning Algorithm

Following the basic decision tree learning algorithm [13], our method employs a top-down strategy to build the model. Initially, a layer is selected and associated with the root node, using a statistical test to verify how well it classifies all samples. Once a layer has been selected, a node is created and a branch is added for each possible value of the attribute of the layer. Then, the samples are distributed among the descendant nodes and the process is repeated for each subtree.

Algorithm 1 shows a general version of the learning method. Initially, termination conditions are tested to solve trivial configurations (for example, when all samples belong to the same class). The `majority_class(`S`)` is the class where most of the samples in S belong to. The crucial point of the algorithm is the selection of the split layer for the current node. In Section 3.3 a strategy based on the notion of entropy is presented to quantify how well a layer separates samples. Once a layer is selected for a test node, the samples are partitioned according to the layer itself and the intersection spatial relation. In Section 3.2 we show how to compute this partition.

3.2 Splitting Samples

When classifying transactions represented as tuples, we aim at grouping transactions together according to an attribute A. If the attribute for splitting is selected in a proper way, the samples in each subpartition may increase their uniformity (or, in other terms, they reduce their entropy).

In the same way, we aim at grouping spatial samples according to the information found in the other layers. We select a layer L_i and we split the samples in

Algorithm 1: Generate_SDT

Input: A layer S of sample areas;
A list $\mathcal{L}$ of layers;
Output: A spatial decision tree
Create a new node N;
if *samples in S are all of class c* **then**
  label N with c;
  exit;
end
if $\mathcal{L}$ *is empty* **then**
  label N with `majority_class(S)`;
  exit;
end
Select layer best_split from $\mathcal{L}$;
Split S according to layer best_split in $\{S(c_1), \ldots, S(c_p)\}$;
foreach $S(c_i)$, $i = 1, 2, \ldots, p$ **do**
  Let N_i = Generate_SDT($S(c_i)$, $\mathcal{L}\backslash\{$best_split $\}$);
  Create a branch from N to N_i labeled with the selected value;
end

layer S according to this layer. In general, if layer L_i has q possible values then it can split the samples in $q+1$ subsets, i.e. a subset for each value $v_j, j = 1, 2, \ldots, q$, and a special subset correspondig to none of these values (termed $\neg L(C)$). We use $L_i(v_j)$ to refer to all the features in L_i that have value v_j.

The *intersection* relation gives us the possibility to express a quantitative measure on the related objects. In fact, given two polygonal geometries, say g_1 and g_2, we obtain a new geometry by means of the function $intersection(g_1, g_2)$. If the two geometries overlap and their intersection geometry is g_3 then the *area* of g_3 gives the quantitative measure of the relation (under the assumption that g_3 is a polygon; the other cases are discussed below).

Given the subset $L_i(v_j)$, for $j = 1, 2, \ldots, q$, we consider all the samples that intersect any feature in $L_i(v_j)$. We denote this sublayer as $L_i(v_j, C)$. For each class value c_k, we denote with $L_i(v_j, c_k)$ the features in $L_i(v_j, C)$ whose class is c_k. Clearly, when a sample overlaps partially a polygon with value v_k and a polygon with value v_l (this situtation is showed in Figure 3) it is split: first, a portion of the sample is computed by intersecting it with polygon v_k; the remaining part of the sample is related to the other polygon; the possibly remaining portion of the sample is left unclassified. For example, the sample in Figure 3(a) is partitioned into three samples (Figure 3(b)).

For the sake of simplicity, we may think of each feature in $L_i(v_j, c_k)$ as a representative for a tuple (v_j, c_k). While in a tuple-transactions context we use cardinality as a quantitative measure, we adopt the *area* as a quantitative measure for "spatial tuples".

Example In Figure 3(c) the result of splitting the samples according to *Water* Layer is showed. There is just one feature in the layer $L_{Water}(Poor)$. Then all the

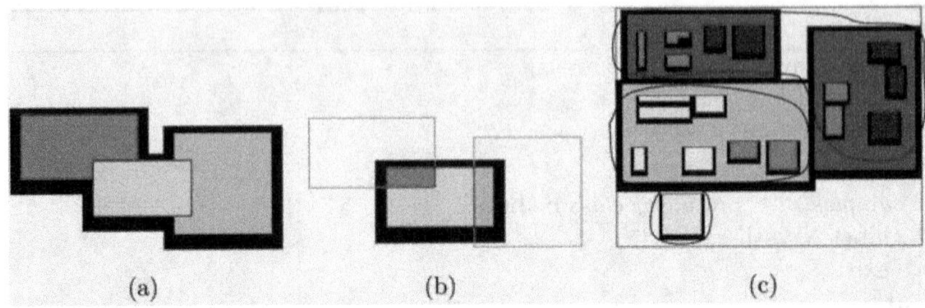

Fig. 3. (a) A sample (at center) overlaps two polygons; (b) the sample after splitting; (c) samples splitted according to the *Water* layer

samples that intersect this object belong to the same subset. The two polygons in $L_{Water}(Rich)$ enclose the other subset of samples. We denote, respectively, with $L_{Water}(Poor, C)$ and $L_{Water}(Rich, C)$ the two subsets of samples. In particular, $L_{Water}(Poor, C)$ is the union of two layers: $L_{Water}(Poor, Corn)$, that contains the five corn polygons on the left, and $L_{Water}(Poor, Tomato)$, that contains the other two polygons.

One of the samples does not overlap any feature either in $L_{Water}(Poor)$ nor in $L_{Water}(Rich)$. Thus, the sample is inserted in $\neg L_{Water}(C)$.

3.3 Selecting Best Split

At each step of the algorithm, we choose one of the candidate layers for growing the tree and for separating the samples. In this section we introduce a statistical measure, the *spatial information gain*, to select a layer that classifies training samples better than the others. The information gain is based on the notion of *entropy*. Intuitively, it is possible to measure the (im)purity of samples w.r.t. their classes by an evaluation of their entropy. Then, candidates for splitting are selected considering the reduction of entropy caused by splitting the samples according to each layer.

Spatial Information Gain. We present now the method to compute the entropy for a layer L, with respect to the class label layer S. First, we evaluate the entropy of the samples, i.e. the information needed to identify the class of a spatial transaction. While in tuple-transaction the frequency of a sample is expressed as a ratio of transaction occurences, we use here the area extent of the samples.

Thus, given a layer L, we denote with $mes(L)$ the sum of the areas of all polygons in L. If S has l distinct classes (i.e. $c_1, c_2, \ldots, c_l$) then the entropy for S is:

$$H(S) = -\sum_{i=1}^{l} \frac{mes(S_{c_i})}{mes(S)} log_2 \frac{mes(S_{c_i})}{mes(S)} \tag{1}$$

If L is a non-class layer with values $v_1, v_2, \ldots, v_q$, we split the samples according to this layer, as showed in Section 3.2. We obtain a set of layers $L(v_i, S)$ for each possible value v_i in L and, possibly, $\neg L(S)$. From equation (1) we can compute the entropy for samples in each sublayer $L(v_i, S)$. The expected entropy value for splitting is given by:

$$H(S|L) = \frac{mes(\neg L(S))}{mes(S)} H(\neg L(S)) + \sum_{j=1}^{q} \frac{mes(L(v_j, S))}{mes(S)} H(L(v_j, S)) \quad (2)$$

The layer $\neg L(S)$ represents the samples that can not be classified by the layer L (i.e. the samples not intersected by the layer L). This scenario may happen, for example, when layer L partially covers the *class label* layer. Thus, when selecting layer L for splitting, we consider the entropy of the samples with empty intersection in the computation of the expected entropy. While the values of layer L may be used to label the edges of the tree, the values of layer $\neg L(S)$ are used as an "handicap" for the layer entropy. For example, consider a layer L containing a single polygon that covers only a small portion of the *class label* layer. The splitting will produce a layer $L(v, S)$ (corresponding to the unique value of layer L) and a layer $\neg L(S)$ larger than the first one. By considering only the layer $L(v, S)$ the resulting entropy would be very low, but a larger part of the sample would remain unclassified. Instead, the entropy measure of layer $\neg L(S)$ gives a measure of the "remaining impurity" left to classify.
The *spatial information gain* for layer L is given by:

$$\text{Gain}(L) = H(S) - H(S|L) \quad (3)$$

Clearly, the layer L that presents the highest gain is chosen as *best split*: we create a node associated with L and an edge for each value of the layer. The samples are splitted among the edges according to each edge value. The selection process is repeated for each branch of the node by considering all the layers except L.

Example Consider the splitting of samples in Figure 1(d) with respect to the *Potassium* layer (Figure 1(b)). The entropy of layers $L_{\text{Potas}}(\text{Rich})$ and L_{Potas} (Poor) is zero. By ignoring the $\neg L_{\text{Potas}}(S)$ the overall entropy of splitting would be zero. Thus, the *Potassium* layer would be selected for splitting. However, so doing, the information of samples in $\neg L_{\text{Potas}}(S)$ is lost: in fact, the $\neg L_{(\cdot)}(S)$ is not used to build the tree.

To clarify the layer selection task, consider the training set in Figure 1(d). Here, the *class label* layer contains polygons whose classes represent crops relative to each area. Following Algorithm 1, we start by building the root node for the tree. One of the available layers has to be selected. So the information gain is computed for each one (i.e. *Water*, *Potassium*, *Climate*).

For instance, we show the information gain computation for the *Water* layer. The splitting of samples according to the *Water* layer is reported in Section 3.2 and it results into three sublayers: $L_{Water}(Poor, C)$, $L_{Water}(Rich, C)$ and

$\neg L_{Water}(C)$. The areas of the samples are reported in Figure 1(f). The entropy of each layer is given by, respectively:

$$H(L_{Water}(Poor,C)) = -\frac{70}{110}log_2\frac{70}{110} - \frac{40}{110}log_2\frac{40}{110} - \frac{0}{110}log_2\frac{0}{110} = 0.9457$$
$$H(L_{Water}(Rich,C)) = -\frac{0}{142}log_2\frac{0}{142} - \frac{47}{142}log_2\frac{47}{142} - \frac{95}{142}log_2\frac{95}{142} = 0.9159$$
$$H(\neg L_{Water}(C)) = -\frac{49}{49}log_2\frac{49}{49} - \frac{0}{49}log_2\frac{0}{49} - \frac{0}{49}log_2\frac{0}{49} = 0.0000$$

From (2), the expected entropy for splitting is:

$$H(S|L_{Water}) = -\frac{110}{301} \times 0.9457 - \frac{142}{301} \times 0.9159 - \frac{49}{301} \times 0.0000 = 0.1830$$

Since the entropy is:

$$H(S) = -\frac{119}{301}log_2\frac{119}{301} - \frac{87}{301}log_2\frac{87}{301} - \frac{95}{301}log_2\frac{95}{301} = 1.5718$$

from (3) we can compute the gain for the *Water* layer (and for the other two layers):

$$Gain(L_{Water}) = 1.3888;\ Gain(L_{Potassium}) = 1.2331;\ Gain(L_{Climate}) = 1.2742;$$

Since the *Water* layer shows the best gain it is selected as root test layer.

3.4 Experiments

We have performed some experiments with real digital maps. In particular, we considered a set of layers from the National Atlas of the United States:

- the *class label* layer contains the information about the average (per farm) market value of sold agricultural products;
- the *ecoregion* layer represents areas that present a common climate;
- the *aquifr* layer shows the distribution of the principal aquifers.
- *cotton, soybean, wheat* layers specify the areas cultivated with cotton, soybeans, wheat for grain respectively;
- *cattle* layers gives the number of cattle and calves per area.

All the layers, but *ecoregion* and *aquifr*, have continuous attributes. We have discretized each attribute by grouping objects into classes with comparable areas.

The spatial operations and indexing are handled by means of the *JTS Topology Suite*[8]. Each layer is indexed with a STRTree [16]. This reduces drastically the execution time for the splitting operation. At each split operation:

- for each polygon in the *class label* layer we compute the intersection with the current layer. The operation is performed in two steps: first, a coarse query is executed on the spatial index; then, the query response is refined and the result is inserted into a new layer;

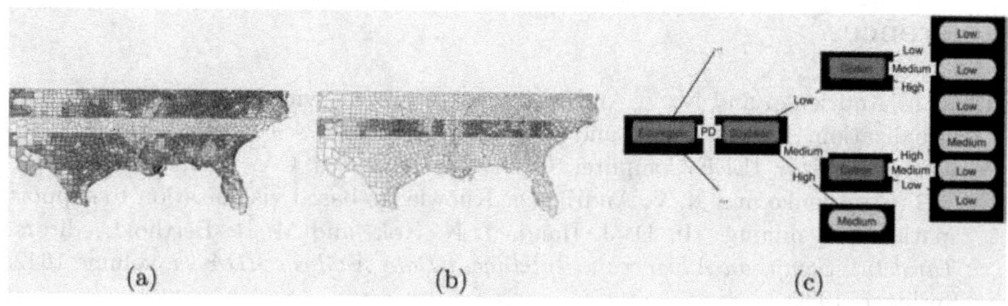

Fig. 4. (a) A portion of a geographical map; (b) the test set considered; (c) the SDT extracted

- the new layer L is partitioned into sublayers $L(v_i)$, for each possible value v_i, to compute the entropy. To speed up this operation, we create an additional hashtable-based index to retrieve each of the sublayer by class.

One of the datasets used for the experiments is showed in Figure 4(a) (the corresponding test set is reported in Figure 4(b)). Continuous layers have been classified into three classes, namely *Low*, *Medium* and *High*. The extracted SDT (Figure 4(c) presents a branch of the whole tree) has the root node associated with the *ecoregion* layer. The accuracy reached on the test set is about 80%: the whole area of the test set is 48.1217; the area of correctly classified polygons is about 39.1.

4 Conclusions and Future Work

We have presented a method for learning a classifier for geographical areas. The method is based on the ID3 learning algorithm and the entropy computation is based on the area of polygons. However, the requirement for polygonal layers may be relaxed and we are currently investigating several possible solutions. It is possible to use line layers (for example, road layer) as well. In this case, the measure of the intersection is computed by considering the *length* of linear geometries. For point layers a suitable measure is the *count* of the occurrences of points. In practice, the entropy measure would be biased to polygonal layers. A good tradeoff is to consider a buffered layer (where each polygon contains all the points within a given distance to an object in the original layer). This solution creates a fresh polygonal layer. However, the choice of the buffer size is not so simple: it can also produce a bias as in the previous scenario. Another solution is to consider the definition of an entropy function *weighted* on each layer. This solution may provide the user with the possibility of promoting some layers during the learning process (i.e. for specifying the interest toward a specific layer). We are also considering some heuristics to improve the quality of data preparation, in particular the discretization of layers with numeric attributes. Another possible direction of research is considering the problem of learning spatially explicit classification models based on intra-layer relations, i.e. topological relations among neighboring spatial objects.

References

1. G. L. Andrienko and N. V. Andrienko. Data mining with c4.5 and cartographic visualization. In N.W.Paton and T.Griffiths, editors, *User Interfaces to Data Intensive Systems*, IEEE Computer Society Los Alamitos, CA, 1999.
2. G. L. Andrienko and N. V. Andrienko. Knowledge-based visualization to support spatial data mining. In D. J. Hand, J. N. Kok, and M. R. Berthold, editors, *Third Int. Symp. on Advances in Intelligent Data Analysis, IDA-99*, volume 1642. Springer, 1999.
3. R. Blazek, M. Neteler, and R. Micarelli. The new GRASS 5.1 vector architecture. In *Open source GIS – GRASS users conference 2002, Trento, Italy, 11-13 September 2002*. University of Trento, Italy, 2002.
4. L. Breiman, J. Friedman, R. Olshen, and C. Stone. *Classification and Regression Trees*. Wadsworth and Brooks, 1984.
5. M. Egenhofer. Reasoning about binary topological relations. In A. P. Buchmann, O. Günther, T. R. Smith, and Y. F. Wang, editors, *Proc. of the 2nd Int. Symp. on Large Spatial Databases (SSD)*, LNCS. Springer-Verlag, 1989.
6. ESRI - Arcview. Available at http://www.esri.com/software/arcview/.
7. M. Ester, A. Frommelt, H. Kriegel, and J. Sander. Spatial data mining: Database primitives, algorithms and efficient DBMS support. *Data Mining and Knowledge Discovery*, 4(2/3):193–216, 2000.
8. Vivid Solutions Inc. - JTS Topology Suite v1.4, 2000.
9. K. Koperski and J. W. Han. Discovery of spatial association rules in geographic information databases. *LNCS*, 951:47–66, 1995.
10. D. Malerba, F. Esposito, and F.A. Lisi. Mining spatial association rules in census data. In *Proc. of the Joint Conf. on "New Techniques and Technologies for Statistcs" and "Exchange of Technology and Know-how"*, 2001.
11. R. T. Ng and J. Han. Efficient and Effective Clustering Methods for Spatial Data Mining. In *Proc. of the 20th Int. Conf. on Very Large Databases*, 1994.
12. Simple feature specification. Available at http://www.opengis.org/specs, 1999.
13. J R Quinlan. Induction of decision trees. *Machine Learning*, 1(1), 1986. QUINLAN86.
14. J. Ross Quinlan. *C4.5: Programs for Machine Learning*. Morgan Kaufmann, 1992.
15. Postgis: Geographic objects for PostgreSQL.
    Available at http://postgis.refractions.net, 2002.
16. Philippe Rigaux, Michel Scholl, and Agnes Voisard. *Spatial Databases: With Application to GIS*. Morgan Kaufmann, 2002.

Digging into Acceptor Splice Site Prediction: An Iterative Feature Selection Approach

Yvan Saeys, Sven Degroeve, and Yves Van de Peer

Department of Plant Systems Biology, Ghent University,
Flanders Interuniversity Institute for Biotechnology (VIB),
K.L. Ledeganckstraat 35, Ghent, 9000, Belgium
{yvan.saeys,sven.degroeve,yves.vandepeer}@psb.ugent.be

Abstract. Feature selection techniques are often used to reduce data dimensionality, increase classification performance, and gain insight into the processes that generated the data. In this paper, we describe an iterative procedure of feature selection and feature construction steps, improving the classification of acceptor splice sites, an important subtask of gene prediction.

We show that acceptor prediction can benefit from feature selection, and describe how feature selection techniques can be used to gain new insights in the classification of acceptor sites. This is illustrated by the identification of a new, biologically motivated feature: the AG-scanning feature.

The results described in this paper contribute both to the domain of gene prediction, and to research in feature selection techniques, describing a new wrapper based feature weighting method that aids in knowledge discovery when dealing with complex datasets.

1 Introduction

During the past decades, feature selection techniques have increasingly gained importance, allowing researchers to cope with the massive amounts of data that have emerged in domains like image processing, text classification and bioinformatics. Within the framework of dimensionality reduction techniques, feature selection techniques are referred to as those techniques that select a (minimal) subset of features with "best" classification performance. In that respect, they differ from other reduction techniques like projection and compression, as they do not transform the original input features, but merely select a subset of them.

The reduction of data dimensionality has a number of advantages: attaining good or even better classification performance with a restricted subset of features, faster and more cost-effective predictors, and the ability to get a better insight in the processes described by the data. An overview of feature selection techniques can be found in [11] and [7].

In many classification problems in bioinformatics, the biological processes that are modelled are far from being completely understood. Therefore, these classification models are mostly provided with a plethora of features, hoping

J.-F. Boulicaut et al. (Eds.): PKDD 2004, LNAI 3202, pp. 386–397, 2004.

that the truly important features are included. As a consequence, most of the features will be irrelevant to the classification task, and will act as noise. This can degrade the performance of a classifier, and obfuscates the interpretation by a human expert, motivating the use of feature selection techniques for knowledge discovery.

An important machine learning task in bioinformatics is the annotation of DNA sequences: given a genome sequence and a set of example gene structures, the goal is to predict all genes on the genome [18]. An important subtask within gene prediction is the correct identification of boundaries between coding regions (exons) and intervening, non coding regions (introns). These border sites are termed splice sites: the transition from exon to intron is termed the *donor* splice site, and the transition from intron to exon is termed the *acceptor* splice site. In higher organisms, the majority of the donor splice sites are characterized by a GT subsequence occurring in the intron part, while the acceptor splice sites have an AG subsequence in the intron part. As a result, each splice site prediction task (donor prediction, acceptor prediction) can be formally stated as a two-class classification task: given a GT or AG subsequence, predict whether it is a true donor/acceptor splice site or not. In this paper we will focus on the prediction of acceptor splice sites in the plant model species *Arabidopsis thaliana*.

The paper is organised as follows. We start by introducing the different classification algorithms and feature selection techniques that were used in the experiments. Then we discuss the application of these techniques to the acceptor prediction task, describing an iterative procedure of feature selection and feature construction steps. We conclude by summarizing the new ideas that emerged from this work, and some suggestions for future work.

2 Methods

It is generally known that there is no such thing as the best classification algorithm. Likewise, there is no best feature selection method, and a comparative evaluation of classification models and feature selection techniques should thus be conducted to find out which combination performs best on the dataset at hand. In this section we discuss the classifiers, feature selection algorithms, and combinations of both techniques that were evaluated in our experiments, followed by a description of the datasets and the implementation.

2.1 Classification Algorithms

For our experiments, we chose two distinct types of classifiers that are widely used in machine learning research : a Bayesian classifier and a linear discriminant function. For the Bayesian classifier, the Naive Bayes method (NBM) was chosen because of its ability to cope with high-dimensional feature spaces and its robustness [5]. For the linear discriminant function, the linear Support Vector Machine (SVM) [3] was chosen because of its good performance on a wide range of classification tasks, and its ability to deal with high-dimensional feature spaces and large datasets.

2.2 Feature Selection Techniques

Techniques for feature selection are traditionally divided into two classes: filter approaches and wrapper approaches [14]. Filter approaches usually compute a feature relevance score such as the feature-class entropy, and remove low-scoring features. As such, these methods only look at the intrinsic properties of the dataset, providing a mechanism that is independent of the classification algorithm to be used afterwards. In the wrapper approach, various subsets of features are generated, and evaluated using a specific classification model. A heuristic search through the space of all subsets is then conducted, using the classification performance of the model as a guidance to find promising subsets. In addition to filter and wrapper approaches, a third class of feature selection methods can be distinguished: embedded feature selection techniques [2]. In embedded methods, the feature selection mechanism is built into the classification model, making direct use of the parameters of the induction model to include or reject features. In our experiments, a representative sample of each type of feature selection techniques was chosen, providing the basis for a comparative evaluation.

Filter Techniques. In general, filter techniques compute a feature relevance score in a single pass over the data, requiring a computational cost that is linear in the number of features. However, this comes at the expense of ignoring feature dependencies, which may result in suboptimal feature sets. To include feature dependencies, more complex filter approaches have been suggested such as the Markov blanket filter [15], or correlation based feature selection [8, 26].

In our work, we used the Markov blanket filter, introduced by Koller and Sahami, as an example of an advanced filter method that deals well with high-dimensional feature spaces [15]. This algorithm eliminates features whose information content is subsumed by some number of the remaining features. An approximate algorithm then solves the task of identifying the Markov blanket for each feature, and - starting from the full feature set - iteratively removes the feature with the "best" Markov blanket. The result of this algorithm is a ranking of all features from least relevant to most relevant. This ranking can then be used by any classification algorithm in a subsequent step. The parameter k of the algorithm determines the size of the Markov blanket, and exponentially increases running time as k gets larger. Typical values for K are {0,1,2}. In our experiments we choose $k = 1$.

Wrapper Techniques. Wrapper based methods combine a specific classification model with a strategy to search the space of all feature subsets. Commonly used methods are sequential forward or backward selection [13], and stochastic iterative sampling methods like genetic algorithms (GA) or estimation of distribution algorithms (EDA) [16, 10]. In our experiments we used an extension of the Univariate Marginal Distribution Algorithm (UMDA, [20]), the most simple EDA. This approach is very similar to the compact GA [9] or to a GA with uniform crossover. However, instead of using the single best feature subset that results from an iterative process like a GA or EDA, we used the frequencies with which the features are present in the final distribution of the UMDA as feature relevance scores. This gives a more dynamic view of the feature selection process, as it allows to derive feature weights. More details about this approach are discussed in [23].

Embedded Techniques. Two embedded feature selection techniques were used in the experiments: a weighted version of the Naive Bayes classifier, to which we will refer as WNBM, and recursive feature elimination using the weights of a linear SVM, further referred to as WLSVM.

The WNBM technique is based on the observation that the Naive Bayes method can be reformulated as a linear classifier when the classification task involves two classes, and all features are binary [5]. In this case, the feature weights can be calculated as

$$w_i = \ln \frac{p_i(1-p_i)}{q_i(1-q_i)} \; \forall \, i = 1, \cdots, n$$

where $p_i = Pr(x_i = 1|c_1)$ and $q_i = Pr(x_i = 1|c_2)$ are the class conditional probabilities of feature x_i being 1. These feature weights can then be sorted, providing a feature ranking. In principle, any classification task can be reduced to a set of two-class classification tasks, and arbitrary features can be converted to binary ones by discretization and sparse vector encoding.

In a similar fashion, the weights w_i of the decision boundary of a linear SVM can be used as feature weights to derive a feature ranking. However, better results can be obtained by recursively discarding one feature at the time, and retraining the SVM on the remaining features, as described in [6]. This is the approach that we adopted. A method, equivalent to this recursive feature elimination (RFE) approach, for feature selection with SVM is described in [24].

2.3 Data Sets and Implementation

The *Arabidopsis thaliana* data set was generated from sequences that were retrieved from the EMBL database, and only experimentally validated genes (i.e. no genes that resulted from a prediction) were used to build the dataset. Redundant genes were excluded, and splice site datasets were constructed from 1495 genes. More details on how these datasets were generated can be found in [4].

Because in real sequences, the number of true acceptor sites is largely outnumbered by the number of false acceptor sites, we chose to enforce a *class imbalance* in our datasets for feature selection. We constructed a dataset of 6000 positive instances and 36,000 negative instances. To obtain stable solutions for feature selection, a 10-fold cross-validation of this dataset was used to test all feature selection methods. This was done by doing 5 replications of a two-fold cross-validation, maintaining the same class imbalance of 1 positive versus 6 negative instances in every partition. For the EDA-based wrapper approach, the internal evaluation of classification performance was obtained by doing a 5-fold cross-validation on the training set.

The methods for feature selection were all implemented in C++, making use of the SVM^{light} implementation for Support Vector Machines [12]. The EDA-based wrapper method is a suitable candidate for parallellization, providing a linear speedup of the selection process. This was done using the MPI libraries, available at `http://www-unix.mcs.anl.gov/mpi/mpich`. However, due to other processes running on our servers, timing comparisons of the different algorithms fall outside the scope of this article.

3 An Iterative Feature Selection Approach

As already discussed in the introduction, many biological processes are still far from being understood. This greatly influences the design of the features that are to be used for a classification model that tries to model this process. In this section we describe an iterative feature construction and feature selection approach, resulting in increasingly more complex features and datasets, the design of which is guided by the feature selection techniques.

We start from the knowledge that the discrimination between true and false acceptor sites is determined by the part of the sequence where the site is located, more precisely the *local context* of the acceptor site. Therefore, the nucleotides A,T,C and G occurring on either side of the acceptor constitute a basic feature set.

3.1 A Simple Dataset: Position Dependent Nucleotides

A local context of 100 nucleotides (50 to the left, 50 to the right) around the acceptor sites was chosen, having at each position one of the four nucleotides {A,T,C,G}. These features were extracted for the positive and negative instances, resulting in a dataset of 100 4-valued features, which were converted into binary format using sparse vector encoding (A=1000,T=0100,C=0010,G=0001). This results in a dataset described by 400 binary features. For this dataset, the following combinations of classifiers and feature selection algorithms were evaluated: the Koller-Sahami filter method (further abbreviated as KS) for both NBM and linear SVM (LSVM), the EDA-based wrapper approach (EDA-R) for both NBM and LSVM, and both the embedded methods WNBM and WLSVM. For EDA-R the distribution size and the number of iterations were tuned to 500 and 20 respectively. For the SVM, the C-parameter was tuned to 0.05, using the full feature set.

A comparative evaluation of the results of our experiments is shown in Table 1. The classifier/selection approach combinations are tabulated row wise, and the results on different feature subsets are shown in the columns. Apart from the results using the full feature set (100%), the results using only 50%, 25%, 10% and 5% of the features are shown. The numbers represent the average F-measure

Table 1. F test comparisons for the dataset of 400 features.

Method	100%	50%	25%	10%	5%
KS NBM	80.87 ± 0.31	$80.85 \pm 0.37^{=}$	78.77 ± 0.45	74.67 ± 0.79	72.14 ± 0.70
EDA-R NBM	80.87 ± 0.31	$82.32 \pm 0.32^{*}$	$80.65 \pm 0.37^{=}$	76.70 ± 0.73	69.49 ± 2.46
WNBM	80.87 ± 0.31	$80.80 \pm 0.42^{=}$	76.84 ± 0.41	67.52 ± 0.52	60.39 ± 1.73
KS LSVM	84.45 ± 0.30	82.75 ± 0.28	80.66 ± 0.49	75.05 ± 0.42	71.00 ± 0.49
EDA-R LSVM	84.45 ± 0.30	$84.17 \pm 0.38^{=}$	81.62 ± 0.46	76.32 ± 0.67	68.73 ± 2.09
WLSVM	84.45 ± 0.30	84.00 ± 0.40	81.87 ± 0.39	76.73 ± 0.43	71.23 ± 0.40

[19] over the 10 cross-validation folds, and the standard deviation. For each of the reduced feature sets, the result was compared to the results on the full feature set, using the combined 5x2 cv F test, introduced in [1]. Statistically significant improvements at confidence intervals of 0.9 and 0.99 were denoted respectively by ‡ and *, statistically equivalent results compared to the full feature set were denoted by $^{=}$ and statistically worse results were not marked.

On a global scale, the only method that achieves better results is NBM combined with EDA-R feature selection, using only half of the features. For NBM, the wrapper method thus seems to produce the best results. The filter method KS produces the second best results, and WNBM performs worst. For the linear SVM, no significant gain in classification performance could be obtained. This can be explained by the fact that the SVM already implicitly uses a feature weighting scheme. For the LSVM, the embedded method WLSVM achieves the best results overall, followed by EDA-R and KS.

For knowledge discovery, the only method in our experiments that is able to derive feature weights is the EDA-R method. Using the results of the EDA-R LSVM combination, we can thus use the derived weights to visualize the relevance of the features. This can be done by color coding the normalized feature weights, as is shown in Figure 1. Part a in this figure shows the results of this color coding, where a gradient ranging from black (unimportant) to white (important) shows the feature weights. For each of the nucleotides (rows), the nucleotide positions (columns) are shown for both parts of the local context, the acceptor site being in the middle. Several patterns can be observed. The nucleotides bordering the acceptor site are of primary importance, representing binding information. Furthermore the nucleotides T in the left part of the context are highly important, representing the well-known pyrimidine-stretch (an excess of nucleotides T and C). A last pattern that can be observed is the three-base periodicity in the right part of the context, especially for nucleotides T and G, capturing the fact that this part of the sequence is the coding part (exon), and that nucleotides in this part are organized in triplets.

3.2 Adding Position Invariant Features

Position dependent features are not the best solutions when trying to capture information like coding potential and composition in the sequence. Therefore, a second dataset was constructed as an extension of the dataset in the previous experiment. In addition to the position dependent nucleotide features, we also added a number of position invariant features. These features capture the occurrence of 3-mers (words of length 3) in the sequence flanking the acceptor. An example of such a feature is the occurrence of the word "GAG" in the left part of the context. This results in another 128 binary features (64 for each part of the context), a 1 decoding the presence, a 0 the absence of the specific word in the context. Together with the position dependent features, this yields a dataset consisting of 528 binary features. The same parameters for EDA-R and SVM were used as with the previous dataset.

Table 2. F test comparisons for the dataset of 528 features.

Method	100%	50%	25%	10%	5%
KS NBM	78.21 ± 0.50	$78.40 \pm 0.50^{=}$	$77.96 \pm 0.64^{=}$	77.26 ± 0.46	74.21 ± 0.58
EDA-R NBM	78.21 ± 0.50	$84.48 \pm 0.30^{*}$	$83.52 \pm 0.36^{*}$	$80.79 \pm 0.57^{\ddagger}$	75.93 ± 0.87
WNBM	78.21 ± 0.50	77.17 ± 0.51	$77.85 \pm 0.37^{=}$	74.06 ± 0.32	67.96 ± 0.68
KS LSVM	87.52 ± 0.49	87.15 ± 0.32	86.03 ± 0.41	82.05 ± 0.46	77.03 ± 0.80
EDA-R LSVM	87.52 ± 0.49	86.72 ± 0.54	85.64 ± 0.59	82.34 ± 0.43	77.02 ± 1.10
WLSVM	87.52 ± 0.49	87.20 ± 0.49	86.40 ± 0.50	84.35 ± 0.48	78.34 ± 0.92

The results of the feature selection experiments on this dataset are shown in Table 2. Comparing the results for NBM to the previous dataset, the classification performance on the full feature set is lower than on the first dataset. However, using feature selection, better classification results than on the first dataset can be obtained. Again, the best results were obtained with the EDA-R wrapper method. Using only 10% of the features, this method still obtains significantly better results than using the full feature set. The KS filter method performs second best, WNBM performs worst. For the SVM, a significant gain of 3% in classification performance was obtained by adding the position invariant features. Similar to the previous dataset, the performance could not be improved using feature selection methods, and the embedded method WLSVM obtains the best results.

The visualization of the feature weights, obtained by the EDA-R LSVM approach, is shown in part b of Figure 1. While the same patterns as in the case of dataset 1 can be observed, it is clear that some information is translated from position dependent features into position invariant features. An example of this is the pyrimidine stretch, which is somewhat shortened compared to the results on the previous dataset, together with the fact that T-rich 3-mers in the left part of the context show up as very important. Another example is the fact that the 3-base periodicity on the coding side is less pronounced, yet some 3-mers are shown to be highly relevant. The results from the feature weighting, combined with the improved classification results explain that indeed position invariant features contribute to the prediction of acceptor sites.

Table 3. F test comparisons for the dataset of 2096 features.

Method	100%	50%	25%	10%	5%
KS NBM	79.21 ± 0.33	79.46 ± 0.30$^=$	79.08 ± 0.39$^=$	79.07 ± 0.57$^=$	78.03 ± 0.97$^=$
EDA-R NBM	79.21 ± 0.33	85.29 ± 0.36*	83.81 ± 0.69*	79.90 ± 0.62$^=$	76.51 ± 0.99
WNBM	79.21 ± 0.33	79.90 ± 0.44‡	79.52 ± 0.34$^=$	77.36 ± 0.50	75.61 ± 0.58
KS LSVM	88.24 ± 0.51	87.56 ± 0.41$^=$	85.62 ± 0.64	83.10 ± 0.49	79.88 ± 1.06
EDA-R LSVM	88.24 ± 0.51	87.90 ± 0.36	86.66 ± 0.43	84.07 ± 0.74	81.73 ± 0.56
WLSVM	88.24 ± 0.51	88.22 ± 0.44$^=$	88.08 ± 0.35$^=$	87.10 ± 0.32	85.86 ±0.34

3.3 Adding More Complex Features: Dependencies Between Adjacent Nucleotides

It is known that correlations exist between nucleotides in the vicinity of splice sites. To detect these dependencies, higher-order (i.e. non-linear) classification methods can be used, like polynomial SVMs. However, these methods have the disadvantage of being quite slow to train, rendering the feature selection process more computationally intensive. Here, we describe another approach to deal with nucleotide dependencies, having the advantage that linear (and thus fast) classification models can be used. We do this by constructing more complex features, capturing the nucleotide dependencies at the feature level. Another important advantage is that the combination with feature selection techniques allows us to select those dependencies that are of primary importance, and visualize them.

To this end, we created complex features that capture dependencies between two adjacent nucleotides. These features are represented as position dependent 2-mers (words of length 2). At each position i of the local context, these features represent the word appearing at position i and $i + 1$. This results in an a set of 1568 binary features (49x16x2). Together with the position dependent nucleotides and the position invariant features, this results in a dataset described by 2096 features. For this dataset, the C-parameter of the SVM was tuned to 0.005.

The results of the feature selection experiments for this dataset are shown in part c of Figure 1. Compared to the results on the previous datasets, similar trends can be observed. The NBM classifier performs worse than dataset 1 on the full feature set, but outperforms the results on dataset 1 and 2, when EDA-R is used with only 50% of the features. For the SVM, an increase in classification performance is noted, compared to dataset 1 and 2. Again, the result on the full feature set cannot be improved using feature selection methods.

Visualizing the weights derived from the EDA-R LSVM combination (Figure 1, part c) reveals some remarkable, new patterns. In addition to the previous patterns, three new patterns, related to the inclusion of dependencies between adjacent nucleotides, can be observed. Firstly, it is observed that nucleotide dependencies immediately neighbouring the acceptor site are of great importance. Furthermore two patterns, related to the 2-mers AG and TG emerge in the left part of the context.

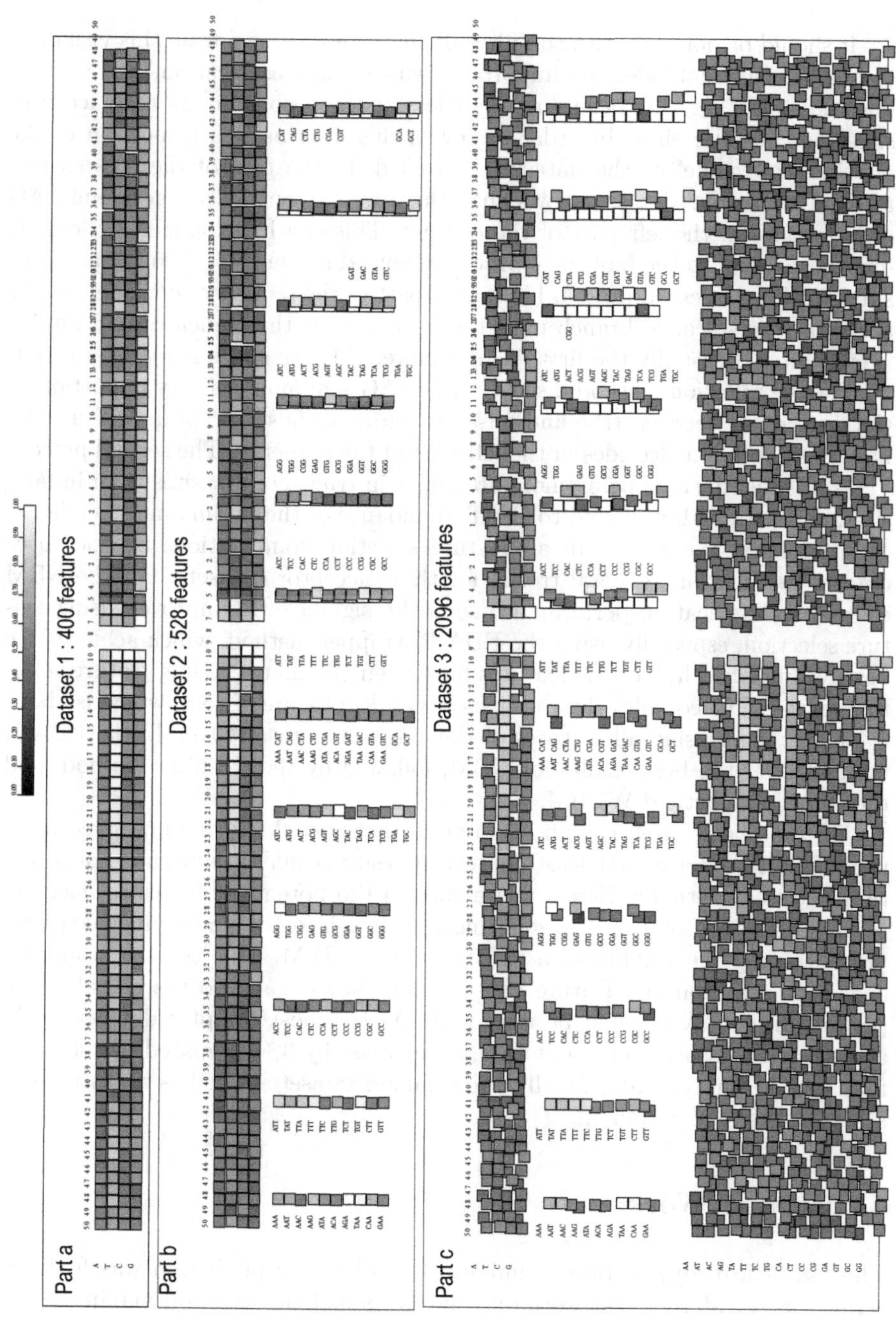

Fig. 1. Visualization of the feature weights for datasets with increasing complexity, obtained by the EDA-R LSVM approach: a simple dataset of 400 binary features (part a), an extended dataset also including position invariant features (part b), and a complex dataset also capturing adjacent feature dependencies (part c).

It should be noted that the only result that can be drawn from this visualization is the fact that these are important features for classification. The method does not tell if e.g. AG occurs more or less at these positions in true acceptor sites than in false sites. In order to reveal this information, inspection of the classification model or the datasets is needed. In the case of the AG-feature, an analysis of our dataset shows that there is a strong selection against AG dinucleotides in the left part of the context. This can be explained by looking into more detail to the biological process involved in the recognition of acceptor sites. In this process, a protein binds to a part of the sequence to the left of the acceptor (the so-called branch point) and then scans the sequence until an AG is encountered (usually the first AG encountered is the splice site). As a result, our feature selection method discovers this "AG-scanning" as very important in the distinction between true and false acceptors, as false acceptors will usually have more AG dinucleotides in the left part of the sequence. The second pattern (TG) was identified as being more abundant in true acceptor sites than in false sites, and is probably related to the T-abundance of the pyrimidine stretch.

Comparing the results of all feature selection combinations on the three dataset reveals some general trends for these acceptor datasets. For the NBM classifier, classification performance could be significantly improved using feature selection, especially using the EDA-R wrapper method, which achieves the best results when half of the features have been eliminated. Using feature selection on the most complex dataset achieves an F-measure of 85%, which is about 5% better than using all features of the simplest dataset. Overall, the EDA-R method gives the best results for NBM, followed by the KS filter method, and the embedded method WNBM.

For the linear SVM, classification performance could not be improved upon using feature selection. At least equivalent results could be obtained using feature selection methods. However, we showed the potential of feature selection techniques as a useful tool to gain more insight into the underlying biological process, using the combination of EDA-R with SVM. The iterative approach of feature selection and feature construction shows that also the classification performance of SVM could be improved. On the most complex dataset, SVM achieves an F-measure of about 88%, an increase by 3% compared to using the most simple dataset. Additionally, the complex dataset allowed us to distinguish a new feature: AG-scanning.

4 Related Work

The use of feature selection techniques for splice site prediction was first described in [4]. More recent work on splice sites includes work on maximum entropy modelling [25], where the authors only consider a very short local context, and the work of Zhang et al. [27] where SVMs are used to model human splice sites. Related work on using EDAs for feature selection can be found in [10, 22], and some recent developments to use feature selection in combination with SVMs are described in [24].

5 Conclusions and Future Work

In this paper, we described an iterative feature selection approach for the classification of acceptor splice sites, an important problem in the context of gene prediction. A comparative evaluation of various feature selection and classification algorithms was performed, demonstrating the usefulness of feature selection techniques for this problem. Furthermore, we proposed a new feature weighting scheme (derived from the EDA-R method) that deals well with datasets described by a large number of features. We showed how this technique can be used to guide the feature construction process, arriving at more complex feature sets with better classification performance. Using these feature weights, and the design of complex features that capture dependencies at the feature level, we demonstrated how important nucleotide correlations could be visualised. Such a visualisation facilitates the discovery of knowledge for human experts, and led to the discovery of a new, biologically motivated feature: the AG-scanning feature.

For the best-scoring method of our experiments, the linear SVM, preliminary experiments with a more complex version of the AG-scanning feature have been designed, incorporating the predictor in a complete gene prediction system. These experiments show promising results with respect to state-of-the-art splice site predictors, like GeneSplicer [21]. Another line of future research is motivated by the use of linear classifiers in combination with more complex features, capturing higher order, or non-adjacent nucleotide dependencies at the feature level, or taking into account secondary structure information. Combining such complex feature sets with linear classifiers and feature selection techniques can be useful to learn and visualise more complex dependencies.

References

1. Alpaydin, E. A Combined 5x2 cv F Test for Comparing Supervised Classification Learning Algorithms. Neural Computation **11(8)** (1999) 1885–1892
2. Blum, A.I., Langley, P.: Selection of relevant features and examples in machine learning. Artificial Intelligence **97** (1997) 245–271
3. Boser, B., Guyon, I., Vapnik, V.N.: A training algorithm for optimal margin classifiers. Proceedings of COLT (Haussler,D. ,ed.), ACN Press (1992) 144–152
4. Degroeve, S., De Baets, B., Van de Peer, Y., Rouzé, P.: Feature subset selection for splice site prediction Bioinformatics **18 Supp.2** (2002) 75–83
5. Duda, R.O., Hart, P.E., Stork, D.G.: Pattern classification. New York, NY, Wiley, 2nd edition (2000)
6. Guyon, I., Weston, J., Barnhill, S., Vapnik, V.N.: Gene Selection for Cancer Classification using Support Vector Machines. Machine Learning **46(1-3)** (2000) 389–422
7. Guyon, I., Elisseeff, A.: An Introduction to Variable and Feature Selection. Journal of Machine Learning Research **3** (2003) 1157–182
8. Hall, M.A., Smith, L.A.: Feature Selection for Machine Learning : Comparing a Correlation-based Filter Approach to the Wrapper. Proc. of the Florida Artificial Intelligence Symposium (1999)
9. Harik, G.R., Lobo, G.G., Goldberg, D.E.: The compact genetic algorithm. Proc. of the International Conference on Evolutionary Computation 1998 (ECEC '98), Piscataway, NJ: IEEE Service Center (1998) 523–528

10. Inza, I.,Larrañaga, P., Sierra. B. Feature Subset Selection by Estimation of Distribution Algorithms. In *Estimation of Distribution Algorithms. A new tool for Evolutionary Computation.* (2001) P. Larrañaga, J.A. Lozano (eds.)
11. Jain, A.K., Duin, R.P.W., Mao, J. Statistical Pattern Recognition: A Review. IEEE Transactions on Pattern Analysis and Machine Intelligence **22(1)** (2000) 4–37
12. Joachims, T.: Making large-scale support vector machine learning practical. B. Schölkopf, C. Burges, A. Smola. Advances in Kernel Methods: Support Vector Machines, MIT Press, Cambridge, MA (1998)
13. Kittler, J.: Feature set search algorithms. In *Pattern Recognition and Signal Processing* (1978) 41–60
14. Kohavi, R., John, G.: Wrappers for feature subset selection. Artificial Intelligence **97(1-2)** (1997) 273–324
15. Koller, D., Sahami, M.: Toward optimal feature selection. Proceedings Thirteenth International Conference on Machine Learning (1996) 284-292
16. Kudo, M., Sklansky, J.: Comparison of algorithms that select features for pattern classifiers. Pattern Recognition **33** (2000) 25–41
17. Larrañaga, P., Lozano, J.A.: Estimation of Distribution Algorithms. A New Tool for Evolutionary Computation. Kluwer Academic Publishers (2001)
18. Mathé, C., Sagot, M.F., Schiex, T., Rouzé, P.: Current methods of gene prediction, their strengths and weaknesses. Nucleic Acids Research **30** (2002) 4103–4117
19. Mladenić, D., Grobelnik, M.: Feature selection on hierarchy of web documents. Decision Support Systems **35** (2003) 45–87
20. Mühlenbein, H., Paass, G.: From recombination of genes to the estimation of distributions. Binary parameters. Lecture Notes in Computer Science 1411 : Parallel Problem Solving from Nature, PPSN IV (1996) 178–187
21. Pertea, M., Lin, X., Salzberg, S.: GeneSplicer: a new computational method for splice site prediction. Nucleic Acids Research **29** (2001) 1185–1190
22. Saeys, Y., Degroeve, S., Aeyels, D., Van de Peer, Y., Rouzé, P.: Fast feature selection using a simple Estimation of Distribution Algorithm: A case study on splice site prediction. Bioinformatics **19-2** (2003) 179–188
23. Saeys, Y., Degroeve, S., Van de Peer, Y.: Feature ranking using an EDA-based wrapper approach. In *Towards a new evolutionary computation: advances in Estimation of Distribution Algorithms* J.A. Lozano et al. (eds), In press
24. Weston, J., Elisseeff, A., Schoelkopf, B., Tipping, M.: Use of the Zero-Norm with Linear Models and Kernel Methods. Journal of Machine Learning Research **3** (2003) 1439–1461
25. Yeo, G., Burge, C.B.: Maximum entropy modelling of short sequence motifs with applications to RNA splicing signals. Proceedings of RECOMB 2003 (2003) 322–331
26. Yu, L., Liu,H.: Feature Selection for High-Dimensional Data: A Fast Correlation-Based Filter Solution. Proceedings ICML 2003 (2003) 856–863
27. Zhang, X., Heller, K., Hefter, I., Leslie, C., Chasin L.: Sequence Information for the Splicing of Human pre-mRNA Identified by Support Vector Machine Classification Genome Research **13** (2003) 2637–2650

Itemset Classified Clustering

Jun Sese[1] and Shinichi Morishita[2]

[1] Undergraduate Program for Bioinformatics and Systems Biology, Graduate School of Information Science and Technology, University of Tokyo
sesejun@cb.k.u-tokyo.ac.jp

[2] Department of Computational Biology, Graduate School of Frontier Sciences, University of Tokyo and Institute for Bioinformatics and Research and Development, Japan Science and Technology Corporation
moris@cb.k.u-tokyo.ac.jp

Abstract. Clustering results could be comprehensible and usable if individual groups are associated with characteristic descriptions. However, characterization of clusters followed by clustering may not always produce clusters associated with special features, because the first clustering process and the second classification step are done independently, demanding an elegant way that combines clustering and classification and executes both simultaneously.

In this paper, we focus on itemsets as the feature for characterizing groups, and present a technique called "itemset classified clustering," which divides data into groups given the restriction that only divisions expressed using a common itemset are allowed and computes the optimal itemset maximizing the interclass variance between the groups. Although this optimization problem is generally intractable, we develop techniques that effectively prune the search space and efficiently compute optimal solutions in practice. We remark that itemset classified clusters are likely to be overlooked by traditional clustering algorithms such as two-clustering or k-means, and demonstrate the scalability of our algorithm with respect to the amount of data by the application of our method to real biological datasets.

1 Introduction

Progress in technology has led to the generation of massive amounts of data, increasing the need to extract informative summaries of the data. This demand has led to the development of data mining algorithms, such as clustering [14, 22, 9, 1, 5, 20], classification [4, 15], and association rules [2, 10, 13]. Recent technological progress in biology, medicine and e-commerce marketing has generated novel datasets that often consist of tuples represented by features and an objective numeric vector. For understanding what causes individual groups of data similar in terms of vectors, it is helpful to associate features with each group. A typical example from molecular biology is association of gene-controlling mechanisms with genes having analogous expression patterns. Such novel data motivate us to develop itemset classified clustering.

J.-F. Boulicaut et al. (Eds.): PKDD 2004, LNAI 3202, pp. 398–409, 2004.

1.1 Motivating Example

We here present a motivating example for showing the difference between traditional approach and our method called "itemset classified clustering."

Consider eight tuples $t_1, ..., t_8$ in Table 1. Each tuple contains feature items $i_1, ..., i_5$ and objective attributes a_1 and a_2. Fig. 1(A) shows objective vectors (a_1, a_2) of the tuples represented by white circles. Each tuple is at regular interval on the same square. For example, tuple t_2 locates at $(1, 2)$.

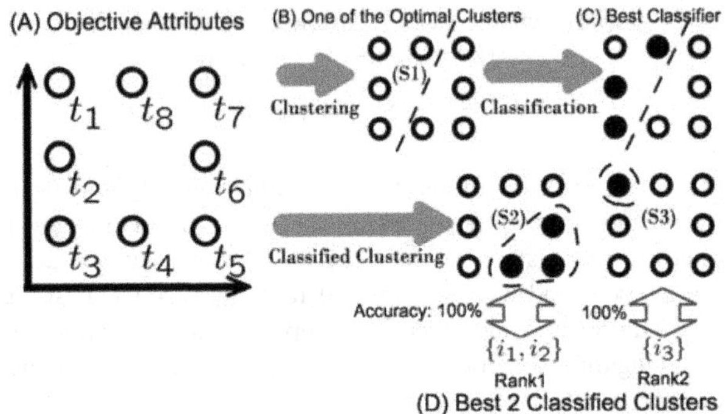

Fig. 1. Motivating Example of Itemset Classified Clustering

Table 1. Example Table

	Feature Items					Objective Attributes	
	i_1	i_2	i_3	i_4	i_5	a_1	a_2
t_1	0	0	1	0	0	1	3
t_2	1	0	0	1	1	1	2
t_3	0	0	0	1	1	1	1
t_4	1	1	0	0	1	2	1
t_5	1	1	0	0	1	3	1
t_6	1	1	0	0	1	3	2
t_7	0	0	0	1	0	3	3
t_8	0	1	0	1	1	2	3

Let us form clusters by dividing the tuples into two groups, S and $\bar{S}$, so that the division optimizes a proper measure. As the measure, one may utilize various values such as the diameter or the connectivity of a group, we here use interclass variance extended to the multi-dimension, which is common measure grounding in statistics to evaluate clusters. For simplicity, we call this multi-dimensional version just *interclass variance* in this paper. Let $c(S)$ denote the centroid of objective vector of S; namely, $\sum_{\boldsymbol{x} \in S} \boldsymbol{x}/|S|$. Interclass variance is defined as: $|S| \left|c(S) - c(S \cup \bar{S})\right|^2 + |\bar{S}| \left|c(\bar{S}) - c(S \cup \bar{S})\right|^2$. One of the solutions that maximize interclass variance is indicated by dotted straight line (S1) in Fig. 1(B). Line (S1) divides the tuples into cluster $S = \{t_1, t_2, t_3, t_8\}$ and cluster $\bar{S} = \{t_4, t_5, t_6, t_7\}$.

To understand the reason why tuples in each cluster are close, traditional clustering-classification approach such as conceptual clustering [12] attempts to find classifiers that are able to exactly classify the clusters using, say, additional feature items $i_1, ..., i_5$. In Table 1, "1" denotes the presence of an item in each tuple, while "0" denotes the absence. For instance, tuple t_2 includes item i_1, i_4 and i_5. In Fig. 1(C), solid black circles indicate tuples that contain itemset

$\{i_4, i_5\}$. Note that three out of four circles on the left of (S1) include the itemset, while none of the four on the right does. From this observation, one may derive the classifier that circles contain $\{i_4, i_5\}$ if and only if they are on the left of (S1), which holds seven out of eight cases, namely 87.5% accuracy. However, use of optimal clustering, such as the division by (S1), may not be able to identify a clustering so informative that each cluster is associated with its special feature items.

For identification of such beneficial clusters, our itemset classified clustering computes optimal clusters under the restriction that allows only splits expressible by a common itemset. We call such clusters *classified clusters.* (S2) in Fig. 1(D) indicates an example of classified cluster because the group $\{t_4, t_5, t_6\}$ is equal to the set of tuples that contain both i_1 and i_2. In other words, classifier $\{i_1, i_2\}$ has 100% accuracy for cluster $\{t_4, t_5, t_6\}$. $\{t_1\}$ is another example of classified cluster because the cluster is associated with the special classifier $\{i_3\}$ ((S3) in Fig. 1(D)). In these classified clusters, the set of tuples whose interclass variance is larger would be better cluster. For example, $\{t_4, t_5, t_6\}$ split by (S2) is better classified cluster than $\{t_1\}$ split by (S3). Note that the classified clusters are overlooked by two-clustering in Fig. 1(B), and the groups would not be found by general clustering algorithms such as k-means clustering.

One may wonder that the itemset associated with any optimal classified cluster is a closed pattern [21], which is a maximal itemset shared in common by transactions including the itemset. However, this claim is not true. For instance, $\{i_1, i_2\}$, which is not a closed pattern, classifies optimal classified cluster $\{t_4, t_5, t_6\}$. Its superset $\{i_1, i_2, i_5\}$ is a closed pattern and it also classifies the same optimal cluster; however, its inclusion of i_5 is superfluous, because neither of its subsets having i_5, namely $\{i_1, i_5\}$ and $\{i_2, i_5\}$, identifies any optimal classified cluster. This observation indicates that $\{i_1, i_2\}$ better classifies the optimal cluster. On the other hand, the closed pattern $\{i_4, i_5\}$ corresponds to non-optimal cluster $\{t_2, t_3, t_8\}$. Consequently, itemsets for optimal classified clusters and closed pattern itemsets are orthogonal notions.

This example reveals us that it is a non-trivial question to compute the optimal classified clusters because of two major problems. First, cluster that maximizes the index such as interclass variance is not always associated with special features. In our example, although cluster segmented by (S1) has the optimal index, the clusters are not associated with special features. Thus, the approach of clustering followed by classification is not effective for deriving classified clusters. Second, the number of combinations of items explodes as the number of items increases.

1.2 Related Work

On clustering-classification approach, refinement of clustering or classification might improve accuracy. Clustering studies have paid a great deal of attention to the choice of a measure that is tailored to the specificity of given data. For example, measures of sub-clustering for gene expression profiles [5, 20] and model-based measures [19] have been proposed. However, in these approaches, feature items are not supposed to be used to output directly constrained clusters.

Improvement of classification would increase the accuracy of the classifier for each cluster. This sophistication, however, could increase the description length of the classifier. For example, the number of nodes in a decision tree such as CART [4] and C4.5 [15] is likely to huge, making it difficult to understand. Moreover, classification methods do not generate classified clusters of similar objects in terms of numeric vectors associated with tuples.

2 Itemset Classified Clustering

In this section, we formalize the itemset classified clustering problem.

Itemset Classified Clustering: Suppose that we classify a tuple by checking to see whether it includes a feature itemset (e.g., $\{i_1, i_2\}$). Compute the optimal classifier that maximizes interclass variance with its corresponding cluster, or list the most significant N solutions.

In the running example, the optimal classifier is the itemset $\{i_1, i_2\}$ and its corresponding cluster is $\{t_4, t_5, t_6\}$. Furthermore, when $N = 10$, the itemset classified clustering problem demands the extraction of ten optimally classified clusters.

Unfortunately, it is difficult to compute an optimal itemset that maximizes the interclass variance, because the problem is NP-hard if we treat the maximum number of items in an itemset as a variable. The NP-hardness can be proved by reduction of the difficulty of the problem to the NP-hardness of finding the minimum cover [8]. The reduction consists of the following three major steps: (1) Treat a tuples as a vertex, and an item as a hyperedge enclosing such vertexes (tuples) that contain the item. (2) A cover is then expressed as an itemset. (3) An optimal itemset is proved to coincide with a minimal cover of vertexes according to the convexity of the interclass variance function [13].

The NP-hardness prompts us to make an effective method to compute the optimal itemset in practice. To compute the itemset classified clustering problem, we present the properties of interclass variance in the next section.

3 Interclass Variance

3.1 Basic Definitions

In this section, we first introduce the index, interclass variance.

Definition 1 Let D be the set of all tuples. Let i_k denote an item. We treat m numerical attributes in the given database as special, and we call these attributes *objective attributes*. Let $a_1, a_2, \ldots, a_m$ denote the objective attributes. Let $t[a_i]$ indicate the value of an objective attribute a_i associated with a tuple t. ■

In Table 1, let $D = \{t_1, t_2, \ldots, t_8\}$ and $i_1, \ldots, i_5$ be items, and a_1 and a_2 be objective attributes. Then, t_2 contains itemset $\{i_1, i_4, i_5\}$ and $t_2[a_1] = 1$ and $t_2[a_2] = 2$.

We divide D into two groups using itemset I, D_I and $\bar{D_I}$. D_I means a set of tuples that include itemset I, and $\bar{D_I}$ is the complement of D_I; namely $D - D_I$. In the running example, when $I = \{i_1\}$, $D_I = \{t_2, t_4, t_5, t_6\}$ and $\bar{D_I} = \{t_1, t_3, t_7, t_8\}$.

Definition 2 Let n be $|D|$ and $x(I)$ be $|D_I|$. Let s_i be $\sum_{t \in D} t[a_i]$, and $y_i(I)$ be $\sum_{t \in D_I} t[a_i]$. We define the interclass variance of itemset I as
$$x(I) \sum_{i=1}^{m} \left(\frac{y_i(I)}{x(I)} - \frac{s_i}{n} \right)^2 + (n - x(I)) \sum_{i=1}^{m} \left(\frac{s_i - y_i(I)}{n - x(I)} - \frac{s_i}{n} \right)^2. \blacksquare$$

Since s_i and n are independent of the choice of itemset I according to the definition of interclass variance, the values of $x(I)$ and $y_i(I)$ uniquely determine interclass variance. Therefore, we will refer to interclass variance as $var(x, y_1, \ldots, y_m)$.

Definition 3 $var(x, y_1, \ldots, y_m) = x \sum_{i=1}^{m} \left(\frac{y_i}{x} - \frac{s_i}{n} \right)^2 + (n - x) \sum_{i=1}^{m} \left(\frac{s_i - y_i}{n - x} - \frac{s_i}{n} \right)^2 . \blacksquare$

In the running example, let $I = \{i_1\}$. $n = 8$, $s_1 = s_2 = 16$, $x(I) = 4$, $y_1(I) = 9$ and $y_2(I) = 6$. Therefore, $var(x(I), y_1(I), y_2(I)) = 2.5$.

When $m = 1$, this measure equals the interclass variance, a well-known statistical measure. Therefore, this index is a multi-dimensional generalization of the interclass variance.

From the definition of interclass variance, we can prove the convexity of $var(x, y_1, \ldots, y_m)$. The convexity is useful for conducting an effective search for significant itemsets.

Definition 4 A function $f(x, y_1, \ldots, y_m)$ is convex if for any $(x, y_1, \ldots, y_m)$ and $(x', y'_1, \ldots, y'_m)$ in the domain of f, and for any $0 \le \lambda \le 1$,
$\lambda f(x, y_1, \ldots, y_m) + (1 - \lambda) f(x', y'_1, \ldots, y'_m) \ge f(\lambda(x, y_1, \ldots, y_m) + (1 - \lambda)(x', y'_1, \ldots, y'_m)) \blacksquare$.

Proposition 1 $var(x, y_1, \ldots, y_m)$ $(0 \le x \le n)$ is a convex function.

Proof *(Omitted)* ■

3.2 Upper Bound

To calculate the set of significant itemsets, it is useful to estimate an upper bound of the interclass variance of any superset J of I because the information allows us to restrict the search space of the itemsets. For example, if an upper bound of itemset $\{i_2\}$ is less than the interclass variance of $\{i_1\}$, then $\{i_2\}$ and its supersets (e.g., $\{i_2, i_3\}$) can be pruned.

To estimate an upper bound, first, we map each itemset $J \supseteq I$ to a tuple $(x(J), y_1(J), \ldots, y_m(J))$, which we call *stamp point* of J. Subsequently, we calculate a hyper-polyhedron that encloses all the stamp points of $J \subseteq I$ for the given itemset I. Finally, we prove that one of the vertexes on the wrapping hyper-polyhedron provides an upper bound. We now present precise definitions and propositions.

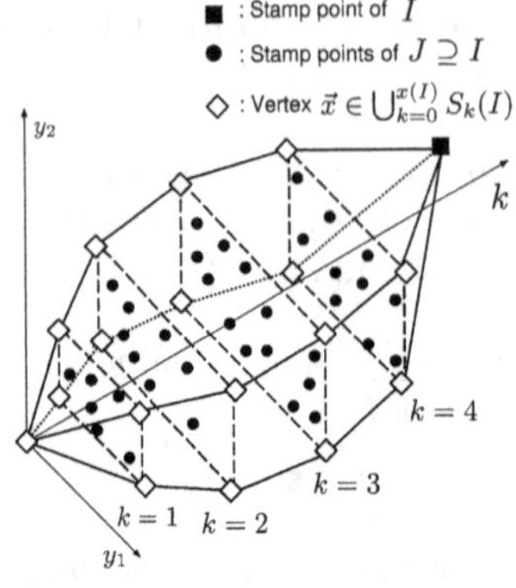

Fig. 2. The hyper-polyhedron surrounding all the stamp points of $J \supseteq I$

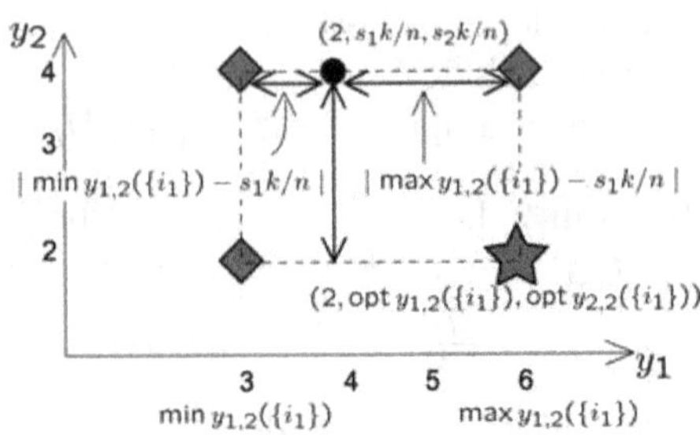

Fig. 3. The point maximizing interclass variance in $S_2(\{i_1\})$

Definition 5 Let $y_{i,k}(I)$ be the multi-set $\{\sum_{t \in D' \subseteq D_I} t[a_i] \mid |D'| = k\}$. Let $S_k(I)$ be $\{(k, z_1, \ldots, z_m) \mid z_i = \max y_{i,k}(I) \text{ or } z_i = \min y_{i,k}(I) \text{ for } i = 1, 2, ..., m\}$, where m is the number of objective attributes. Each element in $S_k(I)$ is a vertex of the wrapping hyper-polyhedron on $x = k$. ■

Here, we describe $y_{i,k}(I)$ as a multi-set, because the multi-set representation will be required later to define the best N solutions.

For example, in Table 1, let I be $\{i_1\}$. Then, $D_I = \{t_2, t_4, t_5, t_6\}$, $y_{1,1}(I) = \{1, 2, 3, 3\}$, $y_{1,2}(I) = \{3, 4, 4, 5, 5, 6\}$. Furthermore, $y_{2,2}(I) = \{2, 3, 3, 3, 3, 4\}$. Therefore, $S_2(I) = \{(2, 3, 2), (2, 3, 4), (2, 6, 2), (2, 6, 4)\}$.

Lemma 1 For any itemset $J \supseteq I$,
$var(x(J), y_1(J), \ldots, y_m(J)) \leq \max_{0 \leq k \leq x(I)} \{var(\boldsymbol{x}) \mid \boldsymbol{x} \in S_k(I)\}$.

Proof It is known that any convex function is maximized at one of the vertexes on the boundary of a convex hyper-polyhedron [1]. From Proposition 1, interclass variance is a convex function. Due to its convexity, it is sufficient to prove that the hyper-polyhedron of $\bigcup_{k=0}^{x(I)} S_k(I)$ encloses all the stamp points of itemsets $J \supseteq I$.

Note that $D_J \subseteq D_I$ for any itemset $J \supseteq I$. Since $y_i(J) \in y_{i,x(J)}(I) (0 \leq i \leq m)$, $\min y_{i,x(J)}(I) \leq y_i(J) \leq \max y_{i,x(J)}(I)$. Therefore, $var(x(J), y_1(J), \ldots, y_m(J)) \leq \max\{var(\boldsymbol{x}) \mid \boldsymbol{x} \in S_{x(J)}(I)\} \leq \max_{0 \leq k \leq x(I)} \{var(\boldsymbol{x}) \mid \boldsymbol{x} \in S_k(I)\}$. ■

According to this lemma, we can estimate an upper bound of the interclass variance of itemset $J \supseteq I$. Fig. 2 illustrates the wrapping strategy. We can confirm that the hyper-polyhedron surrounds all stamp points. Indeed, according to this lemma, we can estimate an upper bound of the interclass variance of any itemset $J \supseteq I$.

However, two problems may appear. First, the wrapping hyper-polyhedron might not be sufficiently tight to form a convex hull of $J \supseteq I$. Second, it could be too costly to calculate an upper bound when m is large because the number of vertices could be $x(I) \times 2^m$ in the worst case.

For the first problem, we show that our wrapping function is tight enough to solve real data in Section 5. To overcome the second problem, we develop a technique that allows us to dramatically reduce the number of vertices to consider in the next subsection.

3.3 Efficient Calculation of the Upper-Bound

We first remark that the vertex farthest away from $s_i k/n$ among the vertices of the hyper-polyhedron on hyper-plain $x = k$ maximizes the interclass variance. From this property, we will devise an efficient algorithm for searching an upper bound.

Definition 6 We denote the vertex farthest away from $s_i k/n$ by $(k, \mathrm{opt}\, y_{1,k}(I), \mathrm{opt}\, y_{2,k}(I), \ldots)$, where

$$\mathrm{opt}\, y_{i,k}(I) = \begin{cases} \min y_{i,k}(I) & \text{if } \mid \min y_{i,k}(I) - s_i k/n \mid > \mid \max y_{i,k}(I) - s_i k/n \mid \\ \max y_{i,k}(I) & \text{otherwise} \end{cases} \blacksquare$$

In the running example, $k = 2$, $n = 8$ and $s_1 = s_2 = 16$. Then, $s_1 k/n = s_2 k/n = 4$. Since $y_{1,2}(\{i_1\}) = \{3, 4, 4, 5, 5, 6\}$, $\min y_{1,2}(\{i_1\}) = 3$ and $\max y_{1,2}(\{i_1\}) = 6$. Therefore, $\mathrm{opt}\, y_{1,2}(\{i_1\}) = \max y_{1,2}(\{i_1\}) = 6$. Similarly, $\mathrm{opt}\, y_{2,2}(\{i_2\}) = \min y_{2,2}(\{i_1\}) = 2$. The arrows in Fig. 3 illustrate the selection of opt.

Lemma 2 $var(k, \mathrm{opt}\, y_{1,k}(I), \ldots, \mathrm{opt}\, y_{m,k}(I)) = \max\{var(\boldsymbol{x}) \mid \boldsymbol{x} \in S_k(I)\}$.

Proof Let $\boldsymbol{c} = (k, s_1 k/n, \ldots, s_m k/n)$. $var(x, y_1, \ldots, y_m) = x \sum_{i=1}^{m} \left(\frac{y_i}{x} - \frac{s_i}{n}\right)^2 + (n-x) \sum_{i=1}^{m} \left(\frac{s_i - y_i}{n-x} - \frac{s_i}{n}\right)^2 = \left(\frac{1}{x} + \frac{1}{n-x}\right) \sum_{i=1}^{m} \left(y_i - \frac{s_i}{n} x\right)^2$. From this equality, on hyper-plain $x = k$, if $|(k, y_1, \ldots, y_m) - \boldsymbol{c}| \geq |(k, y'_1, \ldots, y'_m) - \boldsymbol{c}|$, then $var(k, y_1, \ldots, y_m) \geq var(k, y'_1, \ldots, y'_m)$.

Now, since $\mathrm{opt}\, y_{i,k}(I)$ denotes the value farthest away from $s_i k/n$ among $y_{i,k}(I)$, the point farthest from $\boldsymbol{c}$ on $x = k$ is $(k, \mathrm{opt}\, y_{1,k}(I), \ldots, \mathrm{opt}\, y_{m,k}(I))$. Therefore, $var(k, \mathrm{opt}\, y_{1,k}(I), \ldots, \mathrm{opt}\, y_{m,k}(I)) = \max\{var(\boldsymbol{x}) \mid \boldsymbol{x} \in S_k(I)\}$. $\blacksquare$

In the running example, $(2, \mathrm{opt}\, y_{1,2}(\{i_1\}), \mathrm{opt}\, y_{2,2}(\{i_1\})) = (2, \max y_{1,2}(\{i_1\}), \min y_{2,2}(\{i_1\}))$, and its stamp point is indicated with star in Fig. 3.

Lemma 1 and 2 lead to the following theorem.

Theorem 1 For any itemset $J \supseteq I$,

$$\begin{aligned} & var(x(J), y_1(J), \ldots, y_m(J)) \\ & \leq \max_{0 \leq k \leq x(I)} var(k, \mathrm{opt}\, y_{1,k}(I), \mathrm{opt}\, y_{2,k}(I), \ldots, \mathrm{opt}\, y_{m,k}(I)). \blacksquare \end{aligned}$$

Definition 7

$$u(I) = \max_{0 \leq k \leq x(I)} var(k, \text{opt}\, y_{1,k}(I), \text{opt}\, y_{2,k}(I), \ldots, \text{opt}\, y_{m,k}(I)). \blacksquare$$

$var(2, \text{opt}\, y_{1,2}(\{i_1\}), \text{opt}\, y_{2,2}(\{i_1\})) = var(2, \max y_{1,2}(\{i_1\}), \min y_{2,2}(\{i_1\})) = 5.33$. Similarly, when $k = 1, 3$ and 4, we can calculate the interclass variances as 2.29, 6.93 and 2.5, respectively. Therefore, $u(\{i_1\}) = 6.93$.

Let us consider the effective computation of $u(I)$. We can calculate $\min y_{i,k}(I)$ ($\max y_{i,k}(I)$, resp) for each i by scanning the sorted list of $y_{i,1}(I)$ once. Therefore, the following lemma can be proved, and its pseudo-code used to calculate $u(I)$ is shown in Fig. 4. In the pseudo-code, $y_i^k(I)$ is k-th smallest value in $y_{i,1}(I)$. This pseudo-code confirms the following lemma.

Lemma 3 Let I be an itemset. The time complexity for calculating $u(I)$ is $O(mn \log n)$. $\blacksquare$

4 Itemset Classified Clustering Algorithm

The estimation of an upper bound enables us to design an algorithm to solve the itemset classified clustering problem as a result of the following pruning observation.

```
Calculate-u(Itemset I)
// Preprocessing O(mn log n)
Sort y_{i,1}(I) for each i ∈ [1, m];
u = 0; // u stores upper bound value.
// Select opt y_{i,k}(I). O(mn)
for each k ∈ [1, x(I)] do
  for each i ∈ [1, m] do
    min y_{i,k}(I) := min y_{i,k-1}(I) + y_i^k(I);
    max y_{i,k}(I) := max y_{i,k-1}(I) + y_i^{x(I)-k+1}(I);
    // Select opt y_{i,k}(I) according to Def. 6
    if | min y_{i,k}(I) - s_i k/n |
        >| max y_{i,k}(I) - s_i k/n | then
      opt y_{i,k}(I) := min y_{i,k}(I);
    else opt y_{i,k}(I) := max y_{i,k}(I);
    end
  end
  v := var(k, opt_{1,k}(I), ..., opt_{m,k}(I));
  u := v if v > u; // Update upper bound u
end
Return u;
```

Fig. 4. The pseudo-code used to calculate $u(I)$

```
Itemset Classified Clustering
(Q_1, L) =ICC-init;
// L: list of the best N rules
B_1 := Q_1; k := 1;
repeat until Q_k = φ
  for each B ∈ B_1, Q ∈ Q_k
    st. tail(Q) < head(B) and u(Q) ≥ τ(L)
    // Search only productive Q
    // τ(L) : Nth best value in L
    if u(B) < τ(L)
      // Disposal of unproductive 1-itemsets
      Remove B from B_1; next;
    end
    // Construct new candidate itemset
    // and update Q_{k+1} and L
    (Q_{k+1}, L) := ICC-update(Q ∪ B, Q_{k+1}, L);
  end
  k++;
end
Return L; // the best N rules
```

Fig. 5. The pseudo-code for Itemset Classified Clustering

```
ICC-init
L := φ;
for each I ∈ {J|J is a 1-itemset }
  // Calculate upper-bound
  // and var for each 1-itemset
  (Q_1, L) := ICC-update(I, Q_1, L);
end
Return Q_1 and L;
```

Fig. 6. The pseudo-code for ICC-init

```
ICC-update
(Itemset I, Set of Itemsets Q, List of the best N rules L)
u(I) =Calculate-u(I); // Calculate u(I)
// Update Q and L if necessary (Observation 1)
if u(I) ≥ τ(L)
  Put I into Q;
  if var(x(I), y_1(I), ..., y_m(I)) ≥ τ(L)
    // Update the best list L
    L := list of the best N rules in L ∪ {I};
  end
end
Return Q and L;
```

Fig. 7. The pseudo-code for ICC-update

Table 2. Default parameters

Parameter	Meaning	Default Value
$\|D\|$:	The number of tuples (genes)	4,000
N:	The number of the best classifiers (clusters)	10
L:	The length of the promoter region	300

Table 3. Yeast Gene Expression Profile Dataset

Dataset	# of feature items	# of objective attributes	# of available genes (tuples)
Spellman [8]	86,016	23	4,347
Cho [6]	86,016	17	6,137
DeRisi [7]	86,016	7	5,882

Observation 1 [13] Let us evaluate itemsets using an index satisfying convexity. Let N be the user-specified number of itemset classified clustering rules. Let $\mathcal{L}$ be a list of the best N itemsets, and $\tau(\mathcal{L})$ be the N-th best value in $\mathcal{L}$. For any itemset $J \supseteq I$, since $u(J) \leq u(I)$, J can be pruned when $u(I) < \tau(\mathcal{L})$. ■

This observation enabled us to design the algorithm "itemset classified clustering." To describe the itemset classified clustering, we define the following notation.

Definition 8 Let k-itemset be an itemset containing k items. Let $\mathcal{Q}_k$ and $\mathcal{B}_1$ be a set of k-itemsets and a set of 1-itemsets, respectively. Let us assume that there exists a total order among the items. Let I be an itemset, and head(I)(tail(I), respectively) denote the minimum (maximum) number of items in I. ■

For example, $\{i_1\}$ is 1-itemset and $\{i_1, i_3\}$ is 2-itemset. Assuming that $i_1 \prec i_2 \prec \cdots$ and $I = \{i_2, i_3, i_4\}$. head(I) $= i_2$ and tail(I) $= i_4$.

Fig. 5-7 shows the pseudo-code used in itemset classified clustering. In this pseudo-code, instead of traversing the itemsets over a lattice structure like apriori algorithm [2], we traverse them over a tree structure based on the set enumeration tree [3, 17], which is tailored to computing the best N rules using a statistical measure.

5 Experimental Results

5.1 Dataset

This section presents experimental results examining the effectiveness and performance of itemset classified clustering using yeast gene expression dataset and its DNA sequences. Gene expression dataset includes expression levels of thousands of yeast genes using DNA microarray under distinct conditions and range from 7 to 23 [18, 6, 7]. We consider each level of expression as an objective value, and each gene as a tuple.

Table 2 shows the parameters and their default values used to construct the test data. Table 3 summarizes the three microarray experiments objective values.

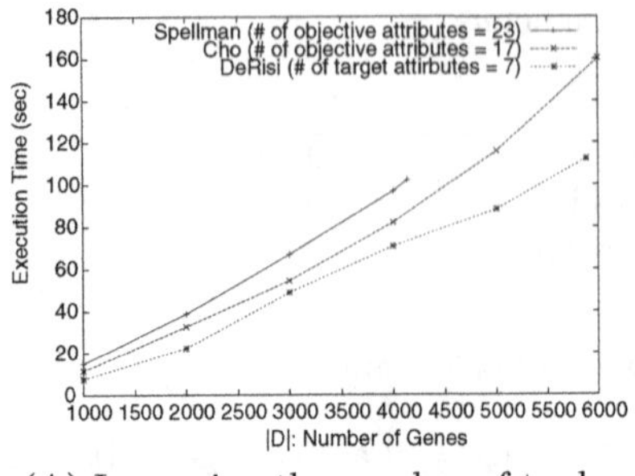

(A) Increasing the number of tuples

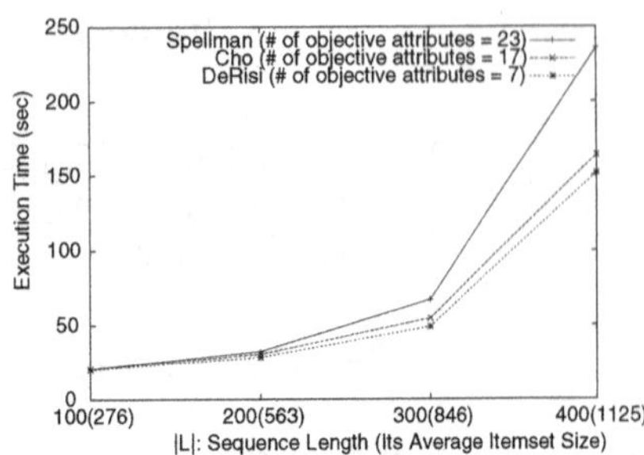

(B) Increasing the average itemset size

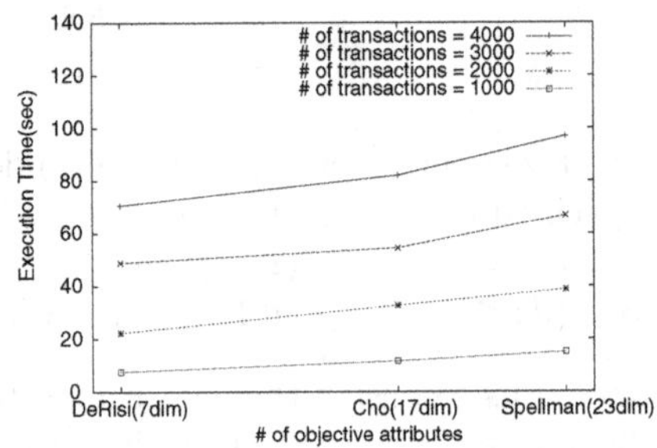

(C) Increasing the number of objective attributes

Fig. 8. Scalability of the performance

The experiments have 7 [7], 17 [6], and 23 [18] objective attributes, respectively. In our experimental results, since gene expression levels are regulated by specific combinations of short subsequences in promoter region (neighbour region of each gene), we use the existence of all the subsequences whose lengths are between six to eight, for instance, "AATGGC" or "AGATCGCC", as feature items. Therefore, the number of items is $4^6 + 4^7 + 4^8 = 86,016$ because a DNA sequence consists of 4 letters, A, C, G, and T. Briefly, we test itemset classified clustering algorithm on a database containing 1,000-6,000 tuples, 7-23 objective attributes, and 86,016 items. As shown in motivating example, itemset classified clustering can extract clusters which are different from clustering-classification approach, we compute best 10 clusters containing more than or equal to 10% of all tuples without any threshold of interclass variance.

We evaluated the overall performance of itemset classified clustering implemented in C with an Ultra SPARC III 900 MHz processor and 1 GB of main memory on Solaris 8.

5.2 Scalability

Two distinct ways of increasing the size of the dataset were used to test the scalability of itemset classified clustering. We represent the scalability in Fig. 8.

Fig. 8(A) illustrates the execution time when we increase the number of tuples $|D|$ from 1,000 to 6,000. The figure shows that the execution time scales almost linearly with the number of tuples for every dataset. Therefore, this figure indicates that our algorithm is scalable for increasing the tuples.

Fig. 8(B) demonstrates the performance of itemset classified clustering when the average number of items in itemsets increases for adding noisy tuples to dataset. Such a dataset can be obtained by increasing promoter length L because the probability of whether each short subsequence appears in the promoter region increases. In Fig. 8(B), L ranges from 100 to 400, while $|D| = 3,000$. This figure shows that the execution time increases quadratically to the average itemset size. This graph shows two types of effectiveness of our itemset classified clustering algorithm. One is that, since our measure, interclass variance, inherits the characteristics of statistical measures, itemset classified clustering effectively neglects noisy tuples. The other is that our upper-bound pruning is effective albeit the search space of itemsets grows more than exponentially according to increase of the average size of itemsets. This dramatic effects of pruning could be observed even when the objective attribute has over twenty dimensions. Indeed, our wrapping hyper-polyhedron of the stamp tuples of all the supersets of an itemset is bigger than their tight convex hull. Nevertheless, these experiments prove that our wrapping upper bound estimation is sufficient for a real dataset.

We converted Fig. 8(A) into Fig. 8(C) to study the effect of the number of objective attributes (dimensions). The figure shows that the execution time also scales almost linearly with the number of objective attributes.

6 Concluding Remarks

This paper presented the demand of the consideration of new data sets consisting of tuples which is represented by a feature itemset and an objective vector. To analyze the data, because traditional clustering-classification method may not always produce clusters associated with a feature itemset, we introduced a new paradigm, itemset classified clustering, which is a clustering that allows only splits expressible by a common feature itemset, and computes the optimal itemset that maximizes the interclass variance of objective attributes, or list the most significant N solutions. This itemset classified clustering can extract clusters overlooked by two-clustering or k-means clustering.

Our experimental results show that the itemset classified clustering has the scalability of performance for tuple size, objective attribute size, and itemset size. Therefore, the method can solve the real molecular biological problem containing 6,000 tuples, more than 80,000 boolean feature items and 23 numerical objective attributes.

Solving the itemset classified clustering problem is applicable to various problems because this problem prompts us to reconsider the results of both clustering and classification analysis. One example is to find the association between patients' gene expressions and their pathological features. [16] Furthermore, the replacement of itemset with other features such as numerical or categorical features might expand the application of clustering and classification algorithms.

References

1. R. Agrawal, J. Gehrke, D. Gunopulos, and P. Raghavan. Automatic subspace clustering of high dimensional data for data mining applications. In *Proc. of ACM SIGMOD 1998*, pages 94–105, 1998.
2. R. Agrawal and R. Srikant. Fast algorithms for mining association rules. In *Proc. of 20th VLDB*, pages 487–499, 1994.
3. R. Bayardo. Efficiently mining long patterns from databases. In *Proc. of ACM SIGMOD 1998*, pages 85–93, 1998.
4. L. Breiman, R. A. Olshen, J. H. Friedman, and C. J. Stone. *Classification and Regression Trees.* Brooks/Cole Publishing Company, 1984.
5. Y. Cheng and G. M. Church. Biclustering of expression data. In *Proc. of the Eighth Intl. Conf. on ISMB*, pages 93–103, 2000.
6. R. J. Cho, M. J. Campbell, et al. A genome-wide transcriptional analysis of the mitotic cell cycle. *Molecular Cell*, 2:65–73, 1998.
7. J. L. DeRisi, V. R. lyer, and P. O. Brown. Exploring the metabolic and genetic control of gene expression on a genomic scale. *Science*, 278:680–686, 1997.
8. M. R. Garey and D. S. Johnson. *Computer and Intractability. A Guide to NP-Completeness.* W. H. Freeman, 1979.
9. S. Guha, R. Rastogi, and K. Shim. Cure: an efficient clustering algorithm for large databases. In *Proc. of ACM SIGMOD 1998*, pages 73–84, 1998.
10. J. Han, J. Pei, and Y. Yin. Mining frequent patterns without candidate generation. In *Proc. of ACM SIGMOD 2000*, pages 1–12, 2000.
11. R. Horst and H.Tuy. *Global optimization: Deterministic approaches.* Springer-Verlag, 1993.
12. R. S. Michalski and R. E. Stepp. *Learning from observation: Conceptual clustering*, pages 331–363. Tioga Publishing Company, 1983.
13. S. Morishita and J. Sese. Traversing itemset lattice with statistical metric pruning. In *Proc. of ACM PODS 2000*, pages 226–236, 2000.
14. R. T. Ng and J. Han. Efficient and effective clustering methods for spatial data mining. In *20th VLDB*, pages 144–155, Los Altos, CA 94022, USA, 1994.
15. J. R. Quinlan. *C4.5: programs for machine learning.* Morgan Kaufmann Publishers Inc., 1993.
16. J. Sese, Y. Kurokawa, M. Monden, K. Kato, and S. Morishita. Constrained clusters of gene expression profiles with pathological features. *Bioinformatics*, 2004. in press.
17. J. Sese and S. Morishita. Answering the most correlated n association rules efficiently. In *Proc. of PKDD'02*, pages 410–422, 2002.
18. P. T. Spellman and other. Comprehensive identification of cell cycle-regulated genes of the yeast *saccharomyces cerevisiae* by microarray hybridization. *Molecular Biology of the Cell*, 9:3273–3297, 1998.
19. J. Tantrum, A. Murua, and W. Stuetzle. Hierarchical model-based clustering of large datasets through fractionation and refractionation. In *Proc. of the KDD '02*, 2002.
20. H. Wang, W. Wang, J. Yang, and P. S. Yu. Clustering by pattern similarity in large data sets. In *Proc. of ACM SIGMOD 2002*, pages 394–405, 2002.
21. M. Zaki and C. Hsiao. Charm: An efficient algorithm for closed itemset mining. In *2nd SIAM International Conference on Data Mining*, 2002.
22. T. Zhang, R. Ramakrishnan, and M. Livny. Birch: an efficient data clustering method for very large databases. In *Proc. of ACM SIGMOD 1996*, pages 103–114, 1996.

Combining Winnow and Orthogonal Sparse Bigrams for Incremental Spam Filtering

Christian Siefkes[1], Fidelis Assis[2],
Shalendra Chhabra[3], and William S. Yerazunis[4]

[1] Berlin-Brandenburg Graduate School in Distributed Information Systems*
Database and Information Systems Group, Freie Universität Berlin
Berlin, Germany
christian@siefkes.net

[2] Empresa Brasileira de Telecomunicações – Embratel
Rio de Janeiro, RJ, Brazil
fidelis@embratel.net.br

[3] Computer Science and Engineering
University of California, Riverside
California, USA
schhabra@cs.ucr.edu

[4] Mitsubishi Electric Research Laboratories
Cambridge, MA, USA
wsy@merl.com

Abstract. Spam filtering is a text categorization task that has attracted significant attention due to the increasingly huge amounts of junk email on the Internet. While current best-practice systems use Naive Bayes filtering and other probabilistic methods, we propose using a statistical, but non-probabilistic classifier based on the *Winnow* algorithm. The feature space considered by most current methods is either limited in expressivity or imposes a large computational cost. We introduce *orthogonal sparse bigrams (OSB)* as a feature combination technique that overcomes both these weaknesses. By combining Winnow and OSB with refined preprocessing and tokenization techniques we are able to reach an accuracy of 99.68% on a difficult test corpus, compared to 98.88% previously reported by the *CRM114* classifier on the same test corpus.

Keywords: Classification, Text Classification, Spam Filtering, Email, Incremental Learning, Online Learning, Feature Generation, Feature Representation, Winnow, Bigrams, Orthogonal Sparse Bigrams.

1 Introduction

Spam filtering can be viewed as a classic example of a text categorization task with a strong practical application. While keyword, fingerprint, whitelist/blacklist, and heuristic–based filters such as SpamAssassin [11] have been successfully

* The work of this author is supported by the German Research Society (DFG grant no. GRK 316).

deployed, these filters have experienced a decrease in accuracy as spammers introduce specific countermeasures. The current best-of-breed anti-spam filters are all probabilistic systems. Most of them are based on Naive Bayes as described by Graham [6] and implemented in *SpamBayes* [12]; others such as the *CRM114 Discriminator* can be modeled by a Markov Random Field [15]. Other approaches such as *Maximum Entropy Modeling* [16] lack a property that is important for spam filtering – they are not *incremental*, they cannot adapt their classification model in a single pass over the data.

As a statistical, but non-probabilistic alternative we examine the incremental *Winnow* algorithm. Our experiments show that Winnow reduces the error rate by 75% compared to Naive Bayes and by more than 50% compared to *CRM114*.

The feature space considered by most current methods is limited to individual tokens (unigrams) or bigrams. The *sparse binary polynomial hashing (SBPH)* technique (cf. Sec. 4.1) introduced by *CRM114* is more expressive but imposes a large runtime and memory overhead. We propose *orthogonal sparse bigrams (OSB)* as an alternative that retains the expressivity of SBPH, but avoids most of the cost. Experimentally OSB leads to equal or slightly better filtering than SBPH. We also analyze the preprocessing and tokenization steps and find that further improvements are possible here.

In the next section we present the Winnow algorithm. The following two sections are dedicated to feature generation and combination. In Section 5 we detail our experimental results. Finally we discuss related methods and future work.

2 The Winnow Classification Algorithm

The Winnow algorithm introduced by [7] is a statistical, but not a probabilistic algorithm, i.e. it does not directly calculate probabilities for classes. Instead it calculates a *score* for each class[1].

Our variant of Winnow is suitable for both binary (two-class) and multi-class (three or more classes) classification. It keeps an n-dimensional weight vector $w^c = (w_1^c, w_2^c, \ldots, w_n^c)$ for each class c, where w_i^c is the weight of the ith feature. The algorithm returns 1 for a class iff the summed weights of all active features (called the score Ω) surpass a predefined threshold θ:

$$\Omega = \sum_{j=1}^{n_a} w_j^c > \theta.$$

Otherwise ($\Omega \leq \theta$) the algorithm returns 0. $n_a \leq n$ is the number of active (present) features in the instance to classify.

The goal of the algorithm is to learn a linear separator over the feature space that returns 1 for the true class of each instance and 0 for all other classes on this instance. The initial weight of each feature is 1.0. The weights of a class

[1] There are ways to convert the scores calculated by Winnow into confidence estimates, but these are not discussed here since they are not of direct relevance for this paper.

are updated whenever the value returned for this class is wrong. If 0 is returned instead of 1, the weights of all active features are increased by multiplying them with a *promotion factor* α, $\alpha > 1$: $w_j^c \leftarrow \alpha \times w_j^c$. If 1 is returned instead of 0, the active weights are multiplied with a *demotion factor* β, $0 < \beta < 1$: $w_j^c \leftarrow \beta \times w_j^c$.

In text classification, the number of features depends on the length of the text, so it varies enormously from instance to instance. Thus instead of using a fixed threshold we set the threshold to the number n_a of features that are active in the given instance: $\theta = n_a$. Thus initial scores are equal to θ since the initial weight of each feature is 1.0.

In multi-label classification, where an instance can belong to several classes at once, the algorithm would predict all classes whose score is higher than the threshold. But for the task at hand, there is exactly one correct class for each instance, thus we employ a *winner-takes-all* approach where the class with the highest score is predicted.

This means that there are situations where the algorithm will be trained even though it did not make a mistake. This happens whenever the scores of both classes[2] are at the same side of the threshold and the score of the true class is higher than the other one – in this case the prediction of Winnow will be correct but it will still promote/demote the weights of the class that was at the wrong side of the threshold.

The complexity of processing an instance depends only on the number of active features n_a, not on the number of all features n_t. Similar to *SNoW* [1], a sparse architecture is used where features are allocated whenever the need to promote/demote them arises for the first time. In sparse Winnow, the number of instances required to learn a linear separator (if exists) depends linearly on the number of relevant features n_r and only logarithmically on the number of active features, i.e. it scales with $O(n_r \log n_a)$ (cf. [8, Sec. 2]).

Winnow is a non-parametric approach; it does not assume a particular probabilistic model underlying the training data. Winnow is a linear separator in the Perceptron sense, but by providing a feature space that itself allows conjunction and disjunction, complex non-linear features may be recognized by the composite feature-extractor + Winnow system.

2.1 Thick Threshold

In our implementation of Winnow, we use a *thick threshold* for learning (cf. [4, Sec. 4.2]). Training instances are re-trained even if the classification was correct if the determined score was near the threshold. Two additional thresholds θ^+ and θ^- with $\theta^- < \theta < \theta^+$ are defined and each instance whose score falls in the range $[\theta^-, \theta^+]$ is considered a mistake. In this way, a large margin classifier will be trained that is more robust when classifying borderline instances.

2.2 Feature Pruning

The feature combination methods discussed in Section 4 generate enormous numbers of features. To keep the feature space tractable, features are stored in an

[2] Resp. two or more classes in other tasks involving more than two classes.

LRU (least recently used) cache. The feature store is limited to a configurable number of elements; whenever it is full, the least recently seen feature is deleted. When a deleted feature is encountered again, it will be considered as a new feature whose weights are still at their default values.

3 Feature Generation

3.1 Preprocessing

We did not perform language-specific preprocessing techniques such as word stemming, stop word removal, or case folding, since other researchers found that such techniques tend to hurt spam-filtering accuracy [6, 16]. We did compare three types of email-specific preprocessing.

- Preprocessing via *mimedecode*, a utility for decoding typical mail encodings (Base64, Quoted-Printable etc.)
- Preprocessing via Jaakko Hyvatti's *normalizemime* [9]. This program converts the character set to UTF-8, decoding Base64, Quoted-Printable and URL encoding and adding warn tokens in case of encoding errors. It also appends a copy of HTML/XML message bodies with most tags removed, decodes HTML entities and limits the size of attached binary files.
- No preprocessing. Use the raw mail including large blocks of Base64 data in the encoded form.

Except for the comparison of these alternatives, all experiments were performed on *normalizemime*-preprocessed mails.

3.2 Tokenization

Tokenization is the first stage in the classification pipeline; it involves breaking the text stream into tokens ("words"), usually by means of a regular expression. We tested four different tokenization schemas:

P (Plain): Tokens contain any sequences of printable characters; they are separated by non-printable characters (whitespace and control characters).

C (CRM114): The current default pattern of *CRM114* – tokens start with a printable character; followed by any number of alphanumeric characters + dashes, dots, commas and colons; optionally ended by a printable character.

S (Simplified): A modification of the CRM114 pattern that excludes dots, commas and colons from the middle of the pattern. With this pattern, domain names and mail addresses will be split at dots, so the classifier can recognize a domain even if subdomains vary.

X (XML/HTML+header-aware): A modification of the **S** schema that allows matching typical XML/HTML markup[3], mail headers (terminated by

[3] Start/end/empty tags: `<tag>` `</tag>` `<br/>`; Doctype declarations: `<!DOCTYPE`; processing instructions: `<?xml-stylesheet`; entity + character references: `—`; attributes terminated by "="; attribute values surrounded by quotes.

Table 1. Tokenization Patterns

<table>
<tr><th>Name</th><th>Regular Expression</th></tr>
<tr><td>P</td><td><code>[^\p{Z}\p{C}]+</code></td></tr>
<tr><td>C</td><td><code>[^\p{Z}\p{C}][-.,:\p{L}\p{M}\p{N}]*[^\p{Z}\p{C}]?</code></td></tr>
<tr><td>S</td><td><code>[^\p{Z}\p{C}][-\p{L}\p{M}\p{N}]*[^\p{Z}\p{C}]?</code></td></tr>
<tr><td>X</td><td><code>[^\p{Z}\p{C}][/!?#]?[-\p{L}\p{M}\p{N}]*(?:["'=;]|/?>|:/*)?</code></td></tr>
</table>

":"), and protocols such as "http://" in a token. Punctuation marks such as "." and "," are not allowed at the end of tokens, so normal words will be recognized no matter where in a sentence they occur without being "contaminated" by trailing punctuation.

The **X** schema was used for all tests unless explicitly stated otherwise. The actual tokenization schemas are defined as the regular expressions given in Table 1. These patterns use Unicode categories – `[^\p{Z}\p{C}]` means everything except whitespace and control chars (POSIX class `[:graph:]`); `\p{L}\p{M}\p{N}` collectively match all alphanumerical characters (`[:alnum:]` in POSIX).

4 Feature Combination

4.1 Sparse Binary Polynomial Hashing

Sparse binary polynomial hashing (SBPH) is a feature combination technique introduced by the *CRM114 Discriminator* [3, 14]. SBPH slides a window of length N over the tokenized text. For each window position, all of the possible in-order combinations of the N tokens are generated; those combinations that contain at least the newest element of the window are retained. For a window of length N, this generates 2^{N-1} features. Each of these joint features can be mapped to one of the odd binary numbers from 1 to $2^N - 1$ where original features at "1" positions are visible while original features at "0" positions are hidden and marked as skipped.

It should be noted that the features generated by SBPH are not linearly independent and that even a compact representation of the feature stream generated by SBPH may be significantly longer than the original text.

4.2 Orthogonal Sparse Bigrams

Since the expressivity of SBPH is sufficient for many applications, we now consider if it is possible to use a smaller feature set and thereby increase speed and decrease memory requirements. For this, we consider only word pairs containing a common word inside the window, and requiring the newest member of the window to be one of the two words in the pair. The idea behind this approach is to gain speed by working only with an *orthogonal* feature set inside the window, rather that the prolific and probably redundant features generated by SBPH.

Instead of all odd numbers, only those with two bits "1" in their binary representations are used: $2^n + 1$, for $n = 1$ to $N - 1$. With this restriction, only $N - 1$ combinations with exactly two words are produced. We call them *orthogonal sparse bigrams (OSB)* – "sparse" because most combinations have skipped words; only the first one is a conventional bigram.

With a sequence of five words, $w_1, \ldots, w_5$, OSB produces four combined features:

*w*4 *w*5

*w*3 <*skip*> *w*5

*w*2 <*skip*> <*skip*> *w*5

*w*1 <*skip*> <*skip*> <*skip*> *w*5

Because of the reduced number of combined features, $N - 1$ in OSB versus 2^{N-1} in SBPH, text classification with OSB can be considerably faster than with SBPH. Table 2 shows an example of the features generated by SBPH and OSB side by side.

Table 2. Features Generated by SBPH and OSB

Number	SBPH	OSB
1 (1)	today?	
3 (11)	lucky today?	lucky today?
5 (101)	feel <*skip*> today?	feel <*skip*> today?
7 (111)	feel lucky today?	
9 (1001)	you <*skip*> <*skip*> today?	you <*skip*> <*skip*> today?
11 (1011)	you <*skip*> lucky today?	
13 (1101)	you feel <*skip*> today?	
15 (1111)	you feel lucky today?	
17 (10001)	Do <*skip*> <*skip*> <*skip*> today?	Do <*skip*> <*skip*> <*skip*> today?
19 (10011)	Do <*skip*> <*skip*> lucky today?	
21 (10101)	Do <*skip*> feel <*skip*> today?	
23 (10111)	Do <*skip*> feel lucky today?	
25 (11001)	Do you <*skip*> <*skip*> today?	
27 (11011)	Do you <*skip*> lucky today?	
29 (11101)	Do you feel <*skip*> today?	
31 (11111)	Do you feel lucky today?	

Note that the *orthogonal sparse bigrams* form an almost complete basis set – by "ORing" features in the OSB set, any feature in the SBPH feature set can be obtained, except for the unigram (the single-word feature). However, there is no such redundancy in the OSB feature set; it is not possible to obtain any OSB feature by adding, ORing, or subtracting any other pairs of other OSB features; all of the OSB features are unique and not redundant.

Since the first term, unigram w_5, cannot be obtained by ORing OSB features it seems reasonable to add it as an extra feature. However the experiments reported in Section 5.4 show that adding unigrams does *not* increase accuracy; in fact, it sometimes decreased accuracy.

5 Experimental Results

5.1 Testing Procedure

In order to test our multiple hypotheses, we used a standardized spam/nonspam test corpus from SpamAssassin [11]. This test corpus is extraordinarily difficult to classify, even for humans. It consists of 1397 spam messages, 250 hard nonspams, and 2500 easy nonspams, for a total of 4147 messages. These 4147 messages were "shuffled" into ten different standard sequences; results were averaged over these ten runs. We re-used the corpus and the standard sequences from [15].

Each test run begins with initializing all memory in the learning system to zero. Then the learning system was presented with each member of a standard sequence, in the order specified for that standard sequence, and required to classify the message. After each classification the true class of the message was revealed and the classifier had the possibility to update its prediction model accordingly prior to classifying the next message[4]. The training system then moved on to the next message in the standard sequence. The final 500 messages of each standard sequence were the *test set* used for final accuracy evaluation; we also report results on an extended test set containing the last 1000 messages of each run and on all (4147) messages. Systems were permitted to train on any messages, including those in the test set, *after* classifying them; at no time a system ever had the opportunity to learn on a message before predicting the class of this message. For evaluation we calculated the *error rate* $E = \frac{\textit{number of misclassifications}}{\textit{number of all classifications}}$; occasionally we mention the *accuracy* $A = 1 - E$.

This process was repeated for each of the ten standard sequences. Each complete set of ten standard sequences (41470 messages) required approximately 25–30 minutes of processor time on a 1266 MHz Pentium III for OSB-5[5]. The average number of errors per test run is given in parenthesis.

5.2 Parameter Tuning

We used a slightly different setup for tuning the Winnow parameters since it would have been unfair to tune the parameters on the test set. The last 500 messages of each run were reserved as test set for evaluation, while the preceding 1000 messages were used as *development set* for determining the best parameter values. The **S** tokenization was used for the tests in the section.

Best performance was found with Winnow using 1.23 as promotion factor, 0.83 as demotion factor, and a threshold thickness of 5%[6]. These parameter values turned out to be best for both OSB and SBPH – the results reported in Tables 3 and 4 are for OSB.

[4] In actual usage training will not be quite as incremental since mail is read in batches.

[5] For SBPH-5 it was about two hours which it not surprising since SBPH-5 generates four times as many features as OSB-5.

[6] In either direction, i.e. $\theta^- = 0.95\,\theta$, $\theta^+ = 1.05\,\theta$.

Table 3. Promotion and Demotion Factors

Promotion	1.35	1.25	1.25	1.23	1.2	1.1
Demotion	0.8	0.8	0.83	0.83	0.83	0.9
Test Set	0.44% (2.2)	0.36% (1.8)	0.44% (2.2)	**0.32% (1.6)**	0.44% (2.2)	0.48% (2.4)
Devel. Set	0.52% (5.2)	0.51% (5.1)	0.52% (5.2)	**0.49% (4.9)**	0.51% (5.1)	0.62% (6.2)
All	**1.26% (52.4)**	1.31% (54.3)	1.33% (55.1)	1.32% (54.7)	1.34% (55.4)	1.50% (62.2)

Table 4. Threshold Thickness

Threshold Thickness	0%	5%	10%
Test Set	0.68% (3.4)	**0.32% (1.6)**	0.44% (2.2)
Development Set	0.88% (8.8)	**0.49% (4.9)**	0.56% (5.6)
All	1.77% (73.5)	**1.32% (54.7)**	1.38% (57.1)

Table 5. Comparison of SBPH and OSB with Different Feature Storage Sizes

	OSB				
Store Size	400000	500000	600000	700000	800000
Last 500	0.36% (1.8)	0.38% (1.9)	**0.32% (1.6)**	0.44% (2.2)	0.44% (2.2)
Last 1000	0.37% (3.7)	0.37% (3.7)	**0.33% (3.3)**	0.37% (3.7)	0.37% (3.7)
All	1.26% (52.3)	1.29% (53.4)	**1.24% (51.4)**	1.26% (52.2)	1.27% (52.5)
	SBPH				
Store Size	1400000	1600000	1800000	2097152 (2^{21})	2400000
Last 500	0.38% (1.9)	**0.36% (1.8)**	0.42% (2.1)	0.44% (2.2)	0.42% (2.1)
Last 1000	0.37% (3.7)	**0.34% (3.4)**	0.38% (3.8)	0.39% (3.9)	0.38% (3.8)
All	1.35% (55.8)	**1.28% (53.1)**	1.30% (54)	1.30% (54)	1.31% (54.2)

5.3 Feature Store Size and Comparison with SBPH

Table 5 compares orthogonal sparse bigrams and SBPH for different sizes of the feature store. OSB reached best results with 600,000 features (with an error rate of 0.32%), while SBPH peaked at 1,600,000 features (with a slightly higher error rate of 0.36%). Further increasing the number of features permitted in the store negatively affects accuracy. This indicates that the LRU pruning mechanism is efficient at discarding irrelevant features that are mostly noise.

5.4 Unigram Inclusion

The inclusion of individual tokens (unigrams) in addition to orthogonal sparse bigrams does not generally increase accuracy, as can be seen in Table 6, showing OSB without unigrams peaking at 0.32% error rate, while adding unigrams pushes the error rate up to 0.38%.

5.5 Window Sizes

The results of varying window size as a system parameter are shown in Table 7. Again, we note that the optimal combination for the test set uses a window size

Table 6. Utility of Single Tokens (Unigrams)

	OSB only	**OSB + Unigrams**	
Store Size	600000	600000	750000
Last 500	**0.32% (1.6)**	0.38% (1.9)	0.42% (2.1)
Last 1000	**0.33% (3.3)**	**0.33% (3.3)**	0.36% (3.6)
All	1.24% (51.4)	**1.22% (50.6)**	1.24% (51.4)

Table 7. Sliding Window Size

Window Size	**Unigrams**	**2 (Bigrams)**	**3**	**4**	**5**	**6**	**7**
Store Size	All (ca.55000)	150000	300000	450000	600000	750000	900000
Last 500	0.46% (2.3)	0.48% (2.4)	0.42% (2.1)	0.44% (2.2)	**0.32% (1.6)**	0.38% (1.9)	0.42% (2.1)
Last 1000	0.50% (5)	0.43% (4.3)	0.39% (3.9)	0.40% (4)	**0.33% (3.3)**	0.38% (3.8)	0.37% (3.7)
All	1.43% (59.2)	1.23% (51.2)	1.24% (51.4)	1.26% (52.2)	1.24% (51.4)	1.28% (53)	**1.22% (50.8)**
Store Size		All (ca.220000)	All (ca.500000)	600000		900000	1050000
Last 500		0.48% (2.4)	0.42% (2.1)	0.42% (2.1)		0.40% (2)	0.46% (2.3)
Last 1000		0.43% (4.3)	0.38% (3.8)	0.38% (3.8)		0.38% (3.8)	0.40% (4)
All		1.24% (51.3)	**1.22% (50.6)**	1.25% (51.8)		1.27% (52.5)	1.25% (51.7)

of five tokens (our default setting, yielding a 0.32% error rate), with both shorter and longer windows producing worse error rates.

This "U" curve is not unexpected on an information-theoretic basis. English text has a typical entropy of around 1–1.5 bits per character and around five characters per word. If we assume that a text contains mainly letters, digits, and some punctuation symbols, most characters can be represented in six bits, yielding a word content of 30 bits. Therefore, at one bit per character, English text becomes uncorrelated at a window length of six words or longer, and features obtained at these window lengths are not significant.

These results also show that using OSB-5 is significantly better then using only single tokens (error rate of 0.46%) or conventional bigrams (0.48%).

5.6 Preprocessing and Tokenization

Results with *normalizemime* were generally better than the other two options, reducing the error rate by up to 25% (Table 8). Accuracy on raw and *mime-decoded* mails was roughly comparable.

Table 8. Preprocessing

Preprocessing	**none**	**mimedecode**	**normalizemime**
Last 500	0.42% (2.1)	0.46% (2.3)	**0.32% (1.6)**
Last 1000	0.37% (3.7)	0.35% (3.5)	**0.33% (3.3)**
All	1.27% (52.5)	1.26% (52.1)	**1.24% (51.4)**

The **S** tokenization schema initially learns more slowly (the overall error rate is somewhat higher) but is finally just as good as the **X** schema (Table 9). **P** and **C** both result in lower accuracy, even though they initially learn quickly.

Table 9. Tokenization Schemas

Schema	X	S	C	P
Last 500	**0.32% (1.6)**	**0.32% (1.6)**	0.44% (2.2)	0.42% (2.1)
Last 1000	**0.33% (3.3)**	**0.33% (3.3)**	0.39% (3.9)	0.38% (3.8)
All	1.24% (51.4)	1.32% (54.7)	1.28% (52.9)	**1.23% (51.1)**

Table 10. Comparison With Naive Bayes and CRM114

	Naive Bayes	**CRM114**	**CRM114**	**Winnow+OSB**
Store Size	All	1048577 ($2^{20}+1$)	All	All
Last 500	1.84% (9.2)	1.12% (5.6)	1.16% (5.8)	**0.46% (2.3)**
All	3.44% (142.8)	2.71% (112.5)	2.73% (113.2)	**1.30% (53.9)**

5.7 Comparison with CRM114 and Naive Bayes

The results for *CRM114* and Naive Bayes on the last 500 mails are the best results reported in [15] for incremental (single-pass) training. For a fair comparison, these tests were all run using the **C** tokenization schema on raw mails without preprocessing. The best reported *CRM114* weighting model is based on empirically derived weightings and is a rough approximation of a Markov Random Field. This model reduces to a Naive Bayes Model when the window size is set to 1. To avoid the different pruning mechanisms (*CRM114* uses a random-discard algorithm) from distorting the comparison, we disabled LRU pruning for Winnow and also reran the *CRM114* tests using all features (Table 10).

5.8 Speed of Learning

The learning rate for the Winnow classifier combined with the OSB feature generator is shown in Fig. 1. Note that the rightmost column shows the incremental error rate on new messages. After having classified 1000 messages, Winnow+OSB achieves error rates below 1% on new mails.

6 Related Work

Cohen and Singer [2] use a Winnow-like multiplicative weight update algorithm called "sleeping experts" with a feature combination technique called "sparse phrases" which seems to be essentially equivalent to SBPH[7]. Bigrams and n-grams are a classical technique; SBPH has been introduced in [14] and "sparse phrases" in [2]. We propose orthogonal sparse bigrams as a minimalistic alternative that is new, to the best of our knowledge.

An LRU mechanism for feature set pruning has been employed by the first author in [10]. We suppose that others have done the same since the idea seems to suggest itself; but currently we are not aware of such usage.

[7] Thanks to an anonymous reviewer for pointing us to this work.

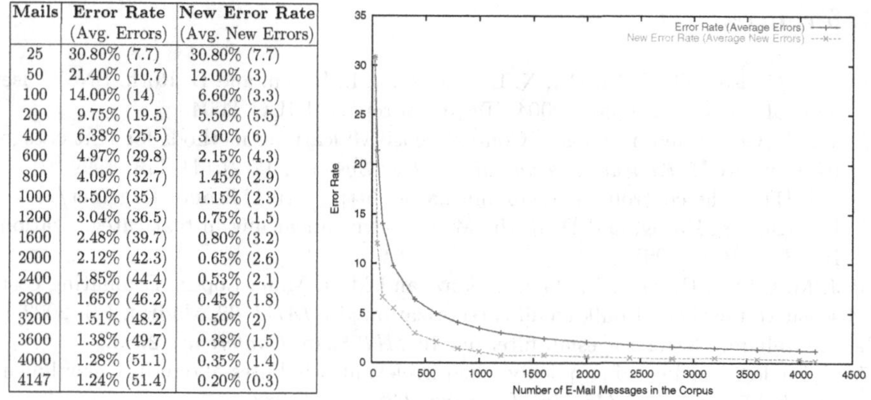

Mails	Error Rate (Avg. Errors)	New Error Rate (Avg. New Errors)
25	30.80% (7.7)	30.80% (7.7)
50	21.40% (10.7)	12.00% (3)
100	14.00% (14)	6.60% (3.3)
200	9.75% (19.5)	5.50% (5.5)
400	6.38% (25.5)	3.00% (6)
600	4.97% (29.8)	2.15% (4.3)
800	4.09% (32.7)	1.45% (2.9)
1000	3.50% (35)	1.15% (2.3)
1200	3.04% (36.5)	0.75% (1.5)
1600	2.48% (39.7)	0.80% (3.2)
2000	2.12% (42.3)	0.65% (2.6)
2400	1.85% (44.4)	0.53% (2.1)
2800	1.65% (46.2)	0.45% (1.8)
3200	1.51% (48.2)	0.50% (2)
3600	1.38% (49.7)	0.38% (1.5)
4000	1.28% (51.1)	0.35% (1.4)
4147	1.24% (51.4)	0.20% (0.3)

Fig. 1. Learning Curve for the best setting (Winnow$_{1.23,0.83,5\%}$ with 1,600,000 features, OSB-5, **X** tokenization)

7 Conclusion and Future Work

We have introduced *orthogonal sparse bigrams (OSB)* as a new feature combination technique for text classification that combines a high expressivity with relatively low computational load. By combining OSB with the *Winnow* algorithm we more than halved the error rate compared to a state-of-the-art spam filter, while still retaining the property of *incrementality.* By refining the preprocessing and tokenization steps we were able to further reduce the error rate by 30% [8].

In this study we have measured the accuracy without taking the different costs of misclassifications into account (it can be tolerated to let a few spam mails through, but it is bad to classify a regular email as spam). This could be addressed by using a cost metric as discussed in [5]. Winnow could be biased in favor of classifying a borderline mail as nonspam by multiplying the spam score by a factor < 1 (e.g. 99%) when classifying.

Currently our Winnow implementation supports only binary features; how often a feature (sparse bigram) appears in a text is not taken into account. We plan to address this by introducing a *strength* for each feature (cf. [4, Sec. 4.3]).

Also of interest is the difference in performance between the LRU (least-recently-used) pruning algorithm used here and the random-discard algorithm used in *CRM114* [15]. When the random-discard algorithm in *CRM114* triggered, it almost always resulted in a decrease in accuracy; here we found that an LRU algorithm could act to provide an *increase* in accuracy. Analysis and determination of the magnitude of this effect will be a concern in future work.

Acknowledgments

We thank the anonymous reviewers for their helpful comments and suggestions.

[8] Our algorithm is freely available as part of the *TiEs* system [13].

References

1. A. J. Carlson, C. M. Cumby, N. D. Rizzolo, J. L. Rosen, and D. Roth. SNoW user manual. Version: January, 2004. Technical report, UIUC, 2004.
2. W. W. Cohen and Y. Singer. Context-sensitive learning methods for text categorization. *ACM Transactions on Information Systems*, 17(2):141–173, 1999.
3. CRM114: The controllable regex mutilator. http://crm114.sourceforge.net/.
4. I. Dagan, Y. Karov, and D. Roth. Mistake-driven learning in text categorization. In *EMNLP-97*, 1997.
5. J. M. Gómez Hidalgo, E. Puertas Sanz, and M. J. Maña López. Evaluating cost-sensitive unsolicited bulk email categorization. In *JADT-02*, Madrid, ES, 2002.
6. P. Graham. Better Bayesian filtering. In *MIT Spam Conference*, 2003.
7. N. Littlestone. Learning quickly when irrelevant attributes abound: A new linear-threshold algorithm. *Machine Learning*, 2:285–318, 1988.
8. M. Munoz, V. Punyakanok, D. Roth, and D. Zimak. A learning approach to shallow parsing. Technical Report UIUCDCS-R-99-2087, Department of Computer Science, University of Illinois at Urbana-Champaign, Urbana, Illinois, 1999.
9. normalizemime v2004-02-04. http://hyvatti.iki.fi/ jaakko/spam/.
10. C. Siefkes. A toolkit for caching and prefetching in the context of Web application platforms. Diplomarbeit, TU Berlin, 2002.
11. SpamAssassin. http://www.spamassassin.org/.
12. SpamBayes. http://spambayes.sourceforge.net/.
13. Trainable Incremental Extraction System. http://www.inf.fu-berlin.de/inst/ag-db/software/ties/.
14. W. S. Yerazunis. Sparse binary polynomial hashing and the CRM114 discriminator. In *2003 Spam Conference*, Cambridge, MA, 2003. MIT.
15. W. S. Yerazunis. The spam-filtering accuracy plateau at 99.9% accuracy and how to get past it. In *2004 Spam Conference*, Cambridge, MA, 2004. MIT.
16. L. Zhang and T. Yao. Filtering junk mail with a maximum entropy model. In *20th International Conference on Computer Processing of Oriental Languages*, 2003.

Asynchronous and Anticipatory Filter-Stream Based Parallel Algorithm for Frequent Itemset Mining*

Adriano Veloso[1], Wagner Meira Jr.[1], Renato Ferreira[1], Dorgival Guedes Neto[1], and Srinivasan Parthasarathy[2]

[1] Computer Science Department, Universidade Federal de Minas Gerais, Brazil
{adrianov,meira,renato,dorgival}@dcc.ufmg.br

[2] Department of Computer and Information Science, The Ohio-State University, USA
srini@cis.ohio-state.edu

Abstract. In this paper we propose a novel parallel algorithm for frequent itemset mining. The algorithm is based on the filter-stream programming model, in which the frequent itemset mining process is represented as a data flow controlled by a series of producer and consumer components (called filters), and the data flow (communication) between such filters is made via streams. When production rate matches consumption rate, and communication overhead between producer and consumer filters is minimized, a high degree of asynchrony is achieved. Following this strategy, our algorithm employs an asynchronous candidate generation, and minimizes communication between filters by transferring only the necessary aggregated information. Another nice feature of our algorithm is a look forward approach which accelerates frequent itemset determination. Extensive evaluation shows the parallel performance and scalability of our algorithm.

1 Introduction

The importance of data mining and knowledge discovery is growing. Fields as diverse as astronomy, finance, bioinformatics, cyber-security are among the many facing the situation where large amounts of data are collected and accumulated at an explosive rate. Analyzing such datasets without the use of some kind of data reduction/mining is getting more and more infeasible and thus there has been an increasing clamor for mining such data efficiently.

The problem is that mining such large and potentially dynamic datasets is a compute-intensive task, and even the most efficient of sequential algorithms may become ineffective. Thus, implementation of non-trivial data mining algorithms in high performance parallel computing environments is crucial to improving response times.

Mining frequent patterns/itemsets is the core of several data mining tasks. Much attention has gone to the development of parallel algorithms for such tasks[2, 7–9, 12, 16]. However, there are yet several challenges as yet unsolved.

First, parallelizing frequent itemset mining can be complicated and communication intensive. Almost all existing algorithms require multiple synchronization points.

* This work has been partially supported by CNPq-Brazil and by CNPq / CT-INFO / PTACS.

J.-F. Boulicaut et al. (Eds.): PKDD 2004, LNAI 3202, pp. 422–433, 2004.

Second, achieving good workload balancing in parallel frequent itemset mining is extremely difficult, since the amount of computation to be performed by each computing unit does not depend only on the amount of data assigned to it. In fact, equal-sized blocks of data (or partitions) does not guarantee equal (nor approximately equal) workloads, since the number of frequent itemsets generated from each block can be heavily skewed[7]. Thus, an important problem that adversely affects workload balancing is sensitivity to data skew.

Third, system utilization is an issue that is often overlooked in such approaches. The ability to make proper use of all system resources is essential in order to provide scalability when mining frequent itemsets in huge datasets.

In this paper we present a new parallel algorithm for frequent itemset mining, based on the filter-stream programming model. Essentially, the mining process is viewed as a coarse grain data flow controlled by a series of components, referred to as *filters*. A filter receives some data from other filters, performs specific processing in this data, and feeds other filters with the transformed/filtered data. Filters are connected via *streams*, where each stream denotes a unidirectional data flow from a producer filter to a consumer filter. This new approach for parallel frequent itemset mining results in interesting contributions, which can be summarized as follows:

- The candidate generation can work in an asynchronous way, yielding a very effective approach for determining frequent itemsets. Also, the algorithm communicates only the necessary aggregate information about itemsets.
- The parallel algorithm is also anticipatory, in the sense that it can look ahead if a candidate is frequent without the necessity of examining it over all partitions first. This ability becomes more effective when the dataset has a skewed itemset support distribution, a common occurrence in real workloads. In some sense, it compensates the general negative impact of data skewness in workload balancing.
- Finally, we demonstrate thought an extensive experimental evaluation that our algorithm utilizes the available resources very effectively and scale very well for huge datasets and large parallel configurations.

2 Definitions and Related Work

DEFINITION 1. [ITEMSETS] For any set $\mathcal{X}$, its size is the number of elements in $\mathcal{X}$. Let $\mathcal{I}$ denote the set of n natural numbers $\{1, 2, ..., n\}$. Each $x \in \mathcal{I}$ is called an item. A non-empty subset of $\mathcal{I}$ is called an itemset. An itemset of size k, $\mathcal{X} = \{x_1, x_2, ..., x_k\}$ is called a k-itemset.

DEFINITION 2. [TRANSACTIONS] A transaction $\mathcal{T}_i$ is an itemset, where i is a natural number called the *transaction identifier*. A transaction dataset $\mathcal{D} = \{\mathcal{T}_1, \mathcal{T}_2, ..., \mathcal{T}_m\}$, is a finite set of transactions, with size $|\mathcal{D}| = m$. The support of an itemset $\mathcal{X}$ in $\mathcal{D}$ is the number of transactions in $\mathcal{D}$ that contain $\mathcal{X}$, given as $\sigma(\mathcal{X}, \mathcal{D}) = |\ \{\mathcal{T}_i \in \mathcal{D}\ | \mathcal{X} \subseteq \mathcal{T}_i\}|$.

DEFINITION 3. [FREQUENT ITEMSETS] An itemset $\mathcal{X}$ is frequent in the dataset $\mathcal{D}$ iff $\sigma(\mathcal{X}, \mathcal{D}) \geq \sigma^{min}$, where σ^{min} is a user-specified minimum-support threshold, with values $0 < \sigma^{min} \leq |\ \mathcal{D}\ |$. The set of all frequent itemsets is denoted as $\mathcal{F}(\sigma^{min}, \mathcal{D})$.

PROBLEM 1. [MINING FREQUENT ITEMSETS] Given σ^{min} and a transaction dataset $\mathcal{D}$, the problem of mining frequent itemsets is to find $\mathcal{F}(\sigma^{min}, \mathcal{D})$.

Several parallel algorithms for frequent itemset mining were already proposed in the literature [2, 7–9, 16]. The majority of the proposed algorithms follow one of the three main parallelizing strategies:

1. COUNT DISTRIBUTION: This strategy follows a data-parallel paradigm in which the dataset is partitioned among the processing units (while the candidates are replicated). One drawback of this strategy is that at the end of each iteration all processing units must exchange local supports, incurring in several rounds of synchronization. We alleviate this problem by presenting algorithms [14] that need only one round of synchronization by employing an upper bound for the global *negative border* [10]. FDM [7] is another algorithm built on the COUNT DISTRIBUTION strategy. It employs new pruning techniques to reduce processing and communication[1]. Several other algorithms [2, 11] also follow this strategy.
2. CANDIDATE DISTRIBUTION: This strategy follows a paradigm that identifies disjoint partitions of candidates. A common strategy is to partition candidates based on their prefixes, and this strategy can incur in poor workload balancing. PARECLAT [16] is an algorithm that follows this strategy.
3. DATA DISTRIBUTION: This strategy attempts to maximize the use of all aggregate main memory, but requires to transfer the entire dataset at the end of each iteration, incurring in very high communication overheads.

Our parallel algorithm distributes both counts and candidates and has an excellent asynchrony. Further, it presents benefits due to the use of a novel anticipation method.

3 The Filter-Stream Programming Model

The filter-stream programming model was originally proposed for Active Disks [1], to allow the utilization of the disk resident processor in a safe and efficient way. The idea was to exploit the extra processor to improve application performance in two ways. First, alleviating the computation demand on the main processor by introducing the extra, mostly idle processor. Second, it was expected that such computation would reduce the amount of data that needed to be brought from the disk to the main memory. The proposed model, introduced the concept of disklets, or filters, which are entities that perceive streams of data flowing in, and after some computation it would generate streams of data flowing out. In a sense, it is very similar to the concept of UNIX pipes. The difference is that while pipes only have one stream of data coming in and one going out, in the proposed model, arbitrary graphs with any number of input and output streams are possible. Later, this concept was extended as a programming model suitable for the Grid environment [5]. A runtime system, called DATACUTTER (DC) was

[1] Other interesting proposal introduced by the same authors is a metric for quantifying data skewness. For an itemset $\mathcal{X}$, let $p_i(\mathcal{X})$ denote the probability that $\mathcal{X}$ occurs in partition i. The entropy of $\mathcal{X}$ is given as $\mathcal{H} = -\sum_i^n p_i(\mathcal{X}) \times log(p_i(\mathcal{X}))$. The skewness of $\mathcal{X}$ is given as $S(\mathcal{X}) = \frac{log(n)-\mathcal{H}(\mathcal{X})}{log(n)}$, where n is the number of partitions. A dataset's total data skewness is the sum of the skew of all itemsets weighted by their supports.

then developed to support such model. Applications from different domains have been successfully implemented using DC [4, 6, 13]. Creating an application in DC consists of decomposing the target application into filters. These filters are than scheduled for execution in the machines comprising a Grid environment.

Streams in DC represent unidirectional pipes. A filter can either write into or read from the stream. The units of communication are fixed size buffers, agreed upon by the two sides. Each stream has a name associated with it and the connecting of the endpoints of a stream is done at execution time. Once it is done, buffers that are written by the sender will eventually be available for reading on the recipient side. Delivery is guaranteed, and there is no duplication.

The instantiation of the filters is performed by the runtime environment. One of the most important concepts in DC is that of transparent filter copies. At execution time, many instances of the same filter can be created. This provides a simple way to express and implement parallelism, reduce the computation time and to balance the time spend on the several stages of the computation. So, while the concept of the decomposition of the application into filters is related to task parallelism, the possibility of having multiple replicas of the same filter, on the other hand, is related to data parallelism. DC nicely integrates both forms of parallelism into one orthogonal environment.

With respect to multiple copies of the same filter, or data parallelism, the difference between one copy and another is the portion of the entire data each copy has seen. If the filter needs to maintain a state, it is vital that data related to the same portion of the entire data to be always sent to the same copy. Moreover, the data buffers being sent onto the streams need to be transported from one copy of the originating filter to one specific copy of the recipient filter.

For most cases, the selection of the actual destination of any given message buffer actually consider the data in the message to be untyped. However, for applications that maintain some state, it is important to have some understanding of the contents of the message as to decide to which copy of the destination filter needs to be delivered. For these cases, DC implements *labelled streams* which extend the notion of the buffer to a tuple $< l, m >$ where l is a label and m is the message. Associated with each stream there is a label domain L and a hash function h which maps the label from L to $h(l)$. The label domain defines valid values for labels in that stream and the hash function defines a domain which may be associated with filter replicas. With the labeled stream, the stream can use a mapping from that value $h(l)$ to the set of replicas to decide to which copy of the filter should the buffer $< l, m >$ be delivered.

4 Filter-Stream Based Parallel Algorithm

In this section we present our parallel algorithm for frequent itemset mining. We start by discussing its rationale and then we raise some implementation issues.

For sake of filter definition, we distinguish three main tasks to determine whether a k-itemset is frequent or not:

1. verify whether its $(k-1)$-subsets are frequent; if so,
2. count its local supports; and,
3. check whether its global support is above the minimum-support.

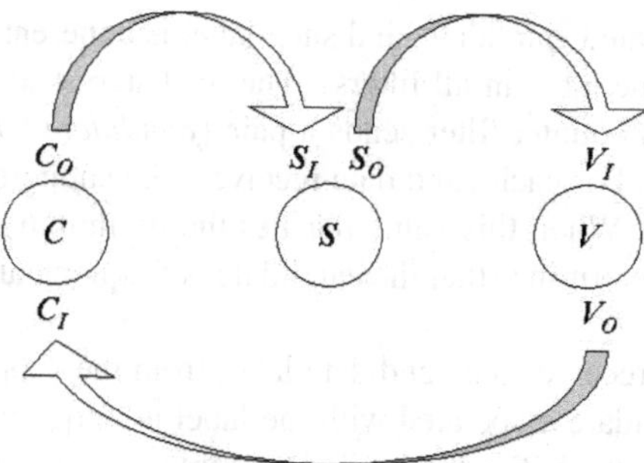

Fig. 1. Algorithm Execution and Data Flow.

The verification filter $\mathcal{V}$ receives as input (represented by the stream $\mathcal{V}i$) the itemsets found to be frequent so far and determines the itemsets that should be verified as being possibly frequent (the candidates), which are the output $\mathcal{V}o$. The counter filter $\mathcal{C}$ receives a candidate itemset through its input stream $\mathcal{C}i$, scans the dataset and determines the support of that candidate itemset, which is sent out using the stream $\mathcal{C}o$. The support checker filter receives the support associated with an itemset through the stream $\mathcal{S}i$, checks whether the counter is above the support threshold and notifies the proper verification filter via the stream $\mathcal{S}o$. We may express the computation involved in determining the frequent itemsets by instantiating the three filters for each itemset, as depicted in Figure 1. In this case, the stream $\mathcal{V}o$ and $\mathcal{C}i$ are connected, as well as the streams $\mathcal{C}o$ and $\mathcal{S}i$. The streams $\mathcal{S}o$ are connected to streams $\mathcal{V}i$ according to the itemset dependence graph[2]. Formally, we have two label domains $\mathcal{I}$ and $\mathcal{T}$ associated with itemsets and transactions, respectively. The streams $\mathcal{S}o$ and $\mathcal{V}i$ are associated with the domain $\mathcal{I}$, while the others with the itemset $\mathcal{T}$.

In the context of filters/streams, there are two dimensions where the parallelization of frequent itemset mining algorithms may be exploited: candidates and counts. We employ both strategies in our algorithm. The verification and counter filters, when created with multiple instances, employ a count distribution strategy, while the support checker filter adopts the candidate division among its instances. This strategy puts together filters from several levels of the dependence graph (that is, filters associated with itemsets of various sizes), using the label concept of our programming model. Although the mapping functions in this case may not be simple, this approach both uses the available plataform efficiently and does not require any replica of the transaction dataset. Further, the granularity of the parallelism that may be exploited is very fine, since we may assign a single transaction or itemset to a filter, without changing the algorithm nor even its implementation.

The execution of the algorithm starts with the counter filters. Each counter filter has access to its local dataset partition, and the first step is to count the 1-itemsets, by scanning its partition and building the *tidsets*[3] of the 1-itemsets. At this point, a label

[2] The vertices in the dependence graph are the itemsets and the edges represent which itemsets that are subsets of a given itemset and must be frequent so that the former may be also frequent.

[3] The set of all transaction identifiers in which a given itemset has occurred. Other data structures, such as diffsets [15], can also be used.

is assigned to each counted candidate, and such label is coherent across all filters (i.e., a candidate has the same label in all filters). The next step is to discover the frequent 1-itemsets, and so each counter filter sends a pair {*candidate label, local support*} to a support checker filter. For each candidate received, the support checker filter simply sums its local supports. When this value reaches the minimum-support threshold, the support checker filter determines that the candidate is frequent, and broadcasts its label to all verifier filters.

Each verifier filter receives the candidate label from the support checker filter and interprets that the candidate associated with the label is frequent. As the labels of frequent 1-itemsets arrive at verifier filters, it is possible to start counting the 2-itemsets that are enumerated from those frequent 1-itemsets. In order to control this process, each verifier filter maintains a prefix-tree that efficiently returns all candidates that must be counted. Note that, because the support checker filter communicates via broadcast and the prefixes are lexicographically ordered, all candidates are verified and counted according to the same order across the filters, being easy to label a candidate, since its label is simply a monotonically increasing number. As soon as a candidate is counted, the counter filter sends another pair {*candidate label, local support*} to the support checker filter, and the process continues until a termination condition is reached and all frequent itemsets were found. From this brief description we distinguish three major issues that should be addressed for implementing our algorithm: performance, anticipation, and termination condition.

Performance: Each filter produces and consumes data at a certain rate. The best performance occurs when the instances of filters are balanced with respect to each other and the communication overhead between the filters is minimized. That is, the data production rates of the producer filters should match the data comsuption rates of the consumer filters. In our case, the number of counter filters must be larger than the number of the other two filters, since counter filters perform a more computational intensive task than the other filters. The optimal number of instances for each filter may vary according to dataset characteristics (i.e., size, density etc.).

Anticipation: A support checker filter does not need to wait for all local supports of a given candidate to determine if it is frequent. In fact, the support checker filter can anticipate this information to the verifier filters, increasing the throughput of the support checker filter and consequently accelerating the whole process. Clearly, data distribution has a major hole in the effectiveness of the anticipation process. In fact, skewed distributions will provide the best gains.

Termination Condition: The execution terminates iff all filters have no more work to be done. However, it is difficult to detect this condition because of the circular dependence among filters. Fortunately, our algorithm has one property that facilitates the termination detection – the candidates are generated in the same order across filters, so that the candidate label may work as a local global clock, which is synchronized when all candidate labels are equal among all filters. At this point, there is no more work to be done.

5 Experimental Evaluation

In this section we present experimental results of our parallel algorithm. Sensitivity analysis on our algorithm was conducted on data distribution, data size, and degree of parallelism. We used both real and synthetic datasets as inputs to the experiments. The real dataset used is called KOSARAK, and it contains click-stream data of a Hungarian on-line news portal (KOSARAK has approximately 900,000 transactions). The synthetic datasets, generated using the procedure described in [3], have sizes varying from 560MB (D3.2MT16I12) to 2.2GB (D12.8MT16I12). To better understand how data distribution (i.e., data skewness) affects the performance of our parallel algorithm, we distributed the transactions among the partitions in two different ways:

- Random Transaction Distribution ($\mathcal{D}_R$): Transactions are randomly distributed among equal-sized partitions. This strategy tends to reduce data skewness, since all partitions have an equal probability to contain a given transaction.
- Original Transaction Distribution ($\mathcal{D}_O$): The dataset is simply splited into blocked partitions, preserving its original data skewness.

We start by analyzing the parallel efficiency of our algorithm. We define the parallel efficiency as: $\mu_{p,q} = \frac{\mathcal{T}_p}{q/p \times \mathcal{T}_q}$, where $\mathcal{T}_p$ is the total execution time when p processors are being employed. A parallel efficiency equals to 1 means linear speedup, and when it gets above 1 it indicates that the speedup is super-linear. Table 1 shows how the parallel efficiency varies as a function of dataset size, transaction distribution and degree of parallelism. Parallel efficiency gets much better when the anticipating procedure is used and dataset is larger. Parallel efficiency continues to be high even for larger degrees of parallelism, reaching 7% of improvement in the best case.

Table 1. Parallel Efficiency.

Dataset	Distribution	Anticipating	$\mathcal{T}_8$ (sec)	$\mu_{8,16}$	$\mu_{16,32}$
D6.4MT16I12	$\mathcal{D}_O$	NO	126.38	1.07	0.88
D6.4MT16I12	$\mathcal{D}_O$	YES	124.11	1.07	1.06
D6.4MT16I12	$\mathcal{D}_R$	NO	94.93	1.02	0.89
D6.4MT16I12	$\mathcal{D}_R$	YES	92.13	1.02	0.99
D12.8MT16I12	$\mathcal{D}_O$	NO	194.74	1.05	0.98
D12.8MT16I12	$\mathcal{D}_O$	YES	188.18	1.07	1.07
D12.8MT16I12	$\mathcal{D}_R$	NO	168.35	1.00	0.97
D12.8MT16I12	$\mathcal{D}_R$	YES	165.96	1.02	1.01
KOSARAK	$\mathcal{D}_O$	NO	642.82	0.96	0.85
KOSARAK	$\mathcal{D}_O$	YES	639.14	1.00	0.95
KOSARAK	$\mathcal{D}_R$	NO	640.87	0.97	0.94
KOSARAK	$\mathcal{D}_R$	YES	639.91	0.99	0.95

We also evaluated our algorithm by means of traditional speedup and scaleup experiments. Figure 2 shows speedup and scaleup numbers obtained from the synthetic

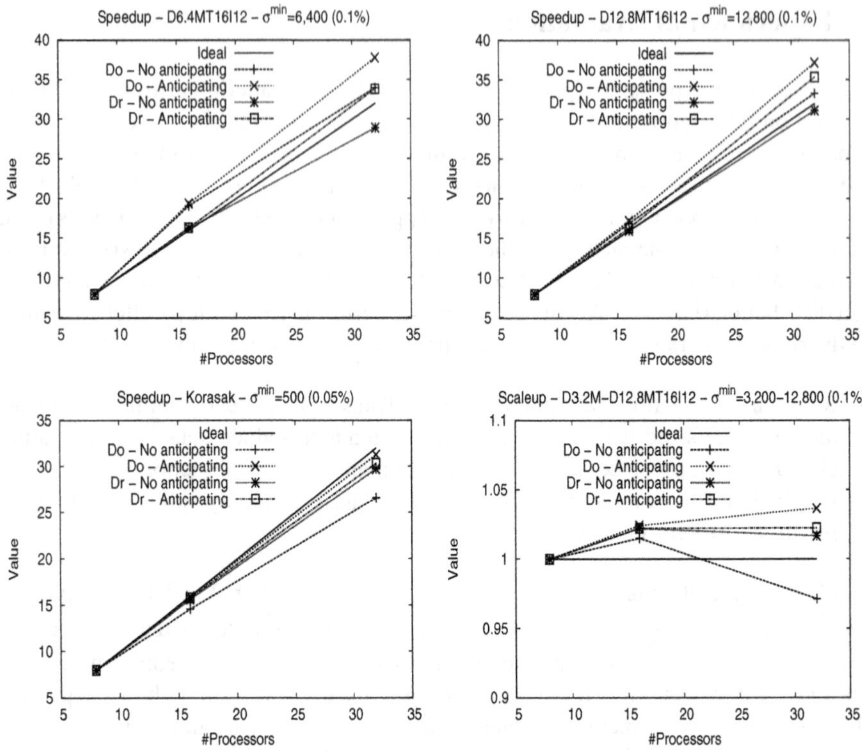

Fig. 2. Speedup and Scaleup Numbers (in relation to 8 processors).

and real datasets. Again, we varied the size, transaction distribution and degree of parallelism. For the speedup experiments with synthetic data we employed datasets with different sizes (6,400,000 and 12,800,000 transactions), and for the two datasets employed we observed superlinear speedups when the anticipation procedure is used. We also observed a superlinear speedup without the anticipation, but in this case the dataset has a random transaction distribution. Further, the speedup number tends to get better for larger datasets, since there is less variability. Impressive numbers were also observed with real data, and the best result was achieved with original (skewed) transaction distribution and using the anticipation procedure.

For the scaleup experiments, we varied dataset size and degree of parallelism in the same proportion. Dataset size ranges from 3,200,000 transactions (with 8 processors) to 12,800,000 transactions (with 32 processors). As we can see in Figure 2, our parallel algorithm also presents ultra scalability when the anticipation procedure is employed, or a random transaction distribution is used. Even when using original transaction distribution without anticipation, our algorithm shows to be very scalable, reaching approximately 95% of scalability.

As seem in both speedup and scaleup experiments, the anticipation is more effective in the presence of data skewness, since the probability of a frequent itemset be sent to the verifier filter using less partitions is higher. Figure 3 shows the average number

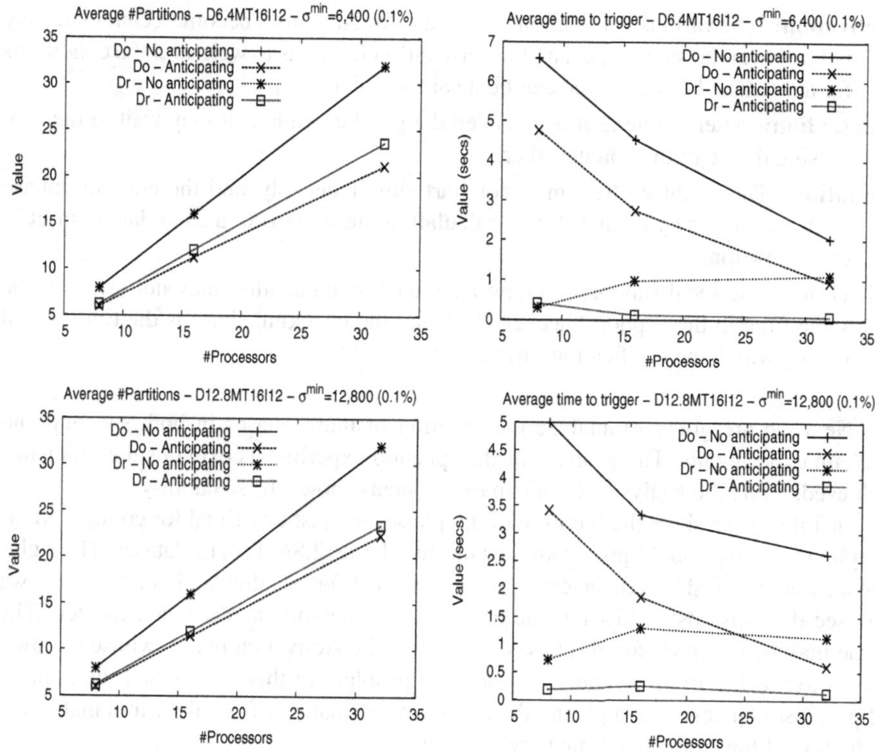

Fig. 3. Effects of Anticipating: Necessary Partitions and Average Time to Trigger.

of partitions necessary to trigger an itemset as frequent. Note that, without any anticipation, all partitions must be analyzed, but when the anticipation procedure is used, the number of partitions that must be analyzed to trigger a frequent itemset is smaller and varies with data skewness. As expected, the same trend is also observed in the average time to trigger a frequent itemset. Further, smaller itemsets usually are detected to be frequent earlier. In Table 2 we present the anticipation gain, that is, how long in advance the itemset is found to be frequent. In a 32-processor configuration using the D12.8MT16I12 dataset the gain varied from 60% to 7%. This observation is particularly interesting because the number of short-sized itemsets is much greater than large-sized itemsets, in particular 2- and 3-itemsets.

Table 2. Anticipation gain for D12.8MT16I12 using 32 processors.

Itemset size	1	2	3	4	5	6
Anticipation gain	60.6%	47.0%	44.5%	16.2%	6.0%	7.2%

In order to better understand the dynamics of the parallel algorithm, we introduce some metrics that quantify the various phases of the algorithm. We may divide the determination of the support of an itemset into four phases:

Activation: The various notifications necessary for an itemset become a candidate may not arrive at the same time, and the verification filter has to wait until the conditions for an itemset be considered candidate are satisfied.

Contention: After the itemset is considered a good candidate, it may wait in the processing queue of the counter filter.

Counting: The counter filters may not start simultaneously, and the counting phase is characterized by counter filters calculating the support of a candidate itemset in each partition.

Checking: The local supports coming from each counter filter may not arrive at the same time in the support checker filter, and the checking phase is the time period during which the notifications arrive.

Next we are going to analyze the duration of these phases in both speedup and scaleup experiments. The analysis of the speedup experiments explains the efficiency achieved, while the analysis of scaleup experiments shows the scalability.

In Table 3 we show the duration of the phases we just described for configurations employing 8, 16, and 32 processors for mining the D12.8MT16I12 dataset. The rightmost column also shows the average processing cost for counting an itemset, where we can see that this cost reduces as the number of processors increase, as expected. The same may be observed for all phases, except for the Activation phase, whose duration seems to reach a limit around 1 second. The problem in this case is that the number of processors involved is high and the asynchronous nature of the algorithm makes the reduction of the Activation time very difficult.

Table 3. Speedup Experiment: Profiling (secs).

Proc	Activation	Contention	Counting	Checking	Processing
8	2.741046	5.564751	9.412093	8.469050	0.001645
16	1.264842	2.058052	4.893773	4.691232	0.000759
32	1.229330	0.273229	1.129718	1.986129	0.000369

Verifying the timings for the scaleup experiments in Table 4, we verify the scalability of our algorithm. We can see that an increase in the number of processors and in the size of the dataset does not affect significantly the measurements, that is, the algorithm implementation does not saturate system resources (mainly communication) when scaled.

Table 4. Scaleup Experiment: Profiling (secs).

Proc	Activation	Contention	Counting	Checking	Processing
8	2.741046	5.564751	9.412093	8.469050	0.001645
16	2.628118	5.538353	9.349371	8.403360	0.001596
32	2.439369	5.021002	10.311501	8.906631	0.001594

6 Conclusion and Future Work

In this paper we proposed an algorithm to conduct parallel frequent itemset mining. The proposed algorithm is based on the filter-stream programming model, where the computation of frequent itemsets is expressed as a circular data flow between distinct components or filters. Parallel performance is optimized, and a high degree of asynchrony is achieved by using the right number of each filter. Further, we propose a very simple anticipation approach, which accelerates frequent itemset determination (specially in the presence of data skew). It was empirically showed that our algorithm achieves excellent parallel efficiency and scalability, even in the presence of high data skewness.

Future work includes the development of parallel filter-stream based algorithms for other data mining techniques, such as maximal/closed frequent itemsets and frequent sequential patterns. Utilization of our algorithms in real applications is also a possible target.

References

1. A. Acharya, M. Uysal, and J. Satlz. Active disks: Programming model, algorithms and evaluation. In *Proc. of the Intl. Conf. on Architectural Support for programming Languages and Operating Systems (ASPLOS VIII)*, pages 81–91. ACM Press, Oct 1998.
2. R. Agrawal and J. Shafer. Parallel mining of association rules. *Transactions on Knowledge and Data Engineering*, 8(6):962–969, 1996.
3. R. Agrawal and R. Srikant. Fast algorithms for mining association rules. In *Proc. of the Intl. Conf. on Very Large Databases (VLDB)*, pages 487–499, SanTiago, Chile, June 1994.
4. M. Beynon, C. Chang, U. Catalyurek, T. Kurc, A. Sussman, H. Andrade, R. Ferreira, and J. Saltz. Processing large-scale multi-dimensional data in parallel and distributed environments. *Parallel Computing*, 28(5):827–859, 2002.
5. M. Beynon, T. Kurc, A. Sussman, and J. Saltz. Design of a framework for data-intensive wide-area applications. In *Proc of the Heterogeneous Computing Workshop (HCW)*, pages 116–130. IEEE Computer Society Press, May 2000.
6. U. Catalyurek, M. Gray, T. Kurc, J. Saltz, and R. Ferreira. A component-based implementation of multiple sequence alignment. In *Proc. of the ACM Symposium on Applied Computing (SAC)*, pages 122–126. ACM, 2003.
7. D. Cheung and Y. Xiao. Effect of data distribution in parallel mining of associations. *Data Mining and Knowledge Discovery*, 3(3):291–314, 1999.
8. E. Han, G. Karypis, and V. Kumar. Scalable parallel data mining for association rules. *Transactions on Knowledge and Data Engineering*, 12(3):728–737, 2000.
9. M. Joshi, E. Han, G. Karypis, and V. Kumar. Efficient parallel algorithms for mining associations. *Parallel and Distributed Systems*, 1759:418–429, 2000.
10. H. Mannila and H. Toivonen. Levelwise search and borders of theories in knowledge discovery. *Data Mining and Knowledge Discovery*, 1(3):241–258, 1997.
11. S. Orlando, P. Palmerini, R. Perego, and F. Silvestri. An efficient parallel and distributed algorithm for counting frequent sets. In *Proc. of the Intl. Conf. on Vector and Parallel Processing (VECPAR)*, pages 421–435, Porto, Portugal, 2002.
12. S. Parthasarathy, M. Zaki, M. Ogihara, and W. Li. Parallel data mining for association rules on shared-memory systems. *Knowledge and Information Systems*, 3(1):1–29, 2001.
13. M. Spencer, R. Ferreira, M. Beynon, T. Kurc, U. Catalyurek, A. Sussman, and J. Saltz. Executing multiple pipelined data analysis operations in the grid. In *Proc. of the ACM/IEEE Conf. on Supercomputing*, pages 1–18. IEEE Computer Society Press, 2002.

14. A. Veloso, M. Otey, S. Parthasarathy, and W. Meira. Parallel and distributed frequent itemset mining on dynamic datasets. In *Proc. of the High Performance Computing Conference (HiPC)*, Hyderabad, India, December 2003. Springer and ACM-SIGARCH.
15. M. Zaki and K. Gouda. Fast vertical mining using diffsets. In *Proc. of the Int. Conf. on Knowledge Discovery and Data Mining (SIGKDD)*. ACM, August 2003.
16. M. Zaki, S. Parthasarathy, M. Ogihara, and W. Li. New parallel algorithms for fast discovery of association rules. *Data Mining and Knowledge Discovery*, 4(1):343–373, December 1997.

A Quantification of Cluster Novelty with an Application to Martian Topography

Ricardo Vilalta[1], Tom Stepinski[2],
Muralikrishna Achari[1], and Francisco Ocegueda-Hernandez[3]

[1] Department of Computer Science, University of Houston
4800 Calhoun Rd., Houston TX 77204-3010, USA
{vilalta,amkchari}@cs.uh.edu
[2] Lunar and Planetary Institute
3600 Bay Area Blvd, Houston TX 77058-1113, USA
tom@lpi.usra.edu
[3] CINVESTAV
López Mateos Sur 590, Guadalajara, Jalisco, C.P. 45090, México
focegued@gdl.cinvestav.mx

Abstract. Automated tools for knowledge discovery are frequently invoked in databases where objects already group into some known classification scheme. In the context of unsupervised learning or clustering, such tools delve inside large databases looking for alternative classification schemes that are both meaningful and novel. A quantification of cluster novelty can be looked upon as the degree of separation between each new cluster and its most similar class. Our approach models each cluster and class as a Gaussian distribution and estimates the degree of overlap between both distributions by measuring their intersecting area. Unlike other metrics, our method quantifies the novelty of each cluster individually, and enables us to rank classes according to its similarity to each new cluster. We test our algorithm on Martian landscapes using a set of known classes called geological units; experimental results show a new interpretation for the characterization of Martian landscapes.

1 Introduction

Clustering algorithms are useful tools in revealing structure from unlabelled data; the goal is to discover how data objects gather into natural groups. Research spans multiple topics such as the cluster representation (e.g., flat, hierarchical), the criterion function (e.g., sum-of-squared errors, minimum variance), and the similarity measure (e.g, Euclidean distance). In real-world applications, however, the discovery of natural groups of data objects is often of limited use; in addition one needs to assess the quality of the resulting clusters against known classifications. An understanding of the output of the clustering algorithm can be achieved by either finding a resemblance of the clusters with existing classes, or if no resemblance is found, by providing an interpretation to the new groups of data objects.

This paper proposes a method to assess the novelty of a set of clusters under the assumption of the existence of a known classification of objects. Most previous metrics output a single value indicating the degree of match between the partition induced

J.-F. Boulicaut et al. (Eds.): PKDD 2004, LNAI 3202, pp. 434–445, 2004.

by the known classes and the one induced by the clusters; approaches vary in nature from information-theoretic [1,4] to statistical [7,3,6]. By averaging the degree of match across all classes and clusters, such metrics fail to identify the potential novelty of single clusters. Moreover, the lack of a probabilistic model in the representation of data distributions precludes inferring the extent to which a class-cluster pair intersect. Our goal is to be able to identify the existence of novel clusters by looking at each of them individually, ranking all classes against each cluster based on their degree of overlap or intersection.

We test our methodology on a database containing images of Mars landscapes produced by the Mars Orbiter Laser Altimeter (MOLA) [9]. Each terrain is characterized through a computational analysis of its drainage networks and represented as a real vector. We apply a probabilistic clustering algorithm that groups terrains into clusters by modelling each cluster through a probability density function; each terrain (i.e., each vector) in the database has a probability of class membership and is assigned to the cluster with highest posterior probability (Section 4). We assess the novelty of the output clusters by applying our proposed methodology using a known classification of Mars surface based on regions known as geological units. The analysis has prompted a new classification of Mars landscapes based on hydrological aspects of landscape morphology.

This paper is organized as follows. Section 2 provides background information and defines current metrics that compare sets of clusters with known object classifications. Section 3 explains our proposed metric. Section 4 describes our domain of study based on a characterization of Mars drainage networks. Section 5 reports our experimental analysis, and provides an interpretation of the output clusters. Lastly, Section 6 gives a summary and discusses future work.

2 Preliminaries: Cluster Validation

We assume a dataset of objects, $\mathcal{D} : \{\mathbf{x}_i\}$, where each $\mathbf{x}_i = (a_1, a_2, \cdots, a_k)$ is an attribute vector characterizing a particular object. We refer to an attribute variable as A_i, and to a particular value of that variable as a_i. The space $\mathcal{X}$ of all possible attribute vectors is called the attribute space. We will assume each attribute value is a real number, $\mathbf{x}_i \in \Re^k$.

A clustering algorithm partitions $\mathcal{D}$ into n mutually exclusive and exhaustive[1] subsets $\mathcal{K}_1, \mathcal{K}_2, \cdots, \mathcal{K}_n$, where $\bigcup_j \mathcal{K}_j = \mathcal{D}$. Each subset $\mathcal{K}_j$ represents a cluster. The goal of a clustering algorithm is to partition the data such that the average distance between objects in the same cluster (i.e., the average intra-distance) is significantly less than the distance between objects in different clusters (i.e., the average inter-distance) [2]. Distances are measured according to some predefined metric (e.g., Euclidean distance) over space $\mathcal{X}$.

We assume the existence of a different mutually exclusive and exhaustive partition of objects, $\mathcal{C}_1, \mathcal{C}_2, \cdots, \mathcal{C}_m$, where $\bigcup_i \mathcal{C}_i = \mathcal{D}$, induced by a natural classification scheme that is independent of the partition induced by the clustering algorithm. Our goal is to

[1] We consider a flat type of clustering (as opposed to hierarchical) where each object is assigned to exactly only cluster.

perform an objective comparison of both partitions. It must be emphasized that the previously known classification is independent of the induced clusters since our main goal is to ascribe a meaning to the partition induced by the clustering algorithm; one may even use multiple existing object classifications to validate the set of induced clusters. When a near-optimal match is found we say the clusters have simply recovered a known class structure.

2.1 Metrics Comparing Classes and Clusters

Several approaches exist attacking the problem of assessing the degree of match between the set $\mathcal{C} = \{\mathcal{C}_i\}$ of predefined classes and the set $\mathcal{K} = \{\mathcal{K}_j\}$ of new clusters. In all cases high values indicate a high similarity between classes and clusters. We divide these approaches based on the kind of statistics employed.

The 2 × 2 Contingency Table

Metrics of a statistical nature usually work on a 2 × 2 table where each entry $\mathcal{E}_{ij}$, $i, j \in \{1, 2\}$, counts the number of object pairs that agree or disagree on the class and cluster to which they belong; $\mathcal{E}_{11}$ corresponds to the number of object pairs that belong to the same class and cluster, similar definitions apply to other entries where $\mathcal{E}_{12}$ corresponds to same class and different cluster, $\mathcal{E}_{21}$ corresponds to different class and same cluster, and $\mathcal{E}_{22}$ corresponds to different class and different cluster. Clearly $\mathcal{E}_{11}$ and $\mathcal{E}_{22}$ denote the number of object pairs contributing to a high similarity between classes and clusters, whereas $\mathcal{E}_{12}$ and $\mathcal{E}_{21}$ denote the number of object pairs contributing to a high degree of dissimilarity. The following statistics have been suggested as metrics of similarity or overlap:

Rand [7]:

$$\frac{\mathcal{E}_{11} + \mathcal{E}_{22}}{\mathcal{E}_{11} + \mathcal{E}_{12} + \mathcal{E}_{21} + \mathcal{E}_{22}} \tag{1}$$

Jaccard [6]:

$$\frac{\mathcal{E}_{11}}{\mathcal{E}_{11} + \mathcal{E}_{12} + \mathcal{E}_{21}} \tag{2}$$

Fowlkes and Mallows [3]:

$$\frac{\mathcal{E}_{11}}{\sqrt{(\mathcal{E}_{11} + \mathcal{E}_{12})(\mathcal{E}_{11} + \mathcal{E}_{21})}} \tag{3}$$

Experiments using artificial datasets show these metrics have good convergence properties (i.e., converge to maximum similarity if classes and clusters are identically distributed) as the number of clusters and dimensionality increase [6].

The $m \times n$ Contingency Table

A different approach is to work on a contingency table defined as follows:

Definition 1. A contingency table $\mathcal{M}$ is a matrix of size $m \times n$ where each row correspond to an external class and each column to a cluster. An entry $\mathcal{M}_{ij}$ indicates the number of objects covered by class $\mathcal{C}_i$ and cluster $\mathcal{K}_j$.

Using $\mathcal{M}$, the similarity between $\mathcal{C}$ and $\mathcal{K}$ can be defined in several forms:

Normalized Hamming Distance [4]:

$$\frac{DH_c(\mathcal{M}) + DH_k(\mathcal{M})}{2|\mathcal{D}|} \tag{4}$$

where $|\mathcal{D}|$ is the size of the dataset (i.e., where $|\mathcal{D}| = \sum_i \sum_j \mathcal{M}_{ij}$) and the directional Hamming distances are defined as follows:

$$DH_c(\mathcal{M}) = \sum_i \max_j \mathcal{M}_{ij} \tag{5}$$

$$DH_k(\mathcal{M}) = \sum_j \max_i \mathcal{M}_{ij} \tag{6}$$

Equation 4 measures accuracy by adding the highest value on each row (conversely column) in $\mathcal{M}$ divided by the total number of objects. Rows and columns are worked out separately since the number of classes and clusters may be different.

Empirical Conditional Entropy [1, 11]:

$$H(C|K) = -\sum_i \sum_j \frac{\mathcal{M}_{ij}}{|\mathcal{D}|} \log_2 \frac{\mathcal{M}_{ij}}{\mathcal{M}_j} \tag{7}$$

where $\mathcal{M}_j$ is the marginal sum $\sum_i \mathcal{M}_{ij}$ and lower values are preferred. Equation 7 measures the degree of impurity of the partitions induced by the clustering algorithm and is biased towards distributions characterized by many clusters; this bias can be adjusted by applying the minimum description length principle [1].

Limitations

All metrics described above output a numeric value according to the degree of match between $\mathcal{C}$ and $\mathcal{K}$. In practice, a quantification of the similarity between classes and clusters is of limited value; any potential discovery provided by the clustering algorithm is only identifiable by analyzing the meaning of each cluster individually. And even when in principle one could analyze the entries of a contingency matrix to identify clusters having little overlap with existing classes, such information cannot be used in estimating the intersection of the probability models from which the objects were drawn, as it is the case with our parametric approach. We address these issues and our proposed novelty metric next.

3 Assessing the Novelty of New Clusters

We start under the assumption that both clusters and classes can be modelled using a multi-variate Gaussian (i.e., Normal) distribution[2]. In this case the probability density function is completely defined by a mean vector μ and covariance matrix $\mathbf{\Sigma}$:

$$f(\mathbf{x}) = \frac{1}{(2\pi)^{k/2}|\mathbf{\Sigma}|^{1/2}} \exp\left[-\frac{1}{2}(\mathbf{x}-\mu)^t \mathbf{\Sigma}^{-1}(\mathbf{x}-\mu)\right] \tag{8}$$

[2] This strong assumption is supported by many real domains where data objects can be seen as random disturbances of a prototype data object.

where $\mathbf{x}$ and μ are k-component vectors, and $|\Sigma|$ and Σ^{-1} are the determinant and inverse of the covariance matrix.

Our goal is simply to assess the degree of overlap between a particular class C_i, modelled as $f_i(\mathbf{x}) : N[\mu_i, \Sigma_i]$, and cluster $\mathcal{K}_j$, modelled as $f_j(\mathbf{x}) : N[\mu_j, \Sigma_j]$. The lower the degree of overlap the higher the novelty of the cluster. Before explaining our methodology (Section 3.3) we introduce two preliminary metrics.

3.1 The Intersecting Hyper-volume

A straightforward approach to measure the degree of overlap between $f_i(x)$ and $f_j(x)$, denoted as $\mathcal{O}(f_i(\mathbf{x}), f_j(\mathbf{x}))$, is to calculate the hyper-volume lying at the intersection of both distributions. This can be done by integrating the minimum of both density functions over the whole attribute space:

$$\mathcal{O}(f_i(\mathbf{x}), f_j(\mathbf{x})) = \int_{\mathbf{x}} \min[f_i(\mathbf{x}), f_j(\mathbf{x})]\, d\mathbf{x} \tag{9}$$

In the extreme case where both distributions have no overlap then $\min[f_i(\mathbf{x}), f_j(\mathbf{x})] = 0$, and hence $\mathcal{O}(f_i(\mathbf{x}), f_j(\mathbf{x})) = 0$. If both distributions are identical or if one distribution is always on top of the other distribution (i.e., if $\forall \mathbf{x}\ f_i(\mathbf{x}) \geq f_j(\mathbf{x})$ or $\forall \mathbf{x}\ f_j(\mathbf{x}) \geq f_i(\mathbf{x})$) then $\mathcal{O}(f_i(\mathbf{x}), f_j(\mathbf{x})) = 1$. Although equation 9 can be approximated using numerical methods the computational cost is expensive; the problem soon turns intractable even for moderately low values of n. In practice, a solution to this problem is to assume a form of attribute independence as explained next.

3.2 The Attribute-Independence Approach

Instead of integrating over all attribute space one may look at each attribute independently. In particular, a projection of the data over each attribute transforms the original problem into a new problem made of two one-dimensional Gaussian distributions (Figure 1 (left)). We represent the two distributions on attribute A_l, $1 \leq l \leq k$, as $f_i^l(x)$ (corresponding to class C_i) and $f_j^l(x)$ (corresponding to cluster $\mathcal{K}_j$). The parameters for these distributions are simply obtained by extracting the corresponding entries on the mean vectors and the diagonal of the covariance matrices.

The computation of the overlap of the two distributions, $\mathcal{O}(f_i^l(x), f_j^l(x))$, is now performed over a single dimension and is thus less expensive (equation 9). To combine the degree of overlap over all attributes we adopt a product approximation:

$$\mathcal{O}(f_i(\mathbf{x}), f_j(\mathbf{x})) = \prod_{l=1}^{k} \mathcal{O}(f_i^l(x), f_j^l(x)) \tag{10}$$

This approach carries some disadvantages. By looking at each attribute independently, two non-overlapping distributions in an n-dimensional space may appear highly overlapped when projected over each attribute. Our challenge lies on finding an efficient approach to estimate $\mathcal{O}(f_i(\mathbf{x}), f_j(\mathbf{x}))$ along a dimension that provides a clear representation of the separation of the two distributions.

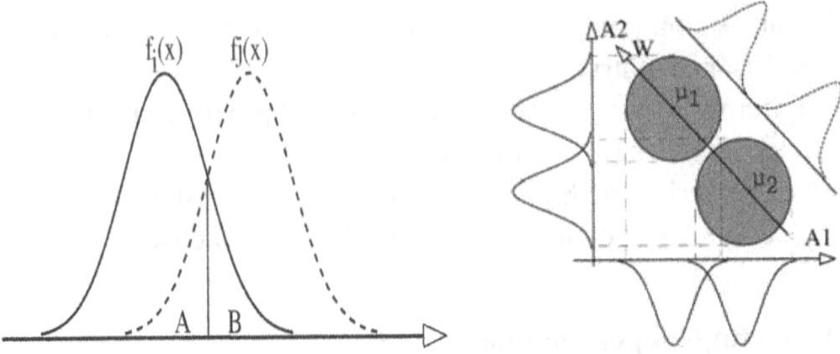

Fig. 1. (left) A measure of the overlap between two distributions; (right) A projection over the difference of the means as a better representation of the separation of the two distributions.

3.3 Projecting over the Difference of the Means

Our proposed solution consists of projecting data objects over a single dimension but in the direction corresponding to the difference of the mean vectors. Specifically, let μ_i be the mean vector of distribution $f_i(\mathbf{x})$ and μ_j be the mean vector of distribution $f_j(\mathbf{x})$; our approach is to project all data objects comprised by class C_i and cluster $\mathcal{K}_j$ into the new vector

$$\mathbf{w} = \mu_i - \mu_j \tag{11}$$

As illustrated in Figure 1 (right), a projection of data objects over vector $\mathbf{w}$ is often a better indicator of the true overlap between both distributions in n dimensions[3]; it captures the dispersion of data objects precisely along the line cutting through the means. In the extreme case where $\mu_i = \mu_j$, we consider $\mathbf{w} = \mu_i = \mu_j$ which is equivalent to considering the origin as one of the means.

Once the projection is done, there is no need to work on each attribute separately (as in equation 10); we simply compute the degree of overlap between the projected distributions along vector $\mathbf{w}$. Thus, our approach efficiently estimates the degree of overlap between two distributions along a single dimension that captures most of the variability[4] of both class C_i and cluster $\mathcal{K}_j$.

Data Projection

To perform the data projection mentioned above we need to compute a scalar dot product

$$x' = \mathbf{w}_0^t \mathbf{x} \tag{12}$$

where $\mathbf{x}$ is an original data point, $\mathbf{w}_0 = \frac{\mathbf{w}}{||\mathbf{w}||}$ is a normalized vector such that $||\mathbf{w}_0|| = 1$, and x' is the (scalar) projection of $\mathbf{x}$ over $\mathbf{w}_0$. We project points over the normalized vector $\mathbf{w}_0$ instead of vector $\mathbf{w}$ simply to give the projection a clear geometrical interpretation (if $||\mathbf{w}|| \neq 1$ the scale of x' is modified).

[3] Alternatively vector $\mathbf{w}$ could be defined as $\mu_j - \mu_i$; both definitions are equally useful.

[4] This is expected to hold as long as C_i and $\mathcal{K}_j$ are close to being hyper-circles, or the direction of the principal axes of the hyper-ellipsoids is close to the direction of vector $\mathbf{w}$.

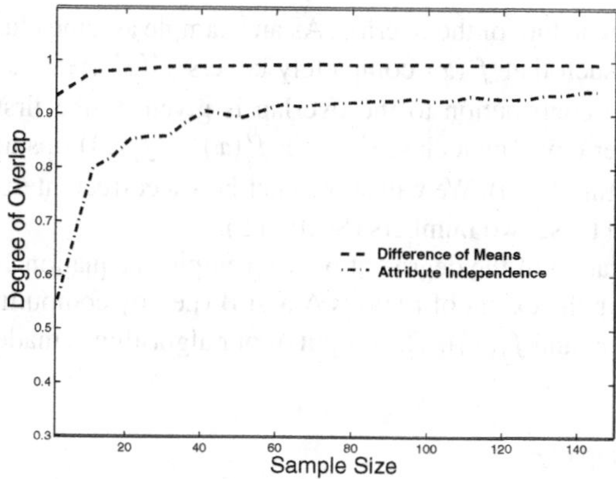

Fig. 2. A comparison of our approach with the attribute independence approach on an artificial dataset where the true overlap is one. Our approach exhibits a faster rate of convergence.

We will refer to the projected density functions over $\mathbf{w}_0$ as $f_i'(x)$ (for class C_i) and $f_j'(x)$ (for cluster $\mathcal{K}_j$). Their parameters can be easily estimated after projecting data objects over $\mathbf{w}_0$. Let μ be the mean of density function $f(\mathbf{x})$, then the projected parameters are defined as

$$\mu' = \mathbf{w}_0^t \mu \qquad \sigma'^2 = \frac{1}{n} \sum (x' - \mu')^2 \tag{13}$$

where μ' and σ'^2 are the projected mean and variance respectively.

In summary, our approach is to quantify the degree of overlap between the two one-dimensional Gaussian distributions $f_i'(x)$ and $f_j'(x)$ obtained after projecting data objects in class C_i and cluster $\mathcal{K}_j$ along vector $\mathbf{w}_0$.

Figure 2 compares our proposed approach with the attribute independence approach (Section 3.2) on an artificial dataset with two Gaussian distributions having the same mean and unit variance. In this experiment there is no cluster novelty, and thus the true overlap (according to equation 9) is one. As shown in Figure 2, both methods tend to stabilize close to one as we increase the sample size, with our proposed approach exhibiting a faster rate of convergence.

A Decomposition of the Degree of Overlap

Until now our measure of overlap has been defined as a function of the intersection of two distributions obtained through a form of data projection. For our purposes we are interested in decomposing the degree of overlap between $f_i'(x)$ and $f_j'(x)$ into two parts:

$$\mathcal{O}(f_i'(x), f_j'(x)) = \int_A f_j'(x) + \int_B f_i'(x) \tag{14}$$

where region A corresponds to all points such that $f_i^l \geq f_j^l$ and region B corresponds to all points such that $f_j^l > f_i^l$ (Figure 1 (left)). The decomposition is important to have an understanding of the nature of the overlap. As an example assume cluster $\mathcal{K}_j$ is a proper subset of class $\mathcal{C}_i$ such that $f_i'(x)$ completely covers $f_j'(x)$ (i.e., $\forall x\ f_i'(x) > f_j'(x)$). In that case all the contribution to the overlap is given by the first integral ($\int_A(\cdot)$). Conversely a cluster covering a class (i.e., $\forall x\ f_j'(x) > f_i'(x)$) tips all the contribution to the second integral ($\int_B(\cdot)$). We will show later how a correct interpretation of cluster novelty depends on these two numbers (Section 5).

Our implementation of the degree of overlap employs equation 14 instead of equation 9 by computing the extent of regions A and B (i.e., by computing the intersecting points between $f_i'(x)$ and $f_j'(x)$). The output of our algorithm is made of three numbers (equation 14):

1. The total overlap $O_\mathcal{T}$
2. The contribution to the overlap by the class ($\int_B(\cdot)$), referred to as $O_\mathcal{C}$
3. The contribution to the overlap by the cluster ($\int_A(\cdot)$), referred to as $O_\mathcal{K}$

4 Drainage Networks in Mars

We now turn to an area of application where our metric for cluster novelty can be tested. Our study revolves around the morphology of Martian landscapes. The goal in this area is to objectively characterize and categorize Martian landscapes in order to understand their origin. Our study uses a dataset characterizing different regions on Mars from the perspective of how they drain. Martian topography based on MOLA data is used to represent landscapes as a series of drainage basins, regardless of the historical presence or absence of actual fluid flow [10]. A drainage network, the part of a basin where the flow is concentrated, is computationally delineated from the basin. Such network has a fractal structure which is described in terms of probability distribution functions of various drainage quantities. Following the method described in [10], the morphology of each network can be encapsulated in a network descriptor or vector of four numbers $\mathbf{x} = (\tau, \gamma, \beta, \rho)$. Briefly, τ, γ, and β are attributes that characterize distributions of contributing areas, lengths of main streams, and dissipated energy, respectively; parameter ρ measures the spatial uniformity of drainage. The network descriptor offers an abstract but very compact characterization of a drainage network.

Our dataset consist of 386 data objects derived from Martian landscapes with a wide range of latitudes and elevations. Our study aims at determining if our characterization of Martian landscapes clusters into natural groups and if there can be a clear interpretation of such clusters. We compare our clusters to a known traditional and descriptive characterization of the Martian surface that divides it into a number of classes called geological units [8]. Division into geological units is based on terrain texture, its geological structure, its age, and its stratigraphy. All this attributes are determined from visual inspection of imagery data. For example unit Hr is described as having "moderately cratered surface, marked by long, linear or sinuous ridges"; unit Npl1 is "highland terrain with high density of craters". Our objects are extracted from surfaces belonging to 16 different geological units representing three major Martian epochs: Noachian, Hesperian, and Amazonian. On the other hand, our clusters group together terrains based on similar drainage patterns. These patterns are obtained from digital topography using a computer algorithm. There is no a priori no clear relation between the two classifications.

5 Empirical Study

We divide our empirical study into two steps: 1) an assessment of the similarity (dissimilarity) of a set of clusters of Mars landscapes with the set of known geological units and 2) an interpretation of the clusters based on hydrological aspects of landscape morphology. We look at each step in turn.

5.1 Cluster Generation

The probabilistic clustering algorithm corresponding to our experiments follows the Expectation Maximization (EM) technique [5]. It groups records into clusters by modelling each cluster through a probability density function. Each record in the dataset has a probability of class membership and is assigned to the cluster with highest posterior probability. The number of clusters is estimated using cross-validation; the algorithm is part of the WEKA machine-learning tool [12].

Applying the EM clustering algorithm directly over the drainage network dataset results in a partition corresponding to nine different clusters. The next step is to assess the degree of match between our nine clusters and the sixteen Martian geological units following our proposed approach (Section 3.3).

5.2 Comparing Clusters to Geological Units

Table 1 shows our results. The first column corresponds to the nine clusters obtained over the drainage network dataset. For each row, the second column corresponds to the class (i.e., geological unit) with highest overlap to that cluster, the third column corresponds to the class with the second highest overlap, and so on. We report on the five classes with highest overlap for each cluster. On each entry we report the total degree of overlap between the cluster and the class ($O_{\mathcal{T}}$) and the corresponding geological unit; within parentheses we show the contribution of the cluster and class to the overlap ($O_{\mathcal{C}}, O_{\mathcal{K}}$).

A first glance at Table 1 may indicate a relatively high overlap between clusters and at least some classes. In some cases this can be explained by looking at the two components of the overlap separately. For example, in some clusters, a high degree of overlap is an artifact of their small size (e.g., cluster C8 with unit Hpl3–a hint is the relatively large value of the second component in the overlap, $O_{\mathcal{K}}$, showing how the unit covers a large portion of the cluster). In other cases, clusters such as C6, C7, and C9, contain many objects and still display a sizable overlap with selected geological units. Using expert knowledge about hydrological properties of clusters and geological properties of classes we may assess whether that overlap is evidence of any significant correlation or if the clusters point to a novel classification of Mars terrains.

Table 1. A measure of the degree of overlap between clusters and classes in the context of Martian topography.

Clusters	Geological Units				
	Most Similar	2nd	3rd	4rd	5th
C1	0.786 Hr (0.309, 0.476)	0.768 Nplr (0.216, 0.552)	0.713 Npld (0.192, 0.522)	0.686 Hnu (0.455, 0.232)	0.669 Npl1 (0.171, 0.498)
C2	0.846 Aoa (0.594, 0.252)	0.533 Aps (0.218, 0.316)	0.438 Hnu (0.141, 0.297)	0.406 Hr (0.0.082, 0.324)	0.384 Npl1 (0.076, 0.308)
C3	0.372 Hvk (0.127, 0.245)	0.362 Hnu (0.070, 0.293)	0.341 Apk (0.064, 0.277)	0.2485 Npld (0.041, 0.208)	0.158 Nh1 (0.023, 0.136)
C4	0.818 Aps (0.260, 0.559)	0.647 Nplr (0.162, 0.485)	0.591 Nh1 (0.142, 0.449)	0.577 Hpl3 (0.139, 0.439)	0.566 Hnu (0.133, 0.433)
C5	0.723 Hh3 (0.248, 0.476)	0.431 Nh1 (0.093, 0.338)	0.353 Hnu (0.069, 0.285)	0.345 Nplr (0.168, 0.177)	0.343 Aps (0.102, 0.241)
C6	0.638 Apk (0.263, 0.375)	0.450 Npld (0.099, 0.351)	0.346 Nh1 (0.066, 0.280)	0.329 Hnu (0.061, 0.268)	0.302 Npl1 (0.054, 0.248)
C7	0.784 Hr (0.246, 0.538)	0.738 Nh1 (0.364, 0.374)	0.696 Nplr (0.182, 0.514)	0.663 Hnu (0.377, 0.287)	0.513 Apk (0.209, 0.304)
C8	0.938 Hpl3 (0.357, 0.581)	0.851 Npl1 (0.290, 0.560)	0.849 Hnu (0.603, 0.247)	0.837 Npld (0.596, 0.241)	0.821 Nplr (0.578, 0.244)
C9	0.488 Nh1 (0.255, 0.233)	0.438 Hh3 (0.113, 0.325)	0.389 Hr (0.232, 0.158)	0.369 Npl1 (0.072, 0.298)	0.278 Nplr (0.048, 0.231)

5.3 Interpretation of Results

The apparent similarity between some clusters and geological units can be explained by observing that basins of all shapes can form in any geological setting. Figure 3 illustrates this point. Four terrains are shown in a 2×2 matrix arrangement. Terrains in the same row belong to the same geological unit, terrains in the same column belong to the same cluster. It is easy to see similarity based on geological unit (look at the distribution of craters across rows), but using drainage networks drawn on top of the terrain, it is also easy to see similarity based on our clustering (look at the shape of the drainage networks across columns).

Cluster C7 describes terrains characterized by drainage basins with widths about the same as their longitudinal extents, these are terrains made of "squared basins". In contrast, cluster C9 groups terrains with narrow basins. The higher overlap of C7 with Hr (0.784) compared to cluster C9 with Hr (0.389) suggests that texture of Hr terrain is such that "squared basins" are preferentially formed (look at the difference in the second component of the overlap). Similar conclusion is also true about the Apk terrain. The most populous cluster, C6, groups terrains that are neither too narrow, nor too squared; it overlaps moderately with a number of geological units. Our interpretation is that *there is no correlation between the classifications based on geology and hydrology*. The texture of all the Martian terrain is such that "squared basins" are simply a common form. This is why a cluster of such objects is the most populous, and this is why it overlaps with many geological units. Other basin shapes are rarer and thus show smaller overlap with

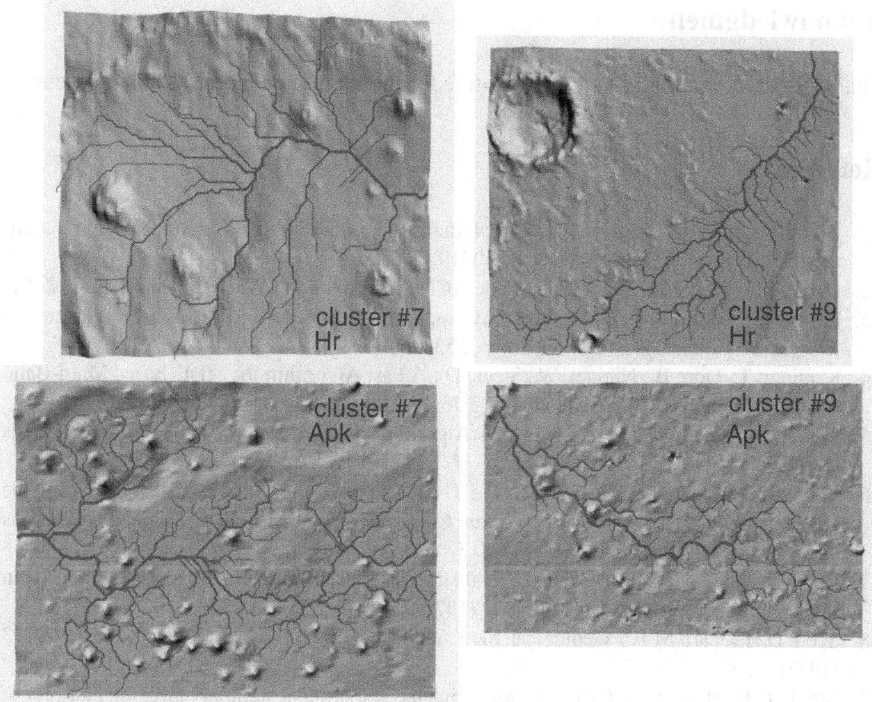

Fig. 3. Four martian terrains from two different geological units and belonging to two different clusters. Drainage networks are drawn on top of the terrain. It is easy to see similarity based on geological unit, but also similarity based on our clustering.

geological units. Some clusters have only a few objects which may happen to belong to particular classes; that results in a false indication of high overlap.

6 Summary and Conclusions

Assessing the degree of match between the partitions induced by a clustering algorithm against those induced by an external classification is normally done through a single numeric response. In this paper we introduce a different approach that quantifies the degree of overlap between two one-dimensional Gaussian distributions obtained after projecting data objects along the vector that cuts through the means.

We test our approach on Martian landscapes by comparing each induced cluster with a set of classes known as geological units. The apparent high overlap between some classes and clusters can be explained by attending to the different nature of the partitions induced by both classifications. Whereas our clusters provide a hydrological view of terrains, the set of existing classes provide a geological view. Both views tend to share terrains whereas in fact there is no correlation among them. In addition, our decomposition of the degree of overlap (Section 3.3) has proved instrumental in giving an interpretation to the similarity (dissimilarity) between clusters and classes.

Acknowledgments

Thanks to the Lunar and Planetary Institute for facilitating data on Martian landscapes.

References

1. Dom Byron: An Information-Theoretic External Cluster-Validity Measure. Research Report, IBM T.J. Watson Research Center RJ 10219 (2001)
2. Duda R. O., Hart P. E., Stork D. G.: Pattern Classification. John Wiley Ed. 2nd Edition (2001)
3. Fowlkes E., Mallows C.: A Method for Comparing Two Hierarchical Clusterings. Journal of American Statistical Association, 78 pp. 553–569 (1983).
4. Kanungo T., Dom B., Niblack W., Steele D.: A Fast Algorithm for MDL-Based Multi-Band Image Segmentation. Image Technology, Jorge Sanz (ed.) Springer-Verlag (1996).
5. McLachlan G., Krishnan T.: The EM Algorithm and Extensions. John Wiley and Sons (1997).
6. Milligan G. W., Soon S. C., Sokol L. M.: The Effect of Cluster Size, Dimensionality, and the Number of Clusters on Recovery of True Cluster Structure. IEEE Transactions on Patterns Analysis and Machine Intelligence, Vol. 5, No. 1pp. 40–47 (1983).
7. Rand W. M.: Objective Criterion for Evaluation of Clustering Methods. Journal of American Statistical Association, 66 pp. 846–851 (1971).
8. Scott D.H., Carr M.H.: Geological Map of Mars. U.S.G.S. Misc Geol. Inv. Map I-1093 (1977).
9. Smith, D.E., et al.: Mars Orbiter Laser Altimeter: Experiment summary after the first year of global mapping of Mars. J. Geophys. Res., Vol. 106, 23,689–23,722 (2001).
10. Stepinski T., Marinova M. M., McGovern P.J., Clifford S. M.: Fractal Analysis of Drainage Basins on Mars. Geophysical Research Letters, Vol. 29, No. 8 (2002).
11. Vaithyanathan S., Dom B.: Model Selection in Unsupervised Learning with Applications to Document Clustering. Proceedings of the Sixteenth International Conference on Machine Learning, Stanford University, CA (2000).
12. Witten I. H., Frank E.: Data Mining: Practical Machine Learning Tools and Techniques with Java Implementations. Academic Press, London U.K (2000).

Density-Based Spatial Clustering in the Presence of Obstacles and Facilitators

Xin Wang, Camilo Rostoker, and Howard J. Hamilton

Department of Computer Science
University of Regina
Regina, SK, Canada S4S 0A2
{wangx,hamilton}@cs.uregina.ca, camilo@scottsdale.ca

Abstract. In this paper, we propose a new spatial clustering method, called DBRS+, which aims to cluster spatial data in the presence of both obstacles and facilitators. It can handle datasets with intersected obstacles and facilitators. Without preprocessing, DBRS+ processes constraints during clustering. It can find clusters with arbitrary shapes and varying densities. DBRS+ has been empirically evaluated using synthetic and real data sets and its performance has been compared to DBRS, AUTOCLUST+, and DBCLuC*.

1 Introduction

Dealing with constraints due to obstacles and facilitators is an important topic in constraint-based spatial clustering. An ***obstacle*** is a physical object that obstructs the reachability among the data objects, and a ***facilitator*** is also a physical object that connects distant data objects or connects data objects across obstacles. Handling these constraints can lead to effective and fruitful data mining by capturing application semantics 69. We will illustrate some constraints by the following example.

Suppose a real estate company wants to identify optimal shopping mall locations for western Canada, shown in the map in Figure 1. In the map, a small oval represents a minimum number of residences. Each river, represented with a light polyline, acts as an obstacle that separates residences on its two sides. The dark lines represent the highways, which could shorten the traveling time. Since obstacles exist in the area and they should not be ignored, the simple Euclidean distances among the objects are not appropriate for measuring user convenience when planning locations of shopping malls. Similarly, since traveling on highways is faster than in urban centers, the length of the highways should be shortened for this analysis. Ignoring the role of such obstacles (rivers) and facilitators (highways for driving) when performing clustering may lead to distorted or useless results.

In this paper, we extend the density-based clustering method DBRS 11 to handle obstacles and facilitators and call the extended method DBRS+. The contributions of DBRS+ are: first, it can handle both obstacles, such as fences, rivers, and highways (when walking), and facilitators, such as bridges, tunnels, and highways (when driving), which exist in the data. Both obstacles and facilitators are modeled as polygons. Most previous research can only handle obstacles. Second, DBRS+ can handle any combination of intersecting obstacles and facilitators. None of previous methods consider intersecting obstacles, which are common in real data. For example, highways or rivers often cross each other and bridges and tunnels often cross rivers. Although the

J.-F. Boulicaut et al. (Eds.): PKDD 2004, LNAI 3202, pp. 446–458, 2004.

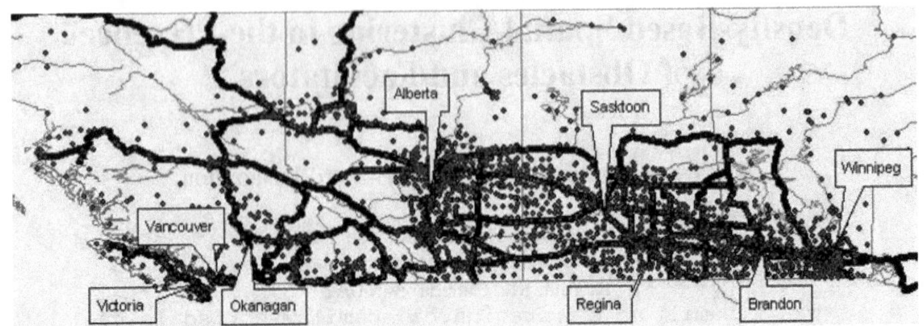

Fig. 1. The Map of Western Population of Canada with Highways and Rivers.

obstacles can be merged in the preprocessing, the resulting polygons cannot be guaranteed to be simple polygons and previous methods do not work on complex polygons. Third, DBRS+ is simple and efficient. It does not involve any preprocessing. The constraints are handled during the clustering process. Almost all previous methods include complicated preprocessing. Fourth, due to capabilities inherited from DBRS, DBRS+ can work on datasets with features such as clusters with widely varying shapes and densities, datasets having significant non-spatial attributes, and datasets larger than 100 000 points.

The remainder of this paper is organized as follows. In Section 2, we briefly discuss three related approaches. Then in Section 3, the DBRS+ algorithm is introduced in detail and its complexity analysis is given. Experimental results are described and analyzed in Section 4. Conclusions are presented in Section 5.

2 Related Work

In this section, we briefly survey previous research on the problem of spatial clustering in the presence of obstacles and facilitators.

COD_CLARANS 10 was the first obstacle constraint partitioning clustering method. It is a modified version of the CLARANS partitioning algorithm 8 adapted for clustering in the presence of obstacles. The main idea is to replace the Euclidean distance function between two points with the ***obstructed distance***, which is the length of the shortest Euclidean path between two points that does not intersect any obstacles. The calculation of obstructed distance is implemented with the help of several steps of preprocessing, including building a visibility graph, micro-clustering, and materializing spatial join indexes. The cost of preprocessing was ignored in the performance evaluation. After preprocessing, COD_CLARANS works efficiently on a large number of obstacles. But, for the types of datasets we described in Section 1, the algorithm may not be suitable. First, the algorithm does not consider facilitator constraints that connect data objects. A simple modification of the distance function in COD_CLARANS is inadequate to handle facilitators because the model used in preprocessing for determining visibility and building the spatial join index would need to be significantly changed. Secondly, as given, COD_CLARANS was not designed to handle intersecting obstacles, such as those present in our datasets. Thirdly, if the dataset has varying densities, COD_CLARANS's micro-clustering approach may not be suitable for the sparse clusters.

AUTOCLUST+ 4 is a version of AUTOCLUST 5 enhanced to handle obstacles. The advantage of the algorithm is that the user does not need to supply parameter values. There are four steps in AUTOCLUST+. First, it constructs a Delaunay diagram 5. Then, a global variation indicator, the average of the standard deviations in the length of incident edges for all points, is calculated to obtain global information before considering any obstacles. Thirdly, all edges that intersect with any obstacles are deleted. Fourthly, AUTOCLUST is applied to the planar graph resulting from the previous steps. When a Delaunay edge traverses an obstacle, the length of the distance between the two end-points of the edge is approximated by a detour path between the two points. However, the distance is not defined if no detour path exists between the obstructed points. As well, the algorithm does not consider facilitator constraints that connect data objects. Since the points connected by facilitators usually do not share a boundary in Voronoi regions, a simple modification of the distance function in AUTOCLUST+ is inadequate to allow it to handle facilitators.

DBCLuC 12, which is based on DBSCAN 3, is the only known previous approach that handles both obstacles and facilitators. Instead of finding the shortest path between the two objects by traversing the edges of the obstacles, DBCLuC determines the visibility through obstruction lines. An ***obstruction line***, as constructed during preprocessing, is an internal edge that maintains visible spaces for the obstacle polygons. To allow for facilitators, entry points and entry edges are identified. The lengths of facilitators are ignored. After preprocessing, DBCLuC is a very good density-based clustering approach for large datasets containing obstacles with many edges. However, constructing obstruction lines is relatively expensive for concave polygons, because the complexity is $O(v^2)$, where v is the number of convex vertices in obstacles. As well, if reachability between any two points is defined by not intersecting with any obstruction line, the algorithm will not work correctly for the example shown in Figure 2. The circle in the figure represents the neighborhood of the central point. Point p and the center are blocked by obstruction lines, but the shortest distance between them is actually less than the radius.

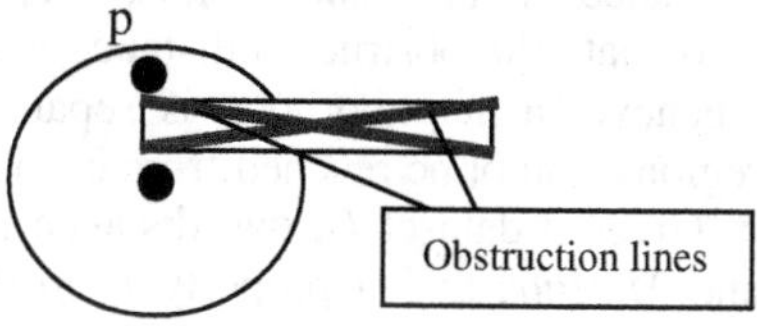

Fig. 2. A Case DBCLuC where may fail.

3 Density-Based Clustering with Obstacles and Facilitators

In this section, we describe a density-based clustering approach that considers both obstacle and facilitator constraints. First we briefly introduce DBRS, then we describe a framework based on DBRS for handling obstacles, then we describe how facilitators are incorporated into this framework, and finally we discuss the neighborhood graph, which provides theoretical support for density-based clustering algorithms.

3.1 Density-Based Spatial Clustering with Random Sampling (DBRS)

DBRS is a density-based clustering method with three parameters, *Eps*, *MinPts,* and *MinPur* 11. DBRS repeatedly picks an unclassified point at random and examines its neighborhood, i.e., all points within a radius *Eps* of the chosen point. The ***purity*** of the neighborhood is defined as the percentage of the neighbor points with the same non-spatial property as the central point. If the neighborhood is sparsely populated ($\leq$*MinPts*) or the purity of the points in the neighborhood is too low ($\leq$*MinPur*) and disjoint with all known clusters, the point is classified as noise. Otherwise, if any point in the neighborhood is part of a known cluster, this neighborhood is joined to that cluster, i.e., all points in the neighborhood are classified as being part of the known cluster. If neither of these two possibilities applies, a new cluster is begun with this neighborhood. The algorithm can identify clusters of widely varying shapes, clusters of varying densities, clusters that depend on non-spatial attributes, and approximate clusters in very large datasets. The time complexity of DBRS is $O(n \log n)$ if an R-tree or SR-tree 7 is used to store and retrieve all points in a neighborhood.

3.2 DBRS+ in the Presence of Obstacles

In DBRS, the distance between two points p and q, denoted as $dist(p, q)$, is computed without considering obstacles. However, when obstacles appear in a neighborhood, the reachability among points can be blocked by these obstacles. So a new distance function, called unobstructed distance, is defined.

Definition 1: The ***unobstructed distance*** between two points p and q in a neighborhood area R, denoted by $dist^R_{ob}(p, q)$, is defined as

$$dist^R{}_{ob}(p,q) = \begin{cases} dist(p,q) & q \text{ and } p \text{ are in the same connected region within R} \\ \infty & \text{otherwise} \end{cases}$$

This definition is based on the observation that in density-based clustering methods, the radius *Eps* is usually set to a small value to guarantee the accuracy and significance of the result. Thus, it is reasonable to use the Euclidean distance to approximate the obstructed distance within a connected neighborhood region. However, whenever a neighborhood is separated into regions, neighbors that are in different regions cannot be reached from each other.

Given a dataset D, two distance functions *dist* and $dist^R_{ob,}$ parameters *Eps*, *MinPts* and *MinPur*, and a property *prop* defined with respect to one or more non-spatial attributes, the definitions of a matching neighbor and reachability in the presence of obstacles are as follows.

Definition 2: The ***unobstructed neighborhood*** of a point p, denoted by $N_{Eps}(p)$, is defined as $N_{Eps}(p) = \{q \in D \mid dist^R_{ob}(p,q) \leq Eps\}$, and its size is denoted as $|N_{Eps}(p)|$.

Definition 3: The ***unobstructed matching neighborhood*** of a point p, denoted by $N'_{Eps\_ob}(p)$, is defined as $N'_{Eps\_ob}(p) = \{ q \in D \mid dist^R_{ob}(p, q) \leq Eps \text{ and } p.prop = q.prop\}$, and its size is denoted as $|N'_{Eps\_ob}(p)|$.

Definition 4: A point p and a point q are ***directly purity-density-reachable*** in the presence of obstacles if (1) $p \in N'_{Eps\text{-}ob}(q)$, $|N'_{Eps\text{-}ob}(q)| \geq MinPts$ and $|N'_{Eps\text{-}ob}(q)| / |N_{Eps}(q)| \geq MinPur$ or (2) $q \in N'_{Eps\text{-}ob}(p)$, $|N'_{Eps\text{-}ob}(p)| \geq MinPts$ and $|N'_{Eps\text{-}ob}(p)| / |N_{Eps}(p)| \geq MinPur$.

Definition 5: A point p and a point q are ***purity-density-reachable (PD-reachable) in the presence of obstacles*** from each other, denoted by $PD_{ob}(p, q)$, if there is a chain of points $p_1,\ldots,p_n$, $p_1=q$, $p_n=p$ such that p_{i+1} is directly purity-density-reachable from p_i in the presence of obstacles.

Definition 6: A ***purity-density-based cluster*** *C* in the presence of obstacles is a non-empty subset of a dataset *D* satisfying the following condition: $\forall p, q \in D$: if $p \in C$ and $PD_{ob}(p, q)$ holds, then $q \in C$.

To determine reachability in the presence of obstacles, we first consider whether a neighborhood contains any obstacles. If not, all points in the neighborhood are reachable from the center. If so, then we say the neighborhood is separated into regions by the obstacles, and we must determine which of these regions contains the center.

When multiple obstacles intersect the neighborhood, not only single obstacles but also combinations of intersecting obstacles need to be considered. Different cases could happen to separate the neighborhood into regions. Perhaps, as shown in Figure 3(a), every obstacle overlaps and intersects the neighborhood, and thus separates a neighbor from the center point. In Figure 3(b), one obstacle (left) divides the neighborhood into two regions. The other obstacle (right) does not divide the neighborhoods by itself. But its combination with the first obstacle generates another region in the neighborhood. In Figure 3(c), each obstacle, does not divide the neighborhood by itself, but its combination with other obstacles does so.

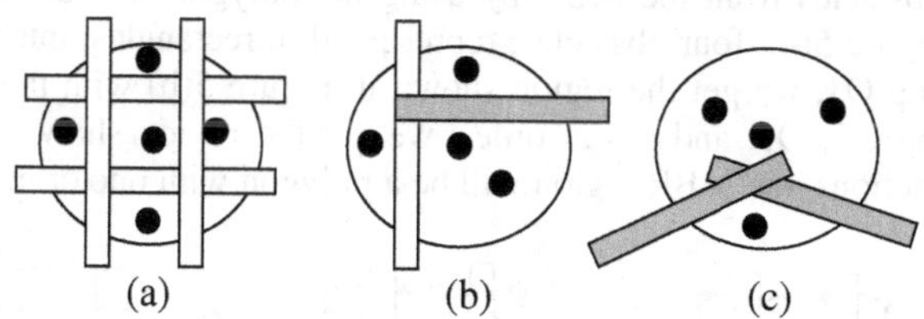

Fig. 3. Multiple Intersecting Obstacles.

To deal with combinations of multiple obstacles, one approach is to merge all obstacles in a preprocessing stage. However, this approach could complicate the problem unnecessarily. First, the result could be a *complex polygon*, i.e., a polygon that consists of an outline plus optional holes. For the case shown in Figure 3(a), the combination of four polygons is a complex polygon with a "#" shape and an internal ring. For a complex polygon, the operations are usually more costly than for a simple polygon. Secondly, the resulting polygon could have more vertices than the original polygon, which could make further operations more costly because intersection points will also need to be treated as vertices. For example, in Figure 3(a), the number of vertices in the original four simple polygons is 16, but the number of vertices in the complex polygon is 32.

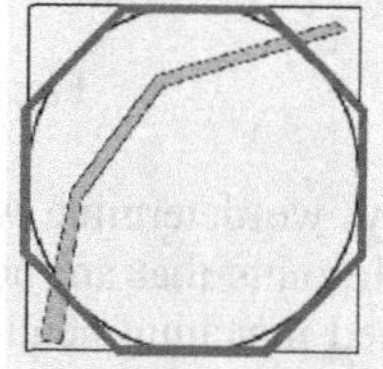

Fig. 4. An obstacle may overlap but not cross the approximation area.

To determine the connected region that contains the central point, we propose a method called ***chop and conquer***. The method subtracts the regions separated by obstacles one by one from the neighborhood, and keeps the simple polygon that includes the center. The process continues until all intersecting polygons have been subtracted.

Chop and conquer consists of four steps. The first step is to create a Minimum Bounding Region (MBR) around the neighborhood. The MBR is a regular polygon with relatively few edges. It is initially created using four edges, but if an obstacle intersects the neighborhood but not the edges of the MBR (as shown in Figure 4), a closer approximation is created by an ***approximation refinement*** process. Each obstacle is checked to see if it intersects the MBR. If so, it is then checked to see if it is completely contained within the MBR. If the obstacle is not completely within the MBR, it is safe to reduce the MBR by performing a geometric subtraction operation with the obstacle. If the obstacle intersects the neighborhood, but does not intersect the MBR, another MBR is created using eight edges (shown as a dark octagon), and the process is repeated with this MBR. The number of edges in the MBR is doubled as necessary until the obstacle intersects both the neighborhood and the MBR. When the algorithm is done, the MBR is topologically equivalent to the neighborhood. The number of edges in the MBR can instead be limited to a specified threshold.

Secondly, each obstacle is checked to see if it overlaps the MBR. If so, a ***local obstacle*** is created by intersecting the obstacle and the MBR. A local obstacle may have fewer edges than the original obstacle, which can save processing time.

Thirdly, after identifying the intersecting local obstacle, DBRS+ subtracts all intersecting local obstacles from the MBR by using the polygon subtraction algorithm 1. As shown in Figure 5(a), four obstacles, represented in rectangles, intersect the MBR. After subtracting O1, we get the region shown in Figure 5(b) with the dark box, and after subtracting O2, O3, and O4 in order, we get the region shown in Figure 5(c). After the subtractions, the MBR region will be a polygon with one or more rings.

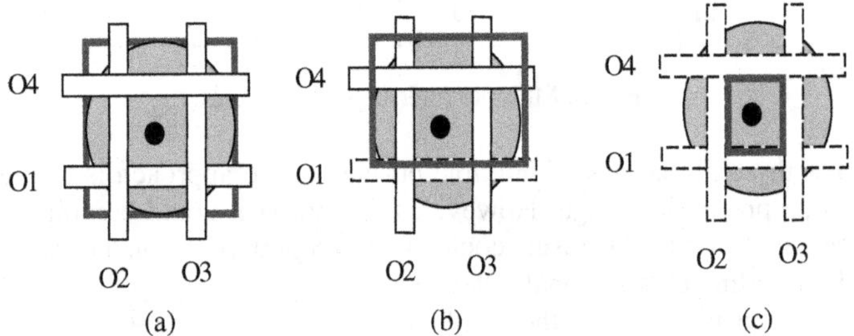

Fig. 5. Subtract obstacles from the MBR.

Fourthly, we determine which ring contains the central point. This ring will also contain all points that are not separated from the central point by obstacles, i.e., the unobstructed matching neighbors. Identifying these unobstructed matching neighbors can be accomplished using a method of determining which point is inside a polygon.

In Figure 6, the DBRS+ algorithm is presented. `D` is a set of points and `O` is a set of obstacles. DBRS+ is equivalent to DBRS except line 4, which calls the `matching-ProperNeighbours` function shown in Figure 7. In Figure 7, we obtain the MBR in line 2 using `ApproximationRefinement`. In lines 3 to 5, we detect obstacles that intersect with the MBR, and add the local obstacles into a list called `localObList`. In lines 8 to 9, the local obstacles in the `localObList` are subtracted from the MBR. In lines 11 to 12, DBRS+ removes all points that are not contained within the same region as the central point.

Determining whether each obstacle intersects with the MBR and obtaining the local obstacles is accomplished in O(log v) time, where v is the total number of vertices of all obstacles. The complexity of subtracting the local obstacles from the MBR is

$O(v'^2)$, where v' is the number of vertices in all local obstacles. Finding the matching neighbors via a region query requires O(log n) time, where n is number of points in the dataset. So the complexity of obtaining the matching neighbors in the presence of obstacles is $O(\log v + v'^2 + \log n)$. Thus the complexity of DBRS+ is $O(nv + nv'^2 + n \log n)$.

```
Algorithm DBRS+ (D, Eps, MinPts, MinPur, O)
    ClusterList = Empty;
 while (!D.isClassified( ))  {
   Select one unclassified point q from D;
   qseeds = D.matchingProperNeighbors(q, Eps, O);
   if (((|qseeds| < MinPts) or (qseed.pur < MinPur)) and
     (qseeds.intersectionWith(ClusterList) == Empty))
      q.clusterID = -1; /*q is noise or a border point */
   else  {
      isFirstMerge = True;
      Ci = ClusterList.firstCluster;
      /* compare qseeds to all existing clusters */
 while (Ci != Empty)  {
          if ( Ci.hasIntersection(qseeds) )
             if (isFirstMerge)  {
                  newCi = Ci.merge(qseeds);
                  isFirstMerge = False; }
             else  {
                  newCi = newCi.merge(Ci);
                  ClusterList.deleteCluster(Ci);}
             Ci = ClusterList.nextCluster;  }
             /*No intersection with any existing cluster */
      if (isFirstMerge)  {
             Create a new cluster Cj from qseeds;
             ClusterList = ClusterList.addCluster(Cj); }
    }  //else
 }  // while !D.isClassified
```

Fig. 6. The DBRS+ Algorithm.

```
SetOfPoints::matchingProperNeighbours(q, Eps, O)
LocalObsList = Empty;
MBR = MBR. ApproximationRefinement(q, Eps,O);
foreach obstacle obi in O
   { localOb = obi.intersect(MBR);
     localObList.addObs(localOb);  }
qseedsRing = MBR;
// subtract all intersecting obstacles from the MBR
foreach localObi in localObList
          qseedsRing = qseedsRing.subtract(localObi)
// only keep points that are in the same ring as the central point
allseeds = D.matchingNeighbours(q, Eps);
qseeds = qseedsRing.removeSeedsNotContained(allseeds);
return(qseeds);
```

Fig. 7. The `matchingProperNeighbours` function.

3.3 DBRS+ in the Presence of Facilitators

When facilitators appear in a neighborhood, the first step is to determine access points for each facilitator. An ***entrance*** refers to a vertex of the facilitator in a neighborhood that is accessible from the central point. An ***exit*** refers to any non-entrance vertex of a facilitator that is reachable from the central point within a distance of *Eps*. All vertices inside the neighborhood are entrances, while only vertices outside the neighborhood may be exits. An entrance that connects directly to an exit is called a ***primary entrance***. In our method, we assume that facilitators are accessed from primary entrances. Figure 8 illustrates a typical facilitator. The vertices of the facilitator are shown as black dots. Each primary entrance, entrance, exit and center is labeled "P", "E", "X", and "C", respectively. In the following, we use the term "entrance" to refer to a "primary entrance" to simplify the description.

Unlike obstacles, facilitators shorten the distance between points. In our method, the ***facilitated distance*** function is defined as follows:

$$dist_{fac}(p,q) = \begin{cases} dist'(p,q) & q \text{ and } p \text{ are in the same connected neighborhood region} \\ Min_f(dist'(p,entrance_f) + length(entrance_f,exit_f) + dist'(exit_f,q)) & \text{otherwise} \end{cases}$$

In the above definition, the *dist'* function could be the Euclidean distance function or the unobstructed distance function. If q is in the same connected neighborhood region with p, they are reachable from each other within the neighborhood. Otherwise, the distance is defined as the shortest distance between them via facilitators. $entrance_f$ and $exit_f$ belong to a facilitator f that makes $dist_{fac}$ have a minimum value.

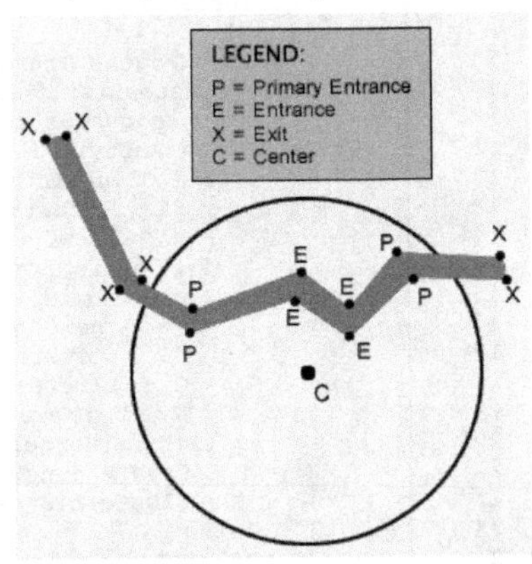

Fig. 8. Entrances, Exits and Primary Entrances.

The length between an entrance and an exit is calculated by accumulating the distances between all the adjacent vertices between them. Thus,

$$length(entrance_f,exit_f) = e\_c * \sum dist(v_i,v_j),$$

where v_i and v_j are adjacent vertices in the same facilitator, including $entrance_f$ and $exit_f$ themselves, and e_c represents the ratio of the facilitated distance to the Euclidean distance. e_c ranges from 0 to 1. For example, if e_c is set to 0.25, the facilitated distance between any two vertices of the same facilitator is only a quarter of the Euclidean distance.

After identifying entrances, the second step is to determine the extra areas reachable from each exit. Since the maximum distance between a neighbor and the central point is *Eps*, the maximum remaining distance from each exit to the extra neighbors, i.e. $dist'(exit, q)$, can be calculated by deducting $dist'(p,entrance_f)$ and $length(entrance_f,exit_f)$ from *Eps*.

The third step in the presence of facilitators is to find the extra neighbors through each exit. Every exit that has a positive remaining distance is used as the center to perform a region query.

After defining the facilitated distance function $dist_{fac}$, facilitated matching neighborhood N'_{Eps_fac}, and reachability PD_{fac} in the presence of facilitators are defined similarly to N'_{Eps_ob} and PD_{ob}, respectively as given in Sec. 3.2, by using $dist_{fac}$ instead of $dist^R_{ob}$.

Figure 9 presents a function for matching neighbors in the presence of obstacles and facilitators. *F* is the set of facilitators present in the dataset. The call to `matching-ProperNeighbors` in line 1 retrieves all matching neighbors in the presence of obstacles. If no obstacles are present, the function works the same way as the function for retrieving matching neighbors in DBRS 11. Lines 3 and 4 determine primary entrances and exits for each facilitator intersecting with the neighborhood. If an entrance appears in the neighborhood, the radius `r_dis` for every exit that leads to extra neighbors is determined in line 7 by deducting from *Eps* the distance from the center `q` to the entrance and from the entrance to the exit. The extra neighbors are retrieved in line 9 by using `exit` as the center and `r_dis` as the radius. The extra neighbors are merged into `qseeds` in line 10.

As discussed in Sec. 3.2, the complexity of the `matchingProperNeighbors` function in line 1 is O (log*n* +log*v* + v'^2), where *n* is the size of the dataset, *v* is the number of vertices in the obstacles and *v'* is the total number of vertices in all local obstacles. If there are entrances for every facilitator in the neighborhood, the method of finding the extra neighbors in line 8, which is the most costly operation, requires O (*f* log *n*) time, where *f* is the number of vertices in the facilitators. Therefore, the worst-case time complexity of DBRS+ in the presence of both obstacles and facilitators is O (*n* log *n*+ *n*log*v*+ nv'^2+ *nf* log *n*) for *n* data points.

```
   SetOfPoints::matchingNeighboursWithObstaclesAndFacilitators(q, Eps, O, F)
1  qseeds = D.matchingProperNeighbours(q, Eps, O);
2  foreach Facilitator fac in SetOfFacilitators F
3       { entranceList = q.getPrimaryEntrances(fac, Eps);
4         exitList = q.getExit(fac, Eps);
5         foreach entrance in entranceList
6             foreach exit in exitList
7                 { r_dis = Eps - dist(q,entrance)-length(entrance,exit);
8                   if (r_dis > 0)
9                     { qseeds_extra = D.matchingNeighbours(exit, r_dis);
10                      qseeds = qseeds.merge(qseeds_extra); }
11                }
12      }
13 return (qseeds);
```

Fig. 9. The matchingNeighborsWithObstaclesAndFacilitators Function.

3.4 The Neighborhood Graph

The *neighborhood graph* for a spatial relation called *neighbor* is a graph *G* = (*V*, *E*) with the set of vertices *V* and the set of directed edges *E* such that each vertex corresponds to an object of the database and two vertices v_1 and v_2 are connected iff *neighbor*(v_1, v_2) holds. As proved in 11, if PD-reachable relation is the neighbor relation, the neighborhood graph generated by DBRS is connected. For DBRS+, the definition of original PD-reachable relation is changed accordingly in the presence of obstacles and facilitators to give the following lemma.

Lemma 1 If the PD-reachable relation in the presence of obstacles and facilitators is the neighbor relation for the neighborhood diagram, the neighborhood graph generated by DBRS+ is connected.

Proof Sketch: Similar to the proof of Lemma 3 in 11.

4 Experimental Results

This section presents experimental results on synthetic and real datasets. All experiments were run on a 2.4GHz PC with 512Mb of memory. To improve the efficiency of region queries, the spatial index was implemented as an R-tree. Nonetheless, the total runtime is linearly related to the number of region queries. For synthetic datasets, each record includes x and y coordinates and one non-spatial property. For all experiments on synthetic datasets (distributed in a 5000x5000 area), *Eps* is 4 or 5 (about 3.14-6E of whole clustering area), *MinPts* is 10, and *MinPur* is set to 0.75. Each reported numeric value represents the average value from 3 runs.

Figure 10 shows a 150k dataset with 10% noise and the clusters found by DBRS+ with and without considering obstacle and facilitator constraints. Figure 10(a) shows the original data with various shapes and different densities. Obstacles analogous to 7 highways or rivers and 2 triangle-like lakes split every cluster. Three facilitators connect three pairs of distant shapes. Figure 10(b) shows the results of 8 clusters found by DBRS+ without considering the presence of obstacles and facilitators. Figure 10(c) shows 28 clusters found in the presence of obstacles but ignoring the facilitators. Figure 10(d) shows the 23 clusters found considering both obstacles and facilitators.

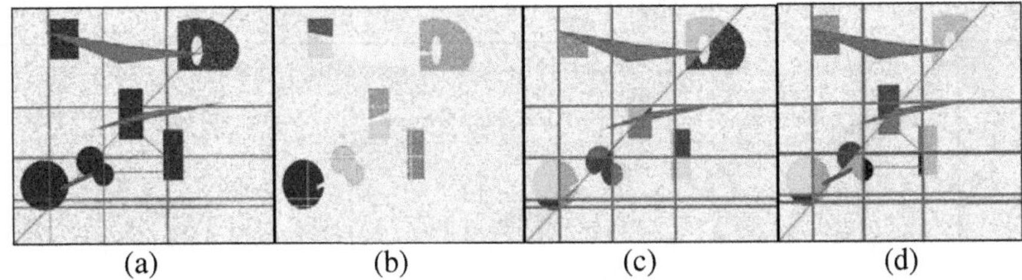

(a) (b) (c) (d)

Fig. 10. Clustering Results with and without Obstacles and Facilitators.

Table 1. Scalability Result in the Presence of Obstacles.

	0 Obstacles (DBRS)		100 Obstacles (10,000 Vertices)		200 Obstacles (20,000 Vertices)		300 Obstacles (30,000 Vertices)	
# of Points	Time (sec)	# of Region Queries	Time (sec)	# of Region Queries	Time (sec)	# of Region Queries	Time (sec)	# of Region Queries
25k	7.17	9523	8.32	9563	9.78	9612	10.98	9551
50k	12.69	11659	14.52	11676	16.10	11668	16.22	11725
75k	25.22	18772	27.33	18858	31.56	18901	33.33	18845
100k	39.92	24426	44.03	24468	46.99	24567	50.54	24520
125k	56.31	30346	62.00	30477	66.91	30402	69.37	30448
150k	76.24	37824	83.54	38049	88.41	37977	92.56	38005
175k	97.67	42914	105.98	42995	111.35	43152	117.06	43059
200k	122.93	49388	132.75	49612	139.60	49583	146.66	49594

Table 1 shows the results of our scalability experiments in the presence of obstacles. All synthetic obstacles and datasets in the following experiments are generated by obGen and synGeoDataGen 13. The size of datasets is varied from 25k to 200k with 10% noise and the number of the obstacles in these datasets is varied from 100 to 300. Each obstacle has 100 vertices, so the total number of obstacle vertices varies from 10,000 to 30,000. The runtime increases with the number of obstacle vertices and with the number of points. Accordingly, the number of region queries also increases with the number of points. We also show the performance of DBRS that does not consider obstacles and facilitators.

We also compared the runtime of DBRS+, AUTOCLUST+, and DBCLuC*. Since AUTOCLUST+, as implemented by its authors, cannot handle 10,000 or more obstacle vertices, we use a smaller obstacle set, including 20 obstacles with 10 vertices for each obstacle, to measure the runtime. We also re-implemented DBCLuC as DBCLuC*, which may differ from the original DBCLuC, because its authors lost their code. For the same input, the three algorithms found identical clusters. Figure 11 shows that AUTOCLUST+ and DBCLuC* are slower than DBRS+. In fairness, transfer time between AUTOCLUST+'s graphical user interface and its algorithm modules may increase its runtime to a certain degree. We did not compare with COD_CLARANS, because the code was unavailable to us. But 4 shows that AUTOCLUST+ has better runtime and clustering results than COD_CLARANS.

Figure 12 shows the runtime comparison with no obstacles, with 30 local obstacles, and with 30 complete obstacles. Local obstacles often have fewer edges than the complete obstacles. For example, for 200k points with no obstacles, the runtime is about 122 seconds; with local obstacles, it is about 145 seconds, but with complete obstacles it is about 158 seconds. Here using local obstacles instead of complete obstacles saves about 30% of the time required to deal with the obstacles.

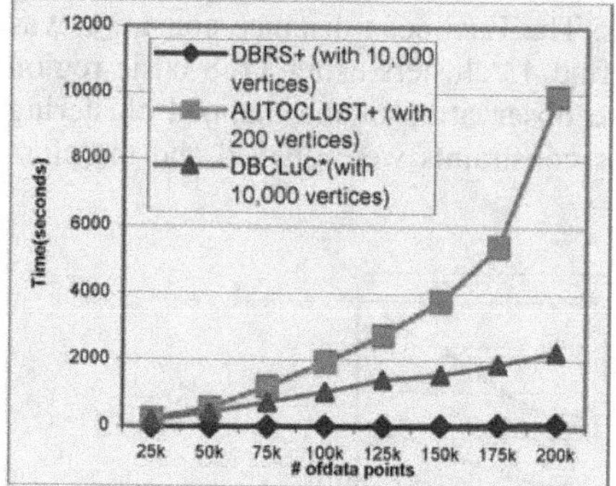

Fig. 11. Runtime Comparisons with AUTOCLUST+ and DBCLuC*.

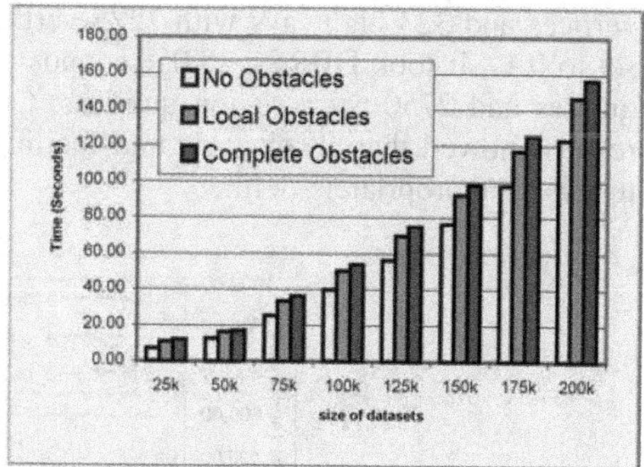

Fig. 12. Runtime Performance of Using Local Obstacles.

Table 2. Scalability Result in the Presence of Facilitators.

	10 facilitators (10,000 vertices)			20 facilitators (20,000 vertices)			30 facilitators (30,000 vertices)		
# of Points	Time (sec)	# of BRQ	# of ERQ	Time (sec)	# of BRQ	# of ERQ	Time (sec)	# of BRQ	# of ERQ
25k	55.96	4070	89155	98.28	4062	156196	143.88	4023	236229
50k	99.32	9097	125902	196.94	9031	254164	275.38	8925	360225
75k	157.03	13963	169216	278.89	13791	312108	393.75	13739	441205
100k	196.20	18472	181509	391.19	18315	412428	555.18	18373	600567
125k	262.33	22959	249421	449.70	22942	442839	668.43	22835	699355
150k	318.88	27198	253960	594.31	26969	599735	800.28	26843	766028
175k	381.97	32515	293909	688.28	32236	608704	975.93	31834	916522
200k	428.37	37473	290443	781.67	36999	621678	1125.74	36875	990377

Table 2 shows the results of our scalability experiments in the presence of facilitators. The size of the datasets is varied from 25k to 200k with 10% noise, and the number of the facilitators in these datasets is varied from 10 to 30. Each facilitator has 1000 vertices, so the number of facilitator vertices varies from 10,000 to 30,000. A ***basic region query (BRQ)*** is a region query on a point from the dataset, and an ***extra region query (ERQ)*** is a region query where a facilitator's exit is treated as a central point. In the experiments, we set e_c to 0.0, and thus every time a facilitator appears in a neighborhood, each exit of the facilitators will be checked for extra neighbors. The total number of region queries increases with the number of facilitator vertices and with the number of points. The runtime also increases accordingly. For datasets with both obstacles and facilitators, the runtime increases both with the number of obstacle vertices and the number of facilitator vertices.

Figure 13 graphs the runtime versus the Euclidean distance constant *e_c*. This experiment is based on datasets with 30 obstacles. When *e_c* is smaller, the facilitated distance is reduced and more exits must be checked for extra neighbors.

We also tested DBRS+ on a real dataset, consisting of the 1996 populations of western provinces of Canada, which are British Columbia, Alberta, Saskatchewan, and Manitoba. There were 189 079 records available for testing. We set *Eps* to 0.15 and *MinPts* as 10.

With no obstacles and facilitators, it took DBRS+ 10.83 seconds to find 65 clusters using 2164 region queries. Then we used rivers as obstacles and highways as facilitators to test DBRS+ with same *Eps* and *MinPts*. There were 175 rivers with 25432 vertices and 323 highways with 7225 vertices. The Euclidean distance const e_c was set to 0.1. It took DBRS+ 28.05 seconds to find 43 clusters using 1488 basic region queries and 2750 extra region queries. Close observation of the mapped clustering results showed that all obstacle and facilitator constraints were obeyed and the clusters were appropriately defined.

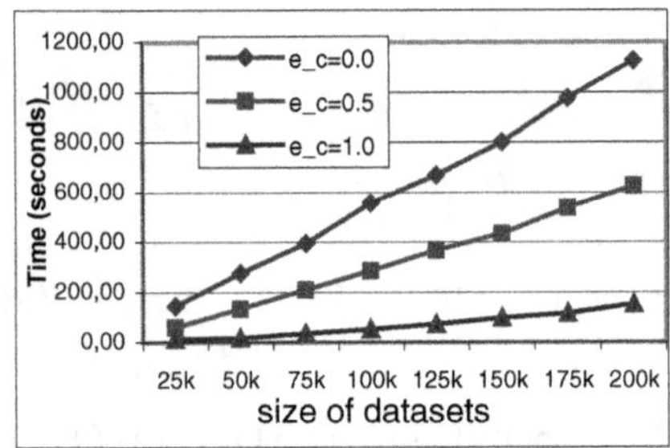

Fig. 13. Euclidean _constant (e_c) Vs. Runtime.

5 Conclusion

We proposed a new spatial clustering method called DBRS+, which aims to cluster spatial data in the presence of both obstacles and facilitators. It can handle datasets with intersected obstacles and facilitators. DBRS+ is faster than other algorithms. It can be used to find patterns in application where obstacles and facilitators appear, such as resource planning, marketing, and disease diffusion analysis.

References

1. Arvo, J. (ed.): Graphics Gems II. Academic Press, Boston (1991)
2. Cormen, T.H., Leiseron, C.E., Rivest, R.L., and Stein, C.: Introduction to Algorithms, The MIT Press, Boston (2001)

3. Ester, M., Kriegel, H., Sander, J., and Xu, X.: A Density-Based Algorithm for Discovering Clusters in Large Spatial Databases with Noise. In: Proc. of 2nd Intl. Conf. on Knowledge Discovery and Data Mining, Portland, OR (1996) 226-231
4. Estivill-Castro, V. and Lee, I. J.: AUTOCLUST+: Automatic Clustering of Point-Data Sets in the Presence of Obstacles. In: Proc. of Intl. Workshop on Temporal, Spatial and Spatio-Temporal Data Mining, Lyon, France (2000) 133-146
5. Estivill-Castro, V. and Lee, I. J.: AUTOCLUST: Automatic Clustering via Boundary Extraction for Mining Massive Point-Data Sets. In: Proc. of the 5th Intl. Conf. On Geocomputation (2000) 23-25
6. Han, J., Lakshmanan, L. V. S., and Ng, R. T.: Constraint-Based Multidimensional Data Mining. Computer 32(8) (1999) 46-50
7. Katayama, N. and Satoh, S.: The SR-tree: An Index Structure for High-Dimensional Nearest Neighbor Queries. In Proc. of ACM SIGMOD, Tucson, Arizona (1997) 369-380
8. Ng, R. and Han, J.: Efficient and Effective Clustering Method for Spatial Data Mining. In: Proc. of 1994 Intl. Conf. on Very Large Data Bases, Santiago, Chile (1994) 144-155
9. Tung, A. K. H., Han, J., Lakshmanan, L. V. S., and Ng, R. T.: Constraint-Based Clustering in Large Databases. In Proc. 2001 Intl. Conf. on Database Theory, London, U.K. (2001) 405-419
10. Tung, A.K.H., Hou, J., and Han, J.: Spatial Clustering in the Presence of Obstacles. In Proc. 2001 Intl. Conf. On Data Engineering, Heidelberg, Germany (2001) 359-367
11. Wang X. and Hamilton, H. J.: DBRS: A Density-Based Spatial Clustering Method with Random Sampling. In: Proc. of the 7th PAKDD, Seoul, Korea (2003) 563-575
12. Zaïane, O. R., and Lee, C. H.: Clustering Spatial Data When Facing Physical Constraints. In Proc. of the IEEE International Conf. on Data Mining, Maebashi City, Japan, (2002) 737-740
13. http://www.cs.uregina.ca/~wangx/synGeoDataGen.html

Text Mining for Finding Functional Community of Related Genes Using TCM Knowledge

Zhaohui Wu[1], Xuezhong Zhou[1], Baoyan Liu[2], and Junli Chen[1]

[1] College of Computer Science, Zhejiang University, Hangzhou, 310027, P.R.China
{wzh,zxz,chenjl}@cs.zju.edu.cn
[2] China Academy of Traditional Chinese Medicine, Beijing 100700, P.R.China
liuby@mail.cintcm.ac.cn

Abstract. We present a novel text mining approach to uncover the functional gene relationships, maybe, temporal and spatial functional modular interaction networks, from MEDLINE in large scale. Other than the regular approaches, which only consider the reductionistic molecular biological knowledge in MEDLINE, we use TCM knowledge(e.g. Symptom Complex) and the 50,000 TCM bibliographic records to automatically congregate the related genes. A simple but efficient bootstrapping technique is used to extract the clinical disease names from TCM literature, and term co-occurrence is used to identify the disease-gene relationships in MEDLINE abstracts and titles. The underlying hypothesis is that the relevant genes of the same Symptom Complex will have some biological interactions. It is also a probing research to study the connection of TCM with modern biomedical and post-genomics studies by text mining. The preliminary results show that Symptom Complex gives a novel top-down view of functional genomics research, and it is a promising research field while connecting TCM with modern life science using text mining.

Keywords: Text Ming, Traditional Chinese Medicine, Symptom Complex, Gene Functional Relationships

1 Introduction

The last decade has been marked by unprecedented growth in both the production of biomedical data and the amount of published literature discussing it. In fact, it is an opportunity, but also a pressing need as the volume of scientific literature and data are increasing immensely. Functional genomics and proteomics have been the foci in the post-genomic life science research. However, the reductionism and bottom-up approach are still the infrastructure of life science research since no holistic knowledge is reached. Traditional Chinese Medicine (TCM) is an efficient traditional medical therapy (e.g. acupuncture and Chinese Medical Formula), which embodies holistic knowledge with thousands of years' clinical practice. Symptom Complex (SC) is one of the core issues studied in TCM, which is a holistic clinical disease concept reflecting the dynamic, functional, temporal and spatial morbid status of human body. Moreover, several bibliographic databases have been curated in TCM institutes and colleges since 90s. One main database is the TCM bibliographic database[1] built by Information

[1] http://www.cintcm.com/index.htm

J.-F. Boulicaut et al. (Eds.): PKDD 2004, LNAI 3202, pp. 459–470, 2004.

Institute of China Academy of TCM, which contains about one half million records from 900 biomedical journals published in China since 1984, and 50% of the records have abstracts. These large amounts of literature storages with high quality will be a good text data sources for text mining.

How to connect TCM with modern life science and using the holistic knowledge of TCM to big biology research is an excited open question, which worthy of large research efforts. The experimental approaches to this objective are even extremely difficult since most TCM concepts are qualitative and systematic complicate. However, automated literature mining offers a yet untapped opportunity to induce and integrate many fragments of information gathered by researchers from multiple fields of expertise, into a complete picture exposing the interrelated relationships of various genes, proteins and chemical reactions in cells, and pathological, mental and intellective states in organisms. Many researches [1,2,3,4,5,6,7,8,9] have focused on the gene or protein name extraction, protein-protein interaction and gene-disease relationship extraction from biomedical literature (e.g. MEDLINE). This paper aims to provide a probing text mining approach to identify the gene functional relationships from MEDLINE using TCM knowledge, which is discovered in TCM literature. A simple but efficient bootstrapping method is used to facilitate the extraction of SC-disease relationships from TCM literature. We obtain the gene nomenclature information from the HUGO Nomenclature Committee[2], which has 17,888 approved gene symbols. The term co-occurrence is used to identify the relationships between disease and gene from MEDLINE. Then we get the SC-gene relationships by one-step inference, that is, to compute the genes and SCs with the same relevant disease. After that, the related gene networks of SCs are computed according to some graph algorithms such as[3]. As the ambiguities and polysemy of terminology using in literature, noise and false positive examples will surely be existed in the simple term match method. Currently, we only use co-occurrence frequency threshold to filter the low frequent relationships, and no other methods be considered for these problems.

The rest of article is structured as follows. To introduce the importance of connecting TCM with modern biomedical research, we give an overview of TCM and biomedical research, and some discussions are also proposed on why and how to do synergistic research of these two fields in Section 2. We review the related research work on biomedical literature mining, which have inspired the work of this paper, in Section 3. Bootstrapping technique and term co-occurrence are two approaches used in text mining process. We introduce them in Section 4. The preliminary text mining results to demonstrate the effective text mining process are proposed in Section 5. Finally, in Section 6, we take a conclusion and give the future work.

2 Modern Biomedical Research and Traditional Chinese Medicine

In the last century, modern biomedical research follows the reductionism and qualitative experimental approach, which is called molecular biology to anatomize the whole

---

[2] http://www.gene.ucl.ac.uk/nomenclature/

[3] http://www.research.att.com/sw/tools/graphviz/

human body to partial organs, tissues, cells and components. Great achievements have been acquired. Furthermore, the working draft of Human Genome sequence has been the crest of molecular biology. However, no other than this huge sequence data prompts the biologists to grasp the life in a reversed approach, which is called holism. System Biology [10] is such a research activity, which is an academic field that seeks to integrate biological data as an attempt to understand how biological systems function. By studying the relationships and interactions between various parts of a biological system (e.g. organelles, cells, physiological systems, organisms etc.) it is hoped that an understandable model of the whole system can be developed. System biology is a concept that has pervaded all fields of science and penetrated into popular thinking. As every biologist knows, there is still a long way to go before understanding biological systems in systematic approaches.

On the other hand, TCM studies the morbid state of human body by clinical practice and holistic quantitative cognitive process. TCM embodies rich dialectical thought, such as that of the holistic connections and the unity of Yin and Yang (two special concepts of TCM, which reflect the two essential states of human body and general material). Other than the disease concept in orthodox medicine, SC reflects the comprehensive, dynamic and functional disease status of live human body, which is the TCM diagnosis result of symptoms (TCM has developed a systematic approach to acquire the symptoms). In clinical practice, there will be several SCs on one specific disease while in different morbid stage. Also, one SC will occur in several different diseases. For example, *kidney YangXu* SC is a basic SC in TCM studies, which will refer to tens to hundreds diseases. Meanwhile, it has been proved that diabetes has several SCs such as *Kidney YangXu*, *YingYangLiangXu* and *QiYingLiangXu* SCs etc.. The therapies (e.g. Acupuncture and Chinese Medical Formula) based on SC are popular and effective in Chinese medical practice. Moreover, the characteristics of SC have much in common with that of genome and proteome such as polymorphism, individuality and dynamics etc.. We believe that SC will provide much more functional and holistic knowledge about the human body, which will largely support the functional genomics and proteomics research since TCM has the thousands of years experience to study the SC status of human body.

There are some studies such as that of Prof. Shen [11] on connecting TCM with molecule biology by experimental approaches. Prof. Shen has spent his fifty years to study the *kidney YangXu* SC at the molecular level. He found that *kidney YangXu* SC is associated with the expression of CRF (C1q-related factor). This paper will have some comparative study with the results of Prof Shen, and produce the novel functional gene relationships plus several novel relevant genes for *kidney YangXu* SC. It is very difficult and a long way to synergize the researches of these two fields in experimental approaches, because still no good approaches to model the dynamic and temporal qualitative organism concepts at molecular level. However, this paper suggests and proposes a synergistic approach of TCM and modern biomedical research using text mining techniques. We take the assumption that since biomedicine and TCM are both focusing on the study of disease phenomenon, we can regard the disease concept as the connecting point of biomedicine and TCM. There is huge literature and clinical bibliographic records on the research of SC-disease relationships in TCM, and the genetic pathology of disease is also intensively studied by modern biomedical research. Therefore, when analyze the relationship between SC and gene or protein through disease concepts, it maybe generate some novel hypothesis knowledge, which is not conceived in TCM or modern biomedical field. To have some probing research with system biology, we focus on connecting SC with gene to get novel gene temporal and spatial interaction information, which cannot be easily acquired by large-scale genomics or proteomics techniques.

3 Related Work on Biomedical Literature Mining

The post-genomic topics have been the main issue in biomedical and life science research. The human genomic sequence and MEDLINE propose the two most important shared knowledge sources, which will greatly contribute to the development and progress of big modern biology research. Knowledge discovery from all kinds of the huge biological data such as genomic sequence, proteomic sequence and biomedical literature (e.g. MEDLINE, the annotations of Swiss-Prot, GenBank etc.) has been the foci of bioinformatics research. Text mining from biomedical literature is one of the most important methods assisting for hypothesis driven biomedical research.

Prof. Swanson is one of the first researchers who use MEDLINE to find novel scientific hypothesis [12]. He proposed the concept of complementary literature probably with innovative knowledge[13] and provided a system named ARROWSMITH to help the knowledge discovery in medical literature of MEDLINE[14]. The work of Swanson gave a set up of knowledge discovery research in medical literature. Gordon and Lindsay got into the literature-based discovery research by applying Information Retrieval (IR) techniques to Swanson's early discoveries [15][16]. Weeber et al proposed the architecture of DAD-system, a concept-based Natural Language Processing system for PubMed citations, to discover the knowledge of drug and food. They claimed that the system could generate and test Swanson's novel hypothesis [17].

Fukuda et al [6] are one of the first researchers using information extraction techniques to extract protein names from biological papers. Since then, the extraction of terminological entities and relationships from biomedical literature is the main research efforts in biomedical literature mining. The research instances include words/phrases disambiguation[1], gene-gene relationships [1][8], protein-protein [3,4,5,7,9], and gene-protein interactions or specific relationships between molecular entities such as cellular localization of proteins, molecular binding relationships[18], and interactions between genes or proteins and drugs[19]. Bunescu et al [20] have a comparative study on the methods such as relational learning, Naïve Bayes, SVM etc. in biomedical information extraction. Hirschman et al [21] have a survey of biomedical literature data mining from natural language processing. They argued that there need a challenge evaluation framework to boost the promising researches. Yandell[22] takes a felicitous discussion of biomedical text mining as a new emerging field-biological natural language processing that will do great help to biology research. Sehgal [23] uses Mesh headings to create concept profiles to compute the similar genes and drugs. Perez-Iratxeta [24] uncovers the relation of inherited disease and gene by Mesh terms of MEDLINE and RefSeq. While, Freudenberg et al [25] got disease clusters according to the fuzzy similarity between phenotype information of disease, which is extracted from OMIM (Online Mendelian Inheritance in Man) then predict the possible disease relevant genes based on the disease clusters.

Most of the recent related work is focus on extraction of terminology concept or concept relationship knowledge, which has been existed in the literature. Moreover, natural language processing techniques (NLP) are preferred since the knowledge is conceived in the sentences. Such analysis is not only computationally prohibitive but also error prone when using NLP, and it is not applicable for large-scale literature mining. We address the large-scale literature analysis by very simple method, which only considers the co-occurrence of terms. Currently, the dictionary-based term extraction method is used. The most similar work is that of Jenssen et al [8] and Wilkinson [26]. They provide the simple approach to build large scale literature network of genes while only considering the gene co-occurrence as the view of gene interaction and dictionary based term extraction method is used. Furthermore, Wilkinson follows the method of [27] to resolve some of the gene terminological ambiguities. However, in this paper we take advantage of TCM knowledge (e.g. SC and the SC-disease relationships) to consider the related genes by a temporal and spatial holistic perspective. The method is based on the assumption that the genes relevant to the same SC will have some temporal or spatial connections since SC reflects the holistic functional state of morbid human body. Currently, there are two aims of the work of this paper. One is to find the relevant genes of SC, and the other is to find the communities of related genes. Bootstrapping technique is used to extract the disease names from TCM literature since no Chinese disease dictionary is available and the irregular using of terminology in clinical literature. The experimental results show that bootstrapping is much suitable to TCM terminology name extraction. The communities of related genes can be modeled as sub-graphs as in [8] and [26]. We believe that SC as a core TCM clinical concept will give novel approach to discovery of gene functional relationships from literature. This paper gives the framework of this approach and some preliminary results.

4 Bootstrapping and Term Co-occurrence

As many modifiers, which reflect the genre, state or characteristic of a specific clinical disease, are used in clinical literature representation, and irregular clinical disease names are also popular, the clinical disease names are far more various than standard disease dictionary such as that of TCM headings terminology. Dictionary based automatic name extraction cannot meet for this situation. We use bootstrapping to extract the disease names from TCM literature, and then extract the relationship of SC and disease from TCM bibliographic records based on metadata of TCM headings and SC standard terminological database, which includes about 790 standard SC names. The term bootstrapping here refers to a problem setting in which one is given a small set of seed words and a large set of unlabeled data to iterated extract the objective patterns and new seeds from free texts. Current work has been spurred by two papers namely that of Yarowsky [28] and Blum [29]. Bootstrapping methods have been used to automatic text classification [30], database curation[31][32] and knowledge base construction[33] from World Wide Web. Next we introduce the bootstrapping method of this article in detail.

Bootstrapping is an iterative process to produce new seeds and patterns when provided with small set of initial seeds. The initial seed information gives the objective semantic type, which bootstrapping technique should extract from the text. In this paper, we aim to extract the terminological name of disease. So some initial disease name seeds are given before the iterative procedure is set up (the initial seeds has 19 disease names). How to define, evaluate and use pattern is one of the core issue of bootstrapping technique. Since TCM literature is written in Chinese, the process of pattern will surely be different from that of bootstrapping used in English like literature. This paper takes a simple but efficient pattern definition and evaluation method, and uses a search-match method without any shallow parsing processing. We define the TCM terminological pattern as a 5-tuple as

$$P =< \text{lPstr}, \text{TermType}, \text{rPstr}, \text{RefCount}, \text{FreqCount} > \quad (1)$$

In which *lPstr* represents the left Chinese string of the seed tuples, *rPstr* represents the right Chinese string of the seed tuples (we only extract two and three Chinese character to form the left and right string respectively called *Two-Left and Three Right pattern* (*TLTR*)), *TermType* represents the semantic type of seed name such as disease and Chinese Medical Formula etc., *RefCount* and *FreqCount* represent the number of seeds produced by pattern and the occurrence of pattern respectively (we also define the corresponding *RefCount* and *FreqCount* of seed).

To keep the bootstrapping procedure be robust in extraction of high quality new seed tuples, we use a dynamic bubble up evaluation method to assure the high quality pattern contribute to the iterative process. Currently, *RefCount* is computed before the next iteration and considered as the only quality criterion to keep the pattern (when *RefCount* is above a pre-assigned threshold) to attend the next step bootstrapping procedure. It is very simple while produce patterns from seeds. Next, we give some descriptions of the procedure of patterns to new seeds. First, the *lPstr* string is used as the search string after the record texts are split by regular punctuations and get the sentences, which have *lPstr* string as their sub-string. Second, the *rPstr* string is used to match the sentences of first step. Finally, we extract the strings between the *lPstr* and *rPstr* string as the objective new seeds. Meanwhile, we refresh the value of *RefCount* and *FreqCount* of pattern. It is experimentally concluded that *TLTR* is very robust and efficient pattern in TCM terminological term extraction as Table 1 shows. The recall of bootstrapping in TCM literature may be due to the iterative threshold, data source quantity and quality. The bootstrapping recall of disease on TCM literature of 2002 (*WX_2002*) is obviously better than that of other years. One reason is that *WX_2002* has high quality data and most of the records have abstracts, but the number of abstracts of the others is small.

Based on the latest version approved gene symbol vocabulary from HUGO in Feb. 2004, and the bootstrapped disease term database, we develop a Perl program to search the title and abstract fields of online MEDLINE to acquire the disease and gene relevant PubMed citations. Before search MEDLINE, Chinese disease name is translated partial automatically to formal English disease name according to the TCM headings database and manually check by TCM terminological expert is needed when no TCM headings of disease exist. Followed we induce the disease and gene term co-occurrences by computing the same PubMed identifiers between each disease and gene term. Currently, no disambiguation method is used and alias gene symbols are not yet considered to search MEDLINE. Fig.1 shows the related data collections used and the related modules for generating novel hypothesis from modern biomedical and TCM literature.

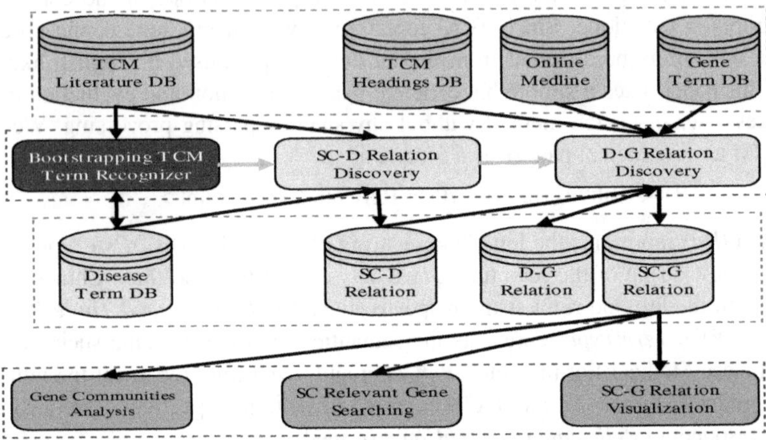

Fig. 1. The relevant data collections and related modules supporting knowledge discovery in biomedical literature. Blue box indicates a bootstrapping term recognizer to extract clinical disease terms. Green boxes represent SC-Disease and Disease-Gene relationships extraction modules. Orange boxes represent the modules that will be used to support biomedical research.

Table 1. Bootstrapping-based TCM terminological name recognition results, the TCM bibliographic database of year 1999, 2001, 2002 and 2003 has (38,937), (43,266), (44,315) and (16,151) records, respectively. P/S is the abbreviation of Pattern/Seed.

Years	Term Type	Iterated P/S Threshold	Iterated P/S Number	Precision	Recall	Pattern Number	Seeds Number
2002	Disease	8/8	8/20	92.6%	48.6%	3153	1018
2002	Disease	7/7	10/20	98.2%	55.5%	3153	1097
2002	Disease	6/6	73/109	90.8%	80.7%	5807	1753
2001	Disease	7/7	9/19	98.9%	30.1%	2430	915
1999	Disease	7/7	4/19	99.4%	35.3%	1459	853
2003	Disease	4/4	27/39	97.6%	21.1%	1684	416

5 Results

We have compiled about two and half million disease relevant PubMed citations, and 1,479,630 human gene relevant PubMed citations in local database. All of the citations are drawn from online MEDLINE by a Perl program. Meanwhile, we get about 1,100

SC-disease relationships from TCM literature (*WX_2002*). Fig.2 gives a picture of the concept associations supported by existed biomedical information. The SC relevant genes are all novel scientific discovery since there are no experiments to study the connections, and obviously, it is vital to boost SC from qualitative sense to quantitative experimental research. Furthermore, the related genes of the same SC will have some functional interactions, which is very different from the literature network built on gene term co-occurrence. And in a way, the appropriate subsets of the specific related genes have some selection of the huge literature gene network conceived in MEDLINE (e.g. the PubGene). To have a demonstration of the work of this article, we take the *kidney YangXu* SC as an instance since it is an important SC involving caducity, neural disease and immunity etc. in TCM, and has been studied by experimental approaches. Table 2 lists the 71 related genes of *kidney YangXu* SC (we exclude the "T" gene symbol because it seems no much information can be given by it). Inspired by the previous studies of Prof. Shen [11], we have an analysis whether the text mining method could find some novel knowledge from MEDLINE compared with experimental approaches. That is, if we can find the relevant genes such as CRF of *kidney YangXu SC*. Because the capacity of our gene symbols vocabulary is limit and alias gene symbols have not been considered (e.g. CRF is an alias name of CRH and is not in our gene symbol vocabulary), we have a confirmation process based on the work of PubGene [8] to have a relative complete view. Before the demonstration, we propose our basic assumption that polygenic etiology or gene interaction network will contribute to the phenotype of SC. We follow the next several steps to verify the efficiency of the text mining research. First, we select some important genes namely CRP (C-reactive protein, pentraxin-related), CRH (corticotropin releasing hormone), IL10 (interleukin 10), ACE (angiotensin I converting enzyme), PTH (parathyroid hormone), MPO (myeloperoxidase) in *kidney YangXu* SC. Second, we search the PubGene for subset network using each of the above genes. We get the subset networks as Fig 3 shows. The third step is analyzing the extracted knowledge. Now suppose that we don't know CRF (literature alias name of CRH) is a relevant gene of *kidney YangXu* SC. By analyzing the six subset networks in the left part and the CRF subset network of Fig 3, we may in a way get the hypothesis that CRF is some relevant to *kidney YangXu* SC, because that the subset network, which reassembled with the gene nodes such as IL10, CRAT, CRF/CRH/ ACE/MPO/PTH etc., constitute possible functional gene communities that contribute to *kidney YangXu* SC. No existed literature reporting the relationship between CRF and *kidney YangXu* SC is used to generate the novel knowledge since CRF is not in our gene vocabulary. It is excited that this simple demonstration has shown the primary text mining results will largely decrease the labor in molecular level SC research. The presented work of this paper gives a tool for the TCM researchers to rapidly narrow their search for new and interesting genes of a specific SC. Meanwhile, we give the specific functional information to the literature networks and divide the large literature networks to functional communities (e.g. the community containing IL10, CRAT, CRH/CRF etc. genes for *kidney YangXu* SC), which cannot be identified in the current PubGene. Moreover, through SC perspective, we can easily have a subset selection or explanation of giant literature based gene network. And it is promising that even the low gene term co-occurrence network is functionally identified by our approach.

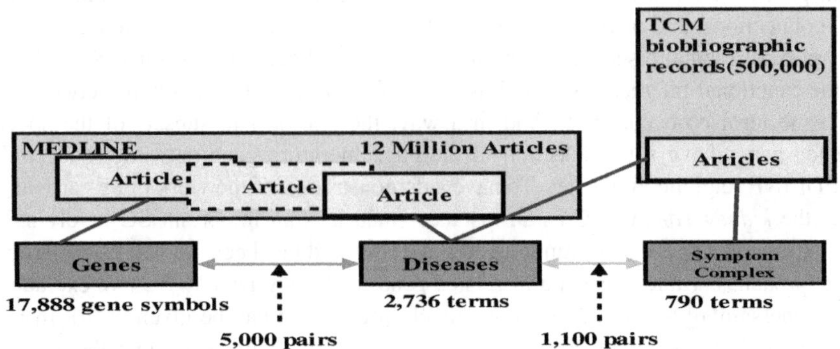

Fig. 2. Components used for identifying associations between genes and Symptom Complex. Green boxes represent databases. Orange boxes indicate the associated concepts and gray arrows represent the association used. Disease is the connecting point concept between gene and Symptom Complex.

Table 2. The 72 relevant genes of *kidney YangXu* SC filtered by total co-occurrence is above 10 or the number of relevant diseases is above 2. NP—526—2 represents the gene NP has co-occurred 526 times with the two diseases in *kidney YangXu* SC. Now there are only 7 relevant diseases of *kidney YangXu* SC, which have related genes in MEDLINE.

ID	Relevant Genes	ID	Relevant Genes	ID	Relevant Genes
1	NP--526--2	25	FCP--18--1	49	GC--11--3
2	IV--137--5	26	TG--18—4	50	IKBKAP--11--1
3	CCR3--61--2	27	MTHFR--18--2	51	EGF--11--4
4	ACE--56--3	28	FAT--18--1	52	AS--9--4
5	CD68--53--4	29	C3--17—6	53	CRP--9--5
6	C5--42--2	30	PON1—17--1	54	IL4--9--3
7	TNF--42--6	31	AHR--16--1	55	MCP--8--4
8	SD--34--5	32	PI3--16—2	56	PCNA--8--4
9	ALK--31--1	33	TAP1--16--1	57	HP--7--4
10	NOS1--31--1	34	HR--15—2	58	TF--7--4
11	SRS--31--2	35	STAT6--15--2	59	VEGF--6--4
12	CD34--30--3	36	STAR—15--1	60	MIP--5--3
13	AA--29--3	37	MPO--15--4	61	NPY--5--3
14	CD28--29--2	38	CD72--14--1	62	SDS--5--3
15	CD4--25--6	39	PTH--14--3	63	PC--5--3
16	CD14--24--3	40	NPHS2--14--2	64	CD2--4--3
17	MLN--24--1	41	LTB--13--3	65	CP--4--3
18	CXCR3--23--1	42	SEA--13--1	66	MIF--4--3
19	CD80--22--1	43	CCR2—12--2	67	DBP--4--3
20	FH--22--2	44	SC--12—4	68	HD--4--3
21	GSTP1--21--2	45	PAX2—12--2	69	EPO--4--3
22	TAT--21--3	46	IL13--12--2	70	CD63--3--3
23	CD86--19--1	47	EGFR—12--2	71	PGM1--3--3
24	ACTN4--18--2	48	CXCR4--11--1	72	T—161--7

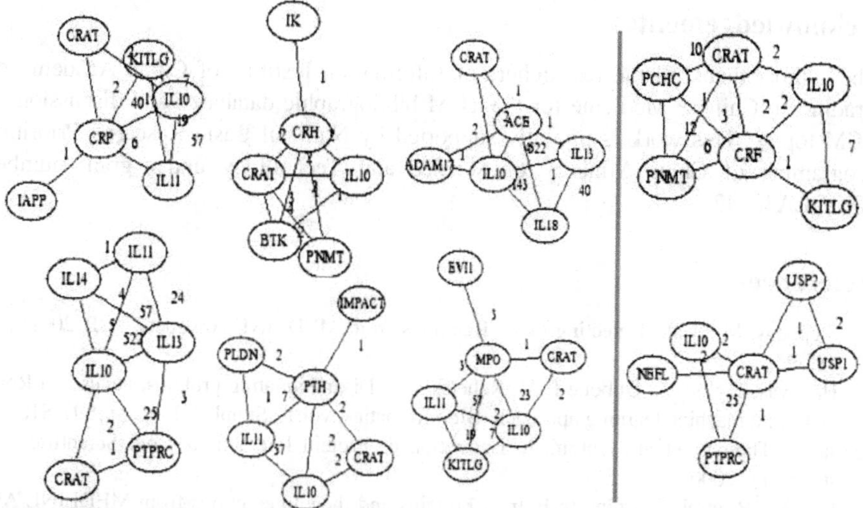

Fig. 3. The six subset networks of each selected genes (left part of the vertical yellow line). Gene CRAT, which in all of the six subset networks, may be a novel relevant gene of *kidney YangXu* SC. The right part gives the subset network of the already known relevant gene CRF, and the subset network of CRAT.

6 Concluding Remarks and Future Work

The exponential growth of MEDLINE and GenBank is rapidly transforming bio-NLP from a research endeavor into a practical necessity. Most of the recent biomedical literature mining studies are focused on information management, and the development of such tools is a necessary and laudable goal. Nevertheless, if bio-NLP (natural language processing) is to achieve its full potential, it will have to move beyond information management to generate specific predictions that pertain to gene function that can be verified at the bench. The synergistic use of sequence and text to extract latent information from the biomedical literature holds much promise in this regard. In this paper, we give a more novel text mining approach to generate large-scale biomedical knowledge by using TCM holistic knowledge. Bootstrapping techniques is used to extract the Chinese disease names from literature, and simple term co-occurrence is used to extract the relationship information. We believe that much more valuable hypothesis will be generated by enhancing the synergy of TCM and modern life science. As protein-protein interactions are central to most biological processes, the systematic identification of all protein interactions is considered a key strategy for uncovering the inner workings of a cell. The future work will focus on studying the protein-protein interactions using SC. Also we will try to use multistrategy learning method to extract more reliable gene-disease or protein-disease relationships from MEDLINE. Because many new terminological names are used in literature, but absent in vocabulary, vocabulary-based name extraction is not applicable to extract the latest relationships. Therefore, we will study on the novel using of bootstrapping techniques to gene or protein name extraction.

Acknowledgements

The authors thank all the researchers of Information Institute of China Academy of Traditional Chinese Medicine for the TCM bibliographic databases and discussion of TCM topics. This work is partially supported by National Basic Research Priorities Programme of China Ministry of Science and Technology under grant number 2002DEA30042.

References

1. Stephens, M. et al., Detecting Gene Relations from MEDLINE Abstracts. PSB 2001, pp: 483-95.
2. Hatzivassiloglou V., Duboue P.A., Rzhetsky A., Disambiguating proteins, genes and RNA in Text: a machine learning approach. Bioinformatics, vol.17 Suppl. 1 2001, pp:S97-S106.
3. James Thomas, et al., Automatic Extraction of Protein Interactions from Scientific Abstracts. psb2000.
4. Bunescu R. et al, Learning to Extract Proteins and their Interactions from MEDLINE Abstracts. Proceedings of ICML-2003 Workshop on Machine Learning in Bioinformatics, Washington DC, August 2003,pp: 46-53.
5. Marcotte E.M., Xenarios L. and Eisenberg D., Mining literature for protein-protein interactions. Bioinformatics,Vol. 17 no.4 2001,pages 359-363.
6. Fukuda K., et al, Toward information extraction: Identifying protein names from biological papers. In Proc. PSB 1998, Maui, Hawaii, January 1998, pp: 707-718.
7. Blaschke M. et al, Automatic Extraction of Biological Information from Scientific Text: Protein-Protein Interactions, Proc. of ISMB'99, pp. 60-67.
8. Jenssen T.-K., et al, A literature network of human genes for high-throughput analysis of gene expression. Nature Genetics 28. 21-28(2001).
9. Humphreys, K., Demetriou, G., & Gaizayskas, R. Two applications of information extraction to biological science journal articles: enzyme interactions and protein structures. Pac. Symp. Biocomput.5. 505-516(2000).
10. Ideker, T., Galitski, T., Hood, L., A new approach to decoding life: Systems Biology. Annu. Rev. Genomics Hum.Genet. 2001,2.
11. Shen Ziyin, The continuation of kidney study. Shanghai, Shanghai scientific & Technical Publishers.1990.3-31.
12. Swanson, D.R., Two medical literature that are logically but not bibliographically connected, Journal of the American Society for Information Retrieval, 1987,38 (4), 228-233.
13. Swanson, D.R, Complementary structures in disjoint science literature. SIGIR-91, pp: 280-289.
14. Swanson, D.R. and Smalheiser, N.R., An interactive system for finding complementary Literature: a stimulus to scientific discovery, Artificial Intelligence, 1997, 91, 183-203.
15. Gordon M.D., Lindsay R.K., Toward discovery support systems: A replication, re-examination, and extension of Swanson's work on literature-based discovery of a connection between Raynaud's and fish oil. J Am Soc Inf Sci 1996; 47 (2):116–128.
16. Lindsay RK, Gordon MD. Literature-based discovery by lexical statistics. J Am Soc Inf Sci 47 (2): 116–128.
17. Weeber, M. et al. Text-based discovery in biomedicine: the architecture of the DAD-system. In: Proceedings of AMIA, November 4-8, 2000, 903-907.

18. Rindflesch, T.C., Rayan, J.V. & Hunter, L. Extracting molecular binding relationships from biomedical text. Association for Computational Linguistics, Seattle, 2000,pp. 188-195.
19. Rindflesch, T.C., et al, EDGAR: Extraction of drugs, genes and relations from the biomedical literature. PSB 2000, 5:514-25.
20. Bunescu R., et al, Comparative Experiments on Learning Information Extractors for Proteins and their Interactions, Special Issue in JAIM on Summarization and Information Extraction from Medical Documents.25, August 2003.
21. Hirschman L., et al., Accomplishments and challenges in literature data mining for biology. Bioinformatics Review.Vol.18 no.12 2002, Pages 1553-1561.
22. Yandell M.D. and Majoros W.H., Genomics and Natural Language Processing. Nature Reviews Genetics, 2002, 3: 601-610.
23. Sehgal A., Qiu X.Y., Srinivasan P., Mining MEDLINE Metadata to Explore Genes and their Connections, SIGIR-03 Workshop on Bioinformatics.
24. Perez-Iratxeta C., Bork P. & Andrade M. A., Association of genes to genetically inherited diseases using data mining, letter to nature genetics, volume 31, july 2002.
25. Freudenberg J. and Propping P., A similarity-based method for genome-wide prediction of disease-relevant human genes. Bioinformatics, Vol. 18 Suppl.2 2002, Pages S110-S115.
26. Wilkinson D. and Huberman B. A., A Method for Finding Communities of Related Genes. Proc. Natl. Acad. Sci. USA, 10.1073/pnas.0307740100.
27. Adamic L.A., et al., A Literature Based Method for Identifying Gene-Disease Connections.Proceedings of the IEEE Computer Society Conference on Bioinformatics, August 14-16, 2002, pp: 109.
28. Yarowsky D., Unsupervised word sense disambiguation rivaling supervised methods. ACL-95, pp. 189–196.
29. Blum A. and Mitchell T., Combining labeled and unlabeled data with co-training. In COLT: Proceedings of the Workshop on Computational Learning Theory. 1998.
30. Jones, R., et al. Bootstrapping for Text Learning Tasks. In IJCAI-99 Workshop on Text Mining: Foundations, Techniques and Applications.
31. Riloff E., Jones R., Learning Dictionaries for Information Extraction by Multi-level Bootstrapping, AAAI-99, pp. 474-479.
32. Brin, S. Extracting Patterns and Relations from the World Wide Web. WebDB Workshop at EDBT-98.
33. Craven, M. et al. Learning to Extract Symbolic Knowledge from World Wide Web. AAAI-98, pp: 509-516.

Dealing with Predictive-but-Unpredictable Attributes in Noisy Data Sources

Ying Yang, Xindong Wu, and Xingquan Zhu

Department of Computer Science, University of Vermont, Burlington VT 05405, USA
{yyang,xwu,xqzhu}@cs.uvm.edu

Abstract. Attribute noise can affect classification learning. Previous work in handling attribute noise has focused on those predictable attributes that can be predicted by the class and other attributes. However, attributes can often be predictive but unpredictable. Being predictive, they are essential to classification learning and it is important to handle their noise. Being unpredictable, they require strategies different from those of predictable attributes. This paper presents a study on identifying, cleansing and measuring noise for predictive-but-unpredictable attributes. New strategies are accordingly proposed. Both theoretical analysis and empirical evidence suggest that these strategies are more effective and more efficient than previous alternatives.

1 Introduction

Real-world data are seldom as perfect as we would like them to be. Except in the most structured environment, it is almost inevitable that data contain errors from a variety of corrupting processes, such as acquisition, transmission and transcription [11, 13]. The corrupted data, namely noise, usually have adverse impact on interpretations of the data, models created from the data, and decisions made based on the data [7]. Since a manual process is laborious, time consuming and itself prone to errors, effective and efficient approaches that automate noise handling are necessary [8].

This paper handles noise in the context of classification learning, which plays an active role in machine learning. In classification learning, data are composed of *instances*. Each instance is expressed by a vector of *attribute* values and a *class* label. We further differentiate attributes into three types: unpredictive, predictive-and-predictable and predictive-but-unpredictable. *Unpredictive* attributes are futile for or irrelevant to predicting the class. They can be discarded by feature selection methods prior to the learning. *Predictive* attributes are useful for predicting the class, among which *predictable* ones can be predicted by the class and other attributes while *unpredictable* ones cannot. We argue that since predictive attributes contribute to classification learning, it is important to handle their noise. However, because predictable attributes and unpredictable attributes have different natures, they require different strategies for noise handling. For convenience of expression, an 'attribute' throughout this paper implies a predictive attribute unless otherwise mentioned.

J.-F. Boulicaut et al. (Eds.): PKDD 2004, LNAI 3202, pp. 471–483, 2004.

Generally, there are two types of noise, attribute noise and class noise. Most previous efforts are engaged in handling class noise, such as for contradictory instances (the same instances with different class labels) or misclassifications (instances labeled with a wrong class). Various achievements have been reported [2–6, 14, 15]. In comparison, much less attention has been paid to attribute noise. Ironically, attribute noise tends to happen more often in the real world. For example, if noise comes from entry mistakes, it is very likely that the class has fewer errors since the people involved know that it is the 'important' value and pay more attention to it [11].

Among few publications on handling attribute noise, an important contribution is LENS [7]. It aims at presenting a 'complete' understanding that helps identify, correct or prevent noise by modelling the generation of clean data, the generation of noise and the process of corruption, under the assumption that the noise generation and the corruption process have learnable underlying structures. If the assumption does not hold, the authors indicate that LENS has limited utility. Different from LENS, another often-cited achievement, polishing [12], deals with attribute noise without knowing its underlying structures. As we will detail later, polishing excels in handling noise for predictable attributes. A predictable attribute is one that can be predicted by the class and other attributes. We suggest that attributes in real-world applications can frequently be otherwise. It is because normally attributes are collected as long as they contribute to predicting the class. Whether or not one attribute itself can be predicted is often not a concern.

Hence, with all due respect to previous achievements, we suggest that the picture of handling attribute noise is incomplete. It is an open question how to address unpredictable attributes when their underlying noise structures are unavailable. These understandings motivate us to explore appropriate strategies that identify, cleanse and measure noise for predictive-but-unpredictable attributes. In particular, Section 2 introduces the background knowledge. Section 3 proposes *sifting* to identify noise. Section 4 discusses *deletion*, *uni-substitution* and *multi-substitution* to cleanse the identified noise. Section 5 studies *literal equivalence* and *conceptual equivalence* to measure noise. Section 6 empirically evaluates our new strategies. Section 7 gives a conclusion and suggests further research topics.

2 Background Knowledge: Polishing

Polishing [12] is an effective approach to handling noise for predictable attributes. When identifying noise for an attribute A, polishing swaps A with the original class, and uses cross validation to predict A's value for each instance. Those values that are mis-predicted are identified as noise candidates. It then replaces each noise candidate by its predicted value. If the modified instance is supported by a group of classifiers learned from the original data, the modification is retained. Otherwise, it is discarded. Polishing assumes that since one can predict the class by attribute values, one can turn the process around and use the class

Table 1. Prediction accuracies of classes and attributes in polishing's data. For instance, mushroom has 100.0% in the 'Class' column indicating that the prediction accuracy for its class is 100.0%. Mushroom has 10 in the column [90%,100%] of 'Attributes' indicating that it has 10 attributes whose prediction accuracies fall into [90%,100%].

Dataset	Class	Attributes			
		[90%,100%]	[70%,90%)	[50%,70%)	[0%,50%)
mushroom	100.0%	10	4	5	3
soybean	91.5%	27	3	3	2
led-24	100.0%	7	0	17	0
vote	96.3%	3	10	3	0
audiology	77.9%	58	9	0	2
promoters	81.1%	0	0	3	54

together with some attributes to predict another attribute's values as long as this attribute is predictive.

However, this 'turn around' can be less justifiable when unpredictable attributes are involved. For example, the often-cited monk's problems from the UCI data repository [1] have 6 attributes $A_1, \ldots, A_6$ and a binary class. One underlying concept is $(A_4 = 1$ and $A_5 = 3)$ or $(A_2 \neq 3$ and $A_5 \neq 4) \Rightarrow C = 1$; otherwise $C = 0$. Accordingly A_5 is a predictive attribute. Although the class can be predicted with 100% accuracy by C4.5trees [9] using 10-fold cross validation, the prediction accuracy for A_5 is only 35% because it has multiple values mapped to a single class. Directed by such a low accuracy, the noise identification has a strong potential to be unreliable. More datasets involving predictive-but-unpredictable attributes will be shown in Section 6.

Polishing has reported favorable experimental results. Nonetheless, they apply to predictable attributes. For each of polishing's datasets, Table 1 summarizes prediction accuracies for its class and attributes[1]. In all datasets except 'promoters', highly predictable attributes dominate the data[2]. In many cases, attributes are even more predictable than the original class. The 'promoters' data, which are not dominated by predictable attributes, produce a less favorable result for polishing.

These observations by no means devalue polishing's key contribution to handling predictable attributes. Nonetheless they raise the concern of polishing's suitability for unpredictable attributes and inspire our further study.

3 Identifying Noise

We propose a *sifting* approach to identifying noise for unpredictable attributes. For a predictable attribute, there exists a value whose probability given an in-

[1] We summarize them because many datasets have too many attributes to be listed individually. For instance, audiology has 69 attributes. Each accuracy results from C4.5trees using 10-fold cross validation, as polishing did.

[2] For the 'vote' dataset, although 17 attributes fall in [50%,70%], these attributes are randomly-added unpredictive attributes.

stance is high enough to dominate alternative values. For an unpredictable attribute, there is often no such dominating value. Instead, multiple values may be valid given an instance.

To identify noise for an attribute, polishing predicts what the clean value is and identify other values suspicious. This strategy is less appropriate if the attribute is unpredictable. For example, we have instances to represent apples and berries. One attribute is color. Apples can be green, yellow or red. Berries can be blue, black or red. If green apples happen to have a slightly higher occurrence, polishing will identify valid instances like $< \cdots, yellow, apple, \cdots >$ and $< \cdots, red, apple, \cdots >$ suspicious since the predicted color of an apple is green.

Neither is it adequate to *individually* identify noise for each attribute, where an attribute is separated from the others, swaped with the class and predicted by other possibly noisy attributes. Suppose an instance to be $< \cdots, black, apple, \cdots >$. When the attribute 'color' is under identification, its value 'black' will be identified suspicious since it should be 'green' given 'apple'. Meanwhile, when the attribute 'fruit' is under identification, its value 'apple' will be identified suspicious since it should be 'berry' given 'black'. Thus both values are identified. However, it is very likely that only one is real noise. The original instance can be a 'black berry' or a 'green apple'. This *compounded suspicion* is caused by identifying noise according to noisy evidence.

Accordingly sifting evaluates whether the pattern presented by a *whole* instance is suspicious instead of judging isolated values. It identifies an instance suspicious *only when* this instance does not satisfy *any* pattern that is learned with certain confidence from the data. We name this strategy 'sifting' because it takes the set of learned patterns as a sifter to sift instances. Instances that match any pattern may go through while the remaining are identified suspicious. By this means, an unpredictable attribute is not forced to comply with a single value. Instead, its multiple valid values can be allowed. For instance, colors of green, yellow and red are all allowable for an apple since we can learn those patterns. A black apple will be identified suspicious since its pattern is very unlikely, and this anomaly is efficiently identified in one go.

Algorithm 1 shows an implementation of *sifting*. Please be noted that we are dealing with situations where the underlying structure of noise (if there is any at all) is not available. This implies that if there is any learnable pattern, it reflects the knowledge of clean data. Furthermore, the learned patterns are in terms of classification rules (with the class, but not any attribute, on the right hand side).

We expect sifting to obtain a subset of instances with a high concentration of noise. We then forward this subset to the next process: cleansing.

4 Cleansing Noise

One should be very cautious when coming to cleanse an instance. For example, the data can be completely noise-free but an inappropriate rule learner is employed. Hence, we suggest that noise cleansing is conducted only when the data's

Algorithm 1: Identifying noise for unpredictable attributes

```
Input: possibly noisy dataset D; a rule learner L;
Output: a set of suspicious instances IS;
Begin
  RS = a set of rules that L learns from D;
  foreach Instance I_i ∈ D
    flag = 0;
    foreach Rule R_j ∈ RS
      if I_i satisfies R_j   {flag = 1; break;}
    if flag == 0   {push I_i into IS;}
End
```

genuine concept is learnable; practically, only when a high prediction accuracy is achievable.

For predictive attributes, it is sensible to only permit changes towards the predicted values, as polishing does. For unpredictable attributes, usually no value can be predicted with such a high accuracy as to exclude alternative values. Accordingly, we study three cleansing approaches, *deletion*, *uni-substitution* and *multi-substitution.*

As delineated in Algorithm 2, deletion simply deletes all identified noisy instances. At first sight, it most likely causes information loss. Nevertheless, this intuition needs to be verified. Uni-substitution cleanses an instance by the rule that minimizes the number of value changes. If there are multiple such rules, uni-substitution filters them by their quality indicators[3] and chooses a single one to cleanse the suspicious instance. Hence uni-substitution maintains the original data amount. Multi-substitution is the same as uni-substitution except at the final stage. If finally several rules are still equally qualified, multi-substitution produces multiple cleansed instances, each corresponding to a rule, and substitutes all of them for the suspicious instance. In this way, it may increase the data amount. But it has a merit that retrieves all valid values.

5 Measuring Noise

It is important to measure how noisy the corrupted data are if compared against its clean version. Otherwise, it is difficult to evaluate the effectiveness of a noise handling mechanism. A common measurement is *literal equivalence* [12, 15]. We believe that it is often of limited utility and propose *conceptual equivalence* instead.

5.1 Literal Equivalence

Suppose CD is a clean dataset and is corrupted into a noisy dataset ND. When measuring noisy instances in ND, literal equivalence conducts a literal comparison. Two common mechanisms are *match-corresponding* and *match-anyone.*

[3] Normally the rule learner attaches quality indicators to each rule. For example, we employ C4.5rules that attaches each rule with *confidence*, *coverage* and *advantage.*

Algorithm 2: Cleansing noise for unpredictable attributes

```
Input: a set of suspicious instances IS identified by a rule set RS in a possibly noisy dataset D;
Output: a cleansed dataset D;
Begin
  if cleanse == deletion
    D = D − IS;
    return D;
  foreach Instance I_i ∈ IS
    foreach Rule R_j ∈ RS
      changes = numbers of attribute values to be changed to make I_i satisfy R_j;
    candidates = set of rules with the smallest changes;
    // Further filter by rules' quality
    if |candidates| > 1  {candidates = rules in candidates with highest confidence;}
    if |candidates| > 1  {candidates = rules in candidates with highest coverage;}
    if |candidates| > 1  {candidates = rules in candidates with highest advantage;}
    if cleanse == uni-substitution
      I' = change I_i to satisfy the first rule^a in candidates;
      D = D − {I_i};
      D = D + {I'};
    if cleanse == multi-substitution
      D = D − {I_i};
      foreach Rule R_c in candidates
        I' = change I_i to satisfy R_c;
        D = D + {I'};
  return D;
End
```

[a] Or a randomly-chosen rule in *candidates*. Since normally the order of the rules implies some overall ranking made by the learner, we here have always chosen the first one.

Match-corresponding. For the ith instance in ND, I_i, match-corresponding compares it with the ith instance I'_i in CD. If I_i matches I'_i, I_i is clean. Otherwise I_i is noisy. Although it is straightforward, match-corresponding has very limited function. The reason is that it is sensitive to the order of instances. As illustrated in Figure 1, (a) is a clean dataset with n instances I_1 to I_n. Suppose that there are no identical instances. Now we shift as in (b) each instance one location left so that I_i takes the location of I_{i-1} for any $i \in [2, n]$ and I_1 takes the location of I_n. The datasets (a) and (b) are the same. But match-corresponding will judge (b) as 100% noise since no instance matches its corresponding one in (a).

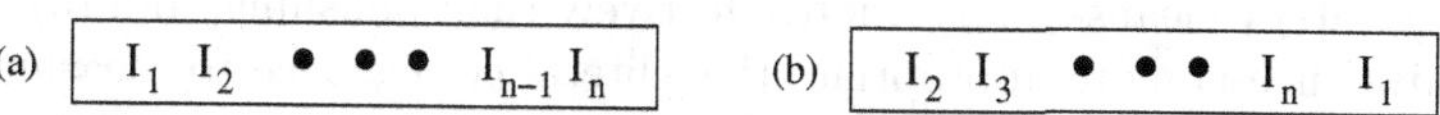

Fig. 1. Match-corresponding is sensitive to the order.

Match-anyone. For an instance I_i in ND, as long as it can match anyone of CD's instances, match-anyone deems it clean. Only if it does not appear in CD at all will I_i be judged noisy. Although it is less censorious, match-anyone can be insensitive to information loss. An extreme example is that an algorithm obtains a rule complying with a single clean instance I_1 and cleanses all other instances into I_1 as in (b) of Figure 2. Although (b) has significantly lost information of (a), match-anyone still judges the cleansing superb with 100% success.

(a) $I_1\ I_2\ \bullet\ \bullet\ \bullet\ I_{n-1}\ I_n$ (b) $I_1\ I_1\ \bullet\ \bullet\ \bullet\ I_1\ I_1$

Fig. 2. Match-anyone is insensitive to information loss.

5.2 Conceptual Equivalence

Other mechanisms of literal equivalence may well exist. However, all measurements involving literal comparison suffer from a problem that they confine the data's legitimacy to the clean dataset at hand. Usually a dataset is only a sample of the whole population. Instances can legitimately exist in the whole population but do not appear in certain samples. This motivates us to propose *conceptual equivalence*, which we expect to be more elegant and capable in measuring noise. Suppose a clean dataset CD is corrupted and is then cleansed into a dataset CD'. Suppose the target concepts learned from CD and CD' are $Concept_{CD}$ and $Concept_{CD'}$ respectively. In order to evaluate how well CD' resembles CD, conceptual equivalence will cross-exam how well CD's data support $Concept_{CD'}$ and how well CD''s data support $Concept_{CD}$. The better both data support each other's concept, the less noise in CD' and the higher the conceptual equivalence. The process is illustrated in Figure 3.

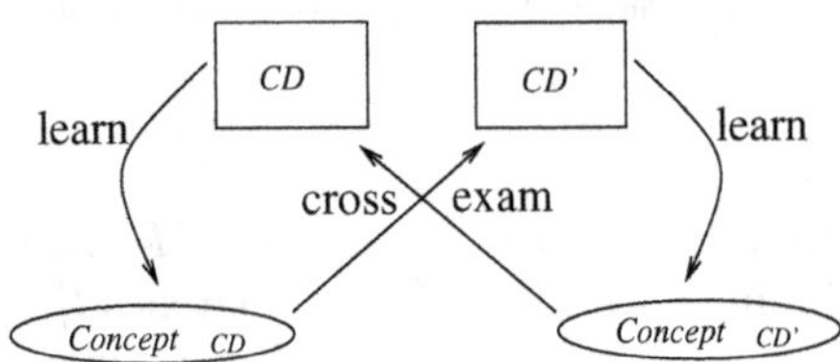

Fig. 3. Conceptual equivalence conducts cross-exam.

6 Experiments

Experiments are conducted to test three hypotheses for unpredictable attributes: (1) our new strategy identifies noise more accurately and efficiently than polishing; (2) if the data's genuine concept is learnable and hence cleansing is allowed, our new strategy cleanses noise more effectively than polishing; and (3) conceptual equivalence is more appropriate than literal equivalence to measure noise. Please be noted that we do not claim that our new strategies outperform polishing for predictable attributes. Instead, the niche of our methodology lies within unpredictable attributes.

6.1 Data and Design

We choose a dataset from the UCI data repository [1] if it satisfies the following conditions. First, its genuine concept is documented. We do not use this information during any stage of identifying or cleansing. It only helps us check the types of attributes and verify our analysis. Second, its genuine concept is learnable, that is, the prediction accuracy is high. Otherwise, we will not be able to differentiate the clean from the noisy. Third, the attributes are predictive but unpredictable. Here we deem an attribute unpredictable if its prediction accuracy is significantly lower than the class. Fourth, there are no numeric attributes. Our approaches can apply to numeric attributes if discretization is employed. However, discretization may introduce extra intractable noise and compromises our understanding of the experimental results. The resulting 5 (3 natural, 2 synthetic) datasets' statistics and prediction accuracies for the class and attributes are in Table 2.

We use C4.5rules [9] as the rule learner for both noise identification and cleansing. Since originally polishing employs C4.5trees, we re-implement polishing using C4.5rules to ensure a fair comparison. Each attribute is randomly corrupted, where each value other than the original value is equally likely to be chosen. A noise level of $x\%$ indicates that $x\%$ instances are corrupted. Each original dataset is corrupted into four levels: 10%, 20%, 30% and 40% respectively.

Table 2. Experimental datasets.

Data set	Size	Class No.	Attribute No.	Prediction accuracy (%)	
				Class	Attributes
car	1728	4	6	92.4	29.3, 30.7, 23.2, 46.8, 36.6, 52.4
monks1	432	2	6	96.5	51.2, 48.1, 41.0, 23.1, 35.9, 44.7
monks3	432	2	6	100.0	23.1, 51.2, 41.2, 28.5, 35.4, 44.2
ttt[a]	958	2	9	85.1	51.0, 48.9, 50.8, 49.6, 62.7, 49.1, 49.8, 49.3, 51.0
nursery	12960	5	8	97.1	42.8, 27.7, 24.2, 25.5, 37.5, 50.8, 34.4, 75.9

[a] The ttt dataset represents the tic-tac-toe dataset.

6.2 Identification

The identification performance is measured by $F1$ *measure* [10], a popular measure used in information retrieval that evenly combines precision and recall of the identification. Precision p reflects the purity of identification. It equals the number of truly noisy instances identified divided by the total number of identified instances. Recall r reflects the completeness of identification. It equals to the number of truly noisy instances identified divided by the total number of truly noisy instances. $F1(p,r)$ equals to $\frac{2pr}{p+r}$, which falls in the range $[0,1]$. The higher a method's $F1$ *measure*, the better this method simultaneously maximizes both precision and recall. The results for all 20 cases (5 datasets, each with 4 levels of corruption) are depicted in Figure 4. No matter whether one uses match-corresponding or uses match-anyone[4], sifting achieves a higher $F1$ *measure* in almost all cases. This indicates that sifting outperforms polishing in identifying noise for unpredictable attributes.

[4] Conceptual equivalence does not apply here since no data have been cleansed yet.

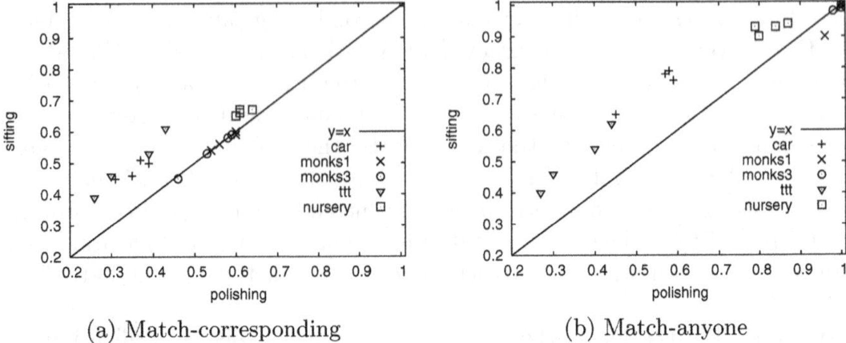

(a) Match-corresponding (b) Match-anyone

Fig. 4. The $F1$ *measure* measures sifting and polishing's identification. Each coordinate represents an experimental case. Its x component represents polishing's $F1$ *measure* and its y component represents sifting's $F1$ *measure*. Hence, any coordinate above the $y = x$ line indicates a case where sifting achieves a higher $F1$ *measure* than polishing does.

6.3 Cleansing

The cleansing performance under literal and conceptual equivalence is studied here.

Under literal equivalence. Figure 5 records *correction accuracy* of uni-substitution and polishing[5] in each of 20 cases, which equals to the number of correctly cleansed instances divided by the total number of identified instances. Although uni-substitution achieves higher correction accuracies more often than not, the results also confirm our belief that literal equivalence tends to improperly reflect the cleansing performance. Correction accuracies under match-corresponding are always low (below 10%) since match-corresponding is sensitive to the instances' order and is over censorious. Correction accuracies under match-anyone are far higher since match-anyone can be insensitive to information loss and has a potential to exaggerate the cleansing efficacy.

However, ttt is an exception that obtains a low accuracy even under match-anyone. Its clean data encode the complete set of legal 3x3 board configurations at the end of tic-tac-toe games and contain 958 instances. But the corruption does not have the legitimacy in mind and has $3^9 = 19683$ configurations at choice. Few can be restored to completely match a clean instance. This raises another type of situation where literal equivalence can not manage well.

Under conceptual equivalence. There can be different ways to calculate conceptual equivalence. Table 3 reports the results of our implementation. We first learn classification rules from clean data and use them to classify

[5] Correction accuracy here does not apply to deletion or multi-substitution since they do not maintain the original data amount.

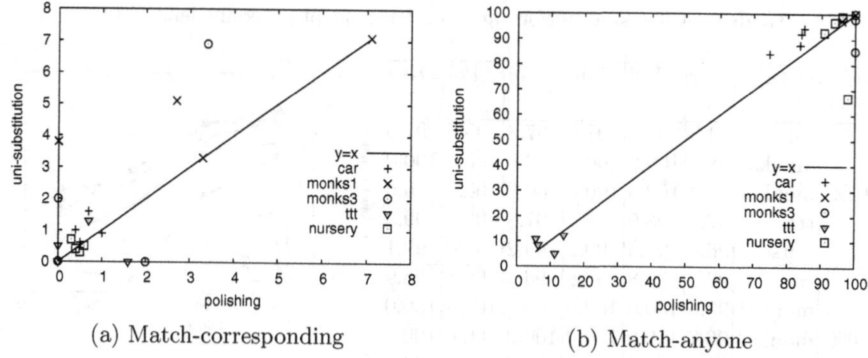

(a) Match-corresponding (b) Match-anyone

Fig. 5. Under literal equivalence, the *correction accuracy*(%) measures polishing and uni-substitution's cleansing. Each coordinate represents an experimental case. Its x component represents polishing's *correction accuracy* and its y component represents uni-substitution's *correction accuracy*. Hence, any coordinate above the $y = x$ line indicates a case where uni-substitution achieves a higher *correction accuracy* than polishing does. The axis value ranges of (a) and (b) are different. The correction accuracies under match-corresponding are far lower than those under match-anyone.

cleansed data, obtaining a classification accuracy acc_1. We then learn classification rules from cleansed data and use them to classify clean data, obtaining another classification accuracy acc_2. Because acc_1 and acc_2 can be associated with different data sizes ($size_1$ and $size_2$ respectively)[6], their *weighted mean* $= \sum_{i=1}^{2}(acc_i \times size_i)/\sum_{i=1}^{2} size_i$ is used to indicate the degree of conceptual equivalence as in (a). Graphs are drawn in (b) and (c) to better reflect the trend of the values. The pattern in (b) corresponds to 'car', which is representative of other three datasets, monks1, monks2 and nursery. The pattern in (c) corresponds to 'ttt', which is a little different from others because of ttt's special nature as we have explained in the previous section.

Compared with corrupted data, all of deletion, uni-substitution and multi-substitution achieve higher conceptual equivalence across all datasets at all noise levels. This suggests that our handling helps improve the quality of the corrupted data. Compared among themselves, there is no significant difference. An interesting observation is that, despite the risk of losing information, deletion works surprisingly well. A closer look reveals that deletion gains this advantage mainly through large datasets like nursery. This suggests that when the available data well exceed the amount that is needed to learn the underlying concept, appropriately deleting suspicious noise may not harm the genuine reflection of the concept while may effectively eliminate the chance of introducing new noise. As for multi-substitution, we have observed that multiple candidates do not often happen. It is because our process of filtering is very fine as given in Algorithm 2. One can make it coarser and may observe more differences between multi-substitution and uni-substitution.

[6] For example, if multi-substitution is used, the size of cleansed data can be bigger than that of clean data.

Table 3. Cleansing performance under conceptual equivalence (%).

Noise level	Data set	Cor	Del	Mul	Uni	Pol	Rev
10%	car	94.8	97.1	97.6	97.8	95.6	97.0
	monks1	98.4	100.0	100.0	100.0	100.0	100.0
	monks3	98.3	100.0	100.0	100.0	100.0	100.0
	ttt	97.3	98.0	97.8	97.8	98.4	99.3
	nursery	96.6	99.3	99.2	99.2	99.0	99.1
20%	car	92.2	96.8	96.8	96.8	95.1	95.9
	monks1	96.6	100.0	100.0	100.0	100.0	100.0
	monks3	96.7	100.0	100.0	100.0	100.0	100.0
	ttt	92.9	94.3	94.1	93.4	93.1	94.7
	nursery	93.7	99.0	98.8	98.8	97.9	98.8
30%	car	86.9	93.9	93.0	93.0	91.6	93.9
	monks1	93.8	100.0	100.0	100.0	100.0	100.0
	monks3	93.9	99.5	99.4	99.4	99.5	99.5
	ttt	87.8	89.9	88.5	88.6	87.6	89.3
	nursery	90.4	98.5	95.5	95.6	96.9	98.0
40%	car	87.4	91.7	92.7	92.5	89.5	91.1
	monks1	91.3	99.2	98.8	98.8	99.7	97.8
	monks3	94.4	99.9	98.4	98.4	100.0	99.9
	ttt	82.3	87.5	86.4	86.4	83.9	85.6
	nursery	87.3	97.3	96.5	96.6	94.3	96.6
Mean		92.6	97.1	96.7	96.6	96.1	96.8
Geomean		92.5	97.0	96.6	96.6	96.0	96.7

(a) summary

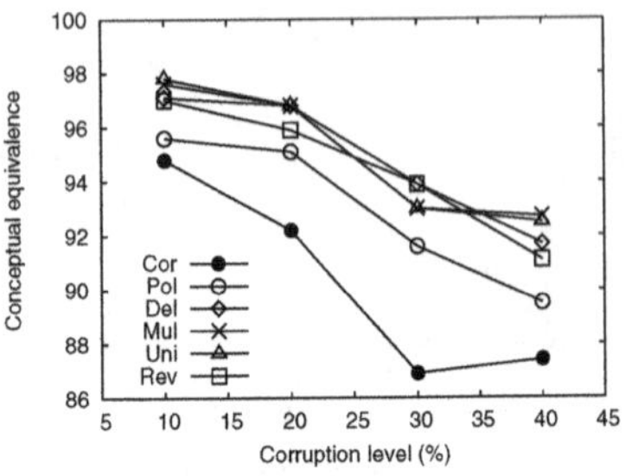

(b) car

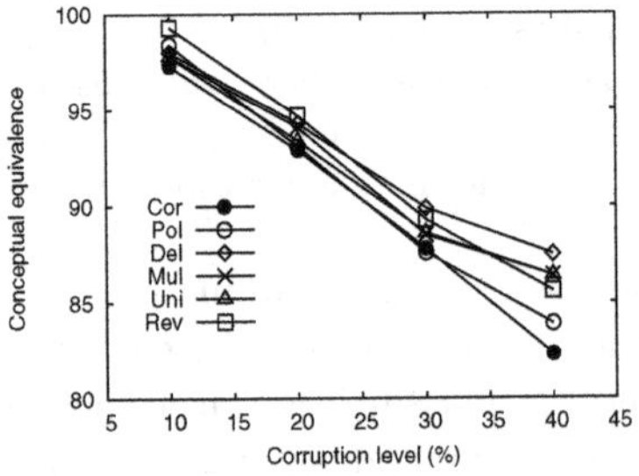

(c) ttt

Note: Each method's conceptual equivalence is calculated between the clean data and the data processed by this method. 'Cor' is corruption; 'Del' is deletion; 'Mul' is multi-substitution; 'Uni' is uni-substitution; 'Pol' is polishing; and 'Rev' is the revised version of polishing that is supplied with sifting's identification. The 'Mean' row and 'Geomean' row record its arithmetic mean and geometric mean across different datasets.

Compared with polishing, all of our cleansing methods achieve higher conceptual equivalence more often than not. Their arithmetic and geometric means are also higher than polishing's. Nevertheless, we do not jump to the conclusion that polishing's cleansing, which is sophisticated, is inferior. It is possible that the disadvantage of its identification is passed on to its cleansing. Hence, we implement a revised version of polishing whose cleansing is supplied with sifting's identification. Thus we can have a pure comparison between the cleansing performances. The experimental results show that revised polishing either improves on polishing or maintains polishing's high conceptual equivalence (like 100%) in 18 out of 20 cases. This from another perspective verifies that sifting is effective.

The revised polishing thus can obtain competitive conceptual equivalence. However, in terms of efficiency, our new strategies are superior to polishing. Suppose the number of instances and attributes to be I and A respectively. Suppose the rule learning algorithm's time complexity and the number of learned rules to be $O(L)$ and R respectively. For each attribute, polishing conducts cross validation to predict its value for each instance. It then recursively tries out different combinations of attribute changes for each instance. It reaches a prohibitive time complexity of $O(ALI) + O(I2^A)$. In comparison, our cleansing needs to learn a rule set for once and match each instance against this rule set. Thus it has a time complexity of $O(L) + O(IR)$. Hence our methods are far more efficient than polishing, which has been verified by the experimental running time and is important in the real world where large data are routinely involved.

7 Conclusion

This paper handles predictive-but-unpredictable attributes in noisy data sources. To identify noise, we have proposed *sifting*. To cleanse noise, we have suggested that unless the genuine concept can be reliably learned, one should be very cautious to modify an instance. When the genuine concept is learnable, we have studied three cleansing approaches, *deletion*, *uni-substitution* and *multi-substitution*. To measure noise, we have argued that *literal equivalence* is often inadvisable and proposed *conceptual equivalence*. Both theoretical analysis and empirical evidence have demonstrated that our strategies achieve better efficacy and efficiency than previous alternatives.

The knowledge acquired by our study, although preliminary, is informative. We expect it to contribute to completing the picture of attribute noise handling. However, a single study seldom settles an issue once and for all. More efforts are needed to further advance this research field. We name three topics here.

First, whether an attribute is predictable or unpredictable is a matter of degree. In our current research, we deem an attribute unpredictable when its prediction accuracy is significantly lower than the class. Further research to work out more sophisticated thresholds or heuristics would be interesting. Second, although we take polishing as straw man, sifting does not claim to outperform polishing for predictable attributes. Instead, sifting and polishing are parallel, each having its own niche to work. Hence it is sensible to combine them. A work frame might be: (1) use feature selection to discard unpredictive attributes; (2) decide whether a predictive attribute is predictable; (3) if it is predictable, use polishing to handle its noise; and if it is unpredictable, use sifting to handle its noise. Lastly, it would be enchanting to extend our research beyond classification learning, such as to association learning where patterns exist but attributes are seldom predictable.

Acknowledgement

This research has been supported by the U.S. Army Research Laboratory and the U.S. Army Research Office under the grant number DAAD19-02-1-0178.

References

1. Blake, C. L., and Merz, C. J. UCI repository of machine learning databases, Department of Information and Computer Science, University of California, Irvine, 1998.
2. Brodley, C. E., and Friedl, M. A. Identifying and eliminating mislabeled training instances. In *Proc. of the 13th National Conf. on Artificial Intelligence* (1996), pp. 799–805.
3. Brodley, C. E., and Friedl, M. A. Identifying mislabeled training data. *Journal of Artificial Intelligence Research 11* (1999), 131–167.
4. Gamberger, D., Lavrac, N., and Dzeroski, S. Noise detection and elimination in data preprocessing: experiments in medical domains. *Applied Artificial Intelligence 14* (2000), 205–223.
5. Gamberger, D., Lavrac, N., and Groselj, C. Experiments with noise filtering in a medical domain. In *Proc. of the 16th International Conf. on Machine Learning* (1999), pp. 143–151.
6. Guyon, I., Matic, N., and Vapnik, V. *Discovering Informative Patterns and Data Cleaning.* AAAI/MIT Press, 1996, pp. 181–203.
7. Kubica, J., and Moore, A. Probabilistic noise identification and data cleaning. In *Proc. of the 3rd IEEE International Conf. on Data Mining* (2003), pp. 131–138.
8. Maletic, J. I., and Marcus, A. Data cleansing: Beyond integrity analysis. In *Proc. of the 5th Conf. on Information Quality* (2000), pp. 200–209.
9. Quinlan, J. R. *C4.5: Programs for Machine Learning.* Morgan Kaufmann Publishers, 1993.
10. Rijsbergen, C. J. v. *Information Retrieval, second edition.* Butterworths, 1979.
11. Schwarm, S., and Wolfman, S. Cleaning data with Bayesian methods, 2000. Final project report for CSE574, University of Washington.
12. Teng, C. M. Correcting noisy data. In *Proc. of the 16th International Conf. on Machine Learning* (1999), pp. 239–248.
13. Teng, C. M. Applying noise handling techniques to genomic data: A case study. In *Proc. of the 3rd IEEE International Conf. on Data Mining* (2003), pp. 743–746.
14. Verbaeten, S. Identifying mislabeled training examples in ILP classification problems. In *Proc. of the 12th Belgian-Dutch Conf. on Machine Learning* (2002), pp. 1–8.
15. Zhu, X., Wu, X., and Chen, Q. Eliminating class noise in large datasets. In *Proc. of the 20th International Conf. on Machine Learning* (2003), pp. 920–927.

A New Scheme on Privacy Preserving Association Rule Mining*

Nan Zhang, Shengquan Wang, and Wei Zhao

Department of Computer Science, Texas A&M University
College Station, TX 77843, USA
{nzhang,swang,zhao}@cs.tamu.edu

Abstract. We address the privacy preserving association rule mining problem in a system with one data miner and multiple data providers, each holds one transaction. The literature has tacitly assumed that randomization is the only effective approach to preserve privacy in such circumstances. We challenge this assumption by introducing an algebraic techniques based scheme. Compared to previous approaches, our new scheme can identify association rules more accurately but disclose less private information. Furthermore, our new scheme can be readily integrated as a middleware with existing systems.

1 Introduction

In this paper, we address issues related to production of accurate data mining results, while preserving the private information in the data being mined. We will focus on association rule mining. Since Agrawal, Imielinski, and Swami addressed this problem in [1], association rule mining has been an active research area due to its wide applications and the challenges it presents. Many algorithms have been proposed and analyzed [2–4]. However, few of them have addressed the issue of privacy protection.

Borrowing terms from e-business, we can classify privacy preserving association rule mining systems into two classes: business to business (B2B) and business to customer (B2C), respectively. In the first category (B2B), transactions are distributed across several sites (businesses) [5,6]. Each of them holds a private database that contains numerous transactions. The sites collaborate with each other to identify association rules spanning multiple databases. Since usually only a few sites are involved in a system (e.g., less than 10), the problem here can be modelled as a variation of secured multiparty computation [7]. In the second category (B2C), a system consists of one data miner (business) and multiple data providers (customers) [8,9]. Each data provider holds only one transaction. Association rule mining is performed by the data miner on the aggregated transactions provided by data providers. On-line survey is a typical example of this type of system, as the system can be modelled as one data miner (i.e., the

* This work was supported in part by the National Science Foundation under Contracts 0081761, 0324988, 0329181, by the Defense Advanced Research Projects Agency under Contract F30602-99-1-0531, and by Texas A&M University under its Telecommunication and Information Task Force Program. Any opinions, findings, conclusions, and/or recommendations expressed in this material, either expressed or implied, are those of the authors and do not necessarily reflect the views of the sponsors listed above.

J.-F. Boulicaut et al. (Eds.): PKDD 2004, LNAI 3202, pp. 484–495, 2004.

survey collector and analyzer) and millions of data providers (i.e., the survey providers). Privacy is of particular concern in this type of system; in fact there has been wide media coverage of the public debate of protecting privacy in on-line surveys [10]. Both B2B and B2C have wide applications. Nevertheless, in this paper, we will focus on studying B2C systems.

Several studies have been carried out on privacy preserving association rule mining in B2C systems. Most of them have tacitly assumed that randomization is an effective approach to preserving privacy. We challenge this assumption by introducing a new scheme that integrates algebraic techniques with random noise perturbation. Our new method has the following important features that distinguish it from previous approaches:

- Our system can identify association rules more accurately but disclosing less private information. Our simulation data show that at the same accuracy level, our system discloses private transaction information about five times less than previous approaches.
- Our solution is easy to implement and flexible. Our privacy preserving mechanism does not need a support recovery component, and thus is transparent to the data mining process. It can be readily integrated as a middleware with existing systems.
- We allow explicit negotiation between data providers and the data miner in terms of tradeoff between accuracy and privacy. Instead of obeying the rules set by the data miner, a data provider may choose its own level of privacy. This feature should help the data miner to collaborate with both hard-core privacy protectionists and persons comfortable with a small probability of privacy divulgence.

The rest of this paper is organized as follows: In Sect. 2, we present our models, review previous approaches, and introduce our new scheme. The communication protocol of our system and related components are discussed in Sect. 3. A performance evaluation of our system is provided in Sect. 4. Implementation and overhead are discussed in Sect. 5, followed by a final remark in Sect. 6.

2 Approaches

In this section, we will first introduce our models of data, transactions, and data miners. Based on these models, we review the randomization approach – a method that has been widely used in privacy preserving data mining. We will point out the problems associated with the randomization approach which motivates us to design a new privacy preserving method, based on algebraic techniques.

2.1 Model of Data and Transactions

Let I be a set of n items: $I = \{a_1, \ldots, a_n\}$. Assume that the dataset consists of m transactions $t_1, \ldots, t_m$, where each transaction t_i is represented by a subset of I. Thus, we may represent the dataset by an $m \times n$ matrix $T = [a_1, \ldots, a_n] = [t_1, \ldots, t_m]'$ [1]. Let $\langle T \rangle_{ij}$ denote the element of T with indices i and j. Correspondingly, for a vector v,

[1] We denote the transpose of matrix T as T'.

Table 1. Transaction Matrix

	a_1	a_2	$\cdots$	a_n
t_1	0	1	$\cdots$	0
$\vdots$	$\vdots$	$\vdots$	$\ddots$	$\vdots$
t_m	1	0	$\cdots$	1

its ith element is represented by $\langle v \rangle_i$. An example of matrix T is shown in Table 1. The elements of the matrix depict whether an item appears in a transaction. For example, suppose the first transaction contains items a_{20} and a_{47}. Then the first row of the matrix has $\langle T \rangle_{1,20} = \langle T \rangle_{1,47} = 1$ and all other elements equal to 0.

An itemset $B \subseteq I$ is k-itemset if B contains k items (i.e., $|B| = k$). The *support* of B is defined as

$$supp(B) = \frac{|\{t \in T | B \subseteq t\}|}{m} \quad (1)$$

A k-itemset B is frequent if $supp(B) \geq min\_supp_k$, where min_supp_k is a predefined minimum threshold of support. The set of frequent k-itemsets is denoted by L_k. Technically speaking, the main task of association rule mining is to identify frequent itemsets.

2.2 Model of Data Miners

There are two classes of data miners in our system. One is *legal data miners*. These miners always act legally in that they perform regular data mining tasks and would never intentionally breach the privacy of the data. On the other hand, *illegal data miners* would purposely discover the privacy in the data being mined. Illegal data miners come in many forms. In this paper, we focus on a particular sub-class of illegal miners. That is, in our system, illegal data miners are *honest but curious*: they follow proper protocol (i.e., they are honest), but they may keep track of all intermediate communications and received transactions to perform some analysis (i.e., they are *curious*) to discover private information [11].

Even though it is a relaxation from Byzantine behavior, this kind of honest but curious (nevertheless illegal) behavior is most common and has been widely adopted as an adversary model in the literatures. This is because, in reality, a workable system must benefit both the data miner and the data providers. For example, an online bookstore (the data miner) may use the association rules of purchase records to make recommendations to its customers (data providers). The data miner, as a long-term agent, requires large numbers of data providers to collaborate with. In other words, even an illegal data miner desires to build a reputation for trustworthiness. Thus, honest but curious behavior is an appropriate choice for many illegal data miners.

2.3 Randomization Approach

To prevent the privacy breach due to the illegal data miners, countermeasures must be implemented in data mining systems. Randomization has been the most common approach for countermeasures. We briefly view this method below.

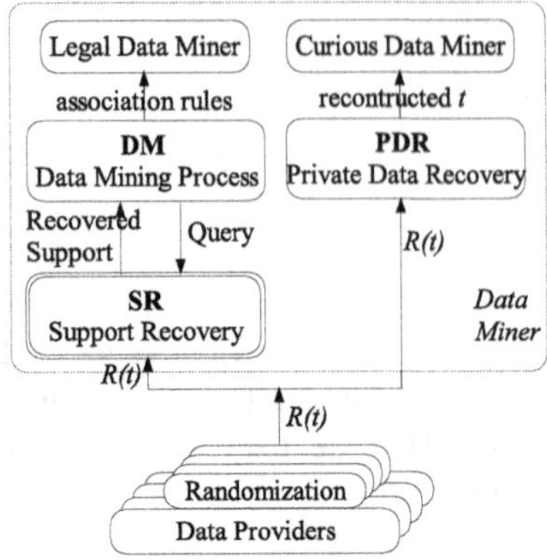

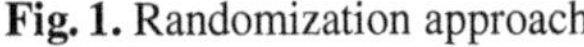
Fig. 1. Randomization approach

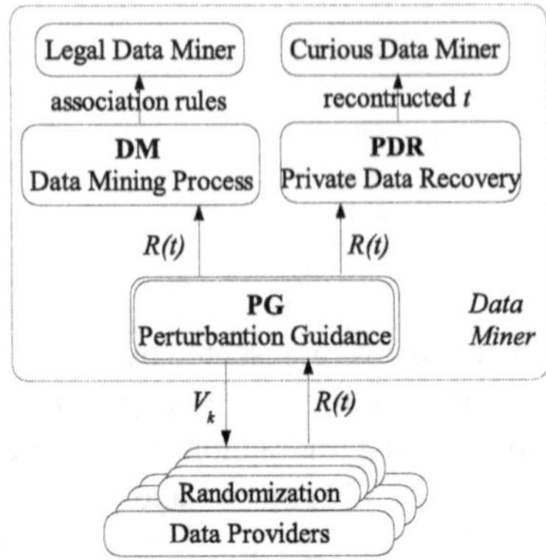

Fig. 2. Our new scheme

We consider the entire mining process to be an iterative one. In each stage, the data miner obtains a perturbed transaction from a different data provider. With the randomization approach, each data provider employs a randomization operator $R(\cdot)$ and applies it to one transaction t which the data provider holds. Fig. 1 depicts this kind of system.

Upon receiving transactions from the data providers, the legal data miner must first perform an operation called *support recovery* which intends to filter out the noise injected in the data due to randomization, and then carry out the data mining tasks. At the same time, an illegal (honest but curious) data miner may perform a particular privacy recovery algorithm in order to discover private data from that supplied by the data providers.

Clearly, the system should be measured by its capability in terms of supporting the legal miner to discover accurate association rules, while preventing illegal miner from discovering private data.

2.4 Problems of Randomization Approach

Researchers have discovered some problems with the randomization approach. For example, as pointed in [8], when the randomization is implemented by a so called *cut-and-paste* method, if a transaction contains 10 items or more, it is difficult, if not impossible, to provide effective information for association rule mining while at the same time preserving privacy. Furthermore, large itemsets have exceedingly high variances on recovered support values. Similar problems would exist with other randomization methods (e.g., MASK system [9]) as they all use random variables to distort the original transactions.

Now, we will explore the reasons behind these problems.

- First, we note that previous randomization approaches are *transaction-invariant*. In other words, the same perturbation algorithm is applied to all data providers. Thus, transactions of a large size (e.g., $|t| > 10$) are doomed to failure in privacy protection by the large numbers of the real items divulged to the data miner. The solution proposed in [8] has ignored all transactions with a size larger than 10. However, a real dataset may have about 5% such transactions. Even if the average transaction

size is relatively small, this solution still prevents many frequent itemsets (e.g., with size of 4 or more) from being discovered.
- Second, previous approaches are *item-invariant*. All items in the original transaction t have the same probability of being included in the perturbed transaction $R(t)$. No specific operation is performed to preserve the correlation between different items. Thus, a lot of real items in the perturbed transactions may never appear in any frequent itemset. In other words, the divulgence of these items does not contribute to the mining of association rules.

Note that invariance of transactions and items is inherent in the randomization approach. This is because in this kind of system, the communication is one-way: from data providers to the data miner. As such, a data provider cannot obtain any specific guidance on the perturbation of its transaction from the (legal) data miner. Consequently, lack of communication between data providers prevents a data provider from learning the correlation between different items. Thus, a data provider has no choice but to employ a *transaction-invariant* and *item-invariant* mechanism.

This observation motivates us to develop a new approach that allows two-way communication between the data miner and data provider. We describe the new approach in the next subsection.

2.5 Our New Approach

Fig. 2 shows the infrastructure of our system. The (legal) data miner S contains two components: DM (data mining process) and PG (perturbation guidance). When a data provider C_i initializes a communication session, PG first dispatches a reference V_k to C_i. Based on the received V_k, the data perturbation component of C_i transforms the transaction t to a perturbed one $R(t)$ and transmits $R(t)$ to PG. PG then updates V_k based on the recently received $R(t)$ and forwards $R(t)$ to the data mining process DM.

The key here is to properly design V_k so that correct guidance can be provided to the data providers on how to distort the data transactions. In our system, we let V_k be an algebraic quantity derived from T. As we will see, with this kind of V_k, our system can effectively maintain accuracy of data mining while significantly reduce the leakage of private information.

3 Communication Protocol and Related Components

In this section, we will present the communication protocol and the associated components in our system. Recall that in our system, there is a two-way communication between data providers and the data miner. While only little overhead is involved, this two-way communication substantially improves performance of privacy preserving discovered association rules.

3.1 The Communication Protocol

We now describe the communication protocol used between the data providers and data miners. On the side of the data miner, there are two current threads that perform the following operations iteratively after initializing V_k:

Thread of registering data provider:

R1. Negotiate on the truncation level k with a data provider;
R2. Wait for a *ready message* from a data provider;
R3. Upon receiving the ready message from a data provider,
   - Register the data provider;
   - Send the data provider current V_k;

R4. Go to Step R1;

Thread of receiving data transaction:

T1. Wait for a (perturbed) data transaction $R(t)$ from a data provider;
T2. Upon receiving the data transaction from a registered data provider,
   - Update V_k based on the newly received perturbed data transaction;
   - Deregister the data provider;

T3. Go to Step T1;

For a data provider, it performs the following operations to transfer its transaction to the data miner:

P1. Send the data miner a ready message indicating that this provider is ready to contribute to the mining process.
P2. Wait for a message that contains V_k from the data miner.
P3. Upon receiving the message from the data miner, compute $R(t)$ based on t and V_k.
P4. Transfer $R(t)$ to the data miner.

3.2 Related Components

It is clear from the above description that the key components of our communication protocol are (a) the method of computing V_k; and (b) the algorithm for perturbation function $R(\cdot)$. We discuss these components in the following. Negotiation is also critical. The details of negotiation protocol can be found in [12].

Computation of V_k. Recall that V_k carries information from the data miner to data providers on how to distort a data transaction in order to preserve privacy. In our system, V_k is an estimation of the eigenvectors of $A = T'T$. Due to space limit, we refer users to [12] about the justification of V_k on providing accurate mining results.

As we are considering dynamic case where data transactions are dynamically fed to the data miner, the miner keeps a copy of all received transactions and need to update it when a new transaction is received. Assume that the initial set of received transactions T^* is empty[2] and every time when a new (distorted) data transaction, $R(t)$, is received, T^* is updated by appending $R(t)$ at the bottom of T^*. Thus, T^* is the matrix of perturbed transactions. We derive V_k from T^*.

In particular, the computation of V_k is done in the following steps. Using singular value decomposition (SVD) [13], we can decompose $A^* = T^{*\prime}T^*$ as (2) where diagonal matrix $\Sigma^* = \text{diag}(s_1^2, \ldots, s_n^2)$ and $s_1^2 \geq \ldots \geq s_n^2$.

$$A^* = T^{*\prime}T^* = V^*\Sigma^*V^{*\prime} \tag{2}$$

V^* is an $n \times n$ unitary matrix composed of the eigenvectors of A^*.

[2] T^* may also be composed of some transactions provided by privacy-careless data providers.

V_k is composed of the first k vectors of V^* (i.e., eigenvectors corresponding to the largest k eigenvalues of A^*). In other words, if $V^* = [v_1, \ldots, v_n]$, then

$$V_k = [v_1, \ldots, v_k] \tag{3}$$

Thus, we call V_k as the k-truncation of V^*. Several incremental algorithms have been proposed to update V_k when a new (distorted) data transaction is received by the data miner [14, 15]. The computing cost of updating V_k is addressed in Sect. 5.

Note that k is a given integer less than or equal to n. As we will see in Sect. 4, k can play a critical role in balancing accuracy and privacy. We will also show that by using V_k in conjunction with $R(\cdot)$, to be discussed next, we can achieve desired accuracy and privacy.

Perturbation Function $R(\cdot)$. Recall that once a data provider receives a perturbation guidance V_k from the data miner, the provider applies a perturbation function, $R(\cdot)$, to its data transaction, t. The result is a distorted transaction that will be transmitted to the data miner. The computation of $R(t)$ is defined as follows. First, for the given V_k, the data transaction, t, is transformed by $\tilde{t} = tV_kV_k'$. Note that the elements in $\tilde{t}$ may not be integers. Algorithm Mapping is employed to integerize $\tilde{t}$. In the algorithm, ρ_{t} is a pre-defined parameter. Finally, to enhance the privacy preserving capability, we need to insert additional noise into $R(t)$. This is done by Algorithm Random-Noise Perturbation.

Algorithm Mapping

```
for every element ⟨t̃⟩_i in t̃ do
  if ⟨t̃⟩_i ≥ 1 − ρ_t then
    ⟨R(t)⟩_i = 1
  else
    ⟨R(t)⟩_i = 0
  end if
end for
```

Algorithm Random-Noise Perturbation

```
for every item a_i ∉ t do
  Choose a real number j uniformly at
  random on [0, 1]
  if j ≥ 1 − ρ_m then
    ⟨R(t)⟩_i = 1
  end if
end for
```

Now, computation of R(t) has been completed and it is ready to be transmitted to the data miner.

We have described our system – the communication protocol and its key components. We now discuss the accuracy and privacy metrics of our system.

4 Analysis on Accuracy and Privacy

In this section, we will propose the metrics of accuracy and privacy with analysis of the tradeoff between them. We will derive a upper bound on the degree of accuracy in the mining results (frequent itemsets). An analytical formula for evaluating the privacy metric is also provided.

4.1 Accuracy Metric

We use the error of support of frequent itemsets to measure the degree of accuracy in our system. This is because general objective of association rule mining is to identify all frequent itemsets with support larger than a threshold min_supp. There are two kinds of errors: *false drops*, which are undiscovered frequent itemsets and *false positives*, which are itemsets wrongly identified to be frequent. Formally, given itemset I_j, let the support of I_j in the original transactions T and the perturbed transactions $R(T)$ be $supp(I_j)$ and $supp'(I_j)$, respectively. Recall that the set of frequent h-itemsets in T is L_h. With these notations, we can define those two errors as follows:

Definition 1. *For a given itemset size h, the error on false drops, ρ_1, and the error on false positives, ρ_2, are defined as*

$$\rho_1 = \max_{I_j \in L_h} (supp(I_j) - supp'(I_j)), \tag{4}$$

$$\rho_2 = \max_{I_j \notin L_h} (supp'(I_j) - supp(I_j)). \tag{5}$$

We define the degree of accuracy as the maximum of ρ_1 and ρ_2 on all itemset sizes.

Definition 2. *The degree of accuracy in a privacy preserving association rule mining system is defined as $\gamma = \max_{h \geq 1} \max(\rho_1, \rho_2)$.*

With this definition, we can derive an upper bound on the degree of accuracy.

Theorem 1. $\gamma \leq 2.618\sigma_{k+1}^2/m$, *where σ_i is the ith eigenvalue of $A = T'T$.*

The proof can be found in [12].

This bound is fairly small when m is sufficiently large, which is usually the case in reality. Actually, our method tends to enlarge the support of high-supported itemsets and reduce the support of low-supported itemsets. Thus, the effective error that may result in false positives or false drops is much smaller than the upper bound. We may see this from the simulation results later.

4.2 Privacy Metric

In our system, the data miner cannot deduce the original t from $\tilde{t} = tV_kV_k'$ because V_kV_k' is a singular matrix with $det(V_kV_k') = 0$ (i.e., it does not have an inverse matrix). Since $t \to \tilde{t} \to R(t)$, t cannot be deduced from $R(t)$ deterministically. To measure the probability that an item in t is identified from $R(t)$, we need a privacy metric.

A privacy metric, *privacy breach*, is proposed in [8]. It is defined by the posterior probability $\Pr\{a_i \in t|t'\}$ that an item could be recovered from the perturbed transaction. Unfortunately, this metric is unsuitable in our system settings, especially to Internet applications. Consider a person taking an online survey of the commodities he/she purchased in the last month. A privacy breach of 50% (which is achieved in [8]) does not prevent privacy divulgence effectively. For instance, for a company who uses spam mail to make advertisement, a 50% probability of success (correct identification of a person who purchased similar commodities in the last month) certainly deserves a try because a wrong estimation (a spam mail sent to a wrong person) costs little.

We propose a privacy metric that measures the number of "unwanted" items (i.e., items not contribute to association rule mining) divulged to the data miner. For an item a_i that does not appear in any frequent itemset (i.e., $a_i \notin \bigcup L_k$), the divulgence of a_i

(i.e., $a_i \in R(t)$) does not contribute to the mining of association rules. Due to survey results in [10], a person has a strong will to filter out such "unwanted" information (i.e., information not effective in data mining) before divulging private data in exchange of data mining results. We evaluate the level of privacy by the probability of an "unwanted" item to be included in the transformed transaction. Formally, the level of privacy is defined as follows:

Definition 3. *Given a transaction t, an item $a_i \in t$ appears in a frequent itemset in t if there exists a frequent itemset I_j such that $a_i \in I_j \subseteq t$. Otherwise we say that a_i is infrequent in t. We define the level of privacy as*

$$\delta = \Pr\{a_i \in R(t) | a_i \text{ is infrequent in } t\} \tag{6}$$

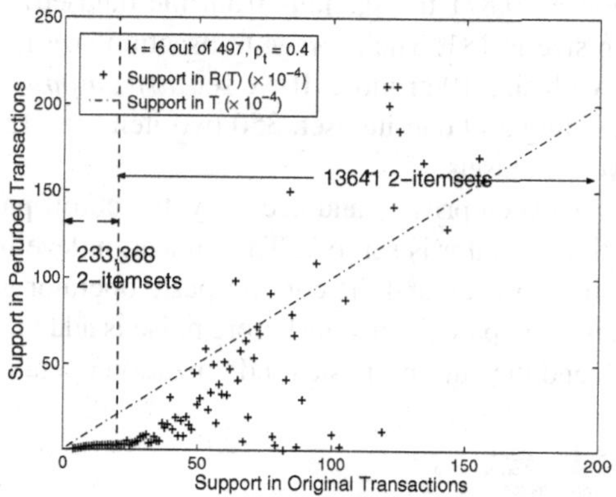

Fig. 3. Comparison of 2-itemsets supports between original and perturbed transactions

Fig. 3 shows a simulation result on all 2-itemsets. The x-axis is the support of itemsets in original transactions. The y-axis is the support of itemsets in perturbed transactions. The figure intends to show how effectively our system blocks the unwanted items from being divulged. If a system preserves privacy perfectly, we should have y equal to zero when x is less than min_supp_2. The data in Fig. 3 shows that almost all 2-itemsets with support less than 0.2% (i.e., $233,368$ unwanted 2-itemsets) have been blocked. Thus, the privacy has been successfully protected. Meanwhile, the supports of frequent 2-itemsets are exaggerated. This should help the data miner to identify frequent itemsets from additional noises.

Formally, we can derive an upper bound on the level of privacy.

Theorem 2. *The level of privacy in our system is bounded by*

$$\delta \leq 1 - \sqrt{\frac{\sigma_{k+1}^2 + \cdots + \sigma_n^2}{\sigma_1^2 + \cdots + \sigma_n^2}}. \tag{7}$$

where σ_i is the ith eigenvalue of $A = T'T$.

The proof can be found in [12].

By Theorems 1 and 2, we can observe a tradeoff between accuracy and privacy. Note that σ_i is sorted in descending order. Thus a larger k results in more "unwanted" items to be divulged. Simultaneously, the degree of accuracy (whose upper bound is in proportion to σ_{k+1}^2) decreases.

4.3 Simulation Results on Real Datasets

We will present the comparison between our approach and the cut-and-paste randomization operator by simulation results obtained on real datasets. We use a real world dataset BMS Webview 1 [16]. The dataset contains web click stream data of several months from the e-commerce website of a leg-care company. It has 59,602 transactions and 497 distinct items.

We randomly choose 10,871 transactions from the dataset as our test band. The maximum transaction size is 181. The average transaction size is 2.90. There are 325 transactions (2.74%) with size 10 or more. If we set $min\_supp = 0.2\%$, there are 798 frequent itemsets including 259 one-itemset, 350 two-itemsets, 150 three-itemsets, 37 four -itemsets and two 5-itemsets.

As a compromise between privacy and accuracy, the cutoff parameter K_m of cut-and-paste randomization operator is set to 7. The truncation level k of our approach is set to 6. Since both our approach and the cut-and-paste operator use the same method to add random noise, we compare the results before noise is added. Thus we set $\rho_\mathrm{m} = 0$ for both our approach and the cut-and-paste randomization operator.

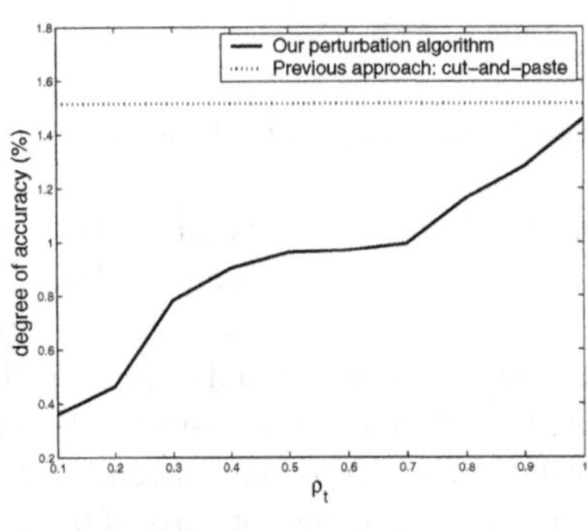

Fig. 4. Accuracy

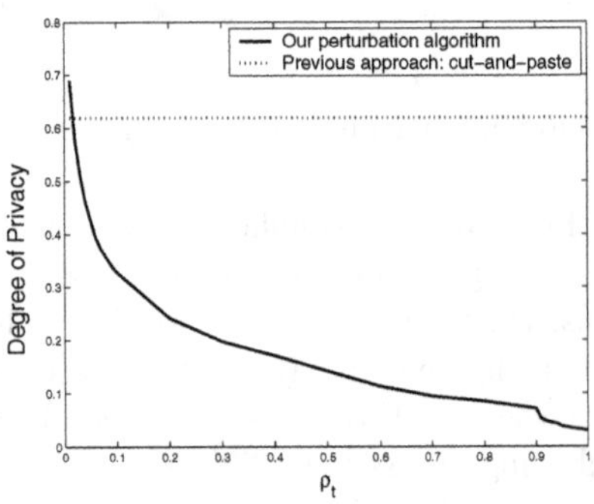

Fig. 5. Privacy

The solid line in Fig. 4 shows the change of degree of accuracy ($\max\{\rho_1, \rho_2\}$) of our approach with the parameter ρ_t. The dotted line shows the degree of accuracy while cut-and-paste randomization operator is employed. We can see that our approach reaches a better accuracy level than the cut-and-paste operator. A recommendation made from the figure is that $\rho_\mathrm{t} \in (0.7, 0.8)$ is suitable for hard-core privacy protectionists while $\rho_\mathrm{t} \in (0.2, 0.3)$ is recommended to persons care accuracy of association rules more than privacy protection.

The relationship between the level of privacy and ρ_t in the same settings is presented in Fig. 5. The dotted line shows the level of privacy of the cut-and-paste randomization operator. We can see that the privacy level of our approach is much higher than the cut-and-paste operator when $\rho_t > 0.1$. Thus our approach is always better on both privacy and accuracy issues when $0.1 \leq t \leq 1$.

5 Implementation

A prototype of the privacy preserving association rule mining system with our new scheme has been implemented on web browsers and servers for online surveys. Visitors taking surveys are considered to be data providers. The data perturbation algorithm is implemented as custom codes on web browsers. The web server is considered to be the data miner. A custom code plug-in on the web server implements the PG (perturbation guidance) part of the data miner. All custom codes are component-based plug-ins that one can easily install to existing systems. The components required for building the system is shown in Fig. 6.

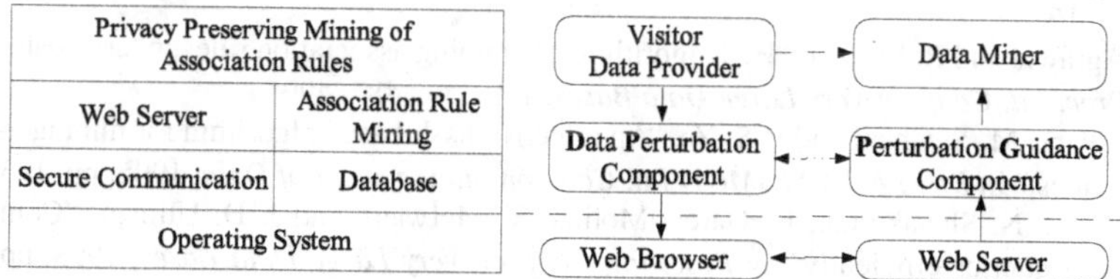

Fig. 6. System implementation

The overhead of our implementation is substantially smaller than previous approaches in the context of online survey. The time-consuming part of the "cut-and-paste" mechanism is on support recovery, which has to be done while mining association rules. The support recovery algorithm needs the partial support of all candidate items for each transaction size, which results in a significant overhead on the mining process.

In our system, the only overhead (possibly) incurred on the data miner is updating the perturbation guidance V_k, which is an approximation of the first k right eigenvectors of $A^* = T^{*\prime}T^*$. Many SVD updating algorithms have been proposed including SVD-updating, folding-in and recomputing the SVD [14, 15]. Since T^* is usually a sparse matrix, the complexity of updating SVD can be considerably reduced to $O(n)$. Besides, this overhead is not on the critical time path of the mining process. It occurs during data collection instead of data mining process. Note that the transfered "perturbation guidance" V_k is of the length kn. Since k is always a small number (e.g., $k \leq 10$), the communication overhead incurred by "two-way" communication is not significant.

6 Final Remarks

In this paper, we propose a new scheme on privacy preserving mining of association rules. In comparison with previous approaches, we introduce a two-way communication mechanism between the data miner and data providers with little overhead. In particular, we let the data miner send a perturbation guidance to the data providers. Using this intelligence, the data providers distort the data transactions to be transmitted to the miner. As a result, our scheme identifies association rules more precisely than previous approaches and at the same time reaches a higher level of privacy.

Our work is preliminary and many extensions can be made. For example, we are currently investigating how to apply a similar algebraic approach to privacy preserving classification and clustering problems. The method of SVD has been broadly adopted to many knowledge discovery areas including latent semantic indexing, information retrieval and noise reduction in digital signal processing. As we have shown, singular value decomposition can be an effective mean in dealing with privacy preserving data mining problems as well.

References

1. R. Agrawal, T. Imielinski, and A. Swami, "Mining association rules between sets of items in large databases," in *Proc. ACM SIGMOD Int. Conf. on Management of Data*, 1993, pp. 207–216.
2. R. Agrawal and R. Srikant, "Fast algorithms for mining association rules in large databases," in *Proc. Int. Conf. on Very Large Data Bases*, 1994, pp. 487–499.
3. J. S. Park, M.-S. Chen, and P. S. Yu, "An effective hash-based algorithm for mining association rules," in *Proc. ACM SIGMOD Int. Conf. on Management of Data*, 1995, pp. 175–186.
4. M. Fang, N. Shivakumar, H. Garcia-Molina, R. Motwani, and J. D. Ullman, "Computing Iceberg queries efficiently," in *Proc. Int. Conf. on Very Large Data Bases*, 1998, pp. 299–310.
5. J. Vaidya and C. Clifton, "Privacy preserving association rule mining in vertically partitioned data," in *Proc. ACM SIGKDD Int. Conf. on Knowledge discovery and data mining*, 2002, pp. 639–644.
6. M. Kantarcioglu and C. Clifton, "Privacy-preserving distributed mining of association rules on horizontally partitioned data," in *Proc. ACM SIGMOD Workshop on Research Issues on Data Mining and Knowledge Discovery*, 2002, pp. 24–31.
7. Y. Lindell and B. Pinkas, "Privacy preserving data mining," *Advances in Cryptology*, vol. 1880, pp. 36–54, 2000.
8. A. Evfimievski, R. Srikant, R. Agrawal, and J. Gehrke, "Privacy preserving mining of association rules," in *Proc. ACM SIGKDD Intl. Conf. on Knowledge Discovery and Data Mining*, 2002, pp. 217–228.
9. S. J. Rizvi and J. R. Haritsa, "Maintaining data privacy in association rule mining," in *Proc. Int. Conf. on Very Large Data Bases*, 2002, pp. 682–693.
10. J. Hagel and M. Singer, *Net Worth.* Harvard Business School Press, 1999.
11. O. Goldreich, *Secure Multi-Party Computation.* Working Draft, 2002.
12. N. Zhang, S. Wang, and W. Zhao, "On a new scheme on privacy preserving association rule mining," Texas A&M University, Tech. Rep. TAMU/DCS/TR2004-7-1, 2004.
13. G. H. Golub and C. F. V. Loan, *Matrix Computations.* Baltimore, Maryland: Johns Hopkins University Press, 1996.
14. J. R. Bunch and C. P. Nielsen, "Updating the singular value decomposition," *Numerische Mathematik*, vol. 31, pp. 111–129, 1978.
15. M. Gu and S. C. Eisenstat, "A stable and fast algorithm for updating the singular value decomposition," Yale University, Tech. Rep. YALEU/DCS/RR-966, 1993.
16. Z. Zheng, R. Kohavi, and L. Mason, "Real world performance of association rule algorithms," in *Proc. ACM SIGKDD Int. Conf. on Knowledge Discovery and Data Mining*, 2001, pp. 401–406.

A Unified and Flexible Framework for Comparing Simple and Complex Patterns*

Ilaria Bartolini[1], Paolo Ciaccia[1], Irene Ntoutsi[2], Marco Patella[1], and Yannis Theodoridis[2]

[1] DEIS – IEIIT/BO-CNR, University of Bologna, Italy
{ibartolini,pciaccia,mpatella}@deis.unibo.it
[2] Research Academic Computer Technology Institute, Athens, Greece
and Department of Informatics, University of Piraeus, Greece
{ntoutsi,ytheod}@cti.gr

Abstract. One of the most important operations involving Data Mining patterns is computing their similarity. In this paper we present a general framework for comparing both simple and complex patterns, i.e., patterns built up from other patterns. Major features of our framework include the notion of structure and measure similarity, the possibility of managing multiple coupling types and aggregation logics, and the recursive definition of similarity for complex patterns.

1 Introduction

Data Mining and Knowledge Discovery techniques are commonly used to extract condensed artifacts, like association rules, clusters, keywords, etc., from huge datasets. Among the several interesting operations on such patterns (modeling, storage, retrieval), one of the most important is that of comparison, i.e., establishing whether two patterns are similar or not [1]. Such operation could be of valuable use whenever we have to measure differences of patterns describing evolving data or data extracted from different sources, and to measure the different behavior of Data Mining algorithms over a same dataset. A similarity operator between patterns could also be used to express similarity queries over pattern bases [5].

In the following we present a general framework for the assessment of similarity between both simple and complex patterns. Major features of our framework include the notion of structure and measure similarity, the possibility of managing multiple coupling types and aggregation logics, and the recursive definition of similarity for complex patterns, i.e., patterns whose structure consists of other patterns. This considerably extends FOCUS [1], the only existing framework for the comparison of patterns, which does not consider complex patterns, neither it allows different matching criteria (i.e., coupling types), since it limits itself to a fixed form of matching (based on a so-called greatest common refinement).

2 A Framework for the Evaluation of Pattern Similarity

We approach the problem of defining a general framework able to guarantee flexibility with respect to pattern types and their similarity criteria and, at the same time, to

* This work was supported by the European Commission under the IST-2001-33058 Thematic Network. PANDA "PAtterns for Next-generation DAtabase systems" (2001-04).

J.-F. Boulicaut et al. (Eds.): PKDD 2004, LNAI 3202, pp. 496–499, 2004.

exploit common aspects of the comparison problem, by starting from the logical model proposed in [3], where each pattern type includes a structure schema *ss*, defining the pattern space, and a measure schema *ms*, describing the measures that quantify the quality of the source data representation achieved by each pattern. A pattern *p* of type *pt* instantiates the structure schema and the measure schema, thus leading to a structure, *p.s*, and a measure, *p.m*. In the basic case, the similarity between patterns is computed by means of a similarity operator, *sim*, which has to take into account both the similarity between the patterns' structures and the similarity between the measures. A pattern is called *simple* if its structure does not include other patterns, otherwise it is called a *complex pattern*. For instance, an Euclidean cluster in a *D*-dimensional space is a simple pattern whose structure is represented by the center (a *D*-dimensional vector) and radius (a real value) of the cluster. Measures for a cluster might include, for example, the average intra-cluster distance and its *support* (fraction of the data points represented by the cluster).

The similarity between two *simple patterns* of the same type *pt* can be computed by combining, by means of an *aggregation function* f_{aggr}, the similarity between both the structure and the measure components:

$$sim(p_1, p_2) = f_{aggr}(sim_{struct}(p_1.s, p_2.s), sim_{meas}(p_1.m, p_2.m)) \quad (1)$$

If the two patterns have the same structural component, then $sim_{struct}(p_1.s, p_2.s) = 1$, and the measure of similarity naturally corresponds to a comparison of the patterns' measures, e.g., by aggregating differences between each measure. In the general case, however, the patterns to be compared have different structural components, thus a preliminary step is needed to "reconcile" the two structures to make them comparable.

The computation of similarity between simple patterns is summarized in Fig. 1. It has to be remarked that the sim_{struct} block could also encompass the use of an underlying domain knowledge. For instance, if we are comparing keywords extracted from textual documents (i.e., the pattern is a keyword), the similarity between them can be computed by exploiting the presence of an ontology, such as WordNet [2].

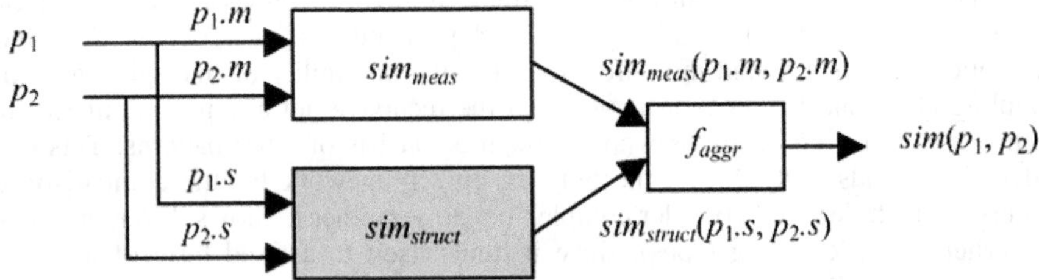

Fig. 1. Assessment of similarity between patterns.

The case of *complex patterns*, i.e., patterns whose structure includes other patterns, is particularly challenging, because the similarity between structures of complex patterns depends in turn on the similarity between component patterns. For instance, a *clustering* pattern is the composition of *cluster* patterns.

Evaluation of similarity between complex pattern follows the same basic rationale shown in Fig. 1 of aggregating similarities between measure and structure components. However, the structure of complex patterns now consists of several other patterns. In our framework, the similarity between the structure of complex patterns is conceptually evaluated in a bottom-up fashion, and can be adapted to specific needs/constraints by acting on two fundamental abstractions, namely the *coupling type*, which is used to establish how component patterns can be *matched*, and the *aggregation logic*, which is used to combine the similarity scores obtained for coupled component patterns into a single overall score representing the similarity between the complex patterns.

Coupling type: Since every complex pattern can be eventually decomposed into a number of component patterns, in comparing two complex patterns, cp_1 and cp_2, we need a way to associate component patterns of cp_1 to component patterns of cp_2. To this end, the *coupling type* just establishes the way component patterns can be matched. Assume without loss of generality that component patterns are given an ordinal number, thus the structure of each complex pattern can be represented as $cp.s = (p^1, p^2, \ldots, p^N)$. Each coupling between cp_1 and cp_2 can be represented by a *matching matrix* $\mathbf{X}_{N \times M} = (x_{ij})$, where each $x_{ij} \in [0,1]$ $(i = 1, \ldots, N; j = 1, \ldots, M)$ represents the (amount of) matching between $p_1{}^i$ and $p_2{}^j$. Different coupling types essentially introduce a number of constraints on the x_{ij} coefficients, e.g.:

- 1–1 *matching*: In this case we accept at most one matching for each component pattern $p_1{}^i$ or $p_2{}^j$.
- *EMD matching*: The Earth Mover's Distance (EMD) [4] is used to compare two distributions and is a particular *N–M* matching. Computing EMD is based on solving the well-known *transportation problem*. EMD has been applied, among others, to compare images by taking into account existing inter-color similarities [4].

Aggregation logic: Among all the feasible matchings, the rationale is to pick the "best" one. To this end, the overall similarity between complex patterns is computed by aggregating similarity scores obtained for matched component patterns, and then taking the maximum over all legal matchings. Formally, each pairing $(p_1{}^i, p_2{}^j)$ contributes to the overall score, as evaluated by the *matching aggregation function* g_{aggr}, with the similarity, $sim(p_1{}^i, p_2{}^j)$, between its matched component patterns:

$$sim_{struct}(cp_1.s, cp_2.s) = \max_{\mathbf{X}}(g_{aggr}((p_1{}^1, p_1{}^2, \ldots, p_1{}^N), (p_2{}^1, p_2{}^2, \ldots, p_2{}^M), \mathbf{X})) \qquad (2)$$

The process of computing the structure similarity between complex patterns can be conceptually summarized as follows. Given a coupling type, any possible legal matching is generated, and the similarity scores between pairs of matched patterns are computed as in Fig. 1. Then, such similarity scores are combined together by means of the matching aggregation function. Finally, the matching attaining the highest overall similarity score is determined. In case of multi-level aggregations, this process has to be recursively applied to component sub-patterns.

The above-described scheme can turn to be highly inefficient. For this reason, when efficient evaluation of $sim_{struct}(cp_1.s, cp_2.s)$ is an issue, we provide efficient algorithms for the solution of the best-coupling problem which do not require going through all possible matchings.

References

1. V. Ganti, J. Gehrke, R. Ramakrishnan, and W.-Y. Loh. A framework for measuring changes in data characteristics. PODS'99, pp. 126-137.
2. G. A. Miller: WordNet: A lexical database for English. CACM, 38(11), 39-41, 1995.
3. S. Rizzi, E. Bertino, B. Catania, M. Golfarelli, M. Halkidi, M. Terrovitis, P. Vassiliadis, M. Vazirgiannis, and E. Vrachnos. Towards a logical model for patterns. ER 2003, pp. 77-90.
4. Y. Rubner, C. Tomasi, and L. J. Guibas. A metric for distributions with applications to image databases. ICCV'98, pp. 59-66.
5. Y. Theodoridis, M. Vazirgiannis, P. Vassiliadis, B. Catania, and S. Rizzi. A manifesto for pattern bases. PANDA Technical Report TR-2003-03, 2003.

Constructing (Almost) Phylogenetic Trees from Developmental Sequences Data

Ronnie Bathoorn and Arno Siebes

Institute of Information & Computing Sciences
Utrecht University
P.O. Box 80.089, 3508TB Utrecht, The Netherlands
{ronnie,arno.siebes}@cs.uu.nl

Abstract. In this paper we present a new way of constructing almost phylogenetic trees. Almost since we reconstruct the tree, but without the timestamps. Rather than basing the tree on genetic sequence data ours is based on developmental sequence data. Using frequent episode discovery and clustering we reconstruct the consensus tree from the literature almost completely.

1 Introduction

One of the big open problems in modern biology is the reconstruction of the course of evolution. That is, when did what (class of) species split off another (class of) species and what were the changes? The ultimate goal is to reconstruct the complete tree of life as correct as possible given the current species and the fossil record. Such a tree is often called an *evolutionary tree* or a *phylogenetic tree*.

Most research in this area uses clustering techniques on genetic data. Using knowledge on, e.g., the rate of change, time-stamps are computed for the different splits. In this paper we show how to reconstruct such trees with a different kind of data, viz., the order of events in the development of an animal [1,2]. Examples of the events are the start of the development of the Heart or of the Eyes. This sequence is fixed for a given species, but varies over species.

2 Method

To construct our tree, we compute the frequent episodes [4,3] over these event sequences in step 1. Next in step 2 we gather the occurence of these episodes in profiles, which we use to calculate the dissimilarity between species using the Jaccard dissimilarity measure in step 3. Finally in step 4, we use agglomerative hierarchical clustering with complete linkage on this dissimilarity matrix.

The resulting tree has no external information on, e.g., the rate of change, we cannot label splits with their probable occurrence in time. Hence we call our tree an *Almost Phylogenetic Tree*.

3 Experimental Results

The experimental results show that clustering results are almost as good as the Phylogenetic trees found in biological literature as can be seen in Figure 2. This result is

J.-F. Boulicaut et al. (Eds.): PKDD 2004, LNAI 3202, pp. 500–502, 2004.

```
BUILD_TREE(data, min_fr, win)
episodes = FindEpisodes(data, min_fr, win)
profiles = makeProfiles(episodes)
dist = makeDistanceMatrix(profiles)
return HierarchicalClustering(dist)
```

Fig. 1. Base algorithm

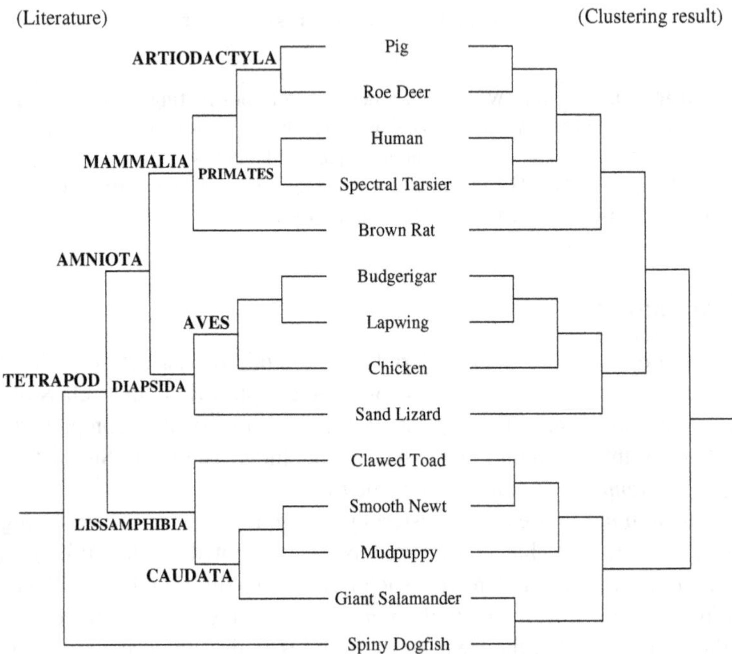

Fig. 2. Comparison of results to the tree from the literature

obtained with a window size of 6 and a minimal frequency of 5%. Lowering the minimal frequency produces trees that become gradualy worse, because the amount of episodes found that have no influence on the distance between species tend to hide the episodes that are key in seperating certain groups of species in the phylogenetic tree. And increasing the window size also produces trees which are worse for the same reason.

Note that the distance in our phylogenetic tree is the distance between species for our distance measure and that it does not mean that the species at the bottom of the tree arose later in evolution. This is because there is no sense of time in these figures. When comparing our experimental result to the Phylogenetic Tree currently accepted in literature we can clearly see that birds and mammals are clustered really well. The amphibians are grouped together but not entirely in the right configuration.

Finally the Spiny Dogfish should have been a group on its own but probably ended up with the amphibians because it was the only creature of its kind in our data set. Which is why episodes being able to separate it from the rest had frequencies which were to low for it to be used in the clustering. That is why we clustered our data with the Spiny Dogfish left out. This gave no changes to the clustering of the rest of the species, giving us reason to believe that patterns able to separate Spiny Dogfish from the rest of the species are not used in the clustering.

4 Conclusions and Future Work

There are a number of reasons that make this application interesting. Firstly, it is a nice illustration of the concept of an *inductive database* [6] given that in the first phase we mine the data for frequent episodes and in the second phase we mine these episodes for the final tree. Moreover, the frequent episodes provide interesting insight in developmental biology by themselves. The second reason is that it illustrates the power of data mining methods. Without expert knowledge we are able to almost reconstruct the consensus tree in the literature, which is based on a lot of biological knowledge. Finally, it shows how much information on evolution is preserved in these developmental event sequences.

This work is part of a larger project that tries to utilize different types of data about the development of different species, such as anatomical data, gene expression data (micro array) and developmental sequence data, e.g., for the construction of evolutionary trees.

References

1. Jonathan E. Jeffery and Olaf R. P. Bininda-Emonds and Michael I. Coates and Michael K. Richardson: Analyzing evolutionary patterns in amniote embryonic development. Evolution & Development Volume 4 Number 4 (2002) 292–302
2. Jonathan E. Jeffery and Michael K. Richardson and Michael I. Coates and Olaf R. P. Bininda-Emonds: Analyzing Developmental Sequences Within a Phylogenetic Framework. Systematic Biology Volume 51 Number 3 (2002) 478–491
3. Mannila, Heikki. and Toivonen, Hannu. and Verkamo, A. Inkeri.: Discovering frequent episodes in sequences. First International Conference on Knowledge Discovery and Data Mining (1995) 210–215
4. Rakesh Agrawal and Ramakrishnan Srikant: Mining Sequential Patterns. International Conference on Data Engineering (1995) 3–14
5. P. W. Holland: The future of evolutionary developmental biology. Nature 402 (1999) 41–44
6. L. De Raedt: A perspective on inductive databases. ACM SIGKDD Explorations Newsletter Volume 4 Issue 2 (2002) 69–77

Learning from Multi-source Data

Élisa Fromont, Marie-Odile Cordier, and René Quiniou

IRISA, Campus de Beaulieu, 35000 Rennes, France
{efromont,quiniou,cordier}@irisa.fr

Abstract. This paper proposes an efficient method to learn from multi source data with an Inductive Logic Programming method. The method is based on two steps. The first one consists in learning rules independently from each source. In the second step the learned rules are used to bias a new learning process from the aggregated data. We validate this method on cardiac data obtained from electrocardiograms or arterial blood pressure measures. Our method is compared to a single step learning on aggregated data.

1 Introduction

In many applications, correlated data are recorded from different sensors observing the same phenomenon from different points of view. We are investigating how to take advantage of this diversity and of these correlations to learn discriminating rules from such data.

Mono source learning consists in learning separately from different sources describing a common phenomenon. When dealing with clear signals, only one channel can be sufficient; when dealing with noisy signals or defective sensors, several sources can be beneficial to ensure a confident recognition. When relations between events occurring on the different sources can be provided, it is interesting to take advantage of the global knowledge. This is achieved by multi source learning. In our case, for instance, relations between events on an blood pressure (ABP) channel and on electrocardiogramm (ECG) channels are explicited to help in characterizing cardiac arrhythmias.

However, in a multi source learning problem, the amount of data and the expressiveness of the language increases compared to mono source learning. These are well known problems in Inductive Logic Programming (ILP). Indeed, the computation time of ILP algorithms grows with the amount of data and so does the size of the hypothesis search space when the expressiveness of the language increases. As pointed out by Fürnkranz [1], novel techniques are still needed to reduce the dimensionality of ILP problems, in particular, the example space and the hypothesis search space. In [2], Quinlan exposes a windowing technique improved later by Fürnkranz [3], to reduce the example space. To reduce the hypothesis search space, many methods have been proposed in ILP, one of them is using a declarative bias [4]. Such a bias aims either at limiting the search space or at deciding in what order hypotheses are to be considered and which hypothesis is better than another.

J.-F. Boulicaut et al. (Eds.): PKDD 2004, LNAI 3202, pp. 503–505, 2004.

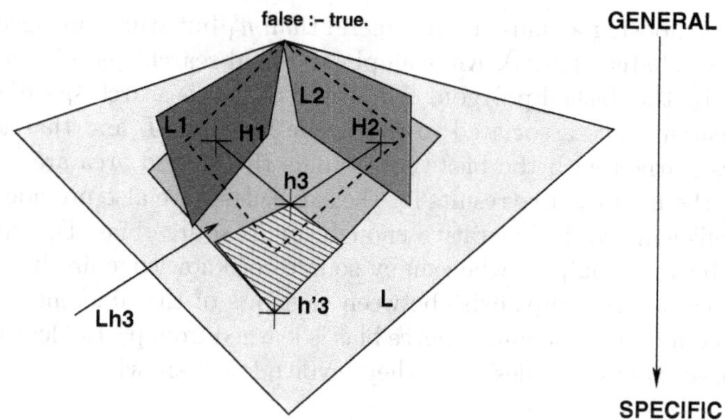

Fig. 1. Hypotheses search space for biased multi source learning

Designing an efficient global bias to cope with the multi source problems is a difficult task. We propose a divide-and-conquer strategy which consists in learning rules independently from each source and then, in using the learned rules to bias a new learning step on the whole dataset.

2 Biased Multi-source Learning

We assume that the reader is familiar with the ILP paradigm (see [5] for details). In order to reduce the multi source learning search space we propose a method to construct a declarative language bias from a bottom clause [5] that bounds this space. In the following, we assume that a language bias can be built from a clause. The method to build such a bias is not detailed in this article.

We consider the case of data coming from two sources. This case can be straightforwardly extended to any number of sources. L_1 (resp. L_2) represents the language of source 1 (resp. 2), and L the multi source language ($L \supseteq L_1 \cup L_2$). A straightforward multi source learning method using the whole language L is complex and takes much computation time. To reduce the search space, and thus the learning complexity, we propose a two step method. Firstly, discriminating rules (H_1 and H_2) are learned on each source independently. The rules are used to select in the language L the literals that are useful to compute the multi source rules. Secondly, another learning step is performed on the whole dataset but with a reduced hypothesis language and so, a reduced search space.

Let L_{h_3} be the language created with the literals of H_1 and H_2 and all literals in L that describe relationships between elements of H_1 and elements of H_2. To design the bias, we focus on two interesting clauses in the multi source search space (cf. Figure 1). The first one is the clause $h_3 = GSS(H_1 \cup H_2)$ (GSS: Greatest Specialization under θ-subsumption [6]). From this clause, the most restrictive bias can be created. However, this clause does not include any relations between the two sources, which can be interesting when some relationship exists. The second interesting clause is h'_3, the most specific clause of L_{h_3}. When knowledge about the expected results is available, the bias can be restricted by choosing a clause more general than h_3' but still more specific than h_3 (in the crosshatched area). An example of a final search space is represented in Figure 1 by the dashed polygon. We proved that this search space is smaller than the search space associated to the whole language L and that the multi source rules learned with the bias that bounds the dashed area are at least as accurate as the mono source results for the same class. We also provide a way to design an efficient, while generative enough, multi source bias. Designing such a bias can be very complex when many sources of knowledge are involved and when complex relationships exist between elements of the different sources. In the presented method, the multi source bias is learned from partial learning steps that use biases simpler to design as they involve fewer knowledge.

3 Preliminary Results

The two step method is compared to the straightforward approach. Experiments were done on five arrhythmias. We focus on two particular arrhythmias: supra-ventricular tachycardia (*svt*) and ventricular extra-systole (*ves*). The predictive accuracy of the learned rules is estimated by a "leave-one-out" cross-validation method because of the relatively small number of positive (about 6) and negative (about 30) examples for each arrhythmia. Rules learned for *esv* using the two methods are sound and complete. Accuracy is 0.973 for the *svt* class with both methods. This is caused by one exceptional negative example misclassified as true positive. The rules learned straightforwardly are less complex (they have about 25 % less literals). This can be explained by the fact that the bias is looser in the straightforward setting than in the biased setting: the shortest clause learned in the former setting has been excluded by the bias constructed from the mono source learned rules. However, the biased learning method reduces computation time very much. The whole computation (mono source then biased multi source) lasted 22.3 CPU seconds for the *ves* class and 49.89 CPU seconds for the *svt* class. These computation times are multiplied respectively by about 10 and 16 for the straightforward learning.

References

1. Fürnkranz, J.: Dimensionality reduction in ILP: A call to arms. In de Raedt, L., Muggleton, S., eds.: Proceedings of the IJCAI-97 Workshop on Frontiers of Inductive Logic Programming, Nagoya, Japan (1997) 81–86
2. Quinlan, J.R.: Learning efficient classification procedures and their application to chess end games. In Michalski, R.S., Carbonell, J.G., Mitchell, T.M., eds.: Machine Learning: An Artificial Intelligence Approach. Springer (1983) 463–482
3. Fürnkranz, J.: Integrative windowing. Journal of Artificial Intelligence Research **8** (1998) 129–164
4. Nedellec, C., Rouveirol, C., Ade, H., Bergadano, F., Tausend, B.: Declarative bias in ILP. In Raedt, L.D., ed.: Advances in Inductive Logic Programming. (1996) 82–103
5. Muggleton, S., De Raedt, L.: Inductive logic programming: Theory and methods. The Journal of Logic Programming **19 & 20** (1994) 629–680
6. Nienhuys-Cheng, S.H., de Wolf, R.: Least generalisations and greatest specializations of sets of clauses. Journal of Artificial Intelligence Research **4** (1996) 341–363

The Anatomy of SnakeT: A Hierarchical Clustering Engine for Web-Page Snippets*

Paolo Ferragina and Antonio Gullì

Dipartimento di Informatica, Università di Pisa
{ferragina,gulli}@di.unipi.it

The purpose of a search engine is to retrieve from a given textual collection the documents deemed *relevant* for a user query. Typically a user query is modeled as a set of keywords, and a document is a Web page, a `pdf` file or whichever file can be parsed into a set of tokens (words). Documents are *ranked* in a flat list according to some measure of *relevance to the user query.* That list contains hyperlinks to the relevant documents, their titles, and also the so called *(page or web) snippets*, namely document excerpts allowing the user to understand if a document is indeed relevant without accessing it.

Even if search engines are successfully used by millions of users to search the web everyday, it is well-known nowadays that the *flat ranked list* is getting obsolete, if not useless, due to the increasing complexity of the web retrieval problem. The key difficulty lies within the definition of what is *"relevant"* to the user issuing a query: Relevance is subjective and may change over the time.

Motivated by these considerations, many IR-tools [1] are currently being designed to help the users in their difficult search task by means of novel ways for reporting the query results. Two success stories are given by Vivisimo and Dogpile that were elected as the best meta-search engines of the last three years by a jury composed by about 500 web users, as reported by SearchEngineWatch.com[1]. These IR-tools add to the flat list of relevant documents a *hierarchy of folders* built on-the-fly over the snippets returned by one (or more) search engine(s). Each folder is *labeled* via a meaningful sentence that captures the "theme" of the snippets (and, thus, of the corresponding documents, or web pages) clustered into it. As a result, users are provided with a concise, but intelligible, picture of numerous query results at various levels of details. Hence the user has no longer to scan the flat list, but may browse the labeled folders according to the "intent behind his/her query". An example is provided in Figure 1.

Various scientific papers have recently investigated this challenging problem [8, 10, 7, 2, 6, 3, 5, 9, 4]. However, none of them achieved results comparable to Vivisimo or DogPile; and none of them provided a web interface with a detailed description of their architecture. The best results to date are "Microsoft and IBM products" [4, 9] not publicly accessible, as the authors communicated to us!

* Partially supported by the Italian MIUR projects ALINWEB and ECD, and by the Italian Registry of `ccTLD.it`.

[1] This IR-tool was introduced in a primitive form by Northernlight, and then improved by many other commercial softwares such as Copernic, iBoogie, Kartoo and Groxis. Unfortunately, very little information is available about their underlying technology!

J.-F. Boulicaut et al. (Eds.): PKDD 2004, LNAI 3202, pp. 506–508, 2004.

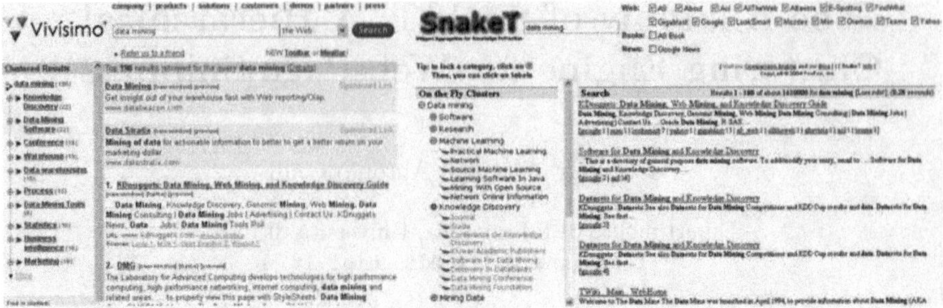

Fig. 1. VIVISIMO and SNAKET tested on the query "*Data Mining*".

In this paper we investigate the *web snippet hierarchical clustering* problem by devising an algorithmic solution, and a software prototype (called SnakeT, `http://roquefort.di.unipi.it/`), that offers some distinguishing features (see Figure 2 for details):

1. It draws the snippets from 16 search engines, the Amazon collection of books (`a9.com`) and the Google News.
2. In response to a user query, it builds *on-the-fly* the folders and labels them with sentences of variable length drawn from the snippets as *non* contiguous sequences of terms. This is crucial to ensure the readability of the labels (a lá Vivisimo), follows the "future directions of research" of Grouper [8], and extends the ideas in [6].
3. It uses some innovative ranking functions based on *two knowledge bases* (KB), built off-line and used at query time. One KB offers some *link and anchor-text information* derived by a crawling of 50 millions of Web pages done via the *Nutch open-source spider*. This information is used to provide "better candidate sentences". The other KB consists of the *Open Directory Project (DMOZ)* and is used to rank the candidate sentences according to their frequency and position of occurrence within the DMOZ's categories.
4. It organizes the folders into a *hierarchy*, and possibly let them overlap. This process is driven by the *set of (approximate) sentences* assigned to each folder and describing its content at various levels of details. The label eventually assigned to a folder will take into account the set of sentences assigned to its children, as well as some coverage and some label distinctiveness measures. As a result, the labels lying over any hierarchy's path are *not necessarily one the substring of the other*, and thus they result much more descriptive. The overall time complexity is linear in the number of candidate labels, and takes few seconds. The picture offered by the hierarchy of labeled folders recalls a sort of knowledge extraction process, which is useful to drive the subsequent user browsing (see the Figure 1).

The web interface available for testing (see Figure 1) and some user studies are detailed in a companion paper published in this proceedings. Many issues remain to be further investigated: We are performing user studies on *free queries*, and on a larger and more variegate set of users; we are investigating the application of SnakeT to other contexts, like *blogs*, news and books; and we are studying the use of SnakeT as a *personalized ranker*, in order to make the flat list of results more effective for resolving the user queries. Check the web page of SnakeT for the latest news.

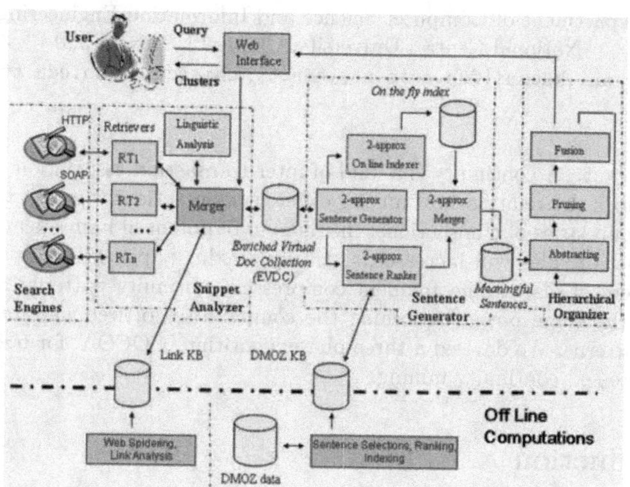

Fig. 2. Architecture of SnakeT.

References

1. CNN.com. Better search results than Google? Next-generation sites help narrow internet searches. Associated Press, January 2004.
2. B. Fung, K. Wang, and M. Ester. Large hierarchical document clustering using frequent itemsets. *SIAM International Conference on Data Mining*, 2003.
3. F. Giannotti, M. Nanni, D. Pedreschi, and F. Samaritani. Webcat: Automatic categorization of web search results. In *SEBD*, pages 507–518, 2003.
4. K. Kummamuru, R. Lotlikar, S. Roy, K. Singal, and R. Krishnapuram. A hierarchical monothetic document clustering algorithm for summarization and browsing search results. In *WWW*, 2004.
5. D. J. Lawrie and W. B. Croft. Generating hiearchical summaries for web searches. In *ACM SIGIR*, pages 457–458, 2003.
6. Y. S. Maarek, R. Fagin, I. Z. Ben-Shaul, and D. Pelleg. Ephemeral document clustering for web applications. Technical Report RJ 10186, IBM Research, 2000.
7. D. Weiss and J. Stefanowski. Web search results clustering in polish: Experimental evaluation of Carrot. In *New Trends in I.I.P. and Web Mining Conference*, 2003.
8. O. Zamir and O. Etzioni. Grouper: a dynamic clustering interface to Web search results. *Computer Networks*, 31:1361–1374, 1999.
9. H. Zeng, Q. He, Z. Chen, and W. Ma. Learning to cluster web search results. In *ACM SIGIR*, 2004.
10. D. Zhang and Y. Dong. Semantic, hierarchical, online clustering of web search results. In *WIDM*, 2001.

COCOA: Compressed Continuity Analysis for Temporal Databases

Kuo-Yu Huang, Chia-Hui Chang, and Kuo-Zui Lin

Department of Computer Science and Information Engineering,
National Central University, Chung-Li, Taiwan 320
{want,kuozui}@db.csie.ncu.edu.tw, chia@csie.ncu.edu.tw

Abstract. A continuity is a kind of inter-transaction association which describes the relationships among different transactions. Since it breaks the boundaries of transactions, the number of potential itemsets and the number of rules will increase drastically. In this paper we consider the problem of discovering frequent compressed continuity patterns, which have the same power as mining the complete set of frequent continuity patterns. We devised a three-phase algorithm, COCOA, for frequent compressed continuity mining.

1 Introduction

Patterns mining plays an important role in data mining areas. Most of the previous studies, such as frequent itemsets and sequential patterns are on mining associations among items within the same transaction. This kind of the patterns describe how often items occur together. In spite the above frequent patterns reflect some relationship among the items, it is not a suitable and definite rule for trend prediction. Therefore, we need a pattern that shows the temporal relationships between items in different transactions. In order to distinguish these two kinds of associations, we call the former task as **intra-transaction association mining**, the latter task as **inter-transaction association mining**. There are three kinds of inter-transaction patterns defined recently, including frequent episodes [2], periodic patterns[4] and inter-transaction associations [3].

An episode is defined to be a collection of events in a specific window interval that occur relatively close to each other in a given partial order. Take Fig. 1 as an example, there are six matches of episode $< AC, BD >$, from E_1 to E_6, in temporal database TD. Unlike episodes, a periodic pattern considered not only the order of events but also the exact positions of events. To form periodicity, a list of k disjoint matches is required to form a contiguous subsequence with k satisfying some predefined minimum repetition threshold. For example, pattern (AC,*,BD) is a periodic pattern that matches P_1, P_2, and P_3, three contiguous and disjoint matches. The notion of inter-transaction association mining is firstly used in [3] by Tung et al. In order to distinguish this kind of inter-transaction patterns from episodes and periodic patterns, we call them as "**continuity**" as suggested in [1]. For example, pattern [AC,*,BD] is a continuity with four matches P_1, P_2, P_3, and P_4 in Fig. 1.

J.-F. Boulicaut et al. (Eds.): PKDD 2004, LNAI 3202, pp. 509–511, 2004.

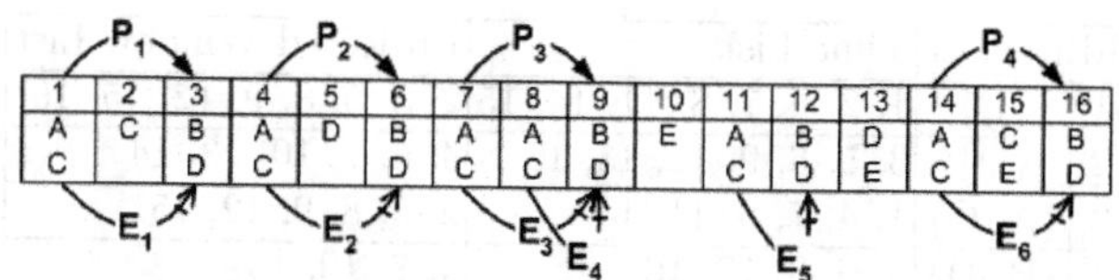

Fig. 1. An example of temporal database TD.

Since inter-transaction associations break the boundaries of transactions, the number of potential itemsets and the number of rules will increase drastically. This reduces not only efficiency but also effectiveness since users have to sift through a large number of mined rules to find useful ones. In this paper, we proposed the mining of frequent **compressed** continuities, which has the same power as mining the complete set of frequent continuities. A compressed continuity is composed of only closed itemsets and the don't-care characters. We devised a three-phase algorithm, COCOA (Compressed Continuity Analysis), for mining frequent compressed continuities from temporal databases with a utilization of both horizontal and vertical formats.

Therefore, the problem is formulated as follows: given a minimum support level *minsup* and a maximum window bound level *maxwin*, our task is to mine all frequent compressed continuities from temporal databases with support greater than *minsup* and window bound less than *maxwin*.

2 COCOA: Compressed Continuities Analysis

This section describes our frequent compressed continuities mining algorithm phase by phase. The algorithm uses both horizontal and vertical database formats. The COCOA algorithm consists of following three phases:

- Phase I: Mining Frequent Closed Itemsets. Since the COCOA algorithm uses a vertical database format, frequent closed itemsets are mined using a vertical mining algorithm, CHARM[5]. For example, there are four closed frequent itemsets in Fig. 1: {AC}, {C}, {BD}, {B}.
- Phase II: Pattern Encoding and Database Recovery. For each frequent closed itemset in TD, we assign a unique number to denote them (see Fig. 2(b)). Next, based on the time lists of the encoded patterns, we construct a recovered horizontal database, RD, as shown in Fig. 3.
- Phase III: Mining Frequent Compressed Continuities. Starting from 1-continuity pattern, we use depth-first enumeration to discover all frequent compressed continuity patterns. For each event (frequent closed itemset) in RD, we calculate its projected window list (PWL, see Fig. 2) from the event's timelist and find all frequent events in its PWL through recovered database RD. We then output the frequent continuity formed by current pattern and the frequent events in the PWL. For each extension pattern, the process is applied recursively to find all frequent continuities until the PWL becomes empty or the window of a continuity is greater than the *maxwin*. Finally, the frequent compressed continuities should be decoded from its primal symbol sets.

Id	itemset	Time List	Projected Window List
[1]	{C}	1, 2, 4, 7, 8, 11, 14, 15	2, 3, 5, 8, 9, 12, 15, 16
[2]	{D}	3, 5, 6, 9, 12, 13, 16	4, 6, 7, 10, 13, 14
[3]	{A, C}	1, 4, 7, 8, 11, 14	2, 5, 8, 9, 12, 15
[4]	{B, D}	3, 6, 9, 12, 16	4, 7, 10, 13

Fig. 2. Frequent closed itemsets for Fig. 1

Time	1	2	3	4	5	6	7	8	9	10	11	12	13	14	15	16
Encoded Pattern	[1] [3]	[1]	[2] [4]	[1] [3]	[2]	[2] [4]	[1] [3]	[1] [3]	[2] [4]		[1] [3]	[2] [4]	[2]	[1] [3]	[1]	[2] [4]

Fig. 3. The recovered horizontal database for Fig. 1

3 Conclusion and Future Work

The performance study shows that the algorithm proposed is both efficient and scalable, and is about an order of magnitude faster than FITI. Readers may consult the authors for the complete paper. For future work, maintaining closed patterns can further reduce the number of redundant patterns and save additional efforts for unnecessary mining. Therefore, more efforts need to be involved in the inter-transaction association mining.

Acknowledgement

This work is sponsored by NSC, Taiwan under grant NSC92-2213-E-008-028.

References

1. K. Y. Huang, C. H. Chang, and Kuo-Zui Lin. Prowl: An efficient frequent continuity mining algorithm on event sequences. In *Proceedings of 6th International Conference on Data Warehousing and Knowledge Discovery (DaWak'04)*, 2004. To Appear.
2. H. Mannila, H. Toivonen, and A. I. Verkamo. Discovering frequent episodes in event sequences. *Data Mining and Knowledge Discovery (DMKD)*, 1(3):259–289, 1997.
3. A. K. H. Tung, H. Lu, J. Han, and L Feng. Efficient mining of intertransaction association rules. *IEEE Transactions on Knowledge and Data Engineering (TKDE)*, 15(1):43–56, 2003.
4. J. Yang, W. Wang, and P.S. Yu. Mining asynchronous periodic patterns in time series data. *IEEE Transaction on Knowledge and Data Engineering (TKDE)*, 15(3):613–628, 2003.
5. M.J. Zaki and C.J. Hsiao. Charm: An efficient algorithm for closed itemset mining. In *Proc. of 2nd SIAM International Conference on Data Mining (SIAM 02)*, 2002.

Discovering Interpretable Muscle Activation Patterns with the Temporal Data Mining Method

Fabian Mörchen, Alfred Ultsch, and Olaf Hoos

Data Bionics Research Group & Department of Sports Medicine
University of Marburg, D-35032 Marburg, Germany

Abstract. The understanding of complex muscle coordination is an important goal in human movement science. There are numerous applications in medicine, sports, and robotics. The coordination process can be studied by observing complex, often cyclic movements, which are dynamically repeated in an almost identical manner. In this paper we demonstrate how interpretable temporal patterns can be discovered within raw EMG measurements collected from tests in professional In-Line Speed Skating. We show how the *Temporal Data Mining Method*, a general framework to discover knowledge in multivariate time series, can be used to extract such temporal patterns. This representation of complex muscle coordination opens up new possibilities to optimize, manipulate, or imitate the movements.

1 Temporal Data Mining Method

The time series Knowledge Discovery framework *Temporal Data Mining Method* (TDM) [1] is a framework of methods and algorithms to mine rules in multivariate time series (MVTS). The patterns are expressed with the hierarchical temporal rule language *Unification-based Temporal Grammar* (UTG) [2]. The whole process is data driven and the resulting rules can be used for interpretation and classification.

Our temporal rule mining framework decomposes the problem of rule discovery into the mining of single temporal concepts. The resulting rules have a hierarchical structure corresponding to these concepts. This gives unique opportunities in relevance feedback during the Knowledge Discovery process and in the interpretation of the results. An expert can focus on particularly interesting rules and discard valid but known rules before the next level constructs are searched.

At each hierarchical level the grammar consists of semiotic triples: a unique symbol (syntax), a grammatical rule (semantic), and a user defined label (pragmatic). The grammatical rule is produced by a mining algorithm for this hierarchy level. An expert is needed to interpret the rule and complete the triple with a meaningful label.

After preprocessing and feature extraction the time series should be grouped into possibly overlapping subsets, called *Aspects*, related w.r.t. the investigated

J.-F. Boulicaut et al. (Eds.): PKDD 2004, LNAI 3202, pp. 512–514, 2004.

problem domain. The remaining TDM steps correspond to the hierarchy levels of the UTG and are described below along with the application.

2 Discovering Movement Patterns

The goal in analyzing a multivariate time series from sports medicine was to identify typical muscle activation patterns during In-Line Speed Skating, which is a cyclic movement with complex inter-muscular coordination pattern. The Knowledge Discovery process started with multivariate electrical signals from kinesiological EMG (3 major leg muscles), angle sensors (hip, knee, ankle), and an inertia switch indicating the ground contact. The analysis resulted in a high level description of the typical movement cycle.

One Aspect per muscle and one for the ground contact were created. The angles formed a multivariate Aspect including the angle velocities. After some preprocessing, e.g. noise filtering, the Aspects needed to be discretized to meaningful state labels, called *Primitive Patterns.*

The multivariate angle Aspect was discretized using spatial clustering with Emergent Self-Organizing Maps (ESOM) [3] and the accompanying visualizations U-Matrix, P-Matrix, and U*-Matrix [4]. The cluster descriptions were found with the rule generation algorithm Sig* [5]. Even though these clusters were found using no time information, the resulting Primitive Patterns were continuous in time. An expert was able to identify and label the clusters as movement phases, e.g. *forward gliding*.

The univariate muscle activity series were discretized using log-normal mixture models. Consulting the expert, 3 states corresponding to *low*, *medium*, and *high* were chosen. The Aspect for the inertia sensor was discretized into two states, namely *on* and *off*.

The union of consecutive Primitive Patterns compose a *Succession* representing the concept of duration. Short, physiologically not plausible, interruptions of otherwise persisting states were filtered out.

The coincidence of several Successions from the different Aspects is an *Event.* The minimum length for Events was chosen to be 50ms, because any co-ordinated muscle activation in a complex movement has at least this duration. The minimum count for an event was set to 10, slightly below the number of movement cycles present in the time series (12-20).

With these setting 5 Events listed in Table 1 were found. The Event labels were given by the expert in order to summarize the set of Succession labels.

Within the Events we searched for typical *Sequences*, representing the concept of order. The minimum length of a sequence was set to 3 and the minimum count to 5. The resulting Sequences are listed in Table 2.

The last TDM step joins several similar Sequences based on the concept of alternative. Due to the small number of Sequences found, and the striking similarity among them, a single Temporal Pattern was constructed manually from the rules describing the 3 Sequences and was labeled as the *Total Movement Cycle in In-line Speed-Skating.*

Table 1. List of interesting Events with statistics and Succession labels.

Symbol	Label	Count	Duration (ms)			Trigger	Move	Muscles		
			Min	Mean	Max			GA	VM	GM
E1	active gliding	29	53	150	476	on	fwd. gliding	medium	high	high
E4	initial gliding	16	119	280	375	on	stabilization	medium	high	high
E5	anticipation	15	217	274	336	off	prep. foot	low	low	low
E8	weight transfer	10	57	91	160	on	fwd. gliding	high	high	high
E10	relaxation	11	51	93	170	off	leg swing	low	low	low

Table 2. List of interesting Sequences with statistics and short UTG notation.

Symbol	Label	Count	Duration (ms)			UTG
			Min	Mean	Max	
S22	contraction & relaxation	12	1378	1848	3310	E1→E4→E5
S24	movement cycle	8	1297	1596	1848	E1→E4→E5→E8
S16	movement cycle +	5	1547	2143	3773	E1→E4→E5→E8→E10

3 Summary

Based on EMG and kinematic measurements the TDM successfully identified functional movement patterns in In-Line Speed Skating. We were able to make the difficult transition from the raw measurement data to interpretable patterns via increasingly complex knowledge representations. Each parameter in the discovery process was carefully selected based on previous results. The ultimate goal of Knowledge Discovery was reached, because the model was interpreted by the domain expert and found novel and interesting. Note that it can also be automatically evaluated on new data. We plan on repeating the method with data from other skaters running at various speeds and investigate possible differences among the resulting patterns.

References

1. Mörchen, F., Ultsch, A.: Discovering temporal knowlegde in multivariate time series. In: GfKl 2004, Dortmund, Germany. (2004)
2. Ultsch, A.: Unification-based temporal grammar. Technical Report 37, Philipps-University Marburg, Germany (2004)
3. Ultsch, A.: Data mining and knowledge discovery with emergent self-organizing feature maps for multivariate time series. In Oja, E., Kaski, S., eds.: Kohonen Maps. (1999) 33–46
4. Ultsch, A.: U*-matrix: a tool to visualize clusters in high dimensional data. Technical Report 36, Philipps-University Marburg, Germany (2004)
5. Ultsch, A.: Connectionistic models and their integration in knowledge-based systems (german) (1991)

A Tolerance Rough Set Approach to Clustering Web Search Results*

Chi Lang Ngo and Hung Son Nguyen

Institute of Mathematics, Warsaw University
Banacha 2, 02-097 Warsaw, Poland
chilang@chilang.com, son@mimuw.edu.pl

Extended Abstract

Two most popular approaches to facilitate searching for information on the web are represented by web search engine and web directories. Although the performance of search engines is improving every day, searching on the web can be a tedious and time-consuming task due to the huge size and highly dynamic nature of the web. Moreover, the user's "intention behind the search" is not clearly expressed which results in too general, short queries. Results returned by search engine can count from hundreds to hundreds of thousands of documents.

One approach to manage the large number of results is clustering. Search results clustering can be defined as *a process of automatical grouping search results into to thematic groups*. However, in contrast to traditional document clustering, clustering of search results are done on-the-fly (per user query request) and locally on a limited set of results return from the search engine. Clustering of search results can help user navigate through large set of documents more efficiently. By providing concise, accurate description of clusters, it lets user localizes interesting document faster.

In this paper, we proposed an approach to search results clustering based on Tolerance Rough Set following the work on document clustering [4, 3]. Tolerance classes are used to approximate concepts existed in documents. The application of Tolerance Rough Set model in document clustering was proposed as a way to enrich document and cluster representation with the hope of increasing clustering performance.

Tolerance Rough Set Model: (TRSM) was developed in [3] as basis to model documents and terms in information retrieval, text mining, etc. With its ability to deal with vagueness and fuzziness, TRSM seems to be promising tool to model relations between terms and documents. In many information retrieval problems, defining the similarity relation between document-document, term-term or term-document is essential.

Let $D = \{d_1, \ldots, d_N\}$ be a set of documents and $T = \{t_1, \ldots, t_M\}$ set of *index terms* for D. TRSM is an approximation space (see [5]) $\mathcal{R} = (T, I_\theta, \nu, P)$ determined over the set of terms T (universe of $\mathcal{R}$) as follows:

* The research has been partially supported by the grant 3T11C00226 from Ministry of Scientific Research and Information Technology of the Republic of Poland.

J.-F. Boulicaut et al. (Eds.): PKDD 2004, LNAI 3202, pp. 515–517, 2004.

1. Let $f_D(t_i, t_j)$ denotes the number of documents in D in which both terms t_i and t_j occurs. The parameterize uncertainty function I_θ is defined as $I_\theta(t_i) = \{t_j \mid f_D(t_i, t_j) \geq \theta\} \cup \{t_i\}$. The set $I_\theta(t_i)$ is called the *tolerance class* of index term t_i.
2. The vague inclusion function is defined as $\nu(X, Y) = \frac{|X \cap Y|}{|X|}$,
3. All tolerance classes of terms are considered as structural subsets: $P(I_\theta(t_i)) = 1$ for all $t_i \in T$.

In TRSM, the lower and upper approximations of any subset $X \subseteq T$ can be determined by

$$\mathbf{L}_{\mathcal{R}}(X) = \{t_i \in T \mid \nu(I_\theta(t_i), X) = 1\}; \quad \mathbf{U}_{\mathcal{R}}(X) = \{t_i \in T \mid \nu(I_\theta(t_i), X) > 0\}$$

By varying the threshold θ (e.g. relatively to the size of document collection), one can control the degree of relatedness of words in tolerance classes. The use of upper approximation in similarity calculation to reduce the number of zero-valued similarities is the main advantage main advantage TRSM-based algorithms claimed to have over traditional approaches. This makes the situation, in which two document are similar (i.e. have non-zero similarity) although they do not share any terms, possible. Let us mention two basic applications of TRSM in text mining area:

1. Enriching document representation: In TRSM, the document $d_i \in D$ is represented by its upper approximation:

$$\mathbf{U}_{\mathcal{R}}(d_i) = \{t_i \in T \mid \nu(I_\theta(t_i), d_i) > 0\}$$

2. Extended weighting scheme for upper approximation: To assign weight values for document's vector, the TF*IDF weighting scheme is used (see [6]). In order to employ approximations for document, the weighting scheme need to be extended to handle terms that occurs in document's upper approximation but not in the document itself. The extended weighting scheme is defined as:

$$w_{ij} = \begin{cases} (1 + log(f_{d_i}(t_j)) * \log \frac{N}{f_D(t_j)} & \text{if } t_j \in d_i \\ \min_{t_k \in d_i} w_{ik} * \frac{\log \frac{N}{f_D(t_j)}}{1 + \log \frac{N}{f_D(t_j)}} & \text{if } t_j \in \mathbf{U}_{\mathcal{R}}(d_i) \backslash d_i \\ 0 & \text{if } t_j \notin \mathbf{U}_{\mathcal{R}}(d_i) \end{cases}$$

The extension ensures that each terms occurring in upper approximation of d_i but not in d_i, has a weight smaller than the weight of any terms in d_i.

The TRC Algorithm: The Tolerance Rough set Clustering algorithm is based primarily on the K-means algorithm presented in [3]. By adapting K-means clustering method, the algorithm remain relatively quick (which is essential for on-line results post-processing) while still maintaining good clusters quality. The usage of Tolerance Space and upper approximation to enrich inter-document and document-cluster relation allows the algorithm to discover subtle similarities not detected otherwise. As it has been mentioned, in search results clustering, the proper labelling of cluster is as important as cluster contents quality.

Since the use of phrases in cluster label has been proven [7] to be more effective than single words, TRC algorithm utilize n-gram of words (phrases) retrieved from documents inside cluster as candidates for cluster description. The TRC

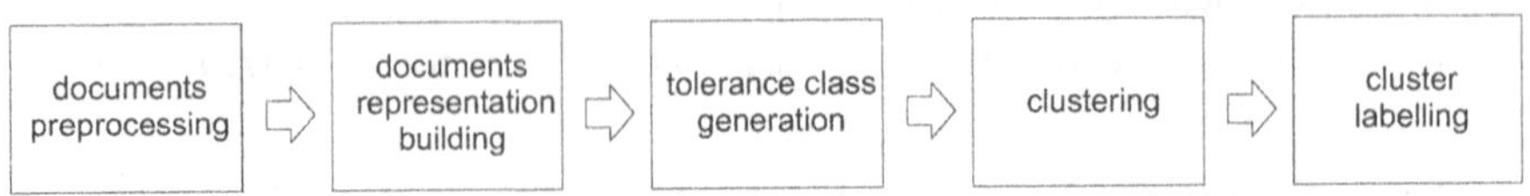

Fig. 1. Phases of TRC algorithm

algorithm consists of five phases (depicted in Fig .1). It is widely known (see [2]) that preprocessing text data before feeding it into clustering algorithm is essentials and can have great impact on algorithm performance. In TRC, the following standard preprocessing steps are performed on snippets: *text cleansing, text stemming, and Stop-words elimination.* As TRC utilizes Vector Space Model for creating document-term matrix representing documents, in *document representation building* step, two main standard procedures: *index term selection and term weighting* are performed.

We have implemented the proposed solution within an open-source framework, *Carrot*^{2}. The implementation of algorithm presented in this paper, including all source codes, will be contributed to the *Carrot*^{2} project and will be available at http://carrot2.sourceforge.net. We hopes that this will foster further experiments and enhancements to the algorithm to be made, not only by the author but also other researchers.

References

1. Baeza-Yates, R., Ribeiro-Neto, B.: Modern Information Retrieval. 1st edn. Addison Wesley Longman Publishing Co. Inc. (1999)
2. Han, J., Kamber, M.: Data Mining: Concepts and Techniques. 1st edn. Morgan Kaufmann (2000)
3. Ho, T.B, Nguyen, N.B.: Nonhierarchical document clustering based on a tolerance rough set model. International Journal of Intelligent Systems **17** (2002) 199–212
4. Kawasaki, S., Nguyen, N.B., Ho, T.B.: Hierarchical document clustering based on tolerance rough set model. In Zighed, D.A., Komorowski, H.J., Zytkow, J.M., eds.: Principles of Data Mining and Knowledge Discovery, 4th European Conference, PKDD 2000, Lyon, France, September 13-16, 2000, Proceedings. Volume 1910 of Lecture Notes in Computer Science., Springer (2000)
5. Skowron, A., Stepaniuk, J.: Tolerance approximation spaces. Fundamenta Informaticae **27** (1996) 245–253
6. Salton, G.: Automatic text processing: the transformation, analysis, and retrieval of information by computer. Addison-Wesley Longman Publishing Co., Inc. (1989)
7. Zamir, O., Etzioni, O.: Grouper: a dynamic clustering interface to web search results. Computer Networks (Amsterdam, Netherlands: 1999) **31** (1999) 1361–1374

Improving the Performance of the RISE Algorithm

Aloísio Carlos de Pina and Gerson Zaverucha

Universidade Federal do Rio de Janeiro, COPPE/PESC, Department of Systems Engineering and Computer Science, C.P. 68511 - CEP. 21945-970, Rio de Janeiro, RJ, Brazil
{long,gerson}@cos.ufrj.br

Abstract. RISE is a well-known multi-strategy learning algorithm that combines rule induction and instance-based learning. It achieves higher accuracy than some state-of-the-art learning algorithms, but for large data sets it has a very high average running time. This work presents the analysis and experimental evaluation of SUNRISE, a new multi-strategy learning algorithm based on RISE, developed to be faster than RISE with similar accuracy.

1 The SUNRISE Algorithm

RISE (Rule Induction from a Set of Exemplars) [2] induces all the rules together. If a generalization of a rule has positive or null effect on the global accuracy, the change is kept. The RISE algorithm is presented in Table 1.

SUNRISE tries to generalize the rules more than once before including them in the rule set and only accepts the changes if the effect on the global accuracy is strictly positive; i.e., a new rule is only added to the rule set if the set achieves a higher accuracy than before its inclusion. The SUNRISE algorithm is presented in Table 2. The fact that SUNRISE does not use Occam's Razor increases the algorithm's speed because it increases the probability of no modification in the rule set after an iteration of the outermost loop Repeat, thus causing the algorithm's stop. Only after k generalizations a rule is evaluated to determine if it must or not belong to the rule set. It makes SUNRISE faster than RISE, since the latter evaluates each generalization made in each rule. The value k is a parameter of the SUNRISE algorithm whose value has to be experimentally determined.

2 Experimental Evaluation

In the experiments, 22 data sets [1] were used to compare the performance of the new algorithm, SUNRISE, to that of the RISE algorithm. The test method used in this research was the paired t test with n-fold cross-validation [5]. To adjust the parameter k, an internal cross-validation was made [5]. The value for k that achieved better performance in most of the data sets was $k \leq 3$. All tests were carried through in a Pentium III 450MHz computer with 64MBytes RAM.

Table 3 presents the running time (training and testing) of each algorithm for each one of the data sets. The two last columns show the results obtained by the SUNRISE

J.-F. Boulicaut et al. (Eds.): PKDD 2004, LNAI 3202, pp. 518–520, 2004.

algorithm when using Occam's Razor, so as to evaluate independently the behavior of the system considering only the effect of the parameter *k*.

Table 1. The RISE algorithm.

```
Input: ES is the training set.
Procedure RISE (ES)
Let RS be ES.
Compute Acc(RS).
Repeat
  For each rule R in RS,
    Find the nearest example E to R not
    covered by it and of the same class of R.
    Let R'=MostSpecificGeneralization(R,E).
    Let RS' = RS with R replaced by R'.
    If Acc(RS') ≥ Acc(RS)
    Then Replace RS by RS',
      If R' is identical to another rule in RS,
      Then delete R' from RS.
Until no increase in Acc(RS) is obtained.
Return RS.
```

Table 2. The SUNRISE algorithm.

```
Input: ES is the training set, k is the SUNRISE
parameter.
Procedure SUNRISE (ES, k)
Let RS be ES.
Compute Acc(RS).
Repeat
  For each rule R in RS,
    Let R' be R.
    Repeat k times
      Find the nearest example E to R' not
      covered by it and of the same class
      of R'.
      R'=MostSpecificGeneralization(R',E).
    Let RS' = RS with R replaced by R'.
    If Acc(RS') > Acc(RS)
    Then Replace RS by RS',
      If R' is identical to another rule in RS,
      Then delete R' from RS.
Until no increase in Acc(RS) is obtained.
Return RS.
```

Table 3. Running times (in seconds).

Data Set	RISE	SUNRISE				SUNRISE-OR	
		$k = 0$	$k = 1$	$k = 2$	$k = 3$	$k = 2$	$k = 3$
Annealing	48.53	3.48	11.83	17.60	19.85	42.82	46.28
Chess endgames	2107.71	261.17	1171.89	1428.81	1744.14	1787.16	1519.58
Credit screening	174.30	5.68	26.94	34.19	42.43	138.04	137.82
DNA promoters	0.69	0.14	0.36	0.36	0.36	0.36	0.36
Echocardiogram	0.72	0.06	0.28	0.28	0.33	0.55	0.55
Glass	2.23	0.31	0.86	0.97	1.24	1.74	1.68
Heart disease	18.97	1.17	4.96	7.00	9.47	18.59	15.40
Hepatitis	5.75	0.20	0.80	0.97	1.24	5.25	3.39
Horse colic	51.34	1.39	6.17	8.26	10.35	37.60	32.82
Iris	0.91	0.09	0.25	0.25	0.31	0.69	0.58
LED	2.05	0.51	1.83	1.83	2.05	2.16	2.32
Liver disease	19.19	1.12	5.57	6.94	9.69	15.46	13.48
Mushroom	4930.79	1258.37	2116.78	2987.38	3868.70	4243.81	4900.24
Pima diabetes	241.17	5.24	33.59	35.51	50.13	182.76	159.74
Post-operative	0.14	0.08	0.08	0.03	0.08	0.08	0.08
Solar flare	4.74	1.39	2.38	2.98	3.42	4.19	4.14
Sonar	11.19	1.74	4.37	4.70	4.81	10.75	6.85
Soybean	6.89	3.09	4.03	4.36	4.91	5.46	5.57
Splice junctions	7043.75	790.24	3522.87	4622.93	5088.37	9785.68	10149.85
Thyroid disease	23767.00	138.92	938.75	1102.65	1461.28	17777.54	16439.19
Wine	4.04	0.31	0.80	1.08	1.30	3.00	2.56
Zoology	0.08	0.03	0.08	0.03	0.08	0.08	0.08

Since SUNRISE presents lower average running time than RISE and achieves good accuracy, the results show that the SUNRISE algorithm seems to be the more indicated choice when working with large data sets. A complete analysis of these experiments can be found in [4].

By adding partitioning [3] to SUNRISE, we expect that the obtained results for large data sets become even better.

3 Conclusions

Besides being faster than RISE, making possible the learning task in data sets intractable with the use of that algorithm, the SUNRISE algorithm showed to be capable to reach an average accuracy as good as that of the RISE algorithm and in some cases superior than that. In 110 tests carried through (not considering $k = 0$), in only 2 of them the SUNRISE algorithm was significantly slower than RISE, achieving a maximum speed-up of 96% in a large data set. In 12 tests the accuracy obtained by SUNRISE was significantly lower than that of RISE, but in 15 tests it was significantly higher.

In an algorithm like RISE, in which the evaluation of the quality of a rule is an expensive process, to generalize a rule more than once and only then to make its evaluation can provide a considerable increase in the algorithm's speed. The use of a more restrictive criterion of acceptance of a generalized rule (turning off the Occam's Razor) provides a final set with less simple rules, but in exchange it can greatly increase the speed of the system. Although these two techniques have been applied to RISE, they could be applied to other bottom-up learning algorithms.

Acknowledgment

We would like to thank Pedro Domingos for providing the source code of the RISE algorithm. The authors are partially financially supported by the Brazilian Research Agency CNPq.

References

1. Blake, C. L., Merz, C. J.: UCI Repository of Machine Learning Databases. Machine-readable data repository. University of California, Department of Information and Computer Science, Irvine, CA (1998) [http://www.ics.uci.edu/~mlearn/MLRepository.html]
2. Domingos, P.: Unifying Instance-Based and Rule-Based Induction. Machine Learning, Vol. 24 (1996) 141–168
3. Domingos, P.: Using Partitioning to Speed Up Specific-to-General Rule Induction. In: Proceedings of the AAAI-96 Workshop on Integrating Multiple Learned Models. AAAI Press, Portland, OR (1996) 29–34
4. de Pina, A. C., Zaverucha, G.: SUNRISE: Improving the Performance of the RISE Algorithm. Technical Report. Federal University of Rio de Janeiro, COPPE/PESC, Department of Systems Engineering and Computer Science, Rio de Janeiro, Brazil (2004) 1–12
5. Mitchell, T. M.: Machine Learning. McGraw-Hill, New York (1997)

Mining History of Changes to Web Access Patterns (Extended Abstract)

Qiankun Zhao and Sourav S. Bhowmick

Nanyang Technological University Singapore, 639798
{pg04327224,assourav}@ntu.edu.sg

1 Introduction

Recently, a lot of work has been done in web usage mining [2]. Among them, mining of frequent Web Access Pattern (WAP) is the most well researched issue[1]. The idea is to transform web logs into sequences of events with user identifications and timestamps, and then extract association and sequential patterns from the events data with certain *metrics*. The frequent WAPs have been applied to a wide range of applications such as personalization, system improvement, site modification, business intelligence, and usage characterization [2]. However, most of the existing techniques focus only on mining frequent WAP from snapshot web usage data, while web usage data is dynamic in real life. While the frequent WAPs are useful in many applications, knowledge hidden behind the historical changes of web usage data, which reflects how WAPs change, is also critical to many applications such as adaptive web, web site maintenance, business intelligence, etc.

In this paper, we propose a novel approach to discover hidden knowledge from historical changes to WAPs. Rather than focusing on the occurrence of the WAPs, we focus on the frequently changing web access patterns. We define a novel type of knowledge, *Frequent Mutating WAP (FM-WAP)*, based on the historical changes of WAPs. The FM-WAP mining process consists of three phases. Firstly, web usage data is represented as a set of WAP trees and partitioned into a sequence of *WAP groups* (subsets of the WAP trees) according to a user-defined *calendar pattern*, where each WAP group is represented as a *WAP forest*. Consequently, the log data is represented by a sequence of WAP forests called WAP history. Then, changes among the WAP history are detected and stored in the *global forest*. Finally, the FM-WAP is extracted by a traversal of the *global forest*. Extensive experiments show that our proposed approach can produce novel knowledge of web access patterns efficiently with good scalability.

2 FM-WAP Mining

Given a WAP forests sequence, to measure the significance and frequency of the changes to the support of the WAPs, two metrics, *S-value* and *F-value*, are proposed. Formally, they are defined as follows.

J.-F. Boulicaut et al. (Eds.): PKDD 2004, LNAI 3202, pp. 521–523, 2004.

Definition (S-Value) Let F_{wi} and $F_{w(i+1)}$ be two WAP forests in the WAP history. Let T_{wk} be a subtree of F_{wi}, $T_{w(k+1)}$ be the new version of T_{wk} in $F_{w(i+1)}$. Then the *S-value* of T_{wi} is defined as

$$S_i(T_{wi}) = \frac{|support(T_{w(k+1)}) - support(T_{wk})|}{max(support(T_{wk}),\ support(T_{w(k+1)}))}$$

Given a threshold α for *S-value*, a WAP tree T_{wi} **changed significantly** from F_{wi} to $F_{w(i+1)}$ if $S_i \geq \alpha$.

Here $support(T_{w(k+1)})$ denotes the support values of $T_{w(k+1)}$ in WAP forest $F_{w(i+1)}$. *S-value* is defined to represent the significance of changes to the support values of a WAP in two consecutive WAP forests. Given a WAP forest sequence, there will be a sequence of *S-values* for a specific WAP tree. It can be observed that *S-value* is between 0 and 1. A larger *S-value* implies a more significant change.

Definition (F-Value) Let H be a WAP history and $\langle F_{w1}, F_{w2}, \cdots, F_{wn} \rangle$. Let $\langle S_1(T_{wk}), S_2(T_{wk}), \cdots, S_{n-1}(T_{wk}) \rangle$ be the sequence of S-values of $T_{wk} \in F_{wj}$, for $1 \leq j \leq n$. Let α be the threshold for the S-value. Then the F-value for T_{wk} is defined as: $\mathrm{F}(T_{wk}, \alpha) = \frac{\sum_{i=1}^{n} f_i}{n}$, where $f_i = 1$, if $S_i(T_{wk}) \geq \alpha$; else $f_i = 0$.

The *F-value* for a tree is defined to represent the percentage of times this tree changed significantly against the total number of WAP forests in the history. The *F-value* is based on the threshold α for *S-value*. It can also be observed that *F-value* is between 0 and 1. Based on the *S-value* and *F-value* metrics, *frequently mutating web access pattern*(FM-WAP) is defined as following.

Definition (FM-WAP) Let H be a WAP history and $\langle F_{w1}, F_{w2}, \cdots, F_{wn} \rangle$. Let α and β be the thresholds for the *S-value* and *F-value* respectively. Then a WAP tree $T_{wj} \in F_{wm}$, where $1 \leq m \leq n$, is a **FM-WAP** if $F(T_{wj}, \alpha) \geq \beta$.

FM-WAPs represent the access patterns that change significantly and frequently (specified by α and β). FM-WAPs can be frequent or infrequent according to existing WAP mining definitions [1]. The difference between frequent WAPs in [1] and the FM-WAPs is that frequent WAPs are defined based on the occurrence of WAPs in the web log data; while FM-WAPs are defined based on the change patterns of web log data. The problem of FM-WAP mining is defined as follows. Given a collection of web usage data, FM-WAP mining is to extract all the FM-WAPs with the user-defined calendar pattern and thresholds of *S-value* and *F-value*.

3 Experimental Results

In the experiments, four synthetic datasets generated by using a tree generator program and a real dataset downloaded from *http://ita.ee.lbl.gov /html/contrib/ Sask-HTTP.html* are used. Figures 1 (a) and (b) show the scalability of our algorithm by varying the size and number of WAP forests. Figure 1 (c) shows that FM-WAP is novel knowledge by comparing with the frequent WAP mining results using WAP-mine[1]. It has been observed and verified that not all FM-WAPs are frequent WAPs and not all frequent WAPs are FM-WAPs.

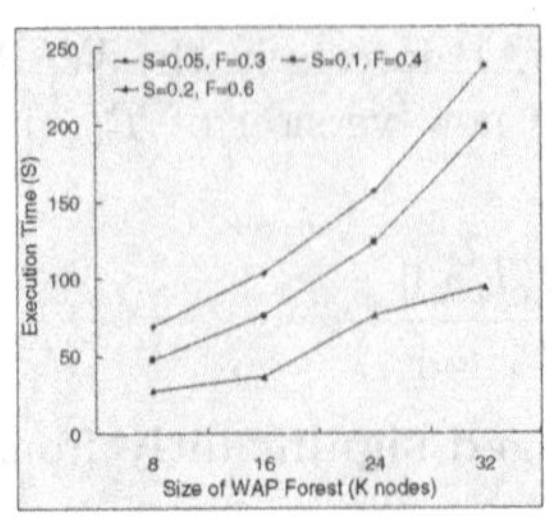

(a) Size of WAP forests

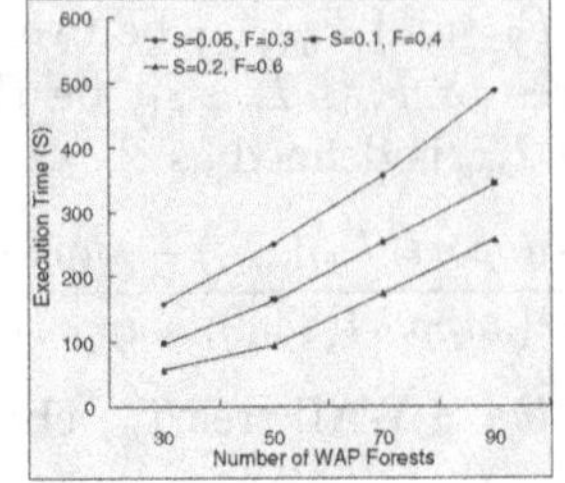

(b) No. of WAP forests

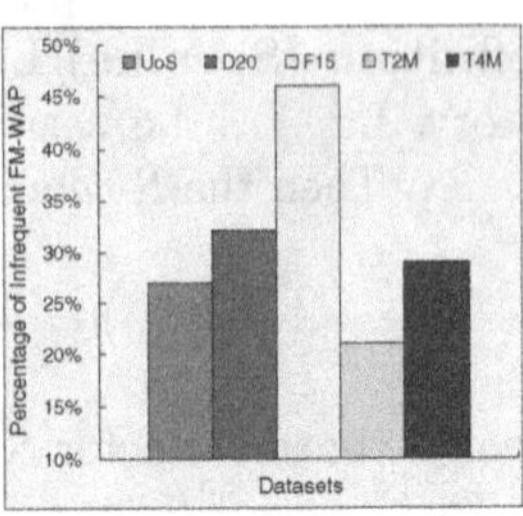

(c) Infrequent FM-WAP

Fig. 1. Experiment Results

4 Conclusion

In this paper, we present a novel approach to extract hidden knowledge from the history of changes to WAPs. Using our proposed data structure, our algorithm can discover the FM-WAPs efficiently with good scalability as verified by the experimental results.

References

1. J. Pei, J. Han, B. Mortazavi-Asl, and H. Zhu. Mining access patterns efficiently from web logs. In *PAKDD*, pages 396–407, 2000.
2. J. Srivastava, R. Cooley, M. Deshpande, and P.-N. Tan. Web usage mining: Discovery and applications of usage patterns from web data. *ACM SIGKDD Explorations*, 1(2):12–23, 2000.

Visual Mining of Spatial Time Series Data

Gennady Andrienko, Natalia Andrienko, and Peter Gatalsky

Fraunhofer Institute AIS,
Schloss Birlinghoven, 53754 Sankt Augustin, Germany
gennady.andrienko@ais.fraunhofer.de
http://www.ais.fraunhofer.de/and

Abstract. CommonGIS is a system comprising a number of tools for visual data analysis. In this paper we demonstrate our recent developments for analysis of spatial time series data.

1 Introduction

Analysis of time series data is one of the most important topics in data mining research. A number of computational methods have been recently developed [1]. Researchers in statistical graphics have also paid a significant attention to this problem, proposing interactive visualisation techniques for time-series [2,3]. These papers introduced basic graphical and interaction primitives for enhancing analytical capabilities of time series plots, or time graphs:

1. Interactive access to values via graphics by pointing on a segment of a line.
2. Tools for overlaying lines for their comparison, with a possibility to distort the lines for better fit.
3. The possibility to select lines with particular characteristics, such as specific values at a given time moment or interval, specific profiles, etc.
4. Dynamic linking between a plot and other displays (scatter-plots, histograms, maps etc.) via highlighting and selection.

However, most of these tools are applicable to analysis of a small number of lines but are not suitable for studying large collections of time series data. Usually such tools are implemented as stand-alone prototypes with a limited number of available complementary displays.

We designed and developed novel methods of time series analysis in our system CommonGIS [4,5], which is unique among both commercial and research software systems as a composition of well-integrated tools that can complement and enhance each other thus allowing sophisticated analyses. The system includes various methods for cartographic visualisation, non-spatial graphs, tools for querying, search, and classification, and computation-enhanced visual techniques. A common feature of all the tools is their high user interactivity, which is essential for exploratory data analysis.

The main features of CommonGIS related to analysis of spatial time series are:

1. A variety of interactive mapping techniques combined with statistical graphics displays and computations.

J.-F. Boulicaut et al. (Eds.): PKDD 2004, LNAI 3202, pp. 524–527, 2004.

2. Animated maps and other time-aware map visualisation techniques [6,7].
3. Novel information visualisation tools (dynamic query, table lens, parallel coordinate plots etc.) dynamically linked to maps and graphics via highlighting, selection, and brushing.
4. Interface to data mining tools [8].

2 Functionality

Let us describe a short example scenario of data analysis that demonstrates some of the new methods. The time graph in Figure 1, upper left, displays results of simulating forest development in 2,600 forest compartments over 100 years with 5-years interval, specifically, the tree biomass in tons per hectare. To eliminate line overplotting, individual lines have been replaced by the median line and 10 percentile lines (10%, 20%, etc.), which form virtual flows, or envelopes (Figure 1, upper right).

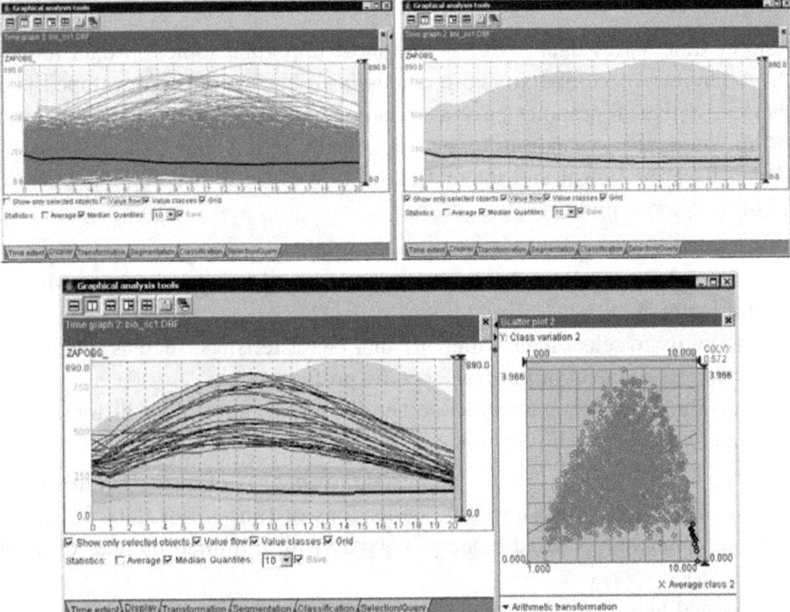

Fig. 1. Individual lines (left), median and 10 envelopes (right), and a scatter plot showing the average envelope and variance, that was used to select particular lines to be shown on the time graph (bottom).

Then, for each individual line, the system has computed the average envelope and the variance. These values have been represented on a scatter plot (Figure 1, bottom right). Using the dynamic link between the scatterplot and the time graph, one may select and explore lines with particular characteristics. Thus, in Figure 1, the lines mostly belonging to the highest envelope have been selected through the scatterplot and, as a result, highlighted on the time graph.

Further analysis can be done using thematic maps. Thus, figure 2 shows a map of relative changes of the tree biomass over the simulated period. The darkest colour shows compartments with biomass increase by more than 10%, the lighter colours correspond to small changes (±10%), decrease (from 10% to 50% of losses) and significant drops (more than 50% of losses). The system allows the user to transform data (e.g. calculate changes, smooth time series, etc.) and to manipulate visualisation properties. Thus, class breaks may be selected interactively on the basis of such criteria as statistical properties of value distribution, the number of objects in classes, or spatial distributional patterns.

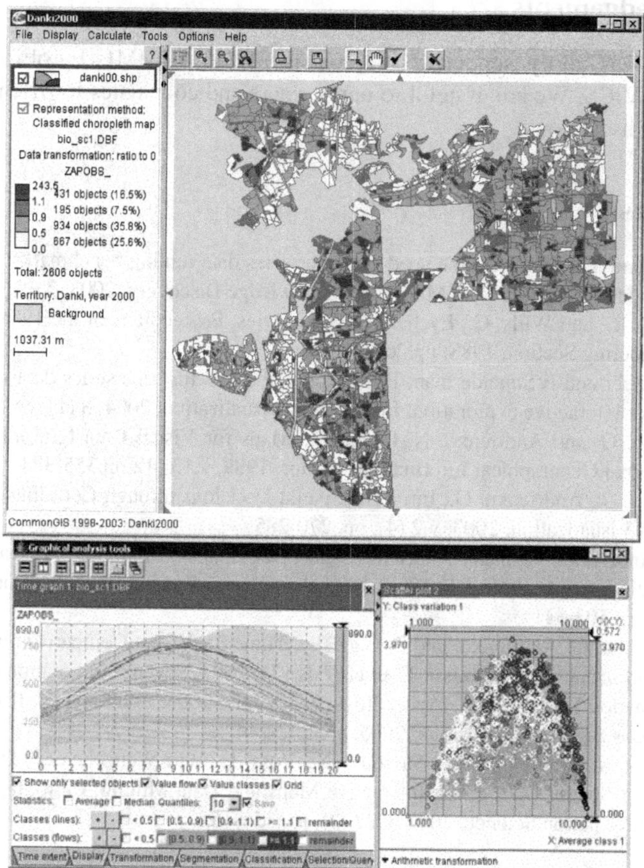

Fig. 2. The map divides forest compartments into classes according to the degree of biomass change. The class colours are projected onto the scatter plot and on the selected lines on the time graph.

3 Discussion and Future Work

We have proposed new interactive methods for analysis of spatial time series data. The methods are scalable with regard to the number of time-series considered simultaneously. However, we see the need in further research for supporting analysis of long time-series and for analysis of multi-attribute time series.

The methods described in this paper will be further developed in a recently approved integrated project OASIS (Open Advanced System for Improved crisiS management, 2004–2008, IST-2003-004677). The focus of further work is the adaptation of the time series analysis for needs and purposes of decision support (analysis of situation, formalization and evaluation of decision strategies, multi-criteria selection and decision sensitivity analysis).

Acknowledgements

This work was partly supported in projects SPIN!, GIMMI, EuroFigures, EFIS, NEFIS, SILVICS. We are grateful to our partners and colleagues for friendly support and fruitful discussions.

References

1. E.Keogh and S.Kasetty, On the need for time series data mining benchmarks: a survey and empirical demonstration, Data Mining and Knowledge Discovery, 2003, 7 (4), 349-371.
2. Unwin, A. R., and Wills, G., Eyeballing Time Series. Proceedings of the 1988 ASA Statistical Computing Section, 1988, pp.263-268
3. H.Hochheiser and B.Shneiderman, Dynamic query tools for time series data sets: Timebox widgets for interactive exploration, Information Visualization, 2004, 3 (1), pp.1-18.
4. Andrienko, G. and Andrienko, N. Interactive Maps for Visual Data Exploration, International Journal Geographical Information Science, 1999, v.13 (4), pp.355-374
5. Andrienko, N., Andrienko, G., Informed Spatial Decisions through Coordinated Views, Information Visualization, 2003, v.2 (4), pp. 270-285
6. Natalia Andrienko, Gennady Andrienko, and Peter Gatalsky, Exploratory Spatio-Temporal Visualization: an Analytical Review, Journal of Visual Languages and Computing, 2003, v.14 (6), pp. 503-541
7. N. Andrienko, G. Andrienko, and P. Gatalsky, Tools for Visual Comparison of Spatial Development Scenarios, In Banissi, E. et al (Eds.) IV 2003. Seventh International Conference on Information Visualization, Proceedings, 16-18 July, 2003, London, UK. IEEE Computer Society, Los Alamitos, California, 2003, pp. 237-244.
8. Andrienko, N., Andrienko, G., Savinov, A., Voss, H., and Wettschereck, D. Exploratory Analysis of Spatial Data Using Interactive Maps and Data Mining, Cartography and Geographic Information Science, 2001, v.28 (3), pp. 151-165

Detecting Driving Awareness

Bruno Apolloni, Andrea Brega, Dario Malchiodi, and Cristian Mesiano

Dipartimento di Scienze dell'Informazione, Via Comelico 39/41, 20135 Milano, Italy
{apolloni,brega,malchiodi,mesiano}@dsi.unimi.it

Abstract. We consider the task of monitoring the awareness state of a driver engaged in attention demanding manoeuvres. A Boolean normal form launches a flag when the driver is paying special attention to his guiding. The contrasting analysis of these flags with the physical parameters of the car may alert a decision system whenever the driver awareness is judged unsatisfactory. The paper presents preliminary results showing the feasibility of the task.

1 The Experiment

We consider persons driving in a very good quality simulator that reproduces some typical phenomena in an interactive environment where they may cross or overpass other cars, follow a queue, face crossroads, etc. This simulator, which is in Queen University of Belfast, is equipped with a Biopac device having some not invasive sensors to record the electrochardiographic signals (ECG), the respiration (RSP), the galvanic skin response (GSR) and the skin temperature (SKT) of the subject at the sampling rate of 200 Hz. A trial is constituted by a road trip that must be followed by the driver with different boundary conditions more or less affecting the driver emotional state. We sharply distinguish between unsoliciting conditions, by default, and soliciting conditions represented by periodically fast direction changes induced by cones in front of the car. The main cinematic parameters of the car, such as speed, acceleration, and wheel angle, are recorded during the trial.

2 A Procedure from Sensory Data to Rules

We may figure the whole procedure hosted by a hybrid multilayer perceptron where first layers compute subsymbolic functions like a conventional neural network, while the latter are arrays of Boolean gates computing formulas of increasing complexity. From an informational perspective the signal flows left to right evolving through representations increasingly compressed where the information is distributed in both the states of the processing elements and their connections. The training of this machinery is reported elsewhere [4]; here we will rather discuss the emerging features.

Since SKT is a constant due to simulation shortness, we extract ten features just from the first three signals, that we avoided to filter as this operation generally destroys relevant informations. We also didn't manage for removing motion artifacts since the features we will consider are insensitive to them. Eight features are conventional, consisting in distances between relevant points of the signals according to medical knowledge [5] (as sketched in Figure 1), while the other two are power spectrum and neural prediction shift. The power spectrum captures global properties of the

J.-F. Boulicaut et al. (Eds.): PKDD 2004, LNAI 3202, pp. 528–530, 2004.

ECG pulse and is computed within every pulse through usual Fast Fourier Transform processing. Finally a "neural" distance is computed between the actual ECG trajectory and the one 10 steps ahead forecasted by a specially featured recurrent neural network. This network is fed by (last line in Figure 1(b)) the current values of the three physiological signals and is embedded with both conventional sigmoidally activated neurons and symbolic ones expressly computing second and fifth order dynamics as baselines of the RSP and ECG signals respectively.

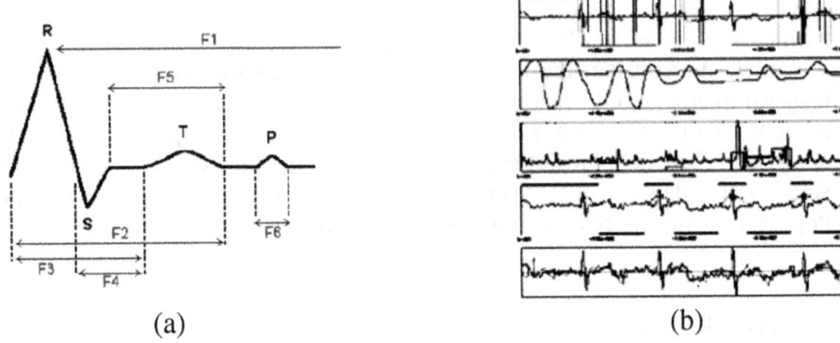

(a) (b)

Fig. 1. A sketch of the extracted features: (a) EGC parameters; (b) coded signals.

The features are compressed into 10 propositional variables through a multilayer perceptron with 80% of pruned connections. To render the symbols sensitive to a short-term dynamics of the features the input layer is made of 30 input nodes, feed by a normalized version of the ten features at time *t*-1 *t*, and *t*+1. The output nodes compute a Boolean vector of ten prpositional variables and the hidden layer has ten nodes as well. The number of connections between neurons are 253 in total. The way of training such a network is not trivial, passing through a RAAM architecture and special entropic error function called "edge pulling" elsewhere [1]. We look for a set of vector of Boolean variables where an almost "one to one" correspondence must exist with the feature vectors, but Boolean assignments may coincide when they code data patterns with the same values of the features of interest to us. We succeed in this target apart for around 20% of examples that met inconsistency, being coded by vectors labeled both by 1 in some examples and by 0 in other. Wether the inconsistency is due to human segmentation or neural classification error is still under investigation.

In order to maintain some understandability of the awareness detection rules, or at least its formal manageability within wider contexts, we required these rules to be symbolic, being expressed on particular as either conjunctive or disjunctive normal forms. We rely on the first level of the PACmeditation procedure discussed in [3] for getting a DNF or a CNF as classifier, and simplify their expressions according to the fuzzy-relaxation procedure discussed in [2]. An example of obtained rule is $x_4x_8x_6 + x_5x_3x_7x_6 + x_{13}x_{10}x_6 + x_{14}x_{11}x_6x_7x_{13} + x_9x_7x_1x_3 + x_7x_{13}x_6x_8 + x_{10}x_3x_6x_1 + x_8x_1x_7x_{12} + x_4x_8x_{10}x_{13}$.

3 Numerical Results and Conclusions

Figure 2 refers to frames of a clip recording a driving simulation experiment. The lowermost graphs report the driving parameters recorded by the simulator, denoting car acceleration, braking strength, wheel angle, distance from the center line. The uppermost graphs show two Boolean variables: dark bars are up when an attention requiring episode occurs according to our above empirical rules; gray bars are up when the system detects a driver attentional state.

The first frame captures the beginning of a foggy tract where the attention of the driver is raised up in spite of absence of particular episodes denoted by the driving parameters. The second one is the crucial instant of a crash due to a "maniac" car suddenly invading the driver's lane. Contemporarily episode and attention bars raise up, but without any chance of avoiding the crash.

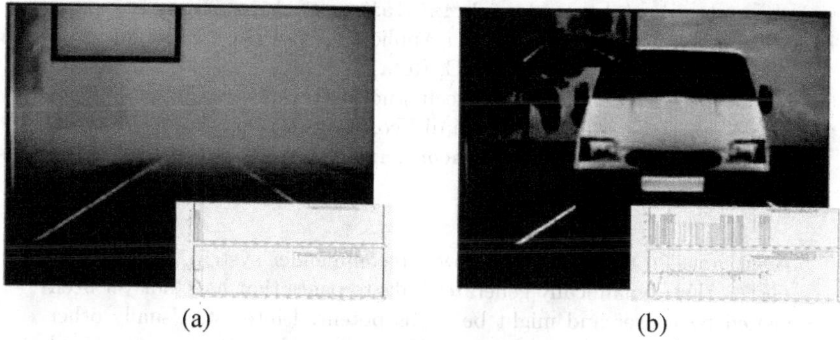

(a) (b)

Fig. 2. Awareness monitoring spots.

The demo will show the entire clip and the companion graphs, denoting how the proposed system detects the awareness or non-awareness state of the driver during various guidance episodes.

References

1. B. Apolloni, S. Bassis, A. Brega, S. Gaito, D. Malchiodi and A. M. Zanaboni, A man-machine human interface for a special device of the pervasive computing world, in Proceedings of DC Tales, CTI Press.
2. B. Apolloni, D. Malchiodi, C. Orovas and A. M. Zanaboni, Fuzzy methods for simplifying a boolean formula inferred from examples, in Proceedings of FSKD'02, Vol. 2, 554-558.
3. B. Apolloni, F. Baraghini and G. Palmas, Pac meditation on Boolean Formulas, *Abstraction, Reformulation and Approximation*, Springer, Berlin: 274-281,2002.
4. B. Apolloni, D. Malchiodi, C. Orovas and G. Palmas. From synapses to rules. Cognitive Systems Research, 3(2): 167-201, 2002.
5. R. Plonsey and D. G. Fleming. *Bioelectric phenomena.* McGraw-Hill, 1969.

An Effective Recommender System for Highly Dynamic and Large Web Sites

Ranieri Baraglia[1], Francesco Merlo[2], and Fabrizio Silvestri[1]

[1] Istituto di Scienze e Tecnologie dell'Informazione (A. Faedo)
ISTI–CNR – Pisa, Italy
{ranieri.baraglia,fabrizio.silvestri}@isti.cnr.it

[2] LIASES – Laboratorio di Informatica Applicata alle Scienze Economiche e Sociali (G. Rota)
Universitá degli studi di Torino
Facoltá di Economia
merlo@econ.unito.it

Abstract. In this demo we show a recommender system, called *SUGGEST*, that dynamically generates links to pages that have not yet been visited by a user and might be of his potential interest. Usually other recommender systems exploit a kind of two-phase architecture composed by an off-line component that analyzes Web server access logs and generates information used by a successive online component that generates recommendations. *SUGGEST* collapse the two-phase into a single online Apache module. The component is able to manage very large Web sites made up of dinamically generated pages by means of an efficient LRU-based database management strategy. The demo will show the way *SUGGEST* is able to anticipate users' requests that will be made farther in the future, introducing a limited overhead on the Web server activity[1].

1 Introduction

The continuous and rapid growth of the Web has led to the development of new methods and tools in the Web recommender or personalization domain [1], [2]. Web Usage Mining (WUM) is the process of extracting knowledge from Web users access data (or clikstream) by exploiting Data Mining (DM) technologies.

In this demo, we present a recommender system, called *SUGGEST*, which is designed to dynamically generated personalized content of potential interest for users of a Web Site. It is based on an incremental personalization procedure, tightly coupled with the Web server. It is able to update incrementally and automatically the knowledge base obtained from historical usage data and to generate a list of page links (*suggestions*). The suggestions are used to personalize *on the fly* the HTML page requested. Moreover, the adoption of a LRU-based algorithm to manage the knowledge base permits us to use *SUGGEST* also on

[1] This work was funded by the Italian MIUR as part of the National Project Legge 449/97, 1999, settore Società dell'Informazione: Technologies and Services for Enhanced Contents Delivery (2002-2004)

J.-F. Boulicaut et al. (Eds.): PKDD 2004, LNAI 3202, pp. 531–533, 2004.

large Web sites made up of dynamically generated pages. Furthermore the system is able to evaluate the importance of the pages composing the underlying Web site by adopting a sort of PageRank estimation based on the maintained usage information. Furthermore, the system proposed was evaluated by adopting the new quality metric we introduced in [3].

2 *SUGGEST*

Schematically, *SUGGEST* works as follows: once a new request arrives at the server, the URL requested and the session to which the user belongs are identified, the underlying knowledge base is updated, and a list of suggestions is appended to the requested page. To catch information about navigational patterns, *SUGGEST* models the page accesses information as a undirected graph $G = (V, E)$. The set V of vertices contains the identifiers of the different pages hosted on the Web server. Based on the fact that the interest in a page depends on its content and not on the order a page is visited during a session [4], we assign to each edge E a weight computed as: $W_{ij} = N_{ij}/max\{N_i, N_j\}$ where N_{ij} is the number of sessions containing both pages i and j, N_i and N_j are respectively the number of sessions containing only page i or page j. Dividing by the maximum between single occurrences of the two pages has the effect of reducing the relative importance of links involving index pages. Such pages are those that, generally, do not contain useful content and are used only as a starting point for a browsing session. Index pages are very likely to be visited with any other page and nevertheless are of little interest as potential suggestions. The data structure we used to store the weights is an adjacency matrix M where each entry M_{ij} contains the value W_{ij} computed according to Formula above. To further reduce the impact of insignificant links we filtered out links whose weigths are under a predetermined threshold *minfreq*. As in the previous version, *SUGGEST* finds groups of strongly correlated pages by partitioning the graph according to its connected components. *SUGGEST* actually uses a modified version of the well known incremental connected components algorithm. After the clustering step, *SUGGEST* has to construct the suggestions list for the current user request. This is done in a straightforward manner by finding the cluster that has the largest intersection with the *PageWindow* related to the current session. The final suggestions are composed by the most relevant pages in the cluster, according to the order determined by a sort of PageRank algorithm applied to the Web site's usage graph.

3 Evaluation and Conclusions

In order to evaluate both the effectiveness (i.e. the quality of the suggestions) and efficiency (i.e. overhead introduced on the overall performance of the Web server) of *SUGGEST* several tests were conducted. All tests were run on a processor Intel Celeron 2,4 GHz with 256 MBytes of RAM, an ATA 100 disk with 30 GBytes, and operating system Linux 2.4.20. The *SUGGEST* effectiveness was evaluated by using a performance parameter that takes into account the distance of the suggestions generated with the actual pages visited during the session. To evaluate the *SUGGEST* effectiveness experimental evaluation was conducted by using three real life access log files of public domains: Berkeley, NASA, USASK [3] For each dataset we measured the quality of the suggestions (Ω) varying the *minfreq* parameter. Figure 1 shows the results obtained. Moreover, we also plotted the curve relative to the suggestions generated by a random suggestion generator (labelled rnd in Fig. 1). As it was expected, the random generator performs poorly and the intersection between a random suggestion and a real session is almost null. On the other hand, suggestions generated by *SUGGEST* show a higher quality, that, in all the datasets, reaches a maximum for *minfreq*=0.2. For low values of the *minfreq* parameter, good values are obtained for

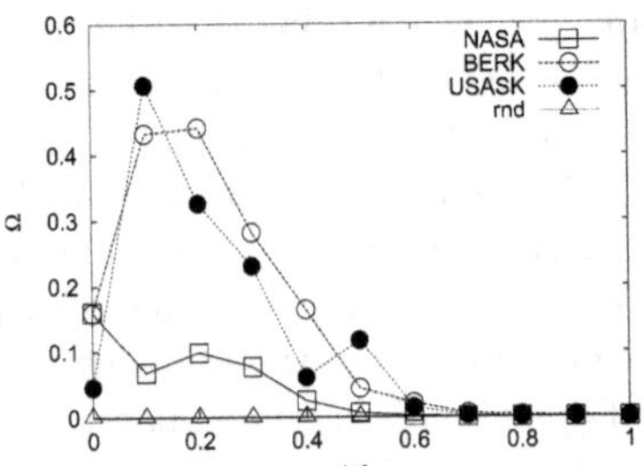

Fig. 1. Coverage of suggestions for the NASA, BERK, USASK access log files, varying *minfreq*.

the quality of the suggestions. In the demo we are going to show the working of *SUGGEST* on a real production Web Site of the Economy Faculty of the University of Turin. Goal of the demo will be showing the ability of *SUGGEST* to generate links to "*useful*" pages.

References

1. Magdalini, E., Vazirgiannis, M.: Web mining for web personalization. ACM Trans. on Internet Technology **3** (2003) 1–27
2. Deshpande, M., Karypis, G.: Item-based top-n recommendation algorithms. ACM Trans. on Information Systems **22** (2004) 143–177
3. Silvestri, F., Baraglia, R., Palmerini, P., M., S.: On-line generation of suggestions for web users. In: Proc. of IEEE Int'l Conf. on Information Technology: Coding and Computing. (2004)
4. Baraglia, R., Palmerini, P.: Suggest: A web usage mining system. In: Proc. of IEEE Int'l Conf. on Information Technology: Coding and Computing. (2002)

SemanticTalk: Software for Visualizing Brainstorming Sessions and Thematic Concept Trails on Document Collections

Chris Biemann, Karsten Böhm, Gerhard Heyer, and Ronny Melz

University of Leipzig, Ifi, Department of Natural Language Processing,
Augustusplatz 10/11, 04109 Leipzig
{biem,boehm,heyer,rmelz}@informatik.uni-leipzig.de
http://www.wortschatz.uni-leipzig.de

Abstract. In this demonstration we introduce a technology to support knowledge structuring processes already at the time of their creation by building up concept structures in real time. Our focus was set on the design of a minimal invasive system, which ideally requires no human interaction and thus gives the maximum freedom to the participants of a knowledge creation or exchange processes. The system captures and displays spoken dialogs as well as text documents for further use in knowledge engineer's tools.

1 Introduction

Our goal is to support the communication and mutual understanding between and within groups in two ways: On one hand we provide a visualisation tool for spoken language, having its application in informal creative and usually highly innovative meetings like brainstorming sessions or open space workshops (cf. [6]. In these scenarios the tool does not only serve as an automatic documentation method by arranging the keywords in a meaningful way, but also provides corpus-based associations in order to enrich the conversation with concepts that are related but might have been forgotten by the participants. On the other hand we use the same visualization engine for displaying single documents as trails on a so-called *semantic map*. The idea of a semantic map is heavily relying on the well known concept of geographical maps, in which visual structuring of interesting locations and the emphasizing of relevant paths between them are the key concepts to provide an orientation for the user. Since the location of the concepts on the map is fixed, users can grasp the contents of a document rapidly and compare documents in a visual way.

Both features are part of a prototype implementation called "SemanticTalk" which has been presented to the public in at the worlds largest IT-exhibition CeBit in Hannover, Germany, in spring 2004 and is described in detail in [3].

2 SemanticTalk – A Tool for Innovation Acceleration

SemanticTalk supports two operation modes: a *brainstorming mode* in which a semantic map will be constructed dynamically from the input and a *red-thread mode* that visualizes a communication trail in a pre-calculated map. The input for the sys-

J.-F. Boulicaut et al. (Eds.): PKDD 2004, LNAI 3202, pp. 534–536, 2004.

tem can be obtained from different sources: it may be loaded from text files, directly typed in using a provided input field or spoken text recorded from a headset.

The SemanticTalk user-interface that represents the different views of the semantic map as well as the controls that modify the behaviour of the tool is show in the figure below.

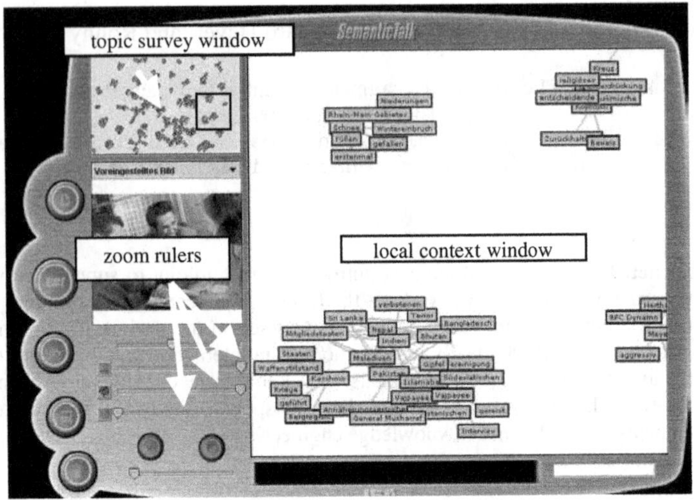

Fig. 1. The *SemanticTalk* user interface. The local context window is a zoomed variant of the topic survey window and can be changed with the three zoom rulers. Other controls: New, Load, Save, Exit, Microphone on/off, Freeze graph, Brainstorming / Red thread Switch, Input text bar and scrolling.

2.1 Visualization of Associations in Brainstorming Mode

For preparation, a text collection of the domain has to be processed as described in [1] to provide the terminology and the global contexts to the system. If existent, ontology or some typological system can be loaded in order to assign types to concepts and relations.

The brainstorming session starts with a blank screen. Words of the conversation that are considered to be important (an easy strategy is to use only nouns of a substantial frequency in the domain corpus) are thrown into the visualization. In addition to this, associations obtained from the text collection are shown. It is possible to include an ontology or alike for typed (colored) nodes and edges.

2.2 Visualization of Concept Trails on Semantic Maps in Red Thread Mode

Semantic maps provide a good overview of a domain by visualizing keywords and their relatedness. While they are extracted from a document collection and serve as model for the whole domain, it is possible to display single documents as paths through them. We visualize the content of the broadcast as trajectory in the pre-calculated semantic map. Words of interest from the document are marked in red and are connected in sequence of their occurrence by directed edges. By using a fixed semantic map and dynamically representing input as indicated, it is possible to examine things like coverage of domain, coverage within (sub-)topics, relatedness of document to the map domain and comparison of different documents. The great advantage of this representation lies in displaying what is described in the document, and the same time, what is *not* being dealt with. By using a fixed representation for the domain, the practiced user can grasp the document's content in a few seconds and then decide whether to read it or not.

3 Knowledge Discovery Aspects

The preliminary text processing (see [1]) is language-independent and fully automatic and learns relations and associations between words. The user is able to discover the knowledge in the text collection by simply talking about the topic he is interested to and watching the associations. Since words and their connections are linked to documents and passages of their origin, the tool can be used for visual Information Retrieval. Other machine learning aspects used for enriching the data are described in [2]: Ways are described how to assign relation types like antonomy, hyponomy and alike to pairs of words by using statistical measures. Training on a set of relations gives rise to rule detection for enhancing the quantity of typed relations automatically.

References

1. Biemann, C., Bordag, S., Heyer, G., Quasthoff, U., Wolff, Chr. (2004): Language-independent Methods for Compiling Monolingual Lexical Data, Springer LNCS 2945
2. Biemann, C., Bordag, S., Quasthoff, U. (2004): Automatic Acquisition of Paradigmatic Relations using Iterated Co-occurrences, Proceedings of LREC2004, Lisboa, Portugal
3. Biemann, C., Böhm, K., Heyer, G., Melz, R. (2004): Automatically Building Concept Structures and Displaying Concept Trails for the Use in Brainstorming Sessions and Content Management Systems, Proceedings of I2CS, Guadalajara, Mexico
4. Faulstich, L., Quasthoff, U., Schmidt, F., Wolff, Chr. (2002): Concept Extractor – Ein flexibler und domänenspezifischer Web Service z. Beschlagwortung von Texten, ISI 2002
5. Jebara, T., Ivanov, Y, Rahimi, A., Pentland, A. (2000): Tracking conversational context for machine mediation of human discourse. In: Dautenhahn, K. (eds.): AAAI Fall 2000 Symposium - Socially Intelligent Agents - The Human in the Loop. Massachusetts: AAAI Press
6. Owen, H. (1998): Open Space Technology: A User's Guide, 1998, Berrett-Koehler Publishers Inc., San Francisco

Orange: From Experimental Machine Learning to Interactive Data Mining

Janez Demšar[1], Blaž Zupan[1,2], Gregor Leban[1], and Tomaz Curk[1]

[1] Faculty of Computer and Information Science, University of Ljubljana, Slovenia
[2] Dep. of Molecular and Human Genetics, Baylor College of Medicine, Houston, USA

Abstract. Orange (`www.ailab.si/orange`) is a suite for machine learning and data mining. For researchers in machine learning, Orange offers scripting to easily prototype new algorithms and experimental procedures. For explorative data analysis, it provides a visual programming framework with emphasis on interactions and creative combinations of visual components.

1 Orange, a Component-Based Framework

Orange is a comprehensive, component-based framework for machine learning and data mining. It is intended for both experienced users and researchers in machine learning who want to write Python scripts to prototype new algorithms while reusing as much of the code as possible, and for those just entering the field who can enjoy in the powerful while easy-to-use visual programming environment. Orange supports various tasks spanning from data preprocessing to modelling and evaluation, such as:

- data management and preprocessing, like sampling, filtering, scaling, discretization, construction of new attributes, and alike,
- induction of classification and regression models, including trees, naive Bayesian classifier, instance-based approaches, linear and logistic regression, and support vector machines,
- various wrappers, like those for calibration of probability predictions of classification models, and those for boosting and bagging,
- descriptive methods like association rules and clustering,
- methods for evaluation and scoring of prediction models, including different hold-out schemes and range of scoring methods and visualization approaches.

1.1 Scripting in Python

As a framework, Orange is comprised of several layers. The core design principle was to use C++ to code the basic data representation and manipulation and all time-complex procedures, such as most learning algorithms and data preprocessing. Tasks that are less time consuming are coded in Python. Python is a popular object-oriented scripting language known for its simplicity and power,

J.-F. Boulicaut et al. (Eds.): PKDD 2004, LNAI 3202, pp. 537–539, 2004.

and often used as a "glue-language" for components written in other languages. The interface between C++ and Python provides a tight integration: Python scripts can access and manipulate Orange objects as if they were implemented in Python. On the other hand, components defined in Python can be used by the C++ core. For instance, one can use classification tree as implemented within Orange (in C++) but prototype a component for attribute selection in Python. For Orange, we took special care to implement machine learning methods so that they are assembled from a set of reusable components one can either use in the new algorithms, or replace them with prototypes written in Python.

Just for a taste, here is a simple Python script, which, using Orange, reads the data, reports on the number of instances and attributes, builds two classifiers and outputs predicted and true class of the first five instances.

```
import orange
data = orange.ExampleTable('voting.tab')
print 'Instances:', len(data), 'Attributes:', len(data.domain.attributes)
nbc = orange.BayesLearner(data)
knn = orange.kNNLearner(data, k=10)
for i in range(5):
    print nbc(data[i]), knn(data[i]), 'vs. true class', data[i].getClass()
```

Another, a bit more complicated script below, implements a classification tree learner where node attributes that split the data are chosen at random by a function randomChoice, which is used in place of data splitting component of Orange's classification tree inducer. The script builds a standard and random tree from the data, and reports on their sizes.

```
import orange, random
def randomChoice(instances, *args):
    attr = random.choice(instances.domain.attributes)
    cl = orange.ClassifierFromVar(whichVar=attr, classVar=attr)
    return cl, attr.values, None, 1

treeLearner = orange.TreeLearner()
rndLearner = orange.TreeLearner()
rndLearner.split = randomChoice

data = orange.ExampleTable('voting.tab')
tree = treeLearner(data)
rndtree = rndLearner(data)
print tree.treesize(), 'vs.', rndtree.treesize()
```

1.2 Visual Programming

Component-based approach was also used for graphical user's interface (GUI). Orange's GUI is made of widgets, which are essentially a GUI wrappers around data analysis algorithms implemented in Orange and Python. Widgets communicate through channels, and a particular set of connected widgets is called a schema. Orange schemas can be either set in Python scripts, or, preferably, designed through visual programming in an application called Orange Canvas.

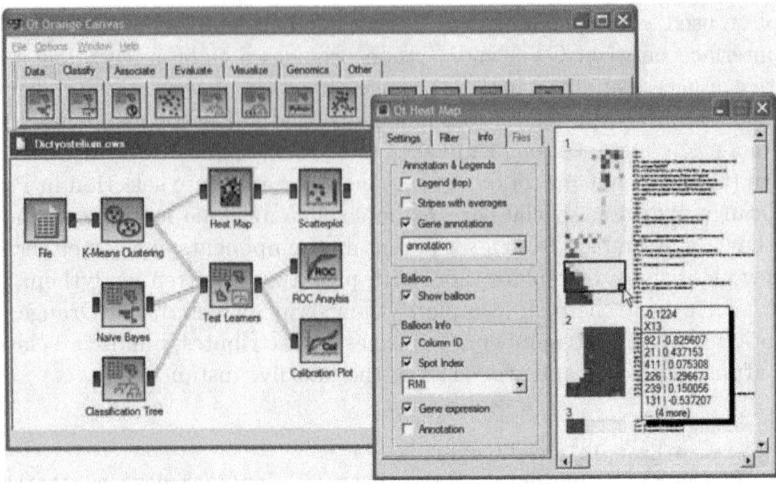

Fig. 1. Snapshot of Orange Canvas with a schema that takes a microarray data, performs k-means clustering, and evaluates the performance of two different supervised learning methods when predicting the cluster label. Clustered data is visualized in the Heat Map widget, which sends any selected data subset to the Scatterplot widget.

Besides ease-of-use and flexibility, data exploration widgets were carefully design to support interaction. Clicking on a classification tree node in the tree visualization widget, for example, outputs the corresponding data instances making them available for further analysis. Any visualization of predictive or visualization models where their elements are associated with particular subsets of instances, attributes, data domains, etc., behave in the similar way. A snapshot of Orange Canvas with an example schema is shown in Fig. 1.

2 On Significance and Contribution

Orange is an open-source framework that features both scripting and visual programming. Because of component-based design in C++ and integration with Python, Orange should appeal to machine learning researchers for the speed of execution and ease of prototyping of new methods. Graphical user's interface is provided through visual programming and carefully designed widgets that support interactive data exploration. Component-based design, both on the level of procedural and visual programming, flexibility in combining components to design new machine learning methods and data mining applications, and user and developer-friendly environment are also the most significant attributes of Orange and those where Orange can make its contribution to the community.

Acknowledgement

This work was supported, in part, by the program and project grants from Slovene Ministry of Science and Technology and Slovene Ministry of Information Society, and American Cancer Society project grant RPG-00-202-01-CCE.

Terrorist Detection System

Yuval Elovici[1], Abraham Kandel[2,3], Mark Last[1], Bracha Shapira[1], Omer Zaafrany[1], Moti Schneider[4], and Menahem Friedman[1,5]

[1] Department of Information Systems Engineering, Ben-Gurion University
Beer-Sheva 84105, Israel
Phone: +972-8-6461397, Fax: +972-8-6477527
{mlast,bshapira,zaafrany}@bgumail.bgu.ac.il,
elovici@inter.net.il

[2] Department of Computer Science and Engineering, University of South Florida,
4202 E. Fowler Ave. ENB 118, Tampa, FL, 33620, USA

[3] Currently at the Faculty of Engineering, Tel-Aviv University, Israel
kandel@csee.usf.edu

[4] School of Computer Science, Netanya Academic College, Netanya, Israel
motis@netanya.ac.il

[5] Department of Physics, Nuclear Research Center – Negev

Abstract. Terrorist Detection System (TDS) is aimed at detecting suspicious users on the Internet by the content of information they access. TDS consists of two main modules: a training module activated in batch mode, and an on-line detection module. The training module is provided with web pages that include terror related content and learns the typical interests of terrorists by applying data mining algorithms to the training data. The detection module performs real-time monitoring on users' traffic and analyzes the content of the pages they access. An alarm is issued upon detection of a user whose content of accessed pages is "too" similar to typical terrorist content. TDS feasibility was tested in a network environment. Its detection rate was better than the rate of a state of the art Intrusion Detection System based on anomaly detection.

1 Introduction

The Internet is an efficient communication infrastructure that is increasingly used by terrorist organizations to safely communicate with their affiliates, coordinate action plans, spread propaganda messages, raise funds, and introduce new supporters into their networks [1]. Governments and intelligence agencies are calling to invest major efforts in development of new methods and technologies for identifying terrorist activities on the web in order to prevent future acts of terror. TDS presents an example to such an effort.

By means of content monitoring and analysis of web pages accessed by a group of web users, it is possible to infer their typical areas of interest [2, 3, 4]. It is also possible to identify users that access specific, potentially illegitimate information on the internet [2, 3, 4]. Using this approach, real time web traffic monitoring may be per-

J.-F. Boulicaut et al. (Eds.): PKDD 2004, LNAI 3202, pp. 540–542, 2004.

formed to identify terrorists as they access typical terror-related content on the internet. Terror Detection System (TDS) described in [2, 3, 4] implemented this approach.

2 Terrorist Detection System (TDS): Overview

TDS is a content-based detection system recently developed to detect users who are interested in terror-related pages on the Web by monitoring their online activities. The reader is referred to [4] for a detailed description of TDS. The system is based on real-time monitoring of internet traffic of a defined group of Web users (e.g. students in a specific University campus). The group is suspected to include hidden individual terrorists and the system aims at detecting them. The current version of TDS refers only to the textual content of the accessed web pages. It consists of two main modules: *a training module* activated in batch, and a real-time *detection module*.

The *training module* receives as input a set of web pages that include terror related content. It applies cluster analysis on the textual representation of each page resulting with a set of vector that efficiently represents typical terrorists' areas of interest.

The *detection module* performs on-line monitoring of all traffic between the users being monitored and the Web. The content of the pages they access is analyzed, transformed to a form of a vector, and added to the vector representing the user profile. The profile for each user is kept during a period of time and number of transactions defined by operative system parameters. Similarity is measured between each user profile and the typical terrorist areas of interests. A consistent high similarity between a specific user and terror-related content would raise an alert about that user. Each user related to the monitored group is identified by a "user's computer" having a unique IP address. In case of a real-time alarm, the detected IP can be used to locate the suspicious computer and hopefully the suspected user who may still be logged on to the same computer. In some intranet environments or cooperative ISPs and according to legal privacy issues users may be identified by their user names to enable fast location upon an alert.

The detection module, being activated in real-time, is required to efficiently capture the textual content of Web pages from the Internet traffic. Actually, the detection efficiency is crucial to TDS effectiveness; skipped pages or inaccurate analysis of pages due to slow handling of traffic might result in unreliable detection. TDS detection rate was compared to performance of ADMIT, a state of the art Intrusion Detection System [5], and obtained higher results as described in [2].

3 TDS Significance

An important contribution of TDS lies in the unique environment of its application. The detection component is planned to run in a real-time wide-area network environment, and should be capable of on-line monitoring of many users. Therefore, a crucial design requirement was high-performance which called for enhancement of the algorithms involved, especially the mining algorithm, to high-performance and scalability. We believe that the adjustment of the algorithm to high-performance might serve many other applications. In addition, TDS is an example for a successful application of data mining and machine learning techniques to international cyber-war effort against world-wide terror.

Acknowledgement

This work is partially supported by the National Institute for Systems Test and Productivity at University of South Florida under the USA Space and Naval Warfare Systems Command Grant No. N00039-01-1-2248 and by the Fulbright Foundation that has granted Prof. Kandel the Fulbright Research Award at Tel-Aviv University, Faculty of Engineering during the academic year 2003-2004.

References

1. Birnhack M. D. and Elkin-Koren, N.: Fighting Terror On-Line: The Legal Ramifications of September 11. Internal Report, The Law and Technology Center, Haifa University. [http://law.haifa.ac.il/faculty/lec_papers/terror_info.pdf]. (2002)
2. Elovici,Y., Shapira, B., Last, M., Kandell, A., and Zaafrany, O.: Using Data Mining Techniques for Detecting Terror-Related Activities on the Web. Journal of Information Warfare, Vol. 3, No.1, (2003) 17-28
3. Shapira, B., Elovici, Y., Last, M., Zaafrany, O., and Kandel, A.: Using Data Mining for Detecting Terror-Related Activities on the Web. European Conference on Information Warfare and Security (ECIW), (2003) 271-280
4. Last, M., Elovici, Y., Shapira, B., Zaafrany, O., and Kandel, A.: Content-Based Methodology for Anomaly Detection on the Web. Advances in Web Intelligence, E. Menasalvas et al. (Editors), Springer-Verlag, Lecture Notes in Artificial Intelligence, Vol. 2663, (2003) 113 – 123
5. Sequeira, K., Zaki, M.: ADMIT: Anomaly-based Data Mining for Intrusions. Proceedings of SOGKDD 02, ACM. (2002) 386-395

Experimenting SnakeT: A Hierarchical Clustering Engine for Web-Page Snippets*

Paolo Ferragina and Antonio Gullì

Dipartimento di Informatica, Università di Pisa
{ferragina,gulli}@di.unipi.it

Current search engines return a ranked list of web pages represented by page excerpts called the *web snippets*. The ranking is computed according to some relevance criterium that takes into account textual and hyperlink information about the web pages (see e.g. [1]). This approach is very well-known and a lot of research is pushing towards the design of better and faster ranking criteria. However, it is nowadays equally known that a flat list of results limits the retrieval of *precise* answers because of many factors. First, the relevance of the query results is a subjective and time-varying concept that strictly depends on the context in which the user is formulating the query. Second, the ever growing web is enlarging the number and heterogeneity of candidate query answers. Third, the web users have limited patience so that they usually just look at the top ten results. The net outcome of this scenario is that the retrieval of the correct answer by a standard user is getting more and more difficult, if not impossible.

It is therefore not surprising that new IR tools are being designed to *boost*, or *complement*, the efficacy of search-engine ranking algorithms. These tools offer new ways of organizing and presenting the query results that are more intuitive and simple to be browsed, so that the users may match their needs faster. Among the various proposals, one became recently popular thanks to the engine Vivisimo (see Figure 1) that got in the last three years the "Best Metasearch Engine Award" by SearchEngineWatch.com.

The goal of this demonstration proposal is to describe the functioning of a Web Hierarchical Clustering engine developed at the University of Pisa, called SnakeT. The algorithmic ideas underlying this IR-tool has been briefly described in a companion paper published in this proceedings. SnakeT offers some distinguishing features with respect to known solutions that make it much closer to VIVISIMO's results: (1) it offers a large coverage of the web by drawing the snippets from 16 search engines (e.g. Google, MSN, Overture, Teoma, and Yahoo), the Amazon collection of books (a9.com) and the Google News, in a flexible and efficient way via I/O Async; (2) it builds the clusters and their labels on-the-fly in response to a user query (without using any predefined taxonomy); (3) it selects on-the-fly the best labels by exploiting a variant of the TF-IDF measure computed onto the whole web directory DMOZ; (4) it organizes the clusters and their labels in a hierarchy, by minimizing an objective function which takes into account various features that abstract some quality and quantitative requirements.

* Partially supported by the Italian MIUR projects ALINWEB and ECD, and by the Italian Registry of ccTLD.it.

J.-F. Boulicaut et al. (Eds.): PKDD 2004, LNAI 3202, pp. 543–545, 2004.

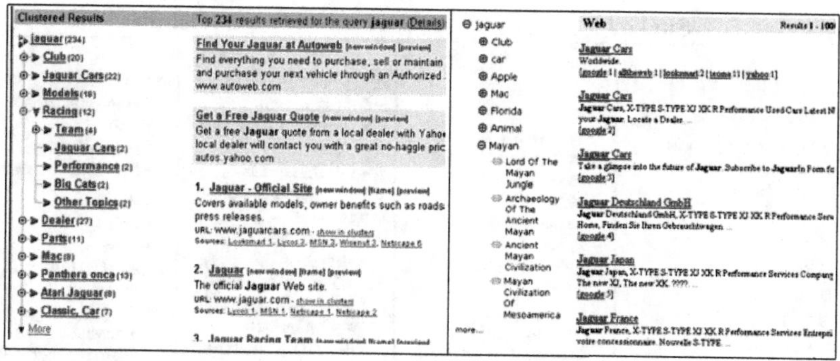

Fig. 1. Two Hierarchical Clustering Engines– VIVISIMO and SnakeT– tested on the ambiguous query "*Jaguar*".

Another specialty of our software is that the labels are *non* contiguous sequences of terms drawn from the snippets. This offers much power in capturing meaningful labels from the poor content of the snippets (see the future works of [7, 6]). The web interface of SnakeT is available at `http://roquefort.di.unipi.it/`.

Here we briefly comment on the latest results [5, 8, 4] for comparison. [5] extracts meaningful sentences from a snippet by using a pre-computed language model, and builds the hierarchy via a recursive algorithm. The authors admit that their hierarchies are often non compact, have large depth and contain some non content-bearing words which tend to repeat. Our work aims at overcoming these limitations by ranking non-contiguous sentences and using a novel covering algorithm for compacting the hierarchies. [8] extracts variable length (contiguous) sentences by combining five different measures through regression. Their clustering is flat, and thus the authors highlight the need of (1) a hierarchical clustering for more efficient browsing, and (2) external taxonomies for improving labels precision. This is actually what we do in our hierarchical engine by developing a *ranker* based on the whole DMOZ. [4] proposes a greedy algorithm to build the hierarchy based on a minimization of an objective function similar to ours. However, their labels are contiguous sentences and usually consist of single words. The best results to date are "Microsoft and IBM products" [4, 8] not publicly accessible, as the authors communicated to us!

In the demo we will discuss the system and comment on some user studies and experiments aimed at evaluating its efficacy. Following the approach of [5], we have selected few titles from TREC Topics and used them to create a testbed enriched with other well-know ambiguous queries. See Figure 2. Following [2, 3] we then prepared two user studies. The first study was aimed at understanding whether a Web clustering engine is an useful complement to the flat, ranked list of search-engine results. We asked to 45 people, of intermediate web ability, to use VIVISIMO during their day by day search activities. After a test period of 20 days, 85% of them reported that using the tool "[..] get a good sense of range alternatives with their meaningful labels", and 72% of them reported that one of the most useful feature is "[..] the ability to produce on-the-fly clusters

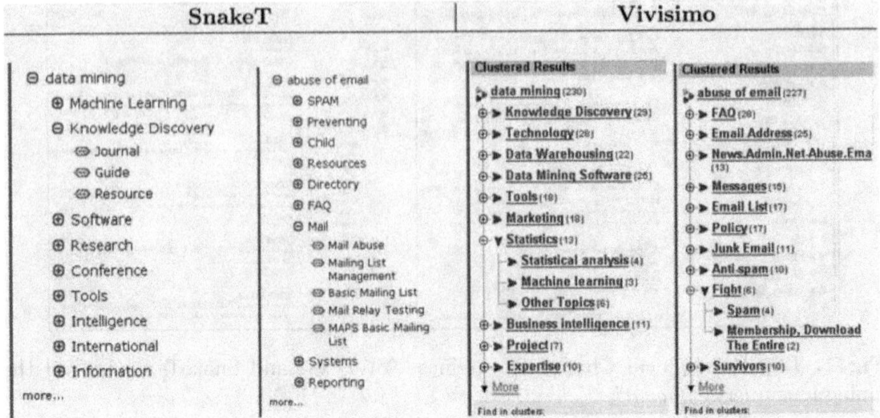

Fig. 2. Two examples of queries: *Data Mining* and *Abuse of Email*, executed on SnakeT and VIVISIMO.

in response to a query, with labels extracted from the text". The second study was aimed at drawing a preliminary evaluation of our software. We selected 20 students of the University of Pisa, each of whom executed 20 different queries drawn from the testbed above. The participants evaluated the answers provided by our hierarchical clustering engine with respect to VIVISIMO. 75% of them were satisfied of the quality of our hierarchy and its labels. This evaluation exercise is going on by increasing the number of users, allowing free queries, and extending its use to other domains, like blogs, news and books.

References

1. S. Brin and L. Page. The anatomy of a large-scale hypertextual Web search engine. *Computer Networks and ISDN Systems*, 30(1–7):107–117, 1998.
2. H. Chen and S. T. Dumais. Bringing order to the web: automatically categorizing search results. In *SIGCHI-00*, pages 145–152, 2000.
3. M. A. Hearst and J. O. Pedersen. Reexamining the cluster hypothesis: Scatter/gather on retrieval results. In *SIGIR-96*, pages 76–84, Zürich, CH, 1996.
4. K. Kummamuru, R. Lotlikar, S. Roy, K. Singal, and R. Krishnapuram. A hierarchical monothetic document clustering algorithm for summarization and browsing search results. In *WWW*, 2004.
5. D. J. Lawrie and W. B. Croft. Generating hiearchical summaries for web searches. In *ACM SIGIR*, pages 457–458, 2003.
6. Y. S. Maarek, R. Fagin, I. Z. Ben-Shaul, and D. Pelleg. Ephemeral document clustering for web applications. Technical Report RJ 10186, IBM Research, 2000.
7. O. Zamir and O. Etzioni. Grouper: a dynamic clustering interface to Web search results. *Computer Networks*, 31:1361–1374, 1999.
8. H. Zeng, Q. He, Z. Chen, and W. Ma. Learning to cluster web search results. In *ACM SIGIR*, 2004.

HIClass: Hyper-interactive Text Classification by Interactive Supervision of Document and Term Labels

Shantanu Godbole, Abhay Harpale, Sunita Sarawagi, and Soumen Chakrabarti

IIT Bombay
Powai, Mumbai, 400076, India
shantanu@it.iitb.ac.in

Abstract. We present the HIClass (Hyper Interactive text Classification) system, an interactive text classification system which combines the cognitive power of humans with the power of automated learners to make statistically sound classification decisions. HIClass is based on active learning principles and has aids for detailed analysis and fine tuning of text classifiers while exerting a low cognitive load on the user.

1 Introduction

Motivated by applications like spam filtering, e-mail routing, Web directory maintenance, and news filtering, text classification has been researched extensively in recent years [1–3]. Most text classification research assumes a simple bag-of-words model of features, a fixed set of labels, and the availability of a labeled corpus that is representative of the test corpus. Many of these assumptions do not hold in real-life.

Discrimination between labels can be difficult unless features are engineered and selected with extensive human knowledge. Often, there is no labeled corpus to start with, or the label set must evolve with the user's understanding. Projects reported routinely at the annual OTC workshops [4] describe applications in which automated, batch-mode techniques were unsatisfactory; substantial human involvement was required before a suitable feature set, label system, labeled corpus, rule base, and system accuracy were attained. Not all commercial systems use publicly known techniques, and few general principles can be derived from them.

There is scope for building learning tools which engage the user in an active dialog to acquire knowledge about features and labels. We present the HIClass system which provides a tight interaction loop for such an active dialog with the expert. *Active learning* has provided clear principles [5–7] and strategies for maximum payoffs from such a dialog. We extend active learning to include feature engineering and multi-labeled document labeling conversations. HIClass is an interactive multi-class multi-labeled text classification system that combines the cognitive power of humans with the power of automated learners to make statistically sound classification decisions (details appear in [8]).

J.-F. Boulicaut et al. (Eds.): PKDD 2004, LNAI 3202, pp. 546–548, 2004.

2 The HIClass Workbench for Text Classification

We present an overview of HIClass in Fig. 1. The lower layer shows the main data entities and processing units. There is a small labeled pool and a large unlabeled pool of documents. The system stores and accesses by name, multiple classifiers with their parameters, for comparative analysis and diagnostics. The upper layer shows main modes/menus of interaction with the system. We outline major components of the system next.

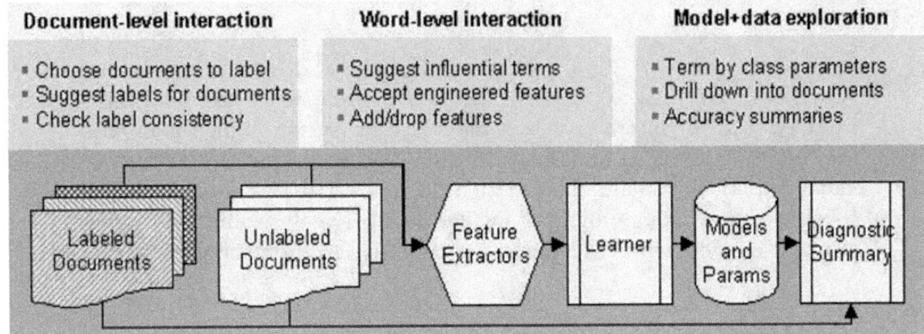

Fig. 1. The architecture of HIClass

Document and classification models: HIClass is designed using a flexible classification model template that (1) suits state-of-the-art automated learners and (2) can be easily interpreted and tuned by the user. HIClass uses **linear additive** classifier models like linear SVMs.

Exploration of data/models/performance summaries: HIClass allows the user to view the trained classifier scores, aggregate and drill-down statistics about terms, documents and classes, and different accuracy measures. An OLAP-like tool enables the human expert to fine tune individual features contributing to various classes with continuous feedback about resultant system accuracy.

Feature engineering: Expert users, on inspection of class-term scores, will be able to propose modifications to the classifiers like adding, removing, or ignoring certain terms for certain classes. They can also provide input about stemming, aggregation, and combination of features.

Document labeling assistant: HIClass maximizes learning rate while minimizing user's cognitive load through various mechanisms: (1) a pool of most uncertain unlabeled documents is selected for user feedback, (2) bulk-labeling is facilitated by clustering documents, (3) the system ranks suggested labels, (4) the system checks labeling conflicts.

Term-level active learning: Initially when bootstrapping from a small corpus, HIClass directly asks users to specify well known trigger features as being positively/negatively associated with classes. This exerts a lower cognitive load on the user compared to reading full documents.

3 Description of the Demonstration

Our demonstration will showcase the detailed working of all aspects of the HIClass system. HIClass consists of roughly 5000 lines of C++ code for the back-end which communicates through XML with 1000 lines of PHP scripts to manage browser-based front-end user interactions [8].

We will allow user interaction along three major modes. First, the user can either bootstrap the classifier by term-based active learning, or engage in traditional document-level labeling. Various labeling aids will minimize cognitive load on the user. The second major mode will be inspection of named learned classifiers in an OLAP-like interface for feature engineering. In the third exploratory mode, various aggregate statistics will draw the user's attention to areas which can benefit by more data and fine tuning. We will present extensive experimental results on benchmark text datasets highlighting the various aspects of the HIClass system.

References

1. T. Joachims. Text categorization with support vector machines: learning with many relevant features. In *Proceedings of ECML 1998.*
2. K. Nigam, J. Lafferty, and A. McCallum. Using maximum entropy for text classification, 1999.
3. J. Zhang and Y. Yang. Robustness of regularized linear classification methods in text categorization. In *SIGIR 2003.*
4. 3rd workshop on Operational Text Classification OTC 2003. At SIGKDD-2003
5. D. A. Cohn, Z. Ghahramani, and M. I. Jordan. Active learning with statistical models. In *Advances in Neural Information Processing Systems*, 1995.
6. Y. Freund, H. S. Seung, E. Shamir, and N. Tishby. Selective sampling using the query by committee algorithm. *Machine Learning*, 28(2-3):133–168, 1997.
7. S. Tong and D. Koller. Support vector machine active learning with applications to text classification. *Journal of Machine Learning Research*, 2:45–66, Nov. 2001.
8. S. Godbole, A. Harpale, S. Sarawagi, and S. Chakrabarti. Document classification through interactive supervision on both document and term labels In *Proceedings of PKDD 2004.*

Balios – The Engine for Bayesian Logic Programs

Kristian Kersting and Uwe Dick

University of Freiburg, Institute for Computer Science, Machine Learning Lab,
Georges-Koehler-Allee 079, 79110 Freiburg, Germany
{kersting,dick}@informatik.uni-freiburg.de

1 Context: Stochastic Relational Learning

Inductive Logic Programming (ILP) [4] combines techniques from machine learning with the representation of logic programming. It aims at inducing logical clauses, i.e, general rules from specific observations and background knowledge. Because of focusing on *logical* clauses, traditional ILP systems do not model uncertainty explicitly. On the other hand, state-of-the-art probabilistic models such as *Bayesian networks* (BN) [5], *hidden Markov models*, and *stochastic context-free grammars* have a rigid structure and therefore have problems representing a variable number of objects and relations among these objects. Recently, various relational extensions of traditional probabilistic models have been proposed, see [1] for an overview. The newly emerging field of *stochastic relational learning* (SRL) studies learning such rich probabilistic models.

2 The Balios Engine

Balios is an inference engine for Bayesian logic programs (BLPs) [3, 2]. BLPs combine BNs with definite clause logic. The basic idea is to view logical atoms as sets of random variables which are similar to each other. Consider the modelling the inheritance of a single gene that determines a person's `P` blood type `bt(P)`. Each person `P` has two copies of the chromosome containing this gene, one, `mc(M)`, inherited from her mother `mother(M, P)`, and one, `pc(F)`, inherited from her father `father(F, P)`. Such a general influence relation cannot be captured within BNs.

Knowledge Representation: Like BNs, BLPs separate the qualitative, i.e., the influence relations among random variables, from the quantitative aspects of the world, i.e., the strength of influences. In contrast to BNs, however, they allow to capture general probabilistic regularities. Consider the BLP shown in Figure 1 modelling our genetic domain. The rule graph gives an overview of all interactions (boxes) among abstract random variables (ovals). For instance, the maternal information `mc/1` is specified in terms of mothers `mother/2`, maternal `mc/1` and paternal `pc/1` information. Each interaction gives rise to a local probabilistic model which is composed of a qualitative and a quantitative part. For instance, rule 2 in Figure 1(a) encodes that

> *"the maternal genetic information* `mc(P)` *of a person* `P` *is influenced by the maternal* `mc(M)` *and paternal* `pc(M)` *genetic information of* `P`*'s mother* `M`.

J.-F. Boulicaut et al. (Eds.): PKDD 2004, LNAI 3202, pp. 549–551, 2004.

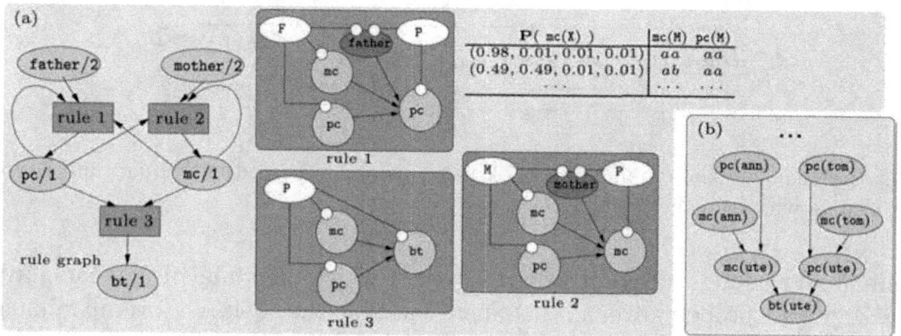

Fig. 1. (a) A graphical BLP. We left out some specification of quantitative knowledge. (b) Parts of the inferred BN specifying the distribution over `bt(ute)`.

Light gray ovals represent abstract random variables such as maternal chromosomes `mc(P)`. Smaller white circles on boundaries denote arguments, e.g., some person `P`. Larger white ovals together with undirected edges indicate that arguments refer to the same person as for `mc(P)` and `mother(M,P)`. To quantify the structural knowledge, *conditional probability distributions* (CPDs) are associated. Some information might be of qualitative nature only, such as `mother(M,P)`. The mother `M` of a person `P` does not affect the CPD but ensures the variable bindings among `mc(P)`,`mc(M)`, and `pc(M)`. Such "logical" nodes are shaded dark gray.

Next to relational probabilistic models, the range of knowledge representation paradigms provided by BALIOS include e.g. BNs (with purely discrete variables, purely continuous variables, or a mix of discrete and continuous variables), hidden Markov models, stochastic grammars, and logic programs.

Inference: To compute the distribution of a finite set of random variables given some evidence, a BN is inferred. To do so, proof techniques from logic programming are employed because the qualitative structure of a BLP corresponds to a logic program. For instance rule 1 in Figure 1(a) corresponds to the clause `pc(P) :- father(F,P),pc(F),mc(F)` . We assume range-restriction, i.e., each variable `P` in the head `pc(P)` also occurs in the body `father(F,P),pc(F),mc(F)`.

To compute the distribution over `bt(ute)`, we first compute `ute`'s paternal `pc(ute)` and maternal `mc(ute)` information due to rule 3. The associated CPD quantifies the influence. Then, in the next iteration, we deduce the influence relations of the chromosomal information of `ute`'s mother (rule 1) and father (rule 2). In this way, we iterate. This yields a BN, see Figure 1(b), if the influence relation is acyclic. In the presence of evidence, e.g., we know that the blood type of `ute`'s sister `nadine` is `a`, we compute the union of the BNs for `bt(ute)` and `bt(nadine)`, and set `a` as evidence for `bt(nadine)` in the resulting network.

Combining Rules and Aggregate Functions: When there are multiple rules firing, a combining rule such as *noisy-or* or *noisy-and* is used to quantify the combined influence. This also allows for aggregate functions such as *median* and *mode*. Consider modelling the ranking of a new book, see Figure 2. The overall ranking depends on the rankings of individual customers who read the book. The

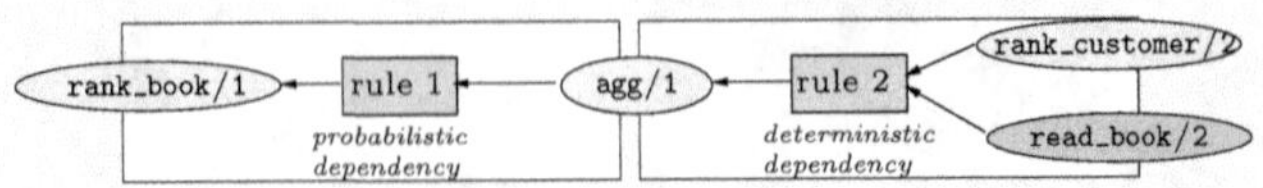

Fig. 2. Modelling the ranking of a book in terms of aggregated individual rankings of customers who read the book.

individual rankings are summarized in `agg/1` which deterministically computes the aggregate property over all customers who read the book. The overall ranking of the book `rank_book/1` probabilistically depends on `agg/1`.

The Engine: BALIOS is written in JAVA. It calls SICSTUS Prolog to perform logical inference and a BN inference engine (e.g. HUGIN or ELVIRA) to perform probabilistic inference. BALIOS features (1) a GUI graphically representing BLPs, (2) compution the most likely configuration, (3) exact (junction tree) and approximative inference methods (rejection, likelihood and Gibbs sampling), and (4) parameter estimation methods (hard EM, EM and conjugate gradient). To the best of the authors' knowledge, BALIOS is the first engine of a turing-complete probabilistic programming language featuring a graphical representation.

3 Demonstration and Concluding Remarks

The demonstration will include examples of Bayesian networks, hidden Markov models, stochastic grammars, and Bayesian logic programs. We will explain the graphical representation, show how to do inference (exact and approximative), and will demonstrate parameter estimation from a database of cases.

At the moment the set of planned future features of BALIOS includes, but is not limited to: effective methods for learning the structure of BLPs, see e.g. [3]; and relational influence diagrams to support decision making.

Acknowledgments: The research was supported by the EU under contract number FP6-508861, *Application of Probabilistic Inductive Logic Programming II.*

References

1. L. De Raedt and K. Kersting. Probabilistic Logic Learning. *ACM-SIGKDD Explorations*, 5(1):31–48, 2003.
2. K. Kersting and L. De Raedt. Adaptive Bayesian Logic Programs. In C. Rouveirol and M. Sebag, editors, *Proceedings of the Eleventh Conference on Inductive Logic Programming (ILP-01)*, volume 2157 of *LNCS*, Strasbourg, France, 2001. Springer.
3. K. Kersting and L. De Raedt. Towards Combining Inductive Logic Programming and Bayesian Networks. In *Proceedings of the Eleventh Conference on Inductive Logic Programming (ILP-2001)*, volume 2157 of *LNCS*. Springer, 2001.
4. S. Muggleton and L. De Raedt. Inductive logic programming: Theory and methods. *Journal of Logic Programming*, 19(20):629–679, 1994.
5. J. Pearl. *Reasoning in Intelligent Systems: Networks of Plausible Inference.* Morgan Kaufmann, 2. edition, 1991.

SEWeP: A Web Mining System Supporting Semantic Personalization

Stratos Paulakis, Charalampos Lampos,
Magdalini Eirinaki, and Michalis Vazirgiannis

Athens University of Economics and Business, Department of Informatics
{paulakis,lampos,eirinaki,mvazirg}@aueb.gr

Abstract. We present SEWeP, a Web Personalization prototype system that integrates usage data with content semantics, expressed in taxonomy terms, in order to produce a broader yet semantically focused set of recommendations.

1 Introduction

Web personalization is the process of customizing a Web site to the needs of each specific user or set of users. Most of the research efforts in Web personalization correspond to the evolution of extensive research in Web usage mining. When a personalization system relies solely on usage-based results, however, valuable information conceptually related to what is finally recommended may be missed. To tackle this problem we propose a Web personalization framework based on semantic enhancement of the Web usage logs and the related Web content. We present SEWeP, a Web Personalization prototype system, based on the framework proposed in [3]. This system integrates usage data with content semantics, expressed in taxonomy terms, in order to produce semantically enhanced navigational patterns that can subsequently be used for producing valuable recommendations.

2 SEWeP Framework

The innovation of the SEWeP prototype system is the exploitation of web content semantics throughout the Web mining and personalization process. It utilizes Web content/structure mining methods to assign semantics to Web pages and subsequently feeds this knowledge to Web usage mining algorithms. Web content is semantically annotated using terms of a predefined domain-specific taxonomy (categories) through the use of a thesaurus. These annotations are encapsulated into C-logs, a (virtual) extension of Web usage logs. C-logs are used as input to the Web usage mining process, resulting in a set of rules/patterns consisting of thematic categories in addition to URIs. Furthermore, the semantically annotated Web pages are organized in coherent clusters (using THESUS system [6]) based on the taxonomy. These clusters are then used in order to further expand the set of recommendations provided to the end user. The whole process results in a broader, yet semantically focused set of recommendations. The system architecture is depicted in Figure 1. The main functionalities of the demonstrated system are described below. A more detailed description can be found in [2,3].

Logs Preprocessing: The system provides full functionality for preprocessing any kind of Web logs, by enabling the definition of new log file templates, filters (including/excluding records based on field characteristics), etc. The new logs are stored in new ("clean") log files.

J.-F. Boulicaut et al. (Eds.): PKDD 2004, LNAI 3202, pp. 552–554, 2004.

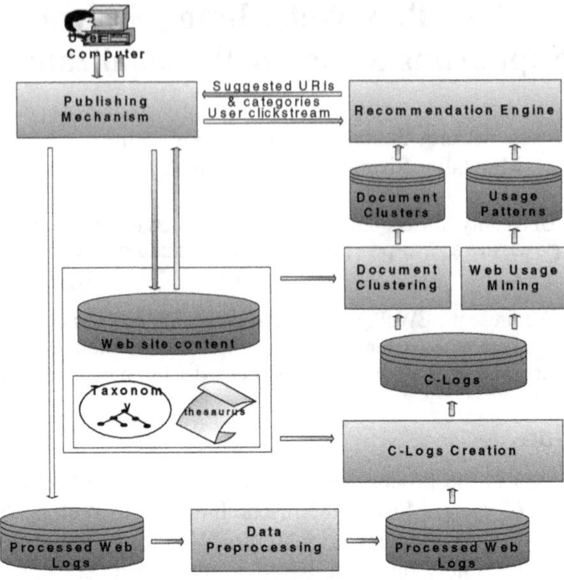

Fig. 1. System architecture.

Content Retrieval: The system crawls the Web and downloads the Web site's pages, extracting the plain text from a variety of crawled file formats (html, doc, php, ppt, pdf, flash, etc.) and stores them in appropriate database tables.
Keyword Extraction and Translation: The user selects among different methods for extracting keywords. Prior to the final keywords selection, all non-English keywords are translated using an automated process (the system currently also supports Greek content). All extracted keywords are stored in a database table along with their relevant frequency.
Keyword – Category Mapping: The extracted keywords are mapped to categories of a domain-specific taxonomy. The system finds the "closest" category to the keyword through the mechanisms provided by a thesaurus (WordNet [7]). The weighted categories are stored in XML files and/or in a database table.
Session Management: SEWeP enables anonymous sessionizing based on distinct IPs and a user-defined time limit between sessions. The distinct sessions are stored in XML files and/or database tables. (Figure 2 includes a screenshot of this module)
Semantic Association Rules Mining: SEWeP provides a version of the apriori algorithm [1] for extracting frequent itemsets and/or association rules (confidence and support thresholds set by the user). Apart from URI-based rules, the system also provides functionality for producing category-based rules. The results are stored in text files for further analysis or use by the recommendation engine.
Clustering: SEWeP integrates clustering facilities for organizing the results into meaningful semantic clusters. Currently SEWEP capitalizes on the clustering tools available in the THESUS system [6].
Recommendations: The (semantic) association rules/frequent itemsets created feed a client-side application (servlet) in order to dynamically produce recommendations to the visitor of the personalized site.

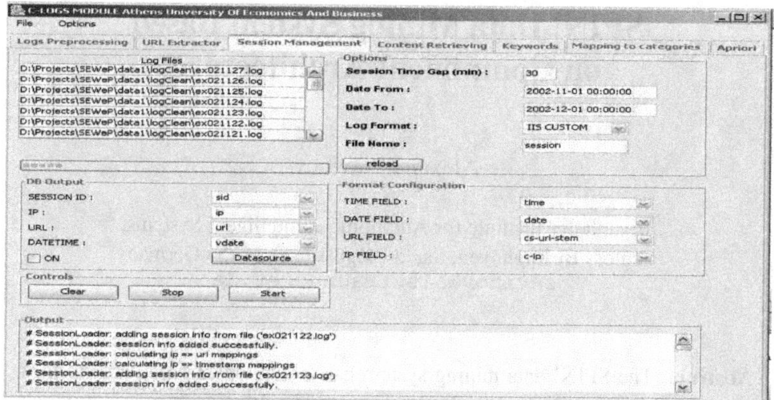

Fig. 2. SEWeP screenshot: Session Management module.

3 Implementation Details

The SEWeP system is entirely based on Java (JDK 1.4 or later). For the implementation of SEWeP we utilized the following third party tools & algorithms: PDF Box Java Library, Jacob Java-Com Bridge Library, and swf2html library (for text extraction); Xerces XML Parser; Wordnet v1.7.1 Ontology; JWNL and JWordnet 1.1 java interfaces for interaction with Wordnet; Porter Stemming Algorithm [4] for English; Triantafillidis Greek Grammar [5]; Apache Tomcat and 4.1 and Java Servlets for recommendation engine; JDBC Library for MS SQL Server.

4 Empirical Evaluation

The main advantage of SEWeP is the involvement of semantics in the recommendation process resulting in semantically expanded recommendations. As long as the system's effectiveness is concerned, we have performed a set of user-based experiments (blind tests), evaluating SEWeP's usefulness, i.e. whether the semantic enhancement results in better recommendations [2,3]. The experiments' results verify our intuitive assumption that SEWeP enhances the personalization process, since users evaluate the system's recommendations as of high quality.

References

1. R. Agrawal, R. Srikant, Fast Algorithms for Mining Association Rules, in Proc. of 20th VLDB Conference, 1994
2. M. Eirinaki, M. Vazirgiannis, H. Lampos, S. Pavlakis, Web Personalization Integrating Content Semantics and Navigational Patterns, submitted for revision at WIDM 2004
3. M. Eirinaki, M. Vazirgiannis, I. Varlamis, SEWeP: Using Site Semantics and a Taxonomy to Enhance the Web Personalization Process, in Proc. of the 9th SIGKDD Conf., 2003
4. M. Porter, An algorithm for suffix stripping, Program (1980)/ Vol. 14, No. 3, 130-137
5. M.Triantafillidis, Triantafillidis On-Line, Modern Greek Language Dictionary, http://kastor.komvos.edu.gr/dictionaries/dictonline/DictOnLineTri.htm
6. I. Varlamis, M. Vazirgiannis, M. Halkidi, B. Nguyen. THESUS: Effective Thematic Selection And Organization Of Web Document Collections Based On Link Semantics, to appear in IEEE TKDE Journal
7. WordNet, http://www.cogsci.princeton.edu/~wn/

SPIN! Data Mining System Based on Component Architecture

Alexandr Savinov

Fraunhofer Institute for Autonomous Intelligent Systems,
Schloss Birlinghoven, Sankt-Augustin, D-53754 Germany
savinov@ais.fraunhofer.de

Abstract. The SPIN! data mining system has a component-based architecture, where each component encapsulates some specific functionality such as a data source, an analysis algorithm or visualization. Individual components can be visually linked within one workspace for solving different data mining tasks. The SPIN! friendly user interface and flexible underlying component architecture provide a powerful integrated environment for executing main tasks constituting a typical data mining cycle: data preparation, analysis, and visualization.

1 Component Architecture

The SPIN! data mining system has a component architecture. This means that it provides only an infrastructure and environment while all the system functionality comes from separate software modules called components. Components can be easily plugged-in into the system thus allowing us to extend its capabilities an importance of which for data mining was stressed in [5]. In this sense it is very similar to such a general purpose environment as Eclipse. Each component is developed as an independent module for solving one or a limited number of tasks. For example, there may be components for data access, analysis or visualization. In order to solve complex problems components need to cooperate by using each other.

All components are implemented on the basis of CoCon Common Connectivity Framework which is a set of generic interfaces and objects in Java and allows components to communicate within one workspace. The cooperation of components is based on the idea that they can be linked by means of different types of connections. Currently there exist three connections: visual, hierarchical and user-defined. Visual connections are used to link a component to its view (similar to Model-View-Controller architecture). Hierarchical connections are used to compose parent-child relationships among components within one workspace, e.g., between folder and its elements or a knowledge base and its patterns. The third and the main type is the user connection, which is used to arbitrary link components in the workspace according to the task to be solved (similar to Clementine). It is important that components explicitly declare their connectivity capabilities, i.e., the system knows how they can be connected and with what other components they can work. In particular, the SPIN! configures itself according to available components by exposing their functions in menu and toolbar.

J.-F. Boulicaut et al. (Eds.): PKDD 2004, LNAI 3202, pp. 555–557, 2004.

2 Workspace Management

Workspace is a set of components and connections among them. It can be stored in or retrieved from a persistent storage like file or database. Workspace appears in two views: tree view and graph view (upper and middle left windows in Fig. 1). In tree view the hierarchical structure of the workspace is visualized with components as individual nodes, which can be expanded or collapsed. In graph view components are visualized as nodes of the graph while user connections are graph edges.

Components can be added to the workspace by choosing them either in menu or in component bar. After a component has been added it should be connected with other relevant components. An easy and friendly way to do this consists in drawing an arrow from the source component to the target one. For example, we might easily specify a data source for an algorithm by linking two components in graph view. While adding connections between components the SPIN! uses information about their connectivity so that only components, which are able to cooperate can be really connected.

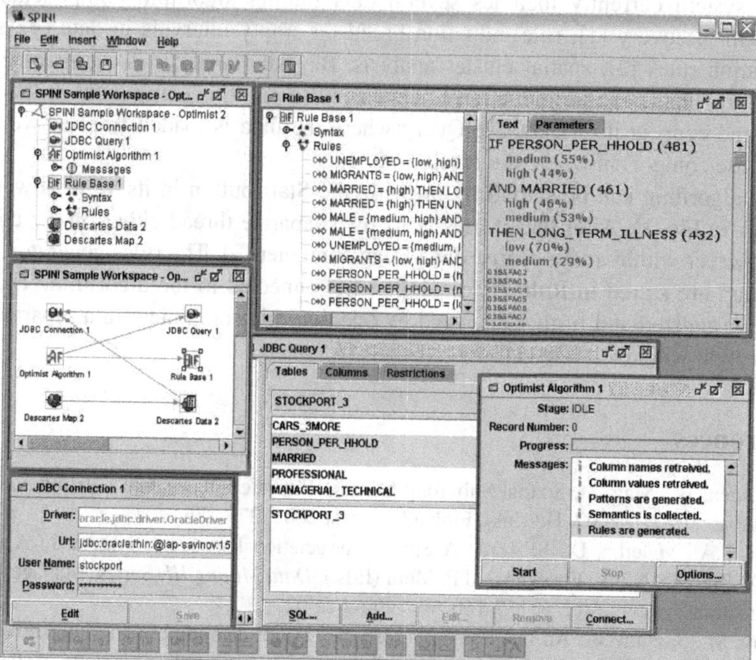

Fig. 1. The SPIN! Data mining system client interface: workspace (upper and middle left windows), rule base (upper right window), database connection (lower left window), database query and algorithm (lower middle and right windows).

Each component has an appropriate view, which is also a connectable component. Each component can be opened in a separate window so that the user can use its functions. When a workspace component is opened the system automatically creates a view, connects it with the model and then displays it within internal window.

3 Executing Data Mining Algorithms

The typical data mining tasks include data preprocessing, analysis and visualization. For data access the SPIN! system includes Database Connection and Database Query components (lower left and middle windows in Fig. 1). The Database Connection represents the database where the data is stored. To use the database this component should be connected to some other component, e.g., via graph view. The Database Query component describes one query, i.e., how the result set is generated from tables in the database. Essentially this component is a SQL query design tool, which allows the user to describe a result set by choosing tables, columns, restrictions, functions etc. Notice also that both Database Connection and Database Query components do not work by themselves and it is some other component that makes use of them. Such encapsulation of functionality and use of connections to configure different analysis strategies has been one of the main design goals of the SPIN! data mining system.

Any knowledge discovery task includes data analysis step where the dataset obtained from preprocessing step is processed by some data mining algorithm. The SPIN! system currently includes several data mining algorithm components, e.g., subgroup discovery [1], rule induction based on empty intervals in data [4], spatial association rules [2], spatial cluster analysis, Bayesian analysis. To use some algorithm, say, Optimist rule induction [4], we need to add this component in the workspace and connect it to Database Query where the data is loaded from as well as to Rule Base component where the result is stored.

The algorithm can be started by pressing the Start button in its view (lower right window in Fig. 1). After that it runs in its own separate thread either on the client or on the server within an Enterprise Java Bean container [3]. The rules generated by the algorithm are stored in Rule Base component connected to the algorithm. The rules can be visualized and further analyzed by opening this component in a separate view (upper right window in Fig. 1).

References

1. Klösgen, W., May, M. Spatial Subgroup Mining Integrated in an Object-Relational Spatial Database, PKDD 2002, Helsinki, Finland, August 2002, 275-286.
2. Lisi, F.A., Malerba, D., SPADA: A Spatial Association Discovery System. In A. Zanasi, C.A. Brebbia, N.F.F. Ebecken and P. Melli (Eds.), *Data Mining III*, Series: Management Information Systems, Vol. 6, 157-166, WIT Press, 2002.
3. May, M., Savinov, A. An integrated platform for spatial data mining and interactive visual analysis, Data Mining 2002, Third International Conference on Data Mining Methods and Databases for Engineering, Finance and Other Fields, 25-27 September 2002, Bologna, Italy, 51-60.
4. Savinov, A.: Mining Interesting Possibilistic Set-Valued Rules. In: Da Ruan and Etienne E. Kerre (eds.), Fuzzy If-Then Rules in Computational Intelligence: Theory and Applications, Kluwer, 2000, 107-133.
5. Wrobel, S., Wettschereck, D., Sommer, E., and Emde, W. (1996) Extensibility in Data Mining Systems. In *Proceedings of KDD'96 2nd International Conference on Knowledge Discovery and Data Mining*. AAAI Press, pp.214-219.

Author Index